Nonequilibrium Superconductivity, Phonons, and Kapitza Boundaries

NATO ADVANCED STUDY INSTITUTES SERIES

A series of edited volumes comprising multifaceted studies of contemporary scientific issues by some of the best scientific minds in the world, assembled in cooperation with NATO Scientific Affairs Division.

Series B. Physics

Recent Volumes in this Series

Volume 58 – Cosmology and Gravitation: Spin, Torsion, Rotation, and Supergravity
edited by Peter G. Bergmann and Venzo De Sabbata

Volume 59 – Recent Developments in Gauge Theories
edited by G. 't Hooft, C. Itzykson, A. Jaffe, H. Lehmann, P. K. Mitter, I. M. Singer, and R. Stora

Volume 60 – Theoretical Aspects and New Developments in Magneto-Optics
edited by Jozef T. Devreese

Volume 61 – Quarks and Leptons: *Cargèse 1979*
edited by Maurice Lévy, Jean-Louis Basdevant, David Speiser, Jacques Weyers, Raymond Gastmans, and Maurice Jacob

Volume 62 – Radiationless Processes
edited by Baldassare Di Bartolo

Volume 63 – Characterization of Crystal Growth Defects by X-Ray Methods
edited by Brian K. Tanner and D. Keith Bowen

Volume 64 – Molecular Electro-Optics: Electro-Optic Properties of Macromolecules and Colloids in Solution
edited by Sonja Krause

Volume 65 – Nonequilibrium Superconductivity, Phonons, and Kapitza Boundaries
edited by Kenneth E. Gray

Volume 66 – Techniques and Concepts of High-Energy Physics
edited by Thomas Ferbel

This series is published by an international board of publishers in conjunction with NATO Scientific Affairs Division

A Life Sciences B Physics	Plenum Publishing Corporation London and New York
C Mathematical and Physical Sciences	D. Reidel Publishing Company Dordrecht, Boston and London
D Behavioral and Social Sciences	Sijthoff & Noordhoff International Publishers
E Applied Sciences	Alphen aan den Rijn, The Netherlands, and Germantown, U.S.A.

Nonequilibrium Superconductivity, Phonons, and Kapitza Boundaries

Edited by

Kenneth E. Gray

US Department of Energy
Argonne National Laboratory
Argonne, Illinois

PLENUM PRESS • NEW YORK AND LONDON
Published in cooperation with NATO Scientific Affairs Division

Library of Congress Cataloging in Publication Data

NATO Advanced Study Institute on Nonequilibrium Superconductivity, Phonons, and Kapitza Boundaries (1980: Acquafredda di Maratea, Italy)
Nonequilibrium superconductivity, phonons, and Kapitza boundaries.

(NATO advanced study institutes series. Series B, Physics; v. 65)
"Published in cooperation with NATO Scientific Affairs Division."
Bibliography: p.
Includes index.
1. Superconductivity. 2. Phonons. 3. Boundary value problems. 4. Solid state physics. I. Gray, Kenneth E. (Kenneth Eugene), 1942- . II. North Atlantic Treaty Organization. Division of Scientific Affairs. III. Title. IV. Title: Kapitza boundaries. V. Series.
QC612.S8N37 1980 537.6'23 81-1858
AACR2

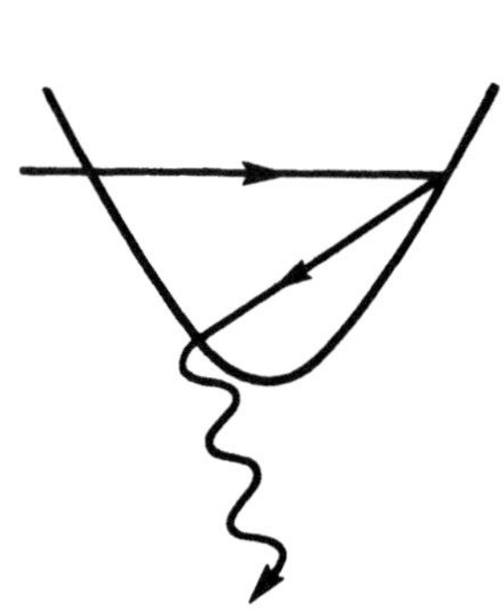

Based on the proceedings of a NATO Advanced Study Institute on Nonequilibrium Superconductivity, Phonons, and Kapitza Boundaries, held August 25–September 5, 1980, in Acquafredda di Maratea, Italy

ISBN-13: 978-1-4684-3937-3 e-ISBN-13: 978-1-4684-3935-9
DOI: 10.1007/978-1-4684-3935-9

Softcover reprint of the hardcover 1st edition 1981
A Division of Plenum Publishing Corporation
233 Spring Street, New York, N.Y. 10013

PREFACE

The importance of phonons has long been recognized by researchers in nonequilibrium superconductivity. Similarly, experimentalists studying phonons at low temperatures have relied heavily on superconductors as sources and detectors. To a large extent this symbiotic relationship has developed with a general mutual awareness; however, to our knowledge these subjects have never been treated together in conferences or study institutes. It was with the hope of further contributing to the awareness and communication between workers in these areas that this NATO Advanced Study Institute (ASI) has been conceived. A second, but equally important, reason for holding this ASI is to fill a void by providing the first general textbook in this important area of physics. Therefore, there was an emphasis on the tutorial nature of the lectures and written contributions to this textbook.

It should not go unnoticed that the experimental and theoretical concepts covered in this textbook are of paramount importance to the various applications of superconductors. Almost by definition, the use of a superconductor implies a nonequilibrium state! For example, phonon conduction to the helium bath is important in devices ranging from microscopic Josephson junctions to large scale magnets and transmission lines. Knowledge of the more fundamental nonequilibrium effects can aid in our understanding of devices as well as provide the potential for entirely new applications.

This book is based on the lectures given at the NATO ASI entitled "Nonequilibrium Superconductivity, Phonons and Kapitza Boundaries" which was held from 25 August to 5 September 1980 at Acquafredda di Maratea in Southern Italy. Individual contributions were coordinated beforehand to avoid undue duplication by allowing the authors to cross-reference between chapters. In addition, standardized notation is used (e.g., the symbol list in the Appendix). Thus this book is suitable as a graduate student textbook, as well as a valuable reference book for researchers in these general areas.

The fields of research covered by this ASI can certainly be considered mature; however there still exist many unresolved problems. In many cases, the lectures and discussion sessions clarified our understanding of these points, and in almost all cases the manuscripts were revised accordingly. It is hoped that these proceedings will provide stimulation for further work aimed at resolving the remaining issues, as well as opening up new research.

The study of the Kapitza boundary resistance provides an excellent example. The first measurements by Kapitza are almost 40 years old. Although the solid-solid boundary is quantitatively described by the acoustic mismatch theory, the solid-liquid helium boundary still has mysteries which are only beginning to be understood. Important new experiments and ideas highlighted the lectures and discussion in this field. The role of the Kapitza resistance is readily seen in measurements of the lifetime of quasiparticle excitations in superconductors. In many experimental situations, what is actually measured is the phonon escape time through this superconductor-substrate boundary.

Superconducting tunneling has been extremely important for the creation and detection of nonequilibrium states as well as phonons. Tunneling techniques are useful for measuring the microscopic properties of energy gaps and distribution functions in superconductors driven away from equilibrium by phonons, microwaves, quasiparticle injection, temperature gradients, transport current, etc.

A particularly exciting result has been the observations of energy gap enhancement in superconductors which have been induced by a perturbation such as tunneling or microwave irradiation. Although this effect seems to defy intuition, it can be understood within the framework of theoretical models.

Instabilities can occur in nonequilibrium superconducting states. These include multigap states and first order transitions into the normal state as a result of perturbations which drive the superconductor out of thermal equilibrium. Both the theoretical and experimental situation seems somewhat unclear at this time, but the stage is set for further developments.

The concept of charge imbalance, that is an exchange of charge between the superconducting pairs and quasiparticles, has received considerable attention recently. Many nonequilibrium states exhibit a charge imbalance, and its relaxation rate back to equilibrium determines the properties of such states. This concept is required to properly describe normal-superconductor boundaries, Josephson devices, thermoelectric effects, phase-slip centers, collective modes and tunnel junction asymmetries. There has been tremendous progress in the theoretical formulation of charge imbalance and its relationship to the various nonequilibrium states of superconductors. Experiments have adequately confirmed many of these ideas.

The motion of magnetic flux (flux flow) was one of the first non-equilibrium states studied, but remains perhaps one of the poorest understood. There is a noticeable lack of microscopic experimental information to compare with theory. On the other hand the phenomenological macroscopic picture is reasonably well understood.

In summary, there are sufficient new and unresolved issues to continue to make these fields interesting and stimulating. The text of the proceedings highlights these in addition to its purpose to provide a tutorial overview of the subject.

The NATO ASI which resulted in the present textbook was completely planned in the space of one calander year. As such, its organization greatly profited from the essential interest and help of many people. At NATO, Dr. M. diLullo provided considerable help and encouragement. Sy Foner kindly provided key insights into the organization based on his experience as a director of three previous NATO Institutes. Herbert Freeman provided information on the hotel in Maratea which was essential and which helped to decide on this location for the ASI. Special mention should go to Donald Langenberg who actively participated in all aspects of the organization until he was prevented from doing so by duties in his new positon as Deputy Director of the National Science Foundation. Additional help in the planning was provided by W. Eisenmenger, V. Narayanamurti, J. Clarke, C. M. Falco, A. Schmid, J. E. Mooij, H. Maris, M. Tinkham and W. Skocpol. Local organization was greatly benefited by the efforts of A. Barone, T. Kester and the hotel management through A. Guzzardi. Those who attended the ASI also relize the invaluable help provided by Susy.

In addition to the support provided by NATO, a significant contribution was made by the U.S. Department of Energy, Division of Basic Energy Sciences, in terms of secretarial support, producing and distributing publicity and the book of abstracts and in providing the ASI director the time to organize these activities. Particular thanks go to Karen Beres and Bonnie Russell for help in preparing the manuscripts for publication.

Further financial support for students and lecturers came from many local sources. Here we mention the U.S. National Science Foundation, which provided complete travel grants for two students from the U.S.

Kenneth E. Gray
Solid State Science Division
Argonne National Laboratory

CONTENTS

THE KAPITZA THERMAL BOUNDARY RESISTANCE BETWEEN TWO SOLIDS

A. C. Anderson

Department of Physics and Materials Research Laboratory
University of Illinois at Urbana-Champaign
Urbana, Illinois 61801

INTRODUCTION

While investigating the transport of heat in superfluid ^{4}He, Kapitza (1941) observed a large temperature difference ΔT between his copper heater and the liquid helium. He deduced that the entire ΔT occurred very close to the copper-helium interface. It is permissible therefore to define a thermal boundary resistance $R_K \equiv \Delta T/\dot{Q}$, where $\dot{Q}$ is the heat flux per unit area of interface. R_K, now called the Kapitza resistance, has been shown to have a temperature dependence of $\approx T^{-3}$.

Initially it was suggested that R_K was related to the superfluid character of liquid ^{4}He at temperatures below 2 K (see the review by Pollack, 1969). However, by 1960 it was apparent that a Kapitza resistance also occurs between two solid materials, and that this phenomenon should be related to the reflection and refraction of thermal phonons at the interface. A theory based on the scattering of phonons at the interface has become known as the acoustic-mismatch model (Little, 1959). This theory successfully accounts for R_K between most solids, but explains only a few of the details in the Kapitza resistance to liquid helium. The present paper discusses the solid-solid boundary resistance; the Kapitza resistance to liquid helium is described in Chapter 2.

THE ACOUSTIC-MISMATCH MODEL

The acoustic-mismatch model will first be developed using several simplifying assumptions. The assumptions will then be removed, noting the effect of each on R_K. The assumptions follow.

(i) The two materials in contact are nonmetallic. Therefore heat is carried only by thermal phonons, not by conduction electrons.

(ii) The temperature T is $\lesssim$ 10 K. Thus, only acoustic phonons are important, and these have a velocity independent of T or of the phonon angular frequency Ω. Three acoustic waves or propagation modes occur, namely one longitudinal and two transverse.

(iii) Both solids are perfect, isotropic continua free of defects. The phonons are scattered only by the interface. The phonon velocity is independent of the direction of propagation and the direction of propagation is parallel to the phonon wave vector. The two transverse modes have the same velocity.

(iv) The interface is ideal. Both displacement and stress are continuous across the interface between the two solids, and the interface is represented by a single plane separating the two materials.

The recipe to calculate R_K is now straightforward. The transmission probability is calculated for each phonon striking the interface, then a summation is performed over the three phonon modes, all directions of incidence, and all phonon frequencies to obtain the net heat flux $\dot{Q}$ carried across the interface.

The phonon transmission probability $\alpha(\Theta)$ is obtained from the boundary conditions of continuum mechanics (see, for example, Kolsky, 1953). The four conditions are that the normal and tangential displacement and the normal and tangential stress must each be continuous across the interface. To satisfy these conditions, an incident wave may give rise to as many as four outgoing waves as shown in Fig. 1. Each wave is assumed to have displacement u of the form

$$u = U\, e^{j\vec{q}\cdot\vec{r} - j\Omega t} \qquad (1)$$

where q is the phonon wave vector. Relations between the angles Θ_i of Fig. 1 are obtained from the requirement that the phase of each wave be the same along the x axis. The result is an acoustic Snell's law for each incident mode,

$$\frac{\sin\Theta_o}{s_o} = \frac{\sin\Theta_{1\ell}}{s_{1\ell}} = \frac{\sin\Theta_{1t}}{s_{1t}} = \frac{\sin\Theta_{2\ell}}{s_{2\ell}} = \frac{\sin\Theta_{2t}}{s_{2t}} \,. \qquad (2)$$

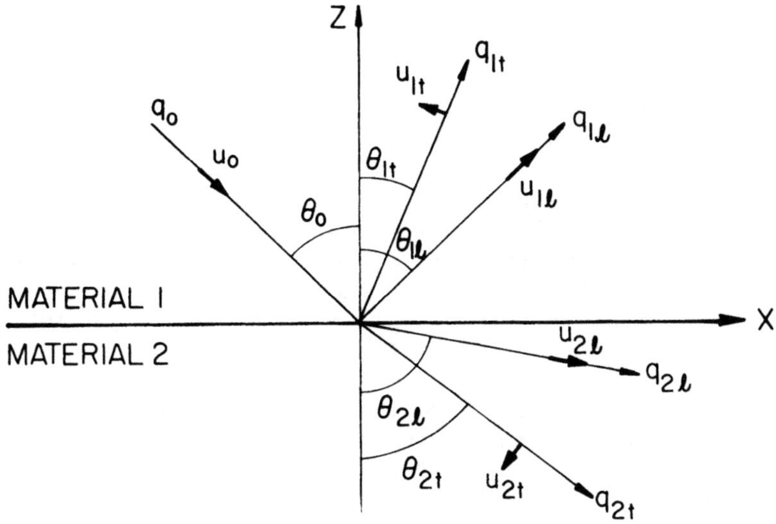

Fig. 1. Waves created at an interface by an incident longitudinal wave. Subscripts on the wave vector q, displacement u and angles Θ have the following meanings: 0-incident wave, 1-reflected wave, 2-refracted wave, ℓ-longitudinal wave, t-transverse wave. A similar behavior occurs for incident transverse waves.

The four boundary conditions plus Eq. 2 lead to four equations between the wave amplitudes U_i. Thus, the amplitude of each outgoing wave may be obtained in terms of the incident amplitude U_o. These relations are written out explicitly in Chapter 2 for the acoustically more simple example of an interface between a solid material and a liquid in which only a longitudinal wave can propagate.

Since the energy flux directed normal to the interface for an acoustic wave is $(1/2)\rho_m s\, U^2\Omega^2 \cos \Theta$, the probability coefficient for the transmission of the phonon energy across the interface is defined as

$$\alpha_i(\Theta) = \frac{\rho_{m2}\, s_{2t}\, U_{2t}^2 \cos \Theta_{2t}}{\rho_{m1}\, s_{1i}\, U_{1i}^2 \cos \Theta_o} + \frac{\rho_{m2}\, s_{2\ell}\, U_{2\ell}^2 \cos \Theta_{2\ell}}{\rho_{m1}\, s_{1i}\, U_{1i}^2 \cos \Theta_o} \qquad (3)$$

for each of the three incident modes i. The transmission is determined largely by the acoustic impedances $\rho_{m1}\, s_1$ and $\rho_{m2}\, s_2$ of the two materials. As an example, for normal incidence application of Eq. 3 and the boundary conditions produces $\alpha(0) = 4(\rho_{m1}\, s_1)(\rho_{m2}\, s_2)$

$[\rho_{m1}\, s_1 + \rho_{m2}\, s_2]^{-2}$. Thus for a large mismatch or acoustic difference between the two materials, α becomes very small.

The transmission coefficient is plotted as a function of angle Θ in Fig. 2 for a particular interface between an epoxy and copper. The coefficient α has been multiplied by $\sin\Theta\cos\Theta$ since this factor appears in the integral for $\dot{Q}$ to be discussed below. The resulting curves in Fig. 2 emphasize the relative importance to $\dot{Q}$ of phonons incident from different angles. Note that total internal reflection occurs for transverse phonons incident at angles greater than 43°. This critical angle is much smaller, $\lesssim 10°$, when one material is liquid or solid helium (see Chapter 2).

The next step in obtaining R_K is to sum over all phonons *striking* the interface from material 1 to find the heat flux $\dot{Q}_{1\to 2}$ transmitted to material 2. The same computation is done for phonons incident from material 2 to obtain $\dot{Q}_{2\to 1}$. The net heat flux $\dot{Q}$ is then simply $\dot{Q} = \dot{Q}_{1\to 2} - \dot{Q}_{2\to 1}$. If $N_{1i}(\Omega,\Theta,\phi,T)$ represents the spectral and directional distribution of phonons of mode i striking the interface from material 1 and $N_{2i}(\Omega,\Theta,\phi,T)$ that for phonons from material 2, then

$$\dot{Q} = \sum_i \int_0^\infty d\Omega \int_0^{\pi/2} d\Theta_{1i} \int_0^{2\pi} d\phi_{1i}\, N_{1i}(\Omega,\Theta,\phi,T)\,\alpha_{1i}(\Theta_{1i})\hbar\Omega s_{1i}\cos\Theta_{1i}\sin\Theta_{1i}$$

$$- \sum_i \int_0^\infty d\Omega \int_0^{\pi/2} d\Theta_{2i} \int_0^{2\pi} d\phi_{2i}\, N_{2i}(\Omega,\Theta,\phi,T)\,\alpha_{2i}(\Theta_{2i})\hbar\Omega s_{2i}\cos\Theta_{2i}\sin\Theta_{2i} \tag{4}$$

The problem now lies in finding a reasonable expression for $N(\Omega,\Theta,\phi,T)$. Little (1959) used an *equilibrium* phonon distribution to represent the spectrum of phonons striking the interface,

$$N(\Omega,\Theta,\phi,T) = N_o(\Omega,T)(4\pi)^{-1}$$

$$= \Omega^2[8\,\pi^3\, s^3\, (e^{\hbar\Omega/k_BT} - 1)]^{-1} \,. \tag{5}$$

Substitution of Eq. 5 into Eq. 4 produces a complicated expression which must generally be integrated on a computer. We represent the result by

$$\dot{Q} = b_1\, \bar{\alpha}_{12}\, T_1^{\,4} - b_2\, \bar{\alpha}_{21}\, T_2^{\,4} \tag{6}$$

where

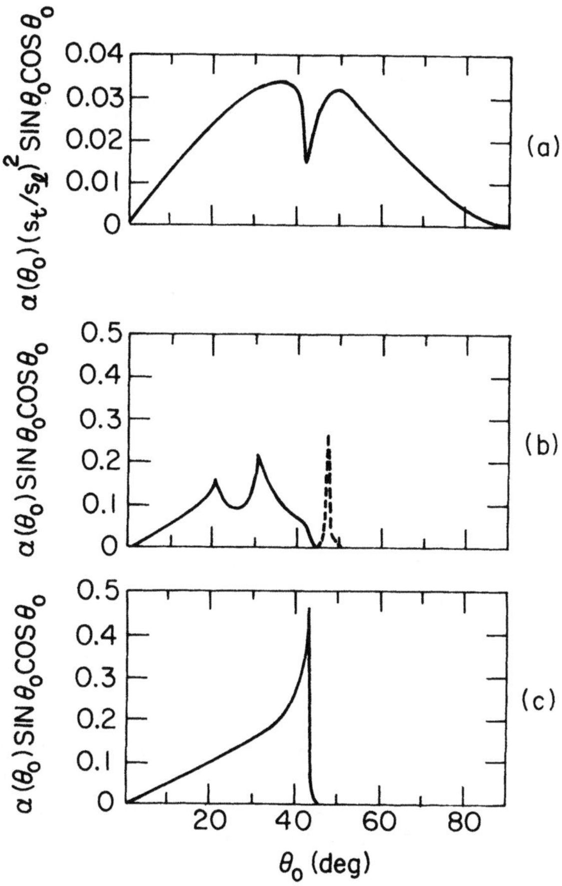

Fig. 2. Relative transmission of energy through an epoxy/copper interface, versus angle of incidence from the epoxy, for (a) longitudinal and (b,c) transverse phonons. The longitudinal phonons contribute only $\approx$ 13% to the total heat flux since the density of longitudinal phonons is smaller by the factor $(s_t/s_\ell)^2$. The dashed curve is explained in the text.

$$b_1 = \frac{\pi^2 k_B^4}{120\hbar^3} \sum_i s_{1i}^{-2} \tag{7}$$

and

$$\bar{\alpha}_{12} = \sum_i \langle\alpha_{1i}\rangle s_{1i}^{-2} / \sum_i s_{1i}^{-2}. \tag{8}$$

Similar expressions are obtained for b_2 and $\bar{\alpha}_{21}$. Here $\langle\alpha_{1i}\rangle$ is the angle-averaged phonon transmission coefficient for each mode of propagation, while $\bar{\alpha}_{12}$ is in addition averaged over all three modes. Under the assumptions listed at the beginning of this section, both b and $\bar{\alpha}$ are independent of temperature.

A simplification occurs for Eq. 6 when it is recognized that, for $T_1=T_2$, $\dot{Q}=0$. Hence $b_1\bar{\alpha}_{12}=b_2\bar{\alpha}_{21}$ and

$$\dot{Q} = b_1\bar{\alpha}_{12} (T_1^4 - T_2^4). \tag{9}$$

Further, if $2(T_1 - T_2)(T_1 + T_2)^{-1} = \Delta T/T << 1$,

$$\dot{Q} = 4 b_1\bar{\alpha}_{12} T^3 \Delta T. \tag{10}$$

Hence

$$R_K = \frac{\Delta T}{\dot{Q}} = (4 b_1\bar{\alpha}_{12} T^3)^{-1}. \tag{11}$$

The magnitude of $\bar{\alpha}$ varies between 0 and 1; a perfect acoustic match corresponds to $\bar{\alpha} = 1$.

It may be helpful to note the similarity between the above calculation and a calculation of the phonon thermal conductivity. The product $4 b_1\bar{\alpha}_{12} T^3$ would be the thermal conductivity of material 1 if $\bar{\alpha}_{12} = 4\Lambda/3L$, where Λ is the phonon mean free path and L is the length of the sample.

Equation 11 is often referred to as the acoustic-mismatch model, although the author prefers to apply this name to the more general expression of Eq. 4. Boundary resistances have been computed for broad ranges of ρ_m and s, and are available in the literature (Cheeke et al., 1976; Kaplan, 1979.)

It would seem a simple matter to experimentally measure the temperature difference $T_2 - T_1 = \Delta T$ resulting from a measured, applied heat flux $\dot{Q}$ across an interface, and to compare the ratio $\Delta T/\dot{Q}$ with the theoretical expression of Eq. 11. In practice one encounters subtleties which are common to many systems not in thermal equilibrium. Consider, for example, an imaginary interface between two identical materials. Experimentally one would expect $\Delta T = 0 = R_K$, but Eq. 11 predicts $R_K = (4b\ T^3)^{-1} \neq 0$. The origin of this apparent paradox was claimed (Simons, 1974; see also Phillips and Sheard, 1976) to lie in the use of an equilibrium phonon distribution for $N(\Omega,\Theta,\phi,T)$ in Eq. 5. If an equilibrium distribution of phonons existed in either material, no heat would flow in that material. To rectify this situation, N was instead approximated by

$$N(\Omega,\Theta,\phi,T) = N_o(\Omega,T)(4\pi)^{-1} + N'(\Omega)(4\pi)^{-1} \cos \Theta \qquad (12)$$

where $N_o(\Omega,T)$ is again the equilibrium distribution as in Eq. 5. The second term now provides a net phonon flux or heat flow, $\dot{Q} = (1/3) \sum_i \hbar s_i \int_0^\infty N'(\Omega)\, \Omega\, d\Omega$, in either material. Note that, instead of considering only those phonons striking the interface, the calculation now focuses on the total distribution of phonons within either material neighboring the boundary.

Substitution of Eq. 12 into Eq. 4 eventually gives (Katerberg et al., 1977)

$$\dot{Q} = b_1 \bar{\alpha}_{12} (1 - \tfrac{1}{2}\bar{\beta}_{12} - \tfrac{1}{2}\bar{\beta}_{21})^{-1} (T_1^4 - T_2^4) \qquad (13)$$

where $\bar{\beta}$ is an integral over angles similar to that in $\bar{\alpha}$ except for an additional factor of cos Θ. β ranges between 0 and 1 depending on the ratio of acoustic impedances of the two materials, with $\bar{\beta} \to 1$ for nearly identical materials. For $\Delta T/T << 1$

$$R_K = (1 - \tfrac{1}{2}\bar{\beta}_{12} - \tfrac{1}{2}\bar{\beta}_{21})(4\, b_1\, \bar{\alpha}_{12}\, T^3)^{-1}. \qquad (14)$$

Now when the two materials are identical, i.e. when $\bar{\alpha}_{12} = \bar{\beta}_{12} = \bar{\beta}_{21} = 1$, Eq. 14 gives $R_K = 0$ as expected. The paradox has been removed. A comparison of Eq. 14 with Eq. 11 is shown in Fig. 3 for a range of acoustic parameters.

While it may appear that Eq. 14 is a better approximation than Eq. 11, in actuality both equations give the same result *provided* the proper thermometers have been used. This statement can be understood qualitatively once it is recognized that the development of Eq. 11 emphasized the distribution of phonons incident on the interface, while the development of Eq. 14 emphasized the total distribution of phonons existing within either material close to the interface.

In order to be more quantitative in discussing the equivalence of Eqs. 11 and 14, we will use the simplified "experiment" shown in Fig. 4. Certain assumptions may at first appear ludicrous, but will be shown later in this chapter to correspond to real physical situations. The following discussion can also serve as an illustration of the problems and subtleties which can arise in a system not in thermal equilibrium. In Fig. 4 material 2 is assumed to be

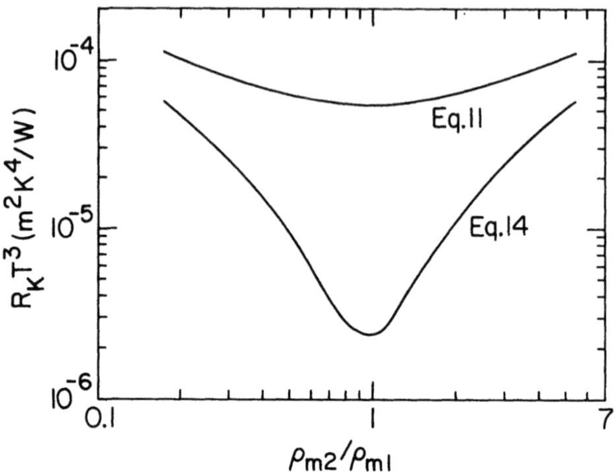

Fig. 3. Thermal boundary resistance calculated from Eqs. 11 or 14 as a function of the ratio of densities of the two materials in contact. Here $s_{2t}/s_{1t} = 1.1$ and $s_{\ell}/s_t = 2$.

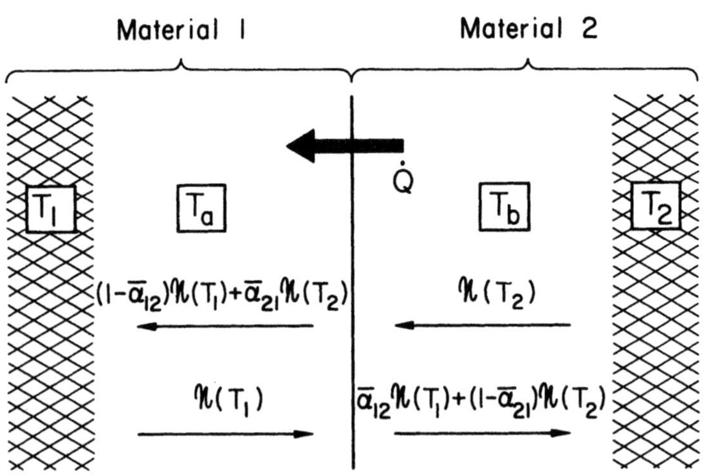

Fig. 4. Interface between materials 1 and 2 with a heat flux directed to the left. The $T_{1,a,b,2}$ represent both the thermometers and the temperatures recorded by those thermometers. No phonon scattering occurs between the interface and the cross-hatched regions.

at a higher temperature than material 1, and so heat flows from 2 to 1. The regions without cross-hatching represents regions free of phonon scattering except at the interface. The cross-hatched areas represent regions which are "black" in an optical sense, i.e. they spew out phonons having a distribution $N(T)$ characteristic of phonon equilibrium at temperature T_1 or T_2. These regions also absorb every phonon incident upon them. The thermometers T_1 and T_2 measure these temperatures. Phonons moving in material 1 toward the interface thus have a frequency distribution $N(T_1)$. Those traveling to the left in material 1 have a distribution $(1-\bar{\alpha}_{21})N(T_1) + \bar{\alpha}_{21}N(T_2)$, where the first term represents phonons reflected by the interface, and the second term represents phonons transmitted across the interface from material 2. For convenience we are taking $\bar{\alpha}_{12} = \bar{\alpha}_{21}$ (and $\bar{\beta}_{12} = \bar{\beta}_{21}$). This is valid as $\bar{\alpha} \rightarrow 1$, and it is in this perfect-acoustic-match limit that the two theories diverge as indicated in Fig. 3.

The thermometers T_a and T_b in Fig. 4 record the apparent temperatures in the regions without cross-hatching, the regions in which phonon scattering does not occur. In steady state, no net energy is exchanged with either of the thermometers, which are internally at temperatures T_a or T_b. Equating the energy exchange on the left-hand side of thermometer T_a with that on its right-hand face gives $T_a^4 - T_1^4 = \bar{\alpha}_{21}\, T_2^4 + (1 - \bar{\alpha}_{12})T_1^4 - T_a^4$ or

$$T_a^4 = T_1^4 + 0.5\,\bar{\alpha}_{12}\,(T_2^4 - T_1^4). \qquad (15)$$

Similarily for thermometer T_b

$$T_b^4 = T_2^4 + 0.5\,\bar{\alpha}_{12}\,(T_1^4 - T_2^4). \qquad (16)$$

We next compute the heat flux $\dot{Q}$ across the interface using both Eq. 9 and Eq. 13. In Eq. 9 we need to know the frequency distribution of phonons *incident* on the interface. From Fig. 4 this would be given by $N(T_1)$ and $N(T_2)$, and therefore thermometers T_1 and T_2 should be used. Thus from Eq. 9

$$\dot{Q}(\text{Eq. 9}) = b_1\,\bar{\alpha}_{12}(T_2^4 - T_1^4). \qquad (17)$$

For Eq. 13 we need to know the actual frequency and directional distribution of all phonons *within* either material, but near the interface. In Fig. 4, the apparent temperatures best representing these phonon distributions are recorded by thermometers T_a and T_b. Thus from Eq. 13, recalling that $\bar{\alpha} \approx \bar{\beta}$ for this problem,

$$\dot{Q}(\text{Eq. 13}) = b_1 \bar{\alpha}_{12}(1 - \bar{\alpha}_{12})^{-1} (T_b^{\,4} - T_a^{\,4}) \qquad (18)$$

To compare this with Eq. 17, the identities of Eqs. 16 and 17 may be substituted to give

$$\dot{Q}(\text{Eq. 13}) = b_1 \bar{\alpha}_{12}(T_2^{\,4} - T_1^{\,4}) \qquad (19)$$

which is identical to Eq. 17. In brief, Eqs. 9 and 13 (or Eqs. 11 and 14) give the *same* result provided the thermometers probe that phonon population which has been assumed by the theoretician in formulating his theory. This lesson must be remembered when comparing theory to experiment. It is also noted that the "paradox", that $R_K \neq 0$ when the two materials are identical, is in fact not a paradox nor a problem if thermometers are properly placed.

We now proceed to remove some of the assumptions made at the beginning of this section. If both materials were metallic, the heat flux carried by phonons would be overwhelmed by that carried by conduction electrons. Thus R_K is important only if one or both of the materials is nonmetallic. The presence of conduction electrons on one side of an interface can provide two additional mechanisms for heat exchange across that interface. These two mechanisms are related, but are conceptually different. Both mechanisms occur only if the metal has larger acoustic velocities than the dielectric with which it is in contact.

The first mechanism involves those phonons incident from the dielectric at angles beyond the critical angle ($\approx 43°$ in Fig. 2). All of these phonons are totally reflected *unless* the metal exhibits an acoustic attenuation Λ^{-1}. The "totally reflected" phonons do strain the metal to a depth of about one wavelength, and the presence of an attenuation can absorb the energy of the phonon. Conduction electrons provide a phonon attenuation, and so energy can be transmitted to the electrons from phonons incident on the interface at angles greater than the critical angle for the dielectric. In the opposite direction, a conduction electron can produce a phonon in the dielectric which leaves the interface at an angle greater than the critical angle.

The second mechanism is similar to the first in that it also involves phonons incident beyond the critical angle. Phonons incident at precisely the correct angle can excite an interface or Stonely wave which propagates along the boundary (Stonely, 1924; Ewing et al., 1957). For a free surface these waves are called Rayleigh waves which are utilized in surface-acoustic-wave (SAW) devices. The Stonely waves can only reexcite phonons in the dielectric unless, again, the second material is a metal in which case the interface waves may be absorbed by the conduction electrons. Thus a phonon from the dielectric is indirectly absorbed by an electron, the Stonely wave serving as an intermediate state in this process. We note that the coupling to attenuating media via a Stonely wave has been developed as a diagonistic tool in the field of applied ultrasonics (Becker and Richardson, 1970).

To incorporate the above two heat-exchange mechanisms into the acoustic mismatch theory requires only that attenuation be added to the plane wave of Eq. 1 (Peterson and Anderson, 1972, 1973). This is done most conveniently by using a complex velocity

$$s' = \Omega/q = s[1 + (j\lambda/4\pi\Lambda)]^{-1} \qquad (20)$$

where, in this expression, λ is the phonon wavelength. Application of the boundary conditions of continuum mechanics, as was done previously, automatically includes the two heat-exchange mechanisms when attenuation Λ^{-1} is present. Since the velocity is complex $\alpha(\Theta)$ is also complex and the computer computations become even more involved. A typical result is shown in Fig. 2. The effect of the first mechanism is seen in the small "foot" in curve b or c near 43°. The contribution of the Stonely wave is shown by the sharp, dashed peak in curve b near 48°. The curves of Fig. 2 were calculated using measured values of ρ and s. The calculation also used the measured phonon mean free path Λ resulting from the electron-phonon interaction in copper (Anderson and Peterson, 1972).

The decrease in R_K as a function of increasing phonon attenuation Λ^{-1} is shown in Fig. 5 for interfaces to several non-superconducting metals. Note that R_K is a function of the ratio λ/Λ (see Eq. 20). The greatest variation in R_K occurs near $\lambda = \Lambda$, reflecting the fact that the strain created by a totally reflected phonon penetrates a distance of only $\approx \lambda$ into the second medium for either of the two heat-exchange mechanisms. If the ratio λ/Λ were a function of either frequency or temperature, then R_K would no longer be proportional to T^{-3}. Note also that R_K is decreased by the two electron-attenuation mechanisms only because the conduction electron which absorbs the phonon rapidly carries that energy away from the interface. The thermal conductance of conduction electrons is large. This is not true for other sources of phonon attenuation such as that caused by lattice defects. Hence, as discussed later, lattice defects may actually appear to increase, rather than reduce, R_K.

Another approximation used in Eq. 11 was that $T < 10$ K. This permitted the integral over Ω in Eq. 4 to be extended to infinity,

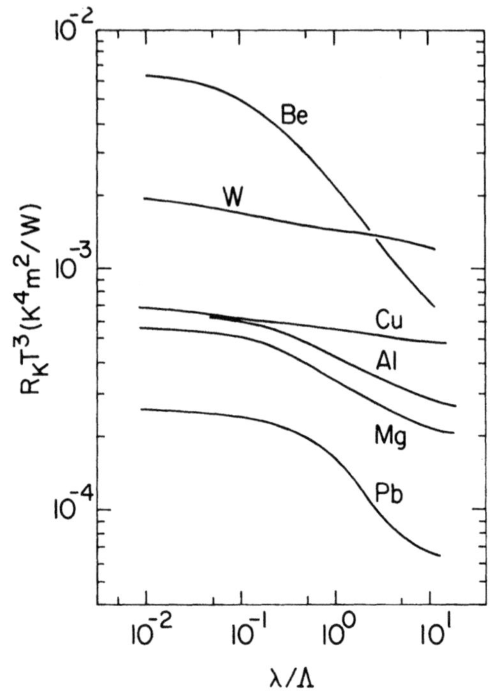

Fig. 5. The thermal boundary resistance of epoxy/metal interfaces as a function of attenuation Λ^{-1} of phonons of wavelength λ. In each case the attenuation of importance would be in the metal, except for Pb for which acoustic velocities are smaller than in the dielectric. The values of λ/Λ for electron-phonon interaction lie in the range 10^{-2} - 10^{-1}.

and the use of frequency-independent s_i. At higher temperatures the related phenomena of phonon dispersion and the cutoff in the phonon spectrum must both be included in the computation. Phonon dispersion occurs when the phonon wavelength approaches the size of atomic spacings. The phonon cutoff or highest phonon frequency occurs when the wavelength equals the atomic spacing. The cutoff is generally approximated by the Debye temperature, $\Theta_D \approx \hbar\, \Omega_{max}/k_B$. Both phenomena cause the product $R_K T^3$ to increase with increasing temperature (Weis, 1969).

Non-isotropic materials greatly complicate the computation of R_K, even though the magnitude of R_K is qualitatively the same as for two isotropic materials (Weis, 1979). In experimental measurements on non-isotropic materials the apparent magnitude of R_K could depend on the geometry of the experiment, that is, the results could depend on the placement of heaters and thermometers. This results from the phenomenon of phonons focussing in which the direction of propagation is not parallel to the phonon wave vector. The phonons are preferentially channeled in certain crystalline directions (Rosch and Weis, 1977). Phonon focusing is discussed in Chapter 4. For the present, we simply note that, once again, the experimentalist must carefully consider the arrangement of thermometers.

In deriving R_K it was assumed that phonons were scattered only by the interface. Defects in dielectric materials can cause phonon scattering and hence a gradient in the apparent phonon temperature within either material. This is troublesome only in the selection of a suitable thermometer arrangement which can probe the pertinent class of phonons. A more difficult problem can arise if the scattering depends on the phonon frequency. Low-frequency phonons usually have a much longer mean free path than high-frequency phonons. At temperatures below $\approx$ 10 K phonon-phonon interactions are weak, which means that little energy is exchanged between high- and low-frequency phonons. The net result is a situation in which high-frequency phonons experience a much larger "temperature" gradient near the interface, and their contribution to $\dot{Q}$ across the interface is small or even zero. The apparent R_K is therefore increased. The word apparent is used here since the increase in the measured $\Delta T/\dot{Q}$ comes about from phonon scattering very close to, but not at, the interface (Reynolds and Anderson, 1975). This behavior will be discussed in more detail below for particular experimental examples.

The final assumption used an ideal interface. It is possible to imagine many non-ideal boundaries, such as an incomplete or weak mechanical bond, a layer of some "dirt", or a microscopically rough or non-planar interface. Most non-ideal boundaries are difficult to treat theoretically. If the acoustic properties of any interleaving dirt layers are known, the acoustic-mismatch model can be extended to include these layers (Opsal and Pollack, 1974; Katerberg et al., 1977).

The "acoustic-mismatch" model has been much maligned in the literature, generally because Eq. 11 has been inappropriately applied to experimental situations for which the various simplifying assumptions were not fulfilled. We have seen in this section, however, that the model can be readily adapted to a broad range of physical situations.

OTHER CALCULATIONS

It is possible to establish a "quantum-mechanical" theory, involving phonon creation and annihilation operators, in a manner similar to that for the tunneling of electrons through metal/insulator/metal junctions (Narnhofer et al., 1970; Sheard et al., 1973). No new information has been obtained from these attempts except to indicate that phonon scattering processes which do not obey Snell's law (Eq. 2) at the interface might significantly reduce R_K. These processes allow phonons having incident angles larger than the critical angle (i.e. $\gtrsim 43°$ in Fig. 2) to partake in energy exchange across the interface. One such process has been discussed above, namely the reduction in R_K contributed from phonon attenuation by conduction electrons.

An alternative theoretical approach is to treat the mechanical response of individual atoms near the interface by solving their equations of motion (Steinbruchel, 1976; Lumpkin et al., 1978). These calculations should be of particular value at higher temperatures where the dominant phonons have wavelengths approaching interatomic distances, and phonon dispersion thus becomes an important factor. Unfortunately, these calculations have been carried out only for a one-dimensional chain of atoms. The results do display the role of dispersion in increasing the product R_KT^3 (Steinbruchel, 1976), and of the possible effect of anharmonicity in reducing R_KT^3 (Lumpkin et al., 1978).

A Boltzmann-equation approach has been used to calculate the spatial dependence of the apparent phonon temperatures within a metal near a metal/dielectric interface, and to determine the effect of this spatial dependence on R_K. This is an attempt to improve upon some of the approximations used in deriving Eq. 11. Apparently, for two solids in contact, any effect on R_K should be small (Levinson, 1977; Perrin, 1978), but these computations are rather difficult to analyze in physical terms.

COMPARISON OF EXPERIMENT AND THEORY

The first measurements on solid-solid interfaces were made in $\gtrsim$ 1955 to obtain engineering information directed to improving the construction of low-temperature apparatus (see Miedema, et. al., 1959, and papers cited therein). Since the measured thermal impedances often contained more than one interface plus the thermal resistances of wires, grease, etc., the results did not correlate well with Eq. 11 of the acoustic-mismatch theory. In later "engineering" measurements, the sample configurations were sufficiently simplified that a Kapitza resistance could be extracted. The measured resistance was $R_KT^3 \approx 10^{-3}$ K m^2/W (Anderson et al.,

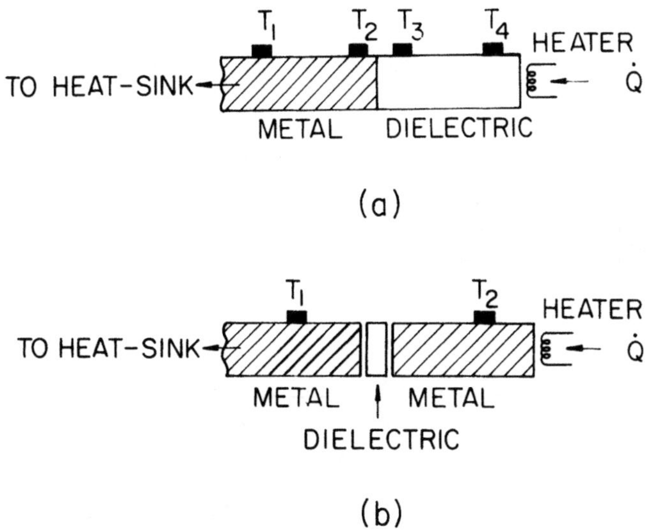

Fig. 6. Two sample arrangements for measuring R_K. Thermometers are represented by T.

1961; Anderson and Peterson, 1970). This is in rough agreement with what would be expected from Eq. 11. However, the acoustic velocities of the engineering materials were not known accurately, which precludes a detailed comparison with theory.

It may be noted that some early papers calculated incorrect or very approximate values of R_K from Eq. 11. The younger reader is reminded that digital computers were, at that time, expensive to use. Numerical integrations were often done with a small density of integration points and thus could miss some of the structure in $\bar{\alpha}$ shown in Fig. 2.

The first measurements explicitly designed to obtain precise Kapitza-resistance data used the arrangement shown in Fig. 6a. The two thermometers on each material allowed temperature gradients to be measured. These gradients were then extrapolated to the interface to obtain the temperatures on the two sides of the boundary. The dielectric could be either a nonmetallic single crystal or a metal in the superconducting state. Early measurements on metal/superconducting interfaces between lead and copper, between indium and copper or between tin and copper gave an R_K only within a factor of ≈ 2 of the expected magnitude and a temperature dependence less than the T^{-3} behavior predicted by Eq. 11 (Barnes and Dillinger, 1966; Suomi et al., 1968; see also

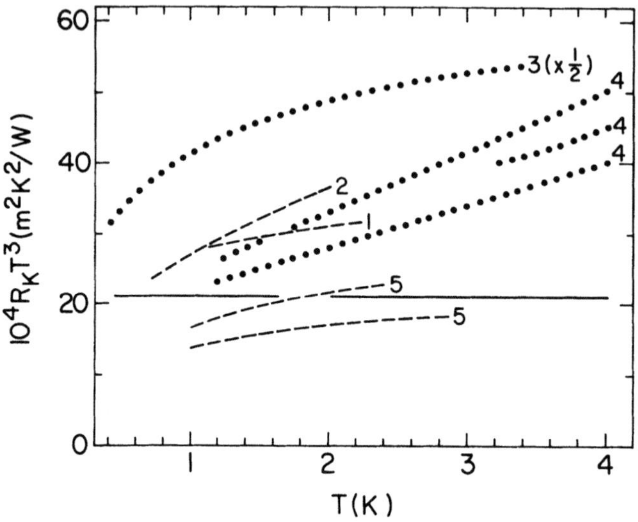

Fig. 7. Measurements of $R_K T^3$ between sapphire and indium in the normal state. 1-Neeper and Dillinger (1964); 2-Park and Narahara (1971); 3-Wolfmeyer et al. (1970) (multiplied by 1/2 to bring the data on scale); 4,5-Schmidt and Umlauf (1976). The solid line is the theory, Eq. 11. The dashed lines represent interfaces made by ultrasonic soldering and casting, the dotted lines by vacuum evaporation and casting.

Schumann et al., 1980). These results do not provide a satisfactory confirmation of the theory. This type of measurement has two inherent problems. First, the interface is undoubtedly non-ideal. Some interdiffusion of the metals probably occurs during the casting process. Second, the extrapolation of the temperature gradient from thermometers T_3 and T_4 of Fig. 6 will give an incorrect temperature at the boundary of a superconductor if there is a greater concentration of phonon-scattering defects near the interface. Dislocations scatter phonons strongly, and it is likely that a larger density of dislocations does occur near the boundary (O'Hara and Anderson, 1974).

These two problems were partly avoided by using metal/sapphire interfaces. Boundaries between indium and sapphire have been investigated by four groups; the results are shown in Fig. 7. Again, the magnitudes (with the exception of curve 3) lie only within a factor of $\simeq$ 2 of the value predicted by Eq. 11, and the temperature dependence is smaller than the T^{-3} expected from the acoustic-mismatch model. Similar measurements were attempted in the author's laboratory using the same techniques (vapor deposition and ultrasonic soldering) for bonding the indium to sapphire. The single-crystal sapphire was in the form of a thin plate, which permitted a visual check of the interface after cycling the sample slowly to liquid nitrogen temperatures. In every case, the interface was observed to have been damaged by differential thermal expansion. These problems may also have occurred for the work represented by Fig. 7.

The availability of ^{3}He - ^{4}He dilution refrigerators has made it a simple matter to extend R_K measurements to temperatures well below 0.1 K. The reduction in temperature has two advantages. First, the dominant phonons in thermal transport have an average frequency of $\Omega/2\pi \simeq 10^{11}T$ Hz K^{-1}. Thus at low temperatures the phonons involved in the Kapitza resistance have longer wavelengths more appropriate to the continuum mechanics utilized in developing the acoustic-mismatch theory. Second, phonon scattering mechanisms generally decrease in magnitude with increasing phonon wavelength (decreasing T) and hence, phonon scattering within the bulk of a material becomes less important relative to scattering at the interface. Thus, at low temperatures one can measure R_K even if one material is amorphous. Amorphous materials at T > 1 K exhibit very strong internal phonon scattering.

An arrangement appropriate for low-temperature measurements is shown schematically in Fig. 6b. The dielectric is sandwiched between two blocks of metal having large thermal conductivity. Each block is therefore sufficiently isothermal at low temperatures that only one thermometer is needed on each. Any phonon scattering occurring within the amorphous dielectric is effectively eliminated by extrapolation of the total measured thermal impedance to T = 0 where only scattering by the interfaces dominates. This extrapolation procedure is demonstrated in Fig. 8 for three sandwiches of magnesium/dielectric/magnesium having different dielectric thicknesses. The extrapolation to T = 0 is independent of the dielectric thickness; the strong phonon scattering which takes place within the amorphous dielectric has been effectively avoided.

The thin layer of dielectric causes one complication at low temperatures. When the phonon mean free path in the dielectric exceeds the thickness of the dielectric layer, it is necessary to calculate the net probability for phonon transmission across two parallel interfaces (Katerberg et al., 1977). The result is that the total thermal impedance across the sandwich (at $T \to 0$) is $(2 - \bar{\alpha})R_K$ rather than 2 R_K as expected for the sum of two isolated interfaces.

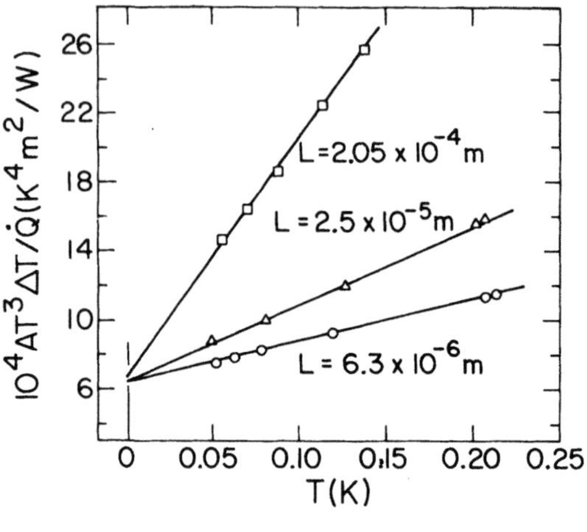

Fig. 8. Total thermal impedance across a Mg/dielectric/Mg sandwich as a function of temperature, for three thicknesses of dielectric. In these measurements the dielectric was Mylar bonded to the Mg with epoxy, but the same results are obtained when only epoxy is present since the low temperature properties of the Mylar and epoxy are so similar (Peterson and Anderson, 1973).

The results obtained from the $T \rightarrow 0$ extrapolation are shown in Fig. 9 as a function of the ratio of the acoustic impedance of the metal to that of the dielectric. The experimental results $(R_K T^3)_E$ have been divided by the value $(R_K T^3)_T$ calculated from theory. Hence the horizontal line indicates perfect agreement between theory and experiment. For points represented by x, $(R_K T^3)_T$ has been calculated assuming that $\bar{\alpha} = 1$, i.e. assuming no phonon scattering at the interface. This has often been called the radiation-limit, and serves to gauge the importance of the Kapitza resistance in reducing the heat flux across the interface. It may be observed from Fig. 9 that scattering at the interface does, in fact, become larger as the acoustic mismatch becomes larger, since the points marked by x deviate more from the solid line on the right of the figure than on the left.

For points represented by the filled circles and squares in Fig. 9, $(R_K T^3)_T$ has been calculated from Eq. 11 using no adjustable parameters. The calculation includes the effect of the electronic attenuation of acoustic waves in the metals. Were this electronic effect to be omitted, the data points would lie 10-20% lower. That is, agreement with the acoustic-mismatch model would be 10-20%

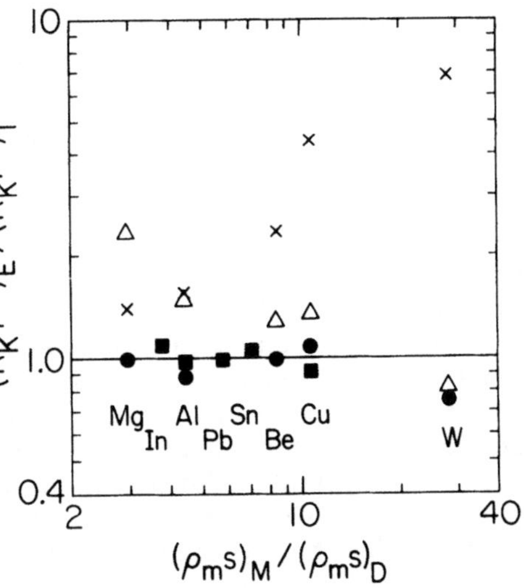

Fig. 9. A comparison of measured and calculated values of R_KT^3 for several dielectric/metal interfaces. See text for details, ●, x, ▲-Peterson and Anderson (1972); ■-O'Hara and Anderson (1974).

less satisfactory. Other than measurements at ultrasonic frequencies, this is the only available evidence that phonon attenuation near an interface does enhance $\dot{Q}$.

The filled circles and squares of Fig. 9 represent two independent measurements. Since R_K can now be measured reproducibly to an accuracy of ≈ 10%, it is worth inquiring whether Eq. 11 or Eq. 14 provides the more appropriate representation of the data. Recall that the difference between Eqs. 11 and 14 was the greatest when the two materials were most identical (see Fig. 13). Hence, in Fig. 9, the predictions of the two equations should show the greatest difference on the left side of the figure. This is in fact true. For points represented by the triangles, $(R_KT^3)_T$ was computed using Eq. 14 with no adjustable parameters. Note that the filled symbols fall closer to the horizontal line by a factor of up to 2.4. Equation 11 provides better agreement with these experimental data than Eq. 14. This implies that the thermometers used in these measurements probe a phonon spectrum which is the same as that striking the interface. They do not probe the total population of phonons both approaching and leaving the interface. Indeed this is true as may be seen from Fig. 6b. The thermometers were located a distance of ≳ 3 x 10^{-3} m from the interface, a distance much larger than the phonon mean free path in copper ($\approx 10^{-4}$ m). Thus the thermometers never detect phonons leaving the interface. Each thermometer detects only phonons created by the conduction electrons, the conduction electrons produce the same spectrum of phonons throughout the copper block, and hence this is the same spectrum of phonons which is incident on the interface.

We have seen that the acoustic mismatch model agrees remarkably well with experimental data in the limit $T \to 0$. Under appropriate experimental conditions, the model can also be tested definitively at finite temperatures. The most successful arrangement has been to measure R_K at the interface between a thin metallic film and a large dielectric single crystal free of internal defects (see Martinon and Weis, 1979, and references cited therein). The thin metallic film avoids the problems caused by differential thermal expansion as occurred for the metal/sapphire arrangement of Fig. 6a and may have been responsible for the large $R_K T^3$ values of curve 3 of Fig. 7. The large single crystal provides a long phonon mean free path limited only by the size of the crystal. If electrical current and thus heat is supplied to the metal film as a sufficiently short pulse, the "hot" phonons characterized by a temperature T_H will propagate away from the interface into the crystal and not return, via scattering, during the pulse interval. Phonons incident on the interface from the crystal continue to be characteristic of the initial temperature T_C before the pulse was applied. The temperature T_H of the heater is not measured with a thermometer, as the response time of most thermometers would be too slow. Instead, the electrical resistance of the film is deduced from the size of the voltage pulse reflected from the film back into the electrical transmission line. From the measured resistance, the temperature T_H is obtained from a predetermined dependence of electrical resistance on temperature. From the measured resistance and voltage, $\dot{Q}$ is calculated. The resistance of the film is temperature dependent only above $\approx$ 10 K, hence this technique is useful only at $T \gtrsim 10$ K.

Since generally $T_H \gg T_C$ in these pulse measurements, the data should be compared with Eq. 9 rather than Eq. 11. The experimental results for a lead/sapphire interface are shown in Fig. 10 as the dotted line. The agreement with Eq. 9 (curve AM) is good to temperatures up to $T_H \approx 40$ K. If T_H is pushed to higher temperatures, however, the data tend to fall above the acoustic-mismatch prediction. This is just the effect expected from phonon dispersion and a cutoff in the phonon spectrum. On the other hand, it is difficult to separate these intrinsic effects from the spurious effect of phonons scattered back to the metal film by defects, impurities, or isotopic mass differences lying within the crystal but close to the interface. This problem will be addressed in greater detail below.

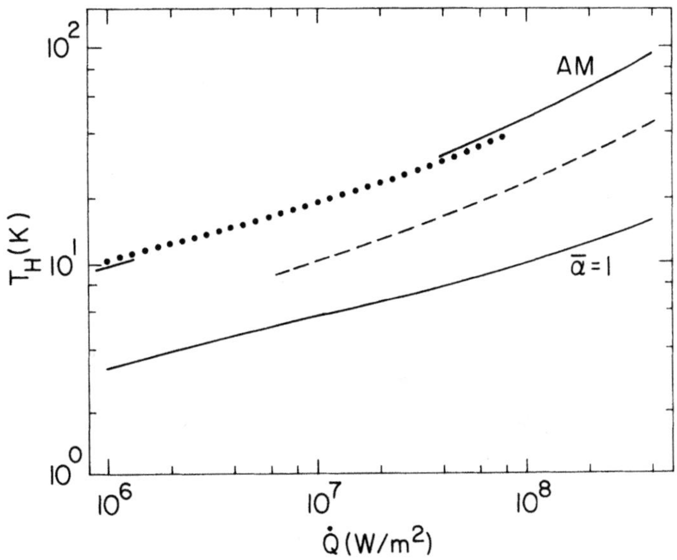

Fig. 10. The temperature T_H of a Pb film on sapphire, from Herth and Weis (1970). (AM)-prediction of acoustic-mismatch model; ($\bar{\alpha}$=1)-acoustic-mismatch model with $\bar{\alpha}$ set = 1, shown for reference; (●●●)-surface cleaned with glow-discharge; (---)-surface not carefully cleaned prior to evaporation of the Pb. The temperature T_C of the sapphire was ≃ 1 K.

Another potential problem of techniques involving thin-film heaters is that the metal film may produce phonons in the crystal which are not characteristic of the mean conduction-electron temperature T_H (Perrin and Budd, 1972). That is, $N(\Omega,\Theta,\phi,T)\neq N_o(\Omega,T_H)/4\pi$ in Eq. 4. This situation can occur because the mean free path of phonons increases with decreasing phonon frequency (Anderson and Peterson, 1972). Only if the film thickness is greater than the phonon mean free path for all frequencies can the emitted phonon distribution be characteristic of the temperature T_H of the electrons. This condition is relaxed somewhat by phonon reflections occurring at the interface. The thickness should be ≈ 1000 Å for T_H ≃ 10 K. A skewing of the emitted phonon spectrum toward higher energy phonons has been reported by Murman and Heber (1977) for effective phonon temperatures of ≈ 30 K.

With each of the experimental techniques discussed above, there arose the possibility of a problem associated with the scattering of phonons within either material but close to the interface. For convenience we refer to this phenomenon as phonon "backscattering". The most severe case of backscattering occurs

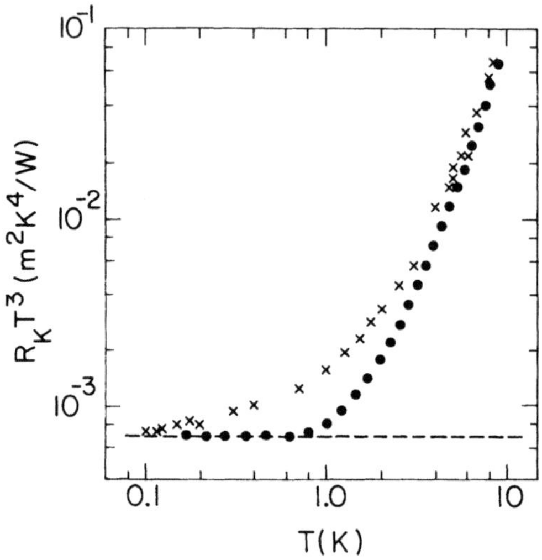

Fig. 11. Apparent $R_K T^3$ of an epoxy/copper interface. (X)-data of Matsumoto et al. (1977); (---)- Eq. 11; (●●●)-calculation as described in text. Similar results have been obtained for other epoxy/metal interfaces by de Araujo and Rosenberg (1975) and by Schmidt (1977), and on an epoxy/sapphire interface by Schmidt (1974).

for amorphous dielectrics, for which R_K can be obtained precisely only by extrapolation to T = 0 as in Fig. 8. If the measurements are instead extended to higher temperatures using the arrangement of Fig. 6b, the results shown by the points marked by x in Fig. 11 are obtained. The bulk thermal impedance d/K of the dielectric has been subtracted, where d is the dielectric thickness and K its thermal conductivity. At 10 K, the experimental data of Fig. 11 deviate by a factor of 100 from Eq. 11 (the dashed line). The explanation lies in the highly frequency-dependent phonon mean free path of this amorphous dielectric, which is shown in Fig. 12. For convenience we approximate the abrupt drop in Λ of Fig. 12 by a vertical line at $\Omega = 8 \times 10^{11}$ sec^{-1}. Phonons of higher frequency cannot propagate away from the boundary into (or from) the epoxy. This effect can be approximated in Eq. 4 by truncating the integral over Ω at $\Omega = 8 \times 10^{11}$ sec^{-1}. The result of this computation is shown by the dotted curve of Fig. 11. Agreement between data and theory is further improved by a more realistic approximation of $\Lambda(\Omega)$ from Fig. 12.

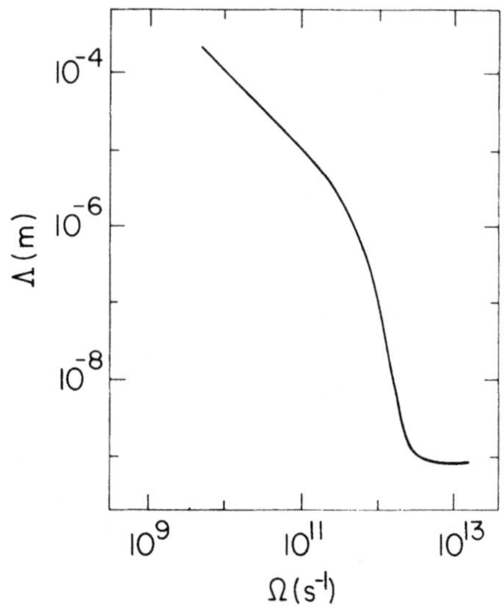

Fig. 12. Frequency dependence of the phonon mean free path in that epoxy used to obtain the data of Fig. 11 (Matsumoto et al., 1977). This frequency dependence is characteristic of amorphous materials.

An equivalent explanation of Fig. 11 is that phonons of $\Omega \gtrsim 8 \times 10^{11}$ sec^{-1} in the amorphous epoxy have a "temperature" at the interface equal to that of the conduction electrons, and a "temperature" gradient away from the interface with a characteristic relaxation length of $\lesssim 10^{-7}$m. Hence the entire "temperature" drop occurs so close to the interface that it cannot be distinguished from a drop ΔT across the interface. Phonons of $\Omega \lesssim 10^{11}$ sec^{-1} have mean free paths similar to the dimensions of the dielectric sample, and the acoustic-mismatch concept can be applied. Note therefore that the phonons in the epoxy next to the interface have a "temperature" of T_H if $\Omega \gtrsim 8 \times 10^{11}$ sec^{-1}, but a "temperature" of $T_H \pm \Delta T$ if $\Omega \lesssim 10^{11}$ sec^{-1}. The phonons in these two frequency regimes are far from thermal equilibrium with each other.

Since the enhanced R_K at temperatures above 1 K in Fig. 11 is related to the phonon mean free path in the amorphous epoxy, R_K at $T > 1$ K should be influenced little by the metal on the opposite side of the interface. This is indeed the behavior observed for a number of normal metals ranging from lead to aluminum (de Araujo and Rosenberg, 1975).

The backscattering of phonons in amorphous materials is probably the most extreme example ever observed. Strong backscattering has apparently been observed at effective temperatures of $\gtrsim$ 20 K in crystals containing large concentrations of intentionally induced defects (Cheeke and Martinon, 1972), and even in good-quality crystals containing isotopic or other defects (Martinon and Weis, 1979; Bron et al., 1979).

Due to the strong scattering of phonons in amorphous materials the thermal conductivity is very small compared to that of the crystalline state. To increase the net thermal conductivity of epoxies, greases and other engineering materials, metallic or other powders are frequently added. With decreasing temperature, however, the thermal boundary resistance at the surface of each particle becomes sufficiently large that the particle is progressively excluded from the thermal path even if the particle is metallic. The particles start to behave like cavities at temperatures below roughly 1 K (Anderson and Rauch, 1970; Garrett and Rosenberg, 1974; Schmidt, 1975). Hence the net thermal conductivity of the composite can be smaller than that of the unfilled amorphous material.

It was mentioned above that no direct experimental evidence has been obtained concerning the role, in R_K, of the electronic attenuation of phonons. However, a number of measurements have been attempted. Generally these have involved superconducting metals. R_K is measured as the metal is transformed from the superconducting to the normal state. This phase transition creates normal conduction electrons and thus turns on the two phonon-attenuation mechanisms of heat exchange. For the measurements on sapphire/indium interfaces of Fig. 7, a decrease in R_K of 10-30% was observed during the superconducting-normal transition. This would be the effect expected from the acoustic-mismatch theory when conduction electrons enhance the transfer of heat across the interface (see Fig. 5). However, this effect should occur only when the metal is the material having the larger acoustic velocities, which is not true for the indium/sapphire system. Hence for this system the acoustic-mismatch model predicts no change in R_K in a normal-superconducting transition, contrary to observation.

The electronic attenuation mechanism should be present for a niobium/epoxy interface. However here it was observed that R_K increases by $\simeq$ 10% when the niobium becomes normal (Schmidt, 1977), the opposite effect from that predicted by the acoustic-mismatch model. Indeed, O'Hara and Anderson (1974) observed both increases *and* decreases in R_K with a normal-superconducting transition, depending on the temperature of the measurement. A theoretical paper (Zelikman and Spivak, 1979) purports to show agreement with these data. However, it must be remembered that the metals in all these measurements are probably strained, that dislocations strongly scatter phonons, and that the determination of an appropriate phonon temperature for the metal in the superconducting state is therefore extremely difficult (O'Hara and Anderson, 1974). It is the author's opinion that the role of conduction electrons in R_K has yet to be demonstrated in a definitive manner.

Another question, which has not been studied in detail, is the possibility of manipulating R_K. The net thermal impedance between two materials might be decreased if a third layer, having intermediate acoustic properties, were inserted between the two. The acoustic-mismatch model may be extended readily to this situation (Opsal and Pollack, 1974). The author is not aware of any systematic experimental investigation of this procedure. (See Chapter 4 by Narayanamurti for related measurements.) However, the effect may be evident in Fig. 10 for a lead/sapphire interface. The dotted curve was obtained for an interface carefully cleaned by a glow-discharge prior to deposition of the lead film. The dashed curve was obtained for an interface not cleaned so carefully. Hence the latter boundary may contain a layer of contamination which has reduced R_K by a factor of $\approx$ 16. It has also been reported (Clarke and Hsiang, 1976) that the R_K between tin films and glass or sapphire is reduced by an intermediate layer of aluminum. In this case, however, the enhanced thermal contact may stem from an improved physical or mechanical bond between the tin film and the substrate.

The question in thermal boundary resistance which has been studied most extensively is depicted in Fig. 13. As the acoustic mismatch between two solid materials becomes larger, the deviation between data and theory also becomes larger. The deviation becomes even larger for liquid ^{3}He or ^{4}He. The reason for this failure of the theory is not yet known. A.F.G. Wyatt discusses the problem in detail in Chapter 2 for liquid helium. Here we summarize only the information available on solid/solid interfaces.

As the ratio of acoustic impedances for the two materials becomes larger, the critical angle for total reflection of incident phonons becomes smaller and $\bar{\alpha}$ therefore becomes smaller. In brief, the energy transmitted across the interface by the acoustic-mismatch mechanism, which is contained in Eq. 11, becomes smaller. If a second mechanism of heat transfer is present, it could become progressively more important in the total heat transfer as the ratio of acoustic impedances increases. Indeed there is some evidence from solid argon/sapphire boundaries that this second mechanism is related to phonons which are *not* scattered at the interface in accordance with the Snell's law of Eq. 2 (Taborek and Goodstein, 1980), and therefore have not been included in the derivation of Eq. 11. There is also evidence from solid helium/

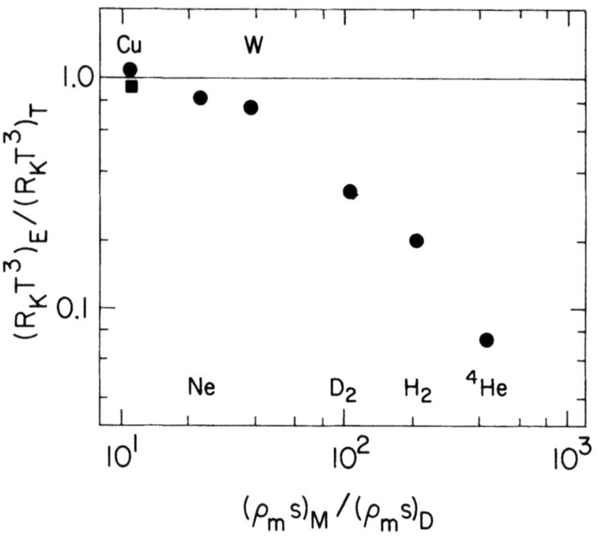

Fig. 13. Experimental data on the boundary resistance between two solids, $(R_K T^3)_E$, compared to the theoretical prediction of Eq. 11, $(R_K T^3)_T$, as in Fig. 9. The Cu/epoxy and the W/epoxy points have been taken from Fig. 9 for comparison. All the solidified-gas interfaces were with Cu in the arrangement of Fig. 6b. (Reynolds and Anderson, 1977).

copper interfaces that this second mechanism decreases in relative importance as temperature is reduced (Folinsbee and Anderson, 1973) This may be related to the increasing wavelength (or decreasing frequency) of the dominant phonons. An understanding of this second mechanism of heat transfer would be beneficial to the goal of controlling R_K in engineering applications.

In summary it may stated that the acoustic-mismatch *concept* has been very successful in accounting for the transport of heat between two different materials, at least one of which is not metallic. The assumptions used in the acoustic-mismatch calculations must be well matched to the experimental situation. The acoustic-mismatch concept does fail to predict the thermal impedance between two materials of very different acoustic properties, where an additional mechanism of heat transfer appears to become dominant. This additional mechanism is the topic of the following chapter.

REFERENCES

Anderson, A. C., and Rauch, R. B., 1970, Low-Temperature Thermal Conductivity of a Suspension of Copper Particles, J. Appl. Phys., 41:3648.

Anderson, A. C., and Peterson, R. E., 1970, Selection of a Thermal Bonding Agent for Temperature Below 1 K, Cryogenics, 10:430.

Anderson, A. C., Salinger, G. L., and Wheatley, J. C., 1961, Transfer of Heat Below 0.15°K, Rev. Sci. Instrum., 32:1110.

Anderson, A. C., and Peterson, R. E., 1972, The Thermal Resistance Between Electrons and Phonons in Copper, Phys. Lett., 38A:519.

Barnes, L. J., and Dillinger, J. R., 1966, Thermal Resistivity at Pb-Cu and Sn-Cu Interfaces between 1.3 and 2.1°K, Phys. Rev., 141:615.

Becker, F. L., and Richardson, R. L., 1970, Ultrasonic Critical Angle Reflectivity, in "Research Techniques in Nondestructive Testing", R. S. Sharp, ed., Academic, New York, Vol. I, p. 91.

Bron, W. E., Patel, J. L., and Schaich, W. L., 1979, Transport of Phonons into Diffusive Media, Phys. Rev. B, 20:5394.

Cheeke, J.D.N. and Martinon, C., 1972, Influence of Surface Defects on High Frequency Phonon Generation, Solid State Commun., 11:1771.

Cheeke, J.D.N., Ettinger, H., and Hebral, B., 1976, Analysis of Heat Transfer Between Solids at Low Temperatures, Can. J. Phys., 54:1749.

Clarke, J., and Hsiang, T., 1976, Low-Frequency Noise in Tin and Lead Films at the Superconducting Transition, Phys. Rev. B, 13:4790.

de Araujo, F.F.T., and Rosenberg, H. M., 1975, The Thermal Boundary Resistance at Epoxy-Resin/Metal Interfaces at Liquid Helium-Temperatures, in "Phonon Scattering in Solids," L. J. Challis, V. M. Rampton and A.F.G. Wyatt, ed., Plenum, New York, p. 43.

Ewing, W. M., Jardetsky, W. S., and Press, F., 1957, "Elastic Waves in Layered Media," McGraw, New York.

Folinsbee, J. T., and Anderson, A. C., 1973, Anomalous Kapitza Resistance to Solid Helium, Phys. Rev. Lett., 31:1580.

Garrett, K. W., and Rosenberg, H. M., 1974, The Thermal Conductivity of Epoxy-Resin/Powder Composite Materials, J. Phys. D, 7:1247.

Herth, P., and Weis, O., 1970, Radiation Temperature of Thermal Phonon Radiators, Z. Angew. Phys., 29:102.

Kapitza, P. L., 1941, The Study of Heat Transfer in Helium II, J. Phys. USSR, 4:181.

Kaplan, S. B., 1979, Acoustic Matching of Superconducting Films to Substrates, J. Low Temp. Phys., 37:343.

Katerberg, J. A., Reynolds, C. L., and Anderson, A. C., 1977, Calculations of the Thermal Boundary Resistance, Phys. Rev. B, 16:673.

Kolsky, H., 1953, "Stress Waves in Solids," Oxford, London.

Levinson, I. B., 1977, Boundary Conditions in Phonon Hydrodynamics, Sov. Phys. JETP, 46:165.

Little, W. A., 1959, The Transport of Heat Between Dissimilar Solids at Low Temperatures, Can. J. Phys., 37:334.

Lumpkin, M. E., Saslow, W. M., and Visscher, W. M., 1978, One-Dimensional Kapitza Conductance: Comparison of the Phonon Mismatch Theory with Computer Experiments, Phys. Rev. B, 17:4295.

Martinon, C., and Weis, O., 1979, α-Quartz as a Substrate in Thermal Phonon Radiation, Z. Phys. B, 32:259.

Matsumoto, D. S., Reynolds, C. L., and Anderson, A. C., 1977, Thermal Boundary Resistance at Metal-Epoxy Interfaces, Phys. Rev. B, 16:3303.

Miedema, A. R., Postma, H., van der Vlugt, N. J., and Steenland, M. J., 1959, Some Experiments on Heat Transfer Below 1°K, Physica, 25:509.

Murmann, H., and Heber, J., 1977, The Phonon Spectrum of High Energy Heat Pulses Generated in Thin Metal Films, Z. Phys. B, 26:137.

Narnhofer, H., Thirring, W., and Sexl, R., 1970, On the Theory of Interfacial Conductivity, Ann. Phys. (N.Y.), 57:350.

Neeper, D. A., and Dillinger, J. R., 1964, Thermal Resistance at Indium-Sapphire Boundaries between 1.1 and 2.1°K, Phys. Rev., 135:A1028.

O'Hara, S. G., and Anderson, A. C., 1974, Thermal Impedance Across Metallic and Superconducting Foils Below 1K, J. Phys. Chem. Solids, 35:1677.

Opsal, J. L., and Pollack, G. L., 1974, Improved Calculations of the Kapitza Resistance: Combined Effects of Phonon Attenuation and Impedance Matching on Kapitza Resistance, Phys. Rev. A, 9:2227.

Park, B. S., and Narahara, Y., 1971, Kapitza Resistance between Dielectrics and Metals in the Normal and Superconducting States, J. Phys. Soc. Jap., 30:760.

Perrin, N., 1978, Determination of the Thermal Boundary Resistance in the Transport Approach, J. Low Temp. Phys., 31:257.

Perrin, N., and Budd, H., 1972, Phonon Generation by Joule Heating in Metal Films, Phys. Rev. Lett., 28:1701.

Peterson, R. E., and Anderson, A. C., 1972, The Transport of Heat Between Solids at Low Temperature, Solid State Commun., 10:891.

Peterson, R. E., and Anderson, A. C., 1973, The Kapitza Thermal Boundary Resistance, J. Low Temp. Phys., 11:639.

Phillips, M. C., and Sheard, F. W., 1976, Path-Integral Formulation of the Theory of Thermal Boundary Resistance, in "Phonon Scattering in Solids," L. J. Challis, V. W. Rampton and A.F.C. Wyatt, ed., Plenum, New York, p. 24.

Pollack, G. L., 1969, Kapitza Resistance, Rev. Mod. Phys., 41:48.

Reynolds, C. L., and Anderson, A. C., 1975, Thermal Boundary Resistance to Noncrystalline Dielectrics, J. Low Temp. Phys., 21:641.

Reynolds, C. L., and Anderson, A. C., 1977, Thermal Boundary Resistance to Solid Helium, Hydrogen, Deuterium, and Neon. II, Phys. Rev. B, 15:5466.

Rosch, F., and Weis, O., 1977, Phonon Transmission from Incoherent Radiators into Quartz, Sapphire, Diamond, Silicon and Germanium within Anisotropic Continuum Acoustics, Z. Phys. B, 27:33.

Schmidt, C., 1975, Influence of the Kapitza Resistance on the Thermal Conductivity of Filled Epoxies, Cryogenics, 15:17.

Schmidt, C., 1977, Thermal Boundary (Kapitza) Resistance at Niobium-Epoxy Interfaces in the Superconducting and Normal States, Phys. Rev. B, 15:4187.

Schmidt, C., 1974, Thermal Boundary Resistance at Interfaces Between Two Dielectrics, Phys. Lett., 50A:241.

Schmidt, C., and Umlauf, E., 1976, Thermal Boundary Resistance at Interfaces Between Sapphire and Indium, J. Low Temp. Phys., 22:597.

Schumann, B., Netsche, F., and Paasch, G., 1980, Thermal Conductance of Metal Interfaces at Low Temperatures, J. Low Temp. Phys., 38:167.

Sheard, F. W., Bowley, R. M., and Toombs, G. A., 1973, Microscopic Theory of the Kapitza Resistance at a Solid-Liquid ^{4}He Interface, Phys. Rev. A, 8:3135.

Simons, S., 1974, On the Thermal Contact Resistance Between Insulators, J. Phys. C, 7:4048.

Steinbruchel, C., 1976, The Scattering of Phonons of Arbitrary Wavelength at a Solid-Solid Interface: Model Calculation and Applications, Z. Phys. B, 24:293.

Stonely, R., 1924, Elastic Waves at the Surface of Separation of Two Solids, Proc. Roy Soc., A106:416.

Suomi, M., Anderson, A. C., and Holmstrom, B., 1968, Heat Transfer Below 0.2°K, Physica, 38:67.

Taborek, P., and Goodstein, D. L., 1980, The Anomalous Kapitza Conductance is not a Quantum Effect, Bull. Am. Phys. Soc., 25:407.

Weis, O., 1969, Thermal Phonon Radiation, Z. Angew, Phys., 26:325.

Weis, O., 1979, Reflection, Transmission and Mode Conversion of Plane Sound Waves at a Plane Interface Between Two Different Anisotropic Solids, Z. Phys. B, 34:55.

Wolfmeyer, M. W., Fox, G. T., and Dillinger, J. R., 1970, An Electron Contribution to the Thermal Conduction Across a Metal-Solid Dielectric Interface, Phys. Lett., 31A: 401.

Zelikman, M. A., and Spivak, B. Z., 1979, Role of Slow Relaxation Processes in the Formation of the Kapitza Jump on the Boundary Between a Superconductor and a Dielectric, Sov. Phys. JETP, 49:377.

ACKNOWLEDGEMENT

The author's research on the thermal boundary resistance has been supported by the U.S. Department of Energy under Contract DE-AC02-76ER01198.

KAPITZA CONDUCTANCE OF SOLID-LIQUID He INTERFACES

A F G Wyatt

Department of Physics
University of Exeter
Exeter UK

INTRODUCTION

The Kapitza conductance problem is concerned with phonon transmission between solids and light atoms in both directions through the interface. The heat transfer across solid-solid interfaces is covered by Anderson in chapter 1. The magnitudes of the conduction between different solids can be well understood in terms of classical elasticity however when ^3He, ^4He, H_2 or D_2 is one of the materials the heat transfer is much greater than that predicted by classical theory. We shall tend to concentrate on ^4He as there is more information available for this material than the others. We shall see that classical transmission does occur with liquid ^4He but that there is another channel in parallel with it which carries most of the heat for temperatures $\gtrsim 0.05K$. Although many of the characteristics of this non-classical channel have been measured the microscopic process still remains a mystery. The subject has been reviewed by Pollack (1969) Snyder (1970) Challis (1974, 1975) Anderson (1976) Kinder et al. (1979) and Wyatt (1979).

The measurements of Kapitza conductance were for many years made by passing a heat current ($\dot{Q}$) down a rod which is in a vacuum except for the end face which is in liquid ^4He. A temperature discontinuity (ΔT) is found across the interface and the Kapitza conductance (Y_K) defined as

$$\dot{Q} = Y_K \, A \, \Delta T$$

where A is the area of the interface. ΔT is kept small compared with the ambient temperature and so the temperature dependence of Y_K can be measured. Clasically we expect $Y_K \propto T^3$ as it just increases with the equilibrium density of phonons, however it is usually found to depend on a higher power of T. Furthermore the magnitude of Y_K is always much greater than the classical value. To show the temperature dependence deviating from T^3 the product $R_K T^3$, where R_K is the Kapitza resistance ($R_K \equiv Y_K^{-1}$), is plotted against temperature. Results for ^3He and ^4He, solid and liquid, measured by Folinsbee and Anderson (1973) are shown in Figure 1. The large change in $R_K T^3$ between 0.1K and 1K can be seen and it is also clear that the measured values are all much less than the acoustic model predictions even at 0.1K. However the elastic properties of the light materials are evident from the fact that the order of the measured values is the same as that of the acoustic model.

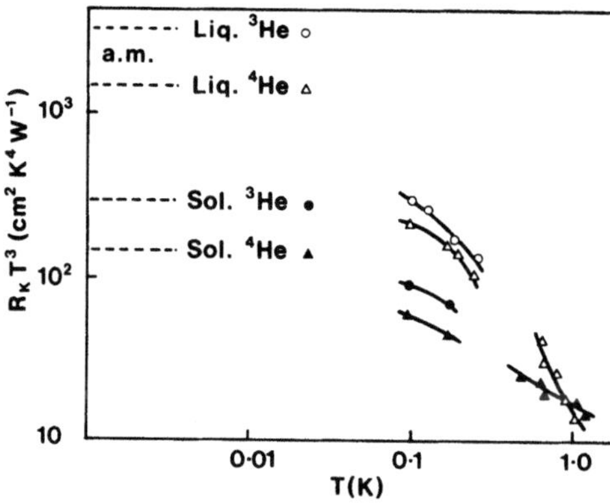

Fig. 1. *The Kapitza resistance as a function of temperature for Copper and ^3He or ^4He, liquid and solid. $R_K T^3$ changes considerably between 0.1 and 1K but at 0.1K there is a trend towards the a.m. values. (Redrawn from Folinsbee and Anderson (1973))*

The microscopic reason for the temperature discontinuity is that phonons on both sides of the interface are only partially transmitted through boundary. Phonons are also reflected at the interface and so different distribution functions for the phonons can be established on the two sides when there is a net phonon flux through the interface. The two phonon distributions correspond to the two temperatures. If the reflectivity of the interface were reduced to zero then the temperature would be continuous at the boundary. In this way it can be seen that the Kapitza conductance depends on the phonon transmission coefficient ($\alpha(\Omega,\theta)$) at the interface. In Figure 2 the frequency dependence is shown schematically but is based on the results of monochromatic phonon experiments. We shall later derive the relationship between R_K and $\alpha(\Omega,\theta)$ but for the present we note that a high α leads to a low R_K.

The classical theory considers an elastic wave incident on the interface and calculates the probability of transmission. The boundary conditions are that normal displacements are continuous, normal components of stress are continuous and stress parallel to the interface is zero as the liquid cannot support a shear wave. Before considering the details we examine the result for longitudinal waves at normal incident. This is the same problem as waves on a string being reflected and transmitted at the join between two different strings. The transmission coefficient $\alpha_{1\to2}$ is given by

$$\alpha_{1\to2} = \frac{4Z_1Z_2}{(Z_1 + Z_2)^2} \tag{1}$$

where the acoustic impedence $Z_i = \rho_{mi}\, s_i$ and ρ_{mi} and s_i are the density and velocity of sound in the ith medium. Eq. 1 is symmetrical in the subscripts so α is the same in both directions of transmission. For a typical solid $\rho_{ms} \sim 3\ 10^{-3}$ kg m^{-3} and s $\sim$ 5 kms^{-1} while for ^{4}He $\rho_m \sim 0.14\ 10^{-3}$ kg m^{-3} and s $\sim$ 0.24 kms^{-1} hence $\alpha \sim 0.01$. This indicates that most of the phonons would be reflected by the interface due to mismatch of the acoustic impedences. As values of α of 0.2 - 0.3 are commonly measured at T $\sim$ 1-2K, the inadequacy of the theory for these surfaces can be seen immediately. However at low phonon frequencies $\lesssim$1 GHz, and hence low temperatures, classical values of α have been found by Keen et al. (1965) between quartz and liquid ^{3}He, and recently by Kinder et al. (1979) have found classical values of reflectivity at high frequencies, up to 290 GHz, for clean and perfect single crystal surfaces.

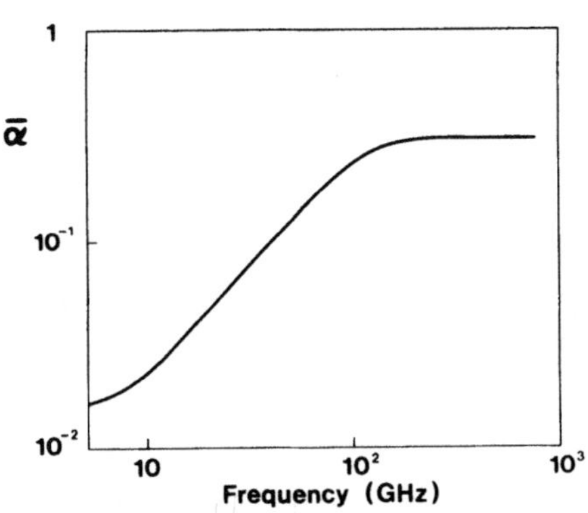

Fig. 2. Schematic drawing of a typical interfacial transmission coefficient $\bar{\alpha}(\Omega)$ as a function of frequency.

KHALATNIKOV'S CALCULATION OF α

The most sophisticated calculation of the classical values of transmission coefficients is due to Khalatnikov (1965). His formulation allows the inclusion of surface waves which if various conditions are satisfied can increase the transmission coefficient by a factor ~2.7. However it should be emphasised that there is no evidence that surface waves do contribute. Other calculations have been made by Little (1959).

Khalatnikov assumes that the solid is isotropic and that the interface with the liquid ^{4}He is planor. As the liquid cannot support shear waves the displacement $\underline{V}$ in the liquid can be written in terms of a scalar potential

$$\underline{V} = \underline{\nabla} \phi \tag{2}$$

while in the solid the more general displacement is represented by

$$\underline{U} = \underline{\nabla}\chi + \underline{\nabla}\wedge\underline{\psi} \tag{3}$$

is required. Cartesian coordinates are chosen so that the interface is in the xy plane at z = 0. As the solid is isotropic, there is axial symmetry about the z axis and we can choose that the displacements are only in the xz plane. This requires $\psi_x = 0 = \psi_z$.

A plane wave incident on the interface from the liquid side give rise to a reflected wave and two transmitted waves, one longitudinal and the other transverse. The potentials have the form:

$$\phi = A_o \exp i\,\{q(x\sin\theta_o - z\cos\theta_o) - \Omega t\} + A_r \exp i\,\{q(x\sin\theta_r + z\cos\theta_r) - \Omega t\} \tag{4}$$

$$\chi = A_L \exp i\,\{q_L(x\sin\theta_L - z\cos\theta_L) - \Omega t\} \tag{5}$$

$$\psi_y = A_T \exp i\,\{q_T(x\sin\theta_T - z\cos\theta_T) - \Omega t\} \tag{6}$$

where A_o, A_r, A_L and A_T are the amplitudes of the incident, reflected, longitudinal transmitted and transverse waves respectively. The q's are wave vectors and Θ's are the angles between the wave vectors and the z axis.

Equations 4 and 5 are substituted into equation 2 and 3 and then the boundary conditions are applied. From the translational invariance in the z = 0 plane we see that $\underline{U} = \underline{V}$ at z = 0 for all x. This requires that the terms in the exponents containing x are all equal, i.e.

$$q_L \sin\theta_L = q_T \sin\theta_T = q\sin\theta_r$$

rearranging using $\Omega = sq$ and writing $\theta = \theta_o = \theta_r$ gives

$$\frac{\sin\theta_L}{s_L} = \frac{\sin\theta_T}{s_T} = \frac{\sin\theta}{s} \tag{7}$$

where the s's are the velocities of sound in the solid and in the liquid ^4He. Equation 7 is just Snell's law which comes from the translational invariance of the interface. There are critical angles for transmission for the longitudinal and transverse waves given by

$$\sin\theta_{cL} = \frac{s}{s_L}\ ; \quad \sin\theta_{cT} = \frac{s}{s_T} \tag{8}$$

Only phonons from the ^4He within θ_{cL} can generate longitudinal phonons in the solid and only those within $\theta_{cT}(>\theta_{cL})$ can generate transverse phonons. For typical solids $\theta_{cT} \sim 6^o$. Conversely phonons from the solid can only be transmitted into the ^4He within the critical angles.

The boundary conditions also require that the normal components of displacement and the components of stress must be continuous. These lead to

$$\frac{A_r}{A_o} = \frac{Z_L\cos^2 2\theta_T + Z_T\sin^2 2\theta_T - Z}{Z_L\cos^2 2\theta_T + Z_T\sin^2 2\theta_T + Z} \tag{9}$$

$$\frac{A_L}{A_o} = \frac{\rho_m}{\rho_{ms}}\ \frac{2Z_L\sin 2\theta_T}{Z_L\cos^2 2\theta_T + Z_T\sin^2 2\theta_T + Z} \tag{10}$$

$$\frac{A_T}{A_o} = \frac{\rho_m}{\rho_{ms}}\ \frac{2Z_T\sin 2\theta_T}{Z_L\cos^2 2\theta_T + Z_T\sin^2 2\theta_T + Z} \tag{11}$$

$$\text{where } Z_L = \frac{\rho_{ms}s_L}{\cos\theta_L}\ ; \quad Z_T = \frac{\rho_{ms}s_T}{\cos\theta_T}\ ; \quad Z = \frac{\rho_m s}{\cos\theta}$$

The reflection coefficient for phonons in the liquid ^4He is $|A_r/A_o|^2$ and so the transmission coefficient α_K is

$$\alpha_K(\theta) = 1 - |A_r/A_o|^2 \tag{12}$$

Now we assume that there is an isotropic and homogeneous distribution of phonons in the liquid ^{4}He so the interface has phonons incident on it from all directions. The value of the transmission coefficient averaged over all angles is a useful quantity, it is defined by

$$\bar{\alpha} = 2\int_0^{\pi/2} \alpha(\theta)\ \cos\theta\ \sin\theta\ d\theta \tag{13}$$

This can be evaluated using equations 9 and 12. However Z_L and Z_T can be imaginary depending on the value of θ. For example $Z_L = \rho_{ms}s_L/\sqrt{(1-\sin^2\theta s_L{}^2/s^2)}$ and if $s_L{}^2/s^2 \sin^2\theta > 1$ then Z_L is imaginary, that is when $\theta > \theta_{cL}$. Khalatnikov evaluates the integral in three regions

i) $0 < \cos\theta < (1 - s^2/s_T{}^2)^{\frac{1}{2}}$ where both Z_T and Z_L are imaginary, this contribution is $\ell_1\ 4\rho_m s/\rho_{ms}s_T$

ii) $(1 - s^2/s_T{}^2)^{\frac{1}{2}} < \cos\theta < (1 - s^2/s^2{}_L)^{\frac{1}{2}}$ where Z_T is real and Z_L is imaginary, this gives contribution $\ell_2\ 4\rho_m s/\rho_{ms}s_T$

iii) $(1 - s^2/s_L{}^2)^{\frac{1}{2}} < \cos\theta < 1$ where both Z_T and Z_L are real, this gives contribution $\ell_3\ 4\rho_m s/\rho_{ms}s_T$. Hence

$$\bar{\alpha}_K = (\ell_1 + \ell_2 + \ell_3)\,\frac{4\rho_m s}{\rho_{ms}s_T} \tag{14}$$

Without further assumptions, the only contributions to $\bar{\alpha}$ are when at least Z_L or Z_T is real, so $\ell_1 = 0$. Khalatnikov argued that under certain circumstances ℓ_1 might contribute to $\bar{\alpha}_K$. The pole in ℓ_1 is given by

$$Z_L \cos^2 2\theta_T + Z_T \sin^2 2\theta_T - Z = 0 \tag{15}$$

and lies only $\rho_m s/\rho_{ms}s_T$ away from the real axis. If the path of integration is distorted to include the pole then ℓ_1 contributes to $\bar{\alpha}$. The physical interpretation is that equation 15, with Z = 0, corresponds to a surface wave at the interface. Normally the bulk phonons in the solid do not couple to the surface wave and so even if phonons in the liquid ^{4}He excite these waves there is no energy transmitted into the solid. However if the bulk phonons in the solid have a finite mean free path due to scattering near the interface, this causes the wave vector to have an uncertainty associated with it, which if large enough could enable the parallel component of wave vector to be conserved in the scattering of bulk modes to surface modes and vice versa.

The values of ℓ_1 and $\ell_2 + \ell_3$ have been calculated by Challis and Cheeke (1968) who define them as F_2 and F_1 respectively. $F_2 \sim 1$ and $F_1 \sim 0.6$ so the value of $\bar{\alpha}_K$ is very similar to α_{12} given in equation 1. The inclusion of ℓ_1 does increase $\bar{\alpha}_K$ by a factor ~2.7. This is not enough to account for the discrepancy between this model and measurements and furthermore we shall see that in experiments which can detect the surface wave effect, no sign of it can be seen. At present it is probably wiser to assume that $\ell_1 = 0$.

The variation of $\alpha_K(\theta)\cos\theta$ is shown in Figure 3. The peak due to surface waves is included to show its effect if it were coupled to the bulk phonons. The parameters are for KCL which is a soft material with low phonon velocities and hence relatively large critical angles.

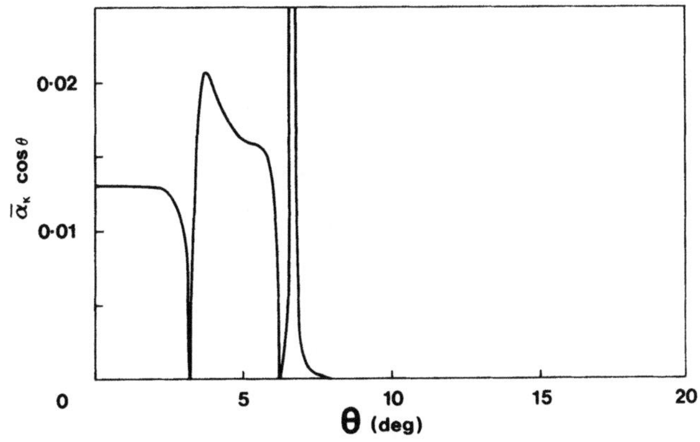

Fig. 3. The angular distribution of phonons emitted from KCl into liquid ^{4}He according to the theory of Khalatnikov. The sharp peak is due to surface waves (Redrawn from Page (1976))

If the phonons near the interface in the solid have an uncertainty in their wave number due to scattering from say dislocations, then not only can there be transitions via the surface waves into the liquid ^{4}He there can also be transmission to angles outside of the critical cone via forced surface oscillations. In reverse, phonons in the ^{4}He with $\theta > \theta_c$ can go into the solid. This can lead to a significant increase in $\bar{\alpha}$. This process was recognised by both Haug and Weiss (1972) and Peterson and Anderson (1972). The effect of this is shown in Figure 4 for various values of the parameter p which is the ratio of the imaginary and real parts of the phonon wave vector. The shorter the mean free path of the phonons in the solid, the larger p is. It can be seen that even with $p = 10^{-2}$ the width of the angular emission is almost doubled and α is increased by a factor ~3. This effect could well be applicable at low temperatures where the phonon wave lengths are longer. However at high temperatures the angular width of the critical cone can be measured quite accurately and the data is consistent with small or zero p. (Wyatt and Page 1978). Moreover this cone breaking effect cannot account for the background channel which has a very different angular dependence. The present situation is that the classical conductance channel is the simplest unmodified form but is in parallel with another channel of different origin and character.

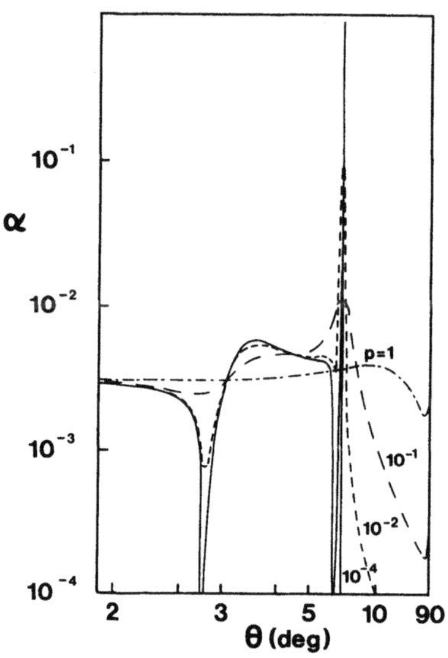

Fig. 4. The calculated angular distribution of phonons emitted from Cu into liquid ^{4}He for several values of the Haug and Weiss damping parameter p. (Redrawn from Haug and Weiss (1972))

RELATIONSHIP BETWEEN α AND Y_K

In general we expect the transmission coefficient α to depend on the phonon frequency and angle of incidence. The value of α averaged over frequency and angle is a useful parameter with which to compare diverse measurements. We adopt the convention that $\bar{\alpha}(\Omega)$ denotes an average over angle for a single frequency and $\alpha(T)$ denotes an average over a Planck distribution of frequencies. $\bar{\alpha}(T)$ denotes both averages.

To derive the relationship between $\bar{\alpha}(T)$ and $Y_K(T)$ we consider an elemental area of interface δA and take this point as the origin of spherical polar coordinates with the polar direction normal to the interface. Consider also a phonon at (r, θ, ϕ) with a random direction of propagation, the probability of it being incident on δA is $\delta A \cos\theta / 4\pi r^2$. The elemental volume at (r, θ, ϕ) is $r^2 \sin\, dr d\theta\, d\phi$ and so with $n(\Omega)F(\Omega)d\Omega$ phonons per unit volume with frequencies between Ω and $\Omega + d\Omega$, and $\alpha(\theta,\Omega)$ the transmission probability, the number of phonons that go through δA per second is

$$n = \int_0^\infty d\Omega \int_0^a dr \int_0^{\pi/2} d\theta \int_0^{2\pi} d\phi \; n(\Omega)F(\Omega)\; \alpha(\theta,\Omega)\, s \frac{\delta A \cos\theta}{4\pi r^2}\, r^2 \sin\theta \quad (16)$$

where s is the velocity of the phonon. For an isotropic Planck spectrum

$$n(\Omega)F(\Omega) = \frac{\Omega^2}{2\pi^2 s^3 \, (e^{\hbar\Omega/kT} - 1)} \quad (17)$$

we can integrate over r and ϕ to get

$$n = \delta A \int_0^\infty d\Omega \int_0^{\pi/2} d\theta \; \frac{\Omega^2 \alpha(\theta,\Omega) \sin\theta \cos\theta}{4\pi^2 s^2 (e^{\hbar\Omega/kT} - 1)} \quad (18)$$

we now define

$$\bar{\alpha}(\Omega) = \frac{\int_0^{\pi/2} d\theta \; \alpha(\theta,\Omega) \sin\theta \cos\theta}{\int_0^{\pi/2} d\theta \; \sin\theta \cos\theta} \quad (19)$$

so

$$n = \delta A \int_0^\infty d\Omega \; \frac{\Omega^2 \bar{\alpha}(\Omega)}{8\pi^2 s^2 (e^{\hbar\Omega/kT} - 1)} \quad (20)$$

As each phonon carries $\hbar\Omega$ of energy the energy flux (Φ) through the interface is given by

$$\Phi = \int_0^\infty d\Omega \; \frac{\hbar\Omega^3 \bar{\alpha}(\Omega)}{8\pi^2 s^2 (e^{\hbar\Omega/kT} - 1)} \quad (21)$$

The transmission coefficient averaged over frequency and angle is

$$\bar{\alpha}(T) = \frac{\displaystyle\int_0^\infty d\Omega \frac{\Omega^3\bar{\alpha}(\Omega)}{(e^{\hbar\Omega/kT}-1)}}{\displaystyle\int_0^\infty d\Omega \frac{\Omega^3}{(e^{\hbar\Omega/kT}-1)}} \tag{22}$$

so

$$\Phi = \bar{\alpha}(T)\int_0^\infty d\Omega \frac{\hbar\Omega^3}{8\pi^2 s^2(e^{\hbar\Omega/kT}-1)} \tag{23}$$

where the integral is just the energy flux incident on the interface. We have assumed that all phonons have the same velocity s and that the temperature is sufficiently low so we can extend the integral to infinity with impunity. Performing the integral we find

$$\Phi = \bar{\alpha}(T)\,\frac{\pi^2}{120}\,\frac{(kT)^4}{\hbar^3 s^2} \tag{24}$$

Phonons will cross the interface in both directions so the net flux from solid to liquid ^{4}He is

$$\Phi_{net} = \frac{\pi^2}{120}\frac{k_B^4}{\hbar^3}\left\{\left[\frac{\bar{\alpha}_L(T_s)}{s_L^2} + \frac{2\bar{\alpha}_T(T_s)}{s_T^2}\right]T_s^4 - \frac{\bar{\alpha}(T)}{s^2}T^4\right\} \tag{25}$$

where T_s and T are the temperatures of the solid and ^{4}He respectively, and s_L, s_T and s are the phonon velocities for the L and T modes in the solid and the L mode in the ^{4}He.

Now when $T_s = T$ there is no net flux and so

$$\frac{\bar{\alpha}_L(T)}{s_L^2} + \frac{2\bar{\alpha}_T(T)}{s_T^2} = \frac{\bar{\alpha}(T)}{s^2} \tag{26}$$

therefore the net energy flux can be written

$$\Phi_{net} = \frac{\pi^2}{120}\frac{k_B^4}{\hbar^3 s^2}\left(\bar{\alpha}(T_s)T_s^4 - \bar{\alpha}(T)T^4\right) \tag{27}$$

if we assume that $\bar{\alpha}(T)$ is a slow function of T and the temperature difference $\Delta T \ll T$ then

$$\Phi_{net} = \frac{\pi^2 k_B{}^4 \bar{\alpha}(T) T^3 \Delta T}{30\hbar^3 s^2} \qquad (28)$$

comparing this equation with the definition of R_K

$$\Phi_{net} = \frac{\dot{Q}}{A} = Y_K \; \Delta T \qquad (29)$$

we see

$$Y_K = \frac{\pi^2 k_B{}^4 T^3 \bar{\alpha}(T)}{30\hbar^3 s^2} \qquad (30)$$

It is clear that if the transmission coefficient from liquid ^4He to the solid depends on the phonon temperature then the Kapitza conductance will not be proportional to T^3. Measurements of $\bar{\alpha}(T)$ by thermal conduction generally show Y_K varying faster than T^3 for temperatures above ~0.1K.

THE CLASSICAL CHANNEL FOR CONDUCTANCE

We expect the theory of elastic wave transmission through interfaces to at least apply in the long wavelength limit. We shall now examine the evidence for its applicability to solid-liquid ^4He boundaries. The theory predicts the amplitude of the transmission, which we have seen is small (few percent), no frequency dependence and a well defined angular distribution with a sharp cut off at the critical angle for transverse polarisations. In particular it will be interesting to see if the theory applies at all in the high frequency limit where we know that the conductance is one to two orders of magnitude higher than this theory predicts.

Angular Distribution

The most outstanding characteristic of the acoustic model (a.m.) is the small critical cone for transmission on the liquid Helium side of the boundary. The half angle of this cone is typically 6°. It can be measured directly, in both directions of transmission, due to the fact that phonons in liquid ^4He can propagate in straight lines over distances of centimetres if the temperature is ~0.1K and the pressure $\gtrsim$18 bar. This allows the angle between the wave vector of the phonon and the surface normal to be measured. The experimental arrangement is shown schematically in Figure 5. The solid is in the form of a small single crystal so i, it approximates to a point source, ii, phonons have mean free paths in the solid which are long compared to the sample dimensions and iii, the faces can be cleaved to give appreciable areas which are atomically flat.

For emission, a thin film metal heater is evaporated onto the back face, which can be pulse-heated to inject high frequency phonons into the crystal. For acceptance a bolometer replaces the heater on the back face and this can detect phonons which enter the crystal through the front face. The sides of the crystal are masked off with an amorphous material which has a short phonon mean free path. The crystal faces a fixed bolometer for the emission measurements and the crystal is rotated about an axis that lies in the plane of the crystal face. For acceptance the crystal is rotated in the same way but a beam of phonons is generated by a thin film heater some 10 mm away.

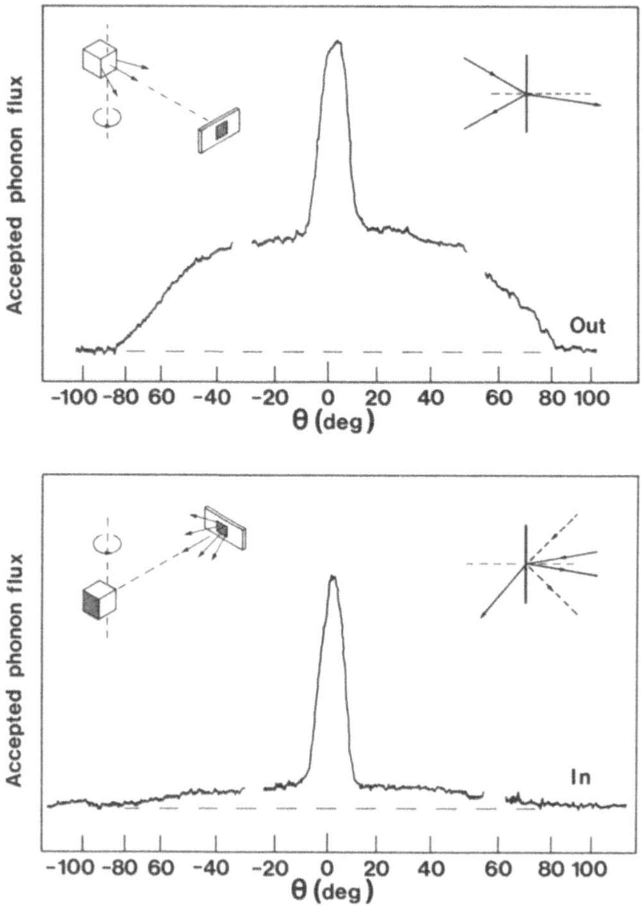

Fig. 5. Measured angular distributions for NaF for emission and acceptance showing the narrow peak and the broad background. (Redrawn from Wyatt and Page (1978))

A typical angular emission (Out) and acceptance (In) data are shown in Figure 5. It is immediately apparent that there is emission and acceptance over a narrow critical cone centred about the face normal, but also there is a slowly changing background emission into the whole of the half space. This background channel in fact carries most of the energy and accounts for the high Kapitza conductance. It will be discussed in detail in the next section. The narrow peak which can be separated off by interpolating the background, can be compared to the one predicted by acoustic theory.

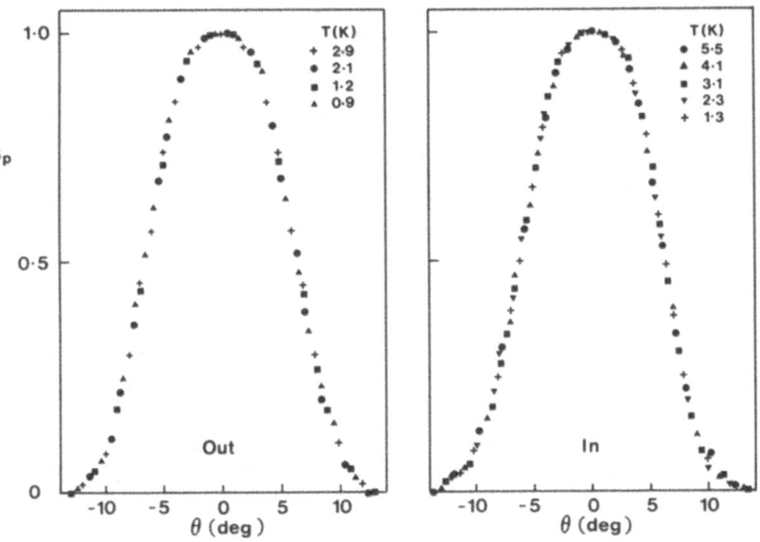

Fig. 6. Measured angular distribution of the peak for NaF at different heater temperatures. The peaks have been normalised and show that the shape is independent of phonon frequency and direction of transmission. (Redrawn from Wyatt and Pàge (1978))

The peak angular distribution measurements are shown in Figure 6 for emission and acceptance for the same NaF crystal sample. The angular width can be seen to be the same in the two directions. In Figure 7a, b is shown a comparison between typical points from Figure 6a and the calculated acoustic theory curve which has been convoluted with the functions that describe the effect of the finite size of the crystal and bolometer. The fit is remarkedly good without any phonon lifetime broadening. There is no sign of any Rayleigh wave contribution so we conclude that Khalatnikov's ℓ_1 parameter is zero and the p parameter of Haug and Weiss is also zero.

Other crystals with different phonon velocities should show different width critical cones. In Figures 7c and 7e we show the measured results for MgO and KCL which are respectively harder and softer than NaF. In Figure 7d and 7f are the calculated angular distributions Page (1977). The agreement is quantitatively good for MgO, and for KCL the intrinsic width of the distribution is wide enough so that the peaks due to the transverse phonons are not lost when the finite sizes of the heater and bolometer are taken into account. The measured transverse phonon peaks are smaller than expected but this could be due to the difficulty in getting a good crystal surface on KCL. In general we see that the shape of the narrow cone for emission and acceptance agrees well with acoustic theory.

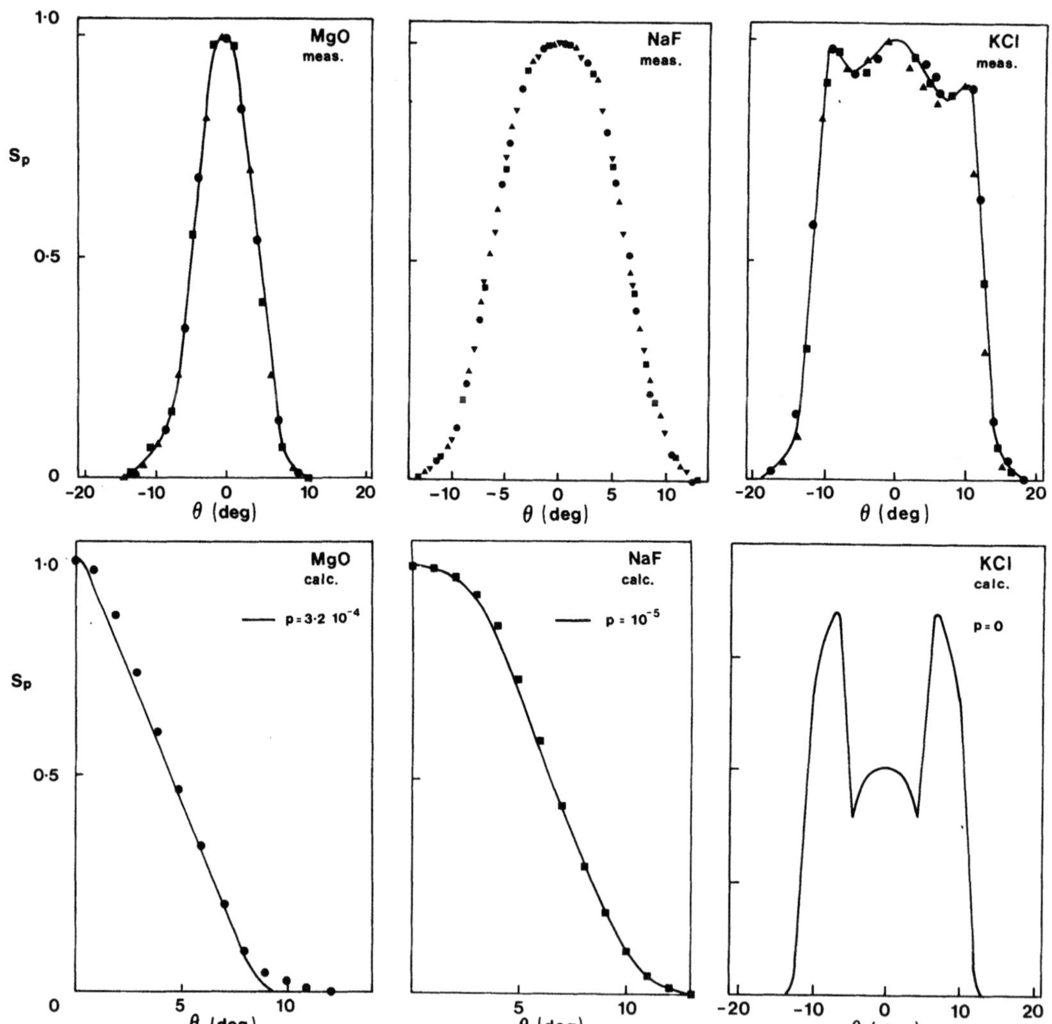

Fig. 7. Comparison of the measured peak angular distributions and those calculated from Khalatnikov's theory modified by Haug and Weiss. The agreement is generally good and only small values of p are necessary. (Redrawn from Page (1977))

Another quite different experiment also gives evidence for a critical cone in emission. Anderson and Sabisky (1971) measured the loss rate of monochromatic phonons from a single crystal of SrF_2 to thin films of liquid ^{4}He coating the crystal's surface. The crystals were doped with Tm^{2+} ions which have magnetic moment. These ions can be excited by microwave radiation and they relax to the ground state by emitting a phonon of the same energy. These phonons also excite the magnetic ions so that there is an equilibrium between the number of excited ions and the phonon number density. The phonon occupation number at this frequency will be a delta function peak on top of the broad thermal spectrum. The difference in occupation of the ground state and excited state can be monitored by the circular dichroism of a photon beam which excites optical transitions from these two states. Because of the coupling between the ions and the phonons this also measures the phonon number density.

The sample is continuously pumped with microwaves and energy is lost from the crystal by phonons escaping to the liquid ^{4}He. In the steady state these two powers balance. The phonon loss rate depends on the product of phonon density and transmission coefficient so that if the transmission coefficient is low then the phonon density is high and vice versa. With no ^{4}He around the crystal the phonon density is high but Anderson and Sabisky found that with a very thin film of helium, perhaps a few atomic layers, the phonon density dropped considerably, then as the film thickness was increased the phonon density oscillated with decreasing amplitude as shown in Figure 8.

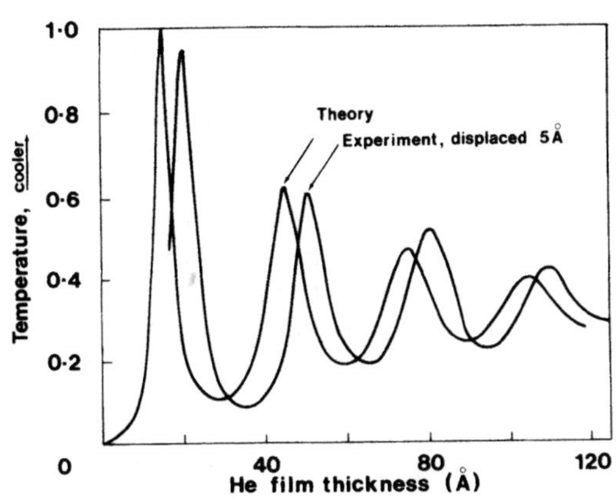

Fig. 8. The Anderson-Sabisky oscillations arising from standing waves in thin ^{4}He films and the result of a model calculation. (Redrawn from Wyatt et al.(1972))

The oscillations are due to multiple beam interference of the phonon in the ^{4}He film. (Wyatt et al. 1972). If the film thickness is an odd multiple of $\lambda/4$ then escape is easier due to the destructive interference of waves going back into the crystal, while for multiples of $\lambda/2$ escape is more difficult. These well defined oscillations would not occur if q_{11} was not conserved across the boundary so this experiment provides excellent evidence for classical acoustic transmission. The oscillations are small changes in phonon density compared to the general decrease that occurs when liquid ^4He coats the crystal. This again can be understood in terms of another more effective channel for phonon escape that does not conserve q_{11}. We would identify this with the background channel.

Magnitude and frequency dependence of the classical channel

At low phonon frequencies ($\lesssim 10^9$ Hz) when the wavelength is long there is good evidence that the only channel for phonon transmission is the classical one. The measurements of $R_K(T)$ show that as the temperature is decreased so R_K increases (See Figure 1). For lower frequencies the phonons can be generated piezo electrically and Keen et al. (1965) have shown that the power lost from a quartz crystal to liquid ^3He at 1 GHz can be accounted for by the acoustic theory. At 10 GHz Sabisky and Anderson (1975) report that they could not detect any gradient in the phonon density in a rod of SrF_2 immersed in liquid ^4He. This indicates that the phonon loss rate is so small that the phonons can come into equilibrium over the whole specimen. They estimate that the minimum value for the reflection coefficient for transverse phonons, over all angles of incidence is 0.98 at 20 GHz. A black body energy density spectrum at 0.2K has its peak at frequency of 10 GHz but conductivity measurements of R_K at this temperature show much lower values than acoustic theory. This apparent conflict with the monochromatic measurements can be resolved if in a sample at 0.2K it is the high frequency tail of the spectrum which escapes into the liquid with a relatively high transmission coefficient and then the spectrum readjusts within the solid by frequency up conversion of the lower frequency phonons.

There are as yet no quantitative measurements of the magnitude of the acoustic transmission when it is in the presence of a large parallel conductance. All the results that can differentiate between the two channels indicate that the acoustic channel is small. An upper limit can be estimated from the angular emission data by taking the total transmission coefficient as unity. This would make the acoustic channel $\lesssim$ 3 times larger than calculated for NaF. However there is evidence that the two channels interact at high temperatures (Wyatt and Page 1978) such that the acoustic channel falls below the calculated value. The monochromatic phonon techniques can also vary the ambient sample temperature and Sabisky and Anderson (1975) report that the total reflectivity of the generated phonons is independent of ambient temperature between 1.4K and 2.0K.

We conclude that the acoustic channel seems well established.

There is good evidence that phonons are transmitted with q_{11} and Ω conserved and that the transmission coefficient is ~1-2%. Measurements still are required above 130 GHz which is the maximum frequency used for thin film resonances (Blackford, 1972), and for detection of the critical cone with superconducting Aℓ tunnel junctions (Wyatt and Crisp 1978). The acoustic channel does not depend strongly on the ambient temperature up to 2K. However at higher temperatures there is evidence that the acoustic channel is decreased by the presence of the background channel.

BACKGROUND CHANNEL

Besides showing that the classical channel operates at high frequencies, the angular distribution measurements show that there is another channel for conductance which carries most of the energy at high frequencies. Because this channel is not restricted by the critical cones and radiates into all directions, it appears as a background distribution underneath the classical peak. The shape of the background channel does not depend on the incident phonon frequency as can be seen in Figure 9, where the backgrounds measured at different heater temperatures have been normalised. There are no systematic differences and also the angular distribution for phonons entering the crystal from the liquid He(In) is very similar to that for emission(Out). The In result shows that phonons at all angles of incidence from the liquid ^{4}He side of the interface, have a non zero probability of transmission into the crystal. This is not a direct corollary from the emitted distribution but is an independent feature of the energy transmission process.

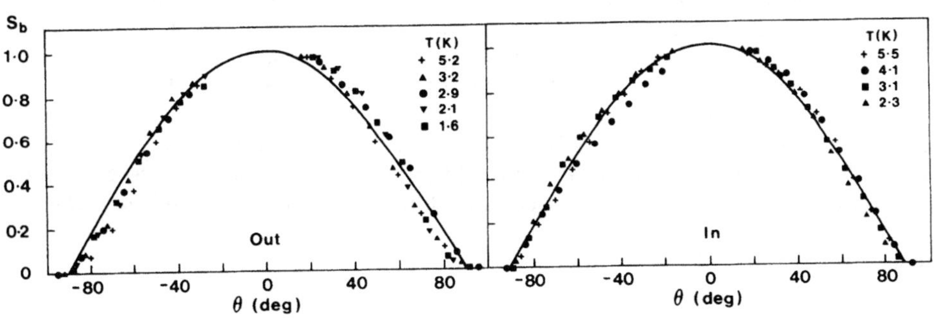

Fig. 9. The background angular distribution for NaF, for emission (Out) and acceptance (In) showing that the form is independent of heater temperature and direction of transmission. The results have been normalised and the solid lines are cosθ. (Redrawn from Wyatt and Page (1978))

The shape of the background shown in Figure 9 follows a cosine distribution fairly closely. If the phonon flux in the ^{4}He at the interface is Φ then the energy entering the crystal is Φ A $\cos\theta$ $(\alpha_{pIn}(\theta) + \alpha_{bIn}(\theta))$ where A is the area of the crystal-He surface. A measured background distribution that varies as $\cos\theta$ implies that α_{bIn} is independent of angle of incidence. A cosine law of emission would follow from an isotropic distribution of phonons in the crystal and no refraction at the interface. Although the first condition is probably correct the second implies a one to one correspondence between phonons in the crystal and those emitted in the background channel which as we shall see later is not the case. More likely the background emission is from the few atomic layers of He at the interface in which the energy is thermalised at a temperature ~0.7K. A cosine emission from such a layer is not unreasonable (Sheard and Fletcher 1979).

The magnitude of the background relative to the peak increases rapidly with heater temperature between 1 and 5K. In this range the energy carried by the background is larger than the peak by a factor between 10 and 10^2. This is just the factor by which the classical theory fails to explain the interfacial conductance.

We leave a discussion on the frequencies emitted in the background channel until we have examined values of $\bar{\alpha}$ from various experiments.

VALUES OF THE TRANSMISSION COEFFICIENT

For many years the study of the Kapitza effect was restricted to conductivity measurements of R_K from which values of $\bar{\alpha}(T)$ solid-He could be determined assuming that there was a Planck distribution at the interface. The dependence of $\bar{\alpha}$ on ambient temperature and sample preparation was found for many solids. The newer techniques have started to differentiate between q vectors and phonon frequencies which are averaged over in conductivity measurements. It is important to know whether these different measurements are mutually consistent and $\bar{\alpha}(T)$ is a useful parameter for comparison. We first briefly look at the values of $\bar{\alpha}(T)$ from R_K measurements which have been extensively reviewed before and then from transmission data.

$\bar{\alpha}$ from R_K

Generally it is found that $\bar{\alpha}(T)$ increases with T in the range 0.3 → 2K for all solids. This is equivalent to $R_K(T)$ increasing faster than T^3. Challis (1974) has reviewed this data and has derived a general T dependence which rises from the acoustic model value at T $\lesssim$ 0.1K to a value $\gtrsim$ 10 higher at ~2K. He draws attention to the lack of evidence at high temperatures to distinguish whether $\alpha(T)$ goes through a maximum or not.

Values of $\bar{\alpha}(T)$ at the same temperature vary between solids but this dependence is not clear due to the wide variation of $\bar{\alpha}(T)$ for the same material. $\bar{\alpha}(T)$ seems particularly dependent on surface purity and strain. Challis (1974) has collated the maximum values of $\bar{\alpha}(T)$ for different materials to emphasize the discrepancy between measured values and those from acoustic theory. He shows that there is no correlation of $\bar{\alpha}(T)$ with values of $\rho_m s$ as this model requires. However since Snyder (1975) and van der Sluijs (1979) have shown that the value of R_K for Cu can be increased by careful surface preparation it would seem that the highest values of $\bar{\alpha}(T)$ merely reflect surface contamination and strain.

Although the detailed work of Snyder (1976) on Cu showed that R_K was reduced in the temperature range 1.6 - 2.2K by electropolishing, Argon ion cleaning and annealing in UHV, there was a large change in the temperature power law for different treatments. One of the "best" surfaces had $Y_K \propto T^{2.4}$ which implies that $\bar{\alpha}(T)$ decreases with T. This has also been seen by Bishop and Van der Sluijs (1980). Other surfaces had $Y_K \propto T^{4.5}$ and a similar absolute value at 1.6K to the "best" surface which implies that at lower temperatures the "best" surface has a higher conductance than ones with more strain or oxide. Measurements down to lower temperatures would help to clarify these issues.

Johnson (1964) has cleaved a single crystal of KCL under liquid ^{4}He, which ensures a clean surface but not necessarily defect free, and found R_K to be the usual low value. The general conclusion appears to be that Y_K increases with temperature up to ~1K to a value an order of magnitude larger than the acoustic model. Above 1K there is a tendency for the "best" surfaces to show a slight decrease in $Y_K(T)$. The notable exception is Ag for which where Bishop and Van der Sluijs (1980) find $\bar{\alpha} \sim 0.025$ at 1.8K rising to 0.03 at 1.2K.

Power sharing technique

Ishiguro and Fjieldly (1973) measured the power sharing of the output of a thin film heater between liquid ^{4}He and Si crystal. Only the power radiated into the crystal was monitored but it was done with and without ^{4}He covering the heater. The drop in signal that was detected, by the bolometer on the crystal face opposite the heater, is assumed to be due to transmission from the heater into the ^{4}He. The results show that the transmission coefficient between the Ni film and liquid ^{4}He increases from 0.15 at heater temperature ~4K, to 0.45 for T and to 0.4 for L polarisation at 6K, the ambient temperature was 1.8K. At higher heater temperatures the emission into the liquid He decreased very rapidly to 0.2 at 15K. Ishiguro and Fjieldly suggest this is in part due to an insulating gas layer on the heater surface. If this decrease is not some artefact of the system then it conflicts with other reliable data.

$\bar{\alpha}(T)$ from second sound pulses

Swaneburg and Wolter (1973) showed that second sound pulses could be generated from phonon pulses which had propagated ballistically through a Si crystal. The phonons were generated in a short pulse from a thin film heater. As the pulse propagated in the crystal the L and T polarisation separated out so that two second sound pulses were created. In this way, they showed that the transmission coefficients for L and T modes are the same over a wide range of heater input powers and hence phonon frequencies.

Buck et al. (1974) showed that the energy in the second sound pulse could be measured using the energy dependence of the velocity which had been measured by Dessler and Fairbanks (1956). Buck et al. report that $\alpha(T) = 0.21$, averaged over L and T polarisations, but they assumed that all the energy from the heater went into the crystal of Si. If they used the power sharing data of Ishiguro and Fjieldly (1973) then $\alpha(T)$ increased to 0.38 at 6K and decreased with heater temperature. As the energy in the phonon pulse at the interface was not measured the quantitative values cannot be relied on.

The technique of Buck et al. (1974) was used by Wolter and Horstman (1977) who made the significant improvement of measuring the energy in the phonon pulse at the interface using an elegant calibration method. This involved a heater in contact with the bolometer on the crystal which could simulate the temperature change, the energy from the heater is the same as that absorbed from the phonons. In this way they measured $\alpha(T)$ averaged over L and T polarisations, for a Si crystal with the liquid ^{4}He at an ambient 1.2K.

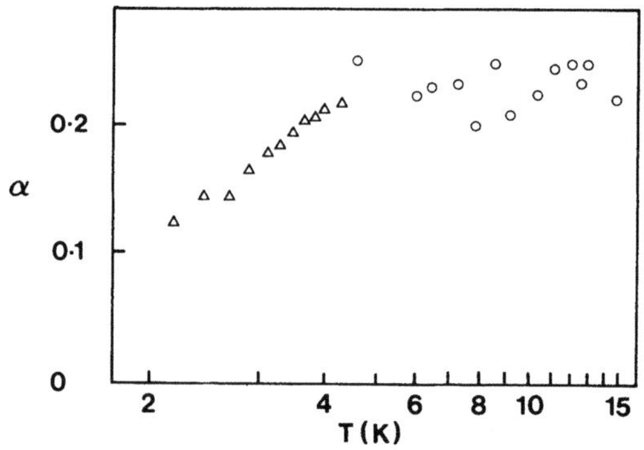

Fig. 10. The transmission coefficient $\alpha(T)$ for a Si-^{4}He interface measured by the second sound technique. (Redrawn from Wolter and Horstman (1977))

Their value of $\alpha(\theta=0)$ as a function of heater temperature is shown in Figure 10. $\alpha(T)$ rises from 0.12 to 2K to 0.23 at 5K and thereafter is constant to 15K, the maximum temperature used. This result is for phonons incident on the Si-He interface at normal incidence, however the surface of the Si was chemically polished so is not necessarily atomically flat nor free from chemical impurities, although it should be defect free. We shall see that this result is in good agreement with other ways of determining $\bar{\alpha}(T)$.

$\bar{\alpha}$ from monochromatic phonons

Sabisky and Anderson (1975) have used their spin-phonon spectrometer to look at the frequency dependence of the overall loss rate of phonons from a single crystal of SrF_2 to bulk liquid He from 15 to 315 GHz. The crystal has phonons injected into it from a thin film heater and the same power is lost to the He. The phonon density in the crystal is therefore higher than ambient (~1.4K) and has a different spectral distribution. The phonon energy density at a given frequency is found from the population of the ground state doublet of Tm^{2+} ions which dope the SrF_2 crystal. The Tm^{2+} levels are split by a magnetic field and they only interact with phonons having the energy of the splitting. The population of the levels is monitored by the circular dichroism (S) of a light beam that passes through the crystal. S is proportional to the difference in population of the two spin levels and is measured as a function of magnetic field so a range of phonon frequencies is considered. At each frequency the phonon energy density $u(\Omega)$ is measured, with the heater on and off. For each polarisation $S = S_o\ m\ \{2(2\pi^2 s^3/\hbar\Omega^3)u+1\}^{-1}$ where m is total number of spins and S_o is a constant. If the spins and phonons are at ambient temperature (T_a) then $S = S_o m\ \{2n(\Omega,T_a)+1\}^{-1}$ where n is the phonon occupation function. When the heater is on, the energy density increases to u' as some fraction (f) of the phonons in the heater come into the crystal. This produces an extra dilute gas with the spectral distribution of the heater, i.e. $u' = (\hbar\omega^3/2\pi^2 s^3)\ (n_a+f'n_{htr})$ where $f' = (s/s_{htr})^3 f$. The size of f' depends on the injection rate and the loss rate to the He. If the transmission to He is low then the phonon density must be high and vice versa. If the injected phonons do not thermalise in the crystal before escape then $f' \propto \alpha^{-1}$ (This is a different interpretation to Sabisky and Anderson (1975) where their $\varepsilon \sim f'$ but they say ε is proportional to the reflection coefficient).

The results of Sabisky and Anderson (1975) show that the transmission coefficient rises between 20 GHz and 150 GHz by a factor 2.5 and at higher frequencies there is no further change. It must be emphasised that the analysis of this experiment depends on details of the phonon transmission from the heater to the crystal and elastic and inelastic scattering within the crystal. These are not sufficiently well known at present and so the results should be treated as a qualitative guide only. For a surface that was contaminated with pump oil condensate the transmission coefficient showed a maximum at 90 GHz. If the surface of the crystal was sandblasted or the crystal coated with a thin layer of glyptal cement the phonon loss was high and frequency independent.

$\bar{\alpha}$ from angular distributions

The angular distribution of emitted phonons gives the relative energy radiated in all directions. To find the transmission coefficient from the crystal we need to know the phonon density inside it by an independent method. As this is unavailable, the peak can be used to normalise the angular distribution if we assume it is given quantitatively by acoustic theory. This is reasonable as the other properties of the acoustic theory have been shown to exist.

At low phonon frequencies the energy in the background (E_b) is not large and we use $\bar{\alpha} = (E_t/E_p)\bar{\alpha}_{am}$ where E_t is the total energy emitted and E_p the energy emitted into the peak. $\bar{\alpha}_{am}$ is averaged over all angles of incidence in the crystal. This formula implies that the emission into the two channels is completely independent and that the frequency distribution in the crystal is constant during the pulse. We would expect this to hold well when both channels have small transmission coefficients. At high heater powers, for some crystals E_t/E_p becomes larger than $\bar{\alpha}_{am}$ so clearly the method breaks down. This is considered to be due to competition between the two channels and to make progress this effect has to be modelled. We assume that a is the probability per unit area of interface of scattering into the background and that the fractional area (1-a) has the possibility of scattering into the peak. The probability of scattering into the peak is then $(1-a)\bar{\alpha}_{am}$ where $\bar{\alpha}_{am}$ is the acoustic model value. If Φ is the energy flux incident on the interface then

$$E_p = (1-a)\bar{\alpha}_{am}\,\Phi$$

$$E_b = a\,\Phi$$

$$E_t = E_p + E_b = \bar{\alpha}\,\Phi$$

this last equation defines $\bar{\alpha}$. Eliminating a and Φ leads to

$$\bar{\alpha} = \bar{\alpha}_{am}\,\{(E_p/E_t)\,(1-\bar{\alpha}_{am}) + \bar{\alpha}_{am}\}^{-1}$$

This reduces to the simpler expression in the limit $E_p/E_t \gg \bar{\alpha}_{am}$ but ensures that $\bar{\alpha} < 1$.

In Figure 11 is shown the ratio of the energy in the peak to the total emitted energy as a function of heater power. This shows that the peak channel carries only a small and decreasing fraction of the energy as the phonon frequency is increased, which explains

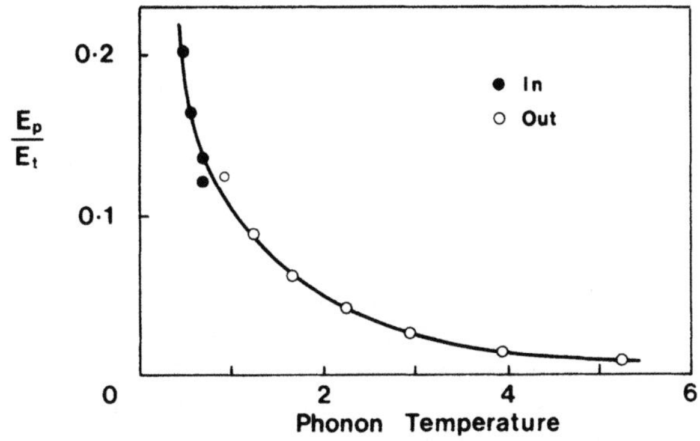

Fig. 11 *The ratio of the energy emitted into the peak and the total energy shown as a function of heater temperature. Temperatures for In points are taken from Fig. 14. Power reference: 0.5 W mm^{-2}. (Redrawn from Wyatt & Page 1978)*

the failure of the acoustic theory to account for measured R_K values. The value of $\bar{\alpha}$, as a function of heater temperature, is shown in Figure 12. $\bar{\alpha}$(T) increases from 0.05 at 1K to 0.25 at 5K. The results are remarkably similar to those for Si Figure 10 and are in general agreement with values $\bar{\alpha}$ from R_K measurements.

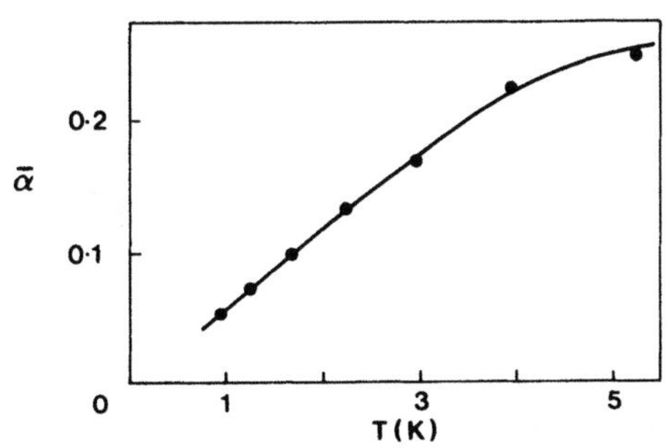

Fig. 12. *The transmission coefficient $\bar{\alpha}$(T) for NaF from the data shown in Figure 11 and the model described in the text. The similarity with Figure 10 is striking. (Redrawn from Wyatt and Page (1978))*

This similarity in the values of α(T) from quite different measurements give confidence that the different techniques are measuring the same quantity. In particular it shows that the broad emission in the background channel can be identified with the anomalously low values of R_K.

FREQUENCIES IN THE BACKGROUND

The classical channel emits with conservation of $\underline{q}_{11}$ and frequency. It has been shown that the background channel certainly does not conserve q_{11} and so there is no reason to expect that Ω is conserved. We now examine the evidence that shows quite clearly that Ω is not conserved.

The most direct evidence comes from measuring the angular distribution emitted from a crystal with both a tunnel junction and a bolometer. A superconducting Aℓ tunnel junction only detects phonons with energy $>2\Delta_{A\ell}$ (~100 GHz) while a bolometer is sensitive to all phonon frequencies. If the background has predominantly lower energy phonons than the peak then the ratio of peak to background will be larger for the tunnel junction than for the bolometer. The measurements are shown in Figure 13 where the peaks have been normalised together. It is quite clear that the background contains lower frequencies than the peak. An analysis by Crisp (1980) of the relative sizes of the background indicates that the background spectrum, if black body, has an equivalent temperature ~1K while the phonons in the crystal are at ~3K. The result also shows that phonons can be transmitted in the classical channel at least up to 100 GHz.

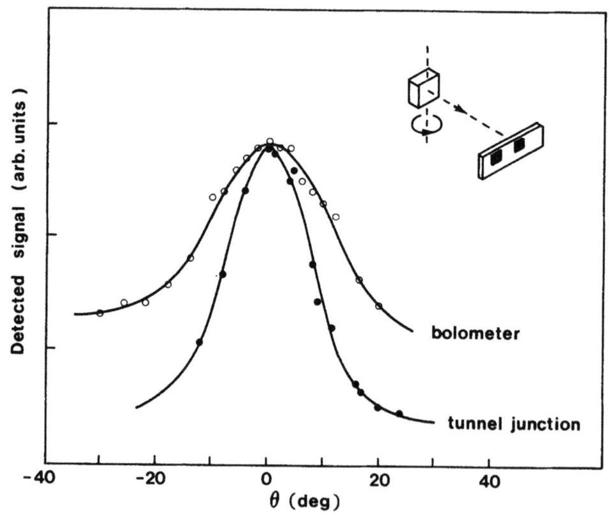

Fig. 13. The angular distribution of the phonons emitted from NaF into liquid ^{4}He measured with a bolometer and superconducting Aℓ tunnel junction ($2\Delta/k_B$=4.2K). The two curves have been normalised at the peak. The relatively small background measured with the tunnel junction indicates that the frequencies in the background are lower than those in the peak. (Redrawn from Wyatt and Crisp (1978))

A thin film heater emitting phonons into liquid ^{4}He is also expected to emit in both channels but because the interface is not plane there is no spatial separation of the peak and background. Instead the peak phonons are emitted over all angles and so at any angle the emission is predominantly background phonons. The phonons radiated from a thin film can be analysed by measuring their dispersion in liquid ^{4}He at 24 bar where the disperions is everywhere normal and has been well documented by neutron diffraction. (Svensson et al 1972). The experiment consists of injecting a relatively long phonon pulse into the liquid. It propagates with the higher frequencies going slower than the lower frequencies, so the pulse spreads out. The effects of the system time constant have to be accounted for before the effects of dispersion alone can be found. The results of a recent measurement are shown in Figure 14 (Wyatt, Sherlock and Allum 1980) where the phonon spectrum in the liquid ^{4}He has been characterised by a single temperature. Also in Figure 14 is shown the calculated temperature of the thin film producing the phonons. It can be seen that this temperature is up to a factor 3 higher than the helium phonon temperature which is a very clear indication that the injection process creates low frequency phonons.

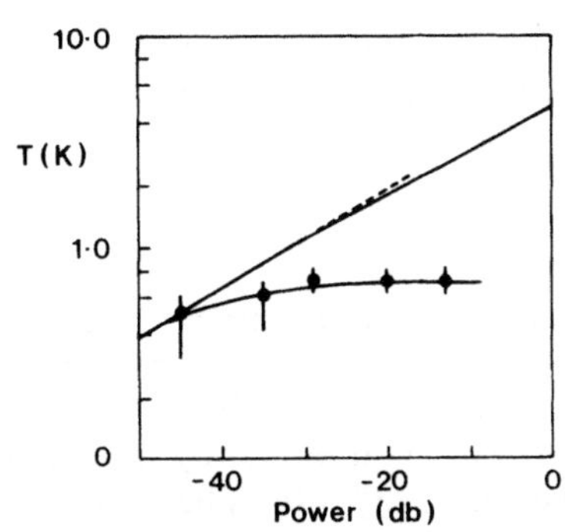

Fig. 14. The heater temperature calculated as a function of heater power and the temperature of the phonons injected into the liquid ^{4}He by the heater. Note the kink in the heater temperature curve which is due to the frequency dependence of α. The low characteristic temperature of the phonons in the ^{4}He compared to the heater indicates frequency down conversion processes at the interface. (From Wyatt, Sherlock and Allum, to be published)

More evidence for this conclusion comes from the In measurements, described in Figure 5, where a phonon beam in the liquid ^{4}He is incident on a crystal which can rotate and so sweep through all angles of incidence. The phonons that enter the crystal are detected by a bolometer on the back of the crystal. The results shown in Figure 5 show that the background is very small compared to the peak, even though the heater temperature of the phonon generator is comparable to that in Out measurements which give easily recognisable backgrounds. This result (Wyatt and Page 1978) was initially very puzzling but can now be readily understood in terms of the temperature of the phonon pulse incident on the crystal. If this pulse is composed of predominantly low frequency phonons and we know from the Out measurements that the background channel is sensitive to frequency and that the transmission coefficient is small at low frequencies, then the background In transmission will be small. The In results are plotted on the E_p/E_t diagram of Figure 11 using the temperatures measured by dispersion in Figure 14 and it can be seen the In and Out results form a continuous curve. This shows that the background channel process appears the same from both sides of the interface which is what we expect from thermodynamic arguments.

From three diverse measurements it is clear that the frequencies emitted by the background channel are quite low compared to those of the phonon that are creating them. Furthermore the transmission coefficient for the background channel increases with the frequency of phonons incident on the interfaces from both sides of the interface. At the moment we have no indication of the frequencies of the phonons created in the solid by phonons incident from the liquid, although it would be surprising if these were not again lower than the incident phonons. These conclusions are consistent with the results of reflection measurements which we consider in the next section.

PHONON REFLECTION

The reflection of phonons from a crystal-liquid ^{4}He interface, first done by Guo and Maris (1972), is a very powerful technique and has given many interesting results. It has shown that only a thin film of He, three atoms thick, is necessary for the Kapitza effect and that this film can "thermalise" the phonons that are transmitted into it. It has allowed crystal surfaces to be coated with a variety of atoms including ^{3}He, H_2, D_2 and Ne both in liquid and solid states, and so confirm the results of R_K measurements which show that the Kapitza effect is common to all light atoms. The frequency dependence of the reflectivity shows that it is constant above 130 GHz and that there are no signs of excitations of either rotons or maxons. Frequency changes have been measured in reflection and it appears that the Kapitza effect is only associated with those processes that give diffuse rather than specular reflection.

The technique consists of generating a short phonon pulse on one side of a single crystal and detecting the phonons that reflect off the opposite side with a detector which is on the same face as the generator. The generators are either thin film heaters or superconducting tunnel junctions which can give a tuneable spectrum. The detectors are either superconducting bolometers or superconducting tunnel junction which have a step function frequency discrimination. The reflection surface is sealed into a cell, separate from the generator-detector face, so that the surface can either be in vacuum or coated with various thicknesses of light atoms. The reflecting surfaces have been mainly polished because Si, Ge and $A\ell_2O_3$ have been used, however there have been measurements on cleaved NaF and LiF with very interesting results when the cleave has been done at low temperatures.

Film Thickness

The reflected signals are normalised to the reflection from the crystal vacuum interface. This signal shows peaks due to longitudinal (L) and transverse (T) polarisations which are specularly reflected, and a peak due to phonons which change their polarisations from L to T or vice versa at the reflecting interface. Usually there are other signals from diffuse reflections at the interface, these can

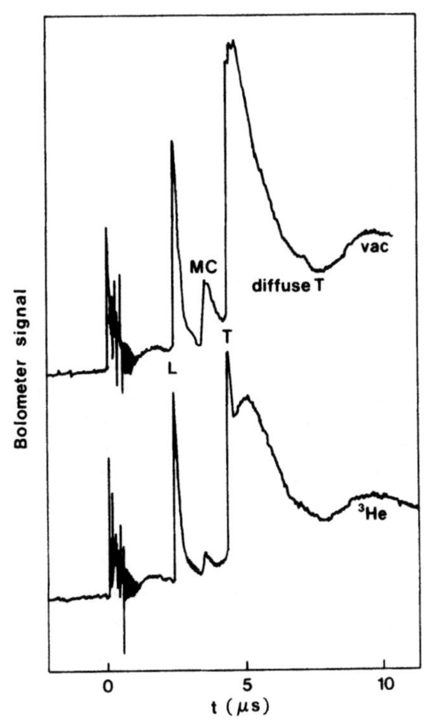

Fig. 15. The reflected signal in a polished Si/^{3}He interface. The large decrease with ^{3}He is typical of non-ideal crystal surfaces. (Redrawn from Follinsbee and Harrison (1978))

appear as tails to the specular signals as can be seen in the results of Folinsbee and Harrison (1978) in Figure 15, or as peaks if the paths are along focussing direction (Horstman and Wolter 1977, Taborek and Goodstein 1979).

As ^4He is added to the reflecting surface the signal changes and so the reflectivity for the various components of the signal can be measured with ^4He film thickness. The results of Kinder and Dietsche (1974) on polished Ge and Long et al. (1974) on cleaved NaF both show that the reflectivity drops rapidly with increasing film thickness and that there is no change after ~3 layers. Typical results are shown in Figure 16. All parts of the reflected signal that change with the introduction of ^4He show that the phonons do not distinguish between ~3 layers or bulk ^4He.

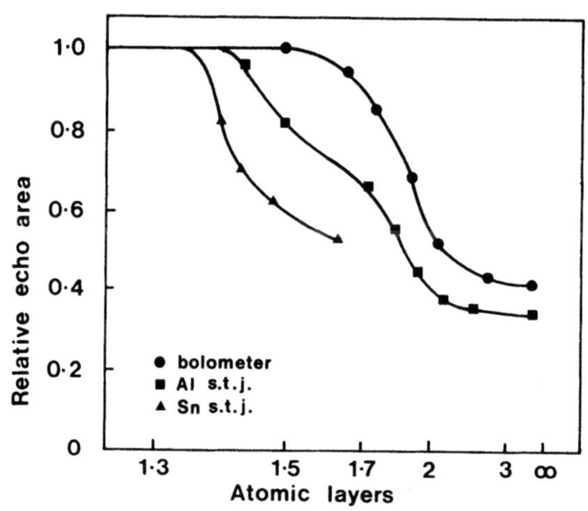

Fig. 16. The decrease in the reflected signal with He coverage for a non-ideal surface measured with bolometer, and Al and Sn superconducting tunnel junctions. The signal ceases to drop by 3 atomic layers. The initial drop is steepest for the Sn junction which indicates that phonons are down converted in frequency on reflection (Redrawn from Kinder et al. (1979))

If the drop in reflectivity is interpreted as loss by transmission into the background channel then this result indicates that the background processes occurs in the second and third ^4He layers.

Magnitude of reflectivity

The magnitude of the change in reflectivity has varied considerably with various experiments which are nominally the same.

Several results have been uncritically collated by Folinsbee and Harrison (1978). Most authors have not distinguished in detail between specular and diffuse reflections but where they have, it is clear that the diffuse reflections are more strongly affected by He. Long et al. (1974) on cleaved NaF, found the diffuse reflectivity was 0.75 for L and 0.55 for T while for the specular it was 0.95 for L and 0.67 for T. Horstman and Wolter (1977) showed that the reflected signal can be dominated by diffusely reflected phonons travelling along focusing directions in the crystal and the reflectivity drops to 0.3 with He for these signals. In the results of Weber et al (1978) on NaF cleaved at 1K shown in Figure 17a, it is clear that the diffusely scattered L tail changes by 20% while the specular L and T modes only change by 3-4%. This point is emphasized by Kinder et al. (1979) and also by Wyatt (1979). Folinsbee and Harrison (1978) using polished Si and liquid ^{3}He, found the diffuse reflectivity is 0.5 while for L it is ~0.95 and for T ~0.8 and again with polished $A\ell_2O_3$ Taborek and Goodstein (1980) clearly show that the specular reflections are hardly changed while the diffuse reflection almost entirely disappears with liquid ^{4}He.

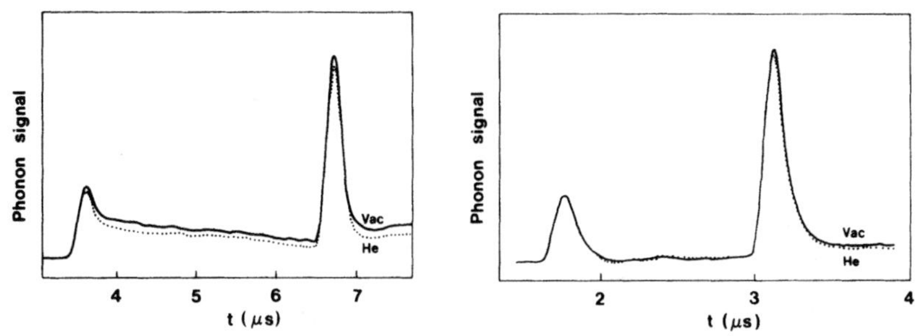

Fig. 17 The reflected signal from a cold cleaved NaF crystal both with and without ^{4}He. The curve on the right is for a more perfect cleave and it shows no diffusive tails to the specular pulses and no He effect. (Redrawn from Weber et al. (1978) and Kinder et al. (1979)

The reflections from NaF have been further refined by Kinder et al. (1979) and they find that their best cleaved surfaces show essentially no diffuse reflections, but only sharp peaks from L, T and mode change reflections. They find that none of these signals change when He is added within the resolution of their measurements as shown in Figure 17b. This result has also been achieved with LiF for a range of angles of incidence and has been shown to be independent of the frequency of the generated phonons.

It appears that as the reflection technique has been improved the specular and diffuse reflections, in the absence of ^{4}He, have been more clearly resolved and so it can be seen that when there are changes, it is the diffusive signal which is strongly affected by the ^{4}He.

The remarkable results of Kinder et al. (1979) show that classical reflectivity (~1) in the presence of He can be achieved if diffusive reflections are eliminated and this can be achieved by obtaining a good and uncontaminated surface. This unexpected and important result will change the emphasis of the Kapitza problem and in future we must try to understand how diffuse reflections occur and how they are associated with transmission into the liquid ^{4}He. However the position is not totally clear as a conventional R_K measurement on KCL cleaved under liquid ^{4}He by Johnson (1964) showed the normal small value of resistance and NaF cleaved under liquid ^{4}He at 0.1K showed a considerable background transmission (Wyatt 1979), see Figure 18, although there was an increase when the surface was thermally cycled to room temperature. However it is possible that these surfaces were not as good as those which were finally used in the reflection experiments of Kinder et al. (1979). The quality of the cleave and the surface damage will have to be monitored if sensible comparisons are to be made in future.

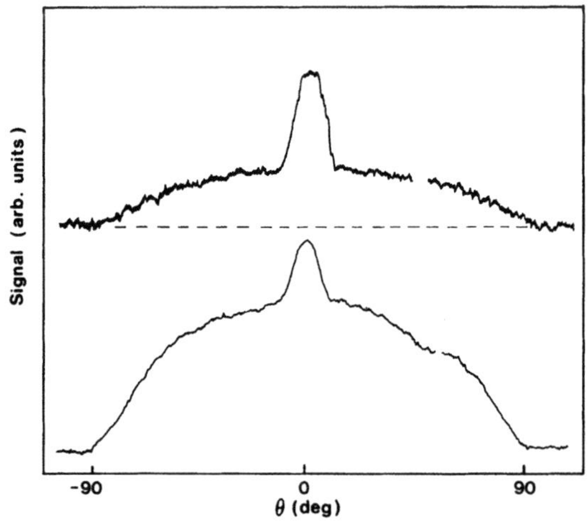

Fig. 18. Angular distribution of phonons emitted from NaF cleaved under liquid ^{4}He. The lower trace shows that the background is increased relative to the peak when the crystal surface is thermally cycled to room temperature. (Redrawn from Wyatt (1979))

One of the main problems with the reflection technique is knowing where the energy goes. Several tacit assumptions are generally made. The reflection is assumed to take place at the crystal-He interface, the energy that is not reflected is assumed to be transmitted into the He, and the angular distribution of the reflected phonons is assumed not to change with He. For example if the effect of adding He was to make reflections that were diffuse, specular then the diffuse signal would vanish because the then specularly reflected phonons would not reach the detector. These points could be clarified by simultaneous transmission and reflection measurements. It must be remembered too that some diffusive signal can always arrive at the same time as a specular signal if it is from the same element of interface and it is probably this which is giving rise to the signal change seen by Weber et al.(1977) for phonons reflected from the bottom of a hole in a Si crystal.

Frequency dependence of reflectivity

The reflectivity has been studied as a function of incident phonon frequency with monochromatic phonons using tunnel junctions (Kinder and Dietsche 1974 and Dietsche and Kinder 1976) and with a black body spectrum generated by a thin film heater (Guo and Maris 1972, 1974, Folinsbee and Harrison 1978). Kinder and Dietsche find that there is no frequency dependence between 130 and 870 GHz while the broad spectrum results show a decrease in reflectivity of 0.3 between 2 and 6K (Guo and Maris 1974) for nominal T phonons in Si. Folinsbee and Harrison (1978) find no change for L and T specular reflection but they find a similar decrease of 0.3 for diffuse reflection between 1.5 and 4K. This apparent difference between the results, using thin film heaters, is probably due to a lack of discrimination between specular and diffuse reflections.

The monochromatic frequency range overlaps the range used by Sabisky and Anderson (1975). They found that the reflectivity changed between 20 and 130 GHz but thereon up to 245 GHz there was no change. So these monochromatic results are in accord. Now 130 GHz is the frequency of the peak in the energy spectrum of a Planck distribution for a temperature of 2.21K. The reflectivity will change with heater temperature until there is a negligible part of the spectrum below 130 GHz. It is clear that this will be around 5K so all the reflection measurements and the results discussed earlier agree with each other.

Frequency changes and time delays

In a series of experiments (Kinder and Dietsche 1974, Dietsche and Kinder 1976 and Weber et al. 1977) it has been shown that when the ^{4}He film covering the crystal is between ~1.4 and ~2.0 layers the reflected phonons have a different frequency spectrum to the transmitted one. Associated with the up and down frequency conversion is a time delay for the down converted phonons which has a maximum value of 0.25 μs at ~1.6 atomic layers. Both these phenomena disappear with bulk ^{4}He and there is the question of whether the frequency converting processes are still operating. They could be occuring but not observable in reflection due to the converted phonons being lost more efficiently to the liquid ^{4}He than being reflected. The time delay, which if associated with low transmission probability between the film and the crystal, would disappear for the same reason. The down conversion is a larger effect than the up conversion both in the energy involved and the shift in frequency. It therefore seems sensible to identify the down converted phonons, which are seen in reflection at low coverages, with the background phonons transmitted at high coverages, as these are down converted too.

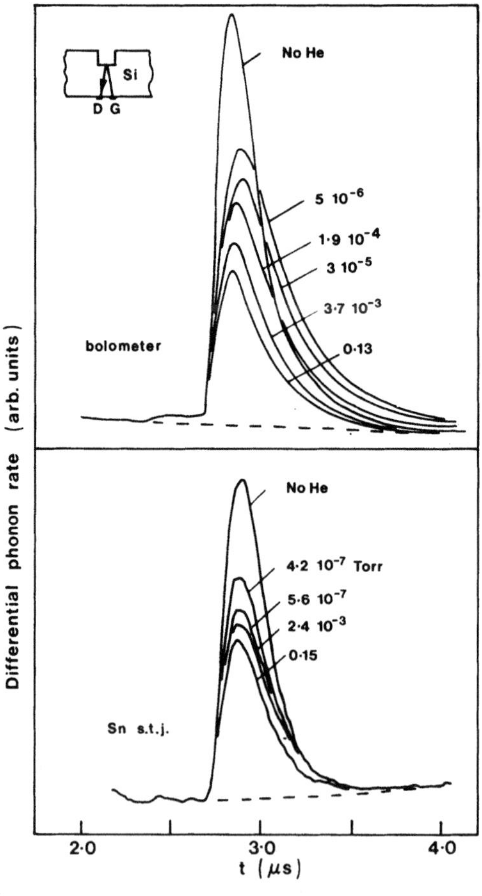

Fig. 19. The reflected signal from a polished Si crystal, at various He coverages, measured with a bolometer and Sn tunnel junction. The bolometer which is sensitive to low frequency phonons shows a broader pulse due to the time delay for energy to re-enter the crystal from the ^{4}He film. (Redrawn from Dietsche et.al. (1976))

The evidence for the down conversion is shown in Figure 19. The transmitted phonons are 290 GHz from the recombination of quasi-particles in a superconducting tunnel junction. The reflected phonons are detected with Aℓ superconducting tunnel junction, which detects $\hbar\Omega > 2\Delta_{A\ell}$ (120 GHz) and a bolometer which is sensitive to all frequencies. As the ^{4}He film grows in thickness the junction response decreases uniformly over the received signal, while the bolometer decreases at the peak but actually increases after the peak. This shows that the thin ^{4}He film increases the "reflected" energy with phonons of frequency < 120 GHz. The increase occurs after the peak showing that the energy is delayed and this is plausibly explained by Weber et al. (1977) as due to the energy leaking back into the crystal from the film with a low transmission probability. The film however has "thermalised" the 290 GHz phonons which were incident on the interface.

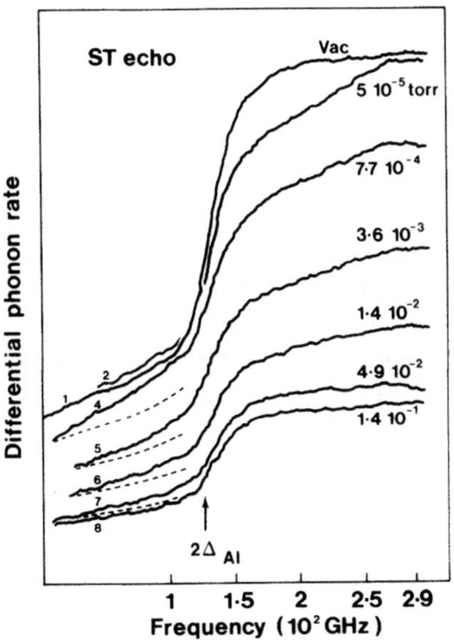

Fig. 20. The differential response of an Aℓ superconducting tunnel junction detecting the reflected phonons from a Ge-^{4}He interface as a function of generator voltage. The box car gate is set on a reflection peak and the generator is biassed with a dc and pulse voltage. Only the signal due to the pulse is recorded. The slope of the signal at $eV_i < 2\Delta_{A\ell}$ is increased by ^{4}He around two monolayers coverage. (Redrawn from Kinder et al. (1974))

The up conversion evidence is shown in Figure 20. The Sn transmitting junction is biassed at $eV_i > 2\Delta_{Sn}$ and a small additional voltage is pulsed. The signal detected by the Aℓ junction, due to the pulses, is measured as a function of the bias voltage. This signal is then due to $2\Delta_{Sn}$ phonons and relaxation phonons with $\hbar\Omega = eV_i - 2\Delta_{Sn}$. If there is no frequency conversion, the detector signal is due only to $2\Delta_{Sn}$ phonons when $eV_i - 2\Delta_{Sn} < 2\Delta_{A\ell}$ so the signal increases slightly with bias voltage due to the generator current increasing slightly faster than linearly with voltage for voltages just above the gap. When $eV_i - 2\Delta_{Sn} = 2\Delta_{A\ell}$ the relaxation phonons can excite the detector and so at this bias there is a step function increase in detected signal. For higher voltages the differential relaxation phonons decrease but the recombination phonons increase so the result is an almost constant detected signal, as shown in Figure 20, trace 1. If the film of ^{4}He were to just decrease the reflectivity at all frequencies, then the detector response would be a scaled down version of trace 1. However it is clear from trace 4 that there is extra signal at $eV_i - 2\Delta_{Sn} < 2\Delta_{A\ell}$ which increases as $2\Delta_{A\ell}$ is approached. An increase in the reflectivity of $2\Delta_{Sn}$ phonons would cause a bias independent increase, so the observed increase is considered to be due to relaxation phonons being up converted to energies $> 2\Delta_{A\ell}$.

Thin layers of He films on crystaline substrates not only have high heat capacities but their temperature dependence shows peaks around 1K, as shown in the results of Pollanco and Bretz (1978) Figure 21. It is thought that there are two dimensional phase changes taking place in the various atomic layers and the latent heat shows up in the heat capacity measurements. Energy thermalising in such a layer would produce a broad spectrum of phonons appropriate to the critical temperature of the phase change. This is possibly the origin of the low frequency phonons seen in reflection and in transmission in the background.

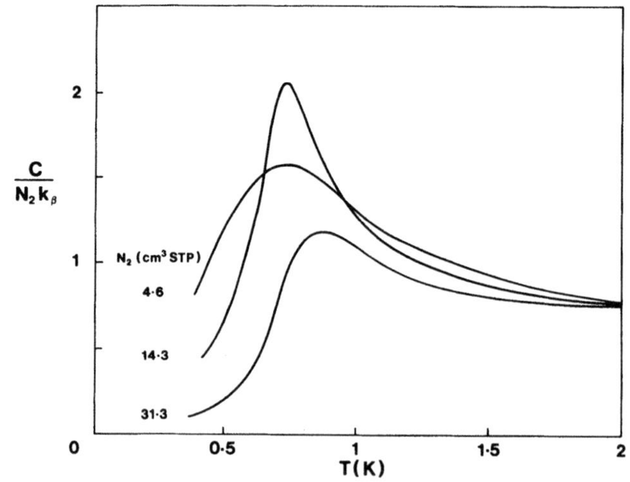

Fig. 21. The heat capacity of thin films of ^{4}He on graphite substrates. (Redrawn from Pollanco and Bretz (1978))

THE QUANTUM MATERIAL

The anomalously high conductance is seen between solids and other light atoms besides ^4He. Reynolds and Anderson (1977) report small R_K values between Cu and solid ^4He, H_2, D_2 and as we have shown in Figure 1, Folinsbee and Anderson (1973) found small R_K values for liquid and solid ^3He and ^4He down to 0.1K although the values were approaching the acoustic model values at the lowest temperature. Using the reflection technique, Buechner and Maris (1975) measured the reflectivity at interface between polished Si and liquid and solid ^4He, solid H_2, D_2 and Ne. Solid Ne behaved most like a classical solid with a significantly reduced reflectivity for the L as well as the T polarisation. All the other materials showed reflectivities around 0.8 for L, 0.67 for T and 0.15 for mode conversion, however no distinction was made between specular and diffuse contributions. The mode conversion reflectivity for Ne was equally small; this value is smaller than any other measurements.

There was no change in reflectivity as the ^4He was solidified as is shown in Figure 22. So this eliminates the liquid state being the important factor in the anomalous conductance. What is in common with these materials is the small mass of the atoms and relatively weak bonding between the molecules. These are just the conditions for zero point motion to be large and quantum effects to show up. Buechner and Maris (1975) point this out and give the values of de Boer's quantum parameter $\Lambda = h/\sigma(me)^{\frac{1}{2}}$ where σ are the range and strength of the inter particle potential and m is the particle mass. Λ values are ^3He:3.08; ^4He:2.68; H_2:1.73 D_2:1.22; Ne:0.59. There would seem to be some correlation with their results but it is not very strong.

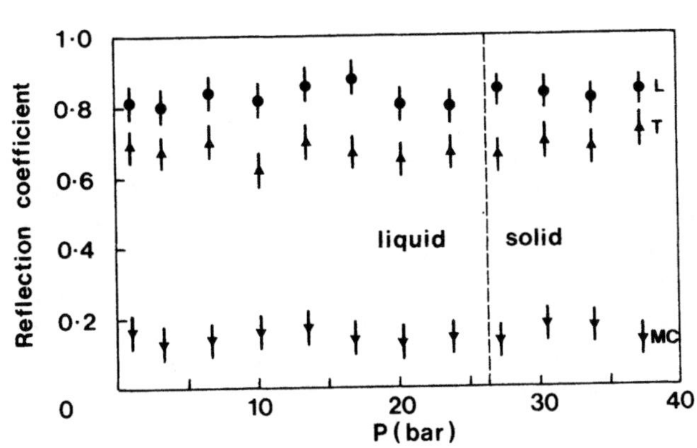

Fig. 22. Reflectivity of Si-^4He interface through the solidification of the ^4He. The results show that the reflectivity is the same for both phases. (Redrawn from Buechner and Maris (1975))

In a transmission experiment in which the L and T phonons could be resolved, Salemink et al. (1978) measured the ratio of the transmission coefficients for L and T polarisations. With a detector on the crystal they measured L/T ratio of 0.31 but with the detector in the liquid ^{4}He, 0.59 mm from the crystal surface, the ratio dropped to 0.20 for heater power <0.8 W/mm^2 and ^{4}He pressure of 24 bar. At lower pressures the ratio decreased to ~0.13 at 1 bar. This probably indicates that the phonons in the liquid ^{4}He generated by the L phonons have a higher frequency distribution than the T ones. These conclusions differ from that of Swaneburg and Wolter (1973) who report that the transmission coefficients are the same for the L and T polarisations. However both groups used polished crystals so the differences are not significant. It would be very interesting to know the transmission coefficient for T polarisations incident normally on an atomically flat crystal because in this case there is no displacement of the crystal into the liquid.

THEORIES

A theoretical treatment of the Kapitza effect would have to consider the i, magnitude and frequency dependence of the transmission and reflection, ii, the two channels, their angular distribution and possible interaction, iii, the low frequencies of the transmitted and reflected phonons, iv, the importance of the first few layers of He, v, generality of behaviour for light atoms vi, ideal, physically non-ideal solid surfaces and chemically contaminated surfaces, vii, diffuse and specular reflections. Many theories have just concentrated on the magnitude and frequency dependence and have assumed that Q_{11} or Ω is conserved in their model. This was quite understandable when the only data available was from conductivity measurements but with our present knowledge we know that these models are not even starting correctly. Many of these theories have been mentioned in the reviews referenced earlier and will not be further discussed.

Recently, theories have been proposed which concentrate on excitations associated with the He at the interface. Namaizawa (1976) suggested that surfons, surface waves in thin layers of He, play the intermediate role between bulk phonons in the crystal and liquid He. A phonon in the solid excites a surfon which then decays to a He phonon. This was the first serious attempt to account for the background channel but the derived angular dependence of $\sin\theta \cos^2\theta$ is quite unlike that measured. In a similar model Guyer (1977) suggests that vacancy waves in the surface layer provide the energy and momentum reservoirs. It is envisaged that a bulk phonon scatters from a vacancy wave, exchanging energy and momentum, and in a virtual process the vacancy wave scatters into a bulk phonon on the other side of the interface. Momentum parallel to the interface is not conserved between the two bulk phonons as momentum can be carried away by the vacancy wave.

Similarly frequency is not conserved. However no details of the predictions of the model are given.

Both surfons and vacancy waves are non-localised excitations which contrast with tunnelling states in the He layer. The possibility of such states being the cause of the Kapitza effect was suggested by Nakayama (1977). With localised states the translational invariance of the interface is destroyed and so there will not be Q_{11} conservation in this model. The tunnelling states are envisaged to arise from an inhomogeneous crystal surface which give double welled potentials for He atoms on the surface. A light atom can tunnel between these wells and so it can exist in one of two energy levels. A phonon from the solid could raise the energy of a surface atom which in turn could relax exciting a phonon in the liquid. As in glasses, it is assumed that there would be a spectrum of two level splittings, extending down to low energies of ~0.1K, so that the phonons typically used in experiments would be scattered. Nakayama shows that with a suitable choice of parameters the conductance increases as measured. However the angular distribution goes as $\cos^4\theta$ which is much faster than measurements suggest. For progress on these ideas, the existence of tunnelling states will have to be demonstrated and even more difficult, their properties measured.

The problem of the angular distribution of the background channel has been considered by Sheard and Fletcher (1979). In a vaguely defined model they envisage non-localised interface states with again a broad energy spectrum. The phonons from the solid excite the interface states but they consider that the surface is not ideal so that Q_{11} is not conserved in this process. This ensures that the critical cone is broken. The interface states are strongly coupled to the He phonons and these are excited with Q_{11} conservation. The angular distribution is then given by the number of phonon modes in the liquid that couple to the interface and this depends on the component of the area in any direction. This very general model shows a way of getting $\cos\theta$ but as the authors point out, phonon frequency is conserved which certainly is not found in practice.

Saslow and Lumpkin (1979) suggest there are nonlinear effects at the interface such that a phonon from the solid creates two phonons in the liquid. A plausibility argument is made that the size of the matrix elements are large enough to achieve the measured conductance values. The model is attractive because it can qualitatively account for many of the properties of reflection and transmission.

The recent reflection results on cold cleaved crystals suggest that there is no anomalous conductance for ideal interfaces. If this is so then models must incorporate non idealities explicitly and theories of ideal surfaces should give the low acoustic values of conductance as has recently been shown again by Maris (1979).

CONCLUSION

The new experiments that have been performed with phonons on the boundary between solids and liquid He have produced a detailed and consistent picture. Phonons between ~10 and 130 GHz have an increasing probability of transmitting energy into liquid He. The phonons produced in the He have a lower frequency than those producing them and are emitted into all angles. Above 130 GHz the probability remains constant and below 10 GHz the acoustic theory gives the correct qualitative behaviour. It is clear that both L and T phonons are effective in exciting phonons in the He. The high conductance is not limited to liquid ^{4}He but other light atoms in both the liquid and solid state show higher transmission than acoustic theory.

It is generally felt that the explanation for the Kapitza effect lies in the behaviour of the first few layers of light atoms which are under the van der Waals attraction of the dense solid. Various suggestions for the interface states have been mooted and localised and non-localised states seem equally possible. It is clear that the answer to the problem does not lie in the light atom layers alone as the surface preparation of the solid has a very considerable influence, with chemical and physical defects increasing the conductance.

The latest reflection measurements indicate that it is possible to produce a surface on a single crystal for which the reflectivity is essentially unity for high frequency phonons. This is prima facie evidence that the extra conductance, above the acoustic model, is due to surface impurities interacting with the first few layers of light atoms. If this is the case it makes the elucidation of the microscopic processes even more difficult and requires the experimentalist to characterise his surface in a detailed way. Such ideal surfaces have not been produced for transmission experiments, as those crystals that have been cleaved under liquid He have shown enhanced conductance. It is clearly desirable for a reflection and transmission measurement to be made simultaneously on an 'ideal' surface. A measure of the ideality of the surface is the diffuse reflectivity which is absent for 'ideal' surfaces. The diffusely reflected signal is strongly decreased by He on the surface and it would be useful to positively show that the cause of the diffuse signal also produced the background phonons seen in transmission.

The increase in our knowledge of the Kapitza effect, which has occurred in the last eight years, far from solving the problem has shown it will require even more imaginative theories and experiments to find the reason for the high conductivity.

ACKNOWLEDGEMENTS

I should like to acknowledge Drs. R.A. Sherlock, A.R. Long, D.R. Allum, N.A. Lockerbie, N.G. Mills, G.J. Page and G.N. Crisp with whom I have had the pleasure to work with on this topic at various times, and my friends at Nottingham and around the globe with whom I have had many fruitful discussions. I am grateful to these authors who have let me reproduce figures from their publications.

NOTE ADDED AFTER THE SUMMER SCHOOL

New results presented at the summer school suggest that the results of the reflection technique should be treated with great caution. Indeed the new evidence shows that the decrease in reflectivity cannot be assumed to be directly associated with an increase in transmission into the coating material and furthermore the decrease in reflection is not peculiar to ^{4}He but is a general effect of any coating.

Taborek and Goodstein have used their high resolution reflection technique with sapphire to show that the decrease in reflectivity comes from the diffuse rather than the specular reflection when the sapphire is coated. They find that exactly the same quantitative decrease occurs when the crystal is coated with any of the inert gases. Scaramuzzi presented results which showed that the reflectivity is unchanged to 1% when ^{4}He is changed from liquid to solid or from normal to superfluid.* It therefore appears that only the presence of material and not its individual properties on the reflecting surface is sufficient to almost entirely eliminate the diffuse reflection. It is not clear whether this energy is reflected elsewhere in the sapphire or whether it is frequency down converted and not detected by the superconducting bolometer.

Horstman pointed out that he and Wolter (1977) have already reported that the diffuse reflectivity is considerably reduced (30%) by a thin metal film (50 nm constantan) on the reflecting surface of a Si crystal, although liquid He reduced the reflectivity even more (~70%). They used a special detector which had a thermalising metal film between the crystal and the superconducting bolometer in order to avoid frequency dependence in the phonon detection. Even so they could not account for the missing energy.

These results indicate that reflection measurements do not give any directly useful information on the anomalous transmission into liquid ^{4}He. They highlight the importance of accounting for all the energy in this type of measurement.

* See chapter 21.

REFERENCES

Anderson, A.C., 1976, The Thermal Boundary Resistance, in "Phonon Scattering in Solids", Challis, L.J. Rampton, V.W. and Wyatt, A.F.G., eds., Plenum, New York.
Anderson, C.H. and Sabisky, E.S., 1971, Spin Phonon Spectrometer, in "Physical Acoustics", Mason, W.P. and Thurston, R.N., eds., Academic Press, New York, 8:1.
Bishop, J.C. and Van der Sluijs, J.C.A., 1980, The Kapitza Conductance and its Enhanced Power Anomaly for Clean Silver to Helium Four Interfaces at Temperatures between 1 and 2K, J. Low Temp. Phys., to be published.
Blackford, B.L., 1972, Phonon Interference in Thin Helium Films Using Superconducting Tunnel Junctions, Phys. Rev. Lett., 28:414.
Buck, J., Lassmann, K. and Eisenmenger, W., 1974, Transmission of Heater Generated High Frequency Phonons through a Sapphire-HeII Boundary, Phys. Lett., 50A:279.
Buechner, J. and Maris, H.J., 1975, Reflection of Phonons at Interfaces between Silicon and Solid Hydrogen and Deuterium, Phys. Rev. Lett., 34:316.
Challis, L.J., 1974, Kapitza resistance and acoustic transmission across boundaries at high frequencies, J. Phys. C. Solid State Phys., 7:481.
Challis, L.J., 1975, Kapitza Conductance, in "The Helium Liquids", Armitage, J.M.G. and Farquer, I.E., eds., Academic Press, New York, Ch. 10.
Challis, L.J. and Cheeke, J.D.N., 1968, Some comments on the theory of the Kapitza conductance between metals and liquid helium, Proc. Roy. Soc., 304A:479, London.
Crisp, G.N., 1980, Phonon transmission through solid-liquid Helium interfaces, Ph.D. Thesis, University of Nottingham,
Dessler, A.J. and Fairbanks, W.M., 1956, Amplitude Dependence of the Velocity of Second Sound, Phys. Rev. 104:6.
Dietsche, W. and Kinder, H., 1976, Frequency Analysis of Phonons Backscattered from Interfaces Between Solids and Helium, J. Low. Temp. Phys. 23:27.
Folinsbee, J.T. and Anderson, A.C., 1973, Anomalous Kapitza Resistance to Solid Helium, Phys. Rev. Lett., 31:1580.
Folinsbee, J.T. and Harrison, J.P., 1978, Phonon Reflection at a Silicon/^{3}He Interface, J. Low.Temp. Phys., 32:469.
Guo, C.J. and Maris, H.J., 1972, Reflection of Phonons at an Interface between a Solid and Helium, Phys. Rev. Lett., 29:855.
Guo, C.J. and Maris, H.J., 1974, Experimental study of the reflection of phonons at an interface between dielectric crystals and liquid helium, gaseous helium, and solid neon, Phys. Rev. A10:960.
Guyer, R.A., 1977, Vacancy Waves in Two Dimensional Systems, Proc. ULT Hakone Symposium p178.
Haug, H. and Weiss, K., 1972, A Modified Theory of the Kapitza Resistance, Phys. Lett., A40:19.
Horstman, R.E. and Wolter, J., 1977, Reflection of High-Frequency Phonons at a Solid-Helium Interface: The Role of Phonon Focusing, Phys. Lett. 62A:279.

Ishiguro, T. and Fjieldly, T.A., 1973, Emission of Thermal Energy into Liquid Helium, Phys. Lett. 45A:127.

Johnson, R.C., 1964, Kapitza Resistance of a Crystal Cleaved While Immersed in Liquid Helium, Bull. Am. Phys. Soc., 9:713.

Keen, B.E., Matthews, P.W. and Wilks, J., 1965, The Acoustic impedance of liquid helium-3, Proc. Roy. Soc., A284:125.

Khalatnikov, I.M., 1965, Heat Exchange between a Solid and Helium II, in "An Introduction to the theory of superfluidity", W.A. Benjamin inc., New York, Amsterdam.

Kinder, H., Weber, J. and Dietsche, W., 1979, Kapitza Resistance Studies Using Phonon Pulse Reflection, in "Phonon Scattering in Condensed Matter", Maris, H.J., ed., Plenum, New York.

Kinder, H. and Dietsche, W., 1974, Strong Phonon Conversion at the Helium-Solid Interface, Phys. Rev. Lett., 33:578.

Little, W.A., 1959, The Transport of Heat Between Dissimilar Solids at Low Temperatures, Can. J. Phys., 37:334.

Long, A.R., Sherlock, R.A. and Wyatt, A.F.G., 1974, Phonon reflection at a cleaved sodium fluoride-helium film interface, J. Low. Temp. Phys., 15:523.

Maris, H.J., 1979, Phonon Transmission across interfaces and the Kapitza resistance, Phys. Rev. B19:1443.

Nakayama, T., 1977, Tunneling states of helium atoms and the Kapitza thermal resistance, J. Phys. C 10:3273.

Namaizawa, H., 1976, A Model of Inelastic Heat Transfer Mechanism for the Excess Kapitza Conductance, in "Phonon Scattering in Solids", Challis, L.J., Rampton, V.W. and Wyatt, A.F.G., eds, Plenum, New York.

Page, G.J., 1977, Phonon propagation across liquid ^{4}He-solid interfaces, Ph.D. Thesis, University of Nottingham.

Page, G.J., Sherlock, R.A., Wyatt, A.F.G. and Ziebeck, K.R.A., 1976, The Angular Distribution of Phonons Radiated from the Cleaved <100> Faces of NaF, KCℓ, and MgO into Liquid ^{4}He, in "Phonon Scattering in Solids", Challis, L.J., Rampton, V.W. and Wyatt, A.F.G., eds., Plenum, New York.

Peterson, R.E. and Anderson, A.C., 1972, The Kapitza Thermal Boundary Resistance, Phys. Letters 40A:317.

Pollack, G.L., 1969, Kapitza Resistance, Rev. Mod. Phys., 41:48.

Polanco, S.E. and Bretz, M., 1978,Thermal Resistivity of Layered ^{4}He Films on 2YX Graphite Below 2K, Phys. Rev., 17:151.

Reynolds, C.L. and Anderson, A.C., 1977, Thermal boundary resistance to solid, helium, hydrogen, deuterium, and neon II, Phys. Rev. B15:5466.

Sabisky, E.S. and Anderson, C.H., 1975, Reflection Coefficient of Phonons Between 15 and 315 GHz at a Crystal-Liquid-Helium Interface, Solid State Comm. 17:1095.

Salemink, H.W.M., Van Kempen, H., and Wyder, P., 1978, Ballistic-Phonon-Pulse Transmission through a Solid-Liquid He II Interface at T = 0.25K, Phys. Rev. Lett., 41:1733.

Saslow, W.M. and Lumpkin, M.E., 1979, Interfacial Three-Phonon Processes and Anomalous Interfacial Energy Transport, Solid State Comm., 29:395.

Sheard, F.W. and Fletcher, J.R., 1979, Kapitza Conduction Via Non Localized Interface Modes, in "Phonon Scattering in Condensed Matter", Maris, H.J., ed., Plenum, New York.

Snyder, N.S., 1970, Heat transport through helium II:Kapitza conductance, Cryogenics, 10:89

Snyder, N.S., 1976, The Kapitza Conductance of the (100) Surface of Copper, J. Low Temp. Phys.,22:257.

Svensson, E.C., Woods, A.D.B. and Martel, P., 1972, Phonon Dispersion in Liquid Helium under Pressure, Phys. Rev. Lett. 29:1148.

Swanenburg, T.J.B. and Wolter, J., 1973, Transmission of High-Frequency Phonons through a Solid-Liquid Helium Interface, Phys. Rev. Lett., 31:693.

Taborek, P. and Goodstein, D., 1979, Phonon reflection at a sapphire-vacuum interface, J. Phys. C, Solid State Phys., 12:4737.

Taborek, P. and Goodstein, D., 1980, Phonon Focusing Catastrophes, Preprint.

Van der Sluijs, J.C.A. and Pattullo, A.W., 1979, Agreement between Experiment and Theory in the Kapitza Conductance between clean copper samples and ^{4}He from 0.3 to 1.3K, in "Phonon Scattering in Condensed Matter", Maris, H.J., ed., Plenum, New York.

Weber, J. Dietsche, W. and Kinder, H., 1977, Time Delay of Phonon Pulses in Thin Helium Films,Phys. Lett., 64A:202.

Weber, J., Sandemann, W., Dietsche, W. and Kinder, H., 1978, Absence of Anomalous Kapitza Conductance on Freshly Cleaved Surfaces, Phys. Rev. Lett., 40:1469.

Wolter, J. and Horstman, R.E., 1977, Quantitative Measurement of the Phonon Energy Transmission Coefficient for a Silicon-He II Boundary, Phys. Lett., 61A:238.

Wyatt, A.F.G., 1979, Characteristics of Kapitza Conductance, in "Phonon Scattering in Condensed Matter", Maris, H.J., ed., Plenum, New York.

Wyatt, A.F.G. and Crisp, G.N., 1978, Frequency of phonons emitted into liquid He by a solid, J. de Physique 39:C6-244.

Wyatt, A.F.G. and Page, G.J., 1978,The Transmission of phonons from liquid He to crystalline NaF, J. Phys. C: Solid State Phys., 11:4927.

Wyatt, A.F.G., Sherlock, R.A. and Allum, D.R., 1980, to be published.

NONEQUILIBRIUM PHONONS

W. Eisenmenger

Physikalisches Institut der Universität Stuttgart

7000 Stuttgart 80, BRD

1. INTRODUCTION

Nonequilibrium or nonthermal phonon distributions in solids in general result from any kind of energy supply to the system. Experimentally thin superconducting films are especially well suited for studies of nonequilibrium excitation distributions. Quasiparticle and phonon distributions depend on the primary excitation processes, e.g. electron-injection by tunneling, phonon, microwave or optical radiation as well on the properties of the system as quasiparticle and phonon lifetimes, mean free path values and phonon escape conditions from the superconductor. Nonequilibrium in superconductors gives rise to an ample field of different phenomena. This chapter reviews work on nonequilibrium phonons resulting from tunnel injection, and phonon generation and detection in Section 2, experimental probing of nonequilibrium phonon emission in Section 3, applications to phonon absorption spectrocopy in Section 4, quantitative phonon intensity measurements in Section 5, applications to phonon emission spectroscopy in Section 6, and aspects of future work in Section 7.

Work in this field started around 1965. Earlier reviews (Eisenmenger 1969, Dayem 1972, Renk 1972, Kinder 1973, Kinder 1975, Eisenmenger 1976a and Bron 1980) discuss fundamental experiments and theory, the application to phonon spectroscopy and compare also with other techniques, as e.g. heat pulse experiments or optical generation and detection of acoustical phonons.

Elastic single electron tunneling between two superconducting films separated by a thin oxide barrier results in a nonequilibrium energy distribution of quasiparticles. The quasiparticles decay predominantly under phonon emission. This results in a nonequilibrium phonon-energy distribution. The phonon frequencies typically extend from 50 GHz to 1.5 THz. Since the phonon mean free path is comparable to or exceeds the film-thickness (100 Å to 1000 Å) a significant number of phonons can escape into the insulator single crystal substrate without reabsorption or thermalization in the superconductor. The phonon spectrum can be analyzed with a frequency selective superconducting tunneling detector, thus giving information on the nonequilibrium distribution. By the characteristic discontinuities of this spectrum, applications for the spectroscopy with acoustic phonons are possible, especially with respect to resonant phonon absorption in the volume of substrate crystals and also phonon reflection and absorption in transmission across boundaries. In addition, phonon absorption and reabsorption processes in superconductors can be studied. Finally, information on phonon spectra can also be obtained by directly analyzing the quasiparticle distribution in superconducting films resulting from phonon absorption by pairbreaking.

2. QUASIPARTICLE AND PHONON TRANSITIONS, TUNNELING, PHONON DETECTION AND GENERATION

2.1 Quasiparticle and Phonon Transitions

Single quasiparticles in superconductors are characterized by the BCS (Bardeen et al., 1957) energy-wavevector relation E(k), cf. Fig. 1. The dispersion relation has a minimum with $E = \Delta$, the energy gap at the Fermi-wavevector k_F. Electron and holelike excitation branches are symmetrical to k_F for small deviations from k_F. Excited quasiparticles decay under spontaneous or stimulated

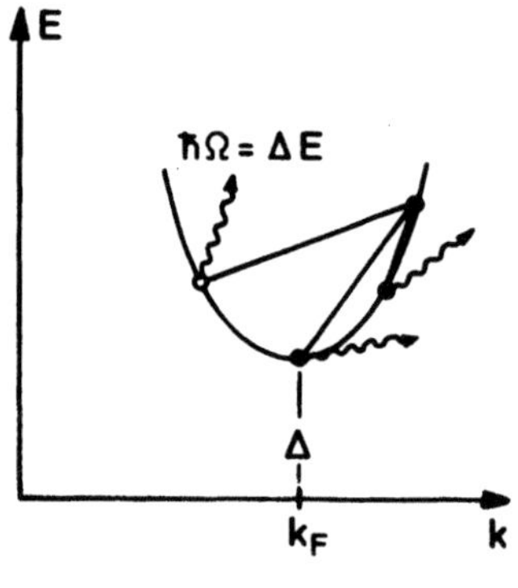

Fig. 1. Phonon emission by quasiparticle relaxation.

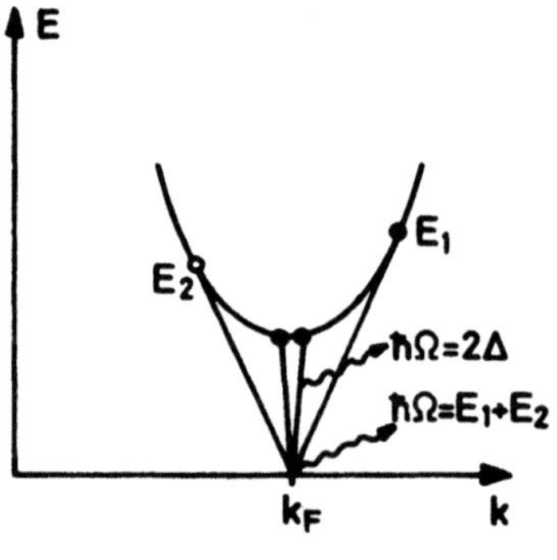

Fig. 2. Phonon emission by quasiparticle recombination.

phonon emission into states of lower energy either within the same branch or changing over to the other branch (Tinkham, 1972). The phonon energies emitted in this relaxation transition (also named phonon "Bremsstrahlung" (Kinder, 1972a)) range from $\hbar\Omega = 0$ to $\hbar\Omega = E - \Delta$ with E the quasiparticle excitation energy. The total spontaneous transition probability τ^{-1} depends on the quasiparticle energy approximately as $\tau^{-1} \propto (E-\Delta)^{3,5}$ (Tewordt, 1962), indicating that relaxation of quasiparticles close at the gap edge is vanishing. This can be understood from the decreasing number of phonon states available for the transition and by destructive interference in the BCS transition probability (Bardeen et al., 1957).

The reverse process of relaxation is phonon reabsorption by excited quasiparticles. This process becomes important at high quasiparticle densities; i.e. at temperatures not small compared to T_c or under conditions of high quasiparticle population by injection, resulting in an inelastic decay of nonequilibrium phonons.

The second process of quasiparticle decay is the recombination of two quasiparticles to a Cooper-pair in the superconducting groundstate under the emission of one recombination phonon of the energy $\hbar\Omega = E_1 + E_2$; E_1 and E_2 are the energies of the recombining quasiparticles. The minimum energy in this process is $\hbar\Omega = 2\Delta$ for quasiparticles at the gap edge. The recombination rate τ_R^{-1} for quasiparticles at the gap edge depends on their number density as $\tau_R^{-1} \propto N(\Delta)$. The reverse process, i.e. phonon absorption by pairbreaking, requires the minimum phonon energy $\hbar\Omega = 2\Delta$. This process is used for phonon detection, and is further important in measurements of the quasiparticle recombination lifetime. Phonons emitted in recombination are reabsorbed via Cooper-pairbreaking with high probability (the phonon mean free path for pairbreaking is of the order of 100 Å to 1000 Å, thus leading to an enhanced effective lifetime $\tau_{eff} \equiv \tau_R^* > \tau_R$ by 2Δ phonon trapping (Rothwarf and Taylor, 1967). Depending on the phonon surface escape and volume decay either by quasiparticle excitation or anharmonic phonon interactions, τ_{eff} in superconducting films is up to two orders of magnitude larger than τ_R (Eisenmenger et al., 1977). The absolute values for τ and τ_R and their dependence on quasiparticle energy and population have been studied theoretically in detail by Kaplan et al. (1976).

Nonequilibrium distributions of quasiparticles and phonons are determined by the processes discussed above. Calculations based on the BCS transition rates make use of coupled kinetic equations (rate equations) of the quasiparticle and phonon system. Important parameters entering such calculations are: The energy distribution of the primary quasiparticle excitation rates as e.g. by tunneling injection, phonon-, microwave- or photon irradiation, the quasiparticle - phonon and the phonon - quasiparticle transition rates, the phonon lifetimes for surface-boundary escape and volume decay, and also elastic scattering rates for quasiparticles and phonons in cases of inhomogeneous distributions; i.e. diffusion. The latter contribution is of importance with respect to high energy phonon escape from superconducting films and the possible occurrence of nonequilibrium instabilities (cf. Langenberg, 1975 and Chapt. 10).

Since the complete numerical treatment of the kinetic equations is complicated, different approximations have been used. With respect to nonequilibrium phonons and quasiparticles at the gap edge a first treatment by Rothwarf and Taylor (1967) showed the important result that the experimental quasiparticle recombination lifetime τ_{eff} is significantly increased as compared to the intrinsic recombination lifetime τ_R. More general models, trying to simplify the nonequilibrium description by introducing quasithermodynamic quantities as an effective chemical potential μ^*, and a "hot quasiparticle" temperature T^* have been proposed by Owen and Scalapino (1972) and Parker (1975). Nonequilibrium phonon emission of tunneling junctions under weak injection has been treated by taking account of only first step relaxation transitions without recombination (Kinder et al., 1970) and including recombination (Welte 1973, cf. Forkel et al., 1973). For the "weak injection (no gap reduction)" situation the detailed balance of high energy quasiparticles and phonons including limited phonon escape has been treated by Dayem and Wiegand (1972). Detailed phonon spectra by relaxation and recombination at finite temperatures and finite tunnel injection voltages taking account of successive relaxation steps have been calculated by Welte (1976). A simple approximation of high energy phonon emission was used by Forkel (1977). Finally strong tunnel injection without phonon reabsorption (Welte 1974, Welte 1976, cf. Welte and Eisenmenger 1980) and with phonon reabsorption and gap reduction under different phonon escape conditions has been treated (Chang and Scalapino, 1978, Chang 1980). In the "weak" tunneling injection case the energy-gap reduction by a stationary increase of the number density of quasiparticles at the gap edge remains below one percent. This condition is typical for the phonon generation with tunneling junctions; i.e. bath temperatures $T < 0.5\ T_c$ and high phonon escape by the condition $d \leq \Lambda$ (d = filmthickness, Λ = phonon mean free path). Although the negligible energy-gap reduction in the weak injection case may appear as characteristic for a minute deviation from thermal equilibrium, the high energy phonon emission spectra in contrast indicate a strong nonequilibrium situation.

Phonon emission spectra of superconducting films, therefore, are very sensitive to deviations from thermal equilibrium. Thus, experimental investigations of phonon emission complement the direct measurements of nonequilibrium quasiparticle distributions by three-layer tunneling experiments discussed in this volume by Gray (1980).

2.2 Tunneling Junctions for Phonon Experiments

In using superconducting film tunneling junctions for phonon generation and detection experiments, elastic single particle tunneling (cf. Gray, 1980) is the most important process to be discussed. The junctions are usually prepared by vacuum deposition (Giaever and Megerle, 1961; Giaever, 1964; Solymar, 1972; Eisenmenger, 1976a) of Al, Sn, or Pb layers with thickness of the order of 1000 Å or less separated by an approximately 10 Å thick oxide barrier which is formed by exposing the first layer to air, oxygen or by glow discharge with oxygen background. In order to obtain a sharp gap structure and an enhanced energy gap, also granular Al-films (Al-evaporation with oxygen background 10^{-5} Torr) or Pb:Bi alloy films are used. The tunneling overlap area is typically of the order of 1 mm^2. As evaporation substrates, especially for phonon experiments, optically polished single crystals; e.g. Al_2O_3, Si, etc.: with long phonon mean free path, typically exceeding 1 cm, are used. In these experiments glass substrates or other materials with high phonon scattering are not well suited. Details for different possible procedures in junction preparation can be found in the literature (cf. Solymar, 1972; Eisenmenger, 1976a).

For most phonon experiments, junctions composed of two identical superconductors are used. The I-V (current - voltage) characteristic (see Fig. 3) can be most conveniently described in terms of the density of states in the semiconductor picture (Giaever and Megerle, 1961), cf. Figs. 4 and 5.

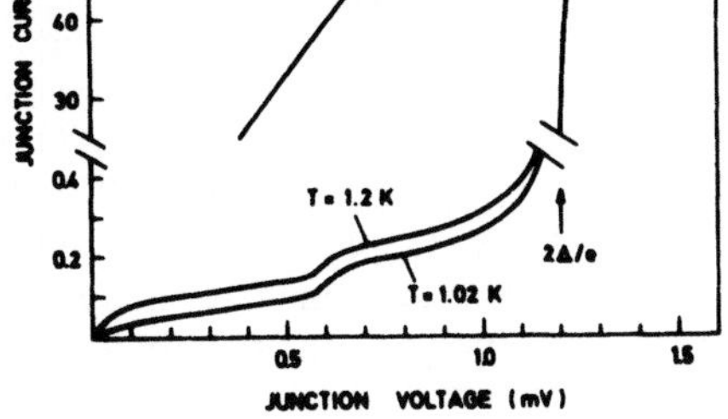

Fig. 3. Current-voltage characteristic of a Sn-I-Sn tunneling junction of 1 mm^2 area, for phonon generation and detection.

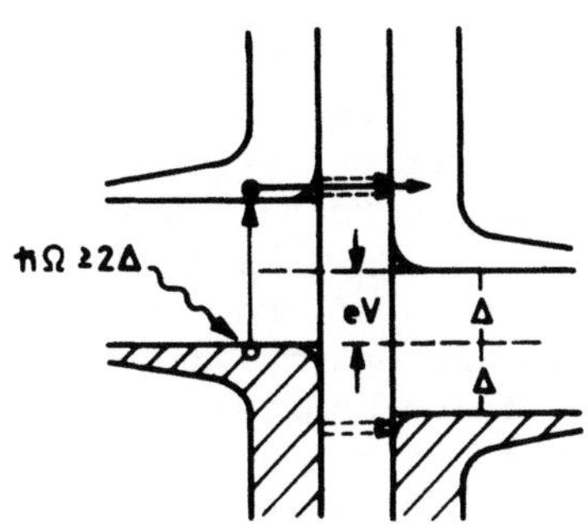

Fig. 4. Phonon detection in the range $0 < eV < 2\Delta$.

For battery voltages $|V| < 2\Delta/e$, with Δ the superconductor energy gap and e the electron charge, only a small temperature dependent current results from thermally excited quasiparticles. Irradiation with phonons of energy $\hbar\Omega > 2\Delta$ leads to additional quasiparticle excitations via Cooper-pairbreaking and a corresponding increase of the tunneling current. Thus, phonons of sufficient energy can easily be detected with a tunneling junction operated in the voltage range $|V| < 2\Delta/e$, as indicated in Figs. 3 and 4.

2.3 Phonon Detection

A quantitative model for phonon detection can be based on parameters which are determined by experiment (cf. Eisenmenger, 1976). Experimentally low ohmic junctions with asymptotic tunneling resistance R_∞ between 10^{-2} Ohms and 10^{-3} Ohms (1 mm²) are preferable since the phonon detection sensitivity increases with tunneling probability.

In order to suppress the DC-Josephson current, a magnetic field of about 20 to 50 Oe parallel to the junction plane (Helmholtz coils) is required. Shielding the earth-magnetic field can increase the electronic sensitivity, since flux trapping in cooling down below T_c reduces the dynamic resistance and τ_{eff} in the biasing point. Instead of magnetic shielding an orientation of the junction plane parallel to the earth-magnetic field is mostly sufficient. The detector is current-biased usually at a voltage $V \cong 1.4\ \Delta/e$, just above the current step at $V = \Delta/e$, caused by two particle tunneling.

Biasing below $V = \Delta/e$ is often difficult because small discontinuities (Fiske steps) and Josephson subharmonic structures limit the range of linearity.

Electronically the detector has the properties of a current generator. Phonons absorbed by pairbreaking increase the quasiparticle population at the gap edge. This results in an almost parallel shift of the I(V) characteristic in the thermal tunneling regime $|V| < 2\Delta/e$ to higher currents corresponding to a differential detector current signal I_d at constant biasing voltage. With constant biasing current instead the resulting open circuit detector voltage signal is $v_d = I_d \cdot R_{dyn}$, with R_{dyn} the dynamic resistance at the biasing point. For good detector junctions R_{dyn} ranges from 0.1 to 1 ohm. An optimal signal-to-noise ration, therefore, requires the use of low-ohmic strip lines and pulse transformation for matching to the typical 50 Ω input impedance of a low-noise-pulse preamplifier. Alternatively a number of small area tunnel detectors in series can be used for matching (cf. Eisenmenger, 1976a).

Detector calibration in terms of the number of phonons absorbed by pairbreaking under stationary conditions is most easily performed using the relation (cf. Eisenmenger, 1976)

$$I_d = 2\dot{N}_\omega \cdot \frac{\tau_{eff}(T)}{\Omega} \cdot \frac{I(T)}{N(T)}$$

with $\dot{N}_\omega$ the number of 2Δ phonons per unit time absorbed by pairbreaking; Ω = the volume of the tunneling junction; τ_{eff} the experimentally observed effective quasiparticle recombination time; $N(T)$ = the equilibrium quasiparticle number density at T and $I(T)$ the tunneling current by thermally excited quasiparticles. $N(T)$ can be calculated according to the approximation valid within 5 % for $T < 0.5\ T_c$:

$$N(T) = 4N(0)\Delta(T)\sqrt{\frac{\pi}{2}\frac{kT}{\Delta(T)}} \cdot e^{-\Delta(T)/kT}$$

with $N(0)$ the one spin density of states at the Fermi level, as known from electronic specific heat data. $I(T)$ is experimentally obtained from the temperature dependence of the tunneling current at constant voltage bias. Plotting this current as function of $\sqrt{T} \cdot e^{-\Delta/kT}$ and extrapolating to $T = 0$ gives the almost temperature independent leakage and two-particle-current contributions which have to be subtracted from the measured current, thus obtaining $I(T)$. $\tau_{eff}(T)$ can be easily obtained in a phonon pulse experiment from the exponential decay of the detector signal depending on temperature in proportion to $T^{-1/2} \cdot e^{\Delta/kT}$. I_d is obtained from the measured electronic signal voltage taking account of the total amplification and the ohmic load of the stripline and pulse-transformer system in parallel to R_{dyn}. Simple calibration of the total electronic system is accomplished by simulating a current pulse signal (introduced from a high ohmic source) at the tunneling junction and measurement of the amplified voltage signal. The detector relation $I_d \propto \dot{N}_\omega$ holds for the linear detection range; i.e. $I_d \ll I(T)$ with increase of the number of quasiparticles by phonon absorption small compared to the number-density of thermally excited quasiparticles. Experimentally, the fulfillment of this condition is easily checked. A further limitation of linearity results from the curvature of the I(V) curve (variation of R_{dyn}) at the detector bias voltage. Mostly a voltage excursion of a few percent of 2Δ can be admitted. The attainable phonon power resolution depends on the noise figure of the preamplifier; minimal phonon powers of 10^{-13} Watts with one second integration time can be detected.

The detector equation, so far, describes the sensitivity with respect to 2Δ-phonons. Phonons with lower energy cannot directly contribute to pairbreaking, but may be weakly absorbed by thermally excited quasiparticles. The energy supplied by this process is partly upconverted to 2Δ-phonons (heating influence). For $T = 0.5\ T_c$ the corresponding detector-signal contribution is, however, at least three orders of magnitude below the sensitivity for 2Δ-phonons. For phonon energies between 2Δ and 4Δ the sensitivity with respect to the number of absorbed phonons remains constant. Increasing the phonon energy above 4Δ results in an almost linear energy dependence, since quasiparticles are excited via pairbreaking also to energies above 3Δ and succeeding relaxation contributes to secondary 2Δ-phonons within the detector. An additional frequency dependence in principle results from the energy dependence of the phonon mean free path. In Sn- and Pb-junctions of at least 2000 Å thickness the 2Δ-phonon mean free path is equal or shorter than this length and all incident 2Δ-phonons are absorbed by pairbreaking. In pure Al-junctions the energy dependence of the phonon mean free path must be taken account of. For further details see (Eisenmenger, 1976a; Wyatt 1976).

2.4 Phonon Generation with Superconducting Tunneling Junctions

At voltages $V \geq 2\Delta/e$ the tunneling current increases discontinuously (Fig. 3) and approaches with higher voltages the linear normal conductor limit. In this regime quasiparticles are excited via pairbreaking by the battery energy leading to nonequilibrium injection into the films. In addition, also tunneling by thermal or injection-excited quasiparticles contributes to the injection distribution which is shown qualitatively in the upper part of Fig. 5. The sharp maxima at the lowest and highest quasiparticle energies result from the singularities in the BCS-densities of state, cf. (Gray, 1980; McMillan and Rowell, 1969). A detailed account of the stationary quasiparticle occupation by tunnel-injection, considering especially branch imbalance and the corresponding relaxation and recombination rates in Al-films, has been given by Kirtley et al. (1980). So far, the possible influence of branch imbalance on phonon experiments has not been studied, and will not be further discussed.

The quasiparticle decay takes place by relaxation and recombination transitions predominantly under phonon emission. Since the available densities of state for photon emission is by several orders of magnitude below the density of phonon states, photon emission can be neglected. The phenomenological similarity between relaxation phonon emission and continuous x-rays has led to the name "phonon Bremsstrahlung" (Kinder, 1972a).

Again as consequence of the singularities in the density of states the quasiparticle relaxation and recombination decay results in characteristic structures (Kinder et al., 1970; Kinder, 1972a; Dayem and Wiegand, 1972; Welte et al., 1972; Welte, 1973; Forkel et al., 1973; Forkel, 1974; Welte, 1976; Forkel, 1977; Chang and Scalapino, 1978) of the phonon spectrum, as shown in the lower part of Fig. 5. The most significant structures are:

a) $\hbar\Omega = 2\Delta$, minimum energy of the recombination spectrum.

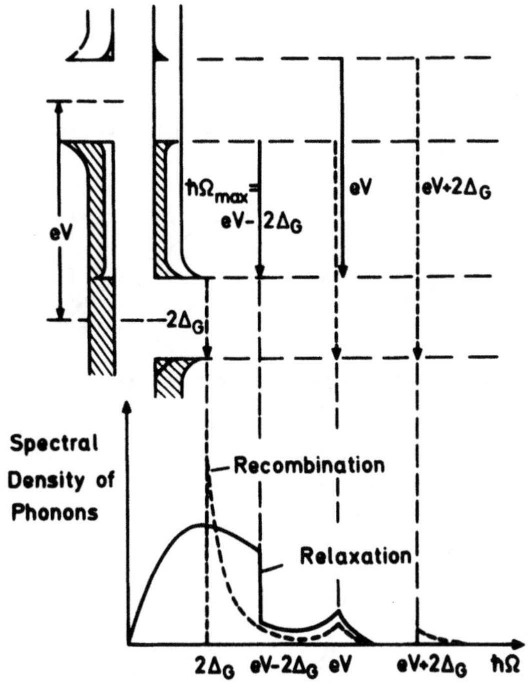

Fig. 5. Phonon emission spectra for relaxation and recombination at finite temperature. Transitions corresponding to density of state- or quasiparticle population singularities give rise to characteristic spectral structures, as discussed in the text.

b) $\hbar\Omega$ = eV, direct recombination of injected quasiparticles with thermal quasiparticles.
c) $\hbar\Omega$ = eV + 2Δ, minimum phonon energy for the direct recombination of tunnel-injected thermal quasiparticles with thermal excitations.
d) $\hbar\Omega$ = eV - 2Δ, upper edge of the relaxation spectrum; the shape and voltage dependence of this structure provides the possibility of voltage tunable phonon spectroscopy.
e) $\hbar\Omega$ = eV, relaxation of tunnel-injected thermal quasiparticles.

In numerical calculations of these spectra, different approximations have been used, cf. Sect. 2.1. In the simplest form which for low injection and low temperatures already reveals the most significant spectral contributions, only first step processes are considered. An example for the calculated first step relaxation spectrum at T > 0 and eV > 2Δ_G is given in Fig. 6 (Welte 1973), indicating the upper discontinuity in the relaxation phonon spectrum at $\hbar\Omega$ = eV_G - 2Δ_G and the presence of the thermal peak structure at $\hbar\Omega$ = eV_G. The height of the relaxation edge decreases with in-

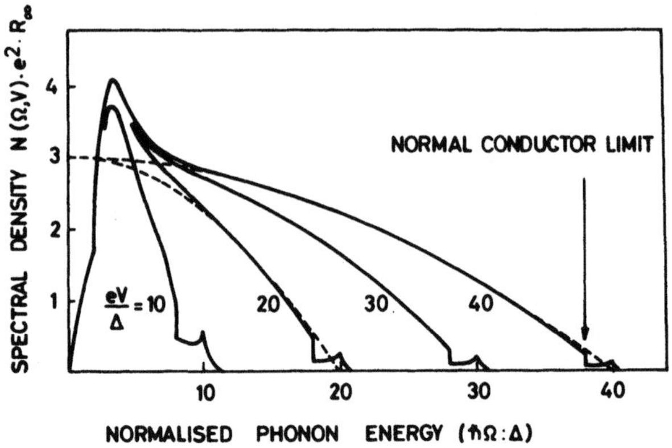

Fig. 6. Calculated relaxation phonon spectra for high generator voltages at finite temperature and weak injection; first step decay; no reabsorption; dotted line: normal conductor limit for T = 0; (Welte, 1973; Forkel et al., 1973).

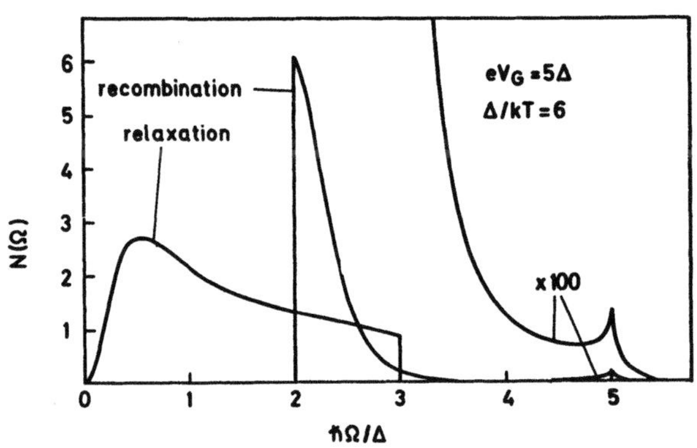

Fig. 7. Calculated relaxation and recombination phonon spectrum at finite temperature and weak injection, no reabsorption, successive quasiparticle decay included (Welte, 1976).

creasing V_G and temperature. For comparison also the phonon spectrum for the normal conducting limit at T = 0 is introduced.

Since quasiparticles decay into all available unoccupied energy levels, a higher degree of approximation must take account of secondary and higher order decay steps. Fig. 7 shows an example for weak injection without phonon reabsorption (Welte, 1976) concerning the relaxation and recombination spectrum at higher temperatures. As seen from this figure, the inclusion of secondary and higher order quasiparticle decay steps does not change the qualitative spectral shape. Recombination between injected quasiparticles becomes important under strong injection as well as reabsorption of phonons by pairbreaking and quasiparticle excitation. The influence of di-

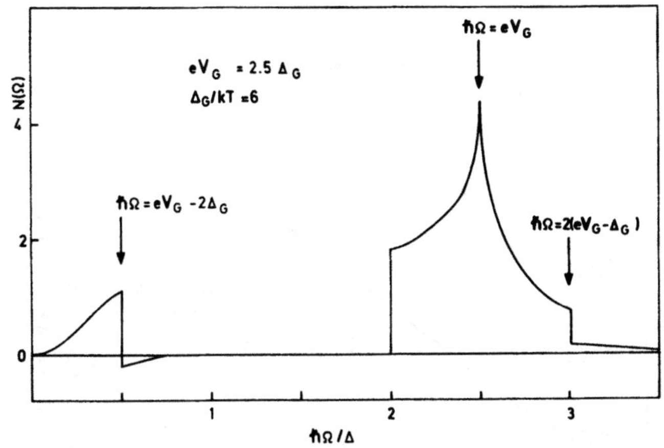

Fig. 8. Calculated relaxation and recombination phonon spectrum at finite temperature and finite tunnel injection, indicating the contribution of the direct recombination of high energy quasiparticles. (Welte, 1976)

rect recombination without reabsorption in the case of strong injection at low temperatures is shown in Fig. 8 (Welte, 1976) by the occurence of an upper spectral edge in the recombination spectrum at $\hbar\Omega = 2(eV-\Delta_G)$, a strong increase of the peak structure at $\hbar\Omega = eV_G$ and by the high population of quasiparticles at the gap edge. Even without taking account of the direct recombination of injected quasiparticles, high injection results in an increased bandwidth of the recombination phonons as consequence of the population of quasiparticle states at and above the gap edge. In this situation a proper account of phonon escape at the film boundaries and the anharmonic phonon decay at the surface or within the volume of the films is necessary. High injection nonequilibrium under limited phonon escape will be discussed in this volume, Chapter 9, by Chang (1980).

3. EXPERIMENTAL EVIDENCE FOR NONEQUILIBRIUM PHONON EMISSION

In the first observation of nonequilibrium phonon emission (Eisenmenger and Dayem, 1967) Sn-I-Sn-junctions as phonon generator and detector on Al_2O_3 as substrate crystal have been used. In this experiment, a nearly linear increase of the detected phonon signal amplitude, as function of the generator current up to the generator voltage of $V_G = 4\Delta_G/e$ is found, cf. Fig. 9. With higher generator current again a linear signal increase is observed but with a slope nearly three times larger than before; i.e. the dependence of the detector signal on the generator current develops a sharp bend at $V_G = 4\Delta_G/e$. This can be easily understood in terms of the onset of the contribution of relaxation phonons with maximum energy $\hbar\Omega = eV_G - 2\Delta_G$ at the voltage $V_G = 4\Delta_G/e$. Below this generator voltage only recombination phonons can contribute to the detector signal.

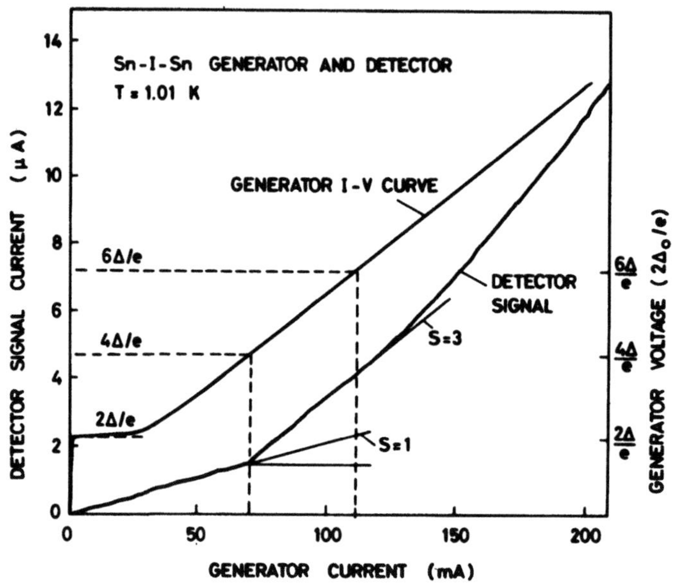

Fig. 9. Detector signal as function of the generator current (voltage: right scale) for Sn-I-Sn generator and detector, T = 1.01 K. For comparison with the characteristic generator voltages 4Δ/e and 6Δ/e, the generator I-V curve is introduced (Eisenmenger and Dayem 1967; Kinder et al., 1970).

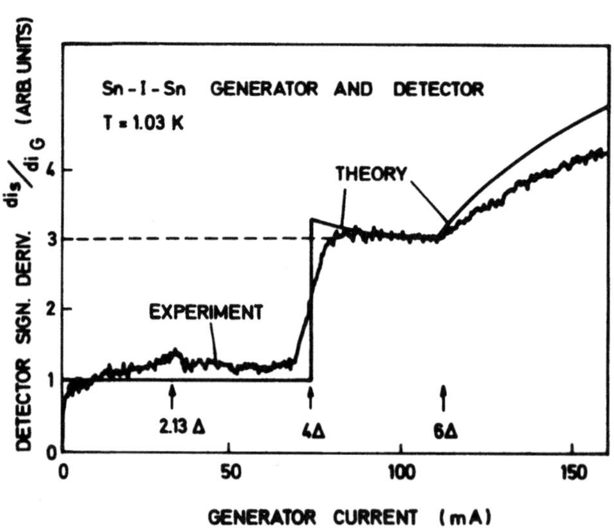

Fig. 10. Derivative of the detector signal with respect to the generator current. Sn-I-Sn generator and detector. T = 1.03 K; cf. Fig. 9 (Eisenmenger and Dayem 1967; Kinder et al., 1970).

This result clearly indicates nonequilibrium phonon emission under weak injection conditions; i.e. strongly impeded phonon thermalization in thin superconducting films. By superimposing small current pulses on the D.C.-generator current it is observed that the differential detector signal is roughly constant with a step at $V_G = 4\Delta_G/e$, and with the onset of a linear increase at $V_G = 6\Delta_G/e$, cf. Fig. 10. This indicates further contributions to the number of detected 2Δ-phonons by repeated phonon reabsorption and reemission in the generator or detector (Kinder et al., 1970); i.e. a splitting or primary relaxation phonons with the energy $\hbar\Omega = 4\Delta$ into two phonons with the energy $\hbar\Omega = 2\Delta$. The theoretical change of slope at $eV_G = 4\Delta_G$ in Fig. 9 and the corresponding step in Fig. 10 for $T = 0$ in the weak injection limit amounts to 3.1 (Kinder et al.,1970; Long and Adkins, 1973). The experiment in general exhibits a somewhat smaller value, especially for low ohmic generator junctions. In this case the high quasiparticle population at the gap edge leads to an enhancement of the direct recombination of high energy quasiparticles and the corresponding reduction of relaxation transitions. The equivalent situation holds for elevated temperatures.

Further experimental evidence for the emission of 2Δ nonequilibrium phonons from Sn-I-Sn-generators was provided by Dynes et al. (1971), using the mechanical stress tunable phonon transition in Sb-doped Ge-substrates as phonon spectrometer.

The first demonstration of the feasibility of nonequilibrium phonon emission from superconducting Sn-I-Sn tunneling junctions for tunable phonon absorption spectroscopy, cf. Sect. 5, was given by Kinder (1972a), who used the voltage dependence of the upper edge of the relaxation phonon spectrum at $\hbar\Omega = eV_G - 2\Delta_G$ for investigating the phonon absorption of V^{3+} in Al_2O_3. In this case relaxation phonons with energy $\hbar\Omega = eV - 2\Delta_G < 2\Delta_G$ at $T < 0.5\ T_c$ are not reabsorbed in the superconducting film and can be radiated into the substrate essentially without further interaction. The upper frequency limit of this method results from the fact that relaxation phonons with the energy $\hbar\Omega \geq 2\Delta_G$ are strongly reabsorbed already in the Sn-films of the generator. This holds also for strong coupling superconductors as Pb. In contrast for thin films of the weak coupling superconductor Al a finite escape of relaxation phonons with $\hbar\Omega > 2\Delta_G$ is to be expected, see Fig. 11, since the 2Δ-phonon mean free path is of the order of 1000 Å (cf. Kaplan et al., 1977).

For the escape of relaxation phonons from the generator junctions in addition to the phonon mean free path the following influences have to be taken into account (Welte et al., 1972; cf. Eisenmenger, 1976a), as depicted in Fig. 11.

i) quasiparticles injected by tunneling have a sharp angle distribution normal (maximum momentum) to the barrier. By momentum conser-

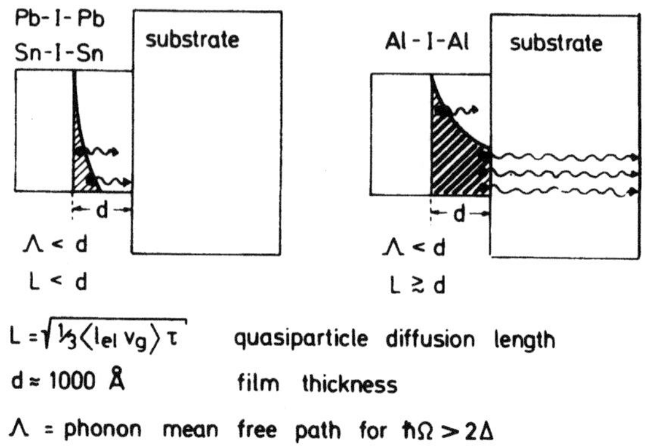

Fig. 11. Phonon escape conditions for $\hbar\Omega > 2\Delta$. The large quasiparticle diffusion length and phonon mean free path Λ in Al-films increase the phonon escape probability, as compared to Pb and Sn-films.

vation, phonon emission in the primary relaxation decay is perpendicular to the quasiparticle momentum and therefore confined to directions in the plane of the junction with little escape probability into the substrate, even for phonon mean free path values exceeding the film thickness. The above-mentioned momentum conditions become less stringent for real metal Fermi surfaces.

ii) This situation changes by elastic scattering of quasiparticles before relaxation, corresponding to quasiparticle diffusion in the generator film. Also boundary scattering at the film surfaces result in a wide angle distribution of quasiparticle momenta. Thus, relaxation phonons are generated omnidirectionally and more close to the substrate (cf. Fig. 11) with increased escape probability. This process becomes important if the characteristic decay time for relaxation is long compared to the elastic scattering time of quasiparticles; i.e. again in the case of weak coupling superconductors.

iii) A simple model (Welte et al., 1972; cf. Eisenmenger, 1976a) taking account of these influences and using literature data for the relaxation time, elastic quasiparticle scattering time and the phonon mean free path, indicates a high phonon escape probability of primary relaxation phonons with $\hbar\Omega > 2\Delta_G$ into the substrate for thin Al-films. Thus, Al-junctions can be used preferably for studies of the nonequilibrium phonon emission in a wide frequency range with little phonon reabsorption and also a weak "feedback" coupling of the phonon system to the quasiparticle system. This makes Al-films also well-suited for direct studies of primary nonequilibrium quasiparticle distributions (cf. Kirtley et al., 1980).

With respect to the phonon spectrum, as depicted in Fig. 5, and the calculated spectra, see Figs. 6, 7 or 8, an experimental analysis is possible by using tunneling detectors with different energy gaps, such as Al-junctions evaporated with oxygen background (Welte, 1974, 1976). Using instead a Sn-I-Sn junction as detector, the voltage-dependent contributions of the Al-generator spectrum with $\hbar\Omega \geq 2\Delta_{Sn}$ can be investigated (Welte et al., 1972; Forkel, 1974; Forkel, 1977) applying modulation techniques.

Since the detector is only sensitive to phonons with energies exceeding the detector gap $2\Delta_D$, the detector signal corresponds to the integral of the emitted generator phonon spectrum with a lower cut-off energy at $2\Delta_D$. Modulation of the generator voltage results in a detector signal modulation arising mainly from the changing generator spectral contribution at the energy $\hbar\Omega = 2\Delta_D$. Thus, modulation is equivalent to differentiating the integral generator spectrum at the lower cut-off energy $\hbar\Omega = 2\Delta_D$. The detector modulation signal, therefore, corresponds to the spectral intensity at $2\Delta_D$. By changing the generator bias voltage it is, therefore, possible to scan the voltage dependent structures of the generator spectrum. The experimental arrangement is shown in Fig. 12. Generator, Al-I-Al and detector, Sn-I-Sn are prepared on the opposing flat surfaces of a high quality Si single crystal cylinder (Forkel, 1974 and 1977). The observed detector modulation signal, as function of the generator bias voltage, is shown in Fig. 13, clearly revealing the characteristic spectral structures depicted in Fig. 5 (with inverted frequency scale). This demonstrates the strong deviation of the phonon-, and indirectly also of the quasiparticle population from thermal equilibrium, even under weak injection conditions. The experiment further gives evidence that phonons with $\hbar\Omega > 2\Delta_G$ escape from Al-I-Al generators into the substrate only with little reabsorption by pairbreaking. Corresponding experiments with a Sn-generator, using as reference the phonon absorption line of ^{16}O in Si

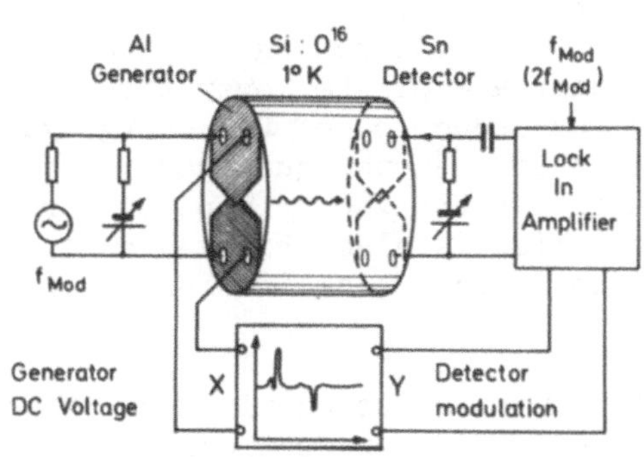

Fig. 12. Experimental arrangement for the analysis of high energy phonon spectra emitted by Al-tunneling junctions as phonon generator. The same measuring technique is used for high energy phonon spectroscopy (Forkel, 1974).

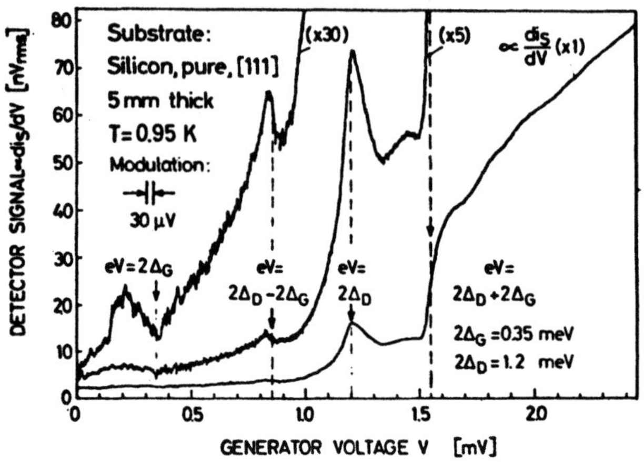

Fig. 13. High energy phonon spectrum emitted by an Al-generator, Sn-I-Sn detector, showing the different spectral structure discussed in Fig. 5 (voltage scale inverted). (Forkel, 1975, see Eisenmenger 1976b).

at 878 GHz, cf. Sect. 4, in contrast demonstrated strong reabsorption (Forkel, 1977). It should be noted that in these experiments with weak injection the strong deviation from equilibrium result from the excitation of quasiparticles by the battery-voltage to energies by far exceeding kT; i.e. the injection of "hot" electrons.

Experiments have been also performed with respect to nonequilibrium phonon emission under strong injection (Welte, 1974, 1976, 1980). These demonstrate the increasing width of the recombination phonon spectrum, the decrease of the lower threshold of 2Δ-recombination spectrum by the reduction of the energy gap, the influence of direct recombination of injected quasiparticles by the enhanced quasiparticle population at the energy gap, thus reducing the relaxation contribution which is also observed at increased temperatures, and finally the direct recombination of injected quasiparticles.

No experimental results on nonequilibrium phonon emission under very high injection with gap reductions up to 50 % are presently known. Such experiments would be valuable with respect to the degree of realization of the μ* and T* quasiparticle distributions in comparison to the results of Willemsen and Gray, (1978) and Kirtley et al., (1980).

4. VOLTAGE-TUNABLE MONOCHROMATIC PHONON SPECTROSCOPY

The upper edge of the phonon spectrum emitted in relaxation of quasiparticles at the energy $\hbar\Omega = eV - 2\Delta_G$ can be used for phonon spectroscopy by differentiation (Kinder, 1972a); i.e. modulation techniques. Applying a small sinusoidal or pulse voltage current, in addition to the constant battery voltage V_G to the generator or operating the generator with amplitude modulated pulses, results in a modulation of the emitted spectrum mainly at the energy $\hbar\Omega = eV - 2\Delta_G$ (Fig. 14 b). The detector modulation signal in this case is mainly caused by nearly monochromatic phonons at $\hbar\Omega = eV - 2\Delta_G$ with a linewidth determined by the modulation amplitude. For a substrate crystal containing defects or impurities with resonant phonon scattering simply a variation of the generator voltage allows tunable monochromatic phonon absorption spectroscopy. This technique meanwhile has found a wide area of applications (cf. Proceedings of Phonon Scattering Conferences 1972, 1976, 1980, and reviews: Kinder, 1973; Kinder, 1975; Eisenmenger, 1976a, and Bron, 1979), and is especially useful in studying phonon selection rules, radiationless transitions, phonon resonance backscattering, phonon propagation in insulators, metals, superconductors, semiconductors, amorphous materials, and phonon transmission and reflection at interfaces as e.g. in the Kapitza problem.

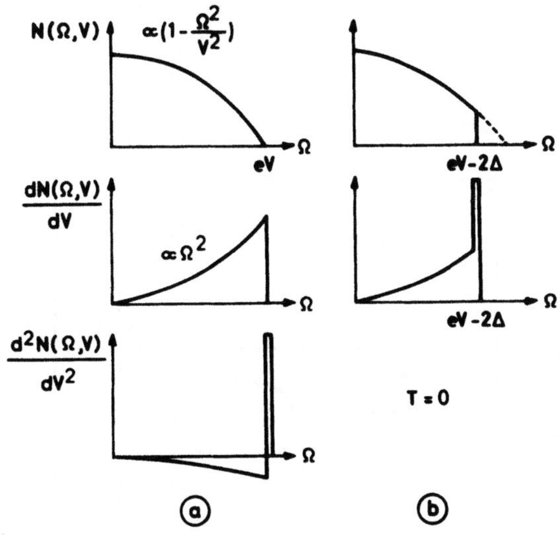

Fig. 14. Modulation of the relaxation spectrum and resulting effective phonon bands: a) approximation for normal conductors or superconductors at energies very large compared to the energy gap, a monochromatic phonon band corresponds to the second derivative (2 ω modulation signal). b) Superconductor relaxation spectrum and monochromatic phonon band of the first derivative (ω modulation signal). (Forkel, 1974)

In phonon spectroscopy two slightly different experimental techniques are used. Avoiding reabsorption in the generator, Kinder (1972a, 1973b) used a Sn-junction as generator for tunable relaxation (Bremsstrahlung) phonons and an Al-junction as detector. In his famous experiment, he demonstrated for the first time monochromatic spectroscopy with nonequilibrium phonons. As a resonant phonon scatterer the Jahn-Teller system $Al_2O_3:V^{3+}$, known from far-infrared and heatconductivity measurements, could be investigated with respect to the selection rules depending on the phonon polarization, see Fig. 15. Also resonant phonon fluorescence in backscattering was demonstrated for the first time. The high phonon-intensity, available in strong coupling superconducting phonon generators, makes this system especially useful for the time of flight separation of longitudinal and transverse phonon contributions.

The upper limiting frequency of 280 GHz for tin junctions can be extended to 870 GHz by using Pb:Bi-junctions as phonon generator (Kinder and Dietsche, 1975). Spectroscopy at higher frequencies may be possible with Nb-junctions.

An alternative way for phonon spectroscopy in the 1 THz-frequency range is the use of Al-generators (Welte et al., 1972, Forkel et al., 1973), since phonon reabsorption for energies $\hbar\Omega > 2\Delta_G$

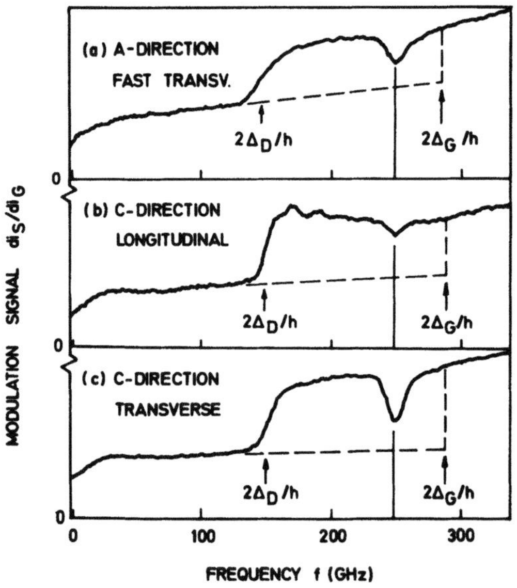

Fig. 15. Phonon-resonant absorption (scattering) in $Al_2O_3:V^{3+}$.
a) fast transverse mode absorption in a-direction.
b) longitudinal mode absorption in c-direction.
e) transverse mode absorption in c-direction.
(Kinder, 1972a).

is comparatively weak (Welte et al., 1972), cf. Sect. 3. The upper frequency limits in Al, resulting from the maxima of phonon dispersion at the Brillouin zone, are 9 THz for longitudinal phonons and 5 THz for transverse phonons. An independent limit results by the momentum conservation between phonons and electrons with $k_{ph\,max} = 2q_F$ at the Fermi surface. Since at higher voltages the relaxation spectrum (Fig. 6) approaches the normal conductor spectrum with a reduction of the amplitude step at $\hbar\Omega = eV - 2\Delta_G$, a second derivative technique (see Fig. 14a) using the second harmonic of the detector signal modulation must be applied. An increase of the available signal power is possible by using very thin Al-films of 100 Å to 200 Å thickness with enhanced energy gap. The high sensitivity and frequency resolution of this technique in the 1 THz range was demonstrated by the resonant phonon absorption (878 GHz) of interstitial oxygen in silicon (Forkel et al., 1973; Forkel, 1977). In the measurement Fig. 16 the resonance absorption of the ^{18}O isotope with the natural concentration of 10^{15} cm^{-3}

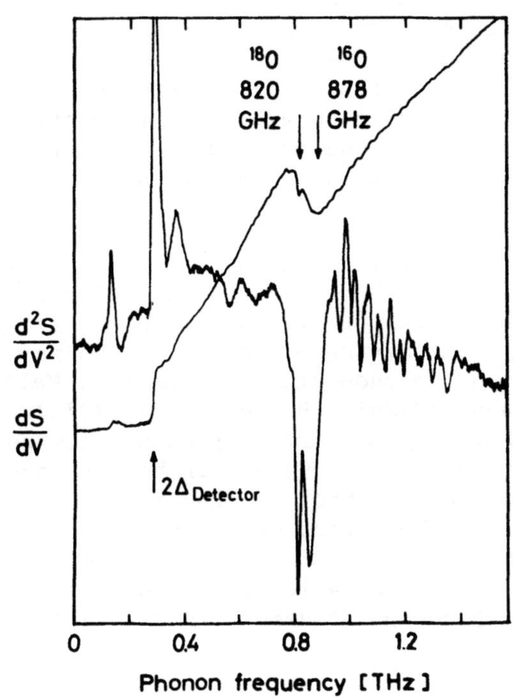

Fig. 16. High frequency phonon absorption of O_i in Si. (Forkel, 1977)

could be detected. The sensitivity is by three orders of magnitude higher than in corresponding FIR measurements and the attainable frequency resolution amounts to 6 GHz. In Fig. 16 also a number of satellite resonances extending to 1.25 THz are observed. These are possibly related to oxygen neighbors with a statistical concentration of 10^{14} cm^{-3}. The main resonance of ^{16}O (concentration 10^{18} cm^{-3}) is "instrumentally" broadened. At lower concentrations the ^{16}O absorption line can be used to study the influence of the generator film-thickness and temperature on the intensity of high frequency phonon emission. Also film-thickness resonances are observable from which the sound velocity and information on phonon reflection at the oxide barrier are obtained. Further spectroscopic examples in the THz-range concern the resonance absorption of the Jahn-Teller systems $Al_2O_3:Ti^{3+}$ at 1.2 THz and $Al_2O_3:V^{4+}$ (Forkel, 1975; Kinder and Dietsche, 1976; Forkel, 1977).

One of the potentialities of acoustic phonon spectroscopy lies in the direct investigation of radiationless transitions, unobservable in FIR-spectroscopy. One of the first examples was the observation of sharp phonon resonance lines in undoped Verneuil Al_2O_3 (Kinder, 1972c; Welte et al., 1972; Kinder and Dietsche, 1975) which are still unexplained. A second example is the phonon absorption in Si:In at 4.2 meV. Ultrasonic absorption measurements (Schad and Lassmann, 1976; Lassmann and Zeile, 1980) indicated that Si:In represents a Jahn-Teller system with a ground-level splitting of 4.2 meV. This could directly be verified by phonon spectroscopy (Schenk et al., 1978). The observed resonance energy and the large linewidth (see Fig. 17) could be explained theoretically (Sigmund and Lassmann, 1980). Corresponding results of ultrasonic absorption and phonon spectroscopy were obtained for the Jahn-Teller system Si:B (cf. Lassmann and Zeile, 1980).

Resonant phonon scattering in solids has been also investigated in paraelectric systems with respect to selection rules under level splitting by mechanical stress (cf. Windheim and Kinder, 1975a,b) in ferrimagnetic systems giving evidence for, so far, unobservable sharp resonant magnon - phonon interactions (cf. Mattes et al., 1978); in $MgO:Cr^{2}$ and $MgO:Mn^{3+}$ (Hasan et al., 1978); in $Al_2O_3:Cr^{3+}$ indicating the possibility of virtual phonon exchange coupling between neighbors (Hasan et al., 1979a); in highly doped $Al_2O_3:V^{3+}$ also possibly indicating neighbor interactions (Hasan et al., 1978b); and in $Al_2O_3:Fe^{2+}$ with the direct determination of the groundstate zero field splitting (Smith et al., 1978). Especially the observation of neighbor-interactions which is often inhibited by the low concentration of neighbor pairs demonstrates the high sensitivity of tunable phonon spectroscopy.

Further, this technique provided significant evidence for the anomalous phonon dispersion in liquid ^4He in the range from 0 to 300 GHz (cf. Dynes and Narayanamurti, 1974, and Narayanamurti and

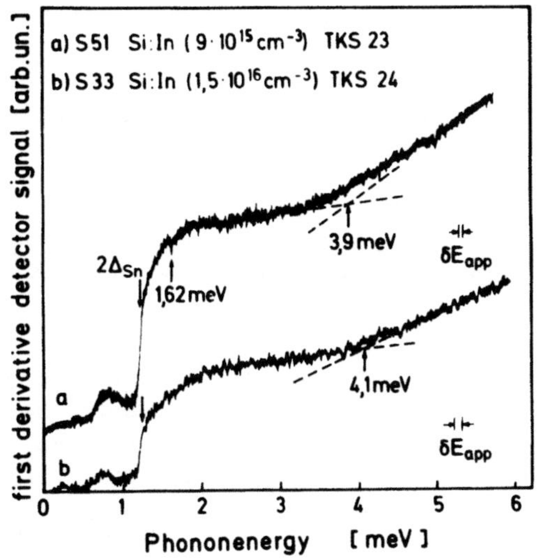

Fig. 17. Phonon absorption in In-doped Si by Jahn-Teller splitting of the In-ground state (Schenk et al., 1978).

Dynes, 1976) and also for inelastic phonon scattering (phonon energy decay) in the anomalous high phonon transmission (Kapitza-anomaly) at the boundary between solids and liquid ^4He (Dietsche and Kinder, 1976), (Wyatt and Crisp, 1978). The present state of knowledge on the Kapitza-anomaly which has been significantly improved by phonon experiments with superconducting tunneling junctions is discussed by Kinder et al. (1980) and Wyatt (1980). Using tunnel junction detectors also the inelastic scattering of two phonon beams in liquid ^4He resulting in frequency up-conversion was observed (Korczynskyj and Wyatt, 1978).

Phonon and quasiparticle propagation in bulk Pb single crystals was studied by Hu et al. (1977), and Narayanamurti et al., (1978), using tunnel junctions and bolometers as detector. Depending on temperature, ballistic phonon propagation and combined diffusion of phonons and quasiparticles could be observed. In similar experiments with bulk Sn single crystals and with tunnel junction phonon detection, Pannetier et al. (1977) were able to determine the temperature-dependent attenuation of ballistic phonons by quasiparticles; also the onset of strong absorption by pairbreaking could be observed.

The first dispersion reduced ballistic phonon propagation in solids was reported by Huet et al. (1972) for InSb in 100 direction. In these experiments superconducting tunneling detectors were used for frequency selection. For degenerately doped n-InSb with different carrier concentrations Huet and Maneval (1974) observed the reduction of longitudinal phonon absorption for $k_{ph} > 2q_F$; i.e. the condition of the Kohn anomaly, using a Sn-I-Sn detector for frequency selection. In addition, the detector was magnetically tuned, (cf. Narayanamurti and Dynes, 1971). Recently also resonant phonon absorption by a superlattice of GaAs/AlGaAs could be demonstrated by Narayanamurti et al. (1979), using the tunnel junction technique. Also other applications in the study of phonon emission and propagation in semiconductors are promising (cf. Reupert et al., 1976; Narayanamurti, 1980).

In addition to phonon propagation in single crystals with dimensions in the mm and cm range also the transverse or longitudinal phonon velocities and scattering mean free path values in thin films have been determined. This was possible by observing standing wave film-thickness resonances (Kinder, 1972) in granular Al-films indicating that the sound velocities do not deviate from the bulk values but that the phonon mean free path is reduced. Additional information by measuring the effective recombination lifetime of quasiparticles in these films indicates an increased electron - phonon interaction enhancement. Other film resonance studies (Forkel, 1977) also indicate an unchanged bulk phonon velocity up to 800 GHz in thin Al-films. Phonon mean free path determinations by direct attenuation in thin Al- and Cu-films have been reported by Long (1973). A very recent measurement of phonon resonances in amorphous Ga-films (Dietsche et al., 1980) showed a reduction of the transverse sound velocity by a factor of 2.7, as compared to the crystalline phase, indicating the breakdown of shear stiffness in amorphous Ga. Applications to inelastic phonon scattering in other amorphous systems; e.g. glasses; will be discussed in Sect. 6.

Further literature on the application of tunable spectroscopy with nonequilibrium phonons can be found by the references given in the introduction of Sect. 4.

5. QUANTITATIVE PHONON INTENSITY MEASUREMENTS

For the analysis of nonequilibrium phonon distributions and also for phonon spectroscopy a quantitative comparison of experiment and theory is desirable. With respect to the phonon emission and detection by tunneling junctions this has been investigated mainly with respect to phonon emission by recombination (cf. Trumpp and Eisenmenger, 1977). For tunneling junctions with equal energy gap as generator and detector the calculation of detector signal amplitudes is straightforward on the basis of a simple ballistic model (cf. Eisenmenger, 1976a).

Within the phonon energy range $2\Delta_D < \hbar\Omega = 4\Delta_D$ the absolute number of phonons absorbed in the detector is obtained by the calibration procedure described in Sect. 2.3. The number of $2\Delta_G$-phonons

$\dot{N}_{\omega G}$ created per unit time in the generator at voltages $2\Delta_G \leq eV_G < 4\Delta_G$ is given by

$$\dot{N}_{\omega G} = \frac{I_G}{e}$$

with I_G the single particle tunneling current, since each tunnel electron corresponds to the breaking up of one Cooper-pair which in recombination delivers one phonon of $2\Delta_G$.

From this generation rate the number of phonons reaching the detector and absorbed by pairbreaking are given (cf. Trumpp and Eisenmenger, 1977) by

$$\dot{N}_{\omega d} = \frac{I_G}{e} \cdot A_d \cdot \frac{1}{\pi r^2_{GD}} \cdot \overline{(f \cdot \Gamma \cdot p)}^{\sigma,\phi_D}_{\sigma,\phi}$$

with A_d = detector area; r_{GD} = distance generator - detector; f = phonon focussing factor; Γ = phonon transmission probability at the substrate - detector interface; p = pairbreaking absorption probability of phonons in the detector, σ = phonon mode, ϕ = angles of ballistic propagation directions between generator and detector areas.

This equation applies to the weighted average of all contributing phonon modes and directions. For time of flight pulse measurements corresponding equations hold separately for each mode where also the different phonon mode contributions in the escape from the generator into the substrate have to be regarded. It is assumed that all phonons created in the generator escape into the substrate (vacuum condition at the free generator surface). Phonon decay by anharmonic processes in the generator or by scattering of excited quasiparticles is neglected. Experimental support for the validity of this assumption is given by the measurement of τ_{eff} in tin junctions on Si as function of film-thickness and temperature (Eisenmenger et al., 1977). In these measurements τ_{eff} agrees within 15 % with the calculation on the basis of acoustic phonon escape into the substrate.The experimental result, see Fig. 18, shows a linear increase of τ_{eff} with film-thickness in accord with the calculated slope, instead bulk processes would lead to a saturation of τ_{eff} with increasing film-thickness.

High phonon trapping within the generator establishes a detailed balance between longitudinal and transverse phonon densities according to the phonon density of states. The ratio of longitudinal and transverse phonons emitted into the substrate is determined by the phonon occupation in the film and the mode-dependent acoustic boundary transmission (Weis, 1969). The strong anisotropy of the ballistic energy flux by phonon focussing (Taylor et al., 1969;

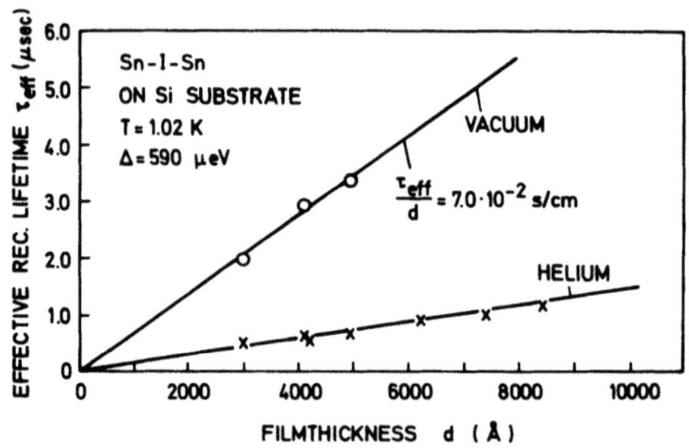

Fig. 18. Effective quasiparticle recombination time τ_{eff} in Sn-I-Sn tunneling junctions of different thickness. The theoretical film thickness dependence (vacuum) results in $\tau_{eff}/d = 6.8 \cdot 10^{-2}$ sec ($T = 1.02^{\circ}$ K). Eisenmenger et al., 1977)

Rösch and Weis, 1976) requires careful integration over all energy propagation angles between generator and detector in evaluating the average phonon focussing factor f. With respect to the transmission factors Γ this is less critical, these factors being close to unity for normal incidence at the Si - Sn boundary. The same holds for the ratio of absorbed and incident phonons. For the Sn-detector filmthickness of at least 2000 Å, taking account of the reflection at the vacuum boundary, the phonon mean free path for 2Δ-phonons of the order of 1000 Å for longitudinal and transverse phonons is short compared to the total propagation path in the film of 4000 Å. The detector has been calibrated according to the procedure outlined in Sect. 2.3. It is important to note that the density of electronic states N(0) at the Fermi-level entering the detector sensitivity must be directly taken from experimental electronic specific heat data. This has been verified by independent measurements of τ_{eff} under pulsed and stationary conditions (cf. Epperlein et al., 1978, and Epperlein and Eisenmenger, 1979). Comparing with the experiment, the observed signal amplitude ranges from 10 % to 15 % of the calculated value for Sn-junctions and generators on Si-substrates. Similar small values are obtained for Sn-junctions on Al_2O_3 and for various propagation directions in Si, see Fig. 19 (Trumpp and Eisenmenger, 1977).

Possible reasons for this discrepancy are: limited phonon emission into the substrate by inelastic volume loss in the Sn-film, boundary decay at the Sn-Si-interface, phonon decay in the bulk of the substrate, or anomalous phonon reflection and phonon decay at the detector - substrate boundary. Since only limited information for distinguishing between these possibilities is available from ballistic propagation, other independent experiments are needed.

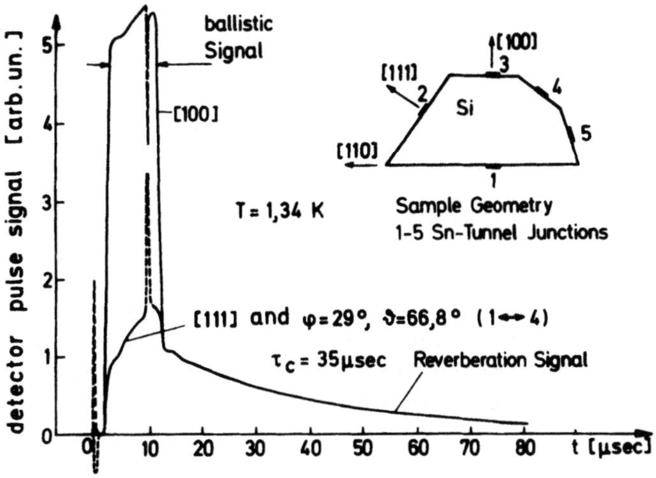

Fig. 19. Detector signal amplitudes for different propagation directions of $2\Delta_{Sn}$ phonons in Si (Sn-I-Sn generator and detector). The reverberation signal amplitude does not depend on the propagation direction in contrast to the ballistic signal. (Trumpp and Eisenmenger, 1977).

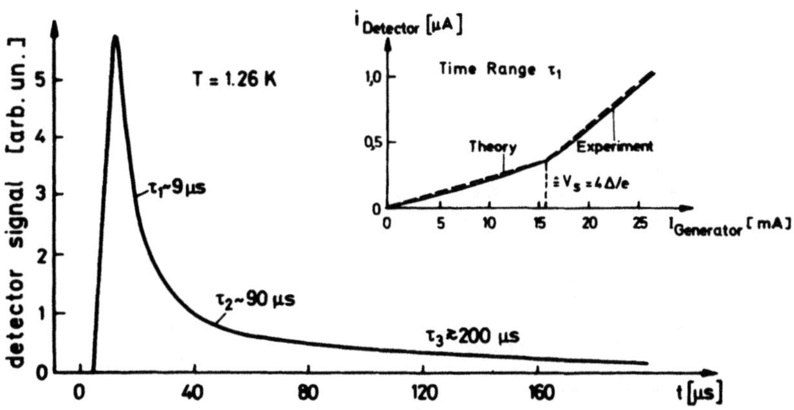

Fig. 20. Frequency check of the reverberation signal by measurement of the detector signal - generator current dependence in the range τ_1. The characteristic bend at eV = 4Δ (insert) is significant for 2Δ phonon reverberation.

One of these is the study of 2Δ-phonon lifetime or reverberation lifetime in the substrate crystal. Fig. 19 shows a slow exponential decay of the detector signal following the ballistic response. Time constant and amplitude agree for different generator positions in contrast to the ballistic signal amplitudes which are strongly influenced by angle-dependent phonon focussing. Measurements of the signal amplitude, as function of the generator voltage, see Fig. 20, show the change of slope at $eV_G = 4\Delta$, indicating that the signal is caused by 2Δ-phonons "reverberating" in the crystal and not by "heat". Experimentally the reciprocal reverberation time depends linearly on the ratio of the crystal surface area covered by the tin film and the crystal volume, see Fig. 21. A simple analysis of the slope of this dependence shows that the phonon absorption coefficient of the Sn-covered area amounts to 87 %. Neglecting boundary reflection, this can be attributed to losses in phonon transmission into the film and after phonon trapping in Sn, losses in backradiation into the substrate; i.e. only 13 % of 2Δ-phonons incident on the film-covered area are backradiated into the Si-crystal after twice traversing the Si - Sn interface. Assuming the absence of bulk losses in the Sn-film as consistent with the τ_{eff} measurements, this can be interpreted as phonon loss or decay in the crystal-superconductor boundary. Since in ballistic experiments phonons also traverse this boundary twice, the results of both measurements are consistent. A similar finding for Al- and In-films on Al_2O_3; i.e. "surface detuning" of phonons; has been reported by Day (1972). As

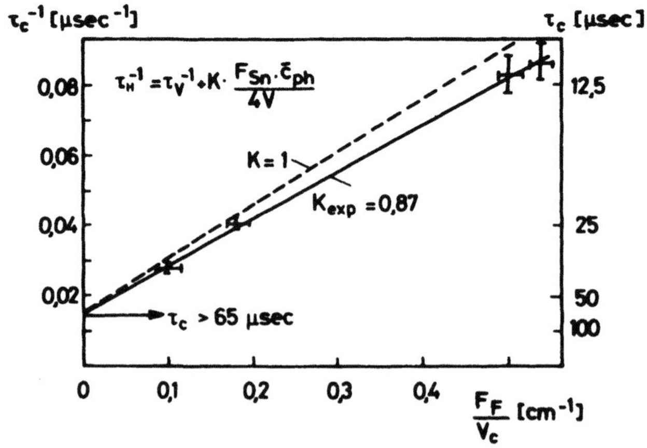

Fig. 21. Dependence of the 2Δ phonon reverberation decay rate τ^{-1} on the Sn coverage surface to volume ratio. The slope K = 1 corresponds to total phonon absorption. (Trumpp and Eisenmenger, 1977).

an additional result of the reverberation measurement of Fig. 21, the extrapolation to vanishing coverages with Sn indicate a phonon lifetime of 65 μs corresponding to a mean free path (dominant transverse phonons) of 25 cm. This indicates that inelastic substrate losses in ballistic experiments and also phonon decay (inelastic scattering) at the uncovered surface are negligible. Since weak residual inelastic scattering at the free crystal surface and also mode conversion to longitudinal phonons with high spontaneous decay rate may contribute to the observed lifetime of 65 μs, the free propagation transverse phonon lifetime may even be longer.

The nature of the interface phonon decay processes is presently unknown but may be related to the nonideal structure of the film-substrate-boundary caused by impurity atoms or other interface defects. These depend on the methods of surface and tunneling junction preparation. In principle, a comparison with the Kapitza anomaly; i.e. the enhanced phonon escape rate at the boundary between solid and liquid ^{4}He is possible. In the latter case, presumably anharmonic processes by the imperfect interface result in enhanced phonon transmission, whereas at the solid-superconductor boundary anharmonic processes result in a decay of high energy phonons.

Phenomenologically the phonon coupling at the insulator - superconductor boundary can be described by an elastic coupling (spring model) parallel to a dissipative or inelastic (dashpot) element, assuming that the interface width is small compared to the phonon wavelength. The elastic contribution represents the binding potential, whereas inelastic processes are possible by multilevel transitions caused by dangling bonds, weakly bound interfacial atoms or other imperfection states. The elastic and dissipative coupling contributions in principle can be determined by a measurement of the elastic phonon transmission and reflection at the interface. For the determination of the elastic reflection and scattering properties the reverberative method has the advantage of simple evaluation. Since it averages over all propagation directions, phonon focussing does not enter, also a quantitative treatment of the phonon generator and detector properties is not necessary. Finally, taking account of the boundary losses, as obtained in reverberation, also the ballistic signal amplitudes are in better agreement with the calculation. This verifies the validity of the theoretical model for ballistic 2Δ-phonon emission and detection if anomalous inelastic and elastic boundary processes are taken into account. If bulk phonon decay in the generator, as for example by magnetic flux trapping or quasiparticle absorption, reduces the phonon signal amplitude, this influence can be experimentally controlled by the measurement of τ_{eff} and the linearity of the signal increase with generator current in the voltage range $0 < eV_G < 4\Delta_G$.

It should be further noted that the reverberation method can be also used for the absolute measurement of the total 2Δ-phonon population in the substrate crystal. For evaluation of the detector signal amplitude it can be either assumed that 2Δ-phonons incident from all directions in the crystal are contributing, or that the quasiparticle and phonon distribution in the detector is in detailed balance equilibrium ($\tau_{eff} << \tau_{crystal}$) with the 2Δ-phonon population of the substrate. In the first case the evaluation results in boundary losses of the same magnitude as determined ballistically, and from the reverberation time constant. In the second case somewhat higher boundary losses were found. This indicates that the assumption of a detailed balance between 2Δ-phonons in the crystal and the detector may be wrong, as interface loss processes represent an irreversible 2Δ-phonon drain.

More specific information on anomalous interface properties can be obtained by phonon reflection and transmission measurements (Marx et al., 1978), as have been successful in studies of the Kapitza-anomaly at the solid He-boundary (cf. Kinder et al., 1980; Wyatt, 1980).

The results of such measurements will be useful for improved crystal surface- and metal film preparation techniques.

It is to be noted that in some interface systems, especially metallic heaters evaporated on insulator crystals, no deviations from the acoustic mismatch model have been observed (Weis, 1969; Weis, 1972; Martinon and Weis, 1979). Also τ_{eff} measurements for Sn-films on silicon show agreement with the acoustic model (Eisenmenger et al., 1977) within 15%. The influence of the observed boundary losses on phonon reflection, therefore, should be comparatively small. In phonon scattering experiments with amorphous layers (Lang et al., 1980) and epitaxial layers (Narayanamurti et al., 1979) no indication of phonon boundary loss was found. This indicates the possibility of avoiding boundary losses by ideal interface conditions.

Experiments with Al-recombination phonons (Trumpp and Eisenmenger, 1977; Welte and Eisenmenger, 1980) showed reduced interface losses as compared to 2Δ-Sn-phonons, indicating a significant but smooth frequency dependence.

Also low energy decay phonons, resulting from interface losses, have been observed (Trumpp and Eisenmenger, 1977; Korczynskji and Wyatt, 1978). With respect to the spectral distribution of phonons resulting from inelastic boundary decay, a high probability with half the original frequency is to be expected from the maximum of available lattice phonon states in a three-phonon interaction. This does not apply to localized boundary states, where a continuous phonon energy distribution or a maximum at low frequencies by multiple decay steps appears possible.

Quantitative evaluation of phonon-transmission experiments have been also reported for relaxation phonons (Welte et al., 1972; Welte, 1976; Forkel, 1977; Welte and Eisenmenger; 1980). The treatment of phonon generation is less simple than in recombination, since primary decay processes and phonon escape are influenced by quasiparticle and phonon scattering in the generator film (see Sect. 3). In addition, it must be expected that this is strongly influenced by the microcrystalline film structure. Model calculations, therefore, remain of approximate nature. The comparison with experimental data resulted in order of magnitude agreement with phonon signal amplitudes within 10 % - 40 % of the calculated values, indicating again phonon interface decay or increased reflection.

An interesting possibility is the observation of changes in the relative amplitude for longitudinal and transverse phonons at the onset of phonon reabsorption by pairbreaking in the generator (Long, 1972; Eisenmenger, 1976a; Welte and Eisenmenger, 1980). Such changes reflect the difference between the relative phonon population by primary relaxation decay in comparison to the equilibrium distribution established by the detailed balance in the case of strong 2Δ-phonon trapping. From observing the relative transverse and longitudinal signal amplitude ratios below and above the generator voltage $eV_G = 4\Delta_G$, in principle conclusions about the strength of the longitudinal and transverse electron - phonon interaction in relaxation decay can be drawn. Such changes of the amplitude ratio are reported by Welte and Eisenmenger (1980). In the quantitative interpretations of these measurements mode conversion of relaxation phonons by elastic bulk and boundary scattering processes, as well as the mode dependence of boundary phonon decay, have to be taken into account.

As a general result it is observed, that primary relaxation decay leads to a dominant transverse phonon contribution caused by the nonspherical Fermi-surface of real metals and the higher densities of transverse phonon states.

Although quantitative evaluation of phonon-transmission experiments are roughly in agreement with simplified model calculations, more information is needed especially with respect to nonideal interface and film-structure properties.

6. EXPERIMENTAL PROBING OF THE QUASIPARTICLE DISTRIBUTION AND APPLICATION TO PHONON SPECTROSCOPY

Tunneling measurements with junctions composed of superconducting films with different energy gaps can be used for an analysis of the energy distribution of excited quasiparticles (Miller and Dayem, 1967; Chang and Scalapino, 1976). The principle of this method, which has been applied to the study of nonequilibrium distributions (Kaplan et al., 1977; Willemsen and Gray, 1978) in superconducting films, is depicted in Fig. 22, showing a three layer Al-I-Al-I-Pb:Bi sandwich tunneling structure. The first and second films are used for quasiparticle injection producing the nonequilibrium distribution. The quasiparticle distribution in the middle film is probed by the third film with the higher energy gap. In varying the voltage between the middle (second) and the third film the tunneling current is mainly determined by excited quasiparticles tunneling into the unoccupied singularity of the density of states of the third film. In the corresponding experiment, Kaplan et al. (1977) were able to analyze the nonequilibrium quasiparticle distribution by weak tunnel injection in Al and to compare their results with theory. Excellent agreement was found including clear evidence for branch imbalance. Recently, the same method has been successfully used to study the quasiparticle distribution (Willemsen and Gray, 1978) in Al-films under high quasiparticle injection. (See Gray, 1980; this volume, Chapter 5). It has been demonstrated that the approximate description of the nonequilibrium distribution by a μ^+ or T^+ model (Owen and Scalapino, 1972; Parker, 1975) depends characteristically on the strength of injection.

As demonstrated by Dietsche (1978), it is also possible to use this method for the analysis of quasiparticle distributions produced by phonon-induced pairbreaking. In this case the first film is not necessary and the tunneling junction, consisting of Al-I-Sn or Al-I-Pb:Bi, is directly used as phonon detector. Phonons absorbed by pairbreaking in the Al-film produce a quasiparticle distribution with an upper energy limit at $E = \hbar\Omega - \Delta_{Al}$ ($\hbar\Omega$ = energy of incident

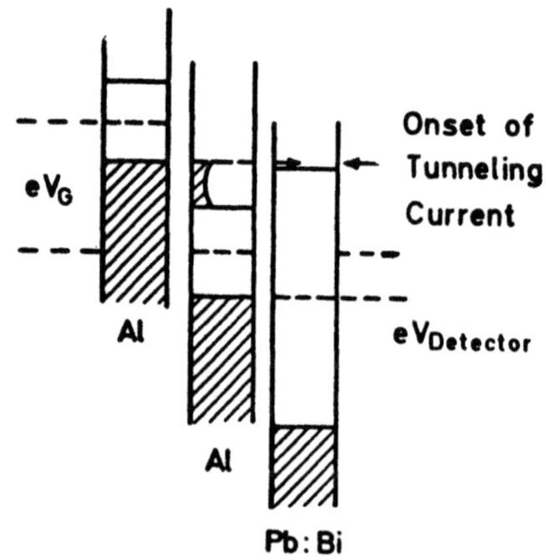

Fig. 22. Direct probing of the quasiparticle distribution in a three layer tunneling sample.

phonons). This can be analyzed in probing the quasiparticle distribution with the higher energy gap film by modulation technique and bias voltage tuning. Using such hetero-junction detectors, it is possible to determine different energies of incident phonons.

This new technique of tunable phonon detection has been successfully applied to the analysis of phonon spectra emitted from Sn-generators (Dietsche, 1978; Lang et al., 1980; Koblinger and Forkel, 1980) and, in addition, to the investigation of inelastic phonon scattering in thin glass films (Dietsche and Kinder, 1979; Lang et al., 1980). By these measurements it was possible to determine the frequency dependence of the phonon mean free path and to give evidence that the dominant scattering process is inelastic.

With respect to nonequilibrium phonons, the generator frequency analysis verified the extreme sharpness of the relaxation phonon edge at $\hbar\Omega = eV_G - 2\Delta_G$, whereas the lower cut-off energy of the recombination spectrum at $\hbar\Omega = 2\Delta_G$ in earlier experiments showed a significant contribution of spectral diffusion extending to energies below 2Δ. In later experiments (Koblinger and Forkel, 1980; Lang et al., 1980) less spectral diffusion was observed. A possible origin of spectral diffusion is magnetic flux trapped in the generator junction which leads to gap smearing and in-gap quasiparticle states which participate in phonon reabsorption and emission thus shifting the 2Δ-phonon spectrum to lower energies.

Tunable phonon detection, using hetero-tunneling junctions or phonon spectroscopy with stress-tunable absorption lines in the substrate, are very promising for investigations of inelastic phonon scattering in solid systems, as e.g. for phonon decay at solid - solid and solid - liquid boundaries.

7. RESULTS AND OPEN QUESTIONS

Nonequilibrium phonon experiments with superconducting film tunneling junctions have demonstrated that phonon spectra produced by nonequilibrium quasiparticle distributions are well in accord with the predictions of theory. This is also demonstrated by the successful application of superconducting tunneling junctions for phonon absorption spectroscopy and recently also for phonon emission spectroscopy.

Open questions are related to the detailed description of the separate contributions of longitudinal and transverse phonon emission and reabsorption (i.e. transverse and longitudinal phonon mean free path data by pairbreaking, quasiparticle scattering and volume decay) in the superconducting films, the influence of branch imbalance, and the properties of film substrate boundaries in the inelastic interface decay of phonons. In addition, more information is needed on quasiparticle and phonon distributions under different - for instance high injection - conditions.

REFERENCES

Bardeen, J., Cooper, L.N., and Schrieffer, J.R., 1957, Theory of Superconductivity, Phys. Rev. 108, 1175.

Bron, W.E., 1980, Spectroscopy of High-Frequency Phonons, Rep. Prog. Phys. 43, 301.

Chang, J.J., and Scalapino, D.J., 1976, Tunneling as a Probe of Nonequilibrium Superconducting States, Phys. Rev. Lett. 37, 522.

Chang, J.J., and Scalapino, D.J., 1978, Nonequilibrium Superconductivity, J. Low Temp. Phys. 31, 1.

Chang, J.J., 1980, chapter 9 this volume, Instabilities of Nonequilibrium Superconducting States.

Day, W., 1972, Phonon Transmission along Sapphire Plates, J. Phys. (France) 33, C4, 65.

Dayem, A.H., 1972, Superconducting Tunnel Junctions as Phonon Sources and Detectors, J. Phys. (France) 33, C4, 15.

Dayem, A.H., and Wiegand, J.J., 1972, Emitted Phonon Spectrum and Its Influence on the Detected Signal in Superconducting Sn Diodes, Phys. Rev. B5, 4390.

Dietsche, W., and Kinder, H., 1976, Frequency Analysis of Phonons Backscattered from Interfaces Between Solids and Helium, J. Low Temp. Phys. 23, 27.

Dietsche, W., 1978, Superconducting Al-PbBi Tunnel Junction as a Phonon Spectrometer, Phys. Rev. Lett., 40, 786.

Dietsche, W., and Kinder, H., 1979, Spectroscopy of Phonon Scattering in Glas, Phys. Rev. Lett. 43, 1413.

Dietsche, W., Kinder, H., Mattes, J., and Wühl, H., 1980, Messung der Schallgeschwindigkeit in amorphen und rekristallisierten Supraleitern, Verhdlg. Deut. Phys. Ges. 3, 402.

Dynes, R.C., Narayanamurti, V., and Chin, M., 1971, Monochromatic Phonon Propagation in Ge:SB Using Superconducting Tunnel Junctions, Phys. Rev. Lett. 26, 181.

Dynes, R.C., and Narayanamurti, V., 1974, Evidence for the Upward or "Anomalous" Dispersion in the Excitation Spectrum of He II, Phys. Rev. Lett. 33, 1195.

Eisenmenger, W., and Dayem, A.H., 1967, Quantum Generation and Detection of Incoherent Phonons in Superconductors, Phys. Rev. Lett. 18, 125.

Eisenmenger, W., 1969, Phonon Generation and Detection by Single-Particle Tunneling in Superconductors, in: "Tunneling Phenomena in Solids", E. Burstein and S. Lundqvist, eds., p. 371, Plenum, New York.

Eisenmenger, W., 1976a, Superconducting Tunneling Junctions as Phonon Generators and Detectors, in: "Physical Acoustics", W.P. Mason and R.N. Thurston, eds., Vol. XII, p. 79-153, Academic Press. New York.

Eisenmenger, W., 1976b, High Frequency Phonon Emission from Superconducting Al-Tunneling Junctions, in: "Phonon Scattering in Solids", L.J. Challis, V.W. Rampton, A.F.G. Wyatt, eds., p. 400, Plenum Press, New York.

Eisenmenger, W., Laßmann, K., Trumpp, H.J., and Krauß, R., 1977, Intrinsic and Experimental Quasiparticle Recombination Times in Superconducting Films, Appl. Phys. 12, 163.

Epperlein, P.W., Laßmann, K., and Eisenmenger, W., 1978, Quasiparticle Recombination Time in Superconducting Tin and Normal Electronic Density of States at the Fermi Surface from Tunnel Junction Experiments, Z. Phys. B 31, 377.

Epperlein, P.W., and Eisenmenger, W., 1979, Measurements of effective Quasiparticle Recombination Times and of Densities of Electronic States at the Fermi Level in Superconducting Al- and PB-Films, Z. Phys. B 32, 167.

Forkel, W., Welte, M., and Eisenmenger, W., 1973, Evidence for 870-GHz Phonon Emission from Superconducting Al Tunnel Diodes through Resonant Scattering by Oxygen in Silicon, Phys. Rev. Lett. 31, 215.

Forkel, W., 1974, Phonon Spectroscopy with Superconducting Aluminum Tunnel Junctions, in: "Microwave Acoustics", E.R. Dobbs and J.K. Wigmore, eds., p. 186, Proc. 8th Internat. Congr. on Acoustics, Inst. Phys. London.

Forkel, W., 1975, see: Eisenmenger, 1976b.

Forkel, W., 1977, Anwendung von Nichtgleichgewichtsprozessen in Supraleiter-Tunneldioden zur Spektroskopie mit akustischen Phononen, Thesis, University of Stuttgart, unpublished.

Giaever, I., and Megerle, K., 1961, Study of Superconductors by Electron Tunneling, Phys. Rev. 122, 1101.

Giaever, I., 1969, Tunneling between Superconductors, in: "Tunneling Phenomena in Solids", F. Burstein and S. Lundquist, eds., p. 255, Plenum Press, New York.

Gray, K.E., 1980, chapter 5 this volume, Tunneling: A Probe of Nonequilibrium Superconductivity.

Hu, P., Dynes, R.C., Narayanamurti, V., Smith, H., and Brinkman, W.F., 1979, Quasiparticle Propagation and Recombination in Bulk, Superconducting Pb, Phys. Rev. Lett. 39, 361.

Huet, D., Maneval, J.P., and Zylbersztejn, A., 1972, Measurement of Acoustic-Wave Dispersion in Solids, Phys. Rev. Lett. 29, 1092.

Huet, D., Maneval, J.P., 1974, Image of the Fermi Surface and Screening Effects in Phonon Attenuation, Phys. Rev. Lett. 33, 1154.

Hasan, F., King, P.J., Murphy, D., Rampton, V.W., 1978, The Phonon Spectroscopy of Doped Magnesium Oxide Using Superconducting Tunnel Junctions, J. Phys. (France) 39, C6, 993.

Hasan, F., King, P.J., Murphy, D., Rampton, V.W., 1979a, Multiple-Ion Phonon Absorption in Al_2O_3:Cr^{2+}, J. Phys. (GB) 12, L431-3.

Hasan, F., King, P.J., Murphy, D., Rampton, V.W., 1979b, An Investigation of V^{3+} Ion Pairs in Al_2O_3 by Phonon Spectroscopy, J. Phys. (GB) 12, L513-16.

Kaplan, S.B., Kirtley, J.R., and Langenberg, D.N., 1977, Experimental Determination of the Quasiparticle Energy Distribution in a Non-Equilibrium Superconductor, Phys. Rev. Lett. 39, 291.

Kaplan, S.B., Chi, C.C., Langenberg, D.N., Chang, J.J., Jafaray, S., and Scalapino, D.J., 1977, Quasiparticle and Phonon Lifetimes in Superconductors, Phys. Rev. B14, 4854; Erratum ibid. 15, 3567.

Kinder, H., Laßmann, K., and Eisenmenger, W., 1970, Phonon Emission by Quasiparticle Decay in Superconducting Tunnel Junctions, Phys. Lett. 31A, 475.

Kinder, H., 1972a, Spectroscopy with Phonons on Al_2O_3:V^{3+} Using the Phonon Bremsstrahlung of a Superconducting Tunnel Junction, Phys. Rev. Lett. 28, 1564.

Kinder, H., 1972b, Use of the "Bremsspectrum" for Phonon Spectroscopy on Al_2O_3:V^{3+} and for a Study of the Phonon Propagation in Granular Aluminum, J. Phys. (France) 33, C4, 21.

Kinder, H., 1972c, Phonon Spectroscopy on the Resonant Scattering of V^{3+} and other Impurities in Al_2O_3 by Tuning the "Bremsspectrum" of a Superconducting Tunnel Junction, Int. Conf. on Phonon Scattering in Solids, H.J. Albany, ed., p. 284, CEN Saclay.

Kinder, H., 1973a, Spektroskopie mit Phononen bei Frequenzen über 100 GHz, Physikertagung, Deutsch-Österr. Phys. Ges., Plenarvortrag Hamburg, 1973, Physik Verlag, Weinheim.

Kinder, H., 1973b, Spin-Phonon Coupling of Al_2O_3:V^{3+} by Quantitative Spectroscopy with Phonons, Z. Phys. 262, 295.

Kinder, H., 1975, Phonon Spectroscopy at Ultrahigh Frequencies, in: "Low Temperature Physics LT14", M. Krusius and M. Vuorio, eds., p. 287, North Holland, Elsevier.

Kinder, H., and Dietsche, W., 1975, Phonon Spectroscopy in Al_2O_3 Doped with Transition Metal Impurities, in: Phonon Scattering in Solids", L.J. Challis, V.W. Rampton, and A.F.G. Wyatt, eds., p. 199, Plenum Press, New York.

Kinder, H., Weber, J., and Dietsche, W., 1980, Kapitza Resistance Studies Using Phonon Pulse Reflection, in: "Phonon Scattering in Condensed Matter", H.J. Maris, ed., p. 173, Plenum Press, New York.

Kirtley, J.R., Kent, D.S., Langenberg, D.N., Kaplan, S.B., Chang, J.J., and Yang, C.C., 1980, Quasiparticle Energy Distribution and Relaxation Times in a Tunnel-Injected Superconductor, Phys. Rev. B, in print.

Koblinger, O., and Forkel, W., 1980, Phononspektroskopie mit Heterodetektoren, Spring Meeting Deutsche Physik. Ges.

Korcynskyj, Y., and Wyatt, A.F.G., 1978, Frequency Up-Conversion in Interacting Phonon Beams in Liquid ^{4}He. J. Phys. (France) 39, C6, 230.

Lang, P., Dietsche, W., and Kinder, H., 1980, Inelastische Phononenprozesse in Gläsern und amorphen Halbleitern, Verhdlg. Deut. Phys. Ges. 3, 234.

Langenberg, D.N., 1975, Nonequilibrium Phenomena in Superconductivity, in: "Low Temperature Physics LT14", H. Krusius and H. Vuorio, eds., Vol. V 223, North Holland, Elsevier.

Laßmann, K., and Zeile, H., 1980, Ultrasonic Spectroscopy of the Acceptor Ground State in Cubic Semiconductors, in: "Phonon Scattering in Condensed Matter", H.J. Maris, ed., p. 369, Plenum Press, New York.

Long, A.R., 1972, Investigations of Electron - Phonon Interactions in Copper and Aluminum, J. Phys. (France) 33, C4, 73.

Long, A.R., 1973, The Attenuation of High Frequency Phonons in metals, J. Phys. F. (GB) Met. Phys. 3, 2023.

Long, A.R., and Adkins, C.J., 1973, Transfer Characteristics of Phonon Coupled Superconducting Tunnel Junctions, Phil. Mag. 27, 865.

McMillan, W.L., and Rowell, J.M., 1969, Tunneling and Strong-Coupling Superconductivity, in: "Superconductivity", R.D. Parks, ed., Marcel Decker, Vol. 1, p. 561.

Martinon, C., and Weis, O., 1979, α-Quartz as a Substrate in Thermal Phonon Radiation, Z. Phys. B32, 259.

Marx, D., Buck, J., Laßmann, K., and Eisenmenger, W., 1978, Reflection of High Frequency Phonons at Free Silicon Surfaces, J. Phys. (France) 39, C6, 1015.

Mattes, J., Berberich, P., and Kinder, H., 1978, Resonant Scattering of Monochromatic Phonons by Magnons in MnF_2 and in YIG, J. Phys. (France), 39, C6, 988.

Miller, B.I., and Dayem, A.H., 1967, Relaxation and Recombination Times of Quasi-Particles in Superconducting Al Thin Films, Phys. Rev. Lett. 18, 1000.

Narayanamurti, V., and Dynes, R.C., 1971, Intense Tunable Phonon Fluorescence in Superconductors, Phys. Rev. Lett. 27, 410.

Narayanamurti, V., and Dynes, R.C., 1976, Roton Propagation and Phonon-Roton Scattering in He II, Phys. Rev. B13, 2898.

Narayanamurti, V., Dynes, R.C., Hu, P., Smith, H., and Brinkman, W.F., 1978, Quasiparticle and Phonon Propagation in Bulk, Superconducting Lead, Phys. Rev. B18, 6041.

Narayanamurti, V., Stormer, H.L., Chin, M.A., Gossard, A.C., and Wiegmann, W., 1979, Selective Transmission of High-Frequency Phonons by a Superlattice: the 'Dielectric' Filter, Phys. Rev. Lett. 43, 2012.

Narayanamurti, V., 1980, Phonon Optics, Carrier Relaxation and Recombination in Semiconductors: Case of GaAs Epitaxial Layers, in: "Phonon Scattering in Condensed Matter", H.J. Maris, ed., p. 385, Plenum Press, New York.

Owen, C.S., and Scalapino, D.J., 1972, Superconducting State under the Influence of External Dynamic Pair Breaking, Phys. Rev. Lett. 28, 1559.

Pannetier, B., Huet, D., Buechner, J., and Maneval, J.P., 1977, Ballistic Propagation of Near-Gap Phonons in Bulk Superconducting Tin, Phys. Rev. Lett. 39, 646.

Parker, W.H., 1975, Modified Heating Theory of Nonequilibrium Superconductors, Phys. Rev. B12, 3667.

Proceedings of Phonon Scattering Conferences:

1972, J. Phys. (France), C33, Conf. Phys. of Very High Frequency Phonons, St. Maxime.

1972, Int. Conf. Phonon Scattering Solids, Paris, H.J. Albany, ed., CEN Saclay.

1976, Phonon Scattering in Solids, L.J. Challis, V.W. Rampton, and A.F.G. Wyatt, eds., Plenum Press, New York.

1980, Phonon Scattering in Condensed Matter, J.M. Maris, ed., Plenum Press, New York.

Renk, K.F., 1972, Phononpulse, Festkörperprobleme, O. Madelung ed., Pergamon, Vieweg, 12, 107.

Reupert, W., Laßmann, K., and de Groot, P., 1976, Anisotropy of Phonon Emission from Hot Electrons in Germanium, in: "Phonon Scattering in Solids", L.J. Challis, V.W. Rampton, and A.F.G. Wyatt, eds., p. 315, Plenum Press, New York.

Rösch, T., and Weis, O., 1976, Geometric Propagation of Acoustic Phonons in Monocrystals within Anisotropic Continuum Acoustics, Z. Phys. B25, 115.

Rothwarf, A., and Taylor, B.N., 1967, Measurement of Recombination Lifetimes in Superconductors, Phys. Rev. Lett. 19, 27.

Schad, H.P., and Laßmann, K., 1976, Ultrasonic Attenuation due to the Neutral Acceptor Indium in Silicon, Phys. Lett. 56A, 409.

Schenk, H., Forkel, W., and Eisenmenger, W., 1978, Phononenspektroskopie mit Supraleiter-Tunneldioden, Verhdlg. Deut. Phys. Ges. 1, 328.

Sigmund, E., and Laßmann, K., 1980, Phonon Scattering at Acceptors with Γ_8 Ground States in Semiconductors, in: "Phonon Scattering in Condensed Matter", H.J. Maris, ed., p. 417, Plenum Press, New York.

Smith, B.S., Wigmore, J.K., Meredith, D.J., 1978, Phonon Spectroscopy of $Al_2O_3:Fe^{2+}$, Phys. Lett. 67A, 416.

Solymar, L., 1972, in: "Superconductive Tunneling and Applications", pp. 45, 234, Chapman and Hall, London.

Taylor, B., Maris, H.J., and Elbaum, C., 1969, Phonon Focusing in Solids, Phys. Rev. Lett. 23, 416.

Taylor, B., Maris, H.J., and Elbaum, C., 1969, Focusing of Phonons in Crystalline Solids due to Elastic Anisotropy, Phys. Rev. B3, 1462.

Tewordt, L., 1962, Lifetimes of Quasi-Particles and Phonons in a Superconductor at Zero Temperature, Phys. Rev. 127, 371

Tinkham, M., 1972, Tunneling Generation, Relaxation, and Tunneling Detection of Hole-Electron Imbalance in Superconductors, Phys. Rev. B6, 1747.

Trumpp, H.J., and Eisenmenger, W., 1977, Sources of Loss Processes in Phonon Generation and Detection Experiments with Superconducting Tunneling Junctions, Z. Phys. B28, 159.

Weis, O., 1969, Thermal Phonon Radiation, Z. Angew. Phys. 26, 325.

Weis, O., 1972, The Solid - Solid Interface in Thermal Phonon Radiation, J. Phys. (France) 33, C4, 48.

Welte, M., Laßmann, K., and Eisenmenger, W., 1972, Emission of High Frequency Relaxation Phonons by Superconducting Aluminum Tunneling Junctions, J. Phys. (France) C33, 4.

Welte, M., 1973, unpublished.

Welte, M., 1974, Phonon Emission from Superconducting Tunnel Junctions at Finite Quasiparticle Densities, in: "Microwave Acoustics", E.R. Dobbs and J.K. Wigmore, eds., Proc. Int. Congr. Acoustics, 8th, p. 183, Inst. Phys. London.

Welte, M., 1976, Spektrale Eigenschaften supraleitender Aluminiumtunneldioden als Phononengeneratoren- und -detektoren, Thesis, University of Stuttgart, unpublished, see also: Welte M., and Eisenmenger, W., 1980.

Welte, M., and Eisenmenger, W., 1980, Analysis of the Phonon Spectrum Emitted by Superconducting Al-Tunneling Junctions at High Quasiparticle Injection Rates, Z. Phys., in print.

Willemsen, H.W., and Gray, K.E., 1978, Distribution Function in Nonequilibrium States of a Superconductor, Phys. Rev. Lett. 41, 812.

Windheim, R., and Kinder, H., 1975a, Phonon Spectroscopy of OH^- Tunneling Levels in NaCl, Phys. Lett. 51A, 475.

Windheim, R., and Kinder, H., 1975b, Phonon Spectroscopy of OH^- and Li^+ Tunneling States in Alkali Halides, in: "Phonon Scattering in Solids", J. Challis, V. Rampton, and A.F.G. Wyatt, eds., p. 220, Plenum Press, New York.

Wyatt, A.F.G., 1976, The Dynamics of Superconducting Phonon Devices, J. Phys. C (GB) 9, 2541.

Wyatt, A.F.G., and Crisp, G.N., 1978, Frequency of Phonons Emitted into Liquid He by a Solid, J. Phys. (France) 39, C6, 244.

Wyatt, A.F.G., 1980, Characteristics of Kapitza Conductance, in: "Phonon Scattering in Condensed Matter", H.J. Maris, ed., p. 181, Plenum Press, New York.

PHONON OPTICS IN SEMICONDUCTORS

V. Narayanamurti

Bell Laboratories
600 Mountain Avenue
Murray Hill, New Jersey 07974, U.S.A.

INTRODUCTION

The primary motivation for high frequency phonon propagation research is its application to a variety of transport problems at thermal frequencies. The strong dependence of anharmonic phonon decay processes, of defect- and electron-phonon scattering and acoustic dispersion on phonon frequency makes it particularly interesting to study high frequency phonon transport in solids. These effects become particularly dominant at frequencies of the order of 10^{12} Hz. During the last decade there has been a great deal of activity in this area. This has been due to the development of thin film generators and detectors such as superconducting tunneling junctions and bolometers. The use of such devices for phonon detection has been reviewed by von Gutfeld (1968) and Eisenmenger (1976, 1980a).

In this paper we will survey the results of phonon propagation experiments in a variety of semiconductors (such as GaAs, InP, InSb, Ge, Si, etc.). Such semiconductors can serve as an excellent arena for the study of a wide range of transport and energy relaxation phenomena. Besides the use of conventional metal heaters or superconducting tunnel junctions as phonon sources one can in the case of the semiconductor study phonon generation by direct optical excitation of e-h pairs, by pn junctions or by layered epitaxial transducers of controlled doping. As shown in Figure 1, the phonons propagate through a thickness of $\sim$ 0.2 to 20 mm of the semiconductor and are detected, after a ballistic time of flight, at the other end by a bolometer or a tunnel junction. In addition controlled photoexcitation can be used to vary the charge state of donors, acceptors or deep

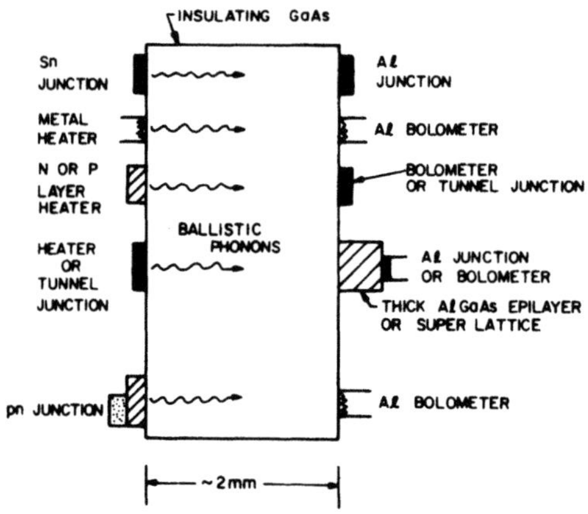

Fig. 1. Schematic of sample arrangement and the variety of transducers which can be used for ballistic phonon experiments in GaAs.

defect states in either the bulk or in the epitaxial layers themselves.

By far the most interesting aspect of phonon transport studies in semiconductors when compared with dielectric solids are the following: 1) Carrier relaxation and e-h recombination processes via multiphonon emission. Even though there has been a great deal of study of relaxation and recombination processes using optical techniques (Ulbrich 1978) almost nothing is known from the phonon side. Yet, as we shall see, phonon relaxation is a dominant process. 2) Study of novel forms of matter such as electron-hole drops. 3) Anisotropic electron-phonon coupling and carrier screening. There is a close analogy here with the case of metals. Semiconductors because of the control over doping concentrations allow a far greater range of studies to be performed than is possible with metals. 4) Because of their technological importance semiconductor crystals of high purity are often available and hence are a valuable testing ground for a variety of ballistic phonon transport studies of the type performed in dielectric solids. One now knows, for example, in great detail the channeling of phonon beams due to elastic anisotropy in crystals such as Ge and GaAs. 5) Again, because of their technological importance one has several exciting possibilities through the use of novel epitaxial transducer materials. One can, for example, by means of molecular beam epitaxy grow superlattice structures with a controlled period. Such structures can be used to build frequency selective phonon filters and reflectors which can in turn be used to manipulate phonon beams in technologically important materials. Such manipulation of phonons with novel transducer materials and the resulting physical studies may best be described as "phonon optics".

BALLISTIC PROPAGATION AND PHONON FOCUSSING

The flow of heat in nonmetallic solids can, as is well known, be described by a kinetic model. The steady state thermal conductivity, K, is given by

$$K = \frac{1}{3} C v \Lambda \tag{1}$$

where C is the phonon specific heat, v an average phonon velocity and Λ the mean free path. At low temperatures $\Lambda \sim d$, where d is the diameter of the sample and $K \propto T^3$ because of the cubic dependence on the temperature, T, of the specific heat. This is the so-called ballistic or Kundsen regime. At higher temperatures scattering due to naturally occurring isotopes and/or impurities and anharmonic phonon-phonon interactions becomes important (Klemens 1958, Carruthers 1961). The thermal conductivity of many tetrahedral semiconductors such as Ge (Geballe and Hull 1958) and III-V semiconductors such as GaAs (Holland 1964) show a characteristic bell-shaped thermal conductivity as a function of T. In Figure 2, I show such a curve for a nominally pure sample of GaAs which shows a well defined T^3 region for temperatures below $\sim$ 10 K for a typical sample cross section $\sim$ few mm. The ballistic window depends strongly on the purity and propagation length.

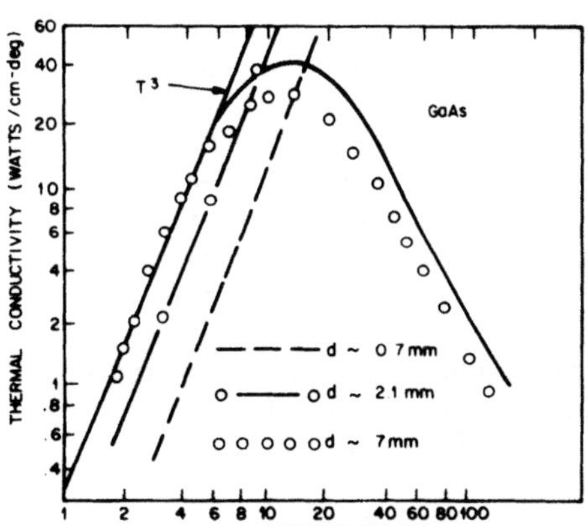

Fig. 2. Thermal conductivity, K, of pure GaAs as a function of temperature. At low temperatures K depends on the diameter, d, and is proportional to the T^3 specific heat.

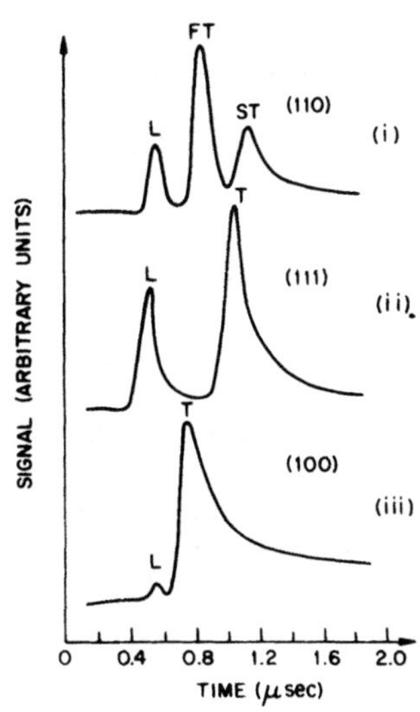

Fig. 3. Ballistic phonons in "pure" GaAs. Heater-bolometer experiment. Low powers ($\ll 1$ W/mm^2). The differences in the mode intensities arises from phonon focussing. After Narayanamurti et al 1978a.

Trace amounts of Cr in the sample, for example, markedly reduces the conductivity at low temperatures due to resonant scattering (Chaudhari et al 1973, Challis and Ramdane 1980).

Such thermal conductivity measurements have served as a very useful way to study phonon transport in solids. However, like other transport measurements, it suffers from the fact that in the analysis one implicitly assumes an isotropic solid and one averages over phonon polarization and frequency. That this can be a gross oversimplification has been strikingly demonstrated by heat pulse type experiments (von Gutfeld 1968) in the ballistic regime. In such experiments, separate pulses are observed due to transverse (T) and longitudinal (L) phonons, these pulses arriving at different times because of the different velocities of the phonons. Taylor, Maris and Elbaum (1969) first showed that even in very pure dielectric solids the intensities of the observed L and T phonons was markedly different from the expected Debye density of states and that the relative intensities depended markedly on the propagation direction. In Figure 3 we show for illustrative purposes ballistic phonon intensities in the (100), (110) and (111) directions of GaAs for almost an identical propagation length $\sim$ 2.5 mm. It is clear that the ratio of the intensities of the T to L pulses changes by more than an order of magnitude depending on the orientation. It is $\sim$ 1 in the (111) direction and $\sim$ 20 in the (100) direction. This focussing of the phonon intensity is due to the elastically anisotropic nature of the solid (Musgrave 1954, Taylor et al 1969, Maris 1971). In an anisotropic medium the energy flow associated with a plane wave does not, in general, coincide with the direction of the wave vector. Thus, even when waves are excited in a given region of a crystal with a uniform angular distribution of wave vectors, the energy flow will be enhanced in some directions and decreased in others with respect to the average. Such focussing of waves in anisotropic media can be calculated on very general grounds from a knowledge of the Gaussian curvature, K (Lighthill, 1960, Lax and Nelson, 1975, 1976; Lax and Narayanamurti 1980) of the so-called slowness surface.

The seminal paper of Maris recognized that the magnification or amplification factor A is determined by the ratio of solid angles in wave vector ($\vec{q}$) space and in ray ($\vec{r}$) space:

$$1/A = d\Omega^r/d\Omega^q \tag{1}$$

Waves uniformly produced in $\vec{q}$ space will be magnified in intensity if they are concentrated in a smaller solid angle $d\Omega^r$ in real space or ray space. The solid angles can be calculated very simply according to the procedure developed by Lax, as follows. In Figure 4 we show the solid angle $d\Omega^q$ associated with a patch of area $dA^q = dudv$ on the surface $\omega(\vec{q}) = \omega$ of free-wave vectors, $\vec{q}$,

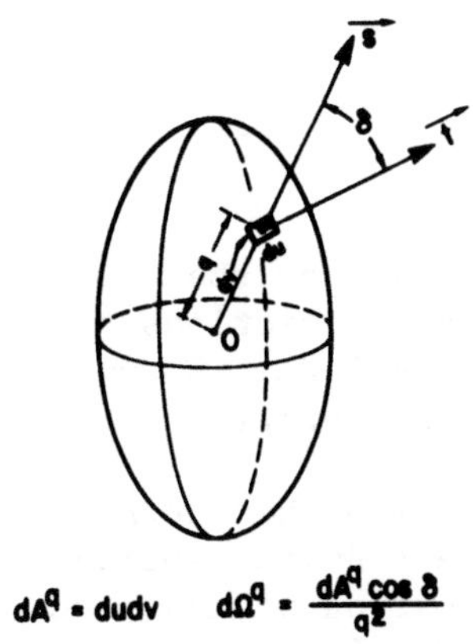

Fig. 4. The solid angle $d\Omega^q$ associated with a patch of area dA^q = dudv on the surface $\omega(\vec{q}) = \omega$ of free wave vectors $\vec{q}$ is given by $d\Omega^q = dA^q \cos\delta/q^2$. After Lax and Narayanamurti 1980.

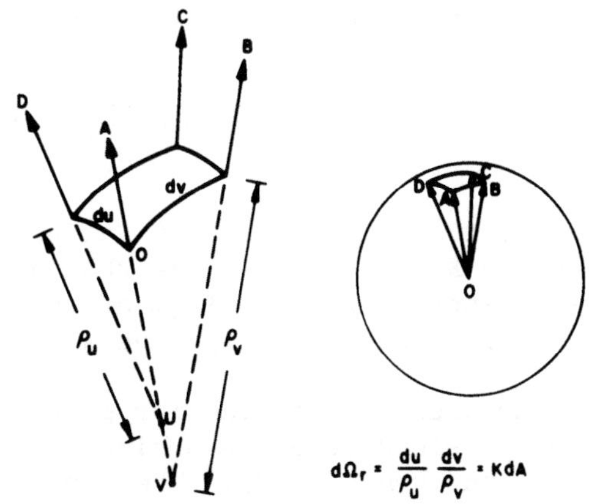

Fig. 5. Gauss' bump theorem relates the area dA = dudv of an element of surface to the area $d\Omega^r$ = ABCD on the surface of a unit sphere subtended by the unit normals $\vec{A}$, $\vec{B}$, $\vec{C}$, $\vec{D}$ to dudv that have been shifted in a parallel manner until their origins coincide at 0, the sphere center. After Lax and Narayanamurti 1980.

is given by $d\Omega^q = \dfrac{dA^q \cos\delta}{q^2}$, where $dA^q\cos\delta$ is the component of the area dA^q normal to $\vec{q}$.

Each point on dA^q possesses a unit normal or ray vector $\vec{t}$ such as the vectors $\vec{A}$, $\vec{B}$, $\vec{C}$ and $\vec{D}$ shown in the left part of Figure 5. If these rays are shifted in a parallel manner until their origins coincide at a point 0 then the area ABCD on the unit sphere is the solid angle subtended by the bundle of rays bounded by dA^q in $\vec{q}$ space. This area is the product of the lengths $\vec{AD}$ and $\vec{AB}$ on the sphere, but these are the angles.

$$\langle AUD = du/\rho_v \quad ; \quad \langle AVB = \frac{dv}{\rho_v} \tag{2}$$

The radii ρ_u and ρ_v are the principal radii of curvature of the surface $\omega(\vec{q})$ at 0 associated with the curved segments du and dv respectively.

The surface area on the unit sphere is given by

$$d\Omega^r = (\langle AUD)\cdot(\langle AVB) = \frac{dudv}{\rho_u\rho_v} = Kdudv \tag{3}$$

Hence, we have the extremely simple relation for the magnification

$$\frac{1}{A} = \frac{d\Omega^r}{d\Omega^q} = \frac{q^2 K}{\cos\delta} \tag{3}$$

Thus the focussing or magnification is determined by the Gaussian curvature K. It is possible that along certain directions K = 0. This depends on the topology of the $\omega(\vec{q}) = \omega$ surface. Whenever, K = 0, one would actually have a focussing singularity.

The presence of singular points, where K = 0, has been recently demonstrated in a high resolution heat pulse experiment (Hensel and Dynes 1979). The angular distribution of measured transverse mode intensities in the (110) plane of Ge is shown in Figure 6(a). The FT mode, e.g., shows sharp jumps in intensities in the vicinity of both the (100) and (110) directions. Figure 6(b) shows theoretically calculated intensities and the agreement with experiment is seen to be excellent. Using novel experimental techniques recently several investigators have directly photographed the focussing of phonon beams in Si and Ge (Northrup and

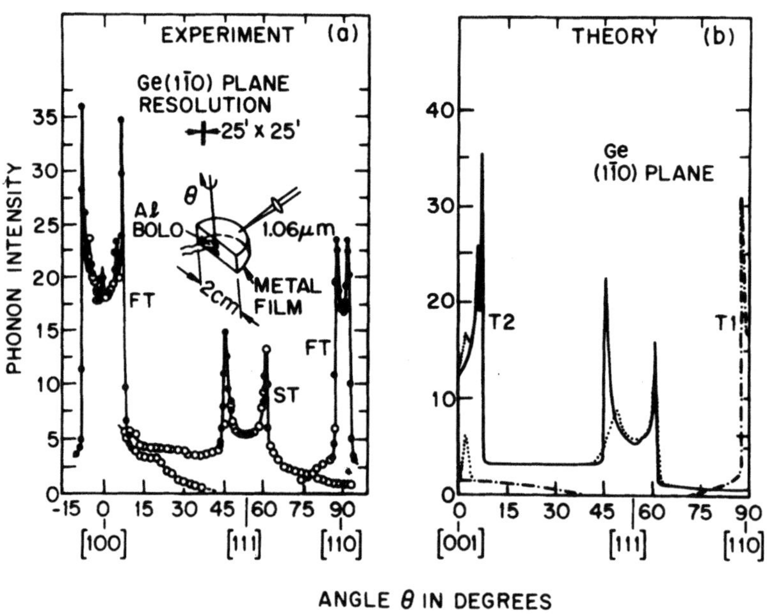

Fig. 6. Angular dependence of phonon focussing in the (110) plane in Ge. (a) Experiment (b) Theory. After Hensel and Dynes (1979).

Wolfe 1979 and Eisenmenger 1980b). The focussing of phonon beams has a direct effect on the matrix elements and coupling of electrons and phonons in all semiconductors, as we shall see below.

PHONON-DEFECT INTERACTIONS IN SEMICONDUCTORS

A great deal of work has been done using the heat pulse and tunnel junction techniques to study donors and acceptors in Si and Ge (see e.g. Pomerantz and von Gutfeld 1968, Dynes and Narayanamurti 1972, Forkel, Welte and Eisenmenger 1973). Work on deep defect states and their symmetries in GaAs and GaAℓAs has been done by Narayanamurti et al 1978a, 1979a. We consider two examples from this list.

Sb in Ge forms a shallow donor whose energy states can be split by the application of uniaxial stress (see inset of Figure 7). The energy splittings lie in the region of 1 meV for moderate stresses. Using superconducting tunnel junctions made of Sn (energy gap $2\Delta \approx 1.2$ meV) as a phonon generator and detector, Dynes, Narayanamurti and Chin (1971) showed a strong resonance absorption at the expected value of stress (Figure 7). Measurements as a function of phonon polarization and energy yielded the symmetry of the states and the strength of the defect phonon coupling (Deformation Potential $E_u \approx 16$ eV). They also provided strong evidence for the basic correctness of the resonant phonon scattering model of Griffin and Carruthers (1963). The resonance absorption of phonons also provided strong evidence for the quantum nature (as distinct from heat) of the recombination process in superconducting junctions.

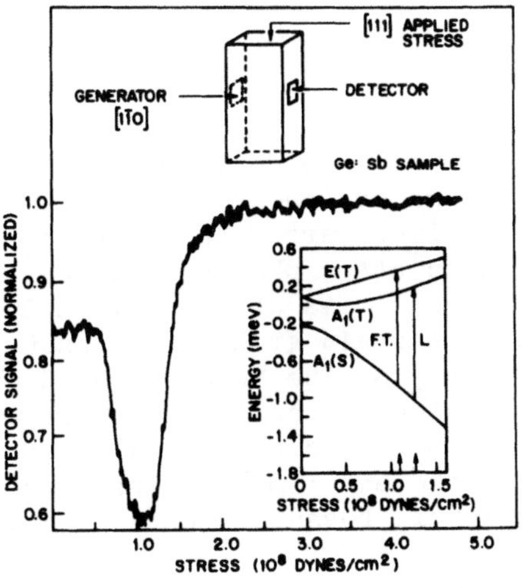

Fig. 7. Transmission of 1.2 meV phonons through a sample of thickness ∿ 4 mm of Ge doped with about 2×10^{15} Sb donors as a function of stress. Insert shows the stress splitting of the donor state. After Dynes, Narayanamurti and Chin 1971.

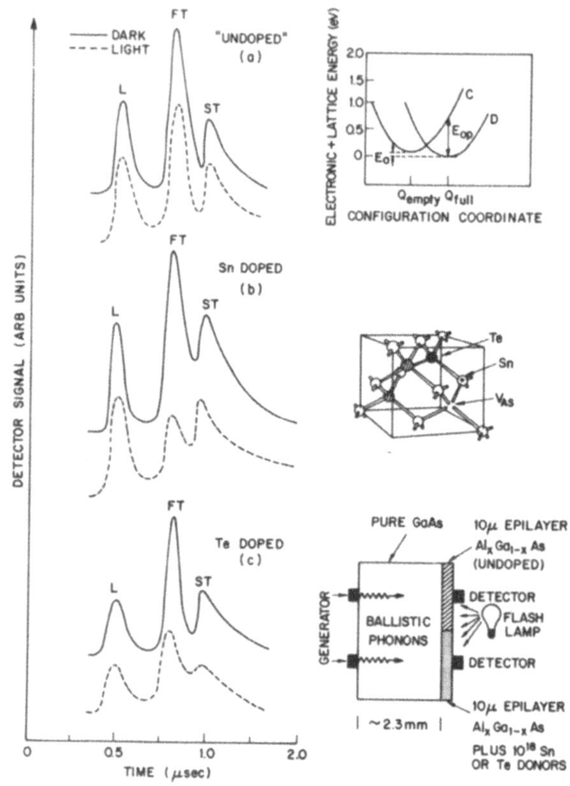

Fig. 3. Ballistic phonon intensities for three different samples of $Al_{0.5}Ga_{0.5}As$: (a) nominally undoped material, (b) with $\sim 10^{18}$ Sn donors per cubic centimeter, and (c) with $\sim 10^{18}$ Te donors per cubic centimeter. Solid lines are data taken in the dark while the dashed lines are taken after photoexcitation. Propagation direction (110). The intensities of the ballistic phonons reveals selective attenuation of certain groups of transverse modes depending upon the symmetry of the donor-related photoconductivity center. The right-hand side of the figure shows a typical configuration-coordinate diagram for such centers. A crystal model of GaAs and a schematic of the experimental arrangement are also shown. After Narayanamurti, Logan and Chin 1979a.

As another example we show the case of deep donor related states in GaAℓAs (Lang et al 1979). Te and Sn donors are believed to form a deep defect complex (DX) but are strongly coupled to the lattice. The strong defect lattice coupling lowers the defect energy (see configuration coordinate diagram of Figure 8) so as to allow several unusual properties such as persistent photoconductivity. Using ballistic phonon techniques, Narayanamurti, Logan and Chin (1979a) were the first to show the symmetry of the centers. Photoexcitation was used to change the occupation of the DX centers in situ in the epitaxial layers which were grown on GaAs substrates. Doped and undoped layers were grown side by side. Preferential attenuation of transverse modes in the (110) direction gave strong evidence for the trigonal and orthorhombic symmetry of the ionized Sn and Te DX centers respectively in AℓGaAs. This has been particularly useful as spin resonance is difficult to perform in this system.

ELECTRON PHONON COUPLING

In a metal or a semiconductor with a spherical Fermi surface, theory (Ziman 1962) predicts that the deformation potential couples free electrons to longitudinal modes only, since the transverse modes result in no dilatation. The maximum momentum of the phonon involved in the scattering corresponds to 2 k_F where k_F is the electron Fermi momentum. This is illustrated schematically in Figure 9. In Figure 10 we show the results of some work by Maneval et al (1971) on the attenuation of ballistic phonons in a degenerately doped sample of InSb. It is clear from this figure

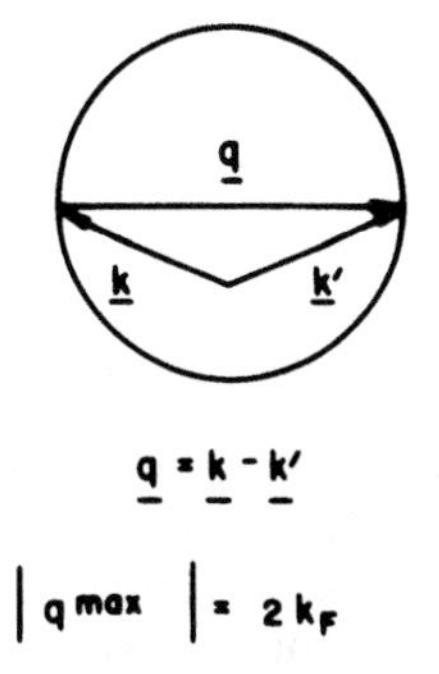

Fig. 9. Phonon-electron scattering for free electrons. The maximum phonon momentum $q^{max} = 2k_F$ where k_F is the radius of the Fermi sphere.

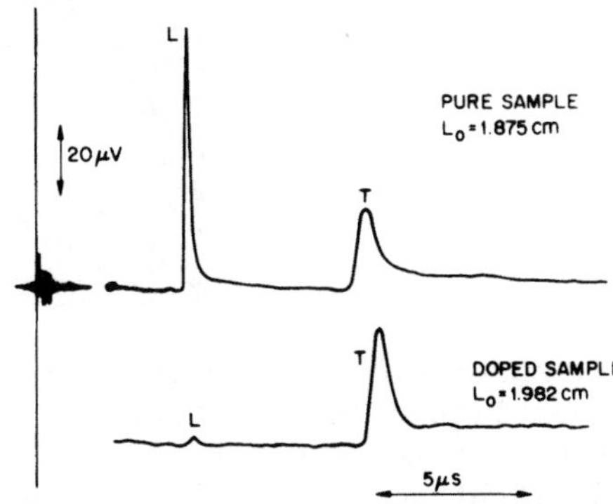

Fig. 10. Heat pulse propagation through nearly 2 cm of bulk InSb. The doped sample had ∿ few×$10^{17}/cm^3$ electrons/cm^3. Propagation direction (111). Note the nearly complete extinction of the L phonons in the doped sample due to the electron phonon interaction. After Maneval et al. (1971).

that L phonons are very strongly attenuated in the doped sample whilst the T phonons are barely effected. Work with superconducting tunnel junctions as sources and detectors of quasimonochromatic phonons has confirmed the $q = 2\ k_F$ nature of the cutoff (Huet and Maneval 1974).

If the Fermi surface is nonspherical, the scattering can be highly anisotropic. A very interesting consequence of this coupling as well as the effects of phonon focussing is the recent work on the interaction of nonequilibrium phonons and e-h drops in Ge. Bagaev et al (1976) and Keldysh (1976) first considered the propulsion and interaction of such drops with phonons and called it the "phonon wind". The e-h drop being a metal absorbs phonons with $q \lesssim 2\ k_F$. Because of the spheroidal nature of the Fermi surface the absorption is highly anistropic. Hensel and Dynes (1977) have applied the heat pulse technique to study this directly. Figure 11 shows their results for a (111) direction of phonon propagation. The e-h drops are produced by volume excitation using a He-Ne laser. In the presence of the drops strong L mode absorption but no T mode absorption along (111) is observed. In other

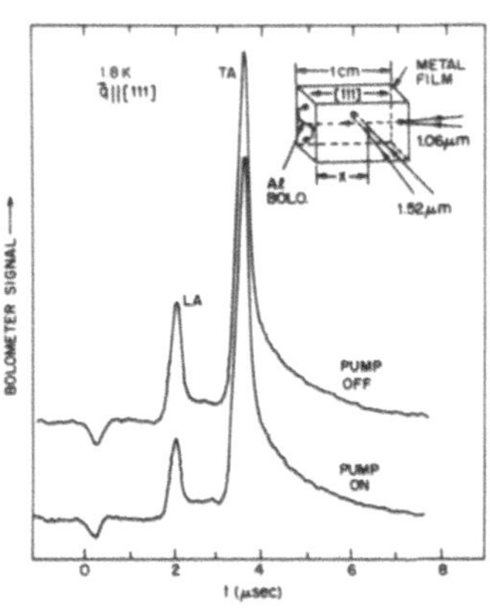

Fig. 11. Absorption of phonons by the e-h liquid in Ge. Traces show phonons time resolved by time of flight across the 1-cm sample length. Excitation: heater, λ=1.06 μm, power=75 mW (0.25 W/mm^2), pulse width=0.3 msec; pump, λ=1.52 μm, power=6 mW cw. The pulse at zero time arises from the laser pulse. The experimental geometry is shown schematically in the inset. After Hensel and Dynes (1977).

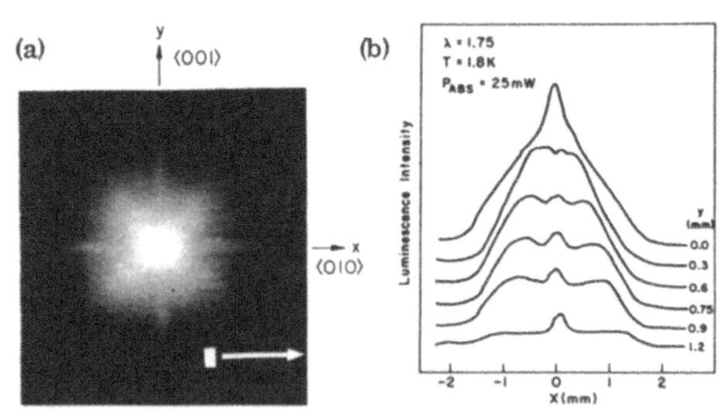

Fig. 12. (a) Photograph of the electron-hole-droplet cloud as viewed through the (100) face opposite the focused excitation. (b) Profiles obtained by scanning the crystal image in x for several y positions of the aperture, showing more quantitatively the structure of the cloud. The structure arises from the highly focused TA phonons which couple to the drops in the (100) direction. After Greenstein and Wolfe (1978).

directions T mode absorption, consistent with symmetry selection rules is observed. Figure 12 shows a photograph of an e-h drop cloud as viewed from the (100) face taken by Greenstein and Wolfe (1978). The flares are believed to arise from the combined effects of phonon focussing (high T mode intensities near the (100) direction) and the allowed coupling in this direction.

So far we have considered deformation potential scattering alone. In piezoelectric semiconductors strong anistropies can arise even for the case of a spherical Fermi surface. This effect has been studied extensively for the case of GaAs (Narayanamurti, Logan, Chin and Lax 1978b,c) doped epitaxial layers. In n-type material the mobility at low temperatures is determined by ionized impurity scattering but the energy relaxation of the carriers under the application of a field takes place via phonon emission. By observing the phonons directly one can determine the strength of the electron phonon scattering and the effects of screening in a very detailed way. In Figure 13 we show some typical phonon generation data at low fields (~ 5 V/cm) for n-layers ($n \sim 2\times10^{17}$ per cm^3) as a function of orientation. The phonons are detected after a ballistic time of flight by a superconducting bolometer. These data should be compared with a constantan metal film generator (see Fig. 3). The data show that the generation rate of L and T phonons is highly anisotropic. For example in the (110) direction we observe only the L and fast transverse (FT) phonons but not the slow transverse (ST) phonons. In the (111) direction the degenerate T modes are extremely weak compared to L but are of the same order in the (100) direction. As discussed above, the deformation potential scattering in GaAs is isotropic. The piezoelectric scattering gives rise to the T mode generation observed. Measurements as a function of carrier concentration show the effects of screening on the piezoelectric scattering. The most interesting case is the (100) direction. Here the piezoelectric scattering is forbidden but the observed T phonons arise from focussing of off-axis piezoelectric phonons onto the detector.

RELAXATION AND RECOMBINATION VIA PHONON EMISSION

We now turn to the case of carrier recombination in semiconductors. This is illustrated schematically in Figure 14. An excited electron loses its energy in several ways. This includes what I would term as 'relaxation' phonon emission within the conduction band. This process is a fast process. In a semiconductor such as GaAs this relaxation process at low energies involves only acoustic phonons whose energy depends on the 'electron temperature'. For high excitation energies relaxation can also take place via longitudinal optic (LO) phonon emission. Carriers at the bottom of the band lose their energy either via resonant phonon emission (shallow donors and acceptors) or via multiphonon emission via deep states (Henry and Lang 1977).

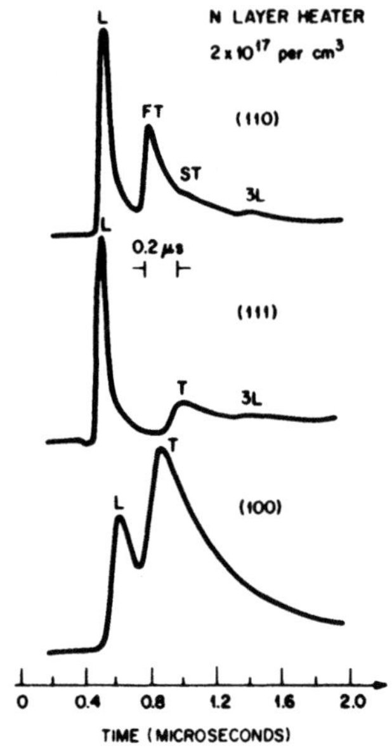

Fig. 13. Ballistic phonons generated by an n epitaxial layer heater containing 2×10^{17} per cm^3 electrons as a function of orientation. In this case, the transverse phonon generation arises through the coupling of the free electron with the phonons via the screened piezoelectric interaction. After Narayanamurti, Logan, Chin and Lax 1978c.

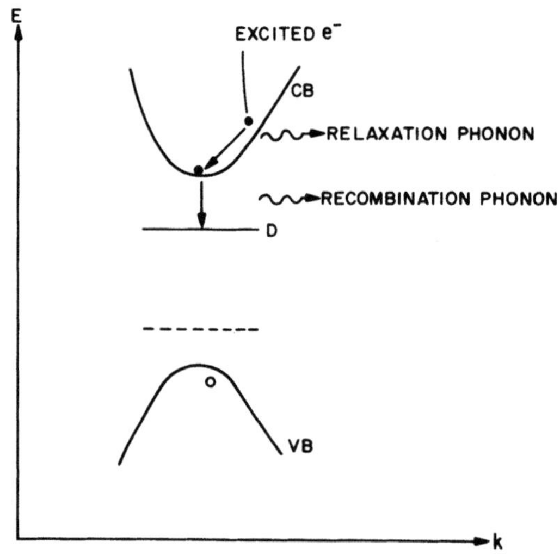

Fig. 14. Band diagram of direct gap semiconductor showing relaxation and recombination phonon generation. Multiphonon recombination processes are important during e-h recombination in the depletion region of pn junctions and in photoexcited semiconductors.

e-h nonradiative recombination processes are of great importance in determining losses in a variety of technologically important luminescent devices. Recent experiments by Narayanamurti, Logan and Chin (1978b) and by Ulbrich, Narayanamurti and Chin (1980) have been designed to study the decay products of multiphonon capture in semiconductors. Figure 15 shows results on phonon generation after photoexcitation in GaAs. The phonon detector was a Pb-oxide-Pb tunnel junction with a threshold in frequency of 0.7×10^{12} Hz. In most directions of propagation these experiments revealed a single broad transverse pulse, whose shape scaled uniformly as a function of Z/t, characteristic of dispersive wave propagation. Such propagation occurred for distances of the order of several mm.

Figure 16 shows a schematic diagram of the phonon dispersion relation in GaAs along the (111) direction after Dolling and Waugh (1964). Also shown is a decay scheme of an optically generated LO phonon into large wave vector TA phonons (Orbach 1966). The

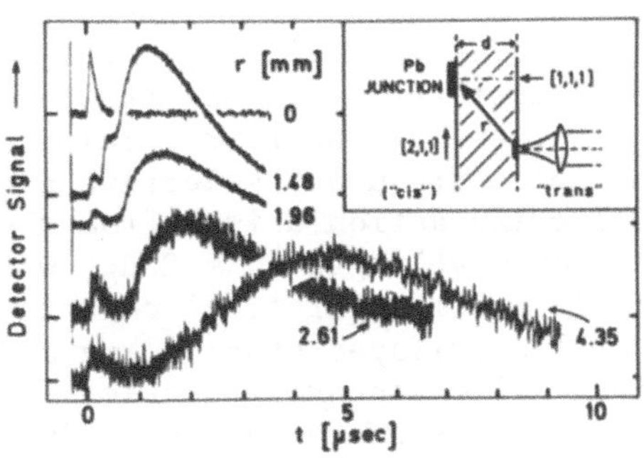

Fig. 15. Phonon propagation in photoexcited GaAs as a function of propagation distance r. Pb junction detector has a threshold $\sim 0.7\times10^{12}$ Hz. Signal at t=0 is due to optical luminescent pickup. After Ulbrich et al. (1980).

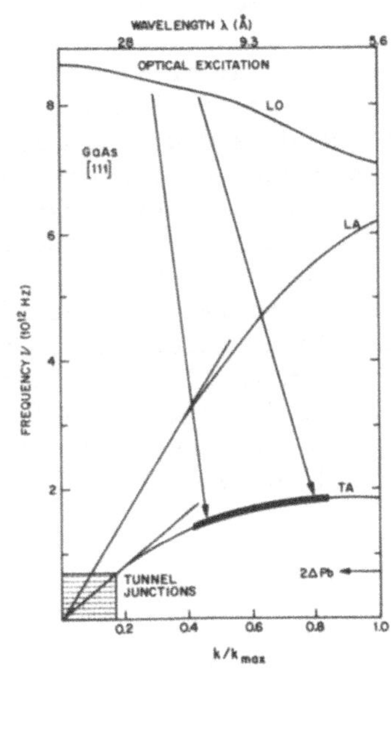

Fig. 16. Dispersion curves for GaAs in (111) direction. Arrows show region of dominant decay to long lived, short wavelength TA phonons of optically excited LO phonons. This is an extremely efficient method for generating high frequency phonons in semiconductors.

shaded area shows the regime of conventional tunnel junction spectroscopy. The data are consistent with propagation of a band of TA phonons with a peak corresponding to phonons well out into the zone. The observed long mean free paths are suggestive of relatively weak impurity scattering in these pure crystals.

Photoexcitation experiments combined with tunnel junctions are a powerful way to study generation, propagation and detection of extremely short wavelength (∿ 10 Å) phonons. The properties of such dispersed phonons (focussing, etc.) are quite different from that of the usual ballistic phonons which correspond to continuum elasticity theory. The high density of states in certain directions (such as (111)) suggests that scattering processes have to be evaluated with care in the large wave vector phonon regime.

SUPERLATTICES AND INTERFACES

Again, because of their technological importance a great deal of effort has been spent in developing ideal epitaxial growth on GaAs using molecular beam techniques (for a recent review of some results see Gossard (1979)). MBE has been used to grow nearly perfect superlattice structures based on the GaAs and AℓGaAs systems. Because of their artificial periodicity such superlattices can serve as frequency selective phonon "mirrors" or "filters". The filtering action arises from the selective reflection of phonons of wavelength $\lambda_o = 2\ d_o$, with d_o the superlattice period. Figure 17 shows theoretical calculations of the reflection properties of a multilayer stack using mathematical procedures described in classical textbooks of optics (Born and Wolfe 1964). The transmission depends on the acoustic impedance ratio Z_1/Z_2 of the 2 media and the number of layers in the stack. 10% disorder in the structure degrades the quality of the filter only slightly.

Recent experiments (Narayanamurti et al 1979b) have shown the phonon filtering action of GaAs:AℓGaAs superlattices directly. Some of the data are shown in Figure 18. Sn tunnel junctions were used as a phonon generator whilst an Aℓ junction served as a detector. In the energy range 0.5 meV to 1.2 meV resonances are observed whose position depended on the superlattice period. The arrows marked E_{sup} show the theoretically expected resonance positions. The lack of interface phonon degradation is suggestive of a very ideal interface. This ideality is quite interesting and surprising in the light of recent Kapitza resistance measurements (Taborek and Goodstein 1980). It seems quite likely that such superlattice structures can find application in phonon transport studies and for phonon optics.

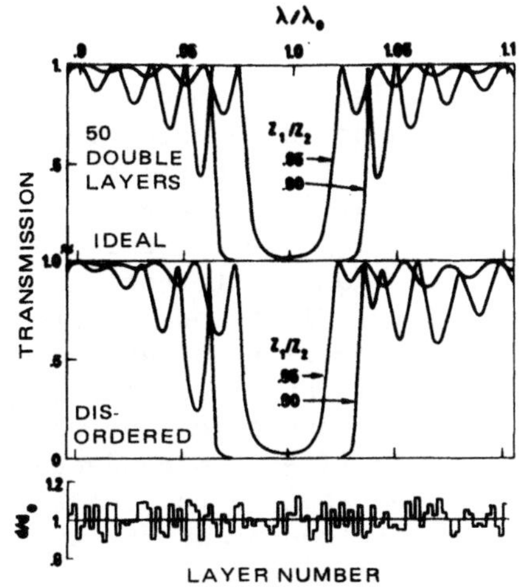

Fig. 17. Theoretical calculation of the transmission of acoustic waves for a multilayer stack of 50 layers and acoustic impedance ratio Z_1/Z_2 of 0.95 and 0.90. Case (a), ideal superlattice. Case (b), superlattice with 10% random disorder in thickness as shown in the trace at bottom. After Narayanamurti et al. 1979b.

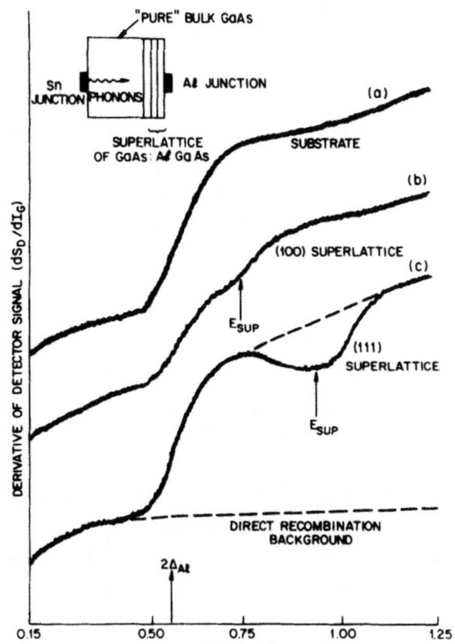

Fig. 18. Phonon transmission through a GaAs:AlGaAs superlattices of periods 93 Å and 122 Å in the (100) and (111) directions respectively. Sn generator, Al detector. The arrows mark the expected positions of the superlattice resonance. After Narayanamurti et al. 1980.

SUMMARY AND OUTLOOK

In summary I have only briefly touched upon some of the many exciting areas of phonon propagation research in semiconductors. Clearly much has been learnt about the details of phonon focussing, electron-phonon interactions, carrier relaxation and recombination and phonon defect interactions in several semiconductors. Much, however, remains to be done. Phonon interactions with excitons, with layered 2 dimensional (or 1 dimensional?) electron gases, with inversion layers and with novel superlattice structures is literally in its infancy. Ideal interfaces can be prepared to study the Kapitza problem. The generation, using the fundamental e-h recombination processes, in III-V semiconductors of large wave vector phonons needs to be applied to a variety of problems at short wavelengths. These include phonon transport (anharmonic 3 phonon processes), energy states of donors and acceptors with small Bohr radii, large lattice relaxation centers and electronic states of electrons and holes in superlattices. Phonon microscopy at 10 Å in technologically important semiconductors appears feasible. Such studies are a powerful way of obtaining stress sensitive information on truly an atomic scale. The field is in its infancy and clearly much needs to be done in the area of phonon optics in semiconductors.

ACKNOWLEDGEMENTS

I would like to acknowledge the close collaboration of several colleagues, particularly in the recent GaAs work. These include Maurice Chin, Mel Lax, Ralph Logan, Rainer Ulbrich, Horst Störmer, Art Gossard, Willie Wiegmann and R. C. Fulton.

REFERENCES

Bagaev, V. S., Keldysh, L. V., Sibeldin, N. N. and Tsvetkov, V. A., 1976, Phonon wind propulsion of excitons and electron-hole droplets, Sov. Phys. JETP 43, 362.

Born, M. and Wolfe, E., 1964, In Principles of Optics (Pergamon, New York).

Carruthers, P. A., 1961, Theory of thermal conductivity of solids at low temperatures, Rev. Mod. Phys. 33, 92.

Challis, L. J., Ramdane, A., 1980, Phonon scattering by chromium ions in GaAs, Proceedings of Int'l. Conf. on Phonon Scattering in Condensed Matter (H. J. Maris, Ed., Plenum Press, New York).

Chaudhuri, N., Wadhwa, R. S., Tiku, P., Sreedhar, A. K., 1973, Thermal conductivity of GaAs at low temperatures, Phys. Rev. 8, 4668.

Dolling, G. and Waugh, J. T., 1964, Normal vibrations in gallium arsenide, In proceedings of the Int'l. Conf. on Lattice Dynamics, Copenhagen, Denmark (Pergamon Press, Oxford).

Dynes, R. C., Narayanamurti, V., Chin, M. A., 1971, Monochromatic phonon propagation in Ge:Sb using superconducting tunnel junctions, Phys. Rev. Lett. 4, 181.

Dynes, R. C., Narayanamurti, V., 1972, Phonon fluorescence in superconductors and the propagation characteristics of high frequency phonons in Ge:Sb and $A\ell_2O_3:V^{3+}$, Phys. Rev. B6, 5143.

Eisenmenger, W., 1976, Tunneling junctions as phonon generators, in Physical Acoustics (W. P. Mason, Ed.), Vol. XII, p. 79 (Academic Press, New York).

Eisenmenger, W., 1980a, Nonequilibrium phonons, Chapter III, this volume.

Eisenmenger, W., 1980b, Phonon detection by the fountain pressure in superfluid ^{4}helium films, Proceedings of Int'l. Conf. on Phonon Scattering in Condensed Matter (H. J. Maris, Ed., Plenum Press, New York).

Forkel, W., Welte, M., Eisenmenger, W., 1973, Evidence for 870-GHz phonon emission from superconducting Aℓ tunnel diodes through resonant scattering by oxygen in silicon, Phys. Rev. Lett. 31, 215.

Geballe, T. H., Hull, G. W., 1958, Isotopic and other types of thermal resistance in germanium, Phys. Rev. 110, 773.

Gossard, A. C., 1979, GaAs/AℓAs layered films, Thin Solid Films 57, 3.

Greenstein, M. and Wolfe, J. P., 1978, Anisotropy in the shape of the electron-hole-droplet cloud in germanium, Phys. Rev. Lett. 41, 715.

Griffin, A. and Carruthers, P., 1963, Thermal conductivity of solids IV: resonance fluorescence scattering of phonons by donor electrons in germanium, Phys. Rev. 131, 1976.

Henry, C. H. and Lang, D. V., 1977, Nonradiative capture and recombination by multiphonon emission in GaAs and GaP, Phys. Rev. B15 , 989.

Hensel, J. C., Dynes, R. C., 1979, Observation of singular behavior in the focussing of ballistic phonons in Ge, Phys. Rev. Lett. 43, 1033.

Hensel, J. C., Dynes, R. C., 1977, Interaction of electron-hole drops with ballistic phonons in heat pulses: the phonon wind, Phys. Rev. Lett. 39, 969.

Holland, M. G., 1964, Phonon scattering in semiconductors from thermal conductivity studies, Phys. Rev. 134, A471.

Huet, D. and Maneval, J. P., 1974, Image of the fermi surface and screening effects in phonon attenuation, Phys. Rev. Lett. 33, 1154.

Keldysh, L. V., 1976, Thermal conductivity and lattice vibrational modes, JETP Lett. 23, 86.

Klemens, P. G., 1958, Thermal conductivity and lattice vibrational modes, Solid State Physics 7, 1.

Lighthill, M. J., 1960, Studies on magneto-hydrodynamic waves and other anisotropic wave motions, Phil. Trans. Roy. Soc. A252, 397.

Lang, D. V., Logan, R. A. and Jaros, M., 1979, Trapping characteristics and a donor-complex (DX) model for the persistent-photoconductivity trapping center in Te-doped $A\ell_xGa_{1-x}As$, Phys. Rev. B19, 1015.

Lax, M., Nelson, D. F., 1975, Radiance theorem and optical invariants in anisotropic media, J. Opt. Soc. Am. 65, 668.

Lax, M., Nelson, D. F., 1976, Imaging through a surface of an anisotropic medium with application to light scattering, J. Opt. Soc. Am. 66, 694.

Lax, M., Narayanamurti, V., 1980, Phonon magnification and the gaussian curvature of the slowness surface in anisotropic media: detector shape effects with application to GaAs, (in press).

Maris, H. J., 1971, Enhancement of heat pulses in crystals due to elastic anisotropy, J. Acoust. Soc. Am. 50, 812.

Maneval, J. P., Zylberstejn, A., and Huet, D., 1971, Direct observation of electron-phonon interaction, Phys. Rev. Lett. 27, 1375.

Musgrave, M. J. P., 1954, On the propagation of elastic waves in aelotropic media. I. General Principles, Proc. Roy. Soc. A226, 337.

Narayanamurti, V., Chin, M. A., and Logan, R. A., 1978a, Direct determination of symmetry of Cr ions in semi-insulating GaAs substrates through anisotropic ballistic phonon propagation and attenuation, Appl. Phys. Lett. 33, 481.

Narayanamurti, V., Logan, R. A., and Chin, M. A., 1978b, Direct observation of phonons generated during nonradiative capture in GaAs p-n junctions, Phys. Rev. Lett. 40, 63.

Narayanamurti, V., Logan, R. A., Chin, M. A., and Lax, M., 1978c, Anisotropic phonon generation in GaAs epilayers and pn junctions, Sol. St. El. 21, 1295.

Narayanamurti, V., Logan, R. A., and Chin, M. A., 1979a, Symmetry of donor-related centers responsible for persistent photoconductivity in $A\ell_xGa_{1-x}As$, Phys. Rev. Lett. 43, 1536.

Narayanamurti, V., Störmer, H. L., Chin, M. A., Gossard, A. C., and Wiegmann, W., 1979b, Selective transmission of high-frequency phonons by a superlattice: the "dielectric" phonon filter, Phys. Rev. Lett. 43, 2012.

Northrup, G. A., Wolfe, J. P., 1979, Ballistic phonon imaging in solids - a new look at phonon focusing, Phys. Rev. Lett. 43, 1424.

Orbach, R., 1966, Nonlinear phonon generation, Phys. Rev. Lett. 16, 15.

Pomerantz, M. and von Gutfeld, R. J., 1968, Heat pulse studies of phonon scattering by impurities in Si and Ge, Proceedings of the Int'l. Conf. on the Physics of Semiconductors, Moscow, (Nanka Publishing House, Leningrad, U.S.S.R.) 2, 690.

Taborek, P. and Goodstein, D., 1980, Diffuse reflection of phonons and the anomalous Kapitza resistance, Phys. Rev. (in press).

Taylor, B., Maris, H. J., Elbaum, C., 1969, Phonon focusing in solids, Phys. Rev. Lett. 23, 416.

Ulbrich, R. G., 1978, Optical studies of hot electrons, Proceedings of 14th Int'l. Conf. on Physics of Semiconductors, Edinburgh, Inst. Phys. Conf. Ser. No. 43, 11.

Ulbrich, R. G., Narayanamurti, V. and Chin, M. A., 1980, Propagation of large wave vector phonons in semiconductors, Proceedings of 15th Int'l. Semiconductor Physics Conf., Kyoto, Japan and Phys. Rev. Lett. (submitted).

von Gutfeld, R. J., 1968, Heat pulse transmission, in Physical Acoustics (W. P. Mason, Ed.), Vol. V, p. 233 (Academic Press, New York).

Ziman, J. M., 1962, in Electrons and Phonons (Clarendon, Oxford, England).

TUNNELING: A PROBE OF NONEQUILIBRIUM SUPERCONDUCTIVITY

Kenneth E. Gray

Solid State Science Division
Argonne National Laboratory
Argonne, IL 60439 USA

INTRODUCTION

This Chapter will begin with a review of the BCS theory of superconductivity and basic tunneling theory, which will form the basis for much of the discussion in other chapters. Many important nonequilibrium experiments have resulted from the use of two tunnel junctions: one to create a nonequilibrium state and another to measure it. The remainder of this Chapter will discuss examples of superimposed tunnel junctions, i.e., in which one film is common to both junctions.

BCS THEORY

The purpose of this section is to provide an overview of the BCS theory of superconductivity (Bardeen et al, 1957). A more comprehensive review can be found in Tinkham (1975). The key idea of BCS is that the conduction electrons condense into a ground state composed of Cooper pairs which consist of electrons in equal and opposite momentum and spin states, i.e., $(k\uparrow,-k\downarrow)$. This condensation occurs when the attractive interaction between electrons due to phonons exceeds the repulsive Coulomb interaction. The physical idea of the phonon interaction is that electrons polarize the medium by attracting positive ions, which in turn attract other electrons. For the ground state wave function, BCS chose

$$\left|\psi_G\right\rangle = \prod_k \left(u_k + v_k\, c^+_{k\uparrow}\, c^+_{-k\downarrow}\right)\left|\psi_o\right\rangle, \qquad (1)$$

where $\left|\psi_o\right\rangle$ is the vacuum state with no particles present, c_k^+ is a creation operator for the electron state k and $u_k^2+v_k^2=1$. This form implies that the probability of a pair state$(k\uparrow,-k\downarrow)$ being occupied is v_k^2. The value of v_k can be calculated by minimizing the free energy with respect to v_k using a suitable hamiltonian. BCS uses the so-called "reduced hamiltonian" which includes only the kinetic energy and the interaction energy of the pairs:

$$H_{BCS} = \sum_k \varepsilon_k n_k + V_{BCS} \sum_{k\ell} c^+_{k\uparrow} c^+_{-k\downarrow} c_{-\ell\downarrow} c_{\ell\uparrow} . \tag{2}$$

Here ε_k is the Bloch energy for an electron in state k, n_k is the number operator $c_k^+ c_k$, and V_{BCS} is the pair interaction potential. It is convenient to define the zero of kinetic energy at the Fermi energy E_F and define $\xi_k=\varepsilon_k-E_F$. Finally, one obtains:

$$v_k^2 = (1 - \xi_k/E_k)/2, \tag{3}$$

$$E_k = (\Delta^2 + \xi_k^2)^{1/2}, \tag{4}$$

$$\Delta = V_{BCS} \sum_k u_k v_k (1-2 f_k), \tag{5}$$

where E_k and Δ are the quasiparticle excitation energy and energy gap respectively, and f_k is the quasiparticle occupation probability or distribution function. For thermal equilibrium, further minimization of the free energy with respect to f_k results in the condition that it is given by the Fermi-Dirac distribution f^o. The variation of Δ with reduced temperature t $(=T/T_c)$ is shown in Fig. 1, where T_c is the transition temperature into the normal state and $\Delta(0) =1.76\ kT_c$. Also shown in Fig. 1 are v_k^2 (the pair occupation probability) and $u_k v_k$. The latter product enters in the expression for Δ and underlines the importance of k states very close to k_F in determining Δ. Note that at T=0, the electron occupation probability in a superconductor does not change discontinuously from 1 to 0 at k_F as in a normal metal. The excitation spectrum of quasiparticles (Eq. 4) is shown schematically in Fig. 2 (note that $\Delta \ll E_F$) along with the reduced superconducting density of states $\rho(E)$. The Cooper pairs reside at the Fermi energy and there is a gap Δ between the pairs and the first excited quasiparticle state. The total number of quasiparticles is given by:

$$N = 2N(0) \int_0^\infty [f_>(E) + f_<(E)]\ \rho(E)dE \tag{6}$$

where N(0) is the density of states at the Fermi energy in the normal metal and $f_>(f_<)$ is the distribution function for the quasiparticle branch with $k>k_F(k<k_F)$. In thermal equilibrium these both equal f^o.

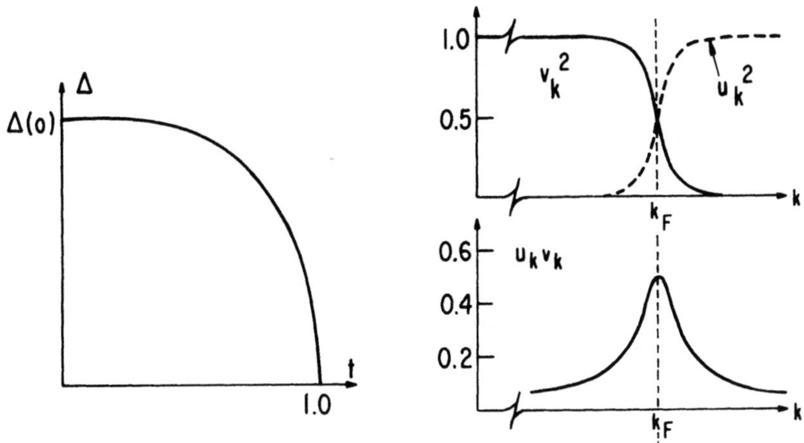

Fig. 1. Parameters of the BCS theory: the pair occupation probability is v_k^2, while $u_k^2 = v_k^2 - 1$; the product $u_k v_k$ which enters the BCS gap equation (Eq. 5); and the temperature dependent energy gap.

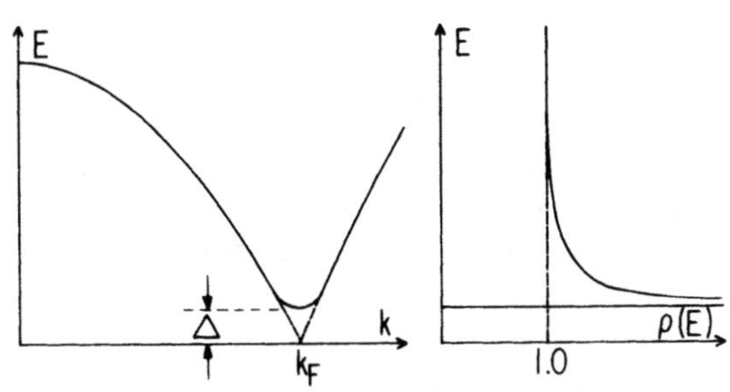

Fig. 2. Left: The quasiparticle excitation spectrum of (Eq. 4); note that the structure near k_F is magnified since $\Delta << E_F$. Right: Reduced density of states (normalized to the normal metal).

It is often useful to work with quasiparticle creation and annihilation operators (γ^+,γ) rather than the electron operators (c^+,c). Bogoliubov (1958) and Valatin (1958) gave the correct transformations:

$$c_{k\uparrow} = u_k\gamma_{k\uparrow} + v_k\gamma^+_{-k\downarrow}$$

$$c^+_{-k\downarrow} = - v_k\gamma_{k\uparrow} + u_k\gamma^+_{-k\downarrow}, \qquad (7)$$

where u_k and v_k are the coherence factors defined above.

TUNNELING

Quasiparticle tunneling, as distinct from Josephson tunneling, has become one of the most important methods for studying the microscopic properties of superconductors outlined above. Quasiparticle tunneling was first reported by Giaever (1960) and the purpose of this section is to review its general features.

Tunneling is a classic quantum mechanical problem. For the rectangular barrier shown in Fig. 3, it is a simple matter to solve the time-independent Schroedinger equation for an electron impinging from the left. In the barrier the wave function decays exponentially with distance so that the amplitude on the right is diminished. The transmission probability therefore depends exponentially on the width of the barrier relative to a characteristic distance which depends on the difference between the barrier height and the energy of the incoming electron. Typical values give a transmission of 10^{-6} for a width of about one nm.

Another approach to tunneling is to use time-dependent perturbation theory to calculate the transition from a state on the left of the barrier to one on the right. The "golden rule" tells us that this rate will depend on the tunneling matrix element $T_{kk'}$, which is determined by the overlap of the wave functions shown in Fig. 4,

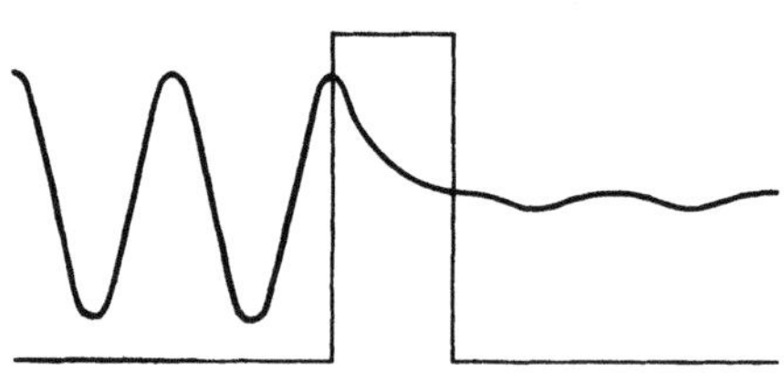

Fig. 3. The solution of the time-dependent Schroedinger equation for an electron impinging a forbidden barrier from the left. The transmitted wave amplitude is reduced.

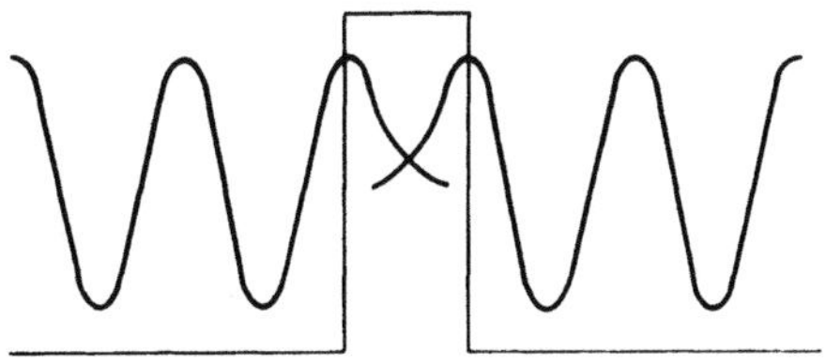

Fig. 4. Wave functions for electrons on each side of the barrier. The coupling is proportional to the overlap of the exponential tails in the barrier and determines the transition rate.

and is hereafter assumed to be constant for the processes of interest. This approach is particularly useful because the hamiltonian can be written in terms of the creation operators of the previous section:

$$H_T = \sum_{kk'} T_{kk'} \, c^{+}_{k'\uparrow} \, c_{k\uparrow} + \text{hermitian conjugate}, \tag{8}$$

where k and k' refer to states on different sides of the barrier. The first term transfers and electron across the barrier in one direction while the h.c. is the inverse transition. The current is 2e (to account for both spins) times the transition probability:

$$2e(2\pi/\hbar) \sum_{kk'} |\langle k'|H_T|k\rangle|^2 \, f_k(1-f_{k'}) \, \delta(E_k + eV - E_{k'}).$$

Here f_k is the probability that the initial state k is occupied and $(1-f_{k'})$ is the probability that the final state k' is unoccupied. The delta function is a statement of energy conservation. For the inverse process, only the distribution functions must be changed, into $f_{k'}(1-f_k)$, and upon subtracting, the net current is

$$I = (4e\pi/\hbar) \sum_{kk'} |\langle k'|H_T|k\rangle|^2 \, (f_k - f_{k'}) \, \delta(E_k + eV - E_{k'}).$$

By converting the summations to an integral, it can be shown that the normal conductance of the junction is $G_{nn}=4\pi e|T_{kk'}|^2 N(0)N'(0)/\hbar$.

If either or both of the metals are superconducting, Eq. 7 must be used to express the electron operators(c^+,c) in terms of the quasiparticle operators (γ^+,γ). In the case of tunneling between two superconductors, there are four terms and their hermitian conjugates. Since the initial or final states are different for each of these terms, no cross products appear when the matrix element is squared to

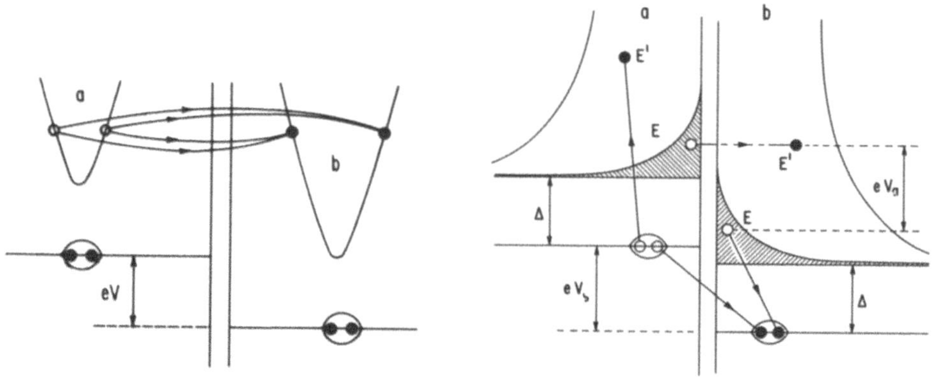

Fig. 5. Left: The four quasiparticle tunneling transitions at one energy due to the term $u_{k}'u_{k}\,\gamma^{+}_{bk'\uparrow}\,\gamma_{ak\uparrow}$. Right: The four transitions on the left are shown as one process in which $(E, a\rightarrow E', b)$ and the branches are suppressed. Also shown is the other process, which involves the pairs, and is due to the term $v_{k}v_{k'}\gamma_{b-k'\downarrow}\gamma^{+}_{a-k\downarrow}$. In thermal equilibrium with equal gaps, the rates of these processes are identical.

get the transition probability. Therefore each term can be considered separately.

Consider first the term $u_{k'}u_{k}\gamma^{+}_{bk'\uparrow}\gamma_{ak\uparrow}$ which creates a quasiparticle in side b and destroys one in a. For given quasiparticle energies E_k and $E_{k'}$ there are four possible transitions depending on which branch of the excitation spectrum (Fig. 2) is involved. These are shown on the left hand side of Fig. 5. The current due to each transition <u>and</u> its inverse is

$$G_{nn}\int_{0}^{\infty} dE_{k}\,\rho(E_{k})\,\rho'(E_{k'})\,u^{2}_{k'}u^{2}_{k}\,(f_{k}-f_{k'}), \tag{9}$$

where $E_{k'}=E_{k}+eV$. Now ρ and ρ' are the same for each branch, as are f_k and $f_{k'}$ in thermal equilibrium, so that the net current is given by summing $u^{2}_{k'}\,u^{2}_{k}$ for the four transitions. This sum can be shown to equal one and the coherence factors drop out of the net current. As a result, it is sometimes useful to suppress the branches, and show these four transitions as a single process. This is shown in the right hand side of Fig. 5, along with the process corresponding to the term $v_{k}v_{k'}\gamma_{b-k'\downarrow}\gamma^{+}_{a-k\downarrow}$ which involves the pairs. Although this second process looks more complicated, the matrix elements for both processes are identical for thermal equilibrium and with equal gaps in a and b. These two processes will play an important role in several examples of nonequilibrium superconductivity to be discussed later in this Chapter.

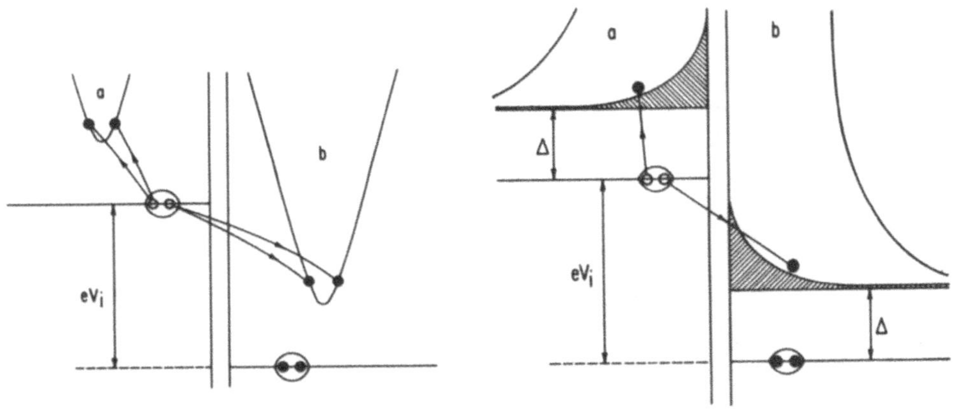

Fig. 6. Left: The four transitions which break pairs in a and inject a quasiparticle into each film. Right: Same as left, but with the dependence on branches suppressed.

There is another very important tunneling process which breaks Cooper pairs. It is represented by the term $u_{k'}v_k\gamma^{+}_{bk'\uparrow}\gamma^{+}_{a-k\downarrow}$ and is shown in Fig. 6. The coherence factors also add to one for the four transitions shown on the left in the case of thermal equilibrium. The current due to pair-breaking and its inverse is

$$G_{nn} \int_0^\infty dE_k\ \rho(E_k)\ \rho'(E_{k'})\ (1-f_k-f_{k'}). \tag{10}$$

This process creates quasiparticles on each side of the barrier by breaking a Cooper pair. As such, it requires a minimum voltage of $2\Delta/e$ to overcome the binding energy of the pair. Note that the fourth term in c^+c, i.e., $v_{k'}u_k\gamma_{b-k'\downarrow}\gamma_{ak\uparrow}$ destroys quasiparticles on each side of the barrier and cannot conserve energy.

Experimental current-voltage characteristics, I(V), are shown in Fig. 7 for a junction between two identical superconductors. The pair-breaking process occurs for $V>2\Delta/e$, and is seen as a sharp rise or step in current because of the singularity in $\rho(E)$ at $E=\Delta$ (Fig. 2). Clearly this structure provides a measure of Δ as a function of T, and these measurments generally confirm the BCS prediction (Fig. 1). At lower voltages, only the processes shown in Fig. 5 occur, and the net current is given by twice Eq. 9. At low temperatures and for V not too low, ρ' is approximately one and $f_{k'} \ll f_k$, so that

$$I \cong 2\ G_{nn} \int_0^\infty dE\ \rho(E)\ f(E) = G_{nn}\ N(T)/2N(0), \tag{11}$$

and I is proportional to N(T), the quasiparticle density (Eq. 6).

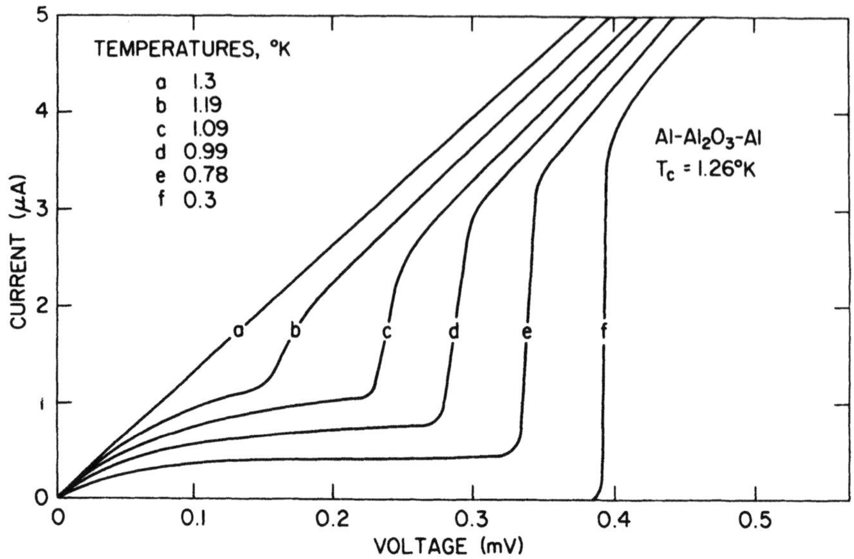

Fig. 7. Experiment I(V) for a tunnel junction between two identical superconductors. The temperature dependence of Δ is seen as a shift in the vertical rising portion of the characteristic which occurs at $V = 2\Delta/e$. Note also the decrease in current at low temperatures for smaller voltages, which indicates a decreasing number of thermally excited quasiparticles.

When only one of the metals is superconducting, the equations simplify, and for thermal equilibrium the current is given by

$$I(V) = G_{nn} \int_0^\infty \rho(E)\,[f(E-eV) - f(E+eV)]\,dE, \tag{12}$$

where f is the distribution function in the <u>normal</u> metal. Taking the derivative of Eq. 12 with respect to V, the quantity in brackets becomes a delta-function in the limit of T=0, so that

$$\frac{dI(V)}{dV} = G_{nn}\,\rho(eV). \tag{13}$$

This demonstrates one of the most important uses of normal-superconducting (SN) junctions - to measure $\rho(E)$. Another important use, which involves the creation and detection of charge imbalance will be covered in detail in Chapter 13.

For tunneling between superconductors with unequal gaps, the evaluation of the terms is only slightly more complex. In particular, the currents due to the two processes shown in Fig. 5 are unequal, and the I(V) has a peak at $|\Delta-\Delta'|/e$, as well as the step at $(\Delta+\Delta')/e$ due to pair breaking (Fig. 17). Thus both gaps are determined. A very important case involves using a superconductor with a large gap Δ' to study the properties of a small gap superconductor ($\Delta \ll \Delta'$). In such a case $f_{k'}$ is negligible and the current for $V<(\Delta+\Delta')/e$ is

$$I(V) = G_{nn} \int_{o}^{\infty} \rho(E)\ \rho'(E + eV)\ f(E)\ dE. \qquad (14)$$

Thus since ρ and ρ' can be independently measured (e.g. by NS tunneling and Eq. 13), the distribution function f(E) can be determined if the integral equation can be inverted. Such an inversion will be discussed later in this Chapter.

Calculations of the currents for all the various cases are given in much greater detail in a review article by Paterson (1978). For a general reference to tunneling, see Solymar (1972).

TUNNELING IN NONEQUILIBRIUM SUPERCONDUCTORS

The previous section has outlined the methods by which tunneling can measure the superconducting properties Δ,ρ,f and N in thermal equilibrium. Results of such measurements in weakly coupled superconductors are in excellent agreement with the BCS theory. The question of whether nonequilibrium superconductors can be probed in the same way must now be addressed. In thermal equilibrium, f is the Fermi function f^o, and Eq. 5 determines Δ self-consistently since u_k and v_k depend on Δ. All the tunneling equations use ρ, u_k and v_k which are determined from the one parameter Δ through Eqns. 3 and 4. In order to proceed, one is forced to assume that these are still determined in the same way for all nonequilibrium situations. In deriving Eqns. 3 through 5, the form of f was not specified, so that this assumption seems reasonable. However due to its importance, this point should be verified experimentally, especially for situations with charge imbalance. Later in this Chapter methods to verify the above assumption will be discussed.

If this assumption is made, it is possible to carry over the formulation of the previous section and measure Δ,ρ,f and N in nonequilibrium states. In the rest of this Chapter as well as others in this book, tunnel junctions will be used to measure the properties of nonequilibrium states created by microwaves, phonons, visible light and tunnel injection of quasiparticles. Because the nonequilibrium properties can be measured easily, tunnel junctions provide excellent detectors, and these are also discussed in other chapters.

A important problem in superconductors is to measure relaxation rates, Γ_x, for various processes. In very general terms, the deviation from equilibrium, δX, of any nonequilibrium state will depend on the ratio of the creation rate, Γ_y, to the relaxation rate. For small deviations from equilibrium, it is possible to determine Γ_x if one can measure δX and Γ_y. Specific examples for δX, Γ_x include: ΔN, the quasiparticle-pair relaxation rate; Q^*, the charge imbalance relaxation rate(Chapter 13); and $\delta f(E)$, the energy-dependent inelastic scattering rate. In all cases one must take precautions to avoid the undesirable effects of heating.

Finally, the use of nonequilibrium tunnel junctions as a source of phonons is discussed in Chapter 3.

DOUBLE JUNCTION EXPERIMENTS

Many important experiments have resulted from the use of two tunnel junctions: one to create a nonequilibrium state and another to measure it. This section will include a compendium of experiments which use two superimposed tunnel junctions, i.e., in which one film is common to both junctions. A cross-sectional schematic view of such a sample is shown in Fig. 8. The injector junction creates the nonequilibrium state in the middle film and the detector junction measures its properties. The middle film, which is always the superconductor to be studied, is usually very thin (30-50 nm) to maximize

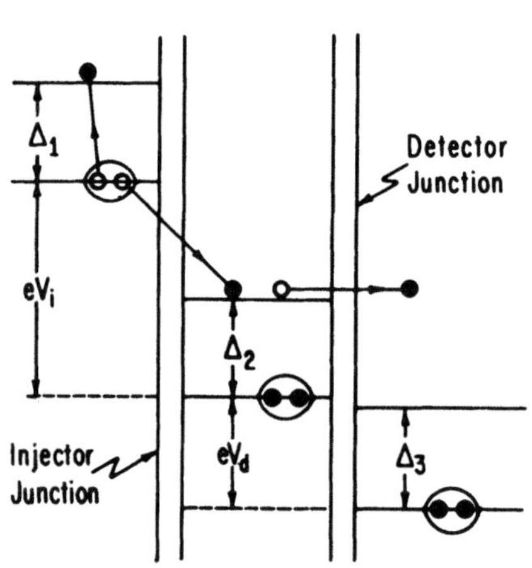

Fig. 8. A schematic cross-sectional view of a double junction sample in which the energy level of pairs and quasiparticle are shown for the three films. Some typical tunneling processes are indicated.

Table 1. A Compendium of Experiments Using Two Superimposed Tunnel Junctions

Experiment	Δ_1	eV_i	Δ_3	eV_d	R_i/R_d
ac Josephson effect	>0	>0	>0	$\sim \Delta_2+\Delta_3$	any
Gap enhancement					
Eliashberg	Δ_2	$<\Delta_1+\Delta_2$	any	$\Delta_2+\Delta_3$	small
Parmenter	$>\Delta_2$	$\simeq\Delta_1-\Delta_2$	any	$\Delta_2+\Delta_3$	small
Instabilities					
Junction	any	$\Delta_1+\Delta_2$	any	$\Delta_2+\Delta_3$	small
Intrinsic	any	$>\Delta_1+\Delta_2$	any	$\Delta_2+\Delta_3$	small
Qp-Pair relaxation	any	$\gg\Delta_1+\Delta_2$	>0	$\sim \Delta_3$	any
Transistor	any	$\gg\Delta_1+\Delta_2$	Δ_2	$\sim \Delta_3$	large
Distribution function	any	$\gg\Delta_1+\Delta_2$	$>\Delta_2$	$<\Delta_3$	small
Inelastic scattering	$\sim \Delta_2$	any	$>\Delta_2$	$<\Delta_3-\Delta_2$	small
Charge relaxation	any	$\neq\Delta_1+\Delta_2$	any	$\sim \Delta_3$	large

the nonequilibrium effects relative to heating. Table 1 summarizes the types of experiments and also indicates the restrictions on the energy gaps, junction resistances and voltages. In virtually all cases, except perhaps charge imbalance relaxation, better results are achieved by using superconductors for all three films.

The ac Josephson effect was clearly demonstrated by Giaever (1965) and will not be discussed here. Experiments on instabilities will be discussed in Chapter 10 and charge imbalance relaxation in Chapter 13. The techniques of gap enhancement by tunneling will be presented in this Chapter, but the relation of the results to other enhancement experiments will be discussed in Chapter 7. The remainder of the experiments will be discussed in the following sections of this Chapter.

QUASIPARTICLE-PAIR RELAXATION

The first experiment to use a double junction was a measurement (Taylor, 1963 and Miller and Dayem 1967) of the quasiparticle-pair recombination rate Γ_R. In this experiment, Γ_R is determined by measuring the excess quasiparticle density ΔN due to tunnel injection at a known rate I_o. The rate equation for N is given by Rothwarf and Taylor (1967) as

$$\frac{\partial N}{\partial t} = I_o + 2\Gamma_B N_\omega - RN^2, \tag{15}$$

$$\frac{\partial N_\omega}{\partial t} = \frac{RN^2}{2} - \Gamma_B N_\omega - \Gamma_{es}[N_\omega - N_\omega(T)] \tag{16}$$

where the N^2 reflects the fact that the quasiparticles must recombine in pairs, and N_ω is the number density of phonons with $\hbar\Omega > 2\Delta$, which break pairs at a rate Γ_B to create new quasiparticles and escape at a rate Γ_{es}. It can be shown that R is the average rate at which a quasiparticle recombines with all others, even though it can only recombine with N/2 of them, i.e., those with opposite spin. In that case $\Gamma_R = RN$, where N is the total density (both spins). Several calculations of R exist, and they are summarized in Kaplan et al. (1976) who also show

$$R = [2\Delta(0)/kT_c]^3/4N(0)\Delta(0)\tau_o$$

$$\tau_o = Z_1(0)\hbar/2\pi kT_c \ \alpha^2F(kT_c/\hbar) \tag{17}$$

where $Z_1(0)$is the real part of the Eliashberg renormalization factor at $\omega=0$, $\alpha^2F(\Omega)$ is the electron-phonon coupling constant times the phonon density of states and τ_o is a time constant characterizing the strength of the electron-phonon interaction in a specific material.

In equilibrium, I_o is zero and $2\Gamma_B N_\omega(T) = RN^2(T)$, so that the ratio R/Γ_B is determined from statistics and the phonon and quasiparticle densities of states. Straightforward evaluation shows that this ratio is constant at low temperatures. Define $N=N(T)+\Delta N$; then for small injection, $\Delta N \ll N(T)$, and

$$\frac{\partial \Delta N}{\partial t} = I_o + 2\Gamma_B \Delta N_\omega - 2\Delta N \Gamma_R(T). \tag{18}$$

If the nonequilibrium phonons escape from the film rapidly, ΔN_ω could be neglected, and steady state measurements of I_o and ΔN would give Γ_R. Unfortunately, ΔN_ω is not small and has the same temperature dependence as $N\Delta_R(T)$. Solving Eqns. 15 and 16 together, Rothwarf and Taylor (1967) show that steady state experiments measure an effective recombination rate Γ_R^*, where

$$\Gamma_R^* = \Gamma_R/(1+\Gamma_B/\Gamma_{es}), \tag{19}$$

$$\Gamma_R^* = I_o/2\Delta N, \tag{20}$$

$$R^* = I_o/2N(T)\Delta N, \tag{21}$$

where Γ_{es} is the rate at which phonons escape. Gray (1971) has shown that Eq. 19 is also valid for time dependent experiments. Unfortunately the phonon trapping factor $F_\omega \equiv 1+\Gamma_B/\Gamma_{es}$ is difficult to reliably calculate or measure in most experiments (see Kaplan, 1979).

In the analysis of these experiments, I_o is related to the current I_i in the injector junction by $I_o = I_i/eA_id$, where A_i is the area of the injector junction and d the thickness of the middle film. The detector is biased at $eV_d < \Delta_2+\Delta_3$ so that no pair-breaking occurs. If $\Delta_2=\Delta_3$, Eq. 11 predicts that $\Delta N/N(T)=2\Delta I_d/I_d(T)$, the factor of two arising because direct injection is into only one film of the detector. Note that the factor of two disappears for $\Delta_3 >> \Delta_2$. It is convenient to define a tunneling rate for quasiparticles, Γ_T, which is the ratio of the quasiparticle current (not pair-breaking) divided by e, to the total number of quasiparticles, i.e.,

$$\Gamma_T = I_d(T)/2eA_d dN(T). \tag{22}$$

For identical superconductors in the detector junction, Eq. 11 can be used to evaluate Eq. 22, resulting in

$$\Gamma_T \cong G_{nn}/4eA_d dN(0). \tag{23}$$

This approximation is valid at low temperatures and for $eV > \Delta/2$. Combining these expressions with Eq. 20, the experimental recombination rate Γ_R^* (the reciprocal of the recombination time τ_R^*) is

$$\Gamma_R^* = \Gamma_T A_d I_i/2A_i\Delta I_d. \tag{24}$$

There is also a geometrical factor relating to the junction overlap.

Near thermal equilibrium, i.e., for weak injection $\Delta N << N(T)$, the temperature dependence of Γ_R (and Γ_R^*) is due to N(T) (see Eq. 6). This has been verified in experiments using aluminum double junctions and the results are shown in Fig. 9. The deviation of τ_R from 1/N(T)

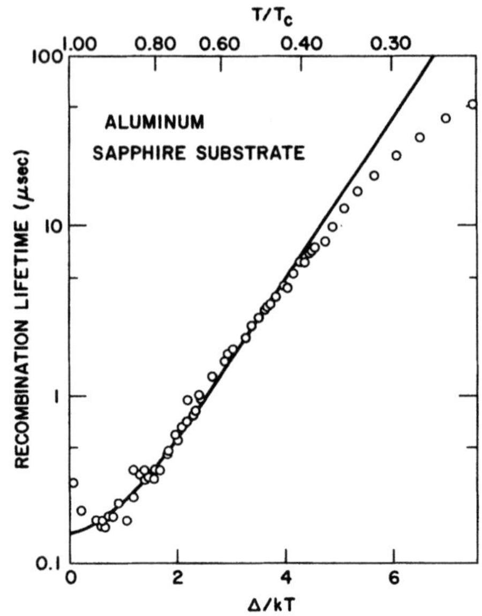

Fig. 9. The experimental recombination lifetime τ_R^* (equal to $1/\Gamma_R^*$) plotted against Δ/kT for aluminum films (Gray et al. 1969). The solid curve is proportional to $1/N(T)$.

at low temperatures is due to residual magnetic fields used to quench the Josephson effect in the junctions, and the increase near T_c will be discussed below. In the opposite limit of large injection, $\Delta N >> N(T)$, Rothwarf and Taylor (1967) showed that

$$R^* = I_o/(\Delta N)^2. \tag{25}$$

In Fig. 10, the data of Gray (1971) is plotted against Eq. 25 with $R^*=1.2\times10^{-12}$ cm^3/s. The excellent agreement, together with the results of Fig. 9 are a strong confirmation that quasiparticle recombination is being observed, and not heating. The value of R^* derived from the fit of Fig. 9 (which is the same sample) is $1.1\times10^{-12} cm^3/s$. This close agreement verifies the measurement of ΔN (see Eqns. 21 and 25), but says nothing about possible systematic errors in the determination of I_o. Therefore it is useful to include time dependent measurements to confirm the magnitude of R^* (see Gray et al., 1969). However, the value of R^* measured in aluminum films is about an order of magnitude smaller than the value of 1.4×10^{-11} cm^3/s calculated by Kaplan et al. (1976) for bulk aluminum. The actual difference is even larger, since the short mean free path in thin films of aluminum enhances the electron-phonon coupling and hence R. The source of this discrepancy is presumably due to phonon trapping, but F_ω is hard to determine.

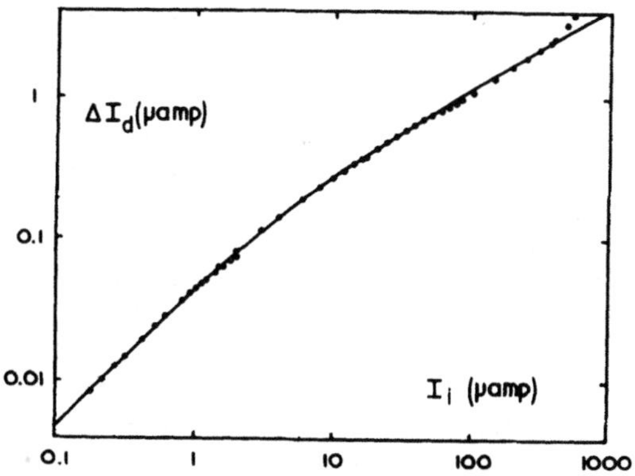

Fig. 10. Comparison of Eq. 25 with the data of Gray (1971) at high injection rates.

The difficulties associated with phonon trapping can be illustrated by considering the limit of Eq. 19 when $\Gamma_{es} \ll \Gamma_B$. In that case the measured Γ_R^* is just $\Gamma_{es}(\Gamma_R/\Gamma_B)$, where the ratio in brackets is wholly determined by statistics and the quasiparticle and phonon densities of states. In other words the information about the strength of the electron-phonon coupling contained in τ_o is common to both terms and cancels out. Therefore Γ_R^* effectively measures Γ_{es} which is unrelated to τ_o. In the most noteworthy attempt to overcome this difficulty, Smith and Mochel (1975) deduced a value for F_ω of about 9 from comparative measurements in vacuum and with a helium film, using similar aluminum samples. However if there is significant electron-electron scattering, this number would be an overestimate. There is a recent critical review of calculations and measurements of F_ω by Kaplan (1979). While alot has been learned about phonons and boundary resistances, F_ω seems to depend on the experimental configuration in ways not fully understood. Therefore, although the study of phonon trapping is important to estimate its effect in practical experimental situations, it is not sufficiently well understood for an accurate determination of τ_o. There is a need for more direct experiments which overcome the uncertainties of phonon trapping. Fortunately this can be done.

A very direct method of measuring the intrinsic electron-phonon coupling time constant, τ_o, is in experiments with charge imbalance. The reason is simply that phonon readsorption cannot produce a charge imbalance (see Chapter 13). A second method involves the intrinsic inelastic scattering time and is described in the last section of this Chapter. Dynes et al. (1978) have recently shown that Γ_R can be deduced from the lifetime broadening of the energy gap measured in a PbBi tunnel junction in equilibrium. Both the temperature dependence and magnitude are in excellent agreement with the predictions of

Kaplan et al. (1976). Unfortunately, this approach is restricted to materials in which $\Gamma_R \sim \Delta(0)/h$.

Finally, measurements of Γ_R near to T_c are not affected by phonon trapping. The simple phonon trapping model which defines a temperature independent phonon trapping factor F_ω is a result of the last two terms in Eq. 18 having the same temperature dependence. Near T_c, this simple model (Eqns. 15 and 16) breaks down because there are no special recombination phonons, N_ω, of energy $2\Delta >> kT$. Phonon and quasiparticle states with energies much greater than 2Δ are occupied. The excess phonons due to the nonequilibrium perturbation (a laser pulse in the following experiment) act only to slightly heat the film, as in the normal state. Thus the trapping of phonons at low temperatures is replaced by the thermal response time of the film near T_c. This time is related to the film's specific heat and Kapitza boundary resistances to the substrate and helium, and is reasonably independent of temperature in the vicinity of T_c. Therefore the intrinsic quasiparticle recombination process and the thermal response of the film are independent and τ_R can be measured, but only if it is longer than the thermal response time.

Schmid and Schön (1975) predicted that Γ_R should decrease to zero at T_c according to

$$\Gamma_R \cong 2.28\ (\Delta/kT)/\tau_o \quad , \qquad (26)$$

so that τ_R diverges and should be longer than the thermal response time close enough to T_c. On the other hand, Kaplan et al., (1976) find Γ_R nearly independent of T near T_c; however they calculate the quasiparticle scattering rate across the Fermi surface, which corresponds to recombination only at low temperatures. Near T_c, the quasiparticles, just as in a normal metal, scatter within a distribution which is about kT wide. However, only those quasiparticles which scatter into momentum states k very close to k_F, such that $\xi_k < \Delta$, can contribute to the condensate (pairs) so that scattering across the Fermi surface is irrelevant. This can be seen by examining the energy gap equation (Eq. 5) and the product $u_k v_k$ shown in Fig. 1. See also the discussion of gap enhancement below and in Chapter 7. Thus if $1/\tau_o$ is a measure of the average scattering rate to all states within kT of E_F, only the fraction Δ/kT of these scattering events can form bound pairs, thereby contributing to the condensate and Γ_R. Thus apart from a numerical factor, Eq. 26 is derived in a simple physical picture. There is a similar argument to explain the charge imbalance relaxation rate which is also proportional to Δ.

Evidence for the predicted divergence of τ_R is seen in Fig. 9, but Schuller and Gray (1976) verified Eq. 26 in detail. In their experiment, the relaxation of the energy gap in an aluminum tunnel junction was measured in real time after a short pulse of laser light.

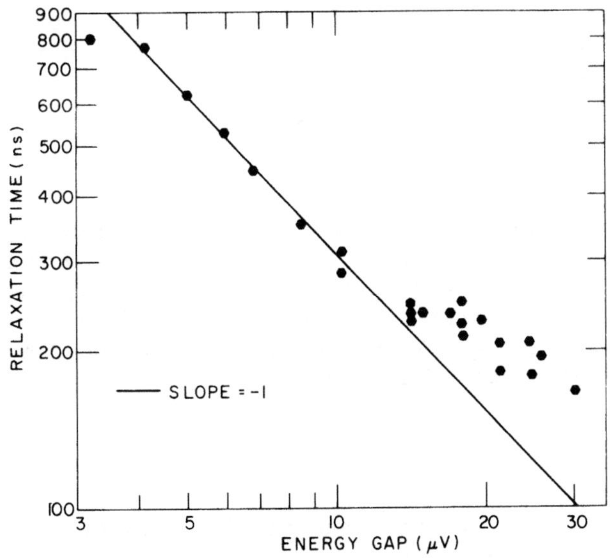

Fig. 11. Results of Schuller and Gray (1976) demonstrating that τ_R is proportional to $1/\Delta$ near T_c.

Their data are shown in Fig. 11, from which a value of 60 ns is derived for τ_0. Taken at face value, these two measurements of Γ_R at low and high temperatures on aluminum junctions would predict a phonon trapping factor of $F_\omega \sim 30$ for the sample shown in Fig. 9. Although this value is larger than $F_\omega \sim 9$ measured by Smith and Mochel (1975) the apparent discrepancy can be understood in the film thicknesses. Rothwarf and Taylor (1967) have shown that F_ω should be proportional to total film thickness which was 2.7 times greater for the sample in which $F_\omega \sim 30$.

Thus the basic features of quasiparticle-pair recombination have been verified at all temperatures. The rate goes to zero at low temperatures because the number density of quasiparticles goes to zero, but also at T_c because the number of states in which pairs can be formed goes to zero. However, at low temperatures, phonon trapping is important when comparing the magnitude of the rates with calculations, and is difficult to determine.

SUPERCONDUCTING TRANSISTOR

The superconducting transistor is a interesting example of additional nonequilibrium effects found in experiments of the last section when the detector resistance is very small. In this case, $\Gamma_T \gtrsim \Gamma_R^*$, so that a tunneling term must be included in the rate equation (Eq. 15) for the middle film. If the third film is assumed to be in thermal equilibrium, then the term to be added is $-\Gamma_T \Delta N$. In steady state, with $\Delta N << N(T)$, the new condition which replaces Eq. 20 would be

$$I_o = (2\Gamma_R^* + \Gamma_T)\,\Delta N. \qquad (27)$$

However, for equal gaps, $\Delta_2 = \Delta_3$, and $\Gamma_T \sim \Gamma_R^*$, the assumption of thermal equilibrium is not valid [i.e., $N_3 \neq N(T)$] and the inverse tunnel process involving the pairs, shown in Fig. 5, must be included. Consider a single quasiparticle state in the middle film in which the injection rate is Γ_i. This rate must be balanced by recombination and tunneling, and if $\Delta N << N(T)$, then

$$\Gamma_i = 2\Gamma_R^*(T) + \Gamma_T, \qquad (28)$$

where the factor of two is the same one found in Eq. 18, and is there because injected quasiparticles mostly recombine with thermally generated quasiparticles. The probability, p_T, that the quasiparticle tunnels into the third film is just Γ_T/Γ_i. Now the second process of Fig. 5 can occur, in which a quasiparticle reappears in the middle film. The probability is p_T^2. Thus a cyclic process continues since the initial tunneling process can occur again with probability p_T^3, etc. For the net detector current, it is necessary to sum over all states <u>and</u> to recognize that <u>each</u> process in Fig. 5 transfers an electron from film 2 to 3. Therefore all the probabilities add, and

$$\Delta I_d = \sum_k e\Gamma_i (p_T + p_T^2 + p_T^3 + \ldots)$$

$$= \sum_k e\Gamma_i p_T/(1-p_T) = (e\Gamma_T/2\Gamma_R^*) \sum_k \Gamma_i \qquad (29)$$

Both Γ_T and Γ_R^* are weakly dependent on the state k, so their average values are used. Since the sum of Γ_i over k states is I_i, Eq. 24 is recovered exactly ($A_d = A_i$ in this calculation) and it is thus shown to be independent of the relative sizes of Γ_R^* and Γ_T.

One important consequence is the possibility of current gain, G, if $\Gamma_T > 2\Gamma_R^*$, since

$$G \equiv \Delta I_d/I_i = \Gamma_T/2\Gamma_R^*. \qquad (30)$$

Such a device, which is analogous to the familiar semiconductor transistor in the common base configuration, has been reported in a double junction composed of three 30 nm thick aluminum films (Gray, 1978a). The results which show a current gain of 4 are summarized in

Fig. 12. Note that the slopes of $\Delta I_d/I_i$ are linear and proportional to $\Gamma_R^*(T)$ indicating the absence of both overinjection and heating. These data indicate that R^* is $5\times10^{-12}cm^3/s$, a value in good agreement with the results of the previous section since F_ω, and hence R^*, are approximately proportional to the total thickness(90 nm in this case compared to 240 nm for the sample in Fig. 9). The thickness dependence of F_ω is because the phonon escape rate Γ_{es} is inversely proportional to total thickness (Rothwarf and Taylor, 1967).

An important operating characteristic of the transistor as a device is the gain-bandwidth product. The bandwidth, B, is limited by the quasiparticle response to $2\Gamma_R^*$, so that $GB=\Gamma_T$. It is therefore advantageous to make Γ_T as large as possible, but unfortunately nonequilibrium effects impose a fundamental limitation. The rate Γ_T can never be greater than the total inelastic scattering rate for quasiparticles, because a second quasiparticle cannot tunnel into an occupied state. This idea is consistent with the experimental observation of a minimum specific resistance, R_nA, for tunnel junctions, which scales with τ_Q. For example, in 30 nm thick aluminum films, when $R_nA<3\times10^{-6}$ Ωcm^2, the junction I(V) departs from ideal behavior. At low temperatures, this R_nAd corresponds to $\Gamma_T\sim10^7$ s^{-1}, which should be compared to the total scattering rate $\Gamma_S+2\Gamma_R^*$, where Γ_S is the energy dependent inelastic scattering rate into another quasiparticle state. Using the results of Kaplan et al. (1976) and a

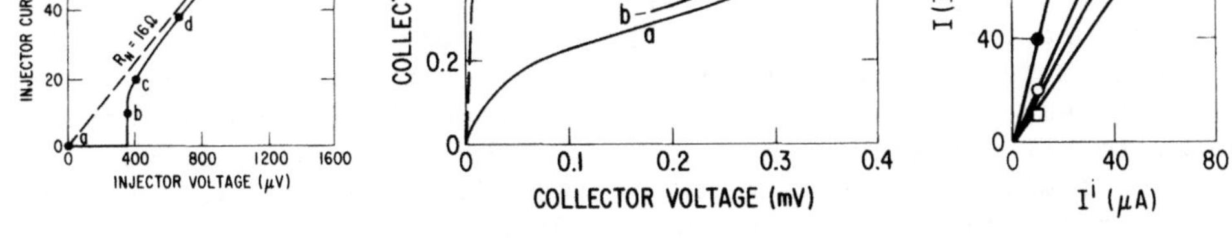

Fig. 12. (a) The two junctions of a superconducting transistor, biased independently. (b) I(V) of injector junction. The onset of pair breaking is at about 360 μV. (c) The I(V) of the detector junction magnified so that the current due to thermally excited quasiparticles as well as injected quasiparticles can be seen. The various labeled curves correspond to points in (b). (d) The change in detector current is plotted against injection current showing a temperature dependent slope greater than one.

typical experimentally determined τ_o of 90 ns, it is found for t<0.8 that $\Gamma_T > \Gamma_S + 2\Gamma_R^*$, and the I(V) should be dominated by details of the energy dependent scattering rates. In the case of tin junctions, a good I(V) is found for R_nA as low as 10^{-7} Ωcm^2, reflecting the smaller τ_o (~2.3 ns), while in niobium, for which τ_o~0.15 ns, Kleinsasser (1980) achieved $R_nA \sim 10^{-8} \Omega cm^2$. Note that this last result indicates a potential GB of greater than 1 GHz for the transistor.

Thus Γ_T (and hence GB) is limited by Γ_S and the maximum gain of the transistor is $\Gamma_S/2\Gamma_R^*$. The ratio Γ_S/Γ_R is predicted by Kaplan et al.(1976) to be about one for typical transistor operation, so the maximum gain is $\sim F_\omega/2$. Thus the gain of 4 shown in Fig. 12 indicates $F_\omega > 8$, and this is consistent with other determinations.

In summary, it is demonstrated that Eq. 24 is not altered by including the loss of quasiparticles by tunneling in the rate equation; however, a more important conclusion is that of a maximum tunneling rate, limited by the inelastic scattering rate.

ENERGY GAP ENHANCEMENT

The general results and implications of the enhancement of superconductivity by microwave irradiation, phonon injection and tunneling are the subject of Chapter 7. The purpose of this section is to outline the methods by which quasiparticle tunneling can create an increased energy gap. The experiments involve the same double tunnel junction described in the previous sections. The detector is chosen to be high resistance and measures changes in Δ and N caused by a quasiparticle(not pair-breaking) current in the injector junction.

In such a configuration, there are two independent mechanisms of gap enhancement: quasiparticle extraction was proposed by Parmenter (1961) and a tunneling analogy to the Eliashberg model of microwave gap enhancement was proposed by Gray (1978b). The Eliashberg model is understood in terms of the BCS theory which discussed in the first section of this Chapter. Because of the factor $u_k v_k$ (see Fig. 1) in the energy gap equation (Eq. 5), the important k states are those closest to k_F. Eliashberg noted that if the thermally excited quasiparticles could be transfered into states further from k_F, they would be less effective at reducing the gap from $\Delta(0)$. Now to do this requires additional quasiparticle energy, as can be seen readily in Fig. 2. Clearly adsorption of microwaves by the quasiparticles will accomplish this. However, if one refers to Fig. 5, the result of the two processes shown on the right is precisely the same. Recall that these processes occur at equal rates if the energy gaps in the two films are identical. In such a case, the net result is to increase the quasiparticles' energies by eV_i in each film at a rate Γ_T (see Eq. 23). In practice the conductance per unit area of the injector junction must be large enough so that Γ_T is a reasonable fraction of the quasiparticle scattering rate Γ_S which tends to restore the equilibrium distribution.

Tunneling enhancement offers two important advantages over microwaves: the injection rate can be precisely measured with an ammeter and the voltage V_i (equivalent to the microwave frequency for experiments with equal gaps) can be varied easily from zero to $2\Delta/e$, at which point pair-breaking reduces the gap. Measurements reported by Gray (1978b) confirm the general picture of gap enhancement including a reduction in the quasiparticle density N. Unfortunately, the requirement of identical gaps is not easily met, and additional enhancement was found if eV_i is approximately equal to the difference in the gaps. With unequal gaps, the balance between the two processes of Fig. 5 is lost and there is a net extraction (injection) of quasiparticles in the film with the smaller (larger) gap. Enhancement due to the extraction process has been studied in detail by Chi and Clarke (1979b). They find enhancements of up to 40%, but at higher voltages, they propose that nonequilibrium phonons, transmitted between the films, limit the size of the enhancement.

Figure 13 shows more recent data for the gap enhancement $\delta\Delta$ as a function of injector voltage in aluminum films for which no difference

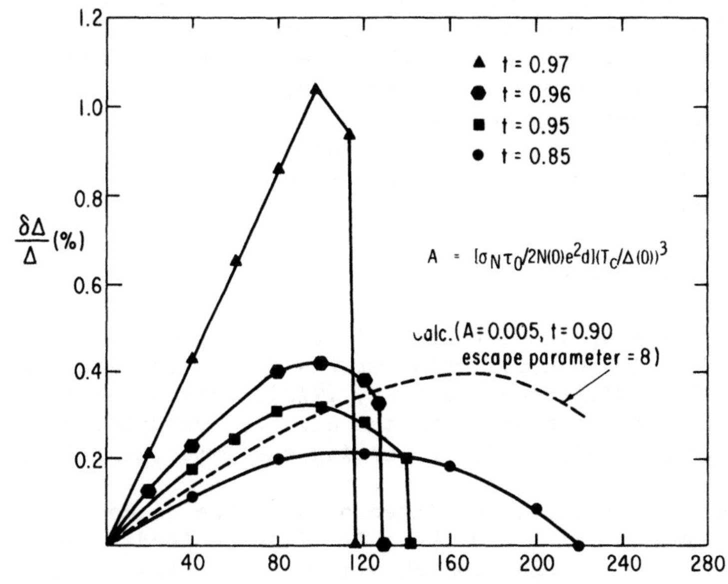

Fig. 13. The enhancement of the energy gap in aluminum films due to quasiparticle redistribution by tunneling between identical superconductors. These results are plotted against tunnel voltage at various temperatures. The dashed curve is the calculations of Chang (1977) scaled to a smaller value of A.

in the equilibrium energy gaps could be detected. The results agree qualitatively with the general features (dashed line) calculated by Chang (1978) who assumed a phonon trapping factor F_ω=9. The only other parameter of the calculation, which relates the strength of Γ_T to Γ_S, agrees with experiment to within 20%, using the "canonical" value of 90 ns for τ_0. Further improvement in the fit at higher voltages is expected if a larger value of F_ω is used in calculations.

This agreement could be regarded as quantitative verification of the Eliashberg model, but it is not the most direct way of doing so. The Eliashberg model really makes the following important assertion: Believe the BCS energy gap equation in nonequilibrium states. The above results rely on an intermediate step, which uses the kinetic equations (Chapter 9) to determine the nonequilibrium quasiparticle distribution function, f(E), with all the problems inherent in phonon escape. A better way is to verify Eq. 5 directly, by measuring f(E) as well as Δ. The following section describes methods to do this.

QUASIPARTICLE DISTRIBUTION FUNCTION

The discussion of tunneling, in the second section of this Chapter, described how the I(V) of a junction between superconductors with unequal gaps (Eq. 14) could be used to determine the distribution function f(E) in the smaller gap electrode. Before discussing the inversion of Eq. 14 to obtain f(E) directly, a two parameter model for the nonequilibrium distribution function will be described. It modifies the Fermi function by including an effective quasiparticle chemical potential, μ^*, first introduced by Owen and Scalapino (1972) and an effective quasiparticle temperature T^* (Parker, 1975) which can be greater than the ambient T:

$$f(E) = 1/[1+\exp(E-\mu^*)/kT^*]. \tag{31}$$

This μ^*,T^* distribution function offers a simple analytical approximation to f(E) in nonequilibrium states due to tunnel injection of quasiparticles (Willemsen and Gray, 1978), however no physical significance should be attached to the μ^*. For example, other expressions may work equally well at fitting the experimental data.

In this experiment, the injector junction is composed of two 30 nm thick aluminum films and the third film is lead. Because of the large gap and strong electron-phonon coupling (i.e., small τ_0) in lead, its distribution function is zero, and Eq. 14 is valid. The gaussian smearing of the aluminum gap and the density of states in the lead are determined in thermal equilibrium. Of the remaining three parameters, T^* and Δ are uniquely determined, respectively, from the slope of the experimental I(V) at small V and the sharp features at $(\Delta'-\Delta)/e$ and $(\Delta'+\Delta)/e$, while μ^* is selected to match the

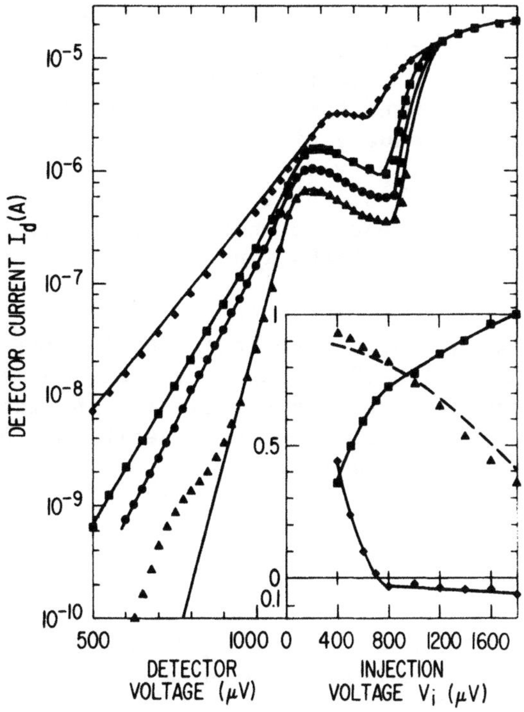

Fig. 14. Comparison of the experimental detector I(V) for quasiparticle injection at various voltages with calculations for a μ^*, T^* distribution. Also included is a fit in thermal equilibrium (squares). Inset, parameters of fits to these and other data (in reduced units): Triangles, $\Delta/\Delta(0)$; squares T^*/T_c; diamonds $\mu^*/\Delta(0)$. The dashed curve is $\Delta/\Delta(0)$ calculated from Eq. 5 using the appropriate μ^*, T^*.

current magnitude at $(\Delta'-\Delta)/e$. The comparison of measured and calculated I(V) in Fig. 14 shows excellent agreement over many orders of magnitude, and the parameters are plotted in the inset against injection voltage. The bump in the lowest curve corresponds to $f(3\Delta)$ and is due to the adsorption of 2Δ recombination phonons by the narrow quasiparticle distribution at $E\sim\Delta$. At higher voltages, both the quasiparticle and phonon distributions broaden and the bump is not discernable. This is consistent with the parameter μ^*. For the curve with the bump, $eV_i \sim 2\Delta(0)$, and quasiparticles are injected at $E\sim\Delta(0)$; the large positive μ^* and small T^* indicate a narrow distribution. As V_i increases, quasiparticles are injected at higher energies and μ^* becomes negative indicating a flat distribution and the bump disappears. At the same time, T^* increases from the ambient T, and the measured value of Δ (determined by the fitting procedure) decreases.

The BCS equation (Eq. 5) can now be used to calculate Δ using Eq. 31, and the results are shown as the dashed line in the inset of Fig. 14. Although the measured Δ are within about 10%, there is a systematic discrepancy, with the best agreement for $\mu^*=0$. Since this could be due to the inadequacy of the μ^*,T^* model, it is necessary to invert Eq. 14 to rigorously verify Eq. 5 in nonequilibrium states. The μ^*,T^* model is still a useful approximation in many cases. However, for small deviations from equilibrium caused by optical illumination, Smith et al. (1980) find $\mu^*=0$ and the T^* model is adequate.

Two approaches have been used to invert Eq. 14: a matrix inversion scheme (Smith et al., 1980) for very small deviations from equilibrium, and an iterative technique valid for the large deviations from equilibrium necessary to adequately verify Eq. 5. The problem is simplified if Eq. 5 is written as

$$\ell n\ \Delta(0)/\Delta = \int_0^\infty dE\ \rho(E)f(E)/E. \qquad (32)$$

Therefore, from Eq. 14, $\rho(E)f(E)$ is required and $\rho'(E)$ can be determined easily with the aluminum films normal.

The iterative method begins with an initial guess of a μ^*,T^* distribution. Starting at a low voltage, V_{d1}, the current is calculated and compared with the measurement. An iterative procedure is used to determine and appropiate constant which is added to $\rho(E)f(E)$ for all $E<\Delta'-eV_{d1}$ and greatly reduces the difference from the measured current. This procedure continues at voltage intervals of 1 to 10 μV until $eV_d=\Delta'$, which corresponds to $\rho(0)f(0)$. The above steps can be repeated indefinitely, using the results of the previous iteration as a new initial guess. One or two iterations are generally sufficient, as the scheme is very insensitive to the initial guess. Examples of $\rho(E)f(E)$ are shown in Fig. 23.

Experiments using aluminum films and a PbBi detector have been analyzed this way. The energy gap is calculated from Eq. 32 using $\rho(E)f(E)$ and compared to the measured value obtained by fitting the peak at $(\Delta'-\Delta)/e$ and step at $(\Delta'+\Delta)/e$, a procedure which is accurate to better than ± 2 μV. In thermal equilibrium, the measured value is within 1% of the BCS value (Fig. 1) and within 1.5% of the calculated value (Eq. 32) for $t<0.8$. The nonequilibrium results are summarized in Fig. 15. The calculated and measured Δ agree to within 3% for reduced gaps greater than 0.45, and the discrepancy is in the same sense as is found in thermal equilibrium. The origin of this discrepancy is the non-zero values of $\rho(E)f(E)$ for $E<\Delta$, that come from the negative resistance part of the I(V) which is difficult to measure accurately. Unfortunately, these uncertainties are amplified by the 1/E factor in Eq. 32.

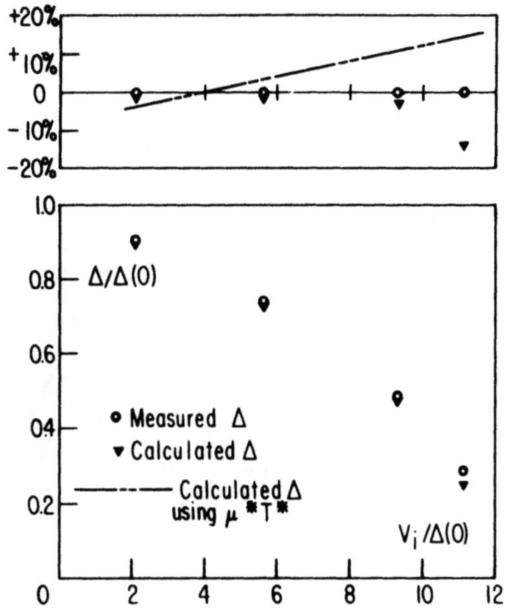

Fig. 15. Bottom: The measured and calculated variation of Δ for quasiparticle injection at a voltage V_i. The calculations are based on Eq. 5 and the measured $\rho(E)f(E)$. Top: Percentage deviations of calculated gap using $\rho(E)f(E)$ and the μ^*, T^* distribution function.

Thus the systematic discrepancy of the μ^*,T^* fit to Eq. 5 is spurious, and the BCS equation is verified for nonequilibrium states without charge imbalance. At higher injection voltages, the gap is smaller and charge imbalance is observed. However in all samples used, gap smearing due to inhomogenieties becomes a reasonable fraction of Δ, and careful comparison is impossible (see highest injection data point in Fig. 15). Therefore, it is an open question about the validity of the BCS equation with significant charge imbalance.

The value of the recombination coefficient R^* can be estimated from the variation of Δ with V_i shown in Fig. 15, using Eq. 25 and the dependence of Δ on N in thermal equilibrium. The resulting value of $4.7\times10^{-12}cm^3/s$ is in excellent agreement with previous estimates.

In Fig. 16, the reduced gap is plotted against T^* for several different resistance injector junctions and for different injector

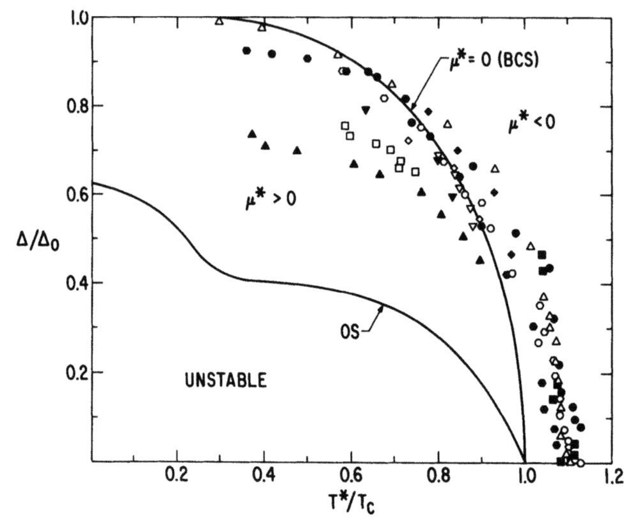

Fig. 16. Compendium of Δ/Δ_0 vs T^*/T_c for several double junction aluminum samples. The solid line (BCS) is the prediction of Eq. 5. The solid line (OS) is the boundary of the predicted first order transition into the normal state (Owen and Scalapino, 1972).

voltages. The solid curve (BCS) is the BCS result shown in Fig. 1, which implies $\mu^*=0$, while outside this curve $\mu^*<0$ and inside $\mu^*>0$. For a given T^*, the smaller resistances correspond to smaller gaps, and the lowest data shown in Fig. 16 is for a junction near the minimum tunnel resistance. The solid curve (OS) represents the first order transition into the normal state predicted by Owen and Scalapino (1972). As the injection level increases, the data follow a line parallel to the OS prediction. Apparently nature is conspiring against us to reach the conditions for this transition (see Chap. 10).

For very small values of Δ, charge imbalance is observed with a different T^* for each branch. However for all samples, the average T^* approaches about 1.1 T_c. The study of distribution functions with a charge imbalance is covered in the next section.

QUASIPARTICLE DISTRIBUTION FUNCTION WITH A CHARGE IMBALANCE

One of the most interesting nonequilibrium properties of superconductors is that of charge or branch imbalance. Branch imbalance (Tinkham and Clarke, 1972) focuses on the relative populations of the branches of the quasiparticle spectrum for k above and below k_F (see Fig. 2). Charge imbalance (Pethick and Smith, 1979) is concerned with the probability that a quasiparticle state is either a hole in or an addition to the pair distribution (v_k^2 in Fig. 1). In the latter description, an imbalance reflects the exchange of charge between the pairs and quasiparticles, but overall charge neutrality is maintained. The difference in these two descriptions results from the fact that the quasiparticle charge does not change sign discontinuously as k crosses k_F, because the distribution of electrons at T=0 is not discontinuous at k_F as in a normal metal.

In experimental situations, the charge imbalance rather that the branch imbalance is most relevant, however it is not obvious which has the more fundamental significance theoretically. Fortunately, these two imbalances can be simply related to each other if the coherence factor u_k or v_k and the quasiparticle distribution f(E) are known as functions of the quasiparticle energy E for the two branches or charges. Within the BCS approximation, u_k and v_k are completely determined by the energy gap, Δ, however the distribution functions must be measured experimentally. There is an open question as to whether the BCS prescription for u_k and v_k is valid with charge imbalance, however for the present it must be assumed so. The transformation for the equilibrium distribution function from branches ($f_>$, $f_<$) to charges (f_e, f_h) is given by the matrix equation:

$$\begin{pmatrix} f_e \\ f_h \end{pmatrix} = \begin{pmatrix} u_>^2 & u_<^2 \\ v_>^2 & v_<^2 \end{pmatrix} \begin{pmatrix} f_> \\ f_< \end{pmatrix} \tag{33}$$

The techniques of the previous section can be extended to include measurments of the distribution functions for the branches or charges. In order to correctly keep track of the charge, the tunneling equations must be considered, before the sum on coherent factors. The results were first worked out for SS junctions by Gray (1969) and for NS junctions by Tinkham and Clarke (1972) and Tinkham (1972). The latter case has been used extensively to measure the branch imbalance relaxation rate and will be covered in Chapter 13.

For the creation of charge imbalance by tunneling between two superconductors, the pair-breaking term (Eq. 10) will be considered here, although the quasiparticle current (Eq. 9) also induces a charge imbalance. Consider the film b on the left hand side of Fig. 6, and sum over the coherence factors in film a, where pair-breaking occurs. Then the coherence factor u_k^2 remains in Eq. 10, and it is more probable that the injected quasiparticle will be on the $k>k_F$ branch. Since $u_k^2=(1+\xi_k/E_k)/2$ and $\rho(E)=E_k/\xi_k$, it follows that the probabilities are

$$P_{k>k_F}(E) = u_>^2 = [1+1/\rho(E)]/2,$$

$$P_{k<k_f}(E) = u_<^2 = [1-1/\rho(E)]/2. \tag{34}$$

These probabilities are reversed for the film on the left, or for a change in the polarity of V_i.

As regards electrons and holes, the situation is more complex. Clearly the battery is pushing electrons through the tunnel barrier and into the film on the right hand side of Fig. 6. There is a conversion of this current into a supercurrent in a short time (Clarke et al., 1979), and the supercurrent leaves the tunneling region to complete the battery circuit and maintain overall charge neutrality. After the conversion process the pairs are again in equilibrium and Eq. 33 correctly describes the probability that the excitation in the state k is a hole in or an addition to the pair distribution. Thus when the probabilities of Eq. 34 are multiplied by u_k^2 the probability of an electron excitation is obtained (and v_k^2 gives the probability for hole excitations):

$$P_e = u_>^4 + u_<^4 = [1+1/\rho^2(E)]/2,$$

$$P_h = u_>^2\, v_>^2 + u_<^2\, v_<^2 = [1-1/\rho^2(E)]/2. \qquad (35)$$

Comparing Eqns. 34 and 35, it is clear that the assymmetry in charge imbalance is greater than for branch imbalance.

The detection of the distribution functions is simplified (as before) by using a detector film with a larger energy gap. Therefore, making the same assumptions for which Eq. 14 is valid, except for charge imbalance, two equations result:

$$I_d(\pm V_d) = G_{nn} \int_0^\infty \rho'(E+eV_d)\, \rho(E)\, f_{e,h}(E)\, dE, \qquad (36)$$

in which $\rho(E)$ is assumed independent of charge. Experimentally, a charge imbalance is induced by tunnel injection into an aluminum film and Fig. 17 shows the resulting I(V) of a detector junction with a PbBi film. The imbalance is seen as an assymmetry in the detector current, from which $\rho(E)f_e(E)$ and $\rho(E)f_h(E)$ are obtained by the methods of the previous section.

The distribution functions for the branches are obtained more easily by taking the sum and difference of Eqns. 36 and applying the transformation in Eq. 34. Thus

$$I_d(+V_d) + I_d(-V_d) = G_{nn} \int_0^\infty \rho'(E+eV_d)\, \rho(E)\, [f_>(E)+f_<(E)]\, dE$$

$$I_d(+V_d) - I_d(-V_d) = G_{nn} \int_0^\infty \rho'(E+eV_d)\, [f_>(E)-f_<(E)]\, dE. \qquad (37)$$

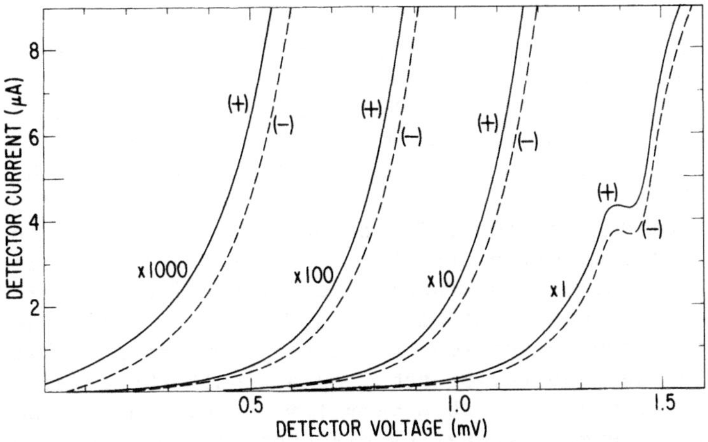

Fig. 17. Detector I(V) for both polarities of V_d for a charge imbalance created by tunnel injection (various magnifications).

Using the gaussian smeared density of states with the appropiate Δ for the aluminum, the procedures of the previous section can be used to determine $f_>$ and $f_<$. Typical results are shown in Fig. 18.

An interesting feature of this data is the different T^* for the two branches. This experiment was conducted at very low temperatures

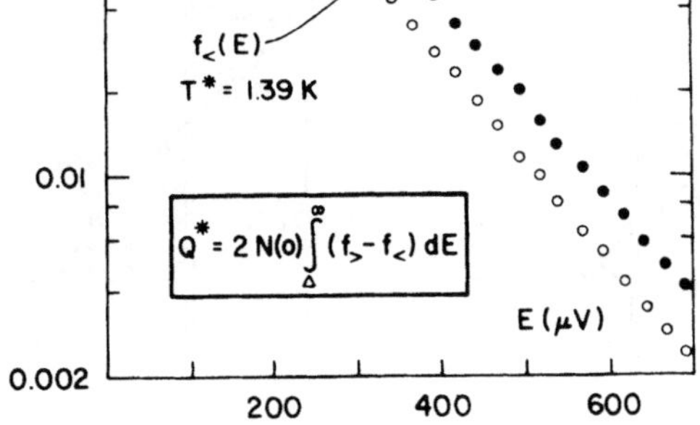

Fig. 18. The quasiparticle distribution functions for the branches $k>k_F$ and $k<k_F$, determined from I(V) data similar to Fig. 17.

so that thermally excited quasiparticles, even in the aluminum, could be neglected. Yet the distribution functions are "thermalized" to some T^*, and are completely unlike the injected distribution shown schematically in Fig. 19. If the thermalization results from inelastic electron-phonon scattering alone, then both branches should have the same T^* since phonons are not expected to distinguish between branches. Likewise, electron-electron scattering should not prohibit scattering between branches. The answer to this dilemma involves the way in which the minority branch is populated. It is well known that scattering from branch to branch occurs predominently for $\xi_k \lesssim \Delta$. Therefore, high energy quasiparticles, which are only injected onto the majority branch, must scatter to very low energies before they can populate the minority branch. As a result the quasiparticles on the minority branch start at a very low energy, and if they then populate the higher energy states by electron-phonon scattering, the T^* of the minority branch should reflect the effective phonon temperature. That this "temperature" should be lower that the T^* for the majority quasiparticle branch follows because the phonons are a subsystem which is "closer" to the heat sink - the substrate or helium.

Returning to Fig. 17, the difference in the currents at opposite polarities increases approximately exponentially with detector voltage. This is because of the cutoff of the integral in Eq. 37 due to the density of states $\rho'(E+eV_d)$. Therefore, unlike NS junction detectors which do not have the cutoff, the full charge imbalance,

$$Q^* = 2N(0) \int_0^\infty (f_> - f_<)\, dE, \tag{38}$$

is not sampled until $V_d = \Delta'$. As a result, the measurement of Q^* is

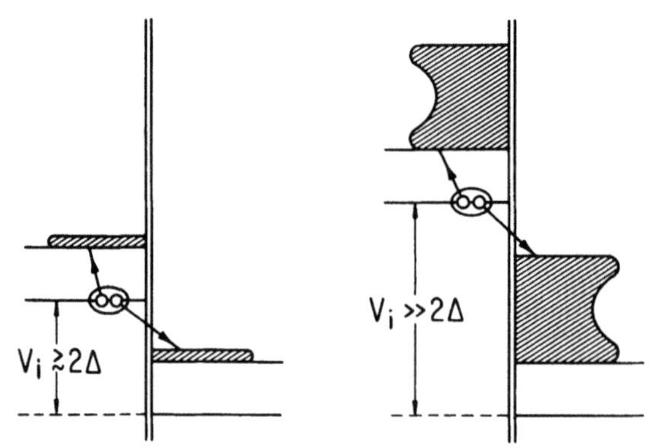

Fig. 19. The schematic distribution of injected quasiparticles by the tunneling process which breaks pairs (Fig. 6). The peaks have been rounded to simulate smearing of the energy gap.

more straightforward with NS junctions. However, the difference in the distribution functions shown in Fig. 18 can be integrated over energy to obtain $Q^*/2N(0)$.

ENERGY DEPENDENT INELASTIC SCATTERING RATE

The total inelastic scattering rate for quasiparticles includes the recombination rate Γ_R and the rate of scattering into other quasiparticle states Γ_S. Both of these have been shown by Kaplan et al. (1976) to be strongly energy and temperature dependent. Previous experiments discussed in this Chapter were concerned with the temperature dependence of this rate after averaging over energy. The first experiment to measure the energy dependence was by Miller and Dayem (1967). In this double junction experiment, the detector current for $V_d<(\Delta'-\Delta)/e$, which is related to the distribution function by Eq. 14, was analyzed as a function of injection current and temperature, using the relaxation time approximation. For this experiment, the injection rate is shown pictorially in Fig. 19, and the relaxation time approximation assumes that quasiparticles only scatter out of these states, and ignores scattering into them. The "scattering in" term is needed to achieve the broad μ^*,T^* distributions discussed earlier in this Chapter for similar experiments. However, on top of the μ^*,T^* distribution there must always be a remanent of the injected distribution:

$$\delta f_k = f_k - f(\mu^*,T^*), \tag{39}$$

and ignoring elastic scattering between branches, the relaxation time approximation correctly predicts

$$\Gamma_R(E_k) + \Gamma_S(E_k) = \Gamma_i(E_k)/\delta f_k, \tag{40}$$

where $\Gamma_i(E_k)$ is the injection rate into the state k and is related to the integrand of Eq. 10.

There are two ways to get δf_k. The first one is for low injection rates and involves sensitive measurements of the second derivative of I(V). In doing so, the step in I_d, which is a result of the top edge of the injected distribution of Fig. 19, is measurable and is shown in Fig. 20. The second method is for high injection rates so that the step is readily visible on the I(V) shown in Fig. 21. Note that in addition to the structure at 0.8 mV caused by the top edge of the injected distribution (due to pair-breaking), the quasiparticle tunneling (inset) also exhibits a charge imbalance which is detected at 0.5 mV. Thus there is a second I(V), of smaller magnitude, superimposed on the main I(V), but shifted to the left by a voltage equal to the injector voltage. It acts like direct tunneling between the

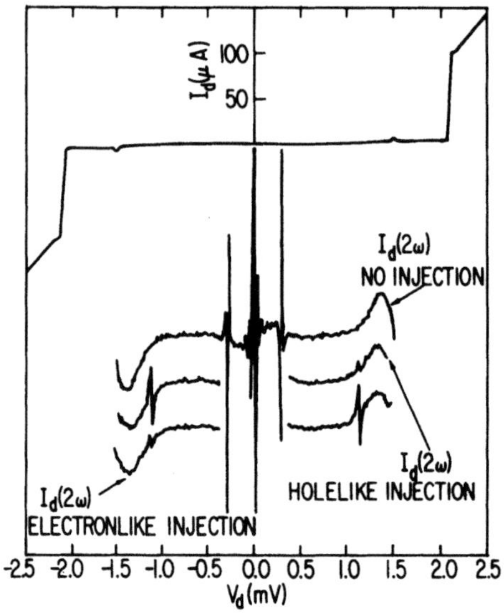

Fig. 20. The I(V) (upper) of the detector junction for weak quasi-particle injection and its second derivative (lower) which shows struture corresponding to the upper edge of the distribution of Fig. 19.

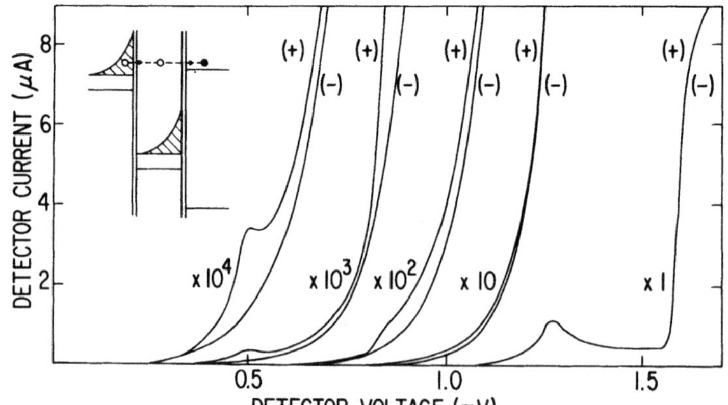

Fig. 21. The I(V) of a detector junction for strong quasiparticle injection showing directly the structure corresponding to the upper edge of the distribution in Fig. 19 due to pair-breaking and also the structure (at 0.5 mV) due to the quasiparticle tunneling shown in the inset diagram.

outside films, but the current is modified by the inelastic scattering rate in the middle film. For example, the structure disappears if the intermediate state is within the gap of the middle film.

In both of the above cases, there is a clear assymmetry with the polarity of V_d, reflecting the fact that injection is predominently onto one branch (charge). In the latter case, it is also clear from Fig. 21 that the charge imbalance in δf_k virtually disappears in the thermalized distribution. This is not true when the aluminum gap and hence the charge imbalance relaxation rate are small (see Fig. 17).

Because of this charge imbalance, Kaplan et al. (1977) pointed out that the elastic branch scattering rate (i.e., $k_> \to k_<$ where $E_k = E_{k'}$) can be measured. In this case, Chang (1977) has generalized Eq. 40 so that measurements of δf_k for both charges (i.e., both polarities of V_d) can be used to calculate both rates. The results for two above methods are shown in Fig. 22. The different materials used (very dirty and clean aluminum) may account for the order of magnitude difference in measured time; otherwise the results are very similar. The upper data are for the clean film and indicate τ_0=100 ns in excellent agreement with other measurements on similar films. On the other hand, the elastic branch scattering time is constant, disagreeing with the theoretical predictions of gap anisotropy (Tinkham

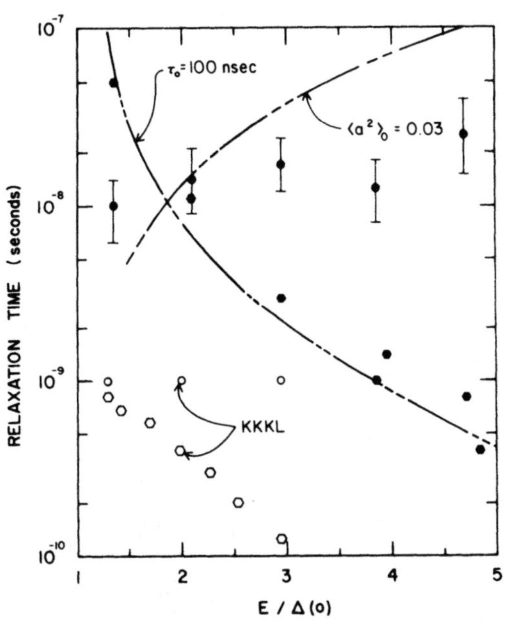

Fig. 22. Quasiparticle relaxation times as a function of energy from data of Figs. 20 and 23. Hexagons, inelastic scattering time. Circles, elastic branch mixing time. The open symbols are from Kirtley et al. (1978). The lines are fits to the solid data.

1972). The anisotropy factor $\langle a^2 \rangle_0 = 0.03$ was representative of values quoted by Chi and Clarke (1979a) for similar aluminum films. This latter result was obtained in both experiments and is not understood.

Using the techniques of the previous section (Eq. 37), the distribution functions can be obtained from the data of Fig. 21, and are shown in Fig. 23. The lower set of data is the branch imbalance distribution function, and when integrated over energy is $Q^*/2N(0)$ from Eq. 38. The charge imbalance distribution function, $f_e - f_h$, is related through Eq. 33:

$$f_> - f_< = \rho(E)\,(f_e - f_h) \tag{41}$$

It is reasonable to define an excitation of a superconductor in which there is one imbalanced charge as a "quon". Dividing the lower curve of Fig. 23 by $\rho(E)$ yields the quon distribution function.

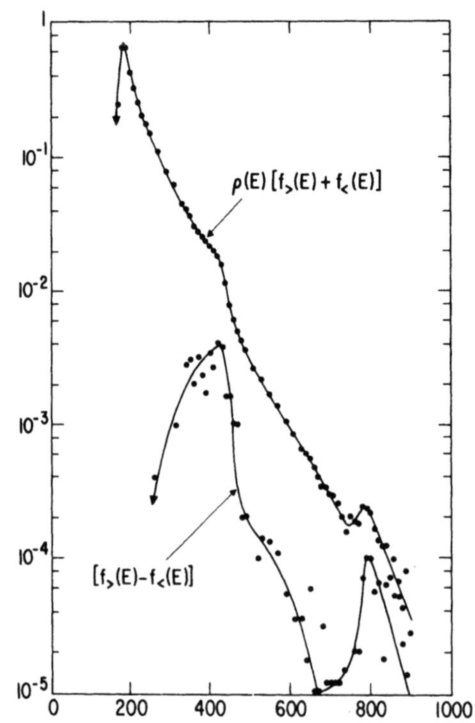

Fig. 23. The quasiparticle distribution functions obtained by inverting I(V) data such as Fig. 21, by the iterative technique.

SUMMARY

Double junction experiments provide a wealth of detailed information on nonequilibrium states. This results from the direct microscopic measurements possible with tunnel junction detectors and from the simplicity and control of perturbations caused by tunnel injection. All the double junction experiments described in this Chapter could be made with one sample by varying the injector and detector biases. In most cases samples are tailored to the specific experiment, but there is a great commonality to them. Nonetheless, one cannot but marvel at the amount of detailed information obtainable and the number of strikingly different phenomena that can be observed. Especially since this is all the result of the simple tunneling hamiltonian (Eq. 8), expanded to include all terms and applied to two interacting junctions.

ACKNOWLEDGEMENTS

Many experimental results reported in this Chapter were due to fruitful collaborations with co-workers. I would like to acknowledge them here for their collective contributions and stimulation. They include I. K. Schuller, J. Zasadzinski and particularly H. W. Willemsen, who has made the most significant contribution to the work reported in this Chapter.

REFERENCES

Bardeen, J., Cooper, L. N., and Schrieffer, J. R., 1957, Theory of Superconductivity, Phys. Rev. 108:1175.

Bogoliubov, N. N., 1958, On a new method in the theory of superconductivity, Nuovo Cimento 7:794.

Chang, J.-J., 1977, Tunneling characteristics of a nonequilibrium double junction, Phys. Rev. Letters 39:1352.

Chang, J.-J., 1978, Gap enhancement in superconducting thin films due to quasiparticle tunnel injection, Phys. Rev. B17:2137.

Chi, C. C., and Clarke, John, 1979a, Quasiparticle branch mixing rates in superconducting aluminum, Phys. Rev. B19:4495.

Chi, C. C., and Clarke, John, 1979b, Enhancement of the energy gap in superconducting aluminum by tunneling extaction of quasiparticles, Phys. Rev. B20:4465.

Clarke, J., Eckern, U., Schmid, A., Schön, G., and Tinkham, M., 1979, Branch imbalance relaxation times in superconductors, Phys. Rev. B20:3933.

Dynes, R. C. Narayanamurti, V., and Garno, J. P., 1978, Direct measurement of quasiparticle-lifetime broadening in a strong-coupled superconductor, Phys. Rev. Letters 41:1509.

Giaever, I., 1960, Energy gap in superconductors measured by electron tunneling, Phys. Rev. Letters 5:147.

Giaever, I., 1965, Detection of the ac Josephson effect, Phys. Rev. Letters 14:904.

Gray, K. E., 1969, Quasiparticle lifetime in superconducting aluminum, Ph.D. Thesis, Cambridge University, unpublished.

Gray, K. E., Long, A. R., and Adkins, C. J., 1969, Measurements of the lifetime of excitations in superconducting aluminum, Philos. Mag. 20:273.

Gray, K. E., 1971, Steady state measurements of the quasiparticle lifetime in superconducting aluminum J. Phys. F: Metal Phys. 1:290.

Gray, K. E., 1978a, A superconducting transistor, Appl. Phys. Letters 32:392.

Gray, K. E., 1978b, Enhancement of superconductivity by quasiparticle tunneling, Solid State Commun. 26:633.

Kaplan, S. B., Chi, C. C., Langenberg, D. N., Chang, J.-J., Jafarey, S., and Scalapino, D. J., 1976, Quasiparticle and phonon lifetimes in superconductors, Phys. Rev. B14:4854.

Kaplan, S. B., Kirtley, J. R., and Langenberg, D. N., 1977, Experimental determination of the quasiparticle energy distribution in a non-equilibrium superconductor, Phys. Rev. Letters 39:291.

Kaplan, S. B., 1979, Acoustic matching of superconducting films to substrates, J. Low Temp. Phys. 37:343.

Kirtley, J. R., Kent, D. S., Kaplan, S. B., and Langenberg, D. N., Experimental determination of the energy and temperature dependences of quasiparticle relaxation times in a nonequilibrium superconductor, J. Physique Supplement, 39: C6-511.

Kleinsasser, A. W., and Buhrman, R. A., 1980, Niobium Josephson junctions with reactive ion beam oxidation, Bull. Am. Phys. Soc. 25: 197.

Miller, B. I., and Dayem, A. H., 1967, Relaxation and recombination times of quasiparticles in superconducting Al thin films, Phys. Rev. Letters 18:1000.

Owen, C. S., and Scalapino, D. J., 1972, Superconducting state under the influence of external dynamic pair breaking, Phys. Rev. Letters 28:1559.

Parker, W. H., 1975, Modified heating theory of nonequilibrium superconductors, Phys. Rev. B12:3667.

Parmenter, R. H., 1961, Enhancement of superconductivity by extraction of normal carriers, Phys. Rev. Letters 7:274.

Paterson, J. L., 1978, Analysis of nonequilibrium double tunnel junctions, J. Low Temp. Phys. 33:285.

Pethick, C. J. and Smith, H., 1979, Relaxation and collective motion in superconductors: a two-fluid description, Annals Phys. (NY) 119:133.

Rothwarf, A., and Taylor, B. N., 1967, Measurement of recombination lifetimes in superconductors, Phys. Rev. Letters 19:27.

Schmid, A., and Schön, G., 1975, Linearized kinetic equations and relaxation processes of a superconductors near T_c, J. Low Temp. Phys. 20:207.

Schuller, I., and Gray K. E., 1976, Experimental observation of the relaxation time of the order parameter in superconductors, Phys. Rev. Letters 36:429.

Smith, A. D., Skocpol, W. J., and Tinkham, M., 1980, Quasiparticle energy distribution in optically illuminated Al-PbBi superconducting tunnel junctions. Phys. Rev. B (submitted).

Smith, L. N., and Mochel, J. M., 1975, Phonon and quasiparticle dynamics in superconducting aluminum tunnel junctions, Phys. Rev. Letters 35:1597.

Solymar, L., 1972, "Superconductive Tunnelling and Applications," Chapman and Hall Ltd., London.

Taylor, B. N., 1963, A study of some electron tunneling phenomena in superconductors and magneto-tunneling in InAs Esaki diodes, Ph.D. Thesis, University of Pennsylvania, unpublished.

Tinkham, M., and Clarke, J., 1972, Theory of pair-quasiparticle potential difference in nonequilibrium superconductors, Phys. Rev. Letters 28:1366.

Tinkham, M., 1972, Tunneling generation, relaxation and tunneling detection of hole-electron imbalance in superconductors, Phys. Rev. B6: 1747.

Tinkham, M., 1975, "Introduction to Superconductivity," McGraw-Hill, New York.

Valatin, J. G., 1958, Comments on the theory of superconductivity, Nuovo Cimento, 7:843.

Willemsen, H. W., and Gray, K. E., 1978, Distribution function in nonequilibrium states of a superconductor, Phys. Rev. Letters 41:812.

THE EFFECT OF MICROWAVES ON TUNNEL JUNCTIONS

Ivan K. Schuller

Solid State Science Division
Argonne National Laboratory
Argonne, IL 60439 USA

E. D. Dahlberg

Physics Department
University of Minnesota
Minneapolis, MN 55455

INTRODUCTION

Interest in the effects of microwave radiation incident on tunnel junctions first arose nearly twenty years ago in the early 1960's. Even though research, both theoretical and experimental, has been extensive in this area, it has not been exhaustive. Recent renewed interest has been motivated by theoretical predictions and experimental observations relating to redistribution of quasiparticles and their relationship to the "gap enhancement" phenomenon. In the present paper we will describe the effect of microwaves on tunnel junctions excluding specifically "gap enhancement," which will be described in Chapter 7 by J. E. Mooij.

Microwaves incident on a superconductor-superconductor tunnel junction have the effect of opening up new tunneling channels which manifest themselves as additional structure in the I-V characteristics and their derivatives. The evolution of this structure as a function of frequency and microwave power has been theoretically calculated and is in good agreement with experimental observations. Other effects include microwave induced Josephson steps and microwave steps related to subharmonic gap structures in tunnel junctions. It is interesting to note that a number of similar effects are present in Esaki tunnel diodes where structures are observed at characteristic phonon frequencies due to phonon assisted tunneling.

In the next section we will describe the theory of photon assisted tunneling. We present the experimental techniques in Section 3 and the experimental results relating to photon assisted tunneling in Section 4. Section 5 is dedicated to the effect of microwaves on phonon structures and in the final section our conclusions are summarized.

2. THEORY OF PHOTON ASSISTED TUNNELING

The original theory for photon assisted tunneling which we will follow was originally developed by Tien and Gordon (1963). Independently, Cohen, Falicov and Phillips (1962) found a similar result using manybody techniques. An electric field perpendicular to a tunnel junction has the effect of setting up a time varying potential difference

$$V_{rf} \cos\omega t \qquad (1)$$

between the two films A and B. If the potential in film A is used as a reference, the effect of the electric field is to add an electrostatic potential to the electrons in film B. The quasiparticle wave function in film B in the absence of microwaves can be written as

$$\Psi(x,y,z,t) = f(x,y,z)e^{-iEt/\hbar} \qquad (2)$$

The new hamiltonian in the presence of microwaves is

$$H = H_o + eV_{rf}\cos\omega t \qquad (3)$$

where H_o is the unperturbed hamiltonian. Since the spatial part of Ψ is not changed by the perturbation, the new wave function is given by

$$\Phi(x,y,z,t) = \Psi(x,y,z,t)\left\{\sum_{n=-\infty}^{\infty} B_n\, e^{in\omega t}\right\} \qquad (4)$$

After substituting Eq. (4) into the time dependent Schrödinger equation one finds that the coefficients B_n obey Bessel function recursion relations so that

$$\Phi(x,y,z,t) = f(x,y,z) \sum_{n=-\infty}^{\infty} J_n(\alpha)\, e^{-it(E+n\hbar\omega)/\hbar} \qquad (5)$$

with

$$\alpha = eV_{rf}/\hbar\omega \qquad (6)$$

It should be noted that the properties of the Bessel functions are such that the new wave function is also normalized. Physically, Eq. (5) shows that an electron of energy $(E + n\hbar\omega)$ has the probability $J_n(\alpha)$ of being in that particular state. If the unperturbed reduced density of states is $\rho(E)$ the density of states in the presence of microwaves is given by

$$\rho'(E) = \sum_{n=-\infty}^{\infty} J_n^2(\alpha)\, \rho(E+n\hbar\omega) \tag{7}$$

It is important to notice that the calculation presented is equivalent to the adiabatic approximation so the frequency ω should be smaller than frequencies corresponding to interstate transition times.

The tunneling current can be calculated from

$$I_{AB}(V_o) = G_{nn} \int_{-\infty}^{\infty} [f(E-eV_o) - f(E)]\rho_A(E-eV_o)\rho_B(E)dE \tag{8}$$

where f is the Fermi function, ρ_A and ρ_B are the reduced densities of states in superconductor A and B, V_o is the applied D.C. voltage and G_{nn} is the normal conductance of the tunnel junction (Chap. 5).

Using Eq. (7) for the density of states in electrode B we find in the presence of microwaves

$$I'_{AB}(V_o) = G_{nn} \sum_{n=-\infty}^{\infty} J_n^2(\alpha) \int_{-\infty}^{\infty} [f(E-eV_o) - f(E+n\hbar\omega)]$$

$$\rho_A(E-eV_o)\, \rho_B(E+n\hbar\omega)\, dE \tag{9}$$

and the conductivity σ' at voltage V_o is

$$\sigma'(V_o) = \sum_{n=-\infty}^{\infty} J_n^2(\alpha)\, \sigma(V_o + \frac{n\hbar\omega}{e}) \tag{10}$$

If the microwave magnetic field is parallel to the junction and if the films are thin enough so that the microwave field penetrates through them, it is found from a similar calculation that Eqs. (9) and (10) are still valid with

$$\alpha = (2\pi)^{3/2} (2e\vec{A}_{rf} \cdot \vec{k}/\omega mc) \tag{11}$$

where $\vec{A}_{rf}$ is the vector potential associated with the microwave field and $\vec{k}$ is the wave vector of an electron. Of course $\vec{A}_{rf} \cdot \vec{k}$ must be averaged over all k before this a is used in the preceding equations.

Cohen, Falicov and Phillips (1962) obtained an equivalent result using a golden rule calculation for the tunneling current and the BCS formulation for the quasiparticle states. The present calculation has been criticized (Cook and Everett, 1967) because both films of the junction are not treated on the same footing although a later (Buttner and Gerlach, 1968) microscopic calculation has shown to give the same results even in a symmetric treatment of the junction.

It is easy to get a pictorial view of photon assisted tunneling in the semiconductor model. Figure 1(a) shows the process by which a quasiparticle absorbs a quantum of energy $\hbar\omega$ and thereby increases the tunneling current at a voltage $E - \hbar\omega$. Stimulated emission (Fig. 1(b)) produces the same effect at the voltage $E + \hbar\omega$. Therefore, equally spaced steps will appear in the I-V characteristics at voltages equal to $E \pm n\hbar\omega$ as predicted formally

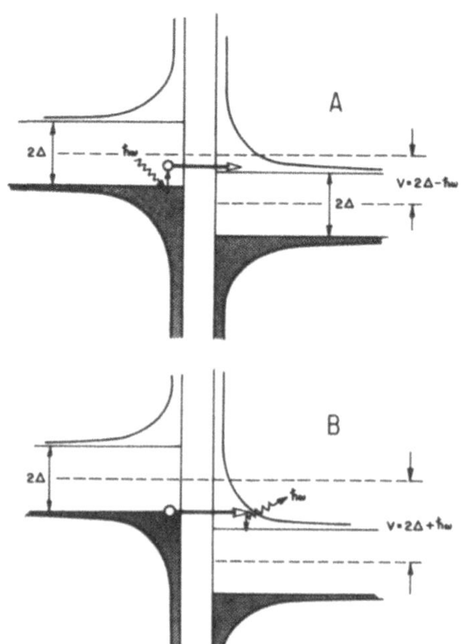

Fig. 1. Photon assisted tunneling by: a) stimulated absorption and b) stimulated emission.

by the analysis presented earlier. Equations (9) and (10) show that the steps have the shape of the 2Δ step in the I-V characteristic of the tunnel junction in the absence of microwaves (i.e., the microwave steps have a finite width which depends on the properties of the electrodes). This implies that there will be two qualitatively distinct regimes depending on the width of the step compared to the separation in between steps. In the "quantum limit" ($\hbar\omega$ > step width) well defined peaks should appear in the tunnel conductance. The amplitude of the steps vary as $J_n^2(\alpha)$ as shown in Eq. (10). In the "classical limit" ($\hbar\omega$ < step width) all the steps overlap and no sharp structure emerges in the conductance as a result of the photon assisted tunneling. Hamilton and Shapiro (1970) have shown using standard asymptotic approximations for $J_n(\alpha)$ that in the small frequency limit (α large) Eq. (9) takes the form

$$I'(V_o) = \pi^{-1} \int_{-\pi/2}^{\pi/2} I_o(V_o + V_{rf} \sin\theta)d\theta \tag{12}$$

where $I_o(V_o)$ is the DC current at voltage V_o in the absence of microwaves. Equation (12) shows that the intuitive idea of the "classical limit" is indeed correct and that no sharp structure appears in the conductance due to photon induced tunneling.

Computer simulations of Eq. (12) indicate that the effect of the microwaves is to split the $\Delta_1 + \Delta_2$ structure into two structures at $V = \Delta_1 + \Delta_2 \pm V_{rf}$.

We should point out that the above argument assumes that the microwave field is a voltage source. The validity of this assumption depends on the impedance of the microwave field coupling to the tunnel junction and the microwave impedance of the junction. The effects of the microwave source being treated as a current source has been investigated theoretically and computer modeled by Tulin (1976) in an effort to determine if gap enhancement could be observed by measuring structures in tunnel characteristics with the application of microwave radiation. Though experimentally these effects have not been investigated, there might be novel effects in high quality low resistance junctions. In this case, when the bias voltage is less than V_o, the junction resistance is very high and the microwave field will be a voltage source. When the bias voltage is greater than V_o, the junction resistance will be low and the microwave field will be a current source. Thus the splitting of $\Delta_1 + \Delta_2$ in this case may not be symmetric or there may be the absence of one peak.

It has been shown by Werthamer (1965) that the tunnel junction distorts considerably the field inside a waveguide or cavity so that the absolute magnitude of the microwave field on the junction might be considerably larger than across an equivalent section of waveguide in the absence of the junction. In addition, it has been shown (Swihart, 1961) that inductive loads due to the junction considerably reduce the wave velocity to

$$v = \frac{C}{[\varepsilon\{1+\lambda/\ell\ (1+\coth d/\lambda)\}]^{1/2}} \tag{13}$$

where C is the velocity in the absence of the junction ε is the effective dielectric constant of the oxide ($\varepsilon \sim 5$ for Al_2O_3), λ is the penetration depth, d and l are the thicknesses of the superconductor and oxide, respectively. This has the effect of setting up standing waves inside the junction which in turn modify the Bessel function dependence as will be shown in Section 4.

3. EXPERIMENTAL TECHNIQUES

In applying microwaves to tunnel junctions several experimental details have to be seriously considered. This involves microwave hardware and junction geometries, substrates, temperature control, tunability of the microwave source for frequency dependent measurements, etc. Several of these techniques are straightforward, although as we will show later, there are pitfalls which are quite subtle and not obvious on a superficial analysis.

The simplest geometry is shown in Fig. 2. It consists of a solid coax and an unmatched loop which is located in physical proximity to the junction. The clear advantage of such a geometry is the ease of construction as well as the ease with which it can be introduced into a dilution refrigerator for low temperature experiments. It also allows the application of microwaves in wide a frequency range. Unfortunately, this geometry does not allow a unique determination of the microwave field and, also, does not confine the microwaves only to the junction. Extreme care must be taken especially when the temperature control and measurement is performed electronically.

An improvement over the solid coax geometry is the use of a wave guide with or without a tunable plunger (Fig. 3). In this geometry the junction is most easily introduced on the side of the wave guide through a removable wall. The DC leads to the junction can be taken out of the guide through ferrite beads which reduce the leakage of microwaves. A tuning plunger operated from room temperature allows changing the field configuration during the experiment. In the TE_{111} propagation mode a perpendicular electric field or a parallel magnetic field can be applied to the junction

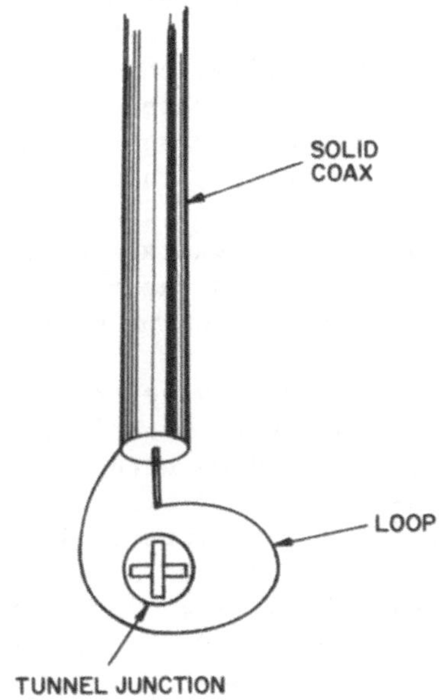

Fig. 2. Solid coax geometry

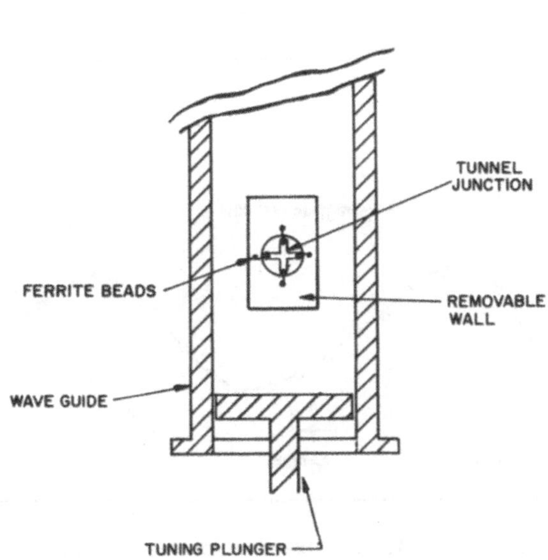

Fig. 3. Waveguide with tunable plunger

by changing the distance between plunger and junction. It was found (Dahlberg, Orbach and Schuller, 1978) that the coupling of the microwave field into superconducting films changes if the field configuration is changed. This is best illustrated by the effect the field has on the enhancement of the critical current of an aluminum film, shown in Fig. 4. It is interesting to note that the coupling seems to be larger when the magnetic field is such that it induces currents through the film. The biggest disadvantage of this geometry is the narrow frequency range (~ 2 GHz) and the potential leakage of microwaves through couplings and around the plunger.

Several configurations of microwave cavities have been successfully used. Figure 5 shows a cavity used in the frequency range 3 - 4.3 GHz (Sweet and Rochlin, 1970) with the electric field oriented perpendicular to the sample. A circular resonator in the TE_{111} mode has been used (Cook and Everett, 1967) at 36 GHz (Fig. 6) to study the effect of microwaves at high powers. A stripline cavity (Dahlberg and Dodds, 1980) (Fig. 7) has been designed recently for use in the frequency range 1-2 GHz. Cavities with high Q can couple large microwave powers although their construction is complicated and they operate in a very restricted frequency range.

Quite recently (Hall et al., 1980), an interesting microstrip transmission line has been constructed where one of the junction electrodes is part of the transmission line (Fig. 8). In order to match impedances the junction geometry in this kind of a setup is somewhat restricted, and it requires the additional formation of ground planes under the junction.

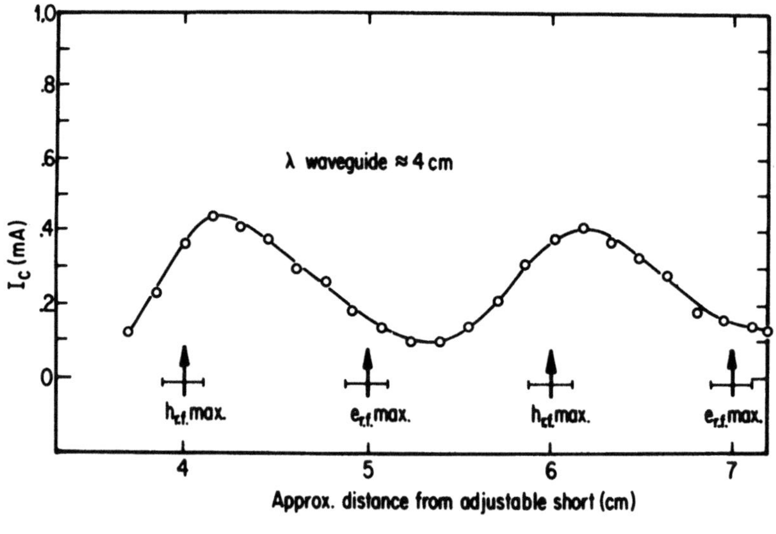

Fig. 4. Enhancement of critical current in an aluminum film.

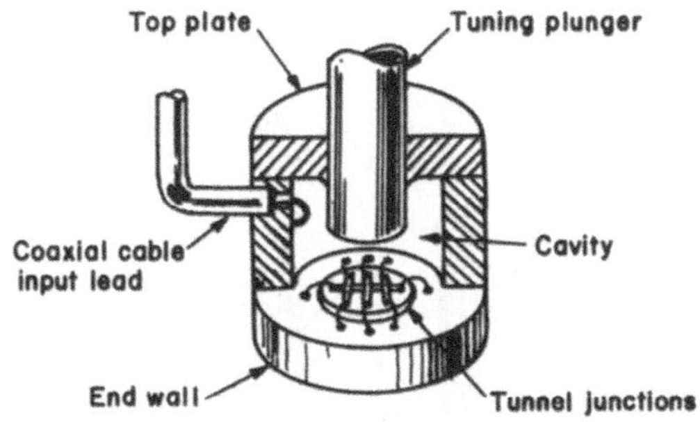

Fig. 5. Tunable microwave cavity

4. EXPERIMENTAL RESULTS

The dependence of the photon induced steps has been extensively studied in a variety of different junctions, frequencies and geometries. A rather large (non-exhaustive) list of experiments in chronological order are summarized in Table I. To our knowledge no experiments have been performed in (i) high T_c superconductors, (ii) at frequencies above 80 GHz or (iii) for superconductors in proximity.

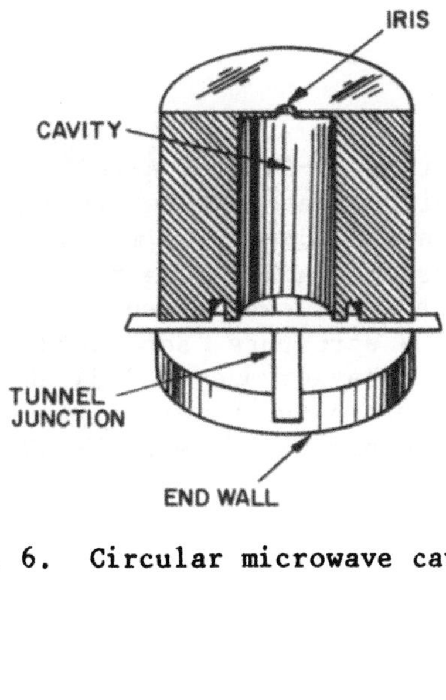

Fig. 6. Circular microwave cavity

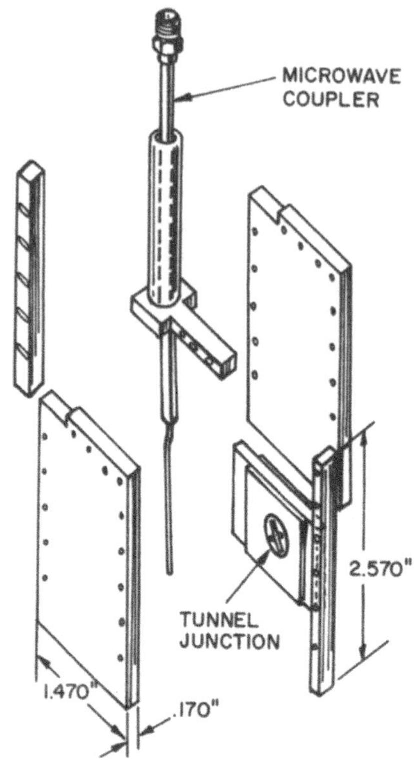

Fig. 7. Low Frequency (1-2 GHz) Stripline Cavity.

Quantum Limit

Typical results for an Al-Al (1000Å-1000Å) tunnel junction in the presence of 2 GHz microwaves is shown in Fig. 9. The location where the peaks occur in the absence of microwaves corresponds to $\Delta_1 \pm \Delta_2$ where Δ_1 and Δ_2 are the gaps in the two electrodes. As the microwave power is increased (lower attenuation) more and more steps appear around the $\Delta_1 \pm \Delta_2$ structure, separated by a voltage $\pm n\hbar\omega/e$. Figure 10 shows similar behavior due to 9.34 GHz radiation ($\hbar\omega/e \sim 39\ \mu V$): again as the microwave power is increased the number of steps increases and the separation now is 39 μV, corresponding to the microwave frequency.

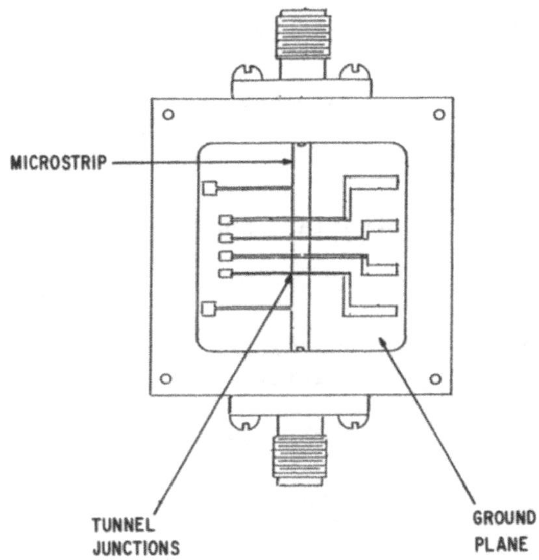

Fig. 8. Microstrip transmission line. One of the electrodes of the junction is part of the stripline.

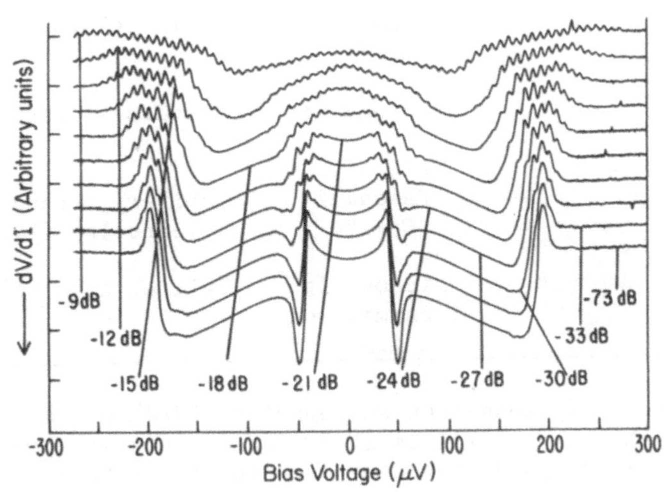

Fig. 9. Microwave induced structure due to 2 GHz radiation as a function of attenuation.

Table I

Junction	Frequency (GHz)	Geometry	Reference
Al-Pb Al-In Al-Sn	38.3	Resonant Cavity	Dayem & Martin (1962)
Al-Pb	0.065	Coaxial Line	Shapiro & Janus (1963)
Al-In	24.2,33,55		Tien & Gordon (1963)
Al-In	36	Resonant Cavity	Cook & Everett (1967)
Sn-Nb	10,37	Waveguide	Bonnet & Rabenhorst (1968)
Sn-Sn	3.93	Resonant Cavity	Sweet & Rochlin (1970)
Sn-Pb	10,35,67	Waveguide & Resonant Cavity	Teller & Kofoed (1970)
Sn-Sn	8,20,80	Waveguide	Hamilton & Shapiro (1970)
Sn-Sn		Waveguide	Hoffmann & Sorensen (1974)
Pb-Pb		Resonant Cavity	Tulin (1976)
Al-Al	10	Waveguide	Kommers & Clarke (1977)
Al-Al Al-Sn	2,10	Waveguide & Resonant Cavity	Dahlberg, Orbach & Schuller (1979)
Al-Al	0.05-18	Microstrip Transmission Line	Hall et al. (1980)

A detailed study (Hamilton and Shapiro, 1970) of the power dependence of the step amplitude at 20 GHz for a small (50 μm) Sn/Sn junction is in good agreement with the theoretical prediction outlined earlier. Figure 11 shows a comparison of experiment and theory where the only adjustable parameter is the absolute magnitude of the microwave power in the junction. The agreement is remarkably good in view of the fact that the dependence of several steps are quantitatively explained using Eq. (10). If the junction size is increased (0.1 mm square) the Bessel junction oscillations predicted

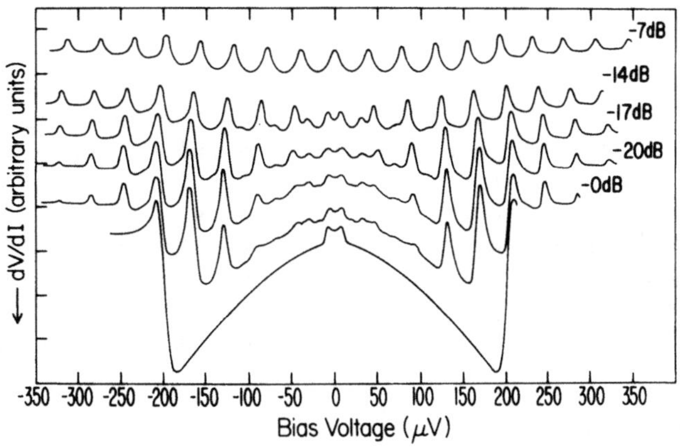

Fig. 10. Microwave induced structures due to 9.34 GHz as a function of attenuation.

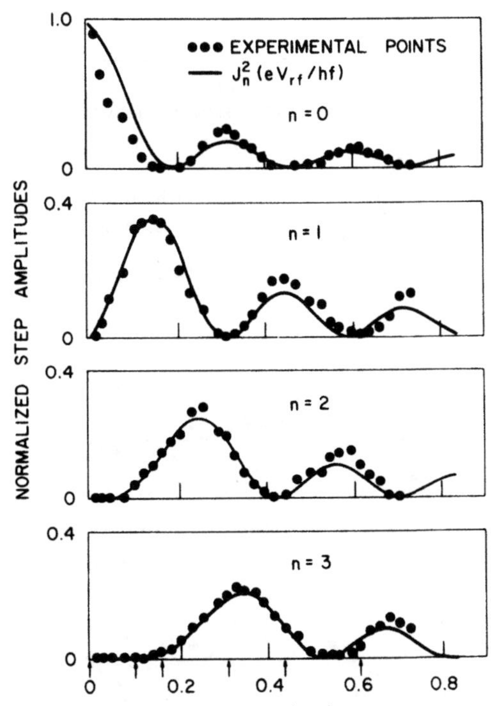

Fig. 11. Dependence of the microwave step amplitude as a function of microwave voltage in a small junction. Solid lines are predictions of the Tien and Gordon (1963) theory.

by Eqs. (9) and (10) are absent as shown in Fig. 12.

This is related to problems relating to standing waves in the junction as suggested in Section 2. Due to the large impedance mismatch between the junction and the wave guide, there will be reflections at the boundaries, thus establishing a standing wave in the junction. As a consequence, the microwave field will no longer be uniform and the power dependence of the steps is modified. In the simplest model it is predicted that the power dependence of the steps can be calculated by simply averaging the voltage across the junction. The power dependence of the step amplitude is given by

$$\frac{1}{L} \int_{o}^{L} J_n^2(eV(x)/\hbar\omega)dx \tag{14}$$

where L is the length of the junction in direction x. Assuming a Voltage Standing Wave Ratio = 5 the agreement between experiment and theory is illustrated in Fig. 12.

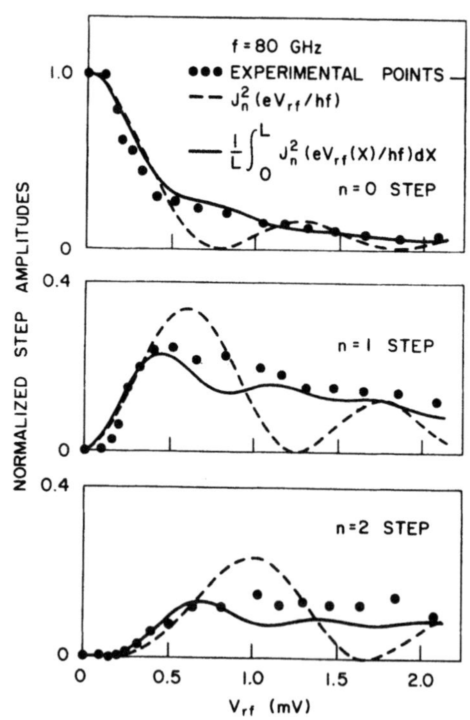

Fig. 12. Evolution of the microwave induced steps as a function of microwave voltage. Solid lines are theoretical predictions as explained in the text.

Classical Limit

If the edges of the junction are not masked or the films are made granular, the width of the gap structure can be larger than the separation of the microwave induced steps. This has the effect of introducing a gradual change on the tunneling curves given by Eq. (12). Figure 13 shows the effect of 2.0 GHz on an Al-Sn junction. The $\Delta_1 \pm \Delta_2$ peaks split into two distinct peaks which separate as the microwave power is increased. Note that if the ac modulation for the derivative is large it is possible to smear the low voltage peak so that only the higher voltage peak is observed. Figure 14 shows similar behavior for an Al-Al junction. The linear dependence of the peak position (Fig. 15.) on the square root of the microwave power shows that this is indeed a microwave voltage induced effect. It is possible to simulate this "classical rectification" by superimposing a "slow" 20 kHz voltage across the junction. A comparison of Fig. 14 and Fig. 16 shows that the two situations have the same qualitative features, and hence this is clearly not a gap enhancement effect.

A study of the intermediate regime has recently been performed (Hall et al., 1980) although not yet published in detail. This work indicates that the Tien and Gordon (1963) calculation can explain quantitatively the results even in the intermediate regime between classical and quantum effects. Of course, for this limit no simple physical picture can be given, and the experimental results have to be compared with numerical calculations based on Eq. (10).

Subharmonic Structures

In addition to the usual photon induced tunneling where the current steps appear at voltages of $\pm n\hbar\omega/e$ measured from the gap

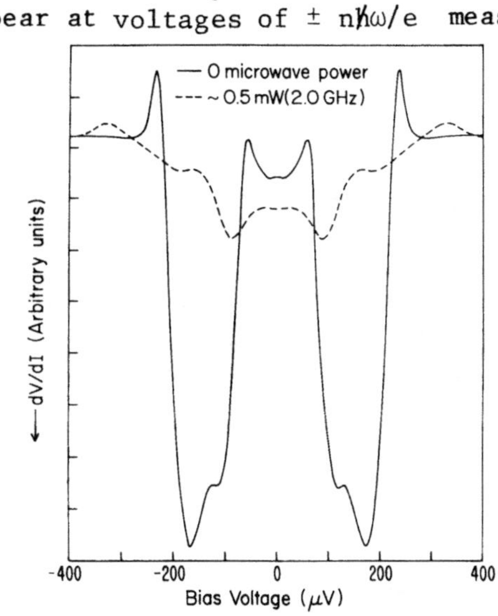

Fig. 13. Microwave induced effects in the classical limit of 1.96 GHz for a Sn-Al junction.

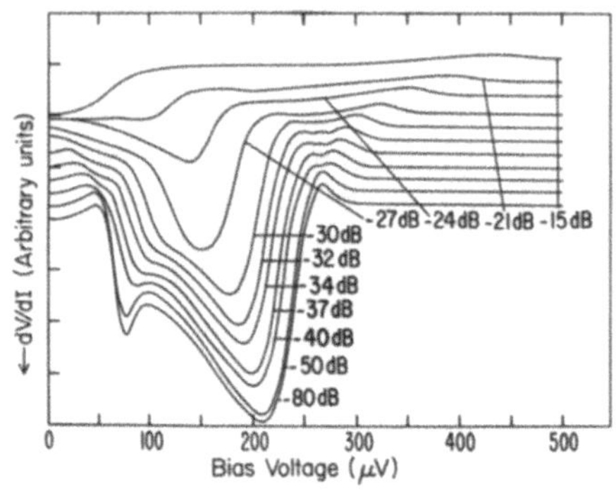

Fig. 14. Evolution of the derivative curves as a function of microwave power for an Al-Al junction in the "classical limit".

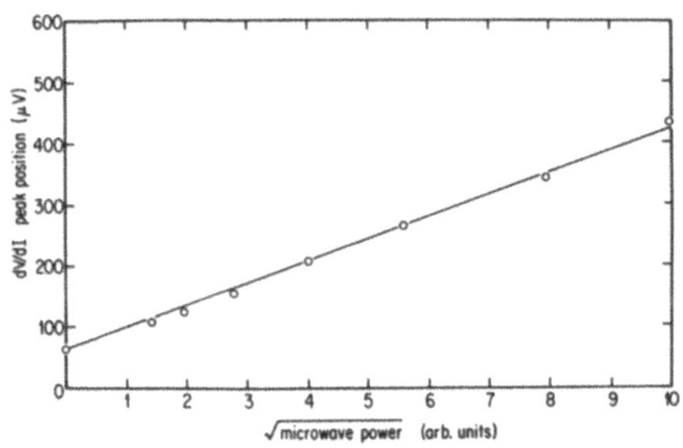

Fig. 15. Evolution of the peak position as a function of microwave voltage.

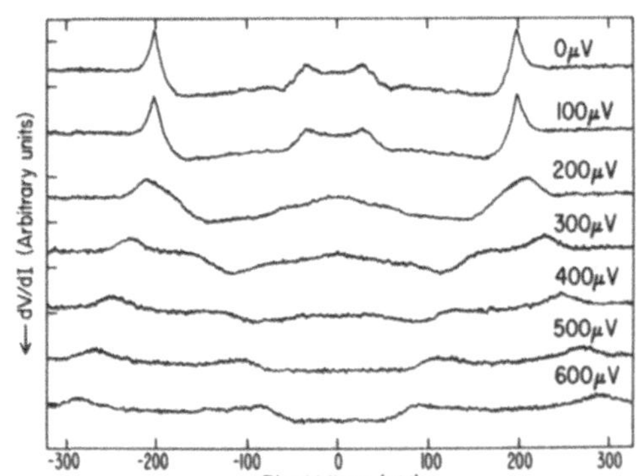

Fig. 16. Simulated microwave experiment in the "classical limit" using 20 kHz ac modulation.

structure in the tunneling characteristic, it is also possible to observe subharmonics of this voltage. Thus abrupt changes are observed in the tunneling current at voltages corresponding to

$$V = \frac{2\Delta \pm n\hbar\omega}{me} \tag{15}$$

where n and m are integers. The magnitude of the changes of the tunneling current are proportional to $J_n^2\left(\frac{meV_{rf}}{\hbar\omega}\right)$ where J_n is the ordinary Bessel function of order n and V_{rf} is the microwave voltage developed across the sample. It should be noted that as shown in the above equation, if n = 0 one will still observe this subharmonic structure. Figure 17 (Hoffman-Soerensen, 1974) shows the differential conductance as a function of bias voltage for different incident microwave power levels. This sample is a Sn-Sn tunnel junction at 1.6 K with the microwave radiation at a frequency of 66 GHz. Figure 18 shows the intensities of the subharmonic structures as a function of incident microwave voltage for two sets of data. The solid lines show that the fit to the Bessel function dependence mentioned earlier (Eq. (10)) is quite good.

5. PHONON STRUCTURE

One of the problems that received least attention in this field is the effect of microwaves on the phonon structure as observed by tunneling. It has been shown (Scalapino, 1969) that the phonon density of states manifests itself as structure in the I-V characteristic at voltages corresponding to the energies of the phonons. If only one phonon peak (of frequency ν_0) is present, one can have a simple

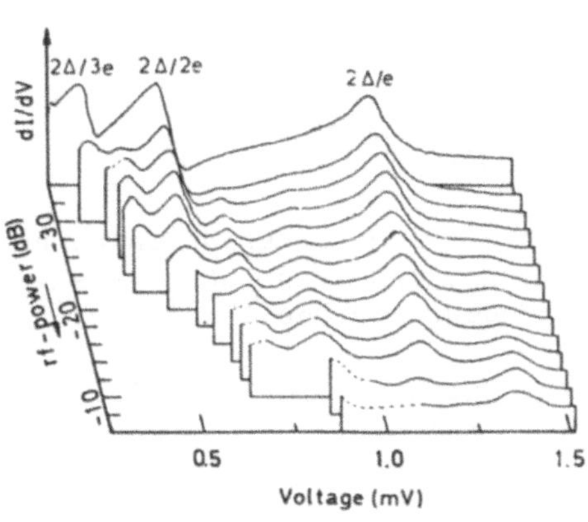

Fig. 17. Evolution of the derivative curves as a function of microwave power showing subharmonic structure.

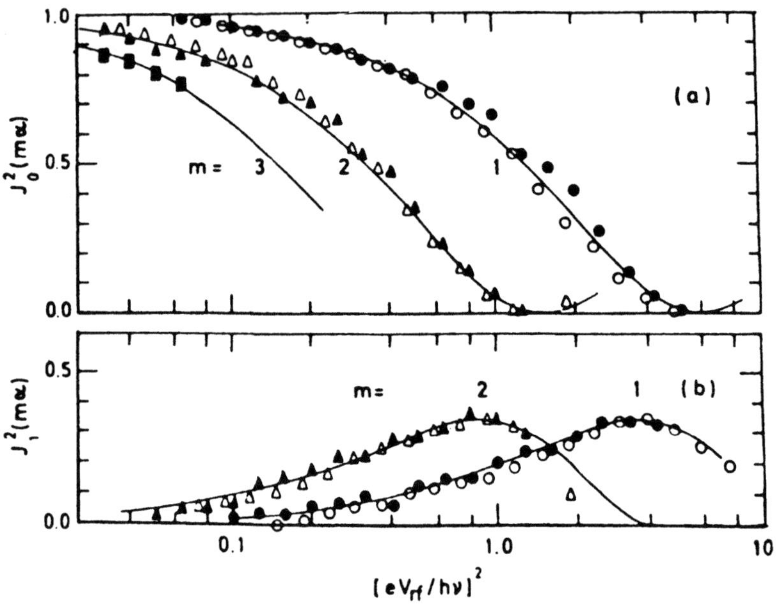

Fig. 18. Evolution of the subharmonic microwave induced steps as a function of power. The solid lines are theoretical fits to experiment.

physical understanding how the phonon structure arises in the tunneling curves. At frequencies ν below ν_0 most of the phonons that are exchanged have larger frequency than ν. This means that the positive ions are underdriven which gives rise to an attractive electron-electron interaction. Above ν_0 the situation is the opposite; the lattice is overdriven, giving rise to an effective repulsive electron-electron interaction. Because of this the complex energy gap changes sign at voltages corresponding to the phonon frequency. Since the quasiparticle density of states is given by

$$\frac{N_T(\omega)}{N_0} = \mathrm{Re}\left\{\frac{\omega}{[\omega^2 - \Delta^2(\omega)]^{1/2}}\right\} \tag{15}$$

structure appears in the I-V characteristics at energies corresponding to the phonon frequency (shifted out by the gap). The amplitude of the phonon peak dependents on the magnitude of the energy gap.

It has been shown (Chaikin et al., 1977) that the magnitude of the phonon peak can be changed by increasing the gap by the proximity effect. The amplitude of the longitudinal Al phonon in an Al/Sn superconductor proximity sandwiches increases markedly if the gap in the backing superconductor is increased. This is quite nicely illustrated in Fig. 19 for an Al/Pb sandwich which shows a consider-

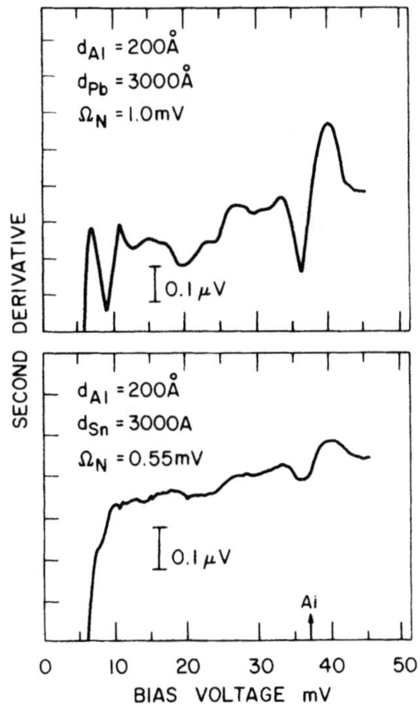

Fig. 19. Phonon structure of an Al/Sn and Al/Pb proximity sandwich. Notice that the amplitude of the Al phonon peak increases roughly quadratically with the induced gap.

ably larger Al phonon structure than the equivalent structure in an Al/Sn sandwich. In fact, the amplitude of the phonon structure should scale with the square of the gap as it can be seen from an expansion of Eq. (15) for $\Delta \ll \omega$.

Very preliminary experiments performed by the authors are shown in Fig. 20 where the second derivative for an Al/Sn junction on sapphire with and without the application of 2 GHz microwaves is displayed.

An expanded graph (Fig. 21) around voltages corresponding to the longitudinal phonon peaks of aluminum does not show large changes in the phonon structure. Clearly more experimental work is needed to clarify this point. To our knowledge, no other experiments have been performed to study changes induced on the I-V characteristics around phonon peak voltages due to microwave irradiation. Since the amplitude of the phonon peaks is proportional to the square of the superconducting gap, large changes in this structure should be observed in "gap enhancement" experiments. This experiment will be particularly interesting in view of the fact that changes in amplitudes are observed as opposed to shifts of structures along the voltage axis.

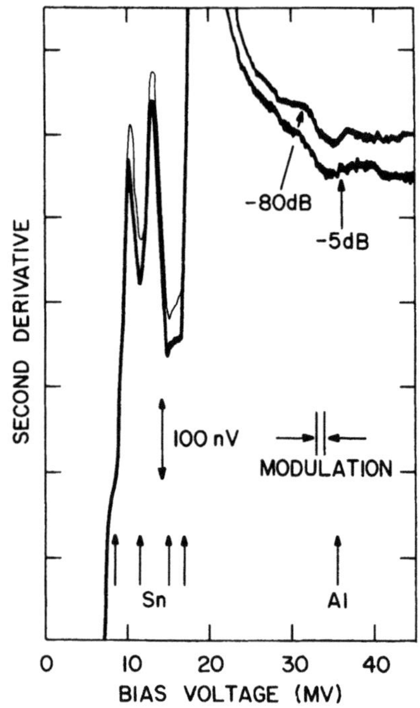

Fig. 20. Second derivative curves for an Al-Sn junction for two different microwave powers.

6. CONCLUSIONS

We have shown that the effect of microwaves on tunnel junctions is to open new tunneling channels for quasiparticles. A theoretical calculation performed in the adiabatic limit is in good quantitative agreement with experiments. It is possible to distinguish two distinct regimes where: the microwave step width > interstep separation ("quantum") and in the opposite case the "classical" limit. The theoretical picture is able to account for the experimental facts in all regimes, even the intermediate one. The study of the effect of microwaves on the phonon structure is suggested for future work.

We would like to acknowledge useful conversations with P. Chaikin, J. Clarke, K. Gray, C. Falco, R. Orbach and A. Goldman.

This work was partially supported by the U.S. Department of Energy, NSF grant DMR 76-82347 and DMR 80-06959 and ONR N0014-75-0245 P4.

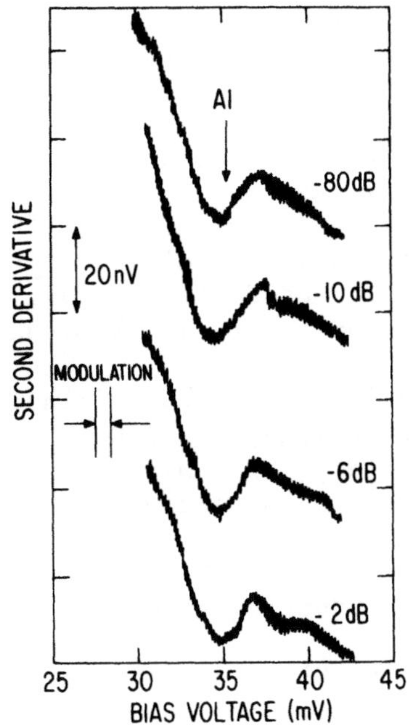

Fig. 21. Evolution of the Al phonon peak as a function of microwave power at 2 GHz and $T/T_c = 0.963$.

REFERENCES

Bonnet, D., and Rabenhorst, H., 1968, Microwave Assisted Tunneling between Superconducting Tin and Bulk Niobium, Phys. Lett. 26A: 174.

Büttner, H., and Gerlach, E., 1968, Microwave Assisted Tunnelling between Superconductors, Phys. Lett. 27A:226.

Chaikin, P. M., Arnold, G., and Hansma, P. K., 1977, Phonon Structure in the Tunneling Characteristics of Thin Proximity Effect Sandwiches, J. Low Temp. Phys., 26:229.

Cohen, M. H., Falicov, L. M., and Phillips, J. C., 1962, Superconducting Tunnelling, Proc. Int. Conf., Low Temp. Phys., 8:178.

Cook, F. C., and Everett. G. E., 1967, Microwave Photon Interaction with Superconducting Tunneling Currents, Phys. Rev., 159:374.

Dahlberg, E. D., Orbach, R. L., and Schuller, I. K., 1979, Microwave-Induced Effects on Superconductors, J. Low. Temp. Phys., 36: 367.

Dayem, A. H., and Martin, R. J., 1962, Quantum Interaction of Microwave Radiation with Tunneling between Superconductors, Phys. Rev. Lett., 8:246.

Falco, C. M., and Lee, T. W., 1980, Josephson Effects in Ultra-Short Mean Free Path Superconductors, to be published.

Hall, J. T., Holden, L. B., and Soulen, R. J., 1980, Photon-Assisted Quasiparticle Tunneling in the Region Intermediate between the Classical and Quantum Limits, Bull. Am. Phys. Soc., 25:411.

Hall, J. T., Holden, L. B., and Soulen, R. J., 1980, Microwave Enhancement of Superconductivity in Aluminum Tunnel Junctions, to be published.

Hamilton, C. A., and Shapiro, S., 1970, rf-Induced Effects in Superconducting Tunnel Junctions, Phys. Rev., B2:4494.

Hoffmann-Sorensen, O., et al, 1974, Microwave Dependence of Subharmonic Gap Structure in Superconducting Junctions, Phys. Rev., 9:3746.

Kommers, T., and Clarke, J., 1977, Measurement of Microwave-Enhanced Energy Gap in Superconducting Aluminum by Tunneling, Phys. Rev. Lett., 38:1091.

Scalapino, D. J., 1969, The Electron-Phonon Interaction and Strong Coupling Superconductors, in: "Superconductivity," R. D. Parks, ed., Marcel Dekker, Inc., New York.

Shapiro, S., and Janus, A. R., 1963, R.F. Detection by Electron Tunnelling between Superconductors, Proc. Int. Conf., Low Temp. Phys., 8:321.

Sweet, J.N., and Rochlin, G. I., 1970, Microwave-Photon-Assisted Tunneling in Sn-I-Sn Superconducting Tunnel Junctions, Phys. Rev., B2:656.

Swihart, J. C., 1961, Field Solution for a Thin-Film Superconducting Strip Transmission Line, J. Appl. Phys., 32:461.

Teller, S., and Kofoed, B., 1970, Microwave-Photon-Assisted Tunneling between Superconductors, Sol. St. Comm., 8:236.

Tien, P. K., and Gordon, J. P., 1963, Multiphoton Process Observed in the Interaction of Microwave Fields with the Tunneling between Superconductor Films, Phys. Rev., 129:647.

Tulin, V. A., 1976, Some Data on Tunneling in the Presence of Microwave Photons, JETP-J. Low Temp. Phys., 2:741.

Werthamer, N. R., 1966, Nonlinear Self-Coupling of Josephson Radiation in Superconducting Tunnel Junctions, Phys. Rev., 147:255.

ENHANCEMENT OF SUPERCONDUCTIVITY

J.E. Mooij

Department of Applied Physics
Delft University of Technology
Delft, The Netherlands

INTRODUCTION

The term 'enhancement of superconductivity' can have different meanings. A materials scientist will use it when he alters the composition or structure of a metal to obtain a material with a higher critical temperature. We will, on the other hand, devote our attention to physical processes in which the superconducting properties of a given material are enhanced by external influences.

It has been found that electromagnetic radiation, acoustic waves and tunneling injection currents may, under certain conditions, cause an increase of the gap in the excitation spectrum, the order parameter, the critical current and the critical temperature. Originated as a puzzling experimental phenomenon that contradicted all intuitive expectations, enhancement later turned out to be one of the interesting consequences of non-equilibrium processes in superconductors. The effects are small and so far of no practical significance except for a good understanding of some weak link devices. However, its paradoxical qualities make the subject fascinating from a physical point of view.

Historical Outline

In the years following the discovery of the Josephson effects in tunnel junctions, other types of junctions or weak links were developed. Anderson and Dayem fabricated constrictions in thin films of tin, with a narrowest width of several microns. When subjected to microwave radiation, these microbridges showed current steps in their current-voltage characteristics similar to those observed in tunnel junctions. It was therefore to be expected that

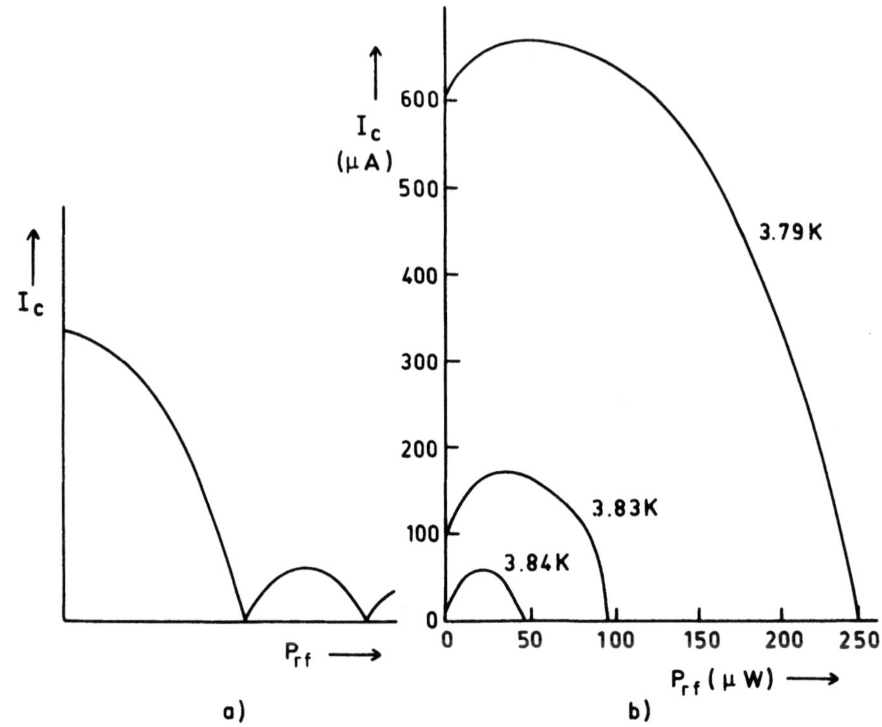

Fig. 1. Dependence of critical current of microbridge on RF power.
a. theoretical expectation
b. experimental observation of Wyatt et al. (1966).

the dependence of the critical current on microwave power would qualitatively be the same as for a tunnel junction. In figure 1[a] such dependence is sketched.

However, by two different groups at approximately the same time, it was found that I_c showed an initial increase when the microwave power was turned on. In figure 1[b] the result of Wyatt, Dmitriev, Moore and Sheard (1966) is given. Similar, more extensive results were obtained by Dayem and Wiegand (1967). This critical current enhancement depended strongly on temperature (most prominent near T_c) as well as microwave frequency. It was later observed in a variety of superconducting weak links, but never in Josephson tunnel junctions. The effect became known as the Dayem-Wyatt effect. Already in the first reports hints are found of apparent small shifts of the critical temperature to higher values. In some later experiments this T_c enhancement was more prominent, but in none of these weak links the enhanced effective T_c exceeded the highest critical temperature of the materials contained in the structure. Several theoretical explanations were put forward which were all based on the weak link character of the sample. None of those theories seemed to fit to all available experimental evidence.

In the years 1970-73 Eliashberg and co-workers published a theory of gap enhancement due to a non-equilibrium distribution of quasiparticles created by a microwave field. This theory was not connected with a weak link structure and predicted enhancement of Δ, I_c and T_c in a homogeneous superconducting film. It took several years for the implications of this theory to be appreciated, but when Tredwell and Jacobsen (1975) observed enhancement of the I_c of junctions by acoustic waves and Latyshev and Nad' (1976) as well as Klapwijk and Mooij (1976) reported on I_c and T_c enhancement by microwaves in long homogeneous strips, the Eliashberg theory was rather generally accepted as the explanation for the Dayem-Wyatt effect. The direct observation of gap enhancement by Kommers and Clarke (1977) provided the final confirmation.

In recent years the theory was extended by Chang and Scalapino (1977) to include effects on the phonon distribution while Schmid (1977) investigated the stability of various states. Experimentally, enhancement by tunneling injection was observed by Gray (1978) and, under different conditions, by Chi and Clarke (1979). Pals and Dobben performed several additional experiments on microwave enhancement.

Lay-out of this chapter

In this chapter most attention will be paid to the Eliashberg theory and its extensions and to those experiments that can be compared best with this theory, i.e. on homogeneous films.
The theory of microwave enhancement will be treated first, followed by a separate discussion of experimental results. Next phonon enhancement, theory as well as experiment, will be treated, followed by enhancement by tunneling. Little attention will be paid to the many experiments on critical current enhancement in weak links by microwaves. Some results are mentioned in the last section together with earlier theories put forward to explain the Dayem-Wyatt effect. Finally some general conclusions are drawn.

MICROWAVE ENHANCEMENT IN THE SPATIALLY UNIFORM CASE; THEORY

Eliashberg mechanism

In the BCS theory there is a fundamental connection between the value of the gap Δ and the occupation of the quasiparticle states. When the temperature is increased from zero, more and more quasiparticles are excited blocking states which were previously available for Cooper pair formation. Ultimately this process leads to the disappearance of superconductivity at the critical temperature. Extraction of quasiparticles leads to enhancement of superconductivity but is difficult to realize. However, there is a second possibility. Electron states near the Fermi surface are more important to Cooper pair formation than states further away from k_F. If a quasiparticle in a state near k_F is removed to a state with higher energy, the energy gap and in general the strength of the superconducting effects increases, although the total number of quasiparticles remains constant. This is the essential basis of the Eliashberg mechanism. In a microwave field, the quasiparticles are 'pumped up', away from the gap. Even if the average energy of the quasiparticles increases, the gap is enhanced. In the following we will give the mathematical formulation of this effect.

It will be assumed that the distribution over the $\underline{k}$ states of the quasiparticles is isotropic, f(E) is the occupation probability. A microwave field has a vector character, but if the superconductor is dirty and an electron is elastically scattered many times in a period of the field, the directional preference is lost. If τ_{el} is the elastic scattering time and ω the microwave frequency it is required that $\tau_{el} \ll \omega^{-1}$. For typical values this condition is well satisfied. In the theory no distinction is made between the quasiparticle branches below and above the Fermi surface. Only the energies are of importance, not hole-like or electron-like character.

The BCS gap equation is:

$$\frac{2}{V_{BCS}} = \sum_k \frac{1 - 2f_k}{E_k} = \sum_k \frac{1 - 2f_k}{(\Delta^2 + \xi_k^2)^{\frac{1}{2}}} \qquad (1)$$

The occupation probability of the quasiparticle state k is f_k, which in equilibrium is equal to the fermi function $f_k^o = [\exp(E_k/k_BT) + 1]^{-1}$. If the quasiparticle distribution is out of equilibrium equation (1) remains valid. The deviation f^1 is defined by:

$$f(E) = f^o(E) + f^1(E) \qquad (2)$$

Near T_c, when $\Delta \ll k_BT_c$, Eq.(1) can be worked out, inserting (2), to give the Ginzburg-Landau equation:

$$\frac{T_c - T}{T_c} - \frac{7\zeta(3)}{8\pi^2}\left(\frac{\Delta}{k_B T_c}\right)^2 - 2\int_{\Delta}^{\infty} \frac{f^1(E)}{(E^2 - \Delta^2)^{\frac{1}{2}}}\, dE = 0 \qquad (3)$$

The last term is the non-equilibrium term. Any $f^1(E) < 0$ leads to enhancement. Those energies close to Δ (k states near the Fermi surface) contribute more to the integral than higher levels of E, for the same f^1. In a series of three papers (Eliashberg, 1971; Ivlev and Eliashberg, 1971; Ivlev, Lisitsyn and Eliashberg, 1973) it was shown how f^1 can be calculated.

The microwave field is represented by a vector potential
(4)

$$\underline{A} = A_\omega \mathrm{Re}(-i\omega t) \qquad (4)$$

which is defined in the London gauge; Δ is assumed to be real. The power of the RF field is expressed as:

$$\alpha_\omega = \frac{2e^2 D}{\hbar c^2} A_\omega^2 \qquad (5)$$

where $D = v_F \ell/3$. The interaction with the RF field leads to a change in the occupation of energy level E by means of three different processes: scattering transitions between levels E and $E + \hbar\omega$, scattering transitions between levels $E - \hbar\omega$ and E and pair breaking leading to creation of quasiparticles at E and $\hbar\omega - E$. The transition rates for each of these processes are determined by the product of several factors: the coupling strength with the field, the density of states $E/(E^2 - \Delta^2)^{\frac{1}{2}}$ of each of the levels involved, the coherence factor, the difference in occupation probability of the levels and one or more θ functions that take into account that no levels $E < \Delta$ are present.
The coherence factors for interaction with microwave radiation are (see Tinkham, 1975, chapter 2.9) for scattering between E and E'

$$(uu' + vv')^2 = \frac{1}{2}\,\frac{EE' + \Delta^2}{EE'} \qquad (6^a)$$

and for pair breaking or recombination:

$$(vu' - uv')^2 = \frac{1}{2}\,\frac{EE' - \Delta^2}{EE'} \qquad (6^b)$$

The transition rate into level E is:

$$I_\omega(E) = 2\hbar^{-1}\alpha_\omega \Bigg[-\frac{E(E+\hbar\omega) + \Delta^2}{(E^2-\Delta^2)^{\frac{1}{2}}\{(E+\hbar\omega)^2-\Delta^2\}^{\frac{1}{2}}} \{f(E)-f(E+\hbar\omega)\}\,\theta(E-\Delta) +$$

$$+ \frac{E(E-\hbar\omega) + \Delta^2}{(E^2-\Delta^2)^{\frac{1}{2}}\{(E-\hbar\omega)^2-\Delta^2\}^{\frac{1}{2}}} \{f(E-\hbar\omega) - f(E)\}\,\theta(E-\hbar\omega-\Delta) +$$

$$+ \frac{E(\hbar\omega-E) - \Delta^2}{(E^2-\Delta^2)^{\frac{1}{2}}\{(\hbar\omega-E)^2-\Delta^2\}^{\frac{1}{2}}} \{1-f(E)-f(\hbar\omega-E)\}\,\theta(E-\Delta)\theta(\hbar\omega-\Delta-E)\Bigg] \quad (7)$$

Only terms to first order in α_ω will be considered. Therefore for $f(E)$ the equilibrium function $f^o(E)$ should be used. If $\hbar\omega/kT_c \ll 1$, to first order in $\hbar\omega/kT_c$, $f(E) - f(E + \hbar\omega)$ is equal to $\hbar\omega/4\,k_BT_c$.

A dynamic equilibrium is reached when the transition rate of Eq. (7) is balanced by the relaxation processes, by means of inelastic scattering with phonons. Eliashberg and co-workers use a simple relaxation time approximation in which the relaxation time does not depend on E and the relaxation rate is proportional to the deviation from equilibrium.

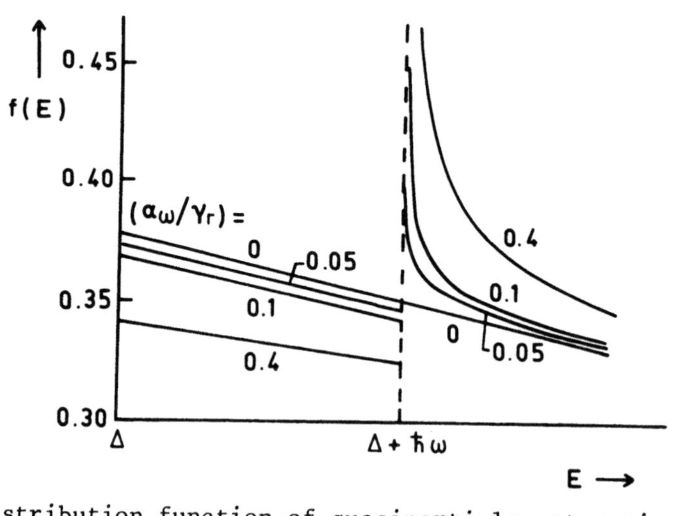

Fig. 2. Distribution function of quasiparticles at various levels of microwave power according to Eqs. (7) and (8). Here T_c = 1.2 K, Δ = 50 μeV and ω corresponds to 3 GHz.

This gives:

$$I_{rel}(E) = 2\hbar^{-1}\gamma_r \frac{E}{(E^2 - \Delta^2)^{\frac{1}{2}}} f^1(E) \tag{8}$$

where $\gamma_r = \hbar/\tau_E$, τ_E is the inelastic collision time at T_c. In the stationary state $I_\omega = I_{rel}$ which directly gives $f^1(E)$. In figure 2 it is sketched what f(E) looks like in this approximation. Between Δ and $\Delta + \hbar\omega$, f^1 is negative, while for $E > \Delta + \hbar\omega$, f^1 is positive. At $\Delta + \hbar\omega$ a root singularity occurs. This singularity is removed when higher order terms (in α_ω) are included. As in the following $f^1(E)$ is integrated, the singularity creates no problems.

We define:

$$\delta_E = -2\int_\Delta^\infty \frac{f^1(E)}{(E^2 - \Delta^2)^{\frac{1}{2}}} dE \tag{9}$$

We see that:

$$\delta_E = \frac{1}{4}\left(\frac{\alpha_\omega}{\gamma_r}\right)\left(\frac{\hbar\omega}{k_B T_c}\right)\Bigg[2\int_\Delta^\infty dE \frac{E(E + \hbar\omega)+\Delta^2}{E\{(E + \hbar\omega)^2-\Delta^2\}^{\frac{1}{2}}(E^2 - \Delta^2)^{\frac{1}{2}}} +$$

$$-2\int_{\Delta+\hbar\omega}^\infty dE \frac{E(E - \hbar\omega)+\Delta^2}{E\{(E - \hbar\omega)^2-\Delta^2\}^{\frac{1}{2}}(E^2 - \Delta^2)^{\frac{1}{2}}} +$$

$$-2\int_\Delta^{\hbar\omega-\Delta} dE \frac{E(\hbar\omega - E)-\Delta^2}{E\ (E - \hbar\omega)^2-\Delta^2\}^{\frac{1}{2}}(E^2-\Delta^2)^{\frac{1}{2}}}\Bigg] \tag{10}$$

The term within brackets has been expressed in terms of complete elliptic integrals of the first kind K and of the third kind Π according to (see Eckern, Schmid, Schmutz and Schön, 1979, appendix C):

$$\delta_E = \frac{1}{4}\left(\frac{\alpha_\omega}{\gamma_r}\right)\left(\frac{\hbar\omega}{k_B T_c}\right)G\left(\frac{\Delta}{\hbar\omega}\right) \tag{11}$$

where

$$G(u) = \begin{cases} 2\pi u(1-u^2)^{-\frac{1}{2}} & u < \frac{1}{2} \\ \frac{2}{u+\frac{1}{2}}\left[K(k)+4u^2\{\Pi(\alpha^2,k)-K(k)\}\right] & u > \frac{1}{2} \end{cases} \tag{12}$$

in this context $k = (u-\frac{1}{2})/(u+\frac{1}{2})$ and $\alpha^2 = (2u+1)^{-2}$. This G function

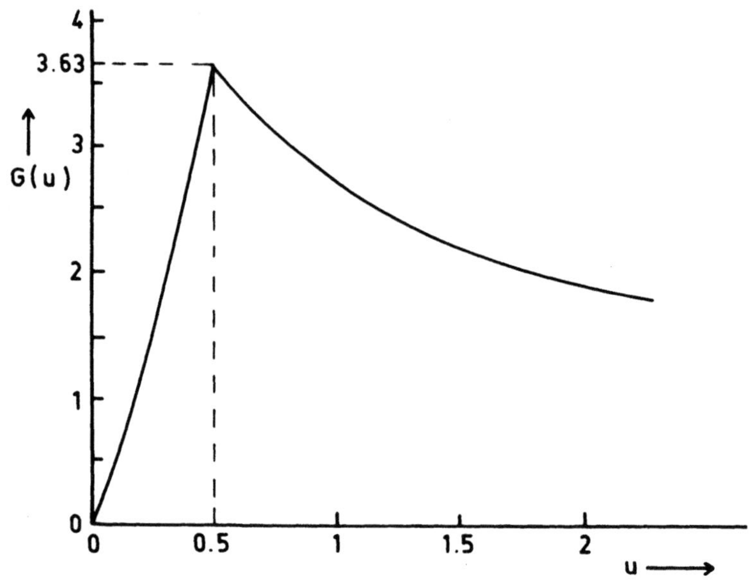

Fig. 3. The function G(u).

has two useful limits:

$$G = \begin{cases} 2\pi u \text{ for } u \to o \\ 2\ln(2.9u)/u \text{ for } u \to \infty \end{cases} \tag{13}$$

For $u = \frac{1}{2}$, $G(\frac{1}{2}) = 3.63$. In figure 3 we give a plot of G(u).

It is now possible to write down the Ginzburg-Landau equation (3) as:

$$\frac{7\zeta(3)}{8\pi^2}\left(\frac{\Delta}{k_BT_c}\right)^2 = \varepsilon + \delta_E = \frac{T_c-T}{T_c} + \frac{1}{4}\left(\frac{\alpha_\omega}{\gamma_r}\right)\left(\frac{\hbar\omega}{k_BT_c}\right) G\left(\frac{\Delta}{\hbar\omega}\right) \tag{14}$$

Before discussing the implications of this result, it is necessary to recall the limitations of the procedure that has been followed so far:

1. Only terms linear in (α_ω/γ_r) have been taken into account.
2. The order parameter suppression that also occurs in a static field has been omitted. If we take the time-averaged value of $\underline{A}(t)^2$. we find an additional pairbreaking term to be added on the righthand side of (14):

$$\delta_{SF} = -\pi\alpha_\omega / 2k_B T_c \tag{15}$$

The effect of this term is small, except at low frequencies.

3. Terms such as $f^o(E) - f^o(E+\hbar\omega)$ were approximated to first order in $\hbar\omega$. It has been calculated by Eckern, Schmid, Schmutz and Schön (1979, appendix F) that the effect of the second order term amounts to another negative extra term in (14):

$$\delta_\omega^2 = -\ 0.17 \left(\frac{\alpha_\omega}{\gamma_r}\right)\left(\frac{\hbar\ \omega}{k_B T_c}\right)^2 \tag{16}$$

at higher frequencies this is an important correction.

4. The relaxation time approximation is only valid for small deviations from equilibrium; τ_E in reality depends on E.
5. The whole derivation is valid for $\Delta/kT_c << 1$ only.
6. It has been assumed that the phonons remain in thermal equilibrium at the measuring temperature T. We will discuss this point, the most serious shortcoming of the Eliashberg theory, later in this chapter.

In table 1 we have collected the values of $\hbar\omega$, $k_B T_c$ and γ_r that occur in the theory, all expressed in µeV, for several materials. For two small values of (α_ω/γ_r) we have illustrated the result of equation (14) in figure 4

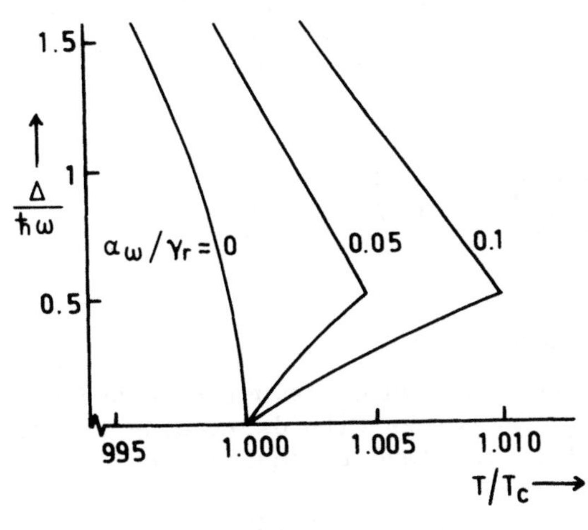

Fig. 4. Solutions to q.(14) at various levels of microwave power.

Table 1. Values of quantities in Eliashberg theory.

Metal	T_c (K)	$\tau_E{}^a$ (s)	$k_B T_c$ (μeV)	γ_r (μeV)
Zn	0.88	$9.3\ 10^{-8}$	76	$7.1\ 10^{-3}$
Al	1.18	$1.3\ 10^{-8}$	102	$5.2\ 10^{-2}$
In	3.41	$9.5\ 10^{-11}$	294	6.9
Sn	3.72	$2.7\ 10^{-10}$	321	2.4
Nb	9.25	$1.8\ 10^{-11}$	798	37

1 GHz corresponds to $\hbar\omega$ = 4.1 μeV.

a from Kaplan et al.(1976), except for Al.

The kinks in the curves for Δ at $\Delta = \hbar\omega/2$ are mathematically connected with the kink of G(u) at u = ½. Physically, at fixed microwave level (α_ω and ω constant) Δ decreases gradually with increasing T until suddenly, when 2 Δ becomes smaller than $\hbar\omega$, pair breaking becomes possible.

The Eliashberg theory predicts a minimal frequency for enhancement. To find this ω_{min}, the correction term δ_{SF}, Eq.(15), has to be taken into account. For enhancement to occur it is necessary that $\delta_E > - \delta_{SF}$ or

$$\frac{1}{4}\left(\frac{\alpha_\omega}{\gamma_r}\right)\left(\frac{\hbar\omega}{k_B T_c}\right) G\left(\frac{\Delta}{\hbar\omega}\right) > \frac{\pi}{2}\frac{\alpha_\omega}{k_B T_c}$$

and consequently

$$\omega_{min} G(\Delta/\hbar\omega_{min}) = 2\pi\tau_E^{-1} \tag{17}$$

ω_{min} is a complicated function of temperature because Δ is temperature dependent. ω_{min} increases with decreasing temperature. The lowest value of ω that can lead to enhancement is found for $\Delta/\hbar\omega$ = ½. This value is:

$$\omega_{min,min} = 1.73\tau_E^{-1} \tag{18}$$

This corresponds to frequencies of 3 MHz for Zn, 20 MHz for Al, 3 GHz for In, 1 GHz for S_n and 15 GHz for Nb.

Stability of states for $T > T_c$

As Fig. 4 shows, in a certain temperature region there are two finite solutions for Δ to eq.(14). Although it is not visible in our formulation, the normal state with $\Delta = 0$ should always be considered as an additional solution to the Ginzburg-Landau equation. The temperature region with two finite solutions is bounded by T_c and T_m, where T_m follows from:

$$\frac{T_m}{T_c} = 1 + 0.91\left(\frac{\alpha_\omega}{\gamma_r}\right)\left(\frac{\hbar\omega}{k_BT_c}\right) - 0.17\left(\frac{\alpha_\omega}{\gamma_r}\right)\left(\frac{\hbar\omega}{k_BT_c}\right)^2 - 0.027\left(\frac{\hbar\omega}{k_BT_c}\right)^2 \quad (19)$$

because at T_m, Δ is equal to $\hbar\omega/2$. The term (16) is included here.

The question now arises which of the three possible states in the region $T_c < T < T_m$ is or are stable. A. Schmid (1977) has investigated this problem. Later a more general paper has appeared on stability of non-equilibrium states (Eckern, Schmid, Schmutz and Schön; 1979). Also chapters 10 and 11 of this book are devoted to this general subject. Schmid generalizes the Ginzburg-Landau free energy density to a non-equilibrium stationary situation with:

$$F(\Delta) = -2N(o)\int_o^\Delta d\Delta'\left[\frac{T_c - T}{T_c} - \frac{7\zeta(3)}{8\pi^2}\left(\frac{\Delta'}{k_BT_c}\right)^2 + \delta_E(\Delta')\right]\Delta' \quad (20)$$

This integral is qualitatively solved in a graphical manner by rewriting (20) as:

$$F(\Delta) = -N(O)\int_o^{\Delta^2} dx\left[\delta_E(x) - \left\{\frac{T - T_c}{T_c} + \frac{0.106}{K_B^2T_c^2}x\right\}\right]$$

where $x = \Delta'^2$ and C is a constant. The two terms of the integrand can both be plotted versus x. $\delta_E(x)$ contains the factor $G(\sqrt{x})$. The intersections of the two curves give the solutions Δ_1^2 and Δ_2^2. The area between the two curves gives F. Because for $0 < \Delta < \Delta_1$ the first term is smaller than the second, F is positive there (with respect to the normal state). Between Δ_1 and Δ_2 the integrand is positive, so that $F(\Delta_2) < F(\Delta_1)$. Whether or not $F(\Delta_2)$ is larger or smaller than zero depends on the areas A_1 and A_2. There are three possibilities, indicated in the Fig. 5 with a, b, and c:

a. T slightly above T_c. A_1 is smaller than A_2, $F(\Delta_2)$ is negative
b. at temperature T_K intermediate between T_c and T_m. A_1 and A_2 are equal, $F(\Delta_2) = o$
c. T close to T_m. A_1 is larger than A_2, $F(\Delta_2)$ is positive.

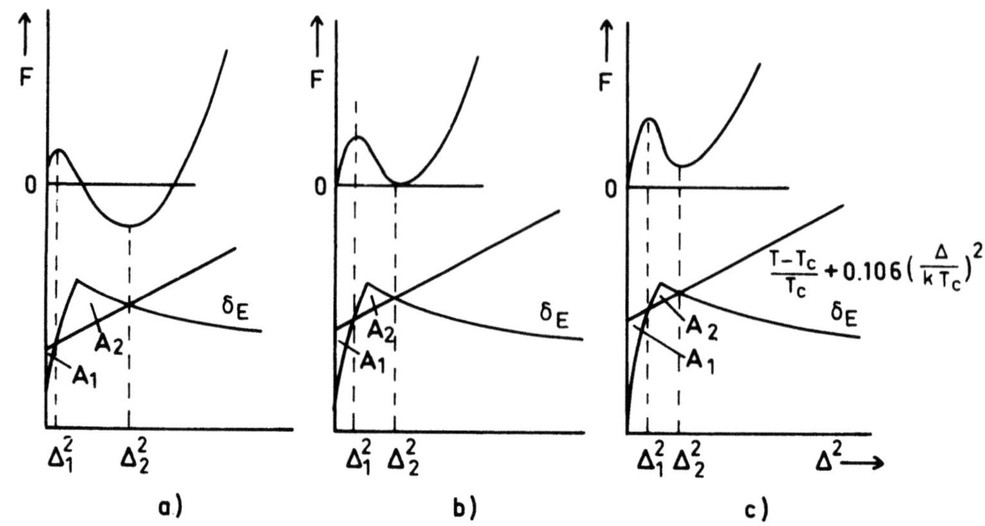

Fig. 5. Graphical evaluation of Eq. (20). The three cases a,b,c correspond to three temperatures T_1, T_2 and T_3 where $T_c < T_1 < T_2 < T_3 < T_m$.

The conclusion from this picture is that for all temperatures the state corresponding to Δ_1 (the solution with $\Delta/\hbar\omega < \frac{1}{2}$) is unstable. For $T_c < T < T_K$ the superconducting state with $\Delta = \Delta_2$ is 'globally' stable while the normal state is metastable. For $T_K < T < T_m$ the superconducting state is metastable and the normal state globally stable. The potential energy barrier between the two stable/metastable states can be found by calculating Δ from (14) and inserting the result in (20). The presence of this potential energy barrier is expected to lead to a hysteretic transition between the superconducting and the normal states.

It is relevant to mention here that the enhanced non-equilibrium state may be spatially unstable. At $T = T_K$ both the superconducting and the normal states have the same free energy. They may therefore co-exist with a boundary layer between the regions. The important length scale in such problems will be the quasiparticle diffussion length $\Lambda_E = (D\tau_E)^{\frac{1}{2}}$. Ivlev (1977) predicts that a homogenous solution is only stable at a certain minimum microwave power.

Calculations of Chang and Scalapino

In all non-equilibrium processes in superconductors, the interaction of the quasiparticles with phonons plays an important role. This interaction provides the process by which relaxation to equilibrium is promoted. With gap enhancement as described in the previous section one should expect that the dynamical equilibrium situation with pumping up and relaxation down of quasiparticles provides a continuous source of non-equilibrium phonons in a specific frequency range. Relaxation of the phonon distribution to equilibrium can occur by way of two processes: interaction with electrons, which will again influence the quasiparticle distribution, and escape from the metal to the surroundings, e.g. the substrate. Only in the ideal case of a very thin metal film in ideally good contact with a non-conducting thermal reservoir, one can expect that the assumption, made by Eliashberg, of a phonon bath in equilibrium at the measuring temperature, is justified.

Chang and Scalapino have investigated the influence of the quasiparticle-phonon interaction on several non-equilibrium processes in superconductors. They considered a set of Boltzmann-like kinetic equations for both the quasiparticles and the phonons. In chapter 9 Chang discusses the general procedure followed and derives the coupled set of linearized kinetic equations. In the specific case of microwave stimulation (Chang and Scalapino; 1977, 1979) the source of non-equilibrium is the interaction with the field, which is introduced with an expression similar to our equation (7). The gap is calculated with the gap equation and follows from the quasiparticle distribution. In the numerical cal-

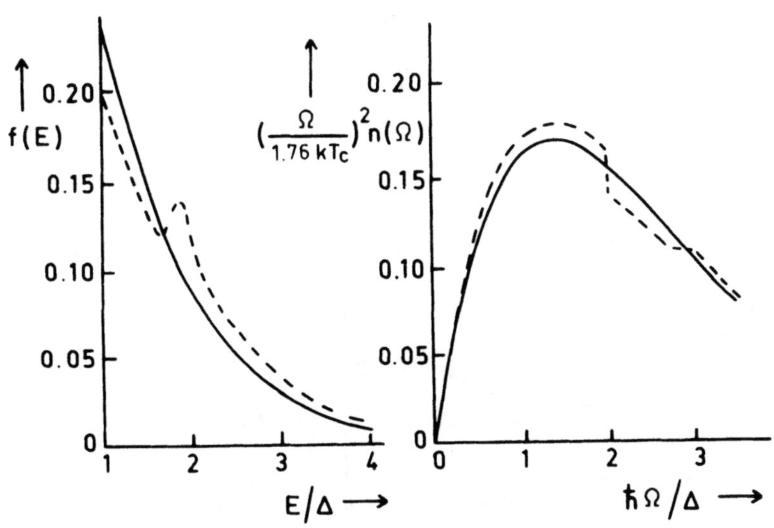

Fig. 6.a. Quasiparticle distribution and b. phonon spectrum as calculated by Chang and Scalapino (1977).
Solid lines: equilibrium, $A_{CS} = 0$, dotted lines : microwave irradiation. $A_{CS} = 0{,}45$, $\omega = 0.46\ k_BT_c/\hbar$, $\tau_{es} = 2\tau_B$. The temperature is $0.9\ T_c$.

culations the whole scheme is run until a stationary solution is obtained which is self-consistent. An important parameter is the coupling of the phonons to the outside world (helium, substrate). Chang and Scalapino use the parameter τ_{es}, the 'escape time', to describe this coupling. This τ_{es} is measured in units of τ_B, the phonon Cooper pair breaking time at zero temperature. A relaxation time approximation is used for the phonons, where the number of phonons leaving the metal is proportional to $\{n(\Omega) - n^o(\Omega)\}/\tau_{es}$. As a whole there are four input parameters: temperature, τ_{es}, microwave frequency and microwave intensity. The last quantity is expressed as A_{CS}, which, when compared with the Eliashberg parameter α_ω in its scattering effect on the quasiparticles corresponds to:

$$A_{CS} = 92 \frac{\alpha_\omega}{\gamma_r} \tag{21}$$

In the work of Chang and Scalapino, the interaction of quasiparticles at energy level E is considered with the phonons of energy $\hbar\Omega$, all separately. There is no assumption of one effective relaxation time. In Figure 6 the quasiparticle distribution function and the phonon spectrum are given in equilibrium and with microwave radiation.

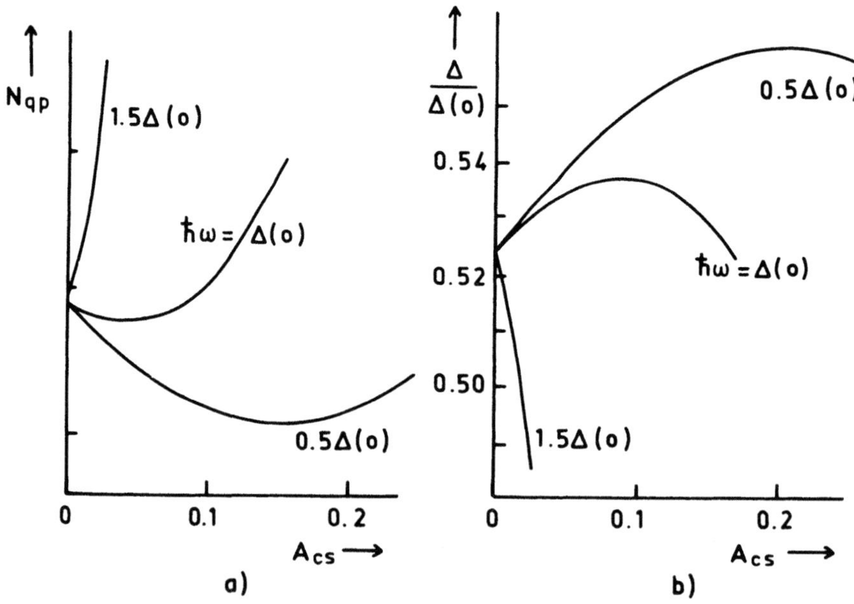

Fig. 7.a. Quasiparticle density N_{qp} and b. energy gap Δ as a function of A_{CS}. The temperature is 0.9 T_c, $\tau_{es}/\tau_B = 8$, while $\hbar\omega$ is 0,5, 1 and 1.5 times $\Delta(o)$, the equilibrium gap at T = 0 (Chang and Scalapino, 1977).

In the Eliashberg theory enhancement is obtained by the redistribution to higher energies while the total number of quasiparticles is conserved if $\omega < 2\Delta/\hbar$. Chang and Scalapino point out that the recombination time for quasiparticles at higher energies is shorter than for quasiparticles near the gap edge. As a secondary result of the redistribution, the effective recombination time is shorter and the total number of quasiparticles decreases. This is an important effect, that in some cases (especially with a short τ_{es}) accounts for up to 70% of the total enhancement.

In figure 7 we reproduce both the density of quasiparticles and the gap as a function of A_{CS}, at $T = 0.9\ T_c$, with $\tau_{es} = 8\ \tau_B$ and for $\omega/\Delta(o) = \frac{1}{4}, \frac{1}{2}$, 1 and 3/2, which roughly corresponds to $\frac{1}{2}$, 1, 2 and 3 times the equilibrium $\Delta(T)$. At the highest frequencies pair breaking dominates, N_{qp} increases and no enhancement occurs. The other curves all show initial enhancement. At higher microwave powers more and more phonons are created with sufficient energy to break pairs. A maximum is found for Δ as a function of A_{CS}.

Simple description of heating effects

In most practical cases of metal films the phonon escape time is much longer than the phonon scattering time. This suggests a simple model for heating in which it is supposed that the phonons are in thermal equilibrium at an effective temperature T_h which is higher than the bath temperature T_b. If the power dissipated in the film per unit area is $\dot{Q}$, T_h is equal to $T_b + Y_K^{-1}\dot{Q}$ where Y_K is the thermal boundary conductance (sum of contributions from two surfaces).

The microwave dissipation depends on the value of Δ. However, when $\Delta \ll k_B T_c$ the dissipation is almost equal to that in the normal state. With $\underline{E} = -c^{-1}\dot{\underline{A}}$ and our definition of α_ω, we find that in a film of thickness d:

$$\dot{Q}_n = d\ \sigma\ \overline{E^2} = 1.5\ d\ \omega^2\ n\hbar\alpha_\omega/E_F$$

E_F is the Fermi energy, T_F to be used later the Fermi temperature. When we define

$$\delta_h = -\frac{T_h - T_b}{T_c} \tag{22}$$

as well as a heating parameter F_h:

$$F_h = 1.5\ \frac{d\ k_B T_c\ n}{Y_k\ \tau_E\ T_F} \tag{23}$$

it follows that

$$\delta_h = - F_h \left(\frac{\alpha_\omega}{\gamma_r}\right)\left(\frac{\hbar\omega}{k_B T_c}\right)^2 \tag{24}$$

This quantity can now be added to the Ginzburg-Landau equation (14) together with the other correction terms. In this way we obtain the full equation which is to be used for comparison with experimental results:

$$\frac{7\zeta(3)}{8\pi^2}\left(\frac{\Delta}{k_B T_c}\right)^2 = \frac{T_c - T_b}{T_c} - \frac{\pi}{2}\frac{\alpha_\omega}{k_B T_c} + \frac{1}{4}\left(\frac{\alpha_\omega}{\gamma_r}\right)\left(\frac{\hbar\omega}{k_B T_c}\right) G\left(\frac{\Delta}{\hbar\omega}\right) +$$

$$- (F_h + 0.17)\left(\frac{\alpha_\omega}{\gamma_r}\right)\left(\frac{\hbar\omega}{k_B T_c}\right)^2 \tag{25}$$

In table 2 we have collected some relevant quantities for several metals. For the thermal boundary conductance Y_K a typical value is given. F_h is given for a film thickness of 100 nm.

Whether or not enhancement occurs will depend on the cooling of the film. The condition is:

$$4(F_h + 0.17)\left(\frac{\hbar\omega}{k_B T_c}\right) < G\left(\frac{\Delta}{\hbar\omega}\right)$$

When Δ becomes large, G gets small and enhancement ceases. However, we do not expect this simple approximation to be relevant for high power levels, where nonlinear terms in (α_ω/γ_r) should be considered.

The above inequality suggests that there will be a maximum frequency for enhancement. As the argument of the G function contains the frequency, this maximum frequency depends on the value of the gap. The highest value of G is reached for $\Delta/\hbar\omega = \frac{1}{2}$. According to the simple heating model, the highest frequency where enhancement will occur for a given film, ω_{max}, is to be found taking $G(\frac{1}{2}) = 3.63$ as the value for G in the inequality. We find that:

$$\hbar\omega_{max} = 0.91\ k_B T_c (F_h + 0.17)^{-1} \tag{26}$$

Table 2. Parameters of importance for heating effects.

metal	T_c (K)	τ_E (s)	T_F (10^4K)	n (10^{22}cm^{-3})	γ_K (W/cm^{2}K) typical	F_h (d=100 nm) typical	$\omega_{max}/2\pi$ (GHz) typical
Zn	0.88	9.3 10^{-8}	11	13	0.02	0.1	62
Al	1.18	1.3 10^{-8}	14	18	0.05	0.5	35
In	3.41	9.5 10^{-11}	10	12	1.1	8	8
Sn	3.72	2.7 10^{-10}	12	15	1.5	2	32
Nb	9.25	1.8 10^{-11}	6	5.6	10	9	19

In table 2 we have included typical values for ω_{max}, calculated for a film thickness of 100 nm. When these values are compared with the minimum frequency from equation (18), it is found that the frequency range for enhancement is very small for some materials, unless much attention is paid to cooling the sample (thin film, good choice of substrate).

MICROWAVE ENHANCEMENT; EXPERIMENTAL RESULTS

In this section, as in the previous one, we limit ourselves to those experiments where the spatial dependence in the sample should not play an important role. There are three types of measurement that fall in this category: measurement of the gap in the excitation spectrum with a tunnel junction, measurement of the critical pairbreaking current in a one-dimensional strip and measurement of the order parameter in a cylindrical film with a squid. We will avoid discussing experimental details, although they are in some cases important for the interpretation of the results obtained.

Microwave coupling

In all experiments the radiation is coupled into a thin film. The classical skin depth in the normal state near T_c at 10 GHz is about 0.25 µm in Al and 0.5 µm in Sn. The penetration depth at zero temperature is of the order of 0.1 µm in dirty films of both materials, near T_c considerably higher. In practical cases with a film thickness around 10 nm, it is reasonable to assume that the field is homogeneous from surface to opposite surface.

In the microwave experiments the power going down into the cryostat P_{rf} is known, but not the field strength α_ω in the sample. The fact that the absolute value of α_ω is not known is a major drawback for the comparison with theory. The impedance of samples is usually low and the size of the samples much smaller than the wavelength. The field might therefore couple to the leads rather than to the sample itself and present an rf current source to the sample. This may lead to an inhomogeneous distribution of the field across the width of the sample or disturb the assumed proportionality between P_{rf} and α_ω.

'Percentages' of enhancement

The experimentally observed enhancement is often expressed in a percentage based on $\delta\Delta/\Delta^o$ or $\delta I_c/I_c^o$. When T is close to T_c this percentage can be very high, above T_c it is not defined. It would be much better if all experiments were interpreted in terms of an apparent relative shift in T_c. We will define an experimental enhancement parameter δ_{exp}, based on the Ginzburg-Landau equation. In the case of gap measurements, δ_{exp} is defined by:

$$0.106 \left(\frac{\Delta}{k_B T_c}\right)^2 = \frac{T_c - T}{T_c} + \delta_{exp} \qquad (27)$$

In this way temperatures above T_c need no special definition. The value of δ_{exp} is in proportion with the magnitude of the basic effect.

Direct measurement of the energy gap

For a comparison with theory a direct measurement of the gap by means of a tunnel junction is clearly the most attractive. At this time only one paper has been published in which gap enhancement was observed. Kommers and Clarke (1977) irradiated $Al\text{-}Al_2O_3\text{-}Al$ junctions with 10 and 12 GHz radiation. In equilibrium the gaps in the two films Δ_1 and Δ_2 were slightly different. They were determined from the structures at $\Delta_1-\Delta_2$ and at $\Delta_1+\Delta_2$ in the plot of dV/dI vs V. When radiation was applied, Δ_1 and Δ_2 both increased. However, immediately additional structure appeared, partly associated with photon-assisted tunneling. The important features at Δ_1 and Δ_2 and $\Delta_1+\Delta_2$ had to be followed closely through successive plots at increasing microwave levels to allow confident identification. The results obtained will be used later in our evaluation. Dahlberg, Orbach and Schuller (1979) tried to repeat the same measurements and compare the results with critical current enhancement in the same film. Surprisingly, they did not observe any gap enhancement, although the critical current did increase. In the experiment structure was observed in dV/dI that moved with P_{rf}, but this structure was associated with 'classical rectification'. For a discussion of the effects of microwaves on tunnel junctions, apart from enhancement in the electrodes, we refer to the previous chapter by Schuller. It is at this time not clear why gap enhancement was seen in one experiment and not in the other. Very recently Hall, Holdeman and Soulen (1980) performed a series of measurements on aluminum tunnel junctions included in a strip line configuration for coupling to microwave radiation. They observed enhancement of the gap at frequencies near 3.7 GHz.

For a good comparison with theory, the experimental values of Δ^2 should be plotted vs T or P_{rf}, rather than Δ. If Δ^o is the equilibrium gap which we will also define above T_c with:

$$\left(\frac{\Delta^o}{k_B T_c}\right)^2 = 9.40\,\frac{T_c - T}{T_c}$$

we see from (27) that

$$\delta_{exp} = \frac{0.106}{(k_B T_c)^2}\,\{\Delta(T,P_{rf})^2 - \Delta^o(T)^2\} \tag{28}$$

Measurement of the critical pair-breaking current

The critical current of a one-dimensional strip is relatively easy to measure and is a good measure of the strength of the superconducting effects. According to equilibrium theory the critical current is:

$$I_c = \Gamma\left(\frac{T_c - T}{T_c}\right)^{3/2}$$

where $\Gamma = 1.03\ A\ H_c(0)(k_B T_c/\hbar\rho_n)^{\frac{1}{2}}$ and A is the cross-sectional area. In equilibrium I_c is proportional to Δ^3, this result is found when the current is included in the Ginzburg-Landau equation for the gap and I_s is maximized within the constraints of that Ginzburg-Landau equation. One might expect that in non-equilibrium a similar procedure could be followed, calculating the depression of Δ by the current from equation (25). However, an extra contribution from the non-equilibrium distribution has to be included. Entin-Wohlman (1980) has recently calculated the critical current in the microwave-enhanced situation. Similar calculations have been performed by Holzmann in Karlsruhe.

For critical current measurements we define δ_{exp} again as the relative shift of T_c needed to obtain the same enhancement of I_c in equilibrium:

$$\delta_{exp} = \Gamma^{-2/3} \{I_c(T,P_{rf})^{2/3} - I_c^o(T)^{2/3}\} \tag{29}$$

Γ is best determined from the experimentally obtained values of $I_c^o(T)$. For a comparison with theory $I_c{}^{2/3}$ is the quantity to be used.

The equilibrium I-V characteristic of a one-dimensional strip is very hysteretic because of heating effects. I_c is usually measured by sweeping the current, reading I_c from an oscilloscope screen. Measuring I_c always involves a transition into the normal or a dynamic intermediate state, which may have consequences. By far the most experimental results on microwave enhancement have been obtained from one-dimensional strips. They were performed by Latyshev and Nad' (1976, 1978), Klapwijk et.al. (1976, 1977, 1978), Pals (1977a, 1977b), Pals and Dobben (1979a, 1979b), Dmitriev and Khristenko (1978, 1979).

Measurement of the order parameter

Pals and Dobben (1980) measured the magnetic flux ϕ contained in a cylindrical aluminum film. Because of fluxoid quantization for a closed contour in the film, there is a direct connection between ϕ, the current density j and the order parameter ψ in the film. A second relation between j and ϕ is given by the inductance of the ring. As a consequence, a change in ψ results in a known change of ϕ. ϕ was measured with a second loop connected to a squid. Microwave radiation was applied to the aluminum film and the resultant variation of ϕ measured. The results indicated that ψ was enhanced. Although a remark to the contrary is made by the authors, the results are in reasonable quantitative agreement with the predictions from the Eliashberg theory. Because few data have been obtained so far, we will not include the results from these measurements in an further analysis. The method seems promising, however, to yield results complementary to the gap measurements with tunnel junctions which should equally well be suitable for comparison with theory.

Dependence on microwave power

In Figure 8a we reproduce results obtained by Kommers and Clarke (1977) with an aluminum tunnel junction, irradiated with 10 GHz radiation. We have taken the average of the two gaps of the films. In Fig. 8b the same data are plotted as δ_{exp} versus P_{rf}, where δ_{exp} was calculated with (28). We will compare δ_{exp} with the predictions of the Eliashberg theory as expressed in the Ginzburg-Landau equation (14). For each value of P_{rf} the experimentally observed Δ can be used to calculate $G(\Delta/\hbar\omega)$, which possibility is a great advantage in the evaluation. The dotted line in Fig. 8b

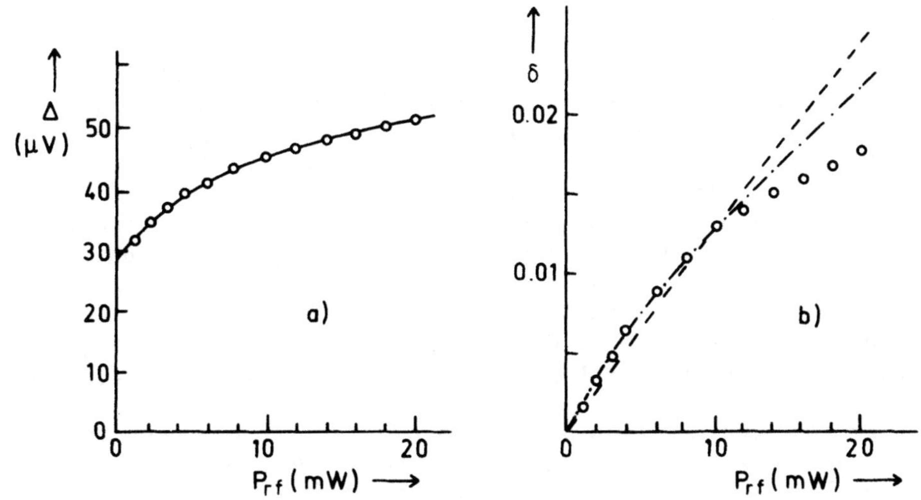

Fig. 8. Results of gap measurements of Kommers and Clarke (1977). a. as reported b. transformed to δ_{exp}. The circles give experimental points. Dashed line: theoretical curve for δ_E with $\alpha_\omega/\gamma_r = 0.05$ at $P_{rf} = 10$ mW; dash-dotted line: theoretical curve for $\delta_E - \delta_{\omega^2} - \delta_h$ with $F_h = 0.9$ and $\alpha_\omega/\gamma_r = 0.15$ at $P_{rf} = 10$ mW.

gives δ_E, assuming that P_{rf} equal to 10 mW corresponds to $(\alpha_\omega/\gamma_r) = 0.05$. We have attempted to obtain a better fit by taking into account the corrections (16) and (24) for second order terms in ω and heating effects respectively. Together they are:
$\delta_{\omega 2} + \delta_h = -(0.17 + F_h)(\hbar\omega/k_BT_c)^2\ (\alpha_\omega/\gamma_r)$. The dash-dotted line was obtained with $F_h = 0.9$ and $(\alpha_\omega/\gamma_r) = 0.15$ at $P_{rf} = 10$ mW. The curve fits well at smaller power levels but still deviates strongly from the experimental points above 10 mW. The value of 0.9 for the heating parameter F_h corresponds to a thermal boundary conductance of 0.04 W/ cm^2K (the total film thickness is 600 nm).

Looking at Figure 8b, it is clear that no definite conclusions can be drawn. The two theoretical curves have been obtained with values of (α_ω/γ_r) which were different by as much as a factor 3. Many of the results obtained in enhancement experiments can be interpreted with either one value for α_ω/P_{rf} or a much higher value for α_ω/P_{rf} together with heating corrections. At lower frequencies heating should be relatively less important. The data obtained with I_c enhancement at relatively low power levels are qualitatively similar.

At high power levels it is very difficult or impossible to determine Δ directly. I_c measurements are possible and have been performed. In figure 9 we give, as an example, results obtained by Pals (1977a) for the critical current of an aluminum strip as a function of P_{rf} (4 GHz radiation). For high values of α_ω, the linear Eliashberg theory cannot be adequate. One should turn to the numerical results of Chang and Scalapino. When we compare figure 9 with figure 7b, obviously good qualitative agreement is obtained (note the logarithmic power scale in figure 9). Enhancement is obtained across four orders of magnitude of P_{rf}. From Pals' data δ_{exp} can be calculated with Eq. (29) by taking into account the values of $I_c^o(T)$ at $P_{rf} = 0$. We calculate that the highest δ_{exp} in the figure is found at the top of the 1.23 K curve, there $\delta_{exp} = 0.067$.

A further confirmation of the importance of phonon non-equilibrium, as suggested by Chang and Scalapino, was obtained by determining $I_c(P_{rf})$ of a film in vacuum and in direct contact with liquid helium. Both Klapwijk et al. (1978) and Pals and Dobben (1979b) found higher maxima for I_c in contact with helium, when the phonon escape time is shorter.

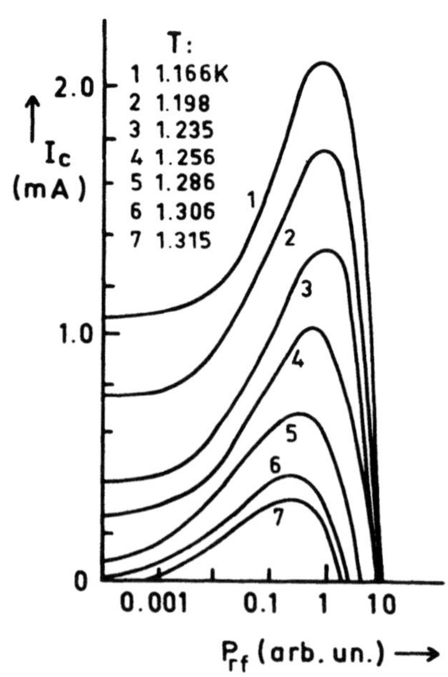

Fig. 9. Critical current of an aluminum strip as a function of microwave power. Frequency 4 GHz (Pals 1977a).

Frequency dependence, materials

Experimentally it is not possible to vary ω while keeping α_ω constant (although P_{rf} can be corrected). One is therefore obliged to use specific features in the plots of Δ or I_c versus T or P_{rf}. In this spirit Klapwijk et al. (1977) give the maximum T_c enhancement obtained in an aluminum sample as a function of frequency. They found the highest enhancement at 3 GHz. However, in later similar experiments in our group on other samples, the optimum frequency turned out to be higher. As heating is quadratic in ω while δ_E is linear, a change in film thickness or cooling conditions may change the frequency dependence.

Pals and Dobben (1979b) plot I_{cm}, the maximum of $I_c(P_{rf})$, as a function of temperature for frequencies between 0.5 and 36 GHz. The sample was an Al strip. Below $T/T_c = 0.94$, I_{cm} keeps increasing with increasing frequency. Nearer T_c, 7 GHz turns out to render the highest I_{cm} values. When the frequency exceeds 7 GHz, I_{cm} decreases. Latyshev and Nad' (1976) give an interesting plot of the frequency limits between which enhancement was observed in several tin strips. The data are reproduced in Figure 10. The lower set of experimental points refers to the minimum frequency. This ω_{min} should, according to the Eliashberg theory, be given by Eq.(17): $\omega_{min}\, G(\Delta/\hbar\omega_{min}) = 2\pi\tau_E^{-1}$. Latyshev and Nad' used the approximation (13) for high values of $(\Delta/\hbar\omega)$ to determine ω_{min}. The result is given in Figure 10. The agreement with experiment is reasonable. The analysis of the high frequency limit is slightly more complicated. Latyshev and Nad' remark that the data points near T_c run parallel to the dash-dotted line that indicates $\hbar\omega = 2\Delta^0$. They conclude that pair breaking dominates in that region (when $\hbar\omega$ exceeds $2\Delta^0$ no continuous enhancement of Δ from zero is possible). At lower temperatures there

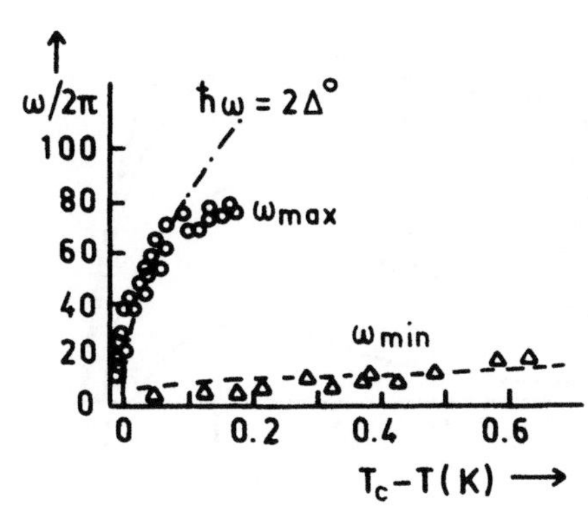

Fig. 10. Minimum frequency ω_{min} and maximum frequency ω_{max} for tin strips (Latyshev and Nad', 1976). Dashed line: theoretical ω_{min}. Dash-dotted line: $\hbar\omega = 2\Delta°$.

is a cut-off at about 75 GHz. In contrast with Latyshev and Nad' we expect that this cut-off is associated with heating. Equation (26) can be used to determine ω_{max}. The samples are about 60 nm thick, Y_k is estimated by the authors to be 1.5 W/cm^{2}K. This gives F_h = 1.3 and the cut-off frequency should be 54 GHz. We consider this value as in good agreement with the experimental value of 75 GHz because Y_k is not well known.

The frequencies that give best results in tin are higher than those for aluminum, as expected from the values of τ_E.As shown by the F_h factors in Table 2, heating will be much more serious in tin. To obtain the same α_ω/γ_r,α_ω has to be 200 times larger. In indium this will be even more pronounced. For indium ω_{max} is about 8Hz, while the minimum frequency is 3GHz. Accordingly, so far no enhancement has been reported on in indium in homogeneous samples. In contrast zinc should be an interesting material to explore.

<u>Temperature dependence, region above T_c</u>

At low temperatures there are few quasiparticles to block states for pairing, and redistribution of those few quasiparticles will have little effect. Pals and Dobben (1979b) have extended their measurement of I_c enhancement in aluminum strips and find that the enhancement decreases with lower T (Figure 11). Never does the maximum enhanced critical current at any temperature or power level exceed the equilibrium I_c at zero temperature, as is to be expected.

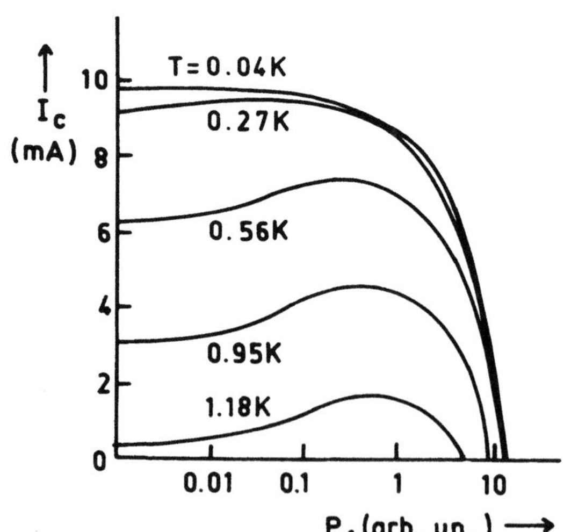

Fig. 11. Critical current of aluminum strip as a function of microwave power (12 GHz). Measurements extended to low temperatures by Pals and Dobben (1979b).

Of particular interest is the temperature region above T_c. As discussed previously, theory predicts that no stable superconducting states exist with Δ values below $\hbar\omega/2$. In the only available measurements of Δ above T_c by Kommers and Clarke (1977), using 10GHz radiation, no values of Δ were seen below 20 μV, in agreement with the theory. However, a more systematic investigation of Δ enhancement above T_c is clearly desirable. As for the critical current, theory predicts that no discontinous behaviour should occur (B. Holzmann, A. Schmidt, to be published). Experimentally, a discontinuous jump is often seen in the critical current but may be associated with hysteretic effects that will be discussed later. Pals and Dobben (1979b) state that in their measurements I_c decreased to zero continuously when T was increased. However, the current resolution in that measurement was limited.

In figure 12 the differential resistance of an aluminum strip is given, measured with a very small current modulation (Mooij, Lambert, Klapwijk; 1980).

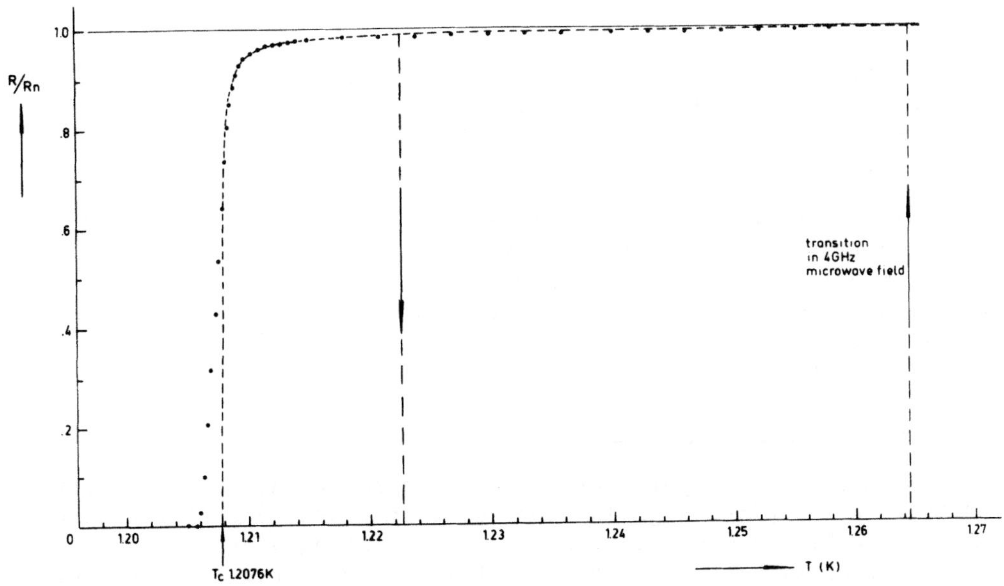

Fig. 12. Superconducting-normal transition of an aluminum strip. The differential resistance at zero bias current was measured with a current amplitude of 10 nA. The black dots give experimental values without microwaves. The dotted line through the dots is a fit to fluctuation theory, used to determine T_c. In the presence of 4 GHz microwave radiation the transition is hysteretic (Mooij, Lambert, Klapwijk; 1980).

Without microwaves there is a continuous, fluctuation-rounded transition. With microwaves a discontinuous hysteretic jump between the superconducting and normal states is observed. This behaviour should be expected from the Eliashberg theory as discussed. At temperatures slightly above T_c (typically 20 mK in Al) it is possible to set the temperature, then turn on the microwaves and obtain superconductivity. At higher temperatures the superconducting state can only be reached by starting in the superconducting state at lower temperatures with the microwaves on and subsequently increasing the temperature. The highest T_c enhancement that has been reported is 100 mK in aluminum, by Pals and Dobben (1979b).

A second explanation is possible for the hysteretic effects. If the dissipation caused by the microwaves is higher in the normal state than in the superconducting state, the effective temperature is higher in the normal state at the same bath temperature. This may lead to instabilities. However, it is certainly not obvious that heating is lower in the superconducting state. If the E field is coupled in, σE^2 will be higher when the sample is superconducting. The hysteretic effects should then strongly depend on the manner in which the rf field is applied. Moreover, in the equilibrium theory at least (Tinkham,1975, chapter 2), there is no discontinuous jump in microwave absorption at T_c, as in the dc resistance. The whole problem calls for additional research (careful analysis of hysteresis at different frequencies where the influence of heating should be different, more measurements of Δ above T_c).

Falco, Werner and Schuller (1980) have questioned the evidence available for T_c enhancement on the basis that the real T_c of samples might be higher than supposed from the $I_c(T)$ plot. In view of all old and new (Mooij, Lambert and Klapwijk, 1980) experimental data we feel safe to state that superconductivity has really been observed above the equilibrium T_c.

Measurement of relaxation times

An interesting experiment has been performed by Pals and Dobben (1979a). They applied signals from two generators, one at frequency ω and the other at $\omega + \delta\omega$, to a sample consisting of an aluminum strip. The critical current of the strip was measured. The applied microwaves can be considered as a field at frequency ω, amplitude modulated with frequency $\delta\omega$. The critical current will be modulated with the same frequency $\delta\omega$. This I_c modulation was measured as a function of modulation frequency. The response turned out to contain two time constants, τ_1 and τ_2, where τ_1 was of the order of 100 ns and τ_2 varied from 1 to 100 ns. The authors interpret τ_1 as the quasiparticle recombination time. This is supported by the fact that τ_1 shows the same dependence on temperature as expected in equilibrium. The nature of τ_2 is unclear. As a whole the theoretical interpretation of both τ_1 and τ_2 requires a further detailed analysis.

ENHANCEMENT BY PHONONS

Theory

The interaction of quasiparticles with photons (microwaves) and with phonons have very similar consequences. The same mechanism of redistribution of quasiparticles to higher energies is possible and enhancement of the gap is obtained. However, the interaction with longitudinal phonons is characterized by different coherence factors in comparison with those that characterize the interaction with a vector field such as electro-magnetic radiation. The coherence factor for scattering of quasiparticles between E and E' for longitudinal phonons is $\frac{1}{2}(1 - \Delta^2/EE')$, to be compared with $\frac{1}{2}(1 + \Delta^2/EE')$ for microwaves. For scattering near the gap the phonon coherence factor is small. The opposite is true for pair breaking processes. The phonon coherence factor is $\frac{1}{2}(1 + \Delta^2/EE')$ whereas it is $\frac{1}{2}(1 - \Delta^2/EE')$ for photons. When the phonon energy $\hbar\Omega$ exceeds 2Δ even stronger effects are expected than with microwaves.

Taking these different coherence factors into account, the Ginzburg-Landau equation can be derived in exactly the same way as for microwaves. We will characterize the phonon power with α_Ω. The Ginzburg-Landau equation for phonon enhancement is (Eckern et.al., 1979):

$$\frac{7\zeta(3)}{8\pi^2}\left(\frac{\Delta}{k_BT_c}\right)^2 = \frac{T_c - T}{T_c} + \frac{1}{4}\left(\frac{\alpha_\Omega}{\gamma_r}\right)\left(\frac{\hbar\Omega}{k_BT_c}\right) G\left(\frac{\Delta}{\hbar\Omega}\right) \qquad (31)$$

where

$$G(u) = \begin{cases} -2\pi u(1-u^2)^{-\frac{1}{2}} & u < \frac{1}{2} \\ \dfrac{2}{u+\frac{1}{2}}\left[K(k) - 4u^2\{\Pi(\alpha^2,k) - K(k)\}\right] & u > \frac{1}{2} \end{cases} \qquad (32)$$

$k = (u - \frac{1}{2})/(u + \frac{1}{2})$, $\alpha^2 = (2u + 1)^{-2}$.

Equation (31) has exactly the same form as Eq.(14), but the G function is different, in particular for $u < \frac{1}{2}$. In figure 13 the result of Eq. (31) is given for one value of α_Ω. The limits for approaching $\Delta = \hbar\Omega/2$ from below and from above are not the same. Between a temperature T_{min} and T_c there is a solution with small Δ. In contrast with the microwave case, for this solution Δ increases with decreasing temperature. The solution is not a priori unstable, but will, according to Eckern et al. (1979), show a strong tendency towards a diffusive spatial instability (see chapter 11). The solution with highest Δ is at least locally stable up to a temperature T_{max}. We see that:

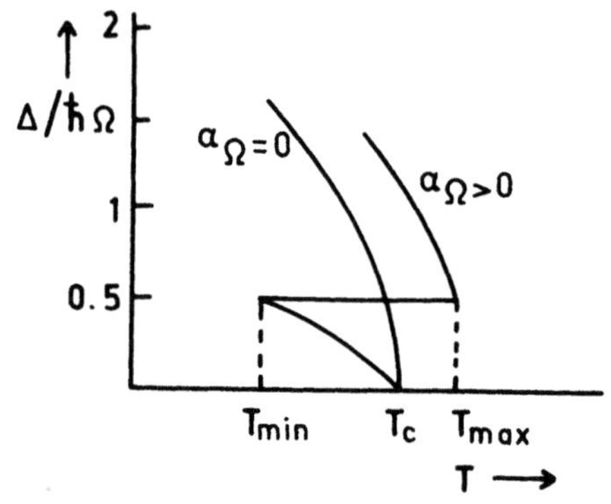

Fig. 13. Gap according to Eq.(31), as a function of temperature.

$$\frac{T_{min}}{T_c} = 1 - 0.027 \left(\frac{\hbar\omega}{k_B T_c}\right)^2 - 0.91 \left(\frac{\alpha_\Omega}{\gamma_r}\right)\left(\frac{\hbar\Omega}{k_B T_c}\right) \quad (33)$$

$$\frac{T_{max}}{T_c} = 1 - 0.027 \left(\frac{\hbar\omega}{k_B T_c}\right)^2 + 0.66 \left(\frac{\alpha_\Omega}{\gamma_r}\right)\left(\frac{\hbar\Omega}{k_B T_c}\right)$$

The stable branch with high Δ for $\Delta/\hbar\Omega$ larger than 0.5 is described by a G function which is considerably smaller than in the case of microwave enhancement. The highest value of G is 2.65 at $\Delta/\hbar\Omega = 0.5$.

The analogy between microwave and phonon enhancement is very strong. Similar correction terms have to be included in the Ginzburg-Landau equation. The only real difference, due to the different coherence factors, is found in the G function. The negative values of G at $\Delta/\hbar\Omega$ below 0.5 lead to a high free energy barrier between the normal and superconducting states above T_c. A Chang and Scalapino-like numerical calculation of enhancement by longitudinal phonons has not been made yet, but is very desirable.

Experiments

Very few experimental results have been obtained on phonon-enhancement. Some authors have seen indications of enhanced gaps in experiments on the absorption of longitudinal and transverse sound in superconductors (Lyall, Meredith and Dobbs, 1976; Fil', Denisenko, Bezuglyi, Priogov, 1976). Tredwell and Jacobsen (1975, 1976) measured the critical current of point contacts and microbridges which were subjected to 10 GHz phonon pulses. The samples consisted of a quartz rod which was placed in a 10 GHz microwave cavity on one end. On to the other end an aluminum film was evaporated. A microbridge was scratched in the film or a tin point pressed to it to form a point contact. The pulsed measuring method made it possible to distinguish phonon effect, which due to the relative low speed of sound arrived later, from microwave effects.

Figure 14 gives results for a microbridge. This particular microbridge had a critical current proportional to (1-t), which suggests that I_c was proportional to Δ^2. Figure 14a gives the data as plotted by Tredwell and Jacobsen. In Figure 14b we plot the same experimental data as δ_{exp} versus 1-t. In this case $\delta_{exp} = (1-t)\ I_c/I_c^o$. We see that δ_{exp} is of the order of 0.003. It decreases with increasing (T_c - T). This is qualitatively in agreement with theory: as Δ increases, $G(\Delta/\hbar\Omega)$ decreases. It is also interesting to note that the equilibrium value of the gap Δ^o equal to $\hbar\Omega/2$ is found about 5 mK below T_c (indicated with a dotted line in Fig.

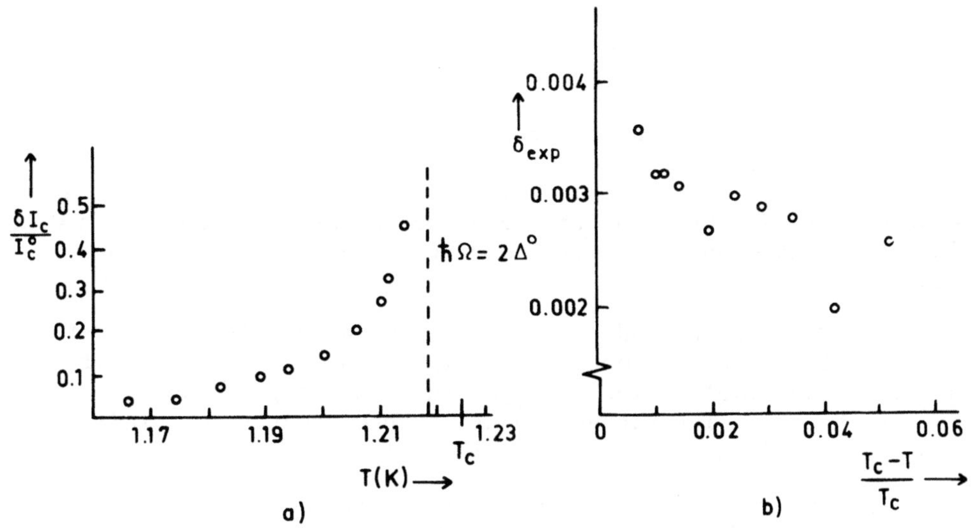

Fig. 14. Critical current enhancement of a microbridge by 10 GHz phonons as observed by Tredwell and Jacobsen (1975). a. Data given by authors. b. Same data expressed as δ_{exp}.

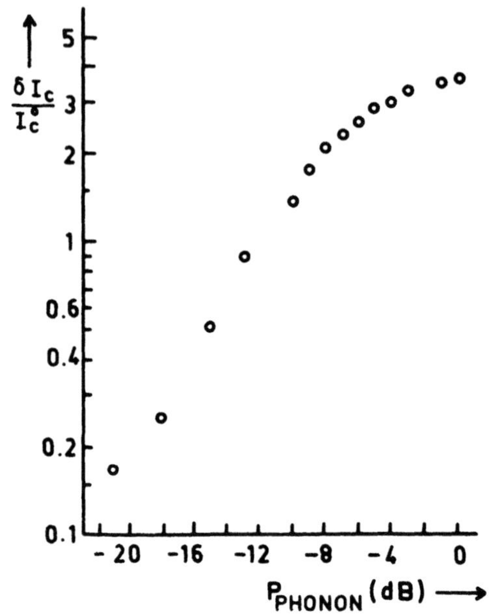

Fig. 15. Enhancement of the critical current of a point contact by 10 GHz phonons as a function of power (Tredwell and Jacobsen, 1976).

14a). Tredwell and Jacobsen never observed enhancement at temperatures above this value. Accordingly, T_c enhancement was never observed. This may be an indication of the more destructive consequences of pair-breaking to phonon-enhancement in comparison with microwave enhancement, associated with the different coherence factors.

In Fig. 15 we reproduce Tredwell and Jacobsen's data for the enhancement of a point contact at $T/T_c = 0.97$ as a function of power. The available power was apparently not high enough to reach saturation. The highest point reached amounts to $\delta I_c/I_c^o = 3.6$. This corresponds to $\delta_{exp} = 3.6(1-t) = 0.108$.

Very recently, preliminary results have been obtained by Seligson and Clarke (private communication), who observed enhancement of the gap of aluminum by 9GHz phonons, using a tunnel junction. It will be of great interest to obtain more experimental data and perform a comparison with the theory.

ENHANCEMENT BY TUNNELING PROCESSES

Tunneling as such and the tunneling mechanisms that may lead to gap enhancement have been discussed in chapter 5. Enhancement is possible by extraction of quasiparticles in an asymmetric junction (S-I-S', the gap in the film with lower T_c is enhanced). This mechanism was first suggested by Paramenter (1961) and later discussed by Kirichenko, Peskovatsky and Seminozhenko (1979). Experimental observations were obtained by Chi and Clarke (1979). Enhancement by tunneling is also possible in symmetric (S-I-S) tunnel junctions by means of the Eliashberg mechanism. Here Peskovatskii and Seminozhenko (1976) calculated the expected enhancement, keeping the phonons in equilibrium. Chang (1978) performed numerical calculations of the enhancement, similar to those of Chang and Scalapino for microwave enhancement, where the phonon distribution was also non-equilibrium. In chapter 9 these calculations are discussed. Experimentally, gap enhancement by tunneling in symmetric junctions was observed by Gray (1978). In view of the discussions in chapters 5 and 9 we will limit ourselves in this chapter to a comparison with the other enhancement processes and a short discussion of the results obtained experimentally.

The role of $\hbar\omega$ and $\hbar\Omega$ in microwave or phonon enhancement is played by eV in tunneling experiments. The conductivity of the tunnel junction, a fixed quantitiy for any sample, decides the strength of the effects. It is a disadvantage with respect to microwave or phonon enhancement that the equivalent of the power cannot be varied. On the other hand the conductivity is well known and the problem of determining α_ω or α_Ω does not occur. This makes tunneling enhancement very attractive for a detailed quantitative comparison with theory.

By Schön and Tremblay (1979) the enhancement by tunneling has been treated in a way completely similar to the one for microwave and phonon enhancement. A Ginzburg-Landau equation can be written down (we have changed the notation slightly to obtain higher similarity):

$$\frac{7\zeta(3)}{8\pi^2}\left(\frac{\Delta}{k_BT_c}\right)^2 = \frac{T_c - T}{T_c} + \frac{1}{4}\left(\frac{\alpha_T}{\gamma_r}\right)\left(\frac{|eV|}{k_BT_c}\right) G\left(\frac{\Delta}{\Delta_G}, \frac{|eV|}{\Delta_G}\right) \tag{34}$$

in which V is the injection voltage, Δ and Δ_G are the gaps in the sample and generator films and α_T is given by:

$$\alpha_T = \frac{\hbar}{4e^2\ d\ R'\ N(0)} \tag{35}$$

d is the thickness of the film, R' the resistance per unit area of the junction. Both the Parmenter and the Eliashberg mechanisms are contained in equation (34). The G function is more complicated here, reflecting the influence of Δ_G. G is zero for eV below $\Delta + \Delta_G$, it contains discontinuities at $\Delta = \pm (\Delta_G \pm eV)$.

It is interesting to demonstrate the similarity with the previously treated other enhancement mechanisms by considering the case where $\Delta = \Delta_G$ at all voltages. Although not usual in actual experiments, this situation would be obtained if the generator and sample films were identical in material and thickness. For this particular case the G function becomes:

$$G(u) = \begin{cases} 0 & u < \frac{1}{2} \\ 2[K(k) - \frac{1}{4}\{\pi(\alpha^2,k) - K(k)\}] & u > \frac{1}{2} \end{cases} \tag{36}$$

where $u = \Delta/|eV|$, $k = (1 - 1/4u^2)^{\frac{1}{2}}$, $\alpha^2 = 1 - 1/2u$. This G function is at all values of u intermediate between the two G functions for microwave and phonon enhancement.

As far as heating is concerned, its influence can be easily calculated with a similar simple model as used previously. If F_h has the same value as in (23), where now d is the thickness of the sample film and Y_K the effective thermal conductance per unit area from the whole packet of three metal films to substrate and helium, the heating correction term is:

$$\delta_h = -2\, F_h \left(\frac{\alpha_T}{\gamma_r}\right)\left(\frac{eV}{k_B T_c}\right)^2$$

The factor 2 reflects the fact that disipation occurs in both the sample film and the generator film.

The experimental arrangement for measurements of gap enhancement by tunneling has been described in chapter 5. When studying the results obtained one should keep in mind that the voltage between the generator and sample films corresponds to the frequency and not to the power in our discussion of enhancement by microwaves and phonons. Gray (1978) reports enhancement $\delta\Delta/\Delta^o$ of up to 10% near T_c where Δ^o is small. In his recent results (chapter 5) Gray finds up to 1% for $\delta\Delta/\Delta^o$ at t = 0.97, 0.4% at t = 0.96, 0.3% at t = 0.95 and 0.2% at t = 0.85. We have calculated the δ_{exp} parameters for these results and find δ_{exp} to be $6\ 10^{-4}$, $3\ 10^{-4}$, $3\ 10^{-4}$ and $6\ 10^{-4}$ for the four temperatures respectively. These are low values in comparison with the results obtained with microwaves and with phonons. This may be due to the following circumstance: once a sample is made there is only one parameter that can be varied: the generator voltage. In contrast, with microwave enhancement both frequency and power can be varied until optimum effects are obtained. The equivalent experiment with tunnel junctions would consist of a set of measurements on a series of samples with varying conductivity of the generator junction.

In their experiments on asymmetric junctions, Chi and Clarke (1979) obtained enhancement $\delta\Delta/\Delta^o$ of up to 40% at a reduced temperature of 0.998. With a critical temperature of 1.35 K this corresponds to $\delta_{exp} = 0.002$. The authors compare their results with a simplified theoretical calculation and find reasonable agreement at low generator voltages, but above $(\Delta_1 - \Delta_2)/e$ the experimental enhancement is significantly smaller than would follow from the theory. Chi and Clarke believe that recombination phonons from the generator film may break pairs in the sample film. Heating effects are also very likely important.

In none of the experiments on tunneling enhancement, stimulated superconductivity above the critical temperature was observed.

MICROWAVE ENHANCEMENT IN WEAK LINKS

Microwave enhancement of the critical current of weak links has been observed in a wide variety of these junctions: constriction microbridges, proximity effect bridges, superconductor -normal-superconductor bridges, point contacts. A detailed discussion is to be found in the review by Dmitriev and Khristenko (1978).

Apart from the Eliashberg mechanism several theoretical explanations have been offered for the observed effects. The first, already suggested by Dayem and Wiegand, was the fluctuation hypothesis. It was supposed that the critical current of the bridge was reduced by fluctuations. The microwaves had a 'synchronizing' effect, suppressing the fluctuations and restoring the critical current to its true value. No explanation was offered for the fluctuation-suppressing action of the microwaves. In view of the detailed analysis of fluctuation effects in superconducting films both below and above T_c that was performed later, the fluctuation hypothesis as an explanation of the Dayem -Wyatt effect seems now of historical interest only.

Notarys, Yu and Mercereau (1973) modified and extended the Aslamazov-Larkin calculation for the time-dependent behaviour of microbridges. Starting with a phenomenological supposition they arrived at a model in which the supercurrent is time dependent in a very complicated way. It turns out that the supercurrent through the bridge in the dynamic state contains a dc component, which accounts for the apparently higher critical current. This explanation does not contain any non-equilibrium elements and it should be possible to confirm the model with a calculation based on time-dependent Ginzburg-Landau equations in the presence of a microwave bias voltage. Such a calculation has not been performed yet. Since the heuristic basis of this model is rather weak it seems reasonable to disregard the Notarys, Yu, Mercereau model for the time being.

Lindelof (1975) proposed a 'smearing' model, based on quasiparticle non-equilibrium. The order parameter (density of Cooper pairs) is smaller in the middle of a microbridge when current is passed through it. In a two-fluid picture the density of quasiparticles must be correspondingly higher in the middle. In an rf field the quasiparticles will be smeared out over a larger region, the pairs will on the average move into the narrow part of the bridge. If the relaxation time is not short in comparison with the rf period a stationary non-equilibrium will result which is strongly position dependent. As pairs move towards the region with high current density, the critical current will be enhanced. The Lindelof model contains interesting elements that are of particular value when weak links are considered with sharp, well-defined boundaries, i.e. proximity effect bridges, S-N-S bridges.

The most generally applicable explanation for the Dayem-Wyatt effect is clearly the Eliashberg mechanism. The gap is enhanced in the film at all positions, which directly gives a higher critical current. Because inevitably some microwave current will be induced through the junction, the critical current will also have a tendency to decrease and to show periodic oscillations as pictured in figure 1a. The product of the function of Fig. 1a and a monotonically increasing intrinsic critical current looks like the picture of Fig. 1b. If the junction character dominates I_c will increase only slightly before going down. If the film character dominates (as it certainly did in the earliest, uniform thickness microbridges) the initial enhancement will be relatively high.

Aslamazov (1979) has calculated the effect of microwaves on a bridge structure where the critical current temperature of the bridge is somewhat lower than the T_c of the banks T_{co}. The quasiparticles with energy below the gap in the banks Δ_o will be Andreev reflected and consequently contained in the bridge while quasiparticles with $E > \Delta$ can move out freely. As a result, the microwave field will not only redistribute the quasiparticles to higher energy levels through the direct action of the electric field (Eliashberg mechanism) but also by the 'trembling of the potential well'. As a result, the effective critical temperature of the bridge may increase to a value near T_{co}. This description is related to that of dynamic enhancement of the critical current, discussed in chapter 8.

An example of an enhancement effect, observed experimentally, that is not readily explained by the Eliashberg theory alone, is given in igure 16. It has been taken from Warlaumont, Brown, Foxe and Buhrman (1979). We see the critical current expressed as I_cR of a lead-copper-lead bridge, 1 μm wide, 2 μm long, 0.15 μm thick. The length of the bridge is equal to 30 times the coherence length in copper at T_c (lead). Consequently, the critical current is extremely small except at very low temperatures (the coherence length is proportional to $T^{-\frac{1}{2}}$). The theoretical expectation is given as a solid line, experimentally Ic is drowned in noise. Irradiated with 10 GHz the experimental points are obtained. For this enhancement in a material with zero pairing potential no gap equation for a homogeneous medium can be adequate. The authors suggest an explanation along the lines indicated by Lindelof or Aslamazov, which we think very reasonable. No quantitative comparison with these theories has been made yet.

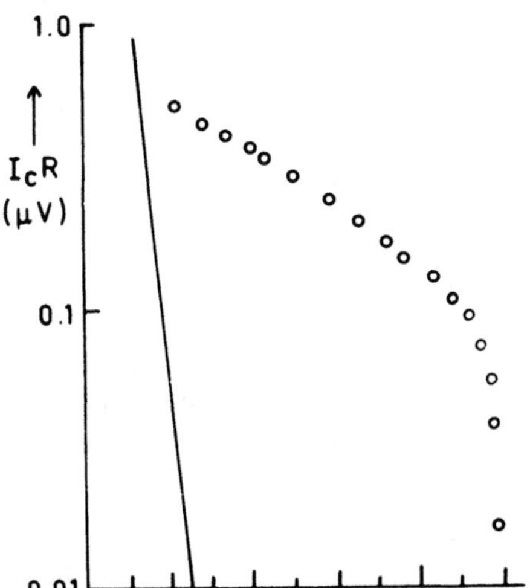

Fig. 16. Critical current (expressed as I_cR, R = 0.13Ω) of a lead-copper-lead bridge. Drawn line: expected value without microwaves (experimentally no I_c could be distinguished). Dots: enhanced values with irradiation by 10 GHz microwaves. (Warlaumont et al.; 1979).

CONCLUSIONS

The Eliashberg theory, as developed by Eliashberg with Ivlev and Lisitsyn, provides a good theoretical framework for the description of gap enhancement. By various other authors the theory has now been extended so far that a good quantitative comparison with experiment should be possible for enhancement by microwave radiation, phonons as well as tunneling injection currents. Heating or non-equilibrium phonon effects are always important, even at low power values. It is likely that at small power levels heating can be accounted for with a simple term in Eliashberg's Ginzburg-Landau equation, so that an analytical formulation can be maintained. For higher power levels the numerical approach of Chang and Scalapino gives a good description. This type of calculation is strongly needed for phonon enhancement as well.

Experimentally a reasonable amount of data has been obtained which are all in satisfactory qualitative agreement with the theory. However, from the experimental side the picture is far from complete. Microwave enhancement has been explored most extensively so far. The frequency dependence and the power dependence have been determined. Both types of measurement are seriously hampered by the fact that the field in the sample is not known absolutely. The stability of the superconducting state above T_c has still to be investigated in more detail. T_c enhancement by microwaves has been clearly established. For a more quantitative evaluation of microwave enhancement more direct measurements of the gap at various frequencies and power levels would be valuable. The maximum enhancement effect that has so far been obtained with microwaves, expressed in the apparent relative shift of the critical temperature δ_{exp}, is about 8%.

With phonons very few, but interesting, experimental results have been obtained. Only preliminary data are available for homogeneous samples, while earlier results, at one frequency only, refer to weak links. No T_c enhancement has been observed, which may be connected with the stronger influence of pair breaking by phonons in comparison with photons.

Enhancement by tunneling is possible by redistribution and by extraction of quasiparticles. Both effects have been observed experimentally. The magnitude of enhancement is relatively small (δ_{exp} = 0.2% with quasiparticle extraction, δ_{exp} = 0.1% with redistribution. No superconductivity has been observed at temperatures above T_c. Tunneling offers the best opportunity for a detailed quantitative comparison with theory. In tunneling the current and the particle energy eV are coupled for one specific junction. To obtain the equivalent of power and frequency dependence with microwaves and phonons, as with tunneling, measurements should be performed on junctions with widely varying conductivities.

All experiments on enhancement in homogeneous films have been limited to aluminum and tin. The 'power' needed to obtain a certain effect is proportional to the inverse relaxation time, whereas the heating effects (phonon escape times) become more serious for materials with lower critical temperatures. It will be valuable to explore other materials. In indium enhancement effects should be smaller, in zinc results similar to those in aluminum should be obtained.

Microwave enhancement in weak links is more difficult to analyze because of various processes occurring simultaneously. A further exploration of enhancement effects in superconductor-normal metal-superconductor bridges, both theoretically and experimentally, will be of interest.

ACKNOWLEDGEMENTS

My understanding of the subject has grown in years of numerous discussions with Dr. T.M. Klapwijk. Correspondence and discussions with Dr. A. Schmid have been extremely valuable, his care and patience in explaining the intricacies of the Eliashberg theory are greatly appreciated. I thank Dr. G. Schön for valuable comments and discussions, and in particular for his verification of the treatment of tunneling enhancement. I gratefully acknowledge useful discussions with Dr. J. Clarke.

REFERENCES

Aslamazov, L.G., 1979, Stimulation of superconductivity in an inhomogeneous bridge in a microwave field, Zh.Eksp.Teor.Fiz., 76: 1775; Sov.Phys. JETP, 49: 902.

Chang, J.J., 1978, Gap enhancement in superconducting thin films due to quasiparticle tunnel injection, Phys.Rev.B, 17:2137.

Chang, J.J. and D.J. Scalapino, 1977, Gap enhancement in superconducting thin films due to microwave irradiation, J. Low Temp.Phys., 29:477.

Chang, J.J. and D.J. Scalapino, 1979, Nonequilibrium superconductivity, J. Low Temp.Phys., 31:1.

Chi, C.C. and J. Clarke 1979, Enhancement of the energy gap in superconducting aluminum by tunneling extraction of quasiparticles, Phys.Rev. B, 20:4465.

Dahlberg, E.D., R.L. Orbach and I. Schuller, 1979, Microwave-induced effects on superconductors, J. Low Temp.Phys., 36:367.

Dayem: A.H. and J.J. Wiegand, 1967, Behavior of thin-film superconducting bridges in a microwave field, Phys.Rev., 155:419.

Dmitriev, V.M. and E.V. Khristenko, 1978, Stimulation and enhancement of superconductivity by external electromagnetic radiation (review), Fiz.Nizk.Temp., 4:821; Sov.J. Low Temp.Phys., 4:387.

Dmitriev, V.M. and E.V. Khristenko, 1979, Phase separation of a homogeneous superconductor in a uhf radiation field, Pis'ma

Zh.Eksp.Teor.Fiz., 29:758; JETP Letters 29:697.
Eckern, U., A. Schmid, M. Schmutz and G. Schön, 1979, Stability of superconducting states out of thermal equilibrium, J. Low Temp.Phys., 36:643.
Eliashberg, G.M., 1970, Film superconductivity stimulated by a high-frequency field, Pis'ma Zh.Eksp.Teor.Fiz., 11:186; JETP Letters, 11:114.
Entin-Wohlman, O., 1980, Effect of microwave radiation on the critical current in dirty superconductors, to be published.
Falco, C.M., T.R. Werner and I.K. Schuller, 1980, T_c enhancement?, Solid state comm., 34:535.
Fil', V.I., Denisenko, P.A. Bezuglyi and A.S. Pirogov, 1976, Nonlinear absorption of sound in superconductors, Fiz.Nizk.Temp. 2:1506; Sov.J.Low Temp., 2:733.
Gray, K.E., 1978, Enhancement of superconductivity by quasiparticle tunneling, Solid State Comm., 26:633.
Hall, J.T., L.B. Holdeman and R.J. Soulen, 1980, Microwave enhancement of superconductivity in aluminum tunnel junctions, Phys.Rev.Letters, 45:1011.
Ivlev, B.I., 1977, Spatial structure in a nonequilibrium superconductor, ZH.Eksp.Teor.Fiz., 72:1197; Sov.Phys.JETP, 45:626.
Ivlev, B.I. and G.M. Eliashberg, 1971, Influence of nonequilibrium excitations on the properties of superconducting films in a high-frequency field, Pis'ma Zh.Eksp.Teor.Fiz., 13:464 ; JETP Letters, 13:333.
Ivlev, B.I., S.G. Lisitsyn and G.M. Eliashberg, 1973, Nonequilibrium excitations in superconductors in high-frequency fields, J. Low Temp.Phys., 10:449.
Kaplan, S.B., C.C.Chi, D.N. Langenberg, J.J. Chang, S. Jafarey and D.J. Scalapino, 1976, Quasiparticle and phonon lifetimes in superconductors, Phys.Rev.B, 14:4854.
Kirichenko, I.K., S.A. Peskovatsky and V.P. Seminozhenko, 1979, Nonequilibrium induction of superconductivity by tunneling injection, Solid State Comm., 31:545.
Klapwijk, T.M., J.N. van den Bergh and J.E. Mooij, 1977, Radiation-stimulated superconductivity, J. Low Temp.Phys., 26:385.
Klapwijk, T.M. and J.E. Mooij, 1976, Microwave-enhanced superconductivity in aluminum films, Physica, 81B:132.
Klapwijk, T.M., H.B. van Linden van den Heuvell and J.E. Mooij, 1978, Phonon effects in microwave-enhanced superconductivity, J. de Physique, 39:C6-525.
Kommers, T.M. and J. Clarke, 1977, Measurement of microwave-enhanced energy gap in superconducting aluminum by tunneling, Phys.Rev. Letters, 38:1091.
Latyshev, Yu.I., and F.Ya. Nad', 1976, Mechanism of superconductivity stimulated by microwave radiation, Zh.Eksp.Teor.Fiz., 71:2158; Sov.Phys. JETP, 44:1136.
Latyshev, Yu.I., and F.Ya. Nad', 1978, Nonequilibrium phenomena in superconducting Al bridges under the microwave radiation, J. de Physique, 39:C6-531.

Lindelof, P.E., 1976, Microwave-enhanced critical currents in superconducting microbridges explained by the electric field induced quasiparticle-pair inequilibrium, Solid State Comm., 18:283.

Lyall, K.R., D.J. Meredith and E.R. Dobbs, 1976, Microwave acoustic studies of the superconducting transition in aluminum, J.Phys.F, 6:807.

Mooij, J.E., N. Lambert and T.M. Klapwijk, 1980, Tc enhancement!, Solid State Comm., to be published.

Notarys, H.A., M.L. Yu and J.E. Mercereau, 1973, Josephson effects at high current density, Phys.Rev. Letters 30:743.

Pals, J.A., 1977a, Microwave-enhanced critical currents in superconducting Al strips with local injection of electrons, Phys.Letters, 61A:275.

Pals, J.A., 1977b, Frequency dependence of microwave-enhanced superconductivity in Al strips with and without local injection of electrons, Phys.Letters, 63A:141.

Pals, J.A. and J. Dobben, 1979a, Measurement of characteristic time constants of microwave-enhanced superconductivity in aluminum films, Phys.Rev.Letters, 42:270.

Pals, J.A. and J. Dobben, 1979b, Measurements of microwave-enhanced superconductivity in aluminum strips, Phys.Rev.B, 20:935.

Pals, J.A. and J. Dobben, 1980, Observation of order-parameter enhancement by microwave irradiation in a superconducting aluminum cylinder, Phys.Rev.Letters, 44:1143.

Parmenter, R.H., 1961, Enhancement of superconductivity by extraction of normal carriers, Phys.Rev.Letters, 7:274.

Peskovatskii, S.A. and V.P. Seminozhenko, 1976, Stimulation of superconductivity by constant tunnel currents, Fiz.Nizk.Temp. 2:943; Sov.J. Low Temp.Phys., 2:464.

Schmid, A., 1977, Stability of radiation-stimulated superconductivity, Phys.Rev. Letters, 38:922.

Schön, G. and A.-M. Tremblay, 1979, Nonequilibrium superconducting states with two coexisting energy gaps, Phys.Rev.Letters, 42:1086.

Tinkham, M., 1975, "Introduction to superconductivity", Mc Graw-Hill, New York.

Tredwell, T.J. and E.H. Jacobsen, 1975, Phonon-induced enhancement of the superconducting energy gap, Phys.Rev.Letters, 35:244.

Tredwell, T.J. and E.H. Jacobsen, 1976, Phonon-induced increase in the energy gap of superconducting films, Phys.Rev. B, 13:2931.

Warlaumont, J.M., J.C. Brown, T. Foxe and R.A. Buhrman, 1979, Microwave-enhanced proximity effect in superconductor-normal-metal-superconductor microjunctions, Phys.Rev.Letters, 43:169.

Wyatt, A.F.G., V.M. Dmitriev, W.S. Moore and F.W. Sheard, 1966, Microwave-enhanced critical supercurrents in constricted tin films, Phys.Rev. Letters, 16:1166.

HEATING AND DYNAMIC ENHANCEMENT IN METALLIC WEAK LINKS

M. Tinkham

Physics Department
Harvard University
Cambridge, MA 02138

INTRODUCTION

In this chapter we treat two contrasting examples of the longitudinal mode of quasiparticle disequilibrium. First, we treat the rather classical problem of heating effects in three-dimensional constriction weak links, which increase the Johnson noise, while reducing the critical current and energy gap. Second, we treat the more subtle nonequilibrium effects generated by a time-varying energy gap. These provide a kind of negative feedback effect which slows down the gap response time from τ_{GL} to τ_{Δ}, which may be orders of magnitude longer, and also provides a type of dynamic enhancement of superconductivity in current-driven metallic weak links.

HEATING EFFECTS IN METALLIC WEAK LINKS

The energy gap frequency sets a limit above which resistive shunting effects progressively degrade the performance of all high-frequency Josephson devices. More restrictive limits may be set by heating effects in metallic weak links. In this section we develop a simplified approximate treatment of these heating effects, followed by a description of experimental observations which confirm the essential correctness of the theoretical results of the model.

Calculation of the Temperature Profile

The essentials of the argument can be outlined as follows: The amplitude of the phase modulation induced in a Josephson device by an rf voltage V_1 at frequency ω_1 is $2eV_1/\hbar\omega_1$. For useful nonlinearity, this must be of order unity, so $V_1 \gtrsim (\hbar\omega_1/2e)$. In a device with resistance R, the associated power dissipation is of order

$$P = (\hbar\omega_1/2e)^2/R \tag{1}$$

This dissipated power raises the temperature of the weak link, and when it is above T_c over a sufficient distance, the Josephson coupling effect is lost and the device fails to operate. In principle, P can be reduced by increasing R, but it is hard to fabricate short bridges with R much above 1Ω, and, in any case, R can not exceed ~100Ω and still have I_c large enough to maintain superconducting coherence against thermal noise. Thus there is a fairly well-defined power level associated by (1) with a given frequency, and to maximize high frequency performance one seeks to reduce the temperature rise resulting from this power dissipation.

It was shown by Skocpol et al. (1974a) that to accomplish this, it is necessary to approach a 3-dimensional cooling geometry in which the nonequilibrium "hot" quasiparticles created by the applied potential can diffuse away in 3-dimensions through the metal. This can be achieved, at least approximately, by use of point contacts or variable thickness bridges (VTB's). Indeed these geometries have demonstrated [Klapwijk and Mooij (1975); Octavio et al. (1977)] performance superior to planar microbridges, in which the diffusive cooling from the microbridge occurs in only 2-dimensions in metal, beyond which heat must be transfered across an interface to the helium bath and/or substrate. One-dimensional cooling is even less effective.

These ideas about dimensionality can be put more quantitatively as follows: If power P must be conducted from r_1 to r_2 through a metal with thermal conductivity K, the temperature gradient and resulting temperature difference is given in the three cases by the following expressions:

$$\text{1-D:} \quad \frac{dT}{dx} = \frac{P}{KA} \qquad \Delta T = \frac{P}{KA}(r_2 - r_1) \tag{2}$$

$$\text{2-D:} \quad \frac{dT}{dr} = \frac{P}{Kd\theta r} \qquad \Delta T = \frac{P}{Kd\theta}\ln\frac{r_2}{r_1} \tag{3}$$

$$\text{3-D:} \quad \frac{dT}{dr} = \frac{P}{K\Omega r^2} \qquad \Delta T = \frac{P}{K\Omega}\left(\frac{1}{r_1} - \frac{1}{r_2}\right) \tag{4}$$

where in 1-D, A is the cross-sectional area of the strip; in 2-D, d is the thickness of the film and θ is its angular width; and in 3-D, Ω is the solid angle of metallic cooling. Note that the 3-D result tends to the finite thermal spreading resistance $(K\Omega r_1)^{-1}$ as $r_2 \to \infty$, while the other two cases require a finite limit on r_2 to hold the heating to finite values. In the 1-D and 2-D cases,

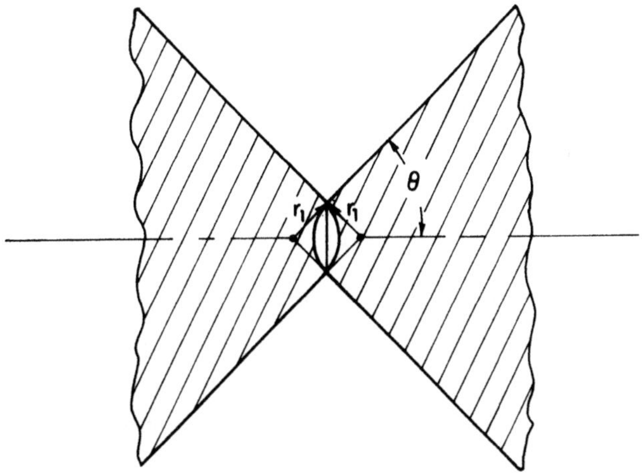

Fig. 1. Longitudinal section of artificial collinear cone neck with apex angle 2θ. The unshaded space inside the radii r_1 is nonphysical.

this limit is set by the thermal healing length $\eta = (KdR_K)^{\frac{1}{2}}$, where R_K is the Kapitza resistance for unit area. This length η, which is the length required to transfer heat across the interface to the substrate, is typically ~5μm. Comparing (3) and (4), we see that the 3-D temperature rise is typically significantly less than that in 2-D, since it is not dependent on surface heat transfer. Accordingly, we henceforth restrict our attention to the 3-D case, which should pertain to VTB's and point contacts.

To allow explicit calculation in the simplest way, we adopt the collinear cone model introduced by Tinkham et al. (1977). In this model, the (imperfectly known) actual geometry is replaced by an artificial one defined by two collinear intersecting cones with apex angle 2θ, each subtending solid angle $\Omega = 2\pi(1-\cos\theta)$, which intersect at a radius r_1 from the apexes, as illustrated in Fig. 1. This r_1 is chosen so that the resistance of the purely radial current flow from r_1 to ∞ in both directions, namely,

$$R = 2\rho/\Omega r_1 \qquad (5)$$

reproduces that of the contact under consideration. Compared to the known resistance of a hyperbolic neck of radius a with the same apex angle 2θ, for the same resistance one has

$$r_1 = \frac{2a}{\pi\sin\theta} \qquad (6)$$

Thus, for typical large values of sin θ, r_1 is not very different from a in magnitude. In any case, it can be thought of qualitatively as the radial distance from the contact center to a point where the cross-section of the constriction has increased by a factor ~4.

Our first step in generalization is to relax the requirement that $\ell << a$ by putting boundary scattering in explicitly, if approximately, by setting

$$\frac{1}{\ell_{eff}} = \frac{1}{\ell} + \frac{1}{2r\sin\theta} \tag{7}$$

so that ℓ_{eff} reduces to the local cone diameter if $\ell \to \infty$. Then

$$\rho_{eff}(r)=\rho_o(1 + \frac{\ell}{2r\sin\theta}) \tag{8}$$

where ρ_o is the bulk resistivity, which we assume is independent of temperature. Using (8), we find the resistance to be

$$R= \frac{2\rho_o}{\Omega r_1}(1 + \frac{\pi\ell}{8a}). \tag{9}$$

Note that in the limiting case of two half-spaces connected through a hole of radius a, $\Omega = 2\pi$ and $r_1 = 2a/\pi$, so that

$$R = \frac{\rho_o}{2a}(1 + \frac{\pi\ell}{8a}) \to \frac{\pi}{16}\frac{\rho_o\ell}{a^2}, \tag{10}$$

where the latter form applies for very small contacts, where $a << \ell$. It is obviously important to use (9) or (10) rather that (5) to infer contact areas from measured resistances in high-resistance (i.e., small-area) contacts. For example, using (9) and taking the typical parameters $\rho_o = 10^{-6}$ Ωcm, $\ell = 10^{-5}$ cm and $\Omega=\pi$, we find a=63 Å for R=10Ω (typical of a Nb point contact) and a=478 Å for R=0.33Ω (typical of a Sn microbridge). If we had used (5) instead, the corresponding values would have been 9 Å and 260 Å. Thus, at least in high-resistance junctions, the size-effect corrections can increase the local resistivity near the neck by almost a factor of 10. For our model and that of Aslamazov and Larkin (1969) to be entirely applicable, it is necessary that a be distinctly less than the coherence length ξ. These typical neck radii fulfill this requirement.

To find the temperature rise for given applied voltage, one must compute the dissipated power and the thermal conductance to the bath. Since we are most interested in assessing high-voltage limits, where the superconductivity is nearly destroyed, we shall simplify by approximating the thermal and electrical resistance properties of the constriction region by those of the normal state of the metal.

The total power dissipated is I^2R. The amount of power dissipated per unit of radius is $dP/dr = I^2\rho_{eff}/\Omega r^2$, which falls very rapidly as r increases from r_1, as is shown in Fig. 2(a). By integration out from r_1, one can compute the radial heat flow P(r) at any radius r. The temperature gradient required to carry this power away is

$$\frac{dT}{dr} = \frac{P(r)}{\Omega K_{eff} r^2} \tag{11}$$

where, by the Wiedemann-Franz law,

$$K_{eff} = (\pi^2/3)(k_B/e)^2 T/\rho_{eff} \tag{12}$$

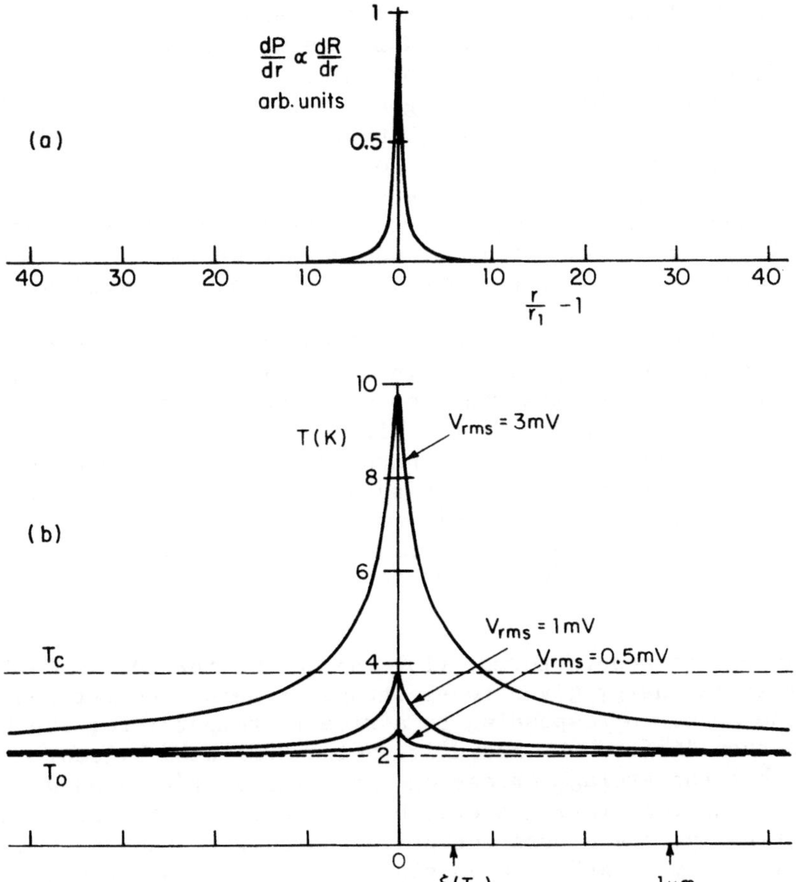

Fig. 2. (a) Radial dependence of contribution to resistance and hence power dissipation. (b) Radial temperature profiles at various applied rms voltages. Parameters used were for a 0.33 ohm tin bridge, with $\ell = 10^{-5}$cm, $\Omega = \pi$, and $r_1 = 350$ Å in collinear cone approximation.

ρ_{eff} being the position-dependent value defined by (8). Integration of (11) then yields the temperature profile T(r). The detailed form of T(r) depends on the ratio (ℓ/r_1) [see Tinkham et al., 1977], but roughly speaking, $T^2(r) - T_o^2$ falls as 1/r, T_o being the asymptotic temperature defined by the bath. A profile for a tin VTB at various voltage levels is shown in Fig. 2(b).

An important special case is the maximum temperature T_m at the center of the constriction, which is given (for all ℓ/r_1)

$$T_m = [T_o^2 + 3(eV/2\pi k_B)^2]^{\frac{1}{2}} \tag{13}$$

This result is not really new. For example, Holm (1967) showed its validity for general 3-D geometries, and a nearly equivalent result was found by Kohlrausch (1900) using the classical theory of electrical and thermal conductivity. Numerically, the second term in (13) is $(3.2V)^2$ if V is in mV. From this, one can conclude that T_m was ~15K in tin VTB's at the highest voltage where they still showed microwave Josephson steps [Octavio et al.,(1979)] and ~70 K in Nb point contacts still showing Josephson steps [McDonald et al., (1974), Weitz et al.,(1978a)]. These examples make clear that we are indeed dealing with large departures from thermal equilibrium at the bath temperature. In fact, the device becomes a sort of SNS weak link because of the high temperature in the neck region.

In assessing (13), it is important to note that all properties of the junction have dropped out, including its resistivity and size except the under-lying requirement of the three-dimensional geometry and the assumption of the Wiedemann-Franz law. The latter holds if the electronic mean free path is the same for electrical and thermal conductivity, which it should be for impurity and boundary scattering as is implicit in our assumed temperature-independent ρ_o. Thus it should apply quite generally, including in the dissipative superconductive state, where the quasiparticle density is nearly normal.

In more microscopic terms, (13) can be interpreted as follows: eV is the extra energy given to electrons on being accelerated through the neck, corresponding to a rise in temperature; the hot electrons then diffuse out radially, being "diluted" by cool electrons so that the average excess energy falls as 1/r, even without individual electrons losing energy by inelastic scattering from phonons; finally, the excess energy is shared with phonons and becomes equilibrium "heat" in a quasiparticle diffusion length $\Lambda=(D\tau_{in})^{\frac{1}{2}}$. Because $\tau_{in}\propto E^{-3}$ or T^{-3}, the appropriate value of Λ here is that corresponding to T_m, which will be much shorter than measured near T_c. For example, in Sn at 15K, $\Lambda\sim0.6\mu m$. Since the electron thermal conductivity usually dominates the phonon contribution, the details of this thermalization are not critical to the argument.

Another point that should be underlined is the importance of the size-effect increase of the local resistivity near the neck. While not upsetting the Wiedemann-Franz relation, this position-dependent resistivity in high-resistance junctions confines the spatial extent of the very high temperature given by (13), which itself is independent of resistivity, thus reducing the effect of heating on superconducting properties.

Experimental Consequences

Because the temperature rise falls off so rapidly away from the center of the link, various measurable characteristics of a junction are sensitive to different aspects of the temperature profile. We now discuss three such experimental consequences of the heating: 1) increased Johnson noise, 2) reduced critical currents with eventual cutoff of microwave Josephson steps, and 3) reduction of the energy gap as measured by observed subharmonic gap structure.

Increased Noise Temperature. If one treats the weak link as containing a series of Johnson noise sources, each of resistance dR and characterized by the local temperature T(r), the suitably averaged noise temperature will be

$$T_N = \frac{1}{R}\int_{r_1}^{\infty} T(r)\,\frac{dR}{dr}\,dr \qquad (14)$$

Although the details of the variation of T(r) and the weighting function dR/dr will depend on (ℓ/r_1) and the voltage level, no large error in the average will result from taking the correctly normalized approximations $T(r)-T_o=(T_m-T_o)(r_1/r)$ and $dR/dr=Rr_1/r^2$. [These r- dependences are asympotically correct when $r >> r_1$, $\ell/r_1 << 1$, and when the fractional temperature rise is small.] With these assumptions, (14) reduces to

$$T_N = \tfrac{1}{2}(T_o + T_m) \qquad (15)$$

That is, the noise temperature should be simply the arithmetic average of the bath temperature and the maximum temperature at the center of the bridge. Since this T_m can be many times as large as T_o, the noise temperature T_N can also be very much increased over T_o.

The noise temperature of a weak link can be estimated experimentally by fitting the shape of the steps induced by high-frequency radiation to the theoretical noise-rounded shape (Stephen, 1969a,b). This was done by Weitz et al. (1978, 1978b) using the steps induced on the I-V curve of a Nb point-contact detector by 496 μm laser radiation. Although the fitting procedure was not highly accurate, the fitted values of T_N did rise from ~5K to ~20K as one considered steps at voltages ranging from ~1 to ~10 mV. This is in very satisfactory agreement with the results predicted by (13) and (15).

Reduction in I_c. In the Aslamazov-Larkin (1969) solution of the Ginzburg-Landau equation for a constriction weak link of size $a << \xi$, one has

$$\psi=\psi_o \left\{f(\underline{r})e^{i\phi_1} + [1-f(\underline{r})]e^{i\phi_2}\right\} \tag{16}$$

where ψ_o is the equilibrium amplitude of the pair wavefunction, $f(\underline{r})$ is a solution of Laplace's equation which varies from 0 to 1 as one goes from one bank to the other, and ϕ is the phase. From this one finds the critical current to be

$$I_c = \pi a^2 (2e\hbar/m^*)|\psi_o|^2\nabla f \tag{17}$$

where ∇f is evaluated at the neck plane of area πa^2. The value of ∇f is geometrically determined, and is $2/\pi a$ for the hyperbolic neck and $(2r_1)^{-1}$ for the collinear cone model. It follows that the critical current density in the neck is larger than the bulk J_c by a factor of order $\xi/a >> 1$. As a result, I_c scales as a rather than a^2 for these small contacts. Since one can show that the electrical resistance of the contact is given by $R=(\pi a^2 \sigma \nabla f)^{-1}$, the I_cR product is universal, independent of all details of the weak link, and near T_c, at least, is given by exactly the same result as for a tunnel junction, namely,

$$I_cR = \frac{\pi\Delta(T)}{2e} \tanh \frac{\Delta(T)}{2k_BT} \tag{18}$$

To take heating into account, one recognizes that ψ is depressed in the neck, but the length scale defining the gradient in (17) will remain the geometrical length a or r_1. In other words, I_c will be reduced by a factor $[2f(0)]^2$, where by symmetry $f(0)=\frac{1}{2}$ in the absence of heating, but it is reduced in the presence of heating. [For details, see Tinkham et al. (1977).]

Since $f(\underline{r})$ can only respond to heating over a length scale of order ξ without excessive gradient energy, for small heating we expect $f(0)$ to reflect the average temperature rise over a volume extending from the contact out to $\sim\xi(T_o)$. Since $I_c \propto (T_c-T)$ this will give a linear decrease in critical current roughly as

$$I_c(P) \simeq I_c(0)\ (1-P/P_o) \qquad (P < P_o) \tag{19}$$

where P_o is the power dissipation required to raise this average temperature from T_o to T_c.

Because the temperature variation occurs over a distance

$a << \xi$, this average temperature procedure is not appropriate for the case of large heating effects, when $T_m >> T_c$. In this case, we resort to a solution of the GL equation using the WKB method to take account (approximately) of the position-dependent GL coefficient α. We start with the GL equation with neglect of vector potential:

$$(\hbar^2/2m^*)\nabla^2\psi + \alpha\psi + \beta|\psi|^2\psi = 0 \tag{20}$$

Specializing to radial solutions without angular dependence, and dropping the nonlinear term since ψ is small in the region above T_c, we have

$$\frac{d^2}{dr^2}(r\psi) = \frac{2m^*\alpha[T(r)]}{\hbar^2}(r\psi) \equiv \frac{1}{\xi^2(r)}(r\psi) \approx \frac{(r\psi)}{\xi^2(0)}[t(r)-1] \tag{21}$$

To simplify as much as possible, we have taken the simple linear variation of α (or ξ^{-2}) with $(T-T_c)$ even well above T_c. Then, neglecting all but the dominant exponential dependence of the WKB approximation, we expect $r\psi$ to be attenuated by a factor

$$2f(0) = \exp(-W), \tag{22}$$

where

$$W = \frac{1}{\xi(0)}\int_{r_1}^{r(T_c)} [t(r)-1]^{\frac{1}{2}} dr. \tag{23}$$

Because the radius $r(T_c)$ of the region with $T > T_c$ will be large compared to r_1 and ℓ for the case of strong heating, we can approximate $T(r)$ using (11), taking $P(r)=\frac{1}{2}P$ (since half the heat goes out on either side) and $K_{eff}=K$, the bulk value. Then, with (12), we obtain

$$T^2 = T_o^2 + \frac{T_c P}{K(T_c)\Omega r} = T_o^2 + \frac{3}{\pi^2}\frac{e^2}{k_B{}^2}\frac{\rho_o}{\Omega}\frac{P}{r} \tag{24}$$

from which it follows that

$$r(T_c) = 3\rho_o P(e/\pi k_B T_c)^2/\Omega(1-t_o^2) \tag{25}$$

where $t_o = T_o/T_c$. For strong heating, $r(T_c) >> r_1$, and one can write (23) approximately as

$$W \simeq [r(T_c)/\xi(0)] \langle [t(r)-1]^{\frac{1}{2}} \rangle_{av} \tag{26}$$

Finally, approximating the average of $(t-1)^{\frac{1}{2}}$ by its value at the average radius, which works out to be $[\frac{1}{2}(1-t_o^2)]^{\frac{1}{2}}$, we obtain

$$I_c(P)/I_c(0) = [2f(0)]^2 = e^{-2W} = e^{-P/Po} \tag{27}$$

where the characteristic power level P_o is

$$P_o = \frac{(1-t_o^2)^{\frac{1}{2}}}{\sqrt{2}} \frac{\pi^2}{3} \frac{k_B^2 T_c^2 \xi(0)\Omega}{e^2 \rho_o} = \frac{(1-t_o^2)^{\frac{1}{2}}}{\sqrt{2}} K(T_c)\, T_c \xi(0)\Omega \quad (28)$$

Note that our low power result (19) is a special case of the result (27), apart from possible numerical factors in the definition of P_o in (19), so (27) should be a good approximation for all power levels although derived for the high power limit.

This P_o forms a figure of merit for avoiding heating limits. It can be maximized for a given material by reducing t_o toward zero and increasing the cooling solid angle Ω toward its limiting value 2π. In comparing materials, one wants to maximize $T_c K(T_c)\xi(0)$ or $T_c^2\xi(0)/\rho_o$. Evidently there is a tendency toward cancellation, since materials with high T_c tend to have short $\xi(0)$ and high ρ_o. In fact, assuming a residual resistance ratio of ~10 in all cases, $\xi(0)T_c^2/\rho_o$ turns out to be about the same (±20%) for Al, Sn, Pb, Nb, but much smaller for Nb_3Sn because its very short coherence length and poor conductivity outweigh the high T_c. Taking $t_o=0$ and $\Omega=\pi$, (28) becomes

$$P_o = \frac{\pi^3}{3\sqrt{2}} \frac{k_B^2}{e^2} \frac{T_c^2\xi(0)}{\rho_o} \approx 10\mu W \quad (29)$$

for the typical simple metallic superconductors. By resorting to inhomogeneous materials, such as Nb-Au bilayer banks with an Au bridge, one can combine to some extent the virtues of high T_c, high K, and high $\xi(0)$, but at the expense of increased fabrication complexity. (van Dover, et al., preprint)

The most direct measure of the critical current I_c under finite voltage conditions is the width of the microwave-induced Josephson steps. In the simple voltage-biased approximation, the width of the step at $V_n=n\hbar\omega_1/2e$ is given by $I_c J_n(2eV_1/\hbar\omega_1)$. Although major corrections to this simple result enter for low-frequency current-biased applications, the simple Bessel function result is quite good at high frequencies and wherever the supercurrent is relatively weak, as it will be near the high-voltage cutoff. Thus, it is a reasonable procedure to compare the cutoff of the measured step widths at high voltage with the theoretical dependence $I_n=I_c e^{-P/P_o} J_n(2eV_1/\hbar\omega_1)$.

The measured results of Octavio et al. (1977) on a 0.33Ω tin VTB irradiated by 10GHz microwaves are shown in Fig. 3, where they are compared with the theoretical expectations for various values of P_o. Fig.3(a), for $P_o=\infty$, corresponds to no heating, and simply shows the Bessel function dependence. The value of V_1 was chosen

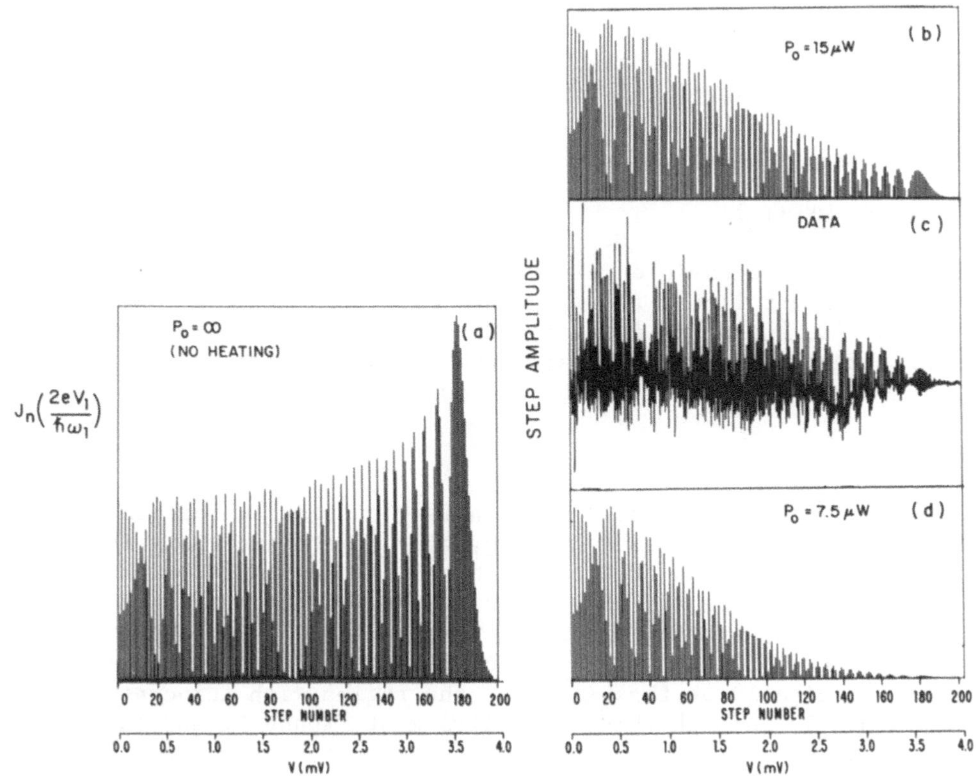

Fig. 3. Theoretical step amplitudes at 10 GHz for different values of P_o compared with experimental amplitudes [panel (c)] for a tin VTB with R=0.33Ω.

to cut off the Bessel function dependence at the right number of steps. P is calculated as V^2/R, where V^2 includes both the observed dc voltage and the inferred V_1. It appears that the best fit occurs for P_o between 10 and 15 μW, in good agreement with our expectation from (29). More generally, this model gives a good account of the high-voltage limit to the observability of microwave-induced steps in both tin VTB's and niobium point contacts. The better performance of the latter seems to be associated more with their higher resistances, giving less dissipation per millivolt, than with their higher T_c and energy gap.

Gap Depression. The dV/dI vs. V curves of metallic weak links show a subharmonic gap structure, with features occurring at voltages $V_n=(2\Delta/e)n^{-1}$. Using the well-known BCS function Δ(T) as a calibration curve, these measured value of V_n can be used to infer an effective temperature. When this is done, one indeed finds a temperature increment ΔT relative to T_o which scales roughly with the dissipated power $P=IV_n$.

However to make a quantitative interpretation of the data, one must identify the position in space to which the inferred temperature rise is associated. In the absence of a definitive theory of the origin of this subharmonic gap structure, we anticipate that the measured ΔT reflects the temperature rise at a radius $\sim\xi$ from the center. Accordingly, from (24) we expect

$$\Delta T \simeq P/2\Omega K\xi(T_o) \tag{31}$$

When the data of Octavio et al. on tin VTB's were plotted in this way, an excellent linear relation between ΔT and P/ξ was found, with a coefficient which corresponded to a value of ΩK which is in good agreement with a priori estimates and with the value inferred from the observed cut off of the microwave-induced steps.

Summary

Measurements of excess Johnson noise, of reduced maximum supercurrent, and of a reduced energy gap in metallic weak links all are consistent with predictions based on the simple model of heating effects presented above. The noise measurement is most sensitive to the high temperature in the center of the constriction where most of the resistance occurs; the reduced supercurrent depends principally on the extent of the region with temperature raised above T_c through which the pairs must "tunnel"; and the reduced gap reflects the rather modest temperature rise at a distance $\xi(T_o)$ out from the center. This overall consistency shows that the sweeping simplifications made in using macroscopic heat flow analysis are remarkably successful in accounting for the major nonequilibrium effects associated with Joule heating. In particular, the importance of using 3-D metallic geometries to facilitate the diffusive escape of non-equilibrium quasiparticles is highlighted.

TIME-DEPENDENT ENERGY GAP AND DYNAMIC ENHANCEMENT

In the first part of this chapter, we discussed the time-average spatially-localized heating associated with the operation of metallic weak links at finite voltages, using rather macroscopic steady-state heat flow analysis. We now turn to more subtle non-equilibrium effects depending on the explicit time variation of the energy gap order parameter Δ.

In dealing with time-dependent superconductivity, a number of characteristic time scales are relevant. The shortest is τ_J, the characteristic time for acceleration of the supercurrent, given by

$$\tau_J = [7\zeta(3)/4\pi^3]\hbar/k_B(T_c - T) \tag{31}$$

where the numerical coefficient is ~ 0.068. The Ginzburg-Landau time τ_{GL} is 5.79 times longer, being given by

$$\tau_{GL} = \pi\hbar/8k_B(T_c - T) \tag{32}$$

Both of these are of order 10^{-12} seconds for $(T_c-T)=1K$. Although it was originally thought that τ_{GL} should govern the time variation of the energy gap, it was later realized that this was true only for the case of gapless superconductors. For superconductors with a gap, Schmid (1968) showed that the gap change should be much slower, the characteristic time scaling with the inelastic electron-phonon time τ_{in}, which ranges from $\sim 10^{-8}$ sec for Al to $\sim 10^{-11}$ sec for Pb. Also the temperature variation of this gap relaxation time τ_Δ, called the "longitudinal relaxation time" $\tau_R^{(L)}$ by Schmid, was different from that of τ_{GL}, namely

$$\tau_\Delta = \tau_R^{(L)} \simeq 1.2\tau_E/(1-t)^{\frac{1}{2}} \tag{33}$$

where τ_E is the inelastic scattering time at the Fermi surface at T_c, and $t=T/T_c$.

What we will do first is show how this result of Schmid can be rederived and given a simple physical interpretation in terms of an effective nonequilibrium temperature change of the quasi-particles induced by the time-dependent energy gap. Then we will indicate a generalization of the method to deal with time-dependence which are too abrupt and rapid to be treated correctly by a simple differential equation of the TDGL sort. The methods will then be applied to the current step in a long filamentary conductor, studied experimentally by Pals and Wolter, and to the short metallic weak link, where we are able to account semiquantitatively for the "foot" structure observed by various workers just above I_c in the I-V curves of microbridges.

Derivation of Ginzburg-Landau Equation for Time-Dependent Gap

The striking contrast between the slow phonon-limited time scale τ_E and the fast intrinsic electronic times τ_J and τ_{GL} makes it useful to treat a simplified model for the nonequilibrium behavior of the system on the time scale τ_E, which treats τ_J and τ_{GL} as vanishingly short by comparison. This "slow motion" regime will be governed by the ordinary static GL equation, but with an effective temperature which reflects the nonequilibrium quasiparticle population. We write this static GL equation in the form

$$[\alpha^* + \beta\,(\Delta^2/T_c^2) - \xi^2(0)\nabla^2]\Delta = 0 \tag{34}$$

where $\alpha^* \equiv (T+\delta T^* - T_c)/T_c = \alpha + \delta\alpha^*$ includes a term $\delta\alpha^*$ to reflect the nonequilibrium population, while $\beta = 7\zeta(3)/8\pi^2 = 0.106$, and $\xi(0)\ (1-T/T_c)^{-\frac{1}{2}}$ is the usual coherence length. With these conventions, the equilibrium gap in the absence of gradients is given by $\Delta_o^2/T_c^2 = -\alpha/\beta = (1-T/T_c)/\beta$.

This nonequilibrium "temperature" shift $\delta\alpha^* = \delta T^*/T_c$ is defined so that in the BCS gap equation for the spatially uniform case, a change δT^* would produce the same change in the equilibrium $\Delta_o(T)$ as would the actual local nonequilibrium population δf_k. Writing the gap equation as

$$\frac{2}{V} = \sum_k \frac{1-2f_k}{E_k} = \sum_k \frac{1-2f_k}{(\Delta^2+\varepsilon_k^2)^{1/2}} \tag{35}$$

we see that a given set of f_k determine a value of Δ. If the f_k are given by the Fermi function $f_o(E/T)$, this defines the BCS form of $\Delta(T)$. However if $\delta f_k \equiv f_k - f_o(E_k/T) \neq 0$, Δ will take on a different value, as specified by (35). Near T_c, where $\Delta/T_c \ll 1$ and the GL equation itself is justified, one can show from (35) that the appropriate relation between the set of δf_k and the $\delta\alpha^*$ which would have the same effect on Δ is

$$\delta\alpha^* = \frac{\delta T^*}{T_c} = \frac{1}{N(0)} \sum_k \frac{\delta f_k}{E_k} = \int_{-\infty}^{\infty} \frac{\delta f}{E}\, d\varepsilon \tag{36}$$

[Note that this $\delta\alpha^* \equiv -\chi$, where χ is the "control function" introduced by A. Schmid (1977). A similar function, denoted by them as ϕ, was also introduced by Aslamazov and Larkin].

For the heating effects in metallic weak links, treated above, the δf_k correspond fairly closely to a Fermi distribution with an increased temperature T^*, which results from an input of dissipated electrical energy. In the present context, however, we will be concentrating on δf_k caused by variation of the energy gap on a time scale rapid enough compared to τ_E that quasiparticle occupation numbers lag behind those that would be correct for thermal equilibrium. To be precise, δf_k is defined as

$$\delta f_k(t) = f_k(t) - f_o[E_k(t)/T] = f_k(t) - f_o\left\{[\Delta^2(t)+\varepsilon_k^2]^{1/2}/T\right\} \tag{37}$$

where $f_k(t)$ is the <u>actual</u> occupation number at time t, when the instantaneous energy gap value is $\Delta(t)$. This T^* value need have no relation to any true thermodynamic temperature; it simply characterizes the effect of the quasiparticles on the strength of the superconductivity.

<u>Slow Gap Variation</u>. To get a feel for the problem, we first consider the case of slow gap variation. Then it is a good approximation to say that the populations lag behind by the relaxation time τ_E, so that

$$f_k(t) \simeq f[E_k(t-\tau_E)/T] \cong f_o[E_k(t)/T] - \tau_E \partial f_o(E_k/T)/\partial t \tag{38}$$

Then

$$\partial f_k = -\tau_E \partial f_o(E_k/T)/\partial t \tag{39}$$

where

$$\frac{\partial f_o(E_k)}{\partial t} = \frac{\partial f_o}{\partial E_k}\frac{\partial E_k}{\partial \Delta}\frac{\partial \Delta}{\partial t} = \frac{\partial f_o}{\partial E_k}\frac{\Delta}{E_k}\frac{\partial \Delta}{\partial t} \tag{40}$$

Restricting attention to $E_k \approx \Delta << T \approx T_c$ as is appropriate in the GL regime, $(\partial f_o/\partial E_k) = -1/4T_c$, so that (39) and (40) lead to

$$\delta f_k = \frac{\tau_E}{4T_c}\frac{\Delta}{E_k}\frac{\partial \Delta}{\partial t} \tag{41}$$

Inserting this in (36), we obtain

$$\delta\alpha^* = \frac{\tau_E}{4T_c}\Delta\frac{\partial \Delta}{\partial t}\int_{-\infty}^{\infty}\frac{d\varepsilon}{\varepsilon^2+\Delta^2} = \frac{\pi}{4T_c}\tau_E\frac{\partial \Delta}{\partial t} \tag{42}$$

so that the GL equation (34) becomes

$$\left[\frac{T-T_c}{T_c} + \beta\frac{\Delta^2}{T_c^2} - \xi^2(0)\nabla^2\right]\Delta = -\frac{\pi}{4T_c}\tau_E\Delta\frac{\partial \Delta}{\partial t} \tag{43}$$

It is convenient to rewrite this in terms of a normalized order parameter $f \equiv \Delta/\Delta_o(T)$. The result is

$$[1 - f^2 + \xi^2(T)\nabla^2]f = 2\tau_\Delta f(\partial f/\partial t) \tag{44}$$

where here τ_Δ is just the Schmid-Schön (1975) longitudinal relaxation time

$$\tau_\Delta = \tau_R^{(L)} = \frac{\pi}{8}\frac{\tau_E}{[\beta(1-t)]^{\frac{1}{2}}} = \frac{1.2\,\tau_E}{(1-t)^{\frac{1}{2}}} \tag{45}$$

To check the appropriateness of this exact definition of τ_Δ, consider a spatially uniform fluctuation δf about the solution $f=1$ to the static GL equation. From (44), one finds that δf is governed by

$$\partial(\delta f)/\partial t = -\tau_\Delta^{-1}(\delta f) \tag{46}$$

so that indeed this τ_Δ gives the exponential damping time for small fluctuations from a uniform solution.

Arbitrary Gap Variation. If the gap variation is too rapid to allow use of the simple approximation (38), one can write the solution of the Boltzmann equation in the relaxation time approximation as

$$\delta f_k = \int_{-\infty}^{t} [-\partial f_o(t')/\partial t]\, e^{-(t-t')/\tau_E}\, dt'$$

$$= \frac{1}{4T_c}\int_{-\infty}^{t} \frac{\Delta'}{E'}\,\frac{\partial \Delta'}{\partial t'}\, e^{-(t-t')/\tau_E}\, dt' \tag{47}$$

using (40) and evaluating $\partial f_o/\partial E'$ near T_c as before. Inserting this in the definition (36) of $\delta\alpha^*$, one has

$$\delta\alpha^* = \frac{1}{4T_c}\iint \frac{\Delta'}{EE'}\,\frac{\partial\Delta'}{\partial t'}\, e^{-(t-t')/\tau_E}\, dt'd\varepsilon \tag{48}$$

The integration over ε has the form

$$\int_{-\infty}^{\infty} \frac{d\varepsilon}{EE'} = \int_{-\infty}^{\infty} \frac{d\varepsilon}{(\varepsilon^2+\Delta^2)^{\frac{1}{2}}\,(\varepsilon^2+\Delta'^2)^{\frac{1}{2}}} \tag{49}$$

Insofar as $\Delta \approx \Delta'$, the integral is elementary and has the value π/Δ'. Its exact value can be written in terms of elliptic integrals, and is very well approximated (to a few percent) by the obvious symmetric interpolation formula $\pi/(\Delta\Delta')^{\frac{1}{2}}$ so long as Δ,Δ' differ by no more than a factor of ~2. (A. Schmid and G. Schön, private communication) This allows us to write (48) as

$$\Delta^{\frac{1}{2}}(t)\ \delta\alpha^*(t) = (\pi/4T_c)\int_{-\infty}^{t} \Delta^{\frac{1}{2}}(t')\ (\partial\Delta/\partial t')\ e^{-(t-t')/\tau_E} dt' \tag{50}$$

It is readily seen that this equation reduces to our earlier result (42) if Δ and $\partial\Delta/\partial t$ vary slowly on the scale of τ_E. However, if $\partial\Delta/\partial t$ varies rapidly, the results are quite different. For example, a small sudden drop of the gap at t=0 from Δ to $\Delta-\delta\Delta$ can be described by $\partial\Delta/\partial t = -(\delta\Delta)\delta(t)$. According to (42), this would cause an infinitely large instantaneous δ-function spike in $\delta\alpha^* = -(\pi\tau_E/4T_c)(\delta\Delta)\delta(t)$. But with (50), the same change in Δ produces a step change $\delta\alpha^*=-\pi(\delta\Delta)/4T_c$, which would then relax over τ_E. This limiting form of behavior reflects the fact that, no matter how rapidly the change in Δ occurs, the maximum departure from the equilibrium population for a given $\delta\Delta$ is that which exists right after the change in gap and before any relaxation of the previously existing f_k toward the new equilibrium has occurred.

Because this model is not capable of ultimate accuracy in any case, it is worthwhile to note that a more transparent version of (50) can be obtained by making the further approximation that $\Delta^{\frac{1}{2}}(t) \approx \Delta^{\frac{1}{2}}(t')$ for $t-t' \lesssim \tau_E$, even when $\partial\Delta/\partial t$ may change rapidly on this time scale. With this approximation, (50) becomes simply

$$\delta\alpha^*(t) = \frac{\pi}{4T_c} \int_{-\infty}^{t} \frac{\partial\Delta}{\partial t'} \, e^{-(t-t')/\tau_E} dt' \tag{51}$$

To conclude this formal development, we now insert the generalized expression (51) for $\delta\alpha^*$ into the GL equation (34), and again express the result in terms of $f = \Delta/\Delta_o$, obtaining the integro-differential equation:

$$[1-f^2 + \xi^2(T)\nabla^2]f = af \int_{-\infty}^{t} (\partial f/\partial t') \, e^{-(t-t')/\tau_E} dt' \tag{52}$$

where $a = 2\tau_\Delta/\tau_E = (\pi/4)\beta^{-\frac{1}{2}} (1-t)^{-\frac{1}{2}} = 2.41(1-t)^{-\frac{1}{2}}$, which is typically large (of order 10). As a check we note that if $\partial f/\partial t$ varies slowly on the scale of τ_E, it can be taken out of the integral, and one recovers the conventional partial differential equation approximation (44).

Transient Superconductivity Above I_c

Pals and Wolter (1979) have recently reported measurements of the time delay t_d between application of a current $I>I_c$ to a superconducting aluminum film strip and the appearance of a resistive voltage. They identify this time delay with the time required for the superconducting order parameter or energy gap Δ to be driven to zero from its zero current equilibrium value Δ_o. Further, they interpret this t_d in terms of a form of time-dependent Ginzburg-Landau equation, and find good agreement with the current-dependent t_d using a relaxation time parameter $\tau \approx 20$ nsec. This value is close to the expected τ_Δ, although longer than τ_{GL} by a factor of $\sim 10^4$; surprisingly, the measured value appears to be independent of temperature over the investigated range, namely, $0.76 < T/T_c < 0.92$. Although this interpretation of the experimental results is certainly reasonable, it is not clear that it is quantitatively reliable. In any case, we now analyze an idealized version of this experimental regime as an illustrative application of the theory outlined above.

We consider a superconductor in which a current $I \gtrsim I_c$ is suddenly applied. For simplicity, we assume transverse dimensions small enough ($d << \lambda, \xi$) to justify a one-dimensional approximation and neglect of magnetic field terms. We also make the seriously restrictive assumption that the solution is spatially homogeneous, i.e., of the form

$$f = |f|e^{iqx} \tag{53}$$

with $|f|$ and the phase gradient $q=\partial\phi/\partial x$ independent of x. The relative stability of such uniform solutions for $I>I_c$ is doubtful, but it provides at least an initial perspective. If we insert this assumed form of f, cancel the factor of f appearing on both sides of the equation, and adopt $\xi(T)$ as the unit of length, (52) becomes

$$1-f^2-q^2 = (2\tau_\Delta/\tau_E)\int_{-\infty}^{t} (\partial f/\partial t')\, e^{-(t-t')/\tau_E} dt' \tag{54}$$

$$\cong 2\tau_\Delta(\partial f/\partial t) \tag{54a}$$

where the second form (54a) is the limiting form of (54) when $\partial f/\partial t$ is slowly varying; it could be obtained directly from (44).

Because $\tau_J << \tau_E$, the initial transient, in which the supercurrent is accelerated to carry the applied current, and the final transient, in which the current is again taken up as a normal current, are both of negligible duration. At all other times, the supercurrent is either essentially the full applied current or zero. Since the supercurrent density $j = f^2q$, using units in which the usual critical supercurrent density is $j_c = 2/3\sqrt{3}$, we can replace q^2 in (54) by j^2/f^4, or by $4I^2/27I_c^2f^4$ if it is desired to normalize explicitly to I_c. For simplicity, we use the first version.

Let us first treat the problem using the approximation (54a) to get an initial orientation. This is essentially the approach used by Pals and Wolter, except that in their work the coefficient of $\partial f/\partial t$ in the right member was taken to be τ/f rather than $2\tau_\Delta$. Although no qualitative difference results, we prefer to use (54a), which has a clearer theoretical foundation in the Schmid-Schön formalism. Then, for imposed supercurrent density j,

$$\frac{\partial f}{\partial t} = \frac{1}{2\tau_\Delta}\left(1-f^2-\frac{j^2}{f^4}\right) \tag{55}$$

which can easily be integrated numerically to find f(t). Results for several representative values of j are shown in Fig. 4. However, additional qualitative insight can be gained by analysis of limiting cases. The initial condition is f = 1, so that just after

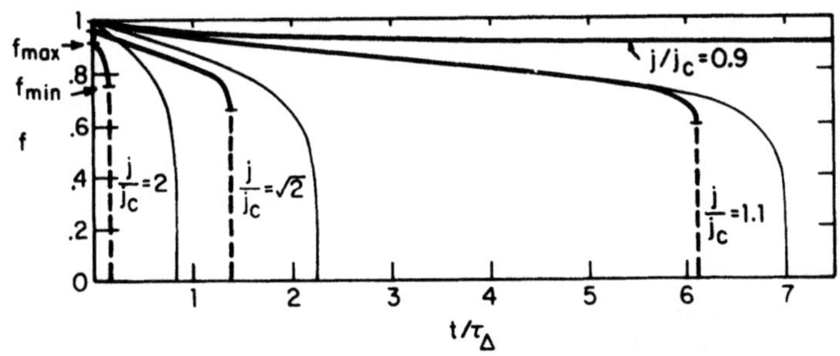

Fig. 4. Time dependence of reduced order parameter $f = \Delta/\Delta_o$ for various current levels. Continuous curves result from differential equation approximation (55); discontinuous curves from integro-differential equation version (61) for $T/T_c=0.9$.

the current step is applied

$$\partial f/\partial t|_o = - j^2/2\tau_\Delta \tag{56}$$

Thus, the initial rate of drop increases as j^2. If this initial drop rate continued, f would collapse to zero in a time $2\tau_\Delta/j^2$. In fact, however, $\partial f/\partial t$ changes as f evolves according to (55), and the possible outcomes fall into two cases, depending on whether or not j exceeds j_c.

If $j<j_c$, f approaches a limiting value, where $\partial f/\partial t = 0$; from (55) we see that this equilibrium value f_{eq} is a solution of the equation

$$j^2 = f^4_{eq} \, (1-f^2_{eq}) \tag{57}$$

As shown in Fig. 6, this f_{eq} drops initially as $1-j^2$, then drops more steeply, reaching a vertical tangent at $f=(2/3)^{\frac{1}{2}}$ when $j=j_c = 2/3\sqrt{3}$. For $j \ll j_c$, the approach of f to f_{eq} is exponential with time constant τ_Δ. However, as j increases, the decay time increases as $\tau_\Delta f_{eq}/(3f^2_{eq}-2)$, which diverges as $f^2_{eq}\rightarrow 2/3$ at j_c. Exactly at j_c, $\delta f \equiv f-f_{eq}$ decays toward zero not exponentially, but only as τ_Δ/t.

If j is slightly above j_c, the rate of decrease of f slows nearly to zero as f passes near $(2/3)^{\frac{1}{2}}$, then speeds up as f falls further, approaching infinity as $f\rightarrow 0$ because of the dominant term $\sim j^2/f^4$ in (55). In fact, if j is substantially above j_c, one can approximate (55) by keeping only this term. Integration is then elementary, and one finds

$$f \simeq (1-5j^2t/2\tau_\Delta)^{1/5} \tag{58}$$

Since this approximation builds in the correct initial and final slopes, it is qualitatively reasonable overall, and predicts a collapse time $t_d = 2\tau_\Delta/5j^2$. This estimate is 5 times shorter than the crude one made from the initial slope because this approximation takes account of the accelerated fall of f as f approaches zero.

Although the approximation (58) is reasonable well above j_c, it fails to capture the very slow drop in f near $(2/3)^{\frac{1}{2}}$ for j just above j_c, mentioned above. This gives rise to a divergence of t_d as $(j-j_c)^{-\frac{1}{2}}$. To see how this comes about, we set $j^2 = j_c^2(1+\varepsilon)$ in (55), find the location $[f_o \simeq (2/3)^{\frac{1}{2}}(1+\varepsilon/6)]$ of the minimum of $df/dt[\simeq -\varepsilon/6\tau_\Delta]$, and expand df/dt about that point to order $(f-f_o)^2$. The result of integrating $(df/dt)^{-1}$ from f = 1 to 0 is

$$t_d = \sqrt{2}\,\pi\,\tau_\Delta\left\{[(j/j_c)^2-1]^{-\frac{1}{2}} - 0.5\right\} \tag{59}$$

The subtractive constant is an estimate of the first departure from the leading asymptotic form.

Although approximations for $j \gg j_c$ and $j \gtrsim j_c$ based on (58) and (59) combine to give the general features of $t_d(j/j_c)$, numerical calculations based on (55) are required for accuracy at all j/j_c. In fact, the result is simply the definite integral

$$t_d = 2\tau_\Delta \int_0^1 \frac{f^4 df}{j^2+f^6-f^4} \tag{60}$$

This result is plotted in Fig. 5 [curve (a)] as a function of j/j_c.

Having sketched the solution as found using the approximation (54a), let us now check its self-consistency; that is, is df/dt always changing slowly compared to τ_E? Since our approximate solution (55) has $df/dt \to \infty$ as $f \to 0$, there is clearly an inconsistency there. Also, at t=0, df/dt changes discontinuously from 0 to the value given by (56), again inconsistent with the approximation. On the other hand, the region in which f is slowly falling, that dominates for $j/j_c \gtrsim 1$, should be well approximated. In short, unless $t_d \gg \tau_E$, (54a) can not be even approximately self-consistent.

Let us now examine how the generalized version (54) deals with the inconsistencies. After replacing q^2 by j^2/f^4 as before, it takes the form

$$1-f^2-\frac{j^2}{f^4} = \frac{2\tau_\Delta}{\tau_E}\int_{-\infty}^{t} \left(\frac{\partial f}{\partial t'}\right) e^{-(t-t')/\tau_E}\, dt' \tag{61}$$

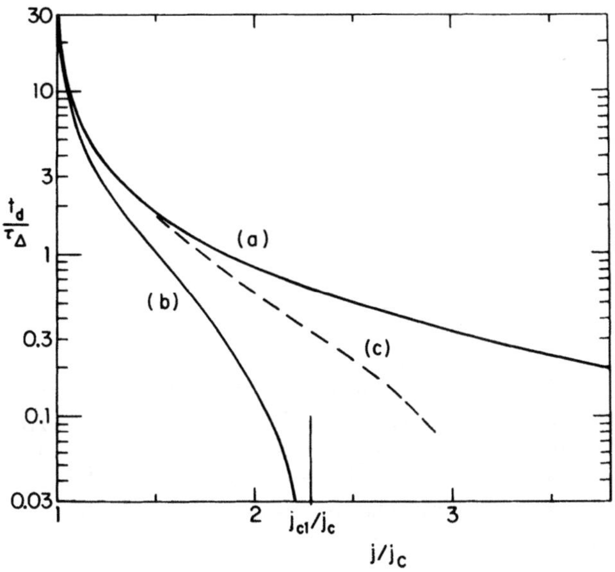

Fig. 5. Calculated delay times for destruction of superconductivity vs. current density. Result of: (a) differential equation approximation (60); (b) integro-differential equation version (64) for $T/T_c=0.9$; (c) Geier calculation, $T/T_c=0.9$.

If we apply this at time t=0+, just after a supercurrent is applied instantaneously (i.e. in time $\sim\tau_J \ll \tau_E$), it is immediately clear that it can not be satisfied without a discontinuous drop in f from its initial value 1; otherwise, the left side is $-j^2$, while the right side is zero. If instead, we assume f drops suddenly (i.e. in much less than τ_E) from 1 to some f_{max} at t=0+ before a continuous further decline, (61) leads to

$$j^2 = f^4_{max}(1-f_{max})(a + 1 + f_{max}) \qquad (62)$$

where, as before, $a = 2\tau_\Delta/\tau_E$. This equation directly gives $j^2(f_{max}, a)$, so that $f_{max}(j,a)$ can be found graphically (See Fig. 6) or numerically. Qualitatively, it is important that the r.h.s. of (62) has a maximum value [when $f_{max} = f_1 \approx 4/5$]. The corresponding maximum value of j is given by $j^{max}_{c1} = 0.29a^{\frac{1}{2}} + \ldots$, or normalized to $j_c = 2/3\sqrt{3}$, and expressing a in terms of (1-t), it is

$$j_{c1}/j_c = 1.15(1-t)^{-\frac{1}{4}}\,[1 + 0.37(1-t)^{\frac{1}{2}} + \ldots] \qquad (63)$$

which is ~2-4 for $0.8 < t < 0.99$. From these considerations, it follows that for $j > j_{c1}$, there is no slow ($\sim\tau_E$) decay of the gap; rather the superconductivity is destroyed in a time of order τ_J. For $j < j_{c1}$, there is an initial sudden drop of f from 1 to $f_{max} \gtrsim 4/5$, the size of the predicted quasi-discontinuous drop approach-

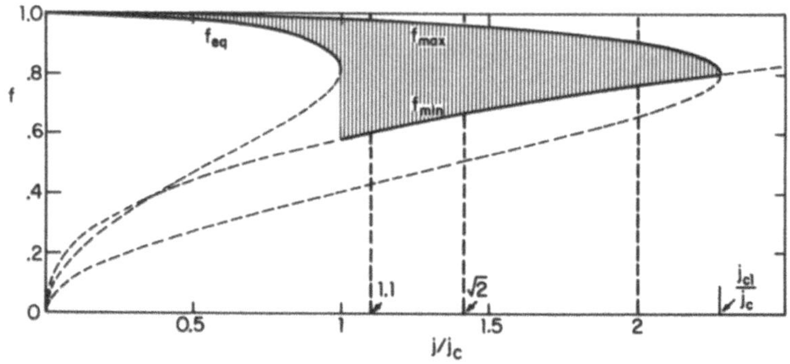

Fig. 6. Plot of f_{eq}, f_{max}, and f_{min} vs. j for $T/T_c = 0.9$. [$f_{eq}(j)$ is independent of T/T_c.] At fixed j, the slow drop in f occurs only between f_{max} and f_{min}, or f_{max} and f_{eq} for $j<j_c$. This slow regime occupies the shaded area. Vertical lines are drawn at the current levels used in Fig. 4. No slow drop occurs for $j>j_{cl}$.

ing zero as $j\to 0$ or $T\to T_c$.

To analyze the continuous change after the initial drop of f to f_{max}, it is convenient to differentiate (61) with respect to t. After some rearrangement, one finds

$$\frac{df}{dt} = \frac{1}{2\tau_\Delta} \frac{1-f^2-j^2/f^4}{1+(\tau_E/2\tau_\Delta)(2f-4j^2/f^5)} \tag{64}$$

For comparison, when we used the slow variation approximation corresponding to (54a), we obtained the same formula except that the denominator was simply 1. This difference is of qualitative significance since (64) gives $df/dt\to\infty$ for a finite value of f, whereas the approximate version (55) gives $df/dt\to\infty$ only as $f\to 0$. This feature of (64) implies a discontinuous drop (continuous on the scale of τ_J) of f to zero from the value f_{min} at which the denominator vanishes. The current dependence of this f_{min} is plotted in Fig. 6 for $T/T_c=0.9$. More physically, one can argue that the strip will act "normal" when the electric field implied by the rapid increase of q to maintain constant $j_s=f^2q$, as f decreases rapidly, is so large as to carry the whole current as a normal current, signaling complete breakdown of the approximation that $j_s=j$. One can show that this condition occurs when

$$\frac{-\partial f}{\partial t} = \frac{1}{2\tau_J} f^3 = \frac{1}{\tau_E} \left(\frac{\tau_E}{2\tau_J}\right) f^3 \tag{65}$$

Since $(\tau_E/2\tau_J)\sim 10^3$, this condition is not far from the condition $\partial f/\partial t\to\infty$ which we use to determine f_{min}.

Having established the limits of the slow phonon-limited reduction of the gap, we can compute the intermediate values of f(t) by integrating the differential equation (64) with the appropriate initial condition f_{max} at $t=0^+$. The total delay time will be given by integration from f_{max} down to f_{min}. Representative curves of f(t) and of $t_d(j)$ are shown in Figs. 4 and 5, respectively.

Discussion. The use of the more correct integro-differential form (54) as opposed to the differential form (54a) has led to the prediction of quasi-discontinuous drops in the energy gap on a time scale $\tau_J << \tau_E$, both when the current step is applied and in the final collapse to the normal state. No direct experimental evidence is yet available to test this qualitative prediction. The principal potentially observable difference in the predictions of the two forms of the theory for t_d is the existence of an "upper critical current" $j_{c1} \simeq j_c/(1-t)^{\frac{1}{4}}\sim 2j_c$, where t_d should drop precipitously with increasing current, as shown in Fig. 5. The value and even the existence of this j_{c1} is, however, called into question by very recent work of Geier, reported at this Institute. He obtained a more rigorous solution by taking into consideration the direct enhancement of j_s by the nonequilibrium quasiparticle population (as well as the indirect enhancement by the effect of the quasiparticles on Δ), an effect discussed in the next section. His result for $t_d(j/j_c)$, sketched in Fig. 5 as curve (c), also shows a drop steeper than j^{-2} at high current levels, but the effect is less dramatic than in our simpler approximation [curve (b)]. Unfortunately, experimental tests of these predictions require work at very short delay times, where measurement difficulties impose limitations. Thus, to date, no data are available which enable one to differentiate between the various predictions.

The absence of any observed temperature dependence in t_d is also puzzling, since all versions of the theory predict that t_d should scale with $\tau_\Delta\sim\tau_E/(1-t)^{\frac{1}{2}}$. This fundamental disagreement, if confirmed in further work, poses the likelihood that the theoretical model treated here is too idealized to incorporate all the relevant physics. For example, phonon disequilibrium is neglected, although it will be large, and the 1-dimensional approximation is at best marginally justified. It also must be emphasized that the measurements of Pals and Wolter were made in a temperature range $[0.76<T/T_c<0.92]$ which excludes the characteristic divergence of τ_Δ near T_c. Moreover, the inequality $\Delta/T << 1$, on which our approximations are based, is only marginally satisfied even at the top of this temperature range. Accordingly, our neglect of Δ^2/T^2 terms seriously limits the quantitative reliability of our results in this temperature range; a more complete theoretical analysis is required to interpret data taken this far below T_c.

Another obvious idealization is the assumption of a spatially uniform system and solution. Any real film strip will have inhomogeneity in critical current, either because of geometrical nonuniformity or variation of T_c. Because the depression of the gap is driven by a term j^2/f^4, this variation of j^2 with position will give rise to a variation of $|f|$ with position which will become accentuated as f drops. This variation is resisted by a term $|f|^{-1}d^2|f|/dx^2$ added to the bracketed quantity in the l.h.s. of (54), but until the variation becomes rapid on a scale of ξ, this will not affect the qualitative behavior. Given that spatial variation, the analysis of the quasiparticle disequilibrium is complicated by the need to include diffusion terms in the Bolzmann equation. These have the effect of reducing the effective τ_E, since the non-equilibrium population is reduced by diffusion away from the region of most rapid generation, i.e., the region where Δ is falling most rapidly. As a result, it seems that even the slightest inhomogeneity will lead to a drop in f which becomes more localized as it drops, and eventually leads to $f\rightarrow 0$ only in the localized region. After this occurs, the resistive region will spread to fill the strip, in agreement with the gradual rise in resistance that is observed by Pals and Wolter to follow after the delay time t_d. However, it must be emphasized that the proper explanation of the temperature-independence of the measured t_d remains an open question at this time.

Dynamic Enhancement in Metallic Weak Links

As a final example, I should like to discuss the implications of nonequilibrium populations which result from the rapid variation of the energy gap at the Josephson frequency in a metallic superconductng weak link. For definiteness, I consider a short link between massive superconductors, an idealized form of the VTB's studied by Octavio et al. and discussed above. As shown by Aslamazov and Larkin (1969), in the limit of a short bridge the solution of the static GL equation for the complex gap is the same as that of Laplace's equation for this geometry, namely a linear interpolation between the boundary values Δ_o at one bank (x=0) and $\Delta_o e^{i\phi}$ at the other (x=L). This implies that

$$|\Delta(x,\phi)|^2=\Delta_o^2[1 - \frac{4x}{L}(1 - \frac{x}{L})\sin^2\frac{\phi}{2}]. \qquad (66)$$

If a voltage V exists between the two banks, the phase ϕ will advance at the Josephson rate $\omega=d\phi/dt = 2eV/\hbar$, and (66) describes an energy gap which cyclically decreases and returns back to Δ_o once each cycle, the greatest decrease being in the middle of the bridge. Insofar as quasiparticle energy levels are defined in such a regime, where Δ is a strong function of position and time, the local values of $E_k = (\Delta^2 + \varepsilon_k^2)^{\frac{1}{2}}$ will also pump up and down periodically in time, along with Δ^2. [See Fig. 7(a).] Only if this variation is very slow compared to the inelastic scattering

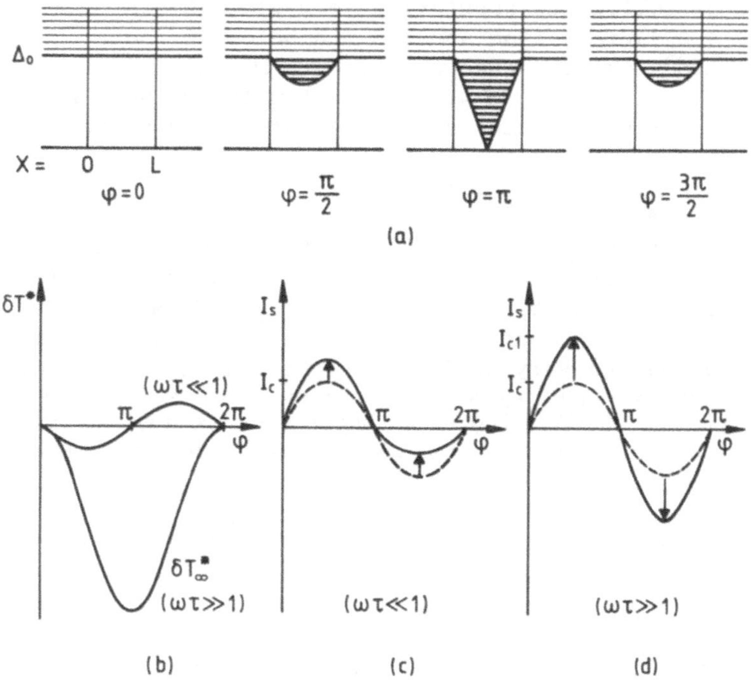

Fig. 7. (a) Cyclic depression of gap as given by Eqn. (66) of text. Quasiparticle levels above Δ_o remain in thermal equilibrium by rapid diffusive contact with the banks. Those in levels below Δ_o (heavy shading) can be out of equilibrium. (b) Effective temperature shifts for rapid and slow modulation of Δ. (c) $I_s(\phi)$ for $\omega\tau < 1$, showing increase in forward current. (d) $I_s(\phi)$ for $\omega\tau > 1$, showing increase of effective I_c to $I_{c1} > I_c$. In (c) and (d), equilibrium supercurrents are shown in dashed curves for comparison.

time τ_E can one assume that the occupation numbers f_k are able to relax and stay in instantaneous thermal equilibrium, so that $f_k = f_o[E_k(t)/T]$. For any finite voltage and hence frequency, there will be departures form equilibrium which imply a periodic instantaneous $\delta T^*(t) \neq 0$.

To find $\delta T^*(t)$, we must know $\delta f_k(t)$. This problem is simplified by noting that for $E_k > \Delta_o$, there is very rapid diffusive contact between the short bridge and the banks, which are assumed to remain in equilibrium. Thus, we take $\delta f_k = 0$ for all $E_k > \Delta_o$, and these levels do not contribute to δT^*. However, since quasiparticles in levels with $E_k < \Delta_o$ are trapped in the bridge, they are not in diffusive contact with a reservoir, and

$\delta f_k \neq 0$. As we follow the population in a given k-state as $E_k = [\Delta^2(t) + \varepsilon_k^2]^{1/2}$ drops from above Δ_o to below, we note that apart from relaxation effects, the occupation numbers will be frozen at the value $f_o(\Delta_o/T)$ they had when contact with the bath was lost. This defines a limiting case of no relaxation for which

$$\delta f_k = \delta f_{k,\infty}(t) \equiv f_o(\Delta_o/T) - f_o[E_k(t)/T], \tag{67}$$

for $E_k = [\Delta^2(x,t) + \varepsilon_k^2]^{1/2} \leq \Delta_o$. Fig. 7(b) shows the corresponding $\delta T^*_\infty(t)$; here the subscript reflects the fact that this limit should hold for $\omega\tau_E \to \infty$. Since $E_k \leq \Delta_o$ in (67), $\delta f_k \leq 0$, and therefore $\delta T^*_\infty(t) \leq 0$. Thus in this limit, there is a cyclic cooling effect, which can be seen to vary qualitatively as $\sin^2(\phi/2)$, being a maximum at $\phi = \pi$ when the gap is most depressed. Since the Josephson current varies as $I_c(T^*)\sin\phi$, such an effect is qualitatively the equivalent of simply an enhancement of I_c to an "effective" value $I_{c1} > I_c$. [See Fig. 7(d).]

The other simple limiting case occurs when $\omega\tau_E << 1$, so that the departures from equilibrium are small. In that case, one can see that

$$\delta f_k = \tau_E \partial(\delta f_{k,\infty})/\partial t = \tau_E(-\partial f_o/\partial E)(\partial E/\partial\Delta)(\partial\Delta/\partial t) \tag{68}$$

From this it follows that

$$\delta T^* \sim \tau_E(\partial\Delta/\partial t) \sim \tau_E[\partial(\delta T^*_\infty)/\partial t] \tag{69}$$

which varies qualitatively as $\sin\phi$, changing sign at $\phi = \pi$; see Fig. 7(b). Keeping track of the sign, $\delta T^* < 0$, giving enhancement of I_c on the forward half cycle of the Josephson current oscillation $(0 < \phi < \pi)$; but $T^* > 0$, giving a reduced reverse current in the second half-cycle $(\pi < \phi < 2\pi)$. [See Fig. 7(c).] The net effect is an algebraic increase in the forward current in both half cycles, the increase being proportional to $\omega\tau_E \propto V$; that is, the average supercurrent acquires a dc component equivalent to adding a normal conductance in parallel with the device. Thus, at low voltages, where $\omega\tau_E << 1$, the nonequilibrium effect is to simulate a lower resistance junction with the same critical current, again a form of enhancement.

When one combines these two regimes, one obtains at least a qualitative explanation for the curious "foot" structure observed by Octavio et al. (1978) on the I-V curves of short microbridges and shown in Fig. 8. Similar, but less complete, observations were reported by Gubankov et al. (1977). Rather than everywhere showing the concave-downward curvature expected from models, such as the resistively shunted junction model, which take no account of nonequilibrium effects, the initial curvature is concave upward. The initial rise in voltage is slow (corresponding to the reduced

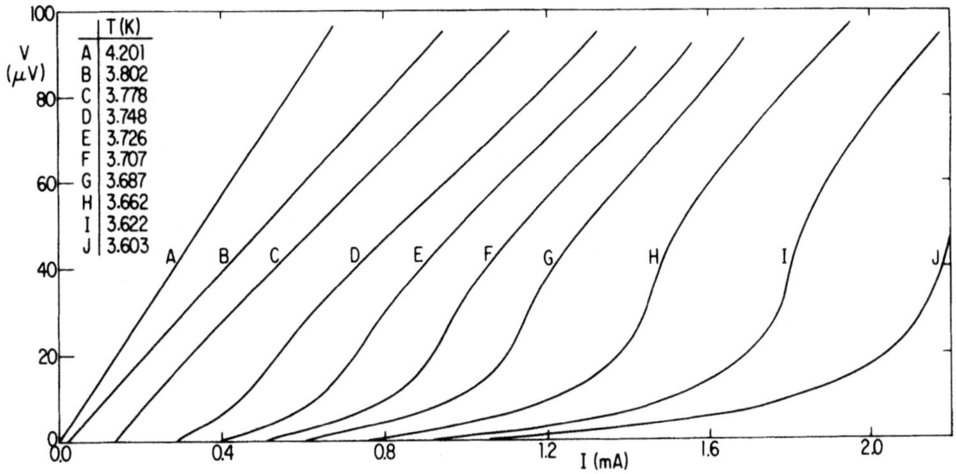

Fig. 8. Low-voltage part of I-V characteristic of tin VTB, studied by Octavio et al. (1978), showing development of "foot" structure as function of temperature.

effective resistance expected for $\omega\tau_E << 1$), while when the voltage at which $\omega\tau_E \sim 1$ is reached, the expected concave-downward shape appears, but extrapolating toward an enhanced critical current $I_{c1} > I_c$, as expected for $\omega\tau_E >> 1$. Thus the qualitative features of the "foot" are accounted for by these nonequilibrium enhancement effects.

Historically speaking, nonequilibrium enhancement effects causing such a "foot" structure were first predicted theoretically by Aslamazov and Larkin (1976) and by Golub (1976). (Their calculations appear to be most applicable in the regimes $\omega\tau_E >> 1$ and $\omega\tau_E << 1$, respectively.) The more physically transparent explanation presented here was developed by the Harvard group, in an effort at interpreting these rather difficult theories, which seemed relevant for understanding the experimental data of Octavio. As presented above and in the work of Octavio et al. (1978), this picture gives only a qualitative account of the data. However, these ideas have more recently been developed in a quantitative fashion by Schmid, Schön, and Tinkham (SST) (1980), using the Schmid-Schön formalism, with attention focussed on the short bridge limit: $L < \xi(0)(1-t)^{-\frac{1}{4}}$. Although it would not be feasible to summarize their entire treatment here, we can highlight a few key points.

A first remark concerns the length dependence. Previous work on the problem had given results which vanish in the limit of zero bridge length. For example, the enhanced critical current I_{c1} of Aslamazov and Larkin scaled linearly with the length L, and the enhanced conductance of Golub scaled as L^4. From the point of view presented above, in which the GL parameter α is modified by a δT^*

term, one would also expect no change to lowest order in L, since the energy gap is given by (66), regardless of the temperature in the bridge, Δ_o being the value of the gap in the banks, where equilibrium is assumed. Rather, one finds only a correction of order $(L^2/\xi^2)(\delta T^*/T_c)$, reflecting the modification of (66) for finite length bridges as first worked out by Likharev and Yakobson (1975). Such length-dependent corrections do indeed occur, but a key finding of SST was that there is a direct contribution to the supercurrent by the nonequilibrium quasiparticle population, in addition to the second-order change which is caused indirectly through an induced change in $\Delta(x,t)$.

Specifically, for a dirty superconductor, SST find the non-equilibrium change in the supercurrent density to be

$$\delta \underline{j}_s = (\sigma/e)\pi\Delta[-\delta f(\Delta)]\underline{q} \tag{70}$$

compared to the equilibrium value

$$\underline{j}_s = (\pi\sigma/4e)(\Delta^2/T)\underline{q} \tag{71}$$

where σ is the conductivity, $\hbar\underline{q}$ is the gauge-invariant superfluid momentum, and $\delta f(\Delta)$ is the (longitudinal mode) nonequilibrium population at $E = \Delta$. Note that if one expresses δf in terms of an equivalent δT^*, as $\delta f(\Delta) = (\Delta/4T^2)\delta T^*$, the result (70) actually would follow directly from (71). A rather similar result for the clean limit can be obtained by considering the expression for the superfluid density in BCS theory:

$$n_s \propto \frac{1}{\lambda_L^2(0)} \left[1-2\int_\Delta^\infty \left(-\frac{\partial f}{\partial E}\right)\frac{E\,dE}{(E^2-\Delta^2)^{\frac{1}{2}}}\right] \tag{72}$$

which gives rise to $n_s \propto \Delta^2/T^2 \propto (1-t)$ in equilibrium. If the effective temperature in the Fermi function and the gap value are decoupled (as is possible in a short bridge, where Δ is constrained by the boundary conditions), this leads to $\delta n_s/n_s = -2\delta T^*/T$, whereas with (71), the ratio is simply $-\delta T^*/T$. The plausibility argument given here for (70) is not entirely limited to nonequilibrium situations describable by a δT^*. The quantity $-\delta f(\Delta)$ in (70) can be written as $\int_\Delta^\infty \delta(\partial f/\partial E)dE$, since $f \to 0$ as $E \to \infty$, so, apart from a weighting function, the nonequilibrium-induced change in the integral in (72) is directly proportional to (70). Accordingly, we can have some confidence that (70) can be given a physical interpretation along these lines.

Given this strong nonequilibrium current even in very short bridges, one is prepared to discuss its magnitude and time dependence in more detail. Let us consider first the low-voltage regime, just above I_{co} in the "foot" region. As we will see, this regime has features which closely resemble those found in the current-induced destruction of superconductivity in a long filament, discussed in the previous section. Upon applying the current, the phase difference ϕ across the link (which is analogous to qL in the filament) moves quickly (i.e., in time $\sim\tau_J$) to a value $(\frac{1}{2}\pi-\delta\phi)$, slightly less than $\pi/2$, where the equilibrium supercurrent $I_{co}\sin\phi$ would be near its maximum value. In the process, the gap is depressed as given by (66), resulting in "cooling" of the quasiparticles, giving a nonequilibrium enhancement of $I_c(T^*)$ to a value essentially equal to the applied current $I>I_{co}$. The rate of continuing phase advance (and the associated tiny voltage and normal current) is then determined by the requirement that $I_c(T^*)$ remain almost equal to I, the time scale being set by the relaxation time τ_E of the nonequilibrium population. There will be a long period of slow phase advance for $\phi\sim\pi/2$, after which the phase will accelerate rapidly when $I_c(T^*)$ can no longer be kept large enough to match I. The electric field associated with the phase acceleration will drive a pulse of normal current while the phase is rapidly sweeping around from $(\frac{1}{2}\pi+\delta\phi)$ through 2π and back to $(\frac{1}{2}\pi-\delta\phi)$ to repeat the cycle. The resemblance to the filamentary case is obvious, but here the strong superconducting banks recreate the superconductivity in the bridge to cause a cyclic Josephson current rather than a one-shot destruction of superconductivity. Since the voltage is essentially negligible during the slow part of the cycle, the time-average voltage $\bar{V}$ arises from the pulses required to swing the phase around at the end of each cycle. By the Josephson relation, $\bar{V} = h/2e(\Delta t)$, where Δt is the duration of the cycle, which is dominated by the slow motion of the phase through the vicinity of $\pi/2$. The numerical results of SST show that over a considerable range of I/I_{co}, $\bar{V}$ rises linearly with a differential resistance $R_{eff}\sim(17\tau_J/\tau_E)R_N$, where $\tau_J = 2\hbar k_B T/\pi\Delta^2 \propto (1-t)^{-1}$. As shown in Fig. 9, the temperature dependence and magnitude of the slope of the foot observed by Octavio in his studies of tin microbridges agrees well with the theory, taking the reasonable value $\tau_E \approx 3\times10^{-10}$ sec.

The "foot" terminates when I reaches the maximum supercurrent that can be carried even when the gap is being depressed rapidly on the scale of τ_E. In the limit of zero bridge length, SST showed that this maximum supercurrent or "upper critical current" I_{c1} was 1.325 I_{co}. More generally, for a finite bridge length L, they found

$$I_{c1}/I_{co} = 1.325 + 0.019\ (\Delta/T)[L/\xi(0)]^2 \tag{73}$$

This result is in satisfactory agreement with the observations of Octavio, considering the considerable scatter in his evaluation of I_{c1} from the smoothly curving experimental curves (Fig. 8). Another way of characterizing the expected limit of the "foot" is to compute $V_{max} = (I_{c1} - I_{co})R_{eff}$, which gives a measure of the

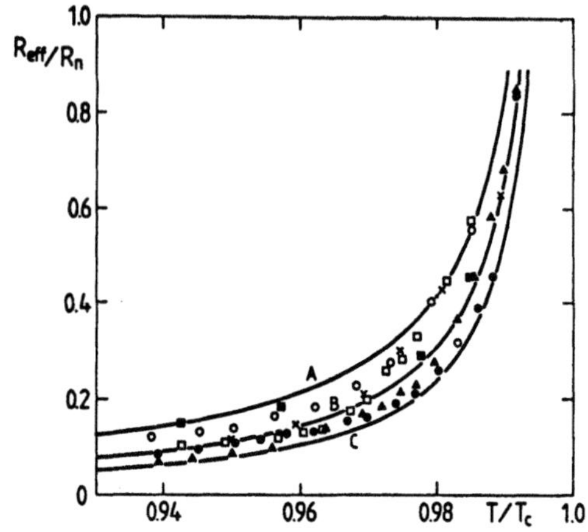

Fig. 9. Normalized differential resistance in the foot region as given by SST theory including finite length corrections for A) L = 0, B) L = $4\xi(0)$, and C) L = $6\xi(0)$. Parameter values: $\tau_E = 2.8 \times 10^{-10}$ sec, $T_c = 3.8$ K. Experimental points are taken from Octavio et al. (1978).

extent of the linear rise in voltage before the limiting enhancement of the supercurrent is reached. In the short bridge limit, the product is simply

$$eV_{max} = 2.76\hbar/\tau_E \qquad (74)$$

For $\tau_E \approx 3 \times 10^{-10}$ sec, V_{max} would be $\sim 7\mu V$, again in satisfactory agreement with Octavio's data. (Fig. 8)

The satisfactory quantitative agreement with theory in the case of the I-V curves of tin VTB's gives evidence in favor of the SST theory. Although the numerical details rely on the use of the full Schmid-Schön formalism, the qualitative picture can be well understood in terms of the modified GL theory with an effective T* to model the nonequilibrium population as presented here.

ACKNOWLEDGEMENT

The work reported here was partially supported by the National Science Foundation, the Office of Naval Research, and the Joint Services Electronics Program.

REFERENCES

Aslamazov, L. G., and Larkin, A. I., 1969, Josephson effect in superconducting point contacts, ZhETF Pis. Red., 9: 150; Engl. trans .JETP Letters, 9: 87.

Aslamazov, L. G., and Larkin, A. I., 1976, Superconducting contacts with nonequilibrium electron distribution functions, Zh. Eksp. Teor. Fiz., 70: 1340; Engl . transl. Sov. Phys. JETP, 43: 698.

Golub, A. A., 1976, Dynamic porperties of short superconducting filaments, Zh. Eksp. Teor. Fiz., 71: 341; Engl. transl. Sov. Phys. JETP, 44: 178.

Gubankov, V. N., Koshelets, V. P., and Ovsyannikov, G. A., 1977, Properties of Josephson thin film variable-thickness microbridges, Zh. Eksp. Teor. Fiz., 73: 1435; Engl. transl. Sov. Phys. JETP, 46: 755.

Holm, R., 1967, "Electric Contacts", Springer-Verlag, Berlin.

Kohlrausch, F., 1900, Ueber den stationären Temperaturzustand eines elektrisch geheizten Leiters, Ann. Phys. (Leipzig), 1: 132.

Klapwijk, T. M., and Mooij, J. E., 1975, Extension of the operating range of superconducting microbridges, IEEE Trans. Magn., MAG-11: 858.

Likharev, K. K., and Yakobson, L. A., 1975, Steady-state properties of superconducting bridges, Zh. Tekh. Fiz., 45: 1503; Engl. transl. Sov. Phys. Tech. Phys., 20: 950.

McDonald, D. G., Kose, V. E., and Evenson, K. M., 1969, Harmonic generation and submillimeter wave mixing with the Josephson effect, Appl. Phys. Lett., 15: 121.

Octavio, M., Skocpol, W. J., and Tinkham, M., 1977, Improved performance of tin variable-thickness superconducting microbridges, IEEE Trans. Mag. MAG-13, 739.

Octavio, M., Skocpol, W. J., and Tinkham, M., 1978, Nonequilibrium-enhanced supercurrents in short superconducting weak links, Phys. Rev. B, 17: 159.

Pals, J. A., and Wolter, J., 1979, Measurement of the order-parameter relaxation in superconducting Al-strips, Phys. Lett., 70A: 150.

Schmid, A., 1968, The approach to equilibrium in a pure superconductor: the relaxation of the Cooper pair denisty, Phys. Kond. Mat., 8: 129.

Schmid, A. and Schön, G., 1975, Linearized kinetic equations and relaxation processes of a superconductor near T_c, J. Low Temp. Phys., 20: 207.

Schmid, A., 1977, Stability of radiation-stimulated superconductivity, Phys. Rev. Lett., 38: 922.

Schmid, A., Schön, G., and Tinkham, M., 1980, Dynamic properties of superconducting weak links, Phys, Rev. B, 21: 5076.

Skocpol, W. J., Beasley, M. R., and Tinkham, M., 1974, Self-heating hotspots in superconducting thin-film microbridges, J. Appl. Phys., 45: 4054.

Stephen, M. J., 1969a, Noise in the ac Josephson effect, Phys. Rev., 182: 539.

Stephen, M. J., 1969b, Noise in a driven Josephson oscillator, Phys. Rev., 186: 393.

Tinkham, M., Octavio, M., and Skocpol, W. J., 1977, Heating effects in high-frequency metallic Josephson devices: Voltage limit, bolometric mixing, and noise, J. Appl. Phys., 48: 1311.

van Dover, R. B., de Lozanne, A., Howard, R. E., McLean, W. L., and Beasley, M. R., preprint, Refractory superconductor S-N-S microbridges,

Weitz, D. A., Skocpol, W. J., and Tinkham, M., 1978, Far-infrared frequency dependence of the ac Josephson effect in niobium point contacts, Phys. Rev., 18: 3282.

Weitz, D. A., Skocpol, W. J., and Tinkham, M., 1978a, Characterization of niobium point contacts showing Josephson effects in the far-infrared, J. Appl. Phys., 49: 4873.

Weitz, D. A., Skocpol, W. J., and Tinkham, M., 1978b, Properties of Josephson point-contact far-infrared detectors, Infrared Phys., 18: 647.

PROPERTIES OF NONEQUILIBRIUM SUPERCONDUCTORS:

A KINETIC EQUATION APPROACH

Jhy-Jiun Chang

Department of Physics
Wayne State University
Detroit, Michigan 48202, U.S.A.

INTRODUCTION

Within the frame work of the BCS theory of superconductivity (Bardeen et al. 1957), a superconductor is a system of three components: The quasiparticles, the phonons, and the paired electrons. In this chapter, we will discuss some of the nonequilibrium properties of superconductors using the kinetic-equation approach. The quasiparticles and phonons will be described by coupled kinetic equations which govern the quasiparticle and phonon distribution functions. The paired electrons are characterized by the energy gap which will be described by a modified gap equation relating the energy gap to the energy spectrum of the quasiparticles. We will use the Golden-Rule approximation in our derivation of the kinetic equations. This approach is simple and brings out clearly the physics involved in the various collision processes. Furthermore, it has been shown that, for all practical purposes, it is equivalent to the more rigorous Green's function approach (Eliashberg, 1970; Entin-Wohlman and Orbach, 1978).

The coupled quasiparticle and phonon kinetic equations were first introduced by Bardeen et al. (1959) to study the thermal conductivity of a superconductor, and were later extended to the strong coupling case by Ambegaokar and Tewordt (1964), and Prange and Kadanoff (1964). Recently, it has been extended even further to include the effects associated with the branch imbalance, the unequal population of the $k>k_f$ and $k<k_f$ branches (Tinkham, 1972; Chang, 1977, 1978, 1979; Paterson, 1978; Pethick and Smith, 1979; Chi and Clarke, 1979a). Here k is the electron wavevector and k_f is its value on the Fermi surface.

In this chapter, only cases in which quasiparticle tunnel injection serves as the driving mechanism will be discussed. We consider cases in which a superconductor is driven into a state far from, as well as close to, equilbrium. In general, in a superconductor far from equilibrium, the relaxation times depend upon the nature of the nonequilibrium state. However, for a system near equilbrium, they are functions of thermodynamic variables and hence would be interesting to consider. We will first introduce the coupled kinetic equations. These equations will then be linearized to discuss the energy-and temperature- dependent quasiparticle and phonon relaxation times associated with the various collision processes. We will also show that different experiments actually probe various weighted averages or combinations of the microscopic lifetimes.

Cases of strongly driven superconductors will be discussed using the full coupled kinetic equations. We will see that when the thermal coupling between the sample and the ambience is weak, a simple analytical model, which is an extension of Parker's T^* model (Parker, 1975; Chang et al., 1979), is adequate. However, when the thermal coupling is good, the nonequilibrium state will reflect the specific signatures of the external driving mechanisms. Many interesting effects result in this case. These include, among other things (See e.g. Chang and Scalapino, 1976, 1978), the enhancement of the energy gap which we will discuss in detail.

THE COUPLED EQUATIONS

We will be interested in homogenous driven steady states. The kinetic equation for the quasiparticle distribution $f(E_k)$ and the phonon distribution $n(\Omega_q)$ have the following forms:

$$\frac{df(E_k)}{dt} = I_{qp}(E_k) + \left.\frac{\partial f(E_k)}{\partial t}\right|_{\text{collision}} \tag{2-1a}$$

and

$$\frac{dn(\Omega_q)}{dt} = I_{ph}(\Omega_q) + \left.\frac{\partial n(\Omega_q)}{\partial t}\right|_{\text{collision}} \tag{2-1b}$$

Here $I_{qp}(E_k)$ and $I_{ph}(\Omega_q)$ are respectively the energy spectra of the quasiparticle and phonon injection rates. The other terms on the right side represent the collision rates associated with the various processes. Since we will consider only tunnel injection of quasiparticles as the driving mechanism, $I_{ph}(\Omega_q)$ is zero. In the following, we first discuss $I_{qp}(E_k)$. Various collision rates will be discussed next.

Tunneling Processes and Quasiparticle Injection Rates

We consider a tunnel junction formed by two superconducting thin films separated by an insulating layer. The tunneling Hamiltonian (Bardeen, 1961, 1962; Cohen et al, 1962) has the form

$$H_T = \sum_{k,q,\sigma} T_{kq} C_{k\sigma}^+ C_{q\sigma} + \text{Hermitian Conjugates} \qquad (2\text{-}2)$$

Here we use k(q) for the wavevector of a quasiparticle in the left (right) film of the junction. In terms of the quasiparticle excitation operators (See e.g. Schrieffer, 1964; Tinkham 1975):

$$\gamma_{eko}^{\ +} = u_k C_{k\uparrow}^+ - v_k S_k^+ C_{-k\downarrow} \qquad (2\text{-}3a)$$

$$\gamma_{ek1}^{\ +} = u_k C_{-k\downarrow}^+ + v_k S_k^+ C_{k\uparrow} \qquad (2\text{-}3b)$$

$$\gamma_{hk}^{\ +} = S_k \gamma_{ek}^{\ +}$$

the Hamiltonian becomes

$$H_T = \sum_{k,q} T_{kq} (u_k \gamma_{eko}^+ + v_k \gamma_{ek1} S_k^+)(u_q \gamma_{eqo} + v_q S_q \gamma_{eq1}^+) + \text{Hermitian Conjugates} \qquad (2\text{-}4)$$

Here $S^+(S)$ is the pair creation (destruction)operator and u_k and v_k are the well known BCS coherence factors. Equation (2-4) is for a single spin only. It is easy to show that terms due to tunneling of different spin do not interfere with each other. The effect of the sum over spin, therefore, gives a factor two in the transition rates and currents calculated using Eq. (2-4).

The four terms, together with their Hermitian conjugates, give the four distinct tunneling processes shown in Fig. 1.

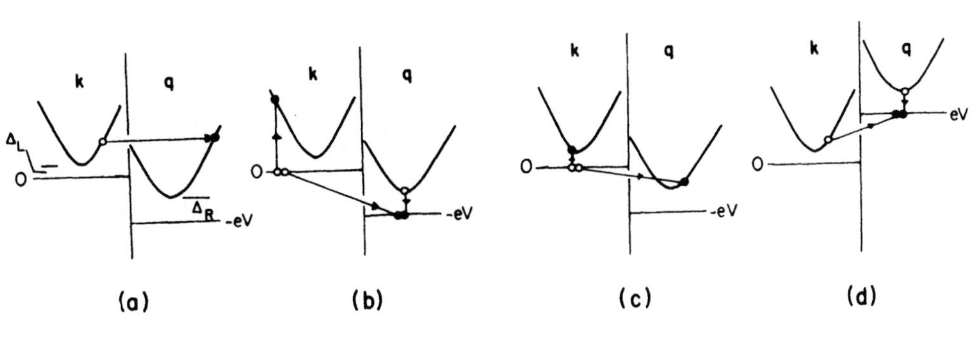

Fig. 1. The four quasiparticle tunneling processes in a superconductor-insulator-superconductor (SIS) junction.

Using the Golden-Rule approximation, the rates of the quasiparticle injection into the right hand film can be shown to be given by

$$
\begin{aligned}
W_q^a &= \frac{2\pi}{\hbar\Omega_R} \sum_k |T_{kq}\ u_k u_q|^2\ (f_k - f_q)\,\delta(E_q - E_k - eV) \\
W_q^b &= \frac{2\pi}{\hbar\Omega_R} \sum_k |T_{kq}\ v_k v_q|^2\ (f_k - f_q)\,\delta(E_q - E_k + eV) \\
W_q^c &= \frac{2\pi}{\hbar\Omega_R} \sum_k |T_{kq}\ v_k u_q|^2\ (1 - f_k - f_q)\,\delta(E_q + E_k - eV) \\
W_q^d &= \frac{2\pi}{\hbar\Omega_R} \sum_k |T_{kq}\ u_k v_q|^2\ (1 - f_k - f_q)\,\delta(E_q + E_k + eV)
\end{aligned}
\tag{2-5}
$$

for the four processes. Here we have defined the junction voltage $V = V_q - V_k$.

These injection rates can be easily expressed in terms of the junction conductance if elastic scattering is strong enough to average out the angular dependence of $\underline{k}$ and $\underline{q}$. We can then convert the sum over the wavevectors $\underline{k}$ and $\underline{q}$ into integrals over the quasiparticle energies E_L and E_R, respectively, for the left and right films. However, due to the different coherence factors involved for states on the $q>q_f$ and $q<q_f$ branches, $f_>(E)$ and $f_<(E)$ are, in general, different in the two junction films. We have

$$
\begin{aligned}
W^a_{\gtrless}(E_R) = \frac{G_{nn}}{2e^2 N_R(0)\Omega_R}\ \rho_L(E_R - eV)\{u_{L>}^2(E_R - eV)\,u_{R\gtrless}^2(E_R)\,[f_{L>}(E_R - eV) \\
- f_{R\gtrless}(E_R)] + u_{L<}^2(E_R - eV)\ u_{R\gtrless}^2(E_R)\,[f_{L<}(E_R - eV) - f_{R\gtrless}(E_R)]\}
\end{aligned}
$$

$W^b_{\gtrless}(E_R) = W^a_{\gtrless}(E_R)$ with the replacements of V by (-V) and u by v.

$$
\begin{aligned}
W^c_{\gtrless}(E_R) = \frac{G_{nn}}{2e^2 N_R(0)\Omega_R}\ \rho_L(-E_R + eV)\{v_{L>}^2(-E_R + eV)\,u_{R\gtrless}^2(E_R)\ [1 - f_{L>}(-E_R + eV) \\
- f_{R\gtrless}(E_R)] + v_{L<}^2(-E_R + eV)\,u_{R\gtrless}^2(E_R)\,[1 - f_{L<}(E_R + eV) - f_{R\gtrless}(E_R)]\}
\end{aligned}
$$

$W^d_{\gtrless}(E_R) = W^c_{\gtrless}(E_R)$ with replacements of V by (-V) and interchange roles of u and v.

(2-6)

Here $G_{nn} = \frac{4\pi e^2}{\hbar} T^2\ N_L(0)\ N_R(0)\ \Omega_L\ \Omega_R$ is the normal state conductance of the junction, T is the effective tunneling matrix element, Ω is the volume under injection, and $\rho_L(E_L) \equiv \theta(E_L-\Delta_L) \times |E_L|/(E_L - \Delta_L)^{\frac{1}{2}}$ is the normalized BCS density of states function for the quasiparticles in the left film. The total quasiparticle injection rate into the right film for the $q \gtrless q_F$ branches are

$$I_{qp\lessgtr}(E_R) = W^a_{\gtrless}(E_R) + W^b_{\gtrless}(E_R) + W^c_{\gtrless}(E_R) + W^d_{\gtrless}(E_R) \qquad (2\text{-}7)$$

The injection spectra $I_{qp\gtrless}(E_R)$ are shown in Fig. 2 for a symmetric junction of $\Delta_L = \Delta_R = \Delta$ and $eV = \Delta$ and 3Δ. The equilibrium distribution function $f(E,T)$ has been used for $f_>$ and $f_<$ on the right side of Eq. (2-6). This approximation introduces error of order G_{nn}^2 or T^4 only. The injection spectrum due to tunneling of quasiparticles from a normal metal can be obtained by setting $\Delta_L = 0$.

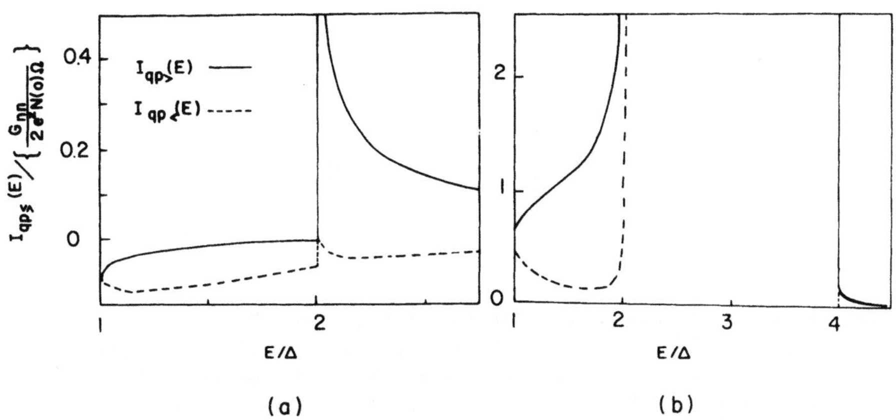

Fig. 2. The energy spectra of the quasiparticle injection rate of a SIS junction with $\Delta_R=\Delta_L=\Delta$. a) $T_a=0.9T_c$ and $eV=\Delta(T)$. b) $T_a=0.5T_c$ and $eV=3\Delta(T)$.

Collision Rates

We will consider relaxation due to the inelastic electron-phonon interaction, the elastic electron-impurity scattering and the "relaxation of phonons" by escaping from the sample film. In general, these are the most important relaxation mechanisms.

(a) Inelastic Electron-Phonon Collision

The Hamiltonian for the electron-phonon interaction has the well known form (See e.g. Schrieffer, 1964):

$$H_{e-ph} = \sum_{pq\sigma\lambda} g_{q\lambda} C^+_{p+q,\sigma} C_{p\sigma} (a_{q\lambda} + a^+_{-q\lambda}) \qquad (2\text{-}8)$$

In terms of the quasiparticle operators [Eq. (2-3)], it becomes

$$H_{e-ph} = \sum_{pq\lambda} g_{q\lambda} \{L(p+q,p) [\gamma^{+}_{e(p+q)0}\gamma_{ep0} + \gamma^{+}_{ep1}\gamma_{e(p+q)1}] + M(p+q,p) [\gamma_{e(p+q)0}\gamma_{hp1} - \gamma_{ep0}\gamma_{h(p+q)1}]\}(a_{q\lambda}+a^{+}_{-q\lambda}) \quad (2\text{-}9)$$

with

$$L(p',p) \equiv u_{p'}u_p - v_{p'}v_p = \tfrac{1}{2}(1 - \frac{\Delta^2-\xi_p\xi_{p'}}{E_pE_{p'}}) \quad (2\text{-}10a)$$

and

$$M(p',p) \equiv u_{p'}v_p + v_{p'}u_p = \tfrac{1}{2}(1 + \frac{\Delta^2-\xi_p\xi_{p'}}{E_pE_{p'}}) \quad (2\text{-}10b)$$

The basic electron-phonon collision processes resulting from this Hamiltonian are illustrated schematicaly in Fig. 3.

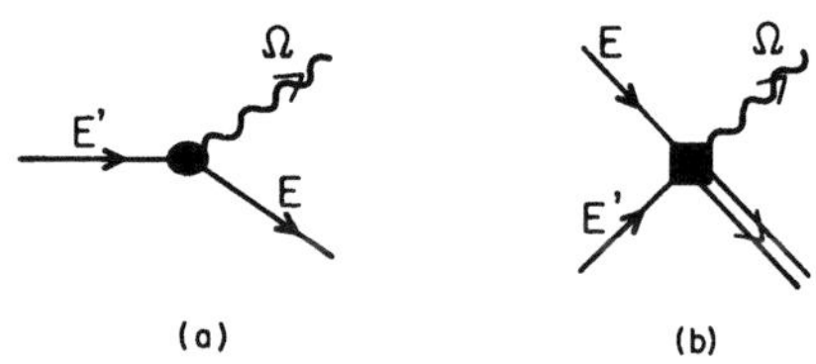

Fig. 3. Basic quasiparticle (solid lines) and phonon (wavy lines) collision processes a) Quasiparticle or phonon scattering process. b) Quasiparticle recombination or phonon pair-breaking process. The double line represents paired electrons.

In the Golden-Rule approximation, these inelastic collisions contribute to the kinetic equation a term:

$$\left.\frac{\partial f(\xi_p)}{\partial t}\right|_{in} = \{ \frac{2\pi}{\hbar} \sum_{q\lambda} |g_{q\lambda}|^2 L^2(p+q,p) [f(\xi_{p+q}) (1-f(\xi_p)) (n(\Omega_{q\lambda}) +1) - (1-f(\xi_{p+q}))f(\xi_p) n(\Omega_{q\lambda})]\delta(\Omega_{q\lambda} + E_p-E_{p+q})\}$$

$$- \{p\leftrightarrow p+q\} - \frac{2\pi}{\hbar} \sum_{q\lambda} |g_{q\lambda}|^2 M^2(p+q,p) [f(\xi_p)f(\xi_{p+q})(n(\Omega_{q\lambda})+1) - (1-f(\xi_p))(1-f(\xi_{p+q}))n(\Omega_{q\lambda})]\delta(E_p+E_{p+q}-\Omega_{q\lambda}) \quad (2\text{-}11a)$$

They also contribute to the phonon kinetic equation a term:

$$\left.\frac{\partial n(\Omega_{q\lambda})}{\partial t}\right|_{in} = -\frac{2\pi}{\hbar}\sum_{p} |g_{q\lambda}|^2 L^2(p+q,p)[f(\xi_p)(1-f(\xi_{p+q}))n(\Omega_{q\lambda})$$

$$- f(\xi_{p+q})(1-f(\xi_p))(n(\Omega_{q\lambda})+1)]\delta(\Omega_{q\lambda} + E_p - E_{p+q})$$

$$- \frac{2\pi}{\hbar}\frac{1}{2}\sum_{p,q} |g_{q\lambda}|^2 M^2(p+q,p) [(1-f(\xi_p))(1-f(\xi_{p+q}))n(\Omega_{q\lambda})$$

$$- f(\xi_p)f(\xi_{p+q})(n(\Omega_{q\lambda}) +1)]\delta(\Omega_{q\lambda} - E_p - E_{p+q}) \tag{2-11b}$$

Again, taking advantage of the strong elastic collision, one can simplify the above expressions by averaging out the angular dependence. Furthermore, using the same treatment as Eliashberg in dealing with strong superconductors (Eliashberg 1960; Scalapino, 1969), one can eliminate $g_{q\lambda}$ in favor of $\alpha^2 F$.

As previously shown, the quasiparticle injection rates for the $k_>$ and $k_<$ branches are different. Therefore in the steady state $f_>(E)$ and $f_<(E)$ may also be different. For convenience in our discussion to follow, we will consider the functions

$$f_+(E) \equiv \tfrac{1}{2}[f_>(E) + f_<(E)] \tag{2-12a}$$

and

$$f_-(E) \equiv [f_>(E) - f_<(E)] \tag{2-12b}$$

which are respectively symmetric and antisymmetric with respect to the $k_>$ and $k_<$ branches.

The collision rates for the distribution functions $f_+(E)$ and $f_-(E)$ can be easily obtained from Eq. (2-11a). We have

$$\left.\frac{\partial f_+(E)}{\partial t}\right|_{in} = -\frac{2\pi}{\hbar}\int_0^\infty dE'\rho(E')\ \{\alpha^2F(|E-E'|)\ [\theta(E-E')\ ((1-\frac{\Delta 2}{EE'})$$

$$\cdot\ (f_+(E)(1-f_+(E'))\ (n(E-E')-(1-f_+(E))f_+(E')n(E-E'))$$

$$+ \frac{|\xi\xi'|}{4EE'} f_-(E)f_-(E')]-[E\rightarrow E' \text{ and } E'\rightarrow E]$$

$$+ \alpha^2F(|E+E'|)\ [(1+\frac{\Delta^2}{EE'})\ (f_+(E)f_+(E')(n(E-E')+1)-(1-f_+(E))\cdot$$

$$(1-f_+(E'))n(E+E')] + \frac{|\xi\xi'|}{4EE'} f_-(E)f_-(E')\} \tag{2-13a}$$

and

$$\left.\frac{\partial f_-(E)}{\partial t}\right|_{in} = -\frac{2\pi}{\hbar}\int_0^\infty dE'\rho(E')\ f_-(E)\ \{\alpha^2F(\ E-E'\ )\cdot((1-\frac{\Delta^2}{EE'})$$

$$[\theta(E-E')\cdot(1+n(E-E')-f_+(E'))+\ \theta(E'-E)\cdot(f_+(E')\ +\ n(E'-E))]$$

$$+\ \alpha^2F(|E+E'|)\ (1+\frac{\Delta^2}{EE'})\ (f_+(E')\ +\ n(E+E'))\}$$

$$+\frac{2\pi}{\hbar}\int_0^\infty dE'\rho(E')\ f_-(E')\ \frac{|\xi\xi'|}{EE'}\ \{\alpha^2F(|E-E'|)$$

$$[\theta(E-E')(f_+(E)\ +\ n(E-E'))+\ \theta(E'-E)\cdot(1+n(E'-E)-f_+(E))$$

$$-\ \alpha^2F(|E+E'|)(f_+(E)+n(E+E'))\} \quad (2\text{-}13b)$$

In terms of $f_+(E)$ and $f_-(E)$, the inelastic collision rate for phonons becomes

$$\left.\frac{\partial n(\Omega)}{\partial t}\right|_{in} = -\frac{2\pi}{\hbar}\frac{4N(0)}{N}\alpha^2\int_0^\infty dE\rho(E)\int_0^\infty dE'\rho(E')$$

$$\cdot\ \{[(1-\frac{\Delta^2}{EE'})(f_+(E)(1-f_+(E'))n(\Omega)-(1-f_+(E))\cdot f_+(E')(n(\Omega)+1))$$

$$+\ \frac{|\xi\xi'|}{4EE'}\ f_-(E)f_-(E')]\delta(\Omega\ +E-E')$$

$$+\ \frac{1}{2}[(1+\frac{\Delta^2}{EE'})((1-f_+(E))\ (1-f_+(E'))n(\Omega)-f_+(E)f_+(E')$$

$$\cdot\ (n(\Omega)+1))+\frac{|\xi\xi'|}{4EE'}\ f_-(E)f_-(E')]\delta(\Omega-E-E') \quad (2\text{-}13c)$$

(b) Elastic Electron-Impurity Collision

Although strong elastic collisions enable us to average out the angular dependence in momentum space, they do not contribute to the branch relaxation if the energy gap is isotropic. This is because the relevant collision matrix has a coherence factor $L(p',p)$ associated with it, and in the case of elastic scattering $L(p',p) = 0$ (See Eqs. (2-9 and 10a)). When the energy gap is anisotropic, it contributes a term

$$\left.\frac{\partial f_-(E)}{\partial t}\right|_{el} = -\ \frac{2f_-(E)}{\tau_{el}(E)} \quad (2\text{-}14)$$

to the kinetic equation (Tinkham, 1972).

(c) Phonon Escape

Through this mechanism the sample gets rid of energy injected by the external source. We are interested in phonons of low energies ($\hbar\Omega\sim 2\Delta<<\hbar\omega_D$) and will use a phenomenological term

$$\left.\frac{\partial n(\Omega)}{\partial t}\right|_{esc} = -\frac{n(\Omega)-n^\circ(\Omega)}{\tau_{es}} \tag{2-15}$$

to take into account this process. Apparently, the phonon escape rate $\tau_{es}{}^{-1}$ depends strongly upon the geometry of the sample and the acoustic match between the sample film and the substrate.

In summary, we have three coupled kinetic equations,

$$\frac{df_+(E)}{dt} = I_{qp}{}^{+}(E) + \left.\frac{\partial f_+(E)}{\partial t}\right|_{in} \tag{2-16a}$$

$$\frac{df_-(E)}{dt} = I_{qp}{}^{-}(E) + \left.\frac{\partial f_-(E)}{\partial t}\right|_{in} + \left.\frac{\partial f_-(E)}{\partial t}\right|_{el} \tag{2-16b}$$

and

$$\frac{dn(\Omega)}{dt} = I_{ph}(\Omega) + \left.\frac{\partial n(\Omega)}{\partial t}\right|_{in} + \left.\frac{\partial n(\Omega)}{\partial t}\right|_{esc} \tag{2-16c}$$

with the quasiparticle injection rates defined to be

$$I_{qp}{}^{+}(E) = \frac{1}{2}\{I_{qp>}(E) + I_{qp<}(E)\} \tag{2-17a}$$

and

$$I_{qp}{}^{-}(E) = I_{qp>}(E) - I_{qp<}(E) \tag{2-17b}$$

Equations (2-8, 14, 15, 16, 17 and 18) form the basis of our investigation.

SYSTEMS DRIVEN NEAR EQUILIBRIUM; MICROSCOPIC RELAXATION TIMES

To obtain the various microscopic relaxation times of the quasiparticles and phonons due to inelastic proceses, we linearize the collision-rate equations (Eqs. (2-13a, b, and c)) with respect to the deviation functions:

$$\delta f_+(E) \equiv f_+(E) - f^\circ(E) \tag{3-1a}$$

$$\delta f_-(E) \equiv f_-(E) \tag{3-1b}$$

and

$$\delta n(\Omega) \equiv n(\Omega) - n^\circ(\Omega) \tag{3-1c}$$

Here $f^\circ(E)$ and $n^\circ(\Omega)$ are respectively the Fermi and Bose functions associated with the nonequilibrium energy gap and the ambient temperature T_a. The zeroth-order terms cancel because of detailed balance in the equilibrium state. In the linear approximation, the rate equations have the form:

$$\left.\frac{\partial f_+(E)}{\partial t}\right|_{in} = -\int_0^\infty dE' 4N(0)\rho(E')\{\delta f_+(E)[K_1(E,E') + K_1(-E,-E')$$
$$+ K_1(E,-E')] + \delta f_+(E')[K_1(E',E) + K_1(-E',-E) - K_1(E',-E)]\}$$
$$- \int_0^\infty d\Omega\ NF(\Omega)\ \delta n(\Omega)\ [K_2(\Omega,E) + K_2(-\Omega,E) - K_2(\Omega,-E)]$$
(3-2a)

$$\left.\frac{\partial f_-(E)}{\partial t}\right|_{in} = -\int_0^\infty dE' 4N(0)\rho(E')\{\delta f_-(E)[K_1(E,E') + K_1(-E,-E') + K_1(E,-E)]$$
$$+ \delta f_-(E')[K_3(E,E') + K_3(-E,-E') + K_3(E,-E')]\} \quad (3\text{-}2b)$$

and

$$\left.\frac{\partial n(\Omega)}{\partial t}\right|_{in} = -\int_0^\infty dE\ 4N(0)\rho(E)\ \delta n(\Omega)\ [K_2(\Omega,E) + \frac{1}{2}K_2(\Omega,-E)]$$
$$- \int_0^\infty dE\ 4N(0)\rho(E)\ \delta f_+(E)\ \frac{4N(0)}{NF(\Omega)}\ [-\rho(E-\Omega)K_1(E,E-\Omega)$$
$$+ \rho(\Omega+E)K_1(-E,-(E+\Omega)) - \rho(\Omega-E)\ K_1(E,-(E-\Omega))] \quad (3\text{-}2c)$$

with the functions K_1, K_2 and K_3 defined to be

$$K_1(E,E') \equiv \frac{2\pi}{\hbar}\frac{\alpha^2F(E-E')}{4N(0)}\left(1-\frac{\Delta^2}{EE'}\right)\theta(E-E')(1+n^\circ(E-E')-f^\circ(E'))$$
$$K_2(\Omega,E) = \frac{2\pi}{\hbar}\frac{\alpha^2F(\Omega)}{NF(\Omega)}\rho(\Omega+E)\left(1-\frac{\Delta^2}{E(\Omega+E)}\right)\theta(\Omega+E-\Delta)(f^\circ(E)-f^\circ(\Omega+E))$$

and

$$K_3(E,E') = \frac{2\pi}{\hbar}\frac{\alpha^2F(\Omega)}{4N(0)}(\rho(E)\rho(E'))^{-1}\theta(E-E')(f^\circ(E)+n^\circ(E-E'))$$

Note that $\delta f_+(E)$ and $\delta f_-(E)$ are decoupled and only $\delta f_+(E)$ appears in the phonon rate equation. The relaxation rates for $\delta f_+(E)$ and $\delta f_-(E)$ are the coefficients of $\delta f_+(E)$ and $\delta f_-(E)$ in Eqs. (3-2a and 2b), and we have

$$[\tau_{qp}^{+}(E)]^{-1} = [\tau_{qp}^{-}(E)]^{-1} = [\tau_s^{-1}(E) + \tau_R^{-1}(E)]. \tag{3-2a}$$

The phonon relaxation rate is

$$[\tau_{ph}(\Omega)]^{-1} = [\tau_s^{ph}(\Omega)]^{-1} + \tau_B^{-1}(\Omega) \tag{3-3b}$$

Here the quasiparticle scattering rate τ_s^{-1}, the quasiparticle recombination rate τ_R^{-1}, the phonon scattering rate τ_s^{ph-1}, and the phonon pair-breaking rate τ_B^{-1} are given by

$$\tau_s^{-1}(E) \equiv \int_0^\infty dE' 4N(0)\rho(E')\ [K_1(E,E') + K_1(-E,-E')]$$

$$\tau_s^{-1}(E) \equiv \int_0^\infty dE' 4N(0)\rho(E')\ K_1(E,-E')$$

$$[\tau_s^{ph}(\Omega)]^{-1} \equiv \int_0^\infty dE' 4N(0)\rho(E')\ K_2(\Omega,E')$$

and

$$\tau_B^{-1}(\Omega) \equiv \int_0^\infty dE' 4N(0)\rho(E')\frac{1}{2}\ K_2(\Omega,-E')$$

Connection With The Macroscopic Relaxation Times

(a) Quasiparticle Density Relaxation Time

The quasiparticle density relaxation time τ_d is defined as

$$\tau_d^{-1} \equiv I_{qp} / \delta N \tag{3-4b}$$

Here $I_{qp}^{+} \equiv \int_0^\infty dE\ 4(N(0)\ \rho(E)\ I^+(E)$ is the total quasiparticle density injection rate, and $\delta N \equiv \int_0^\infty dE\ 4N(0)\rho(E)\delta f_+(E)$ is the total excess quasiparticle density in the nonequilibrium state. To find a relationship between I_{qp}^{+} and δN, we substitute (2-2a) into Eq. (2-16a) and sum over the quasiparticle states. We have

$$I_{qp}^{+} = \int_0^\infty dE\ 4N(0)\delta f_+(E)\ 2\tau_R^{-1}(E) - \int_0^\infty d\Omega\ NF(\Omega)\delta n(\Omega)\ 2\tau_B^{-1}(\Omega)$$

$$\equiv 2\delta N \langle\tau_R^{-1}\rangle_{qp} - 2\delta N_\omega \langle\tau_B^{-1}\rangle_{ph} \tag{3-5}$$

Here $\delta N_\omega \equiv \int_{2\Delta}^{\infty} d\Omega NF(\Omega)\ \delta n(\Omega)$ is the excess density of phonons with energy $\hbar\Omega > 2\Delta$. There are three points worth noting in Eq. (3-5). First, the scattering process which changes the quasiparticle spectrum but not the density does not contribute. Second, only phonons with energy $\hbar\Omega > 2\Delta$ are relevant, because they can break pairs to change the quasiparticle density. Finally, the averages $<\cdots>_{qp}$ and $<\cdots>_{ph}$ are over the excess distributions of quasiparticles and phonons respectively.

To calculate δN_ω we substitute (3-2c) into Eq. (2-16c) and sum over the phonon states. We obtain

$$\delta N_\omega (<1/\tau_s^{ph} + \tau_B^{-1} + \tau_{es}^{-1}>)_{ph} - \delta N(<-\tau_{sE}^{-1} + \tau_{sA}^{-1} - \tau_{es}^{-1}>_{qp}) = 0 \qquad (3\text{-}6)$$

Here $\tau_{sE}^{-1} \equiv \int dE' 4N(0)\rho(E')K_1(E,E')\theta(E-E'-2\Delta)$ and $\tau_{sA}^{-1} \equiv \int dE' 4N(0)\rho(E')\ K_1(-E,-E')\theta(E'-E-2\Delta)$ are the quasiparticle scattering rates associated with, respectively, the emission and absorption of a phonon of energy $\hbar\Omega > 2\Delta$. Eliminating δN_ω from Eqs. (3-5 and 6), we obtain

$$\tau_d^{-1} = 2<\tau_k^{-1}>_{qp} \{1 - \frac{1 + <\tau_{sE}^{-1} - \tau_{sA}^{-1}>_{qp}/<\tau_R^{-1}>_{qp}}{1 + [<1/\tau_s^{ph} + \tau_{es}^{-1}>_{ph}]/<\tau_B^{-1}>_{ph}}\} \qquad (3\text{-}7)$$

If the effects of the scattering processes can be ignored, equation (3-7) reduces to

$$\tau_d^{-1} = 2<\tau_R^{-1}>_{qp}/[1 + <\tau_B^{-1}>_{ph}/<\tau_{es}^{-1}>_{ph}] \qquad (3\text{-}8)$$

which is the same as Eq. (19) in Chapter 5.

Therefore, τ_d^{-1} is a direct measure of the quasiparticle recombination rate averaged over the excess quasiparticle distribution for the phonon escape time $\tau_{es} = 0$. When τ_{es} is finite, the phonon trapping factor $<\tau_B^{-1}>_{ph}/<\tau_{es}^{-1}>_{ph}$ reduces τ_d^{-1} from the average recombination rate.

(b) Quasiparticle Charge Relaxation Time

As will be discussed in Chapters 13 and 14, the quasiparticle charge and its relaxation time are quantities of physical interest. Since a quasiparticle excitation of wavevector k is a combination of u_k^2 electron and v_k^2 hole, the net charge of the excitation, in units of (-e) is

$$q_k = u_k^2 - v_k^2 = \pm \rho^{-1}(E_k) \qquad \text{for } k \gtrless k_F \qquad (3\text{-}9)$$

The quasiparticle charge relaxation rate is defined as

$$\tau_Q{}^{*-1} \equiv I_Q{}^*/Q^* \qquad (3\text{-}10)$$

Here the steady state quasiparticle charge density is

$$Q^* \equiv \sum_k q_k f_k = \int_0^\infty dE\ 2N(0)\ \delta f_-(E) \tag{3-11}$$

while the quasiparticle density injection rate is

$$I_Q^* \equiv \sum_k q_k I(E_k) = \int_0^\infty dE\ 2N(0)\ I_{qp}^-(E)$$

$$\equiv \int_0^\infty dE\ 2N(0)\delta f_-(E)\ [\tau_Q^{*-1}(E) + \tau_{el}^{-1}(E)] \tag{3-12}$$

We have used Eq. (2-16b) to obtain the last equality. The inelastic branch relaxation rate

$$\tau_Q^{*-1}(E) \equiv \frac{2\pi}{\hbar} \int_0^\infty dE'\rho(E')\{[\alpha^2F(|E'-E|)\theta(E'-E)\ (1 - \frac{\Delta^2}{EE'}$$

$$- \rho^{-2}\ (E'))\ (f^\circ(E') + n^\circ\ (E'-E))] + [E\to -E,\ E'\to -E] + [E\to -E]\}$$

(3-13)

is the same as that obtained by Chang (1979) using another method.

In general, it requires a tedious calculation to obtain $\delta f_-(E)$ before one can calculate $\tau_Q^{*\,-1}$ from Eqs. (3-10, 11, and 12). However, when $T \lesssim T_c$, experiment shows that τ_Q^* is long compared to the quasiparticle scattering and recombination times. Therefore one can assume that the quasiparticles are in thermal equilibrium except for a finite amount of excess quasiparticle charge density

$$\sum_k q_k \langle \gamma_{eko}^+ \gamma_{eko} + \gamma_{ek1}^+ \gamma_{ek1} \rangle = \text{constant} \tag{3-14}$$

Using the method of Lagrange multipliers, one can easily show that the nonequilibrium distribution function is given by

$$f(E_k,\ \delta\mu,\ T) = 1/[e^{(E_k - \delta\mu q_k)/kT} + 1] \tag{3-15}$$

Here $\delta\mu$ is determined by the constant in Eq. (3-14). Therefore

$$\delta f_-(E) = \frac{\partial f^\circ(E)}{\partial E}\ |q_k|\ (-2\delta\mu) + 0(\delta\mu)^2 \tag{3-16}$$

Substituting (3-16) into Eq. (3-10) and utilizing the asymptotic form of

$$\tau_Q^{*-1}(E,T) = \tau_o^{-1} \frac{\pi}{2} \left(\frac{E}{kT_c}\right)^2 \left(\frac{\Delta}{kT_c}\right) \coth\left(\frac{E}{2kT_c}\right) \tag{3-17}$$

for $T \to T_c$, derived by Chang (1979), one obtains

$$\tau_Q^{*-1} = \frac{\pi}{4} \frac{\Delta}{kT_c} \tau_{in}^{-1} (\xi=0,T_c) \tag{3-18}$$

This is the same result obtained earlier by Schmid and Schon (1975), and Pethick and Smith (1979).

(c) Measurement of the Microscopic Quasiparticle Relaxation Times

Recently, it has been suggested (Chang, 1977; Chang and Yang, 1978) that the microscopic times $\tau_{qp}(E)$ and $\tau_{el}(E)$ can be measured by the tunneling technique using a double-tunnel-junction structure. To obtain these relaxation times, one studys the structures in the current-voltage (I-V) characteristic associated with the nonequilibrium quasiparticle distribution (See e.g. Kaplan et al., 1977 and Kirtley et al., 1980). Consider an Al/Al/Pb-Bi double junction which they have studied. The low resistance Al/Al junction serves as an injector to drive the center Al film. The driven state is probed by the high resistance Al/Pb-Bi detector junction. When the injector junction is biased at $|V_i| > 2\Delta/e$, four structures in addition to the usual ones at $V_d = \pm(\Delta_{pb} \pm \Delta_{Al})/e$ are seen in the detector d^2I/dV^2 at voltages $V_d = \pm [\Delta_{pb} \pm (\Delta_{Al} - |eV_i|)]/e$ (Fig. 4). Two of them appear at $|V_d| = \pm(\Delta_{pb} + \Delta_{Al} - |eV_i|)/e$ with

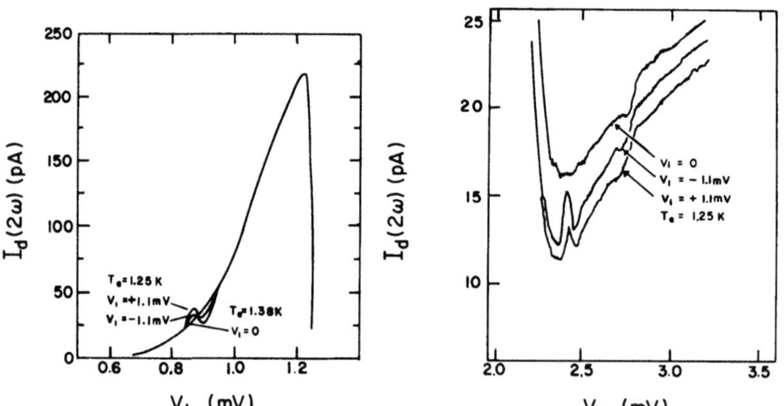

Fig. 4. Second-harmonic current signal vs detector junction voltage. The nonequilibrium structures occur at $V_d=(\Delta_{pb}+\Delta_{Al}-|eV_i|)/e = 0.84$mV and $V_d=(\Delta_{pb}-\Delta_{Al}+|eV_i|)/e = 2.32$mV. The curves for nonzero detector biases have been vertically displaced for $V_d>2$mV for clarity.

$|V_d|$ smaller than the gap-difference voltage. They correspond to small steps in the detector I-V curve. The other two appear at voltages $V_d = \pm(\Delta_{Pb}-\Delta_{Al} + |eV_i|)/e$ with $|V_d|$ greater than the gap-sum voltage. They correspond to small peaks and dips in the detector I-V curve. Furthermore, for each type of structure which occurs at the detector voltages of equal magnitude but opposite sign, the amplitudes are different. This asymmetric effect increases with injection voltage.

It has been shown that these structures result from the sharp structure in the quasiparticle injection rate at energy $E_s \simeq |eV_i| - \Delta_{Al}$ (Fig. 2). Effects due to the structure at $(|eV_i| + \Delta_{Al})$ are too small to be detected. The injected quasiparticles relax and give a steady state distribution which, for the most part, resembles a thermal distribution with an effective temperature $T^* > T_a$. However, the steady -state distribution also contains a small residue from the sharp structure in the injection spectrum. Figure 5 shows schematically the effects of this sharp structure on the detector junction current. The excess quasiparticles enhance the currents contributed by processes (a) and (d) as shown in (i) and (iv). They also block processes (b) and (c), and reduce the currents contributed by them as shown in (ii) and (iii). We note that the most effective enhancements and reductions corresponding to each process occur at a different detector voltage. This results in the four measured structures in the detector d^2I/dV^2 curve shown in Fig. 4, each corresponds to a different process. We also note that, since the magnitudes of the I-V structures depend on the magnitude of the sharp structure at $E_s \simeq |eV_i| - \Delta_{Al}$ in the steady-state quasiparticle distribution which in turn is governed by the relaxation times of the quasiparticles at that energy, the quasiparticle relaxation time at energy E_s can be obtained from a quantitative analysis of the magnitudes of the I-V structures. Furthermore, by sweeping the injector junction voltage, the energy dependence of the relaxation time can be mapped out.

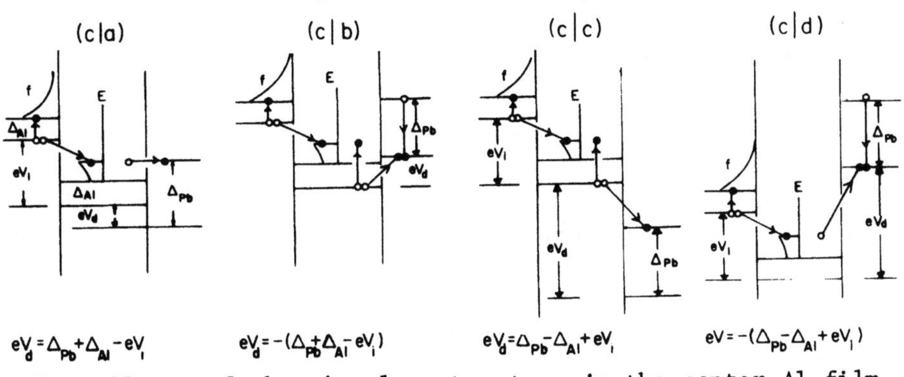

Fig. 5. Effects of the singular structure in the center Al film quasiparticle distribution on the detector junction current contributed by the four tunneling processes.

To obtain a detailed quantitative understanding of the extraction of the quasiparticle relaxation times, we calculate the detector junction current due to the excess quasiparticles in the steady state, particularly, the excess quasiparticle distribution at $E_s \simeq |eV_i| - \Delta_{Al}$ which is responsible for the nonequilibrium I-V structures discussed above.

We apply the results for the quasiparticle injection rate [Eq. (2-6)] and identify the left-(right-) side film to be the Al (Pb-Bi) film. The detector junction current is

$$I_d(V_d) = 2eN_{Pb}(0)\Omega_{Pb} \int_0^\infty dE\ \rho_{Pb}(E)\{[W^a_>(E) + W^a_<(E)] - [W^b_>(E) + W^b_<(E)] + [W^c_>(E) + W^c_<(E)] - [W^d_>(E) + W^d_<(E)]\} \quad (3\text{-}19)$$

the signs for the b- and d- process terms are negative because these processes remove electrons when quasiparticles are added and vise versa.

To make further progress we assume that the Pb-Bi film is not significantly disturbed in the steady state and all the change in the detector current is due to the excess distributions $\delta f_>(E)$ and $\delta f_<(E)$ in the center Al film. In fact, because of the large value of Δ_{Pb}, one can set $f_{R\gtrless} \simeq 0$, while $f_{L\gtrless}(E) = f_\gtrless(E)$.

To obtain $\delta f_\gtrless(E)$ for the injected center Al film, in general, we need to solve the coupled kinetic equations [Eqs. (2-16)]. However, the calculation of $\delta f_\gtrless(E)$ becomes simple if it can be assumed that any quasiparticle having suffered inelastic collision contributes only to the background distribution which is a Fermi function with an elevated effective temperature $T^* > T_a$, and gives no contribution to the sharp structures in the quasiparticle distribution and the detector I-V curve. In this case, the relaxation-time approximation is valid. From Eqs. (2-16a, and b) we have

$$\dot{f}_+(E) = I^+(E) - \delta f_+(E)/\tau_{qp}(E) \quad (3\text{-}20a)$$

and

$$\dot{f}_-(E) = I^-(E) - \delta f_-(E)/\tau_{qp}(E) - 2\delta f_-(E)/\tau_{el}(E) \quad (3\text{-}20b)$$

The steady-state solution for $\delta f_\gtrless(E)$ is

$$\delta f_{\lessgtr}(E) = I^+(E)\tau_{qp}(E) \pm \frac{1}{2} I^-(E)/[2\tau_{el}^{-1}(E) + \tau_{qp}^{-1}(E)] \quad (3\text{-}21)$$

From Eqs. (2-6 and 7, 3-19 and 21), one can show straightforwardly that step structures occur in the I-V curve at voltages $|V_d| = \pm(\Delta_{Pb} + \Delta_{Al} - |eV_i|)/e$, while cusp-like structures occur at voltages $|V_d| = \pm(\Delta_{Pb} - \Delta_{Al} + |eV_i|)/e$, in addition to the usual singular structures associated with an asymmetric junction at $|V_d| = \pm(\Delta_{Pb} \pm \Delta_{Al})/e$. Due to the fact that in reality the energy gap is smeared, the singularities are broadended. The calculated second derivative of the excess detector junction current which is the difference of the steady state value at $V_i \neq 0$ and the equilibrium value at $V_i=0$ is shown in Fig. 6. To obtain Fig. 6, we have neglected branch mixing due to the elastic scattering.

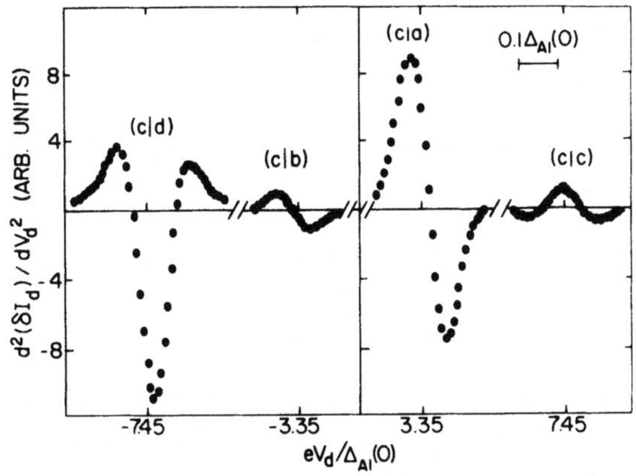

Fig. 6. Theoretical second derivative of the excess detector junction current as a function of detector voltage. We have used parameters $eV_i=3\Delta_{Al}(0)$, $\Delta_{pb}(0)=5.4\Delta_{Al}(0)$, and $T^* = 0.5T_c(Al)$. The half width of the gap smearing is $\delta\Delta_{Al}= \delta\Delta_{pb}=0.05\Delta_{Al}(0)$.

To make a comparison of the calculated curve with the experimental result (Fig.4), one has to realize that due to the electron-hole symmetry, $I_d(V_d,V_i) = -I(-V_d,-V_i)$. Therefore, the detector current for $V_d<0$ and $V_i>0$ has the same value as that for $V_d>0$ and $V_i<0$ except for the sign. We note that theory and experiment agree well in both the position and the qualitative nature (including the shapes and the relative sizes) of the structures.

Although the broadening affects the magnitudes of the peak and dip structures, it should have little or no effect on the magnitudes of the step structure. Therefore, experimental data of the step heights are particularly useful for the extraction of the relaxation times.

The magnitude of the current-step heights at $V_d = \pm (\Delta_{pb} + \Delta_{Al} -$

$|eV_i|)/e \equiv \pm V_d{}^s$ can be readily obtained if one uses a Fermi function $f(E,T^*)$ with an elevated temperature T^* for $f_{\gtrless}(E)$ in Eq. (2-8) to calculate $I^+(E)$ and $I^-(E)$. This approximation introduces an error of order $(G_{nn}{}^i)^2\, G_{nn}{}^d$ in the calculation of the step height which itself is of order $G_{nn}{}^i\, G_{nn}{}^d$. Therefore, the approximation is well justified. It was found that the step heights at $V_d = \pm V_d{}^s$ are

$$\Delta I_d(\pm V_d{}^s) = \frac{\pi\, G_{nn}{}^i\, G_{nn}{}^d\, (\Delta\Delta_{pb})^{1/2}}{8e^2 N(0)\,\Omega}\, \tau_{qp}\,\rho(eV_i-\Delta)$$

$$\left[\, 1- f^\circ(\Delta,T^*) - f^\circ(eV_i-\Delta,T^*)\right] \left[1 \pm \left(1 - \frac{\Delta^2}{(eV_i-\Delta)^2}\right) \left(\frac{\tau_{el}}{2\tau_{qp}+\tau_{el}}\right)\right] \qquad (3\text{-}22)$$

Here, all the unsubscripted functions and parameters such as ρ, Δ, and Ω etc. are for the center Al films. The relaxation times τ_{qp} and τ_{el} are to be evaluated at $E = |eV_i| - \Delta_{Al}$. Clearly, the inelastic quasiparticle relaxation time can be obtained from the sum of the current-step magnitudes for positive and negative detector biases. The elastic branch relaxation time τ_{el} can then be derived from the difference of the two magnitudes. Furthermore, the energy dependence of the relaxation times can be studied by sweeping the injector junction voltage.

Detailed analysis of the experimental data for a dirty Al film has been carried out by Kirtley *et al*. (1980), and for a clean Al film by Moody and Paterson (1980).

We now turn to discuss systems driven far from equilibrium.

SYSTEMS DRIVEN FAR FROM EQUILIBRIUM

Great efforts have been made to determine the energy spectra of the quasiparticle and phonon excitations in nonequilibrium superconductors. Numerical solutions of the coupled kinetic equations for a variety of external drives have been obtained (Chang and Scalapino, 1978; Chang *et al*., 1979). It was found that when the thermal coupling between the superconductor and its environment is strong, the distributions show specific structure associated with the particular drive. However, when the thermal coupling is weak so that the superconductor has a phonon trapping factor $\tau_{es}/\tau_B(\Delta_0)>20$, the numerical results can be well approximated by an analytic model. We first discuss this severe phonon trapping case. An example of good thermal coupling will be discussed later in the context of gap enhancement due to the quasiparticle tunnel injection.

The Extended T^* Model

For simplicity, we will ignore the effects of branch imbalance and set $\delta f_-(E) = 0$. This is a good approximation when the ambient temperature is sufficiently below T_c or the phonon-trapping factor is large so that $\delta f_-(E)$ is much smaller than $\delta f_+(E)$. In the case of our present interest, these conditions are satisfied.

We also need an equation to govern the behavior of the energy gap. It has been shown (Eliashberg, 1970) that Δ in the nonequilibrium state can be determined from the modified BCS gap equation:

$$\mathrm{Ln}(\frac{\Delta}{\Delta_0}) = -\int_{\Delta}^{\infty} dE \, \frac{2f_+(E)}{\sqrt{E^2-\Delta^2}} \tag{4-1}$$

with Δ_0 the zero-temperature energy gap.

We consider a symmetric superconductor-insulator-superconductor tunnel junction. The energy spectrum of the quasiparticle injection rate is given by Eqs. (2-6 and 7) with $\Delta_L = \Delta_R = \Delta$. Equation (4-1) is solved self-consistently with the coupled quasiparticle and phonon kinetic equations [Eqs. (2-16a and c)] using a different set of values for the parameters such as the ambient temperature, the strength of the driving force, the junction voltage, and the phonon trapping factor. Also, we have used an energy independent electron-phonon coupling constant and a quadratic form for the phonon density of states. Typical results are shown in Fig. 7. In (a) the solid line shows the steady-state quasiparticle-distribution function obtained from the numerical calculation. Parameter $A \equiv (G_{nn}\, \tau_0 T_c^3)/(2N(0)\, e^2 \Omega \Delta_0^3)$ is a measure of the strength of the driving force. Here τ_0 is the characteristic quasiparticle lifetime for the film and its value is tabulated by Kaplan _et al_.

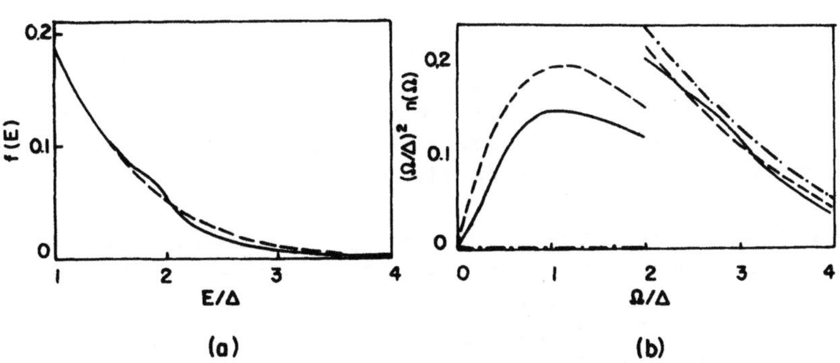

Fig. 7. Nonequilibrium steady-state quasiparticle distribution $f(E)$ and phonon spectrum $(\Omega/\Delta)^2 n(\Omega)$. We have set $eV=2\Delta_0$, $T_a=0$, $A=0.01$ and $\tau_{es}/\tau_B(\Delta_0)=60$ in the calculation. Solid curves are numerical solutions, dashed curves are results from the extended T^* motel, while dash-dot curves are Parker's T^* model results, with $T^*=0.94T_c$.

(1976). The dashed line in Fig. 7a is a Fermi distribution function with an effective temperature $T^* = 0.82\ T_c$ which is related to the calculated gap through the BCS gap equation. Evidently, the fit is good. It is better for higher phonon trapping factors. However, the numerical solution of the phonon distributions shown as the solid line in Fig. 7b cannot be described by a Bose function. In fact, there is a discontinuity in the phonon spectrum at $\hbar\Omega = 2\Delta$ due to the fact that phonons of energy $\hbar\Omega > 2\Delta$ can break a pair and have lifetimes shorter than those of $\hbar\Omega < 2\Delta$. This discontinuity is more significant for a superconductor with a smaller phonon-trapping factor.

In the early stage of the study of nonequilibrium superconductivity, Parker (1975) had proposed a T^* model. In this model, the quasiparticle excitations are described by a Fermi distribution with an effective temperature T^*. The phonon excitations are described by a Bose distribution containing either the ambient temperature T_a or the effective temperature T^* depending on the phonon energy being smaller than or greater than 2Δ. It turns out that if we take the quasiparticle distribution function to be a Fermi distribution at a modified temperature T^*, it is straightforward to solve the kinetic equation for the phonon distribution. We have

$$n(\Omega) = \frac{[1/\tau_s^{ph}(\Omega,T^*) + \tau_B^{-1}(\Omega,T^*)]n(\Omega,T^*) + \tau_{es}^{-1}\ n(\Omega,T_a)}{\tau_s^{ph-1}(\Omega,T^*) + \tau_B^{-1}(\Omega,T^*) + \tau_{es}^{-1}} \tag{4-2}$$

<u>The phonon distribution is a weighted average of the Bose functions with the effective temperature and the ambient temperature.</u> $n(\Omega)$ given by Eq. (4-2) has a discontinuity at $\hbar\Omega = 2\Delta$ because $\tau_B^{-1}(\Omega,\ T=0)$ is zero for $\hbar\Omega<2\Delta$ and has a step singularity at $\hbar\Omega = 2\Delta$. Phonon spectrum obtained from Eq. (4-2) is plotted as the dashed line in Fig. 7b. The fit to the numerical results is good and represents dramatic improvement over the T^* model that predicts zero phonon density for $\hbar\Omega<2\Delta$ in the present case.

Experimental measurements of the quasiparticle distribution in a nonequilibrium superconductor have been carried out using a tunneling technique (See e.g. Willemsen and Gray, 1978; Jaworski and Parker, 1979; Smith <u>et al</u>., 1980). In general, the results support the T^* model prediction.

On the other hand, it is difficult to probe the phonon spectrum directly because of the lack of a good tunable phonon spectrometer. However, it has been pointed out that the μ^* model (Owen and Scalapino, 1972), the extended T^* model (Chang <u>et al</u>., 1979) and the simple heating model predict quite different behaviors for the power dependence of the suppression of the energy gap in a driven film. Therefore, a measurement of $\delta\Delta(P)$ should be able to the determine the best among the three analytic models. Indeed, Rajeevakumar et al. (1979) did the experiment and found that the extended T^* model can best describe their data.

Energy Gap Enhancement by Tunneling

The gap enhancement in superconductors by quasiparticle tunnel injection has been predicted and observed in superconducting thin films with good thermal coupling with the ambience. Parmenter (1961) first suggested that this can occur in films in an asymmetric junction. Later, Peskovatskii and Seminozhenko (1976) and Chang (1978) studied the gap enhancement in a symmetric junction. Experimental studies of the enhancement effects have been carried out by Gray (1978) using a symmetric Al/Al juntion and by Chi and Clarke (1979b) using an asymmetric Al/Al' junction.

The reason that the energy gap in a superconductor can be enhanced by the tunneling of quasiparticles is not difficult to understand. Consider a junction biased at a voltage V smaller than the sum of the left-film gap Δ_L and the right-film gap Δ_R. There is no pair breaking due directly to tunneling processes because the tunneling processes shown in Figs. (1c) and (1d) which break pairs have contributions only for $|V| > (\Delta_L + \Delta_R)/e$. In process (a), a quasiparticle is transferred from the left film to the right film while in process (b) the reverse is true. In both processes there is no change in the total number of paired electrons in the junction. In an asymmetric junction, contributions of processes (a) and (b) to the quasiparticle injection rate are not equal because of the different joint density of states involved. Therefore, a net increase in the quasiparticle density results in one of the films to give a suppressed energy gap in that film, while in the other film, there will be a reduction of the quasiparticle density, and the energy gap in this film may be enhanced. In a symmetric junction, processes (a) and (b) contribute to the quasiparticle injection rate by the same amount [Eq. (2-6)] but with different signs. Therefore, tunneling does not change the quasiparticle density in the individual films. The net effect of the primary tunneling processes (a) and (b) is to shovel low-energy quasiparticles into higher-energy states.

Eliashberg (1970) has suggested in his study of microwave enhancement of superconductivity, that it is the occupation of the low-energy states that most severely interferes with the pair correlation of the electrons. Thus clearing out the low-energy states by driving the quasiparticles into higher states could result into an enhanced energy gap.

It has also been suggested (Chang and Scalapino, 1976) that shoveling of quasiparticles can result in a reduction of quasiparticle density and hence an enhancement of the energy gap. This can be achieved because higher-energy quasiparticles can recombine faster than lower-energy ones. To be more specific, we note first that the phonon density of states in the Debye approximation varies as the phonon energy squared. Thus for phonons emitted by recombination of quasiparticles near the gap edge, the final phonon density of states involved is proportional to $(2\Delta)^2$. However, if the two quasiparticles are shoveled into states of energy $\Delta + eV$ before they recombine into a pair, the recombination phonon would have energy $2(\Delta + eV)$ and the final phonon density of states involved would be proportional to $4(\Delta + eV)^2$. Therefore tunneling excitation of quasiparticles can lead to faster quasiparticle recombination and a net reduction of the quasiparticle density in the steady state.

The above discussion holds only if the phonon trapping factor is small so that the injected energy can be released into the ambience easily. For a sample with severe phonon trapping, "heating" effects will be important and the energy gap will be suppressed instead of enhanced.

Again the energy gap and the distribution functions for quasiparticles and phonons are calculated from the coupled equations (Eq. (2-16)) and Eq. (4-2), with $\delta f_-=0$. The distribution functions have the same features as those for the case of microwave irradiation (Fig. 6 in Chapter 7). The calculated voltage dependence of the energy gap is shown in Fig. 8 for A=0.05 and 0.1, and $T_a=0.9T_c$. The phonon escape time used is $\tau_{es} = 8\tau_B(2\Delta_0)$, typical

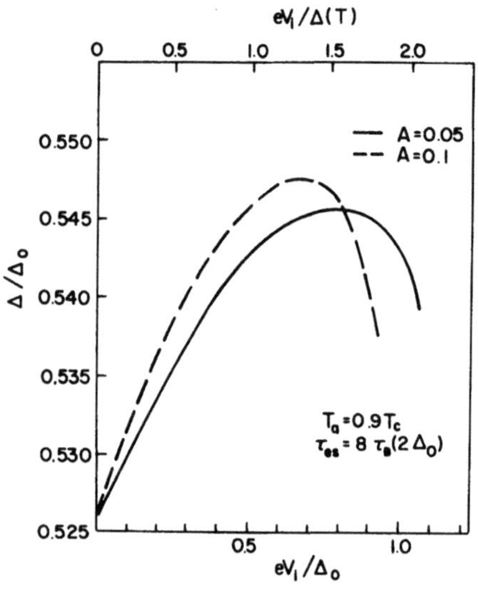

Fig. 8 Bias voltage dependences of the gap enhancement for A=0.05 and 0.01.

for an Al film on sapphire dipped in He. The maximum gap enhancement is slightly less than 4%. It is interesting to note that the energy gap begins to decrease before the bias voltage reaches $2\Delta/\hbar$ when the tunneling processes which break pairs set in. This is because for eV slightly less than 2Δ there will be a significant number of quasiparticles being excited into states of E 3Δ by the tunneling processes if the temperature is high enough as in the present case. When these quasiparticles relax, phonons of energy $\hbar\Omega > 2\Delta$ are generated to break pairs and to reduce the energy gap. This result is different from that obtained by Peskovatskii and Seminozhenko (1976) who assumed phonons to be unperturbed and the energy gap was found to be more enhanced as eV increases and approaches 2Δ.

The size of the gap enhancement depends upon the strength of the driving force, the relaxation time and also the ability of the phonons to escape from the sample which is governed by the acoustic match between the sample and the environment. What is the maximum amount of gap enhancement and whether it is the redistribution of quasiparticles in energy space or the reduction of their density that is more responsible for the observed gap enhancement remain as intriguing questions.

ACKNOWLEDGEMENT

I would like to thank Professor D. J. Scalapino for introducing me to this exciting field of research. Much of the physics reported here is the result of cooperation with my colleagues. I would like to thank them for their valuable contributions and stimulating discussions. Financial support from the Office of Naval Research and the National Science Foundation through the years during which this research was performed are greatly appreciated.

REFERENCES

Ambegoakar, V., and Tewordt, L., 1964, Theory of electronic thermal conductivity of superconductors, Phys. Rev., A134:805.

Bardeen, J., Cooper, L.N., and Schrieffer J.R., 1957, Theory of superconductivity, Phys. Rev., 108:1175.

Bardeen, J., Rickazen, G., and Tewordt, L., 1959, Theory of thermal conductivity of superconductors, Phys. Rev., 113:982.

Bardeen, J., 1961, Tunneling from a many-particle point of view, Phys. Rev. Lett., 6:57; 1962, Tunneling into superconductors, Phys. Rev. Lett., 9:147.

Chang, Jhy-Jiun, and Scalapino, D.J., 1976, Tunneling as a probe of nonequilibrium superconducting states, Phys. Rev. Lett., 37:522.

Chang, Jhy-Jiun, 1977, Tunneling characteristic of a nonequilibrium double junction, Phys. Rev. Lett., 39:1352.

Chang, Jhy-Jiun, 1978, Gap enhancement in superconducting thin film due to quasiparticle injection, Phys. Rev. B, 17:2137.

Chang, Jhy-Jiun, and Scalapino, D.J., 1978, Nonequilibrium superconductivity, J. Low Temp. Phys., 31:1.
Chang, Jhy-Jiun, and Yang, Chi-Chung, 1978, New structure in the tunneling characterictics of a nonequilibrium double junction, Bull. Am. Phys. Soc., 23:263.
Chang, Jhy-Jiun, 1979, Relaxation of the quasiparticle charge imbalance in superconductors, Phys. Rev. B, 19:1420.
Chang, Jhy-Jiun, Lai, W.Y., and Scalapino, D.J., 1979, T* model of the energy distribution of excitations in nonequilibrium superconductors, Phys. Rev. B, 20:2739.
Chi, C.C., and Clarke, J., 1979a, Quasiparticle branch mixing rates in superconducting aluminum, Phys. Rev. B, 19:4495.
Chi, C.C., and Clarke, J., 1979b, Enhancement of the energy gap in superconducting aluminum by tunneling extraction of quasiparticles, Phys. Rev. B, 20:4465.
Cohen, M.H., Falicov, L. M., and Phillips, J. C., 1962, Superconductive tunneling, Phys. Rev. Lett., 8:316.
Eliashberg, G. M., 1960, Interactions between electrons and lattice in a superconductor, Zh. Eksp. Teor, Fiz., 38:966 [Sov. Phys. JETP Lett., 11:696].
Eliashberg, G. M. 1970, Film superconductivity stimulated by a high frequency field, Pisma Zh. Eksp. Teor. Fiz., 11:186 [Sov. Phys. JETP Lett., 11:114].
Entin-Wohlman, O., and Orbach, R., 1978, Mode propagation in superconductors near T_c, Ann. Phys., 116:35; 1979, On the microscopic and Boltzmann equation approaches to nonequilibrium superconductors, Ann Phys., 122:64.
Gray, K. E., 1978, Enhancement of superconductivity by quasiparticle tunneling, Solid State Commun., 26:633.
Jaworski F., and Parker, W.H., 1979, Experimental determination of the quasiparticle energy distribution in optically perturbed superconducting films, Phys. Rev. B, 20:945.
Kaplan, S.B., Chi, C.C., Langenberg, D.N., Chang, Jhy-Jiun, Jafarey, S., and Scalapino, D.J., 1976, Quasiparticle and phonon lifetimes in superconductors, Phys. Rev. B, 14:4854; Errati: ibid., 15, 3567 (1977).
Kaplan, S.B. Kirtley, J.R., and Langenberg, D.N., 1977, Phys. Rev. Lett., 39:291.
Kirtley, J.R., Kent, D.S., Langenberg, D.N., Kaplan, S.B., Chang, Jhy-Jiun, and Yang, Chi-Chung, 1980, Quasiparticle energy distribution and relaxation times in a tunnel injected superconductor, Phys. Rev. B, 22:1218.
Moody, M.V., and Paterson, J.L., 1980, Quasiparticle Relaxation time in clean Al films, Preprint.
Owen, C.S., and Scalapino, D.J., 1972, Superconducting state under the influence of external dynamic pair breaking, Phys. Rev. Lett. 28:1559.
Parker, W.H., 1975, Modified heating theory of nonequilibrium superconductors, Phys. Rev. B, 12:3667.
Parmenter, R.H., 1961, Enhancement of superconductivity by extraction of normal carriers, Phys. Rev. Lett., 7:274.

Paterson, J.L. 1978, Analysis of nonequilibrium double junctions, J. Low Temp. Phys., 33:285.
Peskovatskii, S.A., and Seminozhenko, V.P., 1976, Stimulation of superconductivity by constant tunnel currents, Fiz. Nizk. Temp., 2:943 [Sov. J. Low. Temp. Phys., 2:464].
Pethick, C.J., and Smith, H., 1979, Relaxation and Collective motion in superconductors; a two fluid description, Ann. Phys., 119:133.
Prange, R.E., and Kadanoff, L.P., 1964, Transport theory for electron-phonon interaction in metals, Phys. Rev., A134:566.
Rajeevakumar, T.V., Chang, Jhy-Jiun, and Chen, J.T., 1979, Energy gap suppression and instability in superconducting tin films under strong quasiparticle injection, J. Low Temp. Phys., 37:77.
Scalapino, D.J., 1969, The electron-phonon coupling and strong superconductors, in "Superconductivity", Parks, R.D., ed., Mercel Dekker, New York.
Schmid, A., and Schon, G., 1975, Linearized kinetic equations and relaxation processes of a superconductor near T_c, J. Low Temp. Phys., 20:207.
Schrieffer, J.R. 1964, "Theory of Superconductivity", Benjamin, New York.
Smith, A.D., Skocpol, W.J., and Tinkham, M., 1980, Quasiparticle energy distribution in optically illuminated Al-PbBi superconducting tunnel junctions, Phys. Rev. B, 21:3879.
Tinkham, M., 1975, "Introduction to Superconductivity", McGraw Hill Inc., New York.
Tinkham, M., 1972, Tunneling generation, relaxation, and tunneling detection of hole-electron imbalance in superconductors, Phys. Rev. B, 6:1747.
Willemsen, H.W., and Gray, K.E., 1978, Distribution function in nonequilibrium states of a superconductor, Phys. Rev. Lett., 41:812.

STABILITY OF NONEQUILIBRIUM SUPERCONDUCTING STATES I:

GENERAL PRINCIPLES

A.-M.S. Tremblay

Département de physique
Université de Sherbrooke
Sherbrooke, Québec J1K 2R1, Canada*

ABSTRACT

Some of the concepts necessary to study the stability of non-equilibrium superconductors are introduced including probabilistic methods which account for fluctuations (thermal noise).

INTRODUCTION

When a small steady flow of energy goes through a system, it usually behaves more or less like its equilibrium conterpart except for small quantitative differences: we have seen and will see again at this School, that by injecting quasiparticles or shining micro-waves, light or phonons, the properties of a superconductor (for example its distribution function, gap or critical current) can be modified. However, if the external driving force becomes too "in-tense", more drastic things can happen: for example, the super-conductor may suddenly become normal or it may develop a spatially inhomogeneous structure, or do a variety of other things. In other words, the homogeneous superconducting state can become "unstable" and the new state which becomes stable can be qualitatively diffe-rent from the original state. The analogy with phase transitions has been stressed many times (by now it is almost a "cliché").

The formation of such "order" (dissipative structures) is one of the most striking features of dissipative (or nonequilibrium, or non-thermal) steady states. These dissipative structures have been observed long ago. For example, if a fluid is heated from below

*Part of this work was done at Cornell University, Ithaca, U.S.A.

then at a certain critical value of the rate of heating, the buoyancy force is large enough to overcome the viscous effects and convection sets in: at some places, the hot fluid moves upward to carry the heat away. Cold fluid comes back at other places. This effect is called the Bénard[1] instability (Chandrasekhar, 1968). Various methods have been developed over the years to understand such phenomena.

Most of the concepts we use in nonequilibrium superconductivity to study instabilities, have a long history and are basically the same as those used in other fields. The difference lies mainly in the system of equations used and in the more "microscopic" manifestations of the instabilities we are interested in. In hydrodynamics, the dissipative structures appear even though at the microscopic level the system looks more or less like the original one. In our case, the system can be quite different at the microscopic level: for example, it can become normal, or it can have a different gap (in both cases, this means a different elementary excitation spectrum) or it can have sharp features in the distribution function which are at the origin of the instability. Also in superconductors (or chemical reactions) the scale of the dissipative structures can be determined by internal parameters of the system (coherence length etc...) in contrast with hydrodynamics where the scale is determined mainly by boundary conditions. The microscopic structure can also play an important role in semiconductor instabilities (Gunn effect for example) (Volkov and Kogan, 1968). Note that if $k_B T$ is of the order of the energy of the dissipative structures, we are interested in, thermal fluctuations will be important. (They are not important for fluids or for many of the other known examples.)

In the following pages, we study <u>simplified</u> sets of equations which illustrate the <u>concepts</u> which we need. We take experimental examples only when needed. The theoretical disucssion is simple. We assume that the reader knows from other lectures in this School, how to compute nonequilibrium distribution functions. Here, we insist only on how these can lead to instabilities. My effort has been mainly pedagogical. No attempt has been made to establish priorities or to give a complete bibliography. Comments are sometimes included with the references to help the reader find good introductory papers. The Appendix contains a detailed example of stability analysis for nonequilibrium superconductors which should concretize the concepts introduced in the text. Specific theories and the corresponding experiments are discussed in the following Chapter.

STABILITY OF THE STATIONARY SOLUTIONS OF DETERMINISTIC EQUATION: SIMPLE EXAMPLES

In this Section we define what is meant by "linear" (equivalently "local") stability. We look at the one mode and many modes cases and introduce for example the notions of critical slowing down, phase coexistence, speed of nucleation front and Lyapunov functional. We will give only simple minded examples. (Haken, 1974)

One Mode

Consider a system which can be described by one degree of freedom Δ and obeys an equation of motion of the form,

$$\frac{d\Delta}{dt} = -\frac{1}{\tau}\frac{dV(\Delta)}{d\Delta} . \tag{2.1}$$

A superconductor close to equilibrium obeys an equation similar to (2.1) (the time dependent Ginzburg-Landau equation: TDGL) with[1] $V(\Delta) = \alpha\Delta^2/2 + \beta\Delta^4/4$. This function is sketched in Figure (1a) for $\beta > 0$ and various values of α. A "stationary" (equivalently "time independent" or "steady state") solution of Eq. (2.1) corresponds to an extremum of $V(\Delta)$, i.e. $dV/d\Delta = 0$. One should recognize that $\alpha = 0$ is a "critical value". For $\alpha > 0$ there is only one extremum while for $\alpha < 0$ there are more than one.

In a nonequilibrium superconductor under steady external perturbation, an equation similar to (2.1) holds if it is possible to express the nonequilibrium distribution function in terms of the gap and some external parameters only. For example, in the case of optical illumination one can sometimes write $V_B(\Delta) = \alpha\Delta^2/2 + \beta\Delta^4/4 - B\Delta^3$ where B depends on the absorbed power. In general, this nonequilibrium contribution is not an analytic function of Δ. $V_B(\Delta)$ is sketched in Figure (1b) for $\alpha > 0, \beta > 0$ and various values of B. For $B > B_b$ there are more than one stationary solutions. For $B = B_b + \varepsilon$ the new stationary solution has a value of Δ quite different from that for $B < B_b - \varepsilon$. This is the analog of a "first order phase transition". Figure (1a) ($B = 0$) represents the analog of a "second order phase transition" where the new solution is arbitrarily close to the original one when α is arbitrarily close to $\alpha = 0$. One can draw in the parameter space B, α, the analog of a phase diagram.

To test for the stability of stationary states to small perturbations it suffices to linearize Eq. (2.1) around the stationary value Δ_{st} of the gap:

$$\frac{d\delta\Delta}{dt} = -\frac{1}{\tau}\left.\frac{d^2V}{d\Delta^2}\right|_{\Delta_{st}} \delta\Delta. \tag{2.2}$$

In more general cases we will also be studying linear equations. Hence we will always try solutions of the form

$$\delta\Delta = \delta\Delta_0 \exp(-st) \tag{2.3}$$

[1] For the longitudinal mode in a clean superconductor, τ would also depend on Δ. See Prof. Schmid's lecture: Chapter 14.

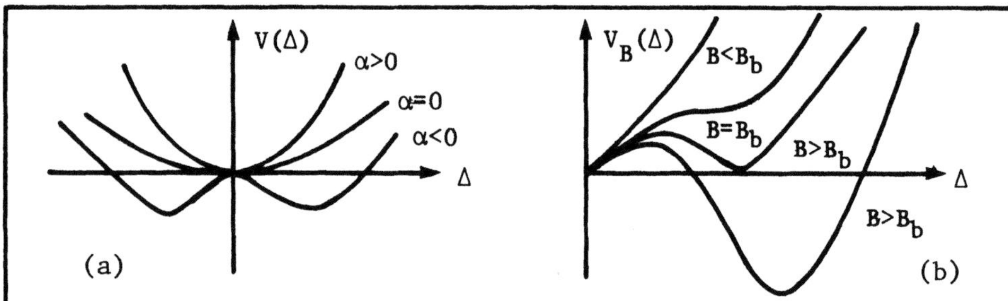

Figure 1: Qualitative sketches for (a) a second order transition, $V(\Delta) = \alpha\Delta^2/2 + \beta\Delta^4/4$; (b) a first order transition, $V_B(\Delta) = V(\Delta) - B\Delta^3$

and solve eigenvalue equations for s. Clearly if $V(\Delta)$ is a minimum, i.e. $(d^2V/d\Delta^2) > 0$ in Eq. (2.2), s is real and positive, hence the system is stable. The system is unstable if $V(\Delta)$ is a maximum.

The case of Figure (1a) where one goes from one stable to two stable solutions is called in mathematics a (Hopf) bifurcation. Note that if we had looked only at the stationary solution $\Delta = 0$ we would have found that as a function of α the system goes continuously from stable $(s > 0)$ to unstable $(s < 0)$. This is analogous to "second order" transitions. As we approach the critical value $\alpha = 0$, we have "critical slowing down", i.e. $s \to 0$ and the relaxation time becomes infinite. When the system becomes unstable $(s < 0)$ one must look for other stable solutions. For a "first-order" transition, a test for local stability on the "old" solution would not reveal that there is a "new" stationary solution elsewhere.

Many Modes

It is somewhat unrealistic to describe a many-body system with only one variable. Very often however, it is possible to eliminate many "microscopic" degrees of freedom of the system which relax fast and to keep only a "few" slow modes. The modes can be slow because of conservation laws or broken symmetries (phase transitions). For example since energy is conserved, it cannot disappear from one part of the system without flowing elsewhere. As a consequence, sudden changes of the energy over large distances (long wavelengths) are not expected. They must occur "slowly". Also, as we have just seen, close to the transition temperature the order parameter relaxes slowly because of "critical slowing down", and there are long range correlations in the system: hence if the order parameter changes over large distances, it must do it slowly. We are interested in the slow modes because they are the ones whose frequency could conceivably go through zero and hence lead to an "instability" towards a new stationary state. The particular mode whose frequency vanishes in "second order" dissipative transitions may itself become a new "order parameter" in the new stable state.

In a superconductor, one sometimes concentrates on the gap equation with gradient terms. Already this leads to an infinite number of degrees of freedom since each point of space is characterized by a value of the gap which depends on the values of the gap at neighbouring points. Another favorite choice is Rothwarf-Taylor type equations where the variables are the local values of the number of excited quasiparticles and phonons.

Stationary solutions: For definiteness we limit ourselves to the equation

$$\frac{d\Delta(\vec{x},t)}{dt} = -\frac{1}{\tau}\left[\frac{dV(\Delta)}{d\Delta} - \xi^2(0)\nabla^2\Delta(\vec{x},t)\right] \tag{2.4}$$

where $\xi^2(0)$ is a microscopic length (T = 0 coherence length in Ginzburg-Landau theory). The equation for the stationary solutions,

$$\xi^2(0)\nabla^2\Delta(\vec{x},t) = \frac{dV(\Delta)}{d\Delta} \tag{2.5}$$

is analogous to the equation of motion for a particle of coordinate Δ in a potential minus $V(\Delta)$ $(-V(\Delta))$ if we restrict ourselves to one dimension and replace $\vec{x}$ by time. The analog of energy conservation then becomes

$$\frac{1}{2}\,\xi^2(0)\left(\frac{d\Delta}{dx}\right)^2 = V(\Delta) - C \tag{2.6}$$

where C is a constant. Depending on the value of C and on the shape of the potential, qualitatively different types of solutions are possible: constant, periodic or localized. This is illustrated in Figure (2c,d,g,h). The potential and the values of C corresponding to the various solutions are in Figure (2a and e). Figure (2b and f) represent the "trajectories in phase space" i.e. $d\Delta/dx$ as a function of Δ. The trajectories are found from the equation

$$\frac{d(d\Delta/dx)}{d\Delta} = \frac{1}{\xi^2(0)}\,\frac{dV/d\Delta}{d\Delta/dx} \tag{2.7}$$

which is derived from Eq. (2.5). Trajectories are popular in the Russian literature. They can be helpful as an intermediate step to determine the character of the solutions. Note that when both $dV/d\Delta$ and $d\Delta/dx$ vanish the right-hand side of Eq. (2.7) is indeterminate. This corresponds to the case where both $d^2\Delta/dx^2$ and $d\Delta/dx$ are zero. Then Δ depends very weakly on position.

Linear stability analysis: To find whether the stationary solutions are stable in time, we linearize Eq. (2.4) and solve for the eigenvalues[1] s.

[1] The following analysis can be found in many references, e.g. Eckern et al. (1979), Appendix B.

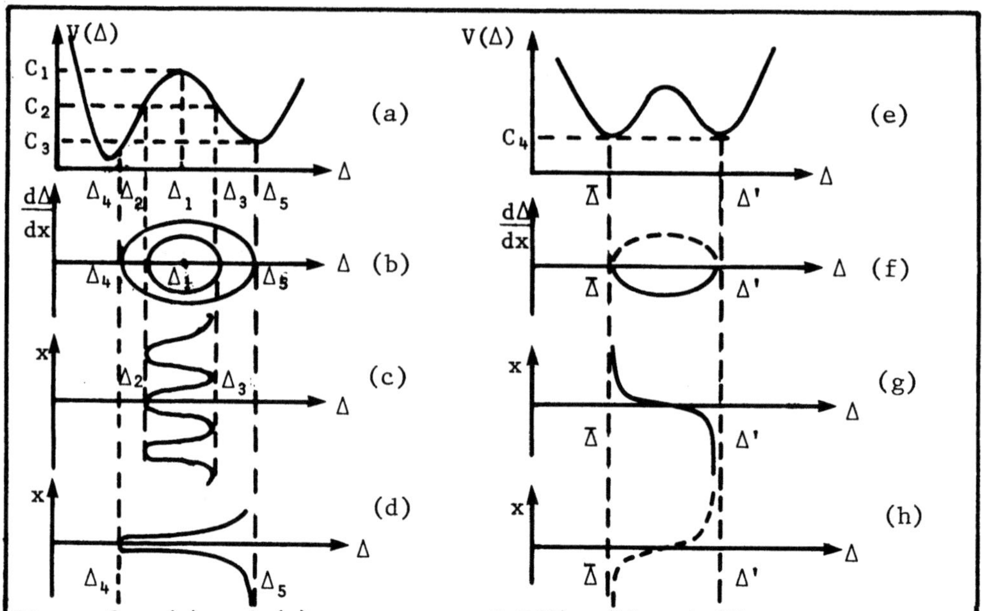

Figure 2: (a) and (e), two types of $V(\Delta)$. (b) and (f), trajectories from (2.7). Solutions $\Delta(x)$ of (2.6) for two $V(\Delta)$ and various values of C. (c) $C = C_2$ oscillating; (d) $C = C_3$ localized; (g) and (h) $C = C_4$ wall. For $C = C_1$, Δ_1 is constant.

$$s\delta\Delta = \frac{1}{\tau}\left[\frac{d^2V}{d\Delta^2}\Big|_{\Delta_{st}} - \xi^2(0)\nabla^2\right]\delta\Delta. \tag{2.8}$$

This is analogous to Schrodinger's equation for a spatially dependent potential $(d^2V/d\Delta_{st}^2)$. A stationary solution is unstable to small perturbations if at least one eigenvalue is negative.

When Δ_{st} is independent of position we can Fourier transform Eq. (2.8) in space to obtain,

$$s_q = \frac{1}{\tau}\left[\frac{d^2V}{d\Delta^2} + \xi^2(0)q^2\right]. \tag{2.9}$$

Hence the system is stable $(s_q > 0)$ to small perturbations of any wavevector if the potential is a minimum while if the potential is a maximum, it is unstable to all perturbations of wavevector $q < \xi^{-1}(0)\ (d^2V/d\Delta^2)^{\frac{1}{2}}$.

To study the stability of periodic (Fig. 2c) and localized (Fig. 2d,g,h) solutions, note that $\delta\Delta = d\Delta_{st}/dx$ is the only eigenfunction of Eq. (2.8) with zero eigenvalue. This can be seen by differentiating Eq. (2.5) with respect to x. For periodic solutions, $d\Delta_{st}/dx$ has infinitely many nodes. The eigenfunctions with a smaller number of nodes have $s_q < 0$, hence periodic solutions are unstable to these modes. Localized solutions of the type of Figure (2d) are also unstable since $d\Delta_{st}/dx$ has one node hence there exist one eigenfunction with s < 0. On the other hand, for localized solutions of the "wall" type (Fig. 2g,h), $d\Delta_{st}/dx$ has no node thus these stationary states are "marginally" stable: the lowest eigenvalue is s = 0.

Walls (Fig. 2g,h) represent the coexistence of two phases. One with $\Delta = \bar{\Delta}$ and one with $\Delta = \Delta'$. Walls exist when $V(\bar{\Delta}) = V(\Delta')$. Hence this becomes the prediction of the deterministic equation of motion (2.4) for the coexistence of two phases. Suppose now the external perturbation is changed so that $V(\bar{\Delta})$ and $V(\Delta')$ differ slightly. We can compute the speed at which the wall will move (Landauer, 1977). Letting $\Delta(x,t) = \Delta\ (x - vt)$ in Eq. (2.4) one finds,

$$- v \frac{d\Delta}{dx} = - \frac{1}{\tau} \left[\frac{dV}{d\Delta} - \xi^2(0) \frac{d^2\Delta}{dx^2} \right]. \tag{2.10}$$

Integrating over Δ, using $d\Delta = (d\Delta/dx)\ dx$ and assuming that $d\Delta/dx$ vanishes at infinity, we find

$$v \int_{-\infty}^{\infty} dx \left(\frac{d\Delta}{dx}\right)^2 = \frac{1}{\tau} \left[V(\Delta(x=\infty)) - V(\Delta(x=-\infty)) \right] . \tag{2.11}$$

To calculate the velocity v to leading order, it suffices to evaluate the integral on the left-hand side from the solution when $V(\bar{\Delta}) = V(\Delta')$. In the mechanical analogy, Eq. (2.11) is the statement that the change in potential energy is dissipated in heat.

One theory suggests (Elesin, 1977) that when a superconductor is illuminated by light, such a situation occurs. At a large enough value of the external power, "nucleation fronts" (walls) move, leaving behind the system in its normal state. When the laser intensity is increased, the "nucleation fronts" move faster. Hence, for a fixed laser pulse duration, the fraction of normal state created depends on the laser power. This is in qualitative agreement with pulsed experiments where beyond a critical value of the absorbed power one sees a gradual rise in the resistance as the laser power is increased (Sai-Halasz, Chi, Denenstein and Langenberg, 1974). Golovashkin et al. (1975) also varied the pulse duration and observed results consistent with the above interpretation.

There is another, sometimes covenient, way to study stability. Defining

$$U(\Delta) = \int dx\ [V(\Delta) + \tfrac{1}{2}\ \xi^2(0) \left(\frac{d\Delta}{dx}\right)^2] \tag{2.12}$$

the equation of motion (2.4) may be rewritten,

$$\frac{d\Delta(\vec{x},t)}{dt} = - \frac{1}{\tau} \frac{\delta U}{\delta\Delta(\vec{x},t)} . \tag{2.13}$$

Hence,

$$\frac{dU}{dt} = \int dx \frac{\delta U}{\delta \Delta} \frac{d\Delta}{dt} = -\frac{1}{\tau} \int dx \left(\frac{\delta U}{\delta \Delta}\right)^2 . \tag{2.14}$$

Thus, U always decreases in time unless $(\delta U/\delta\Delta) = 0$ everywhere. If furthermore $U_L = U - U(\Delta_{st})$ is positive and it plus its first derivative are continuous then U_L is a Lyapunov functional (Haken 1974). The existence of such a functional in a region around Δ_{st} guarantees that Δ_{st} is stable. Since $dU_L/dt < 0$ the system will always evolve towards the stable solutions where $dU_L/dt = 0$. In particular, there are no temporal oscillations (limit cycles) in the system. Note that limit cycles are characteristic of dissipative systems: they cannot occur in conservative systems (Minorski 1947). Phase slip centers can be looked at as some sort of limit cycles. This is a fascinating subject but we will not dwell on it. (See Chapters 8, 18 and 19).

A final remark. When the quasiparticle diffusion length is larger than $\xi(0)$ one sometimes ends up with an equation of the form (2.9) with $\xi^2(0)$ replaced by a negative constant, say $-b^2$. In that case, the uniform system is unstable for all $q > b^{-1}(d^2V/d\Delta^2)^{\frac{1}{2}}$. This kind of instability is sometimes called "diffusive" since the coefficient of q^2 is often a diffusion constant. One must then take into account higher order gradients and Eq. (2.9) is replaced by

$$s_q \, \delta\Delta_q = \frac{1}{\tau} \left(\frac{d^2V}{d\Delta^2} - b^2q^2 + c^4q^4\right) \delta\Delta_q \tag{2.15}$$

where c is a constant. Then there is only a finite range of wavevectors which is unstable because of the stabilizing effect of the c^4q^4 term. The "instability" threshold is at the value of the external parameter where $V(\Delta)$ is such that there is an infinitesimal range of wavevectors (centered around $q \equiv q_c$) which is unstable. The new stable state can be quite hard to find in function space. However, it should have a periodicity characterized by q_c. One can use variational methods, i.e. use a trial function and look for a minimum of the Lyapunov functional defined above (Eckern, Schmid, Schmutz and Schön, 1979). There also exist "finite amplitude perturbation theory". Hida (1978) for example has used such a method to study the diffusive instability of the μ* model found by Scalapino and Huberman (1977). Note that finite amplitude perturbation theory is also used to check for "global stability", i.e. for the possibility of a first-order type transition where "new" solutions appear while the "old" solution is still locally stable.

GLOBAL STABILITY AND THE EFFECT OF NOISE

Most "phase transitions" in nonequilibrium superconductors can be accounted for by deterministic methods. Nevertheless, one should be aware of the basic role of "noise" in determining possible steady states.

Langevin and Fokker-Planck Equations and Generalized Free Energy

We have just seen that the stability analysis of deterministic equations of motion can yield many states which are locally stable at the same time. There may occur "first order" type transitions between these states. To find when the transition occurs, one looks for conditions allowing the existence of marginally stable walls between two phases. If there were no fluctuations (noise) in the system however, it could stay in any locally stable state forever. In equilibrium, we know that to find which state the system will be in, one finds which one minimizes the free energy. We proceed to give an alternate way of formulating that result which can be generalized to nonequilibrium systems.

Up to now we have included in our analysis only a few "slow" modes. The effect of the other "high frequency" modes and of the environment can be represented in the equations of motion by stochastic (random) forces, i.e. we write,

$$\frac{d\Delta}{dt} = -\frac{1}{\tau}\frac{dV(\Delta)}{d\Delta} + \theta_\Delta \tag{3.1}$$

where θ_Δ is a random force, also called Langevin force, with the properties

$$< \theta_\Delta(t) > = 0 \tag{3.2a}$$

$$< \theta_\Delta(t)\, \theta_\Delta(t') > = 2\mathcal{D}\delta(t - t') \tag{3.2b}$$

where $\mathcal{D}$ is a constant to be determined. It is also assumed that the Langevin force is Gaussian, i.e. any higher order correlation function can be expressed as a sum of products of the correlation functions in Eq. (3.2).

Because of (3.1), Δ is now also a random quantity. Its probability distribution can be found from that for θ_Δ if (3.1) can be solved. Since $dV(\Delta)/d\Delta$ is in general a nonlinear function of Δ, this can be quite hard. There is a trick however which consists in solving instead an equation of motion directly for the probability distribution of Δ. To find that equation, define $P_2(\bar{\Delta},\bar{t}|\Delta',t')$ the conditional probability that Δ has the value $\bar{\Delta}$ at time $\bar{t}$ given that it had the value Δ' at time t'. P_2 obeys the following (Smoluchowski) equation[1] :

$$P_2(\Delta,t + \delta t|\Delta',0) = \int d\bar{\Delta}\; P_2(\Delta,t + \delta t|\bar{\Delta},\bar{t})\; P_2(\bar{\Delta},\bar{t}|\Delta',0) \tag{3.3}$$

for any value of $\bar{t}$ between $t + \delta t$ and 0. Eq. (3.3) says that the probability that Δ changes from Δ' to Δ between time 0 and $t + \delta t$

1 This equation is valid only for Markov processes. See Wang and Uhlenbeck (1945). For additional criticism see van Kampen (1965).

equals the probability that it changes from Δ' to $\bar{\Delta}$ between time 0 and $\bar{t}$ and from $\bar{\Delta}$ to Δ between time $\bar{t}$ and $t + \delta t$, summed over all possible intermediate values of $\bar{\Delta}$. We can transform Eq. (3.3) into a differential equation. Let $\bar{\Delta} \equiv \Delta - y$ and $\bar{t} = t$. Then choose a time interval $\delta t \ll \tau$ during which Δ cannot change much. Then we can expand $P_2(\Delta,t + \delta t|\Delta - y,t)\, P_2(\Delta - y,t|\Delta',0)$ in powers of y about $P_2(\Delta + y,t + \delta t|\Delta,t)\, P_2(\Delta,t|\Delta',0)$ to obtain

$$\frac{\partial P_2}{\partial t} \equiv \lim_{\delta t\to 0} \frac{P_2(\Delta,t + \delta t|\Delta',0) - P_2(\Delta,t|\Delta',0)}{\delta t}$$

$$= \sum_{n=1}^{\infty} \frac{(-)^n}{n!} \frac{\partial^n}{\partial \Delta^n} [\, M_n(\Delta)\, P_2(\Delta,t|\Delta',0)\,] \tag{3.4}$$

where

$$M_n(\Delta) = \lim_{\delta t\to 0} \frac{1}{\delta t} \int dy\, y^n\, P_2(\Delta + y,t + \delta t|\Delta,t) \tag{3.5}$$

where we also used $\int dy\, P_2(\Delta + y,t + \delta t|\Delta,t) = 1$. The moments in Eq. (3.5) can be computed from (3.1). Using

$$\Delta(t + \delta t) - \Delta(t) = -\frac{1}{\tau} \frac{dV(\Delta(t))}{d\Delta} \delta t + \int_t^{t + \delta t} dt'\theta_\Delta(t') \tag{3.6}$$

and the properties of the random force (3.2) we find,

$$M_1(\Delta) = \lim_{\delta t\to 0} \frac{< \Delta(t + \delta t) - \Delta(t) >}{\delta t} = -\frac{1}{\tau} \frac{dV(\Delta)}{d\Delta} \tag{3.7}$$

$$M_2(\Delta) = \lim_{\delta t\to 0} \frac{< (\Delta(t + \delta t) - \Delta(t))^2 >}{\delta t} = 2\mathcal{D} \tag{3.8}$$

$$M_n(\Delta) = 0 \text{ for } n > 2. \tag{3.9}$$

Then Eq. (3.4) becomes the Fokker-Planck equation

$$\frac{\partial P_2}{\partial t} = -\frac{\partial}{\partial \Delta} [\, -\frac{1}{\tau} \frac{dV(\Delta)}{d\Delta} P_2] + \frac{1}{2} \frac{\partial^2}{\partial \Delta^2} [2\mathcal{D} P_2]. \tag{3.10}$$

Equation (3.10) also holds for the single time probability $P_1(\Delta,t) \equiv \int d\Delta'\, P_2(\Delta,t|\Delta',0)$ we are interested in.

In equilibrium we know that

$$P_1(\Delta,t) = C \exp [\, - V(\Delta)/k_B T\,] \tag{3.11}$$

where $\mathcal{C}$ is a normalization constant. On the other hand, setting the time derivative equal to zero in (3.10) we find,

$$P_1(\Delta,t) = \mathcal{C} \exp [-V(\Delta)/\mathcal{D}\tau]. \tag{3.12}$$

The integration constant which we have set equal to zero is a probability current as will be clear from part B of this Section. From (3.11) and (3.12) it is clear that $\mathcal{D}=k_BT/\tau$ yields the correct equilibrium distribution function. You should recognize this as the Einstein relation or the fluctuation-dissipation theorem.

Out of equilibrium, $dV(\Delta)/d\Delta$ in general differs from the equilibrium result. However, if fluctuations in the external source can be neglected, it is natural to assume that the random force defined by (3.2) remains unchanged since it represents the effect of molecular chaos (thermal agitation). However one must use $\mathcal{D} = k_B\bar{T}/\tau$ where $\bar{T}$ is the actual temperature. The stationary solution of (3.10) gives the steady-state probability distribution. Written in the form (3.12) we see that the argument of the exponential is a generalization of the free energy density since the most probable state is that which minimizes the argument.

Concerning the assumption that the stochastic force remains unchanged out of equilibrium two more remarks are in order. First, this seems an appropriate assumption when the system can be described quite accurately by a local temperature (Tremblay,Siggia,Arai, 1980). When the distribution function greatly differs from a local equilibrium one, it has been checked explicitly for a simple model that the constant $\mathcal{D}$ differs by factors of order unity from $k_B\bar{T}/\tau$ (Tremblay, Patton, Martin, Maldague, 1979). Second,in clean superconductors, it seems necessary to consider a probability space for both the quasiparticle distribution function f_k and the gap Δ for the Fokker-Planck description to remain valid. Out of equilibrium, the phonons should also be included. The derivation of the steady-state probability distribution for nonequilibrium superconductors is technically more complicated than the above argument suggests but the basic ideas are the same (Schmid 1977).

"First Order Phase Transition" and Nucleation Rates

If Eq. (3.12) is valid then when $V(\Delta_1) = V(\Delta_2)$ there can be a transition from a state with gap Δ_1, to one with gap Δ_2 since $V(\Delta_1)(1 - v) + V(\Delta_2)v$ is a constant for arbitrary values of v the fractional volume ($0 < v < 1$) occupied by the phase Δ_2.

To find the rate at which the new phase nucleates, we must solve a time dependent problem. To understand the method of calculation (Kramers 1940) we rewrite the Fokker-Planck equation in the suggestive form,

$$\frac{\partial P_1}{\partial t} + \partial_\Delta J_{P_1} = 0 \tag{3.13}$$

where J_{P_1} can be thought of as the divergence of the probability current

$$J_{P_1} = -\frac{1}{\tau}\frac{dV(\Delta)}{d\Delta} P_1 - \mathcal{D}\partial_\Delta P_1 \tag{3.14}$$

which contains a deterministic probability flow and a diffusive one. If one computes the probability current flowing between the two minima of V as a function of the difference in occupation probability of each minimum (say δn) then the nucleation rate γ is $J_{P_1} = \gamma\delta n$. The problem can be done in quasi steady state. In reality P_1 depends on infinitely many variables; Δ for each point of space. In that case the most probable nucleation path goes through a saddle point of the generalized free energy (Landauer and Swanson 1961, Langer 1968). The rate turns out to be proportional to the exponential of the difference between the initial and saddle point generalized free energies.

Unusual Features of "Phase Transitions" in Dissipative Systems

The criterion $V(\Delta_1) = V(\Delta_2)$ for a phase transition is the same as that we derived earlier from the requirement that the deterministic equations of motion predict a marginally stable wall between phases Δ_1 and Δ_2. However, if instead of $\xi^2(0)\nabla^2\Delta$ we had had $\xi^2(0)\nabla^2 g(\Delta)$ (where $g(\Delta)$ can be Δ^2 for example) in Eq. (2.4) then the result obtained from the deterministic equations of motion would have differed in general from the Fokker-Planck result. Cases where the deterministic and probabilistic approaches give different results have been found by Eckern et al. (1979) when $(D\tau_E)^{\frac{1}{2}}$, the quasiparticle inelastic relaxation length, is much larger than the coherence length $\xi(T)$ and the superconductor is illuminated by microwaves, phonons or light. The transition temperature found from the two approaches however differed very little from one another. The fact that they do differ is nevertheless an indication that our probabilistic approach was incomplete: in Eqs (3.10) to (3.12) we assumed that Δ did not depend on position while in reality P_1 should be a functional of $\Delta(\vec{x})$. In the latter case, the probability space has many dimensions and the stationary solution to the Fokker-Planck equation may contain non-zero probability currents (whose "divergence" vanishes). When such currents exist, detailed balance is not satisfied. When detailed balance is violated, the probability for a transition from Δ_1 to Δ_2 is not equal to the probability for a transition along the reverse path; as a consequence, to find the relative probability of Δ_1 and Δ_2 one must consider all paths connecting them and that makes the problem quite hard. Nonequilibrium steady states with detailed balance are analogous to equilibrium states and as such quite a bit easier to analyze because in that case to find the relative probability of two spatially independent values of Δ it is sufficient to consider only intermediate states with values of Δ independent of space as we did in Eqs. (3.10) to (3.12).

Sometimes the deterministic analysis can be wrong (Arnold, Horsthemke and Lefever 1977, Landauer 1978). Suppose we have a one dimensional Fokker-Planck equation with a value of $\mathcal{D}$ which depends on Δ . Then the solution of Eq. (3.10) would give $P_1 = C \exp [-\int^{\Delta} d\Delta' \, [(\tau^{-1} \, dV/d\Delta' - d\mathcal{D}/d\Delta')/\mathcal{D}(\Delta')]]$. Clearly the noise $\mathcal{D}$ may now produce large deviations from the deterministic prediction that the stationary states are at extrema of V. No such example is known yet in superconductors. There are also cases in other fields where the noise induces limit cycles which the deterministic analysis does not predict (Hongler and Ryter 1978).

CONCLUSION

It should be clear that not everything is said and done in this field but to emphasize that point, let me mention a few of the things that are left (or may be left!) on the more theoretical side. More concrete conclusions are in the next Chapter.

In this lecture, we have not considered at all convective instabilities, i.e. cases where $\vec{v}_s$ or $\vec{v}_n$ could differ from zero in the inhomogeneous state. Baru and Sukhanov (1975) have included the possibility of a non zero $\vec{v}_s$ in their calculation for the μ^* model. I know of no other theoretical or experimental work on that subject (except that related to phase slip centers). The effect of nonequilibrium phonons on the "generalized free energy density" (Schmid 1977) should also be investigated.

Phenomena which are really characteristic of dissipative systems have been briefly discussed here. For example we have seen there may occur phase transitions which cannot be predicted by the deterministic analysis. This type of behavior has not been observed yet in superconductors. Experimental or theoretical advances in that direction would certainly teach us a great deal about the statistical mechanics of nonequilibrium systems. One may also wonder if limit cycles of a type different from phase slip centers will be found and whether for stronger perturbations a turbulent regime may be reached[1].

---

[1] Huberman, B.A., Crutchfield, J.P., Packard, N.H. (preprint) have found that driven Josephson junctions have a turbulent regime.

The systematic study of stability and instabilities in nonequilibrium superconductors started about three years ago. It is a comparatively new area of research which still contains some interesting physics and may even lead to some surprises.

ACKNOWLEDGMENTS

I should like to thank A. Caillé, L.G. Caron and G. Slater for comments on the manuscript. Most of this work was performed in Sherbrooke and supported by NSERC (Canada) through a grant and through the program "Attaché de recherche CRSNG". The work in Cornell was supported by the U.S. Office of Naval Research.

APPENDIX

LOCAL STABILITY OF UNIFORM NONEQUILIBRIUM SUPERCONDUCTING STEADY STATE

This Appendix assumes a somewhat larger background than the rest of the lecture. A useful exercise to make concrete some of the general remarks of the previous Section and has some experimental relevance is presented. It is drawn from a combination of the work of Aronov and Spivak (1978) and of Eckern, Schmid, Schmutz and Schön (ESSS) (1979).

A superconductor is described by a kinetic equation for the quasiparticles and the phonons, a gap equation and the continuity equation for the current and Maxwell's equations. For simplicity we assume that the phonons remain in equilibrium and that the superfluid stays at rest. We are left with the following equations (Aronov and Gurevich 1974), (Tremblay, Patton, Martin 1980):

$$\frac{\partial f_k}{\partial t} + \frac{\partial E_k}{\partial \vec{k}} \cdot \frac{\partial f_k}{\partial \vec{r}} - \frac{\partial E_k}{\partial \vec{r}} \cdot \frac{\partial f_k}{\partial \vec{k}} = - I\{f_k\} + \dot{f}_D \tag{A.1}$$

$$0 = \frac{\Delta}{N(0)V_{BCS}} - \int_{-\infty}^{\infty} d\,\xi_k \frac{\Delta}{2E_k} (1 - 2 f_k) - \xi^2(0)\nabla^2\Delta. \tag{A.2}$$

I is the collision operator and $\dot{f}_D$ the external perturbation. Explicit expressions for that quantity have been given by other lecturers. In general close to T_c it is a good approximation to use the equilibrium distribution function to evaluate $\dot{f}_D$. Define f_{st} the stationary solution of (A.1) (A.2). To test for stability we linearize around the stationary state.[1] Let

[1] ESSS have a definition of f_1 which differs from Aronov and Spivak and from ours but to the order we are working the results are identical.

$$\Delta = \Delta_{st} + \Delta_1(\vec{r},t), \qquad f = f_{st}(\Delta) + f_1. \tag{A.3}$$

Keeping only terms linear in the small quantities Δ_1 and f_1 in Eq. (A.2) we obtain

$$\frac{\partial f_1}{\partial t} + \frac{d\Delta_1}{dt}\frac{df_{st}}{d\Delta} + \frac{\xi_k}{E_k}\vec{v}_F \cdot \left[\frac{d\Delta_1}{d\vec{r}}\left(\frac{df_{st}}{d\Delta}\right)_E + \frac{df_1}{d\vec{r}}\right] =$$

$$-\frac{1}{\tau_{tr}}\frac{|\xi_k|}{E_k}(f_1 - \bar{f}_1) - \frac{1}{\tau_E} f_1 \tag{A.4}$$

$$df_{st}/d\Delta \equiv (\partial f_{st}/\partial\Delta)_E + (\partial f_{st}/\partial E)_t \, \Delta/E_k \quad . \tag{A.5}$$

$\bar{f}_1$ stands for the average of f_1 over the direction of $\hat{k}$ at constant energy. In the collision operator we kept impurity scattering and used a relaxation time approximation for inelastic scattering by the phonons. The latter approximation is justified when the quasiparticles of interest have energies in the range $\Delta << k_BT_c$. Then since ξ and Δ are small we can also neglect the energy dependence of τ_E. f_D disappeared for Eq. (A.4) because it is independent of f_1 to leading order in the ratio of the excitation rate and the relaxation rate and because for a uniform stationary state,

$$-\frac{1}{\tau_{tr}}\frac{|\xi_k|}{E_k}(f_{st}(\Delta) - \bar{f}_{st}(\Delta)) - \frac{1}{\tau_E} f_{st}(\Delta) + \dot{f}_D(\Delta) = 0. \tag{A.6}$$

To solve the Boltzmann equation (A.4), we look at one Fourier component q of the spatial dependence and assume $f = f^{(i)} + \hat{k} \cdot \hat{q}\, f^{(a)}$ where $f^{(i)}$ and $f^{(a)}$ are now independent of the direction of k. Projecting out the part of the Boltzmann equation which depends on $\hat{k} \cdot \hat{q}$ and that which does not, we find in the limit $\tau_{tr} << \tau_E$ and $df_1/dt << f_1/\tau_{tr}$,

$$f_1^{(a)} = - \operatorname{sgn}[\xi_k] iqv_F \tau_{tr} [\Delta_1 (-\frac{\partial f_{st}}{\partial\Delta})_E + f_1^{(i)}] \tag{A.7}$$

and hence

$$- sf_1^{(i)} - s\Delta_1\frac{df_{st}}{d\Delta} + \frac{|\xi_k|}{E_k} Dq^2[\Delta_1 (\frac{df_{st}}{d\Delta})_E + f_1^{(i)}] = -\frac{f_1^{(i)}}{\tau_E} \tag{A.8}$$

From now on, Δ stands for Δ_{st}. Linearizing the gap equation (A.2),

$$\frac{\Delta_1}{\Delta} = -2\int_0^\infty d\xi_k \Delta_1 \frac{d}{d\Delta}\left(\frac{f_{st}}{E_k}\right) - 2\int_0^\infty d\xi_k \frac{f_1^{(i)}}{E_k} + \frac{\xi^2(0)}{\Delta}\nabla^2\Delta_1, \quad (A.9)$$

we substitute the solution of Eq. (A.8) and obtain,

$$X = -2\int_0^\infty d\xi_k \frac{\frac{\Delta}{E_k}\left(s - \frac{|\xi_k|}{E_k} Dq^2\right)\left.\frac{\partial f_{st}}{\partial \Delta}\right|_E + \frac{\Delta^2}{E_k^2}\, s\, \frac{df_{st}}{dE_k}}{-s + \frac{|\xi_k|}{E_k} Dq^2 + \frac{1}{\tau_E}} - q^2\xi^2(0) \quad (A.10)$$

where

$$X = -\Delta\frac{d}{d\Delta}\left[\int_{-\infty}^{\infty} d\xi_k \frac{1}{2E_k}(1 - 2f_{st}) - \frac{1}{N(0)V_{BCS}}\right]. \quad (A.11)$$

Given a particular f_{st}, one can find the eigenvalue s from Eq. (A.11). This is not an easy task in general. We look at a few limiting cases:

(i) Stability to the $q = 0$ mode. Since we are interested in the case where the eigenvalue s is of the order of the dimensionless parameter characteristic of the external perturbation, to linear order in that parameter, we can replace f_{st} by the equilibrium distribution function when it is multiplied by s. Then we find for $s\tau_E << 1$,

$$s\tau_E\Delta\left(\frac{\pi}{4T_c}\right) = X\ . \quad (A.12)$$

The system is stable to small uniform perturbations ($q = 0$) if $X > 0$. This reduces to a familiar result in equilibrium since then $2N(0)X$ is the second derivative of the free energy density with respect to Δ

(ii) Case $\xi^2(0)q^2 << 1 << (D\tau_E)\, q^2$. Since the smallest allowed value of q is of the order of the inverse system size, L, the last inequality requires an inelastic diffusion length $(D\tau_E)^{\frac{1}{2}}$ larger than L. This is a condition easier to satisfy at low temperature where, incidently, the relaxation time approximation for inelastic collisions (recombination) can also be justified. From (A.10)(A.11), we find,

$$\frac{n_s}{n_e} = -2\int_0^\infty d\xi_k \frac{s}{-s + \frac{|\xi_k|}{E_k} Dq^2}\frac{\partial f_{st}}{\partial E_k}\frac{\Delta^2}{E_k^2} \quad (A.13)$$

where

$$\frac{n_s}{n_e} = 1 + 2 \int_0^\infty d\xi_k \left(- \frac{\Delta^2}{E_k^{\,3}} f_{st} + \frac{\Delta^2}{E_k^{\,2}} \frac{\partial f_{st}}{\partial E_k}\right) = 1 + 2 \int_0^\infty d\xi_k \frac{\partial f_{st}}{\partial E_k} \qquad (A.14)$$

is, in a clean superconductor, the fraction of superconducting electrons. From (A.14) it is clear that the system is stable when $n_s/n_e > 0$. When $n_s/n_e < 0$ the superconductor is in a paramagnetic instead of diamagnetic state and it is unstable to fluctuations in the magnitude of the gap. The condition $n_s/n > 0$ is also required for stability to electromagnetic oscillations (Aronov and Spivak 1975).

At low temperature the μ^* distribution is sometimes appropriate. To find what Eq. (A.14) means for such a distribution assume that N, the density of excited quasiparticles, is what characterizes the external perturbation. For all values of N the gap equation is satisfied. Differentiating that equation with respect to N we find,

$$\frac{d\Delta}{dN} \frac{n_s}{n_e} = - 2 \frac{d\mu^*}{dN} \int_0^\infty d\xi_k \frac{\Delta}{E_k} \left(\frac{\partial f_{st}}{\partial \mu^*}\right)_{E_k} . \qquad (A.15)$$

Intuition and a direct calculation reveal that $d\Delta/dN < 0$ hence the stability condition $n_s/n_e > 0$ becomes $d\mu^*/dN > 0$. This criterion was first formulated by Chang and Scalapino (1974) on the basis of the analogy between μ^* and the first derivative of the free energy density with respect to N. The criterion $n_s/n_e > 0$ is due to Baru and Sukhanov (1975). If the gap equation is differentiated with respect to μ^* instead of N, then the analog of (A.15) reveals that the inequality $n_s > 0$ implies $d\Delta/d\mu^* < 0$. This criterion for stability was formulated in slightly more general terms by Smith (1977).

The possible uniform solutions to the gap equation depend on the details of the distribution function f_{st}. As we can see from this Section and as we will see in the next Chapter, the local stability of the various uniform steady states also depends on the details of f_{st} as well as on new characteristics of the relaxation process e.g. coherence length and quasiparticle diffusion length. We have neglected various other lengths and times associated with the phonons which can in principle be important. The calculations of Scalapino and Huberman (1977) and of Hida (1978) illustrate the potential influence of phonons on the instability threshold.

REFERENCES

Arnold, L., Horsthemke, W., and Lefever, R., 1978, White and coloured external noise and transition phenomena in nonlinear system, Z. Physik B, 29:367.

Aronov, A.G., and Gurevich, V.L., 1974, Response of a pure superconductor to a slowly varying perturbation, Fiz. Tverd. Tela (Leningrad), 16:2656; Engl. transl. 1975, Sov. Phys. Solid State, 16:1722.

Aronov, A.G., and Spivak, B.Z., 1975, Stability of states of a superconductor with a "negative superfluid density", Fiz. Tverd. Tela (Leningrad), 17:2806; Engl. transl. Sov. Phys. Solid State, 17:1874.

Aronov, A.G., and Spivak, B.Z., 1978, Nonequilibrium distributions in superconductors (Review), Fiz. Nizk. Temp., 4:1365; Engl. transl. Sov. J. Low Temp. Phys., 4:641.

Baru, V.G., and Sukhanov, A.A., 1975, New types of instabilities in nonequilibrium excitation of a superconductor, ZhETF Pis. Red., 21:209; Engl. transl. JETP Lett., 21:93.

Chang, J.J., and Scalapino, D.J., 1974, New instability in superconductors under external dynamic pair breaking, Phys. Rev. B, 10:4047.

Chandrasekhar, S., 1968, "Hydrodynamic and hydromagnetic stability," Clarendon, Oxford.

Eckern, U., Schmid, A., Schmutz, M., and Schön, G., 1979, Stability of superconducting states out of thermal equilibrium, J. Low Temp. Phys., 36:643.

Elesin, V.F., 1977, Nonstationary "intermediate" state of nonequilibrium superconductors, Zh. Eksp. Teor. Fiz., 73:355; Engl. transl. Sov. Phys. JETP, 46:185.

Golovashkin, A.I., Mitsen, K.V., and Motulevich, G.P., 1975, Experimental investigation of the nonequilibrium state of superconductors excited by lasers, Zh. Eksp. Teor. Fiz., 68:1408; Engl. transl. 1976, Sov. Phys. JETP, 41:701.

Haken, H., 1975, Cooperative phenomena in systems far from thermal equilibrium and in nonphysical systems, Rev. Mod. Phys., 47:67. Good introduction.

Hida, K., 1978, Phase transitions in nonequilibrium superconductors induced by quasiparticle injection, J. Low. Temp. Phys., 32:881.

Hongler, M.-O., and Ryter, D.M., 1978, Hard mode stationary states generated by fluctuations, Z. Physik B, 31:333.

Kramers, H.A., 1940, Brownian motion in a field of force and the diffusion model of chemical reactions, Physica, VII:284.

Landauer, R., and Swanson, J.A., 1961, Frequency factors in the thermally activated process, Phys. Rev., 121:1668. "Easy" reference on nucleation rates in the case of many variables.

Landauer, R., 1978, Stability in the dissipative steady state, Physics Today, Nov. 78:23. Good introduction to some of the concepts described in this Chapter.

Landauer, R., 1977, The ballast resistor, Phys. Rev. A, 15:2117. Introduces many of the concepts of this Chapter in the context of a simple model.

Langer, J.S., 1968, Theory of nucleation rates, Phys. Rev. Lett., 21:973.

Minorski, N., 1947, "Nonlinear Mechanics," J.W. Edwards, Ann Arbor. Good introduction to the mathematics of stability analysis.

Sai-Halasz, G.A., Chi, C.C., Denenstein, A., and Langenberg, D.N., 1974, Effect of dynamic external pair-breaking in superconducting films, Phys. Rev. Lett., 33:215.

Scalapino, D.J., and Huberman, B.A., 1977, Onset of an inhomogeneous state in a nonequilibrium superconducting film, Phys. Rev. Lett., 39:1365.

Schmid, A., 1977, Stability of radiation-stimulated superconductivity, Phys. Rev. Lett., 38:922.

Smith, L.N., 1977, Diffusive instability in optically excited superconducting films, J. Low Temp. Phys., 28:519.

Tremblay, A.-M., Patton, B., Martin, P.C., and Maldague, P.F., 1979, Microscopic calculation of the nonlinear current fluctuations of a metallic resistor: the problem of heating in perturbation theory, Phys. Rev. A, 19:1721.

Tremblay, A.-M.S., Siggia, E.D., and Arai, M.R., 1980, Fluctuations about hydrodynamic nonequilibrium steady states, Phys. Lett., 76A:57. The last two papers illustrate some general points made in the text about Langevin forces.

Tremblay, A.-M., Patton, B.R., and Martin, P.C., 1980, Kinetic equations in superconductors and the nature and decay rate of a new mode, Ann. Phys. (New York), 124:401.

van Kampen, N.G., 1965, Fluctuations in nonlinear systems, in: "Fluctuation phenomena in solids," R.E. Burgess, ed., Academic Press, New York.

Volkov, A.F., and Kogan, Sh.M., 1968, Physical phenomena in semiconductors with negative differential conductivity, Usp. Fiz. Nauk., 96:633; Engl. transl. 1969, Soviet Physics Uspekhi, 11:881.

Wang, M.C., and Uhlenbeck, G.E., 1945, On the theory of Brownian motion II, Rev. Mod. Phys., 17:323. Reprinted 1954 in: "Selected papers on noise and stochastic processes," N. Wax, ed., Dover, New York.

STABILITY OF NONEQUILIBRIUM SUPERCONDUCTING STATES II:

THEORY AND EXPERIMENT

A.-M.S. Tremblay

Département de Physique
Université de Sherbrooke
Sherbrooke, Québec J1K 2R1, Canada

ABSTRACT

Theories and experiments on the stability of superconductors under the influence of laser illumination, tunnel injection, microwave or phonon irradiation are reviewed.

INTRODUCTION

This Chapter and the preceeding one have quite a different purpose. In Chapter 10 we have seen various theoretical concepts, what they mean and how they could be used to do calculations. The only detailed example is presented in the Appendix. Here instead, we wish to give the results of specific theories and compare their predictions with experiment. Hence these two Chapters can, for the most part, be read independently, according to taste.

A few warnings. This Chapter should have been written by an experimenter. Due to uncontrolable circumstances, I had to replace him at the very last minute. This has two consequences. First, as a theorist my insight into the details of experiments is limited. Second, I had little time to write this contribution because of the publisher's deadline and so the final product is not entirely to my taste. I am consoled only by the thought that this Chapter may soon be outdated: even as I was preparing it I had to change my outlook because new results were constantly coming out in print. In my opinion, this part of nonequilibrium superconductivity is the most unsettled one and as such it is likely to evolve quite rapidly.

As in the preceeding Chapter I have made no attempt to follow an historical line or to establish priorities. I have just tried to be as pedagogical as possible. The bibliography is far from complete and I apologize in advance to many authors whose work I could not discuss for conciseness reasons or because it was slightly out of the main line of discussion. Neither negative nor positive opinions I have expressed about theories or experiments should be taken as authoritative. In such a rapidly evolving field one must remain critical. This is sometimes in conflict with the requirement of being pedagogical. I have tried to arrive at an acceptable compromise.

The text is devided in two main Sections. Ths first one contains theory and experiment on "wide sources", i.e. laser and high voltage tunnel injection,while the second one discusses theory and experiments on "narrow sources", i.e. low voltage tunnel injection and monochromatic phonons and microwaves. The precise reason for that division will become clear shortly. The discussion on "wide sources" is somewhat more lengthy partly because much of the results on "narrow sources" have already been exposed in Chapter 7. The conclusions to the various Sections have been gathered at the end in the Conclusion so that someone already familiar with the field can find the main points in one place. The subject to which the various conclusions refer are clearly identified and the more systematic reader should refer to them immediately after reading the corresponding Section of the main text.

The expression "wide sources" refers to external perturbations (laser, large voltage tunnel injection) which create quasiparticles with an energy much larger than the energy gap Δ . These quasiparticles decay down towards the gap very rapidly. At the slowest stage of the relaxation process (near the gap) we end up with quasiparticles distributed over a large range of energies. Thus, effectively, "wide sources" create quasiparticles over an energy range much larger than the gap. Because there have been many collisions before the quasiparticles reach the gap, one expects that the steady-state distribution functions will assume "quasi-thermal" forms, independently from the precise details of the external perturbation: This is not however completely borne out by experiments with laser illumination and high voltage tunnel injection.

"Narrow sources" on the other hand (threshold (low) voltage tunnel injection and monochromatic microwaves and phonons) create quasiparticles of energy close to the gap (and over an energy range much smaller than the gap). Since quasiparticles close to the gap are quite effective in determining the value of the gap (see for example Chapter 7) then it is understandable that the detailed character of the external perturbation will be important. In other words, in contrast with "wide sources" where quasiparticles suffer many collisions before reaching energies close to the gap, for "narrow sources" the final form of the distribution function will certainly depend on details of the external perturbation (unless there is too strong phonon trapping). By contrast with quasi-thermal models, here instabilities or multiple steady states will occur because of pronounced ("sharp") features in the distribution function.

One may be led to think that the instabilities of the quasi-thermal models are more "intrinsic" to the superconductor than those which are caused by narrow sources but this is only a matter of semantics. In all cases it is the distribution function which determines what happens in the superconductor and in all cases the distribution function is determined by the Boltzmann equation which contains the influence of both the external perturbation and the detailed dynamics of the superconductor. The quasi-thermal models are at best approximate solutions to the Boltzmann equation which may be reasonable for certain kinds of external sources. We will not consider certain types of instabilities as being more "intrinsic" than others.

WIDE SOURCES: LASER AND HIGH VOLTAGE TUNNEL INJECTION

Models for Wide Sources

The two most famous quasi-thermal models are the μ^* (Owen and Scalapino, 1972) and the T* model (Parker, 1975). In the μ^* model it is assumed that quasiparticles thermalize rapidly to the ambient temperature but that there is a "recombination bottleneck" which leaves the quasiparticles out of "chemical" equilibrium with the pairs. Hence the distribution function is taken to be a Fermi function with a chemical potential μ^* different from the pair's chemical potential. The μ^* model is expected to be valid at best only at low temperatures. In the T* model the quasiparticles and phonons of energy larger than 2Δ are in equilibrium at an effective temperature larger than the bath temperature while the phonons of energy smaller than 2Δ remain at the bath temperature. An improvement of that model has been suggested by Chang, Lai and Scalapino (1979). They have shown that if the quasiparticles are assumed to be distributed with a Fermi function at a temperature T* then the kinetic equations can be solved exactly for the phonons leading to a distribution function which is a linear combination of two Bose functions, one at the bath temperature, and the other at T*. The linear combination is different above and below 2Δ .

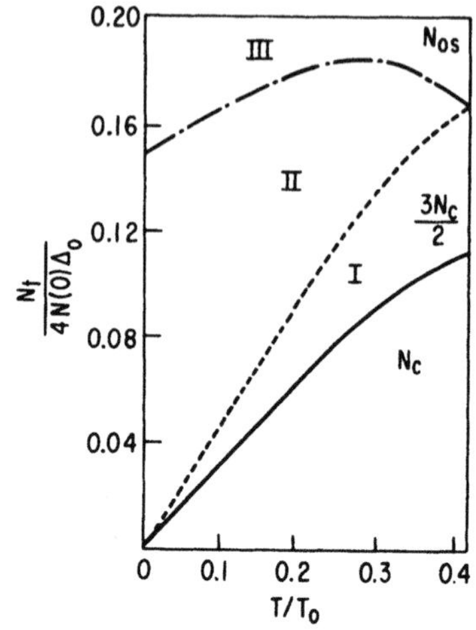

Fig. 1. The μ^* model predicts a first order-like transition when N_t is in region I and a second order-like transition when N_t is in region II. When N_t is in region III the superconductor should undergo the Owen-Scalapino first-order transition when N reaches the dashed-dotted line. (From Hida 1978a)

We are interested in the "instabilities" predicted by these models. The T* model leads to "dull" predictions: The superconductor will behave more or less like an equilibrium superconductor at an effectively higher temperature. In particular, the gap will go smoothly to zero when the pump power is increased. The results of the μ^* model are somewhat more complicated. They are summarized in Figure 1 (Hida 1978a). The superconductor should have a "diffusive" instability (see Chapter 10, discussion of Eq.(2.15)) when the density of excess quasiparticles is near the value N_t (Scalapino and Huberman 1977):

$$N_t \equiv N_c + \left[\frac{2R\xi^2 N(o)T}{D} \frac{\tau_B}{\tau_{es} + \tau_B} \frac{B}{A^2}\right]^{\frac{1}{2}} N(0)\Delta_o \qquad (2.1)$$

where the symbols follow the standard notation of this book except for A and B which are constants of order unity, $2RN_{QP} \equiv \tau_R^{-1}$ with N_{QP} the steady state quasiparticle number and N_c which is the value at which $d\mu^*/dN = 0$[1]. According to Hida (1978a), the order of the transition can be different depending on which of the regions of Fig. 1 N_t falls. If N_t is in region I, the superconductor should go discontinuously (first order transi-

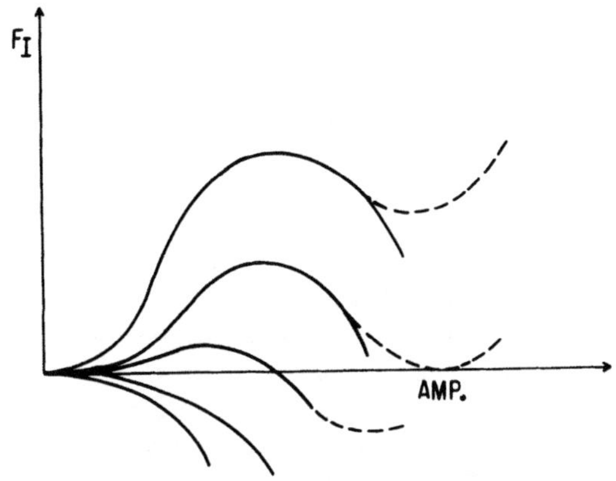

Fig. 2

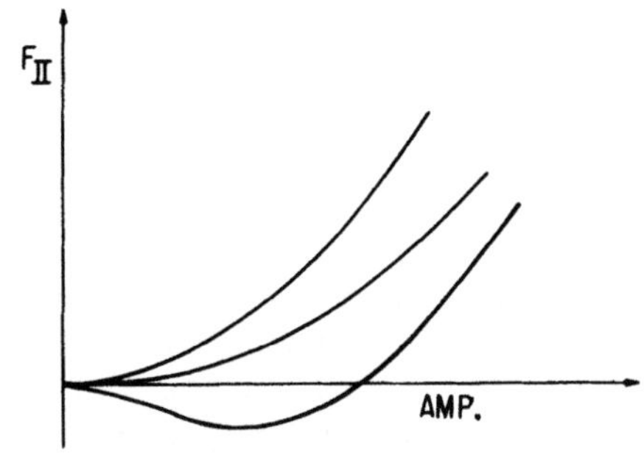

Fig. 3

Fig. 2. "Potential" F_I when N_t is in region I. The horizontal axis is the amplitude Amp. of the unstable mode. The dotted line is infered, not calculated. (From Hida 1978a).

Fig. 3. "Potential" F_{II} when N_t is in region II. Amp. is the amplitude of the unstable mode. (From Hida 1978a).

tion) to an inhomogeneous state whose exact nature is unknown. If N_t falls in region II, there should be a "second order" transition to a state described by a wavevector q_m (Scalapino and Huberman 1977). Finally, when N_t is in region III the superconductor should undergo a first order transition to the normal state as soon as the excess quasiparticle density reaches the dash-dotted line labeled N_{os} . This is the transition predicted in the original Owen-Scalapino paper. The shape of the "potential" F (the analog of $V(\Delta)$ in Chapter 10) as a function of the amplitude of the unstable mode is illustrated in Fig. 2 for the case where N_t is in region I and in Fig. 3 for the case where N_t is in region II of Fig. 1.

The origin of the μ^* model instability can be understood physically from the (disputable) phenomenological formula for the quasiparticle current $J \sim - \nabla\mu^* \sim - (d\mu^*/dN)\nabla N$. If $(d\mu^*/dN) < 0$ then the quasiparticle current flows from regions of low quasiparticle concentration towards regions of high quasiparticle concentration, increasing the size of spontaneously formed inhomogeneities. Regimes when $d\mu^*/dN < 0$ exist in the μ^* model (Chang

[1] Note that as τ_{es} increases (larger phonon trapping), N_t in Eq. (2.1) moves towards N_c, the instability predicted by Chang and Scalapino (1974) by thermodynamic-like arguments (which disregarded the dynamics). See also the Appendix of Chapter 10.

and Scalapino 1974) because at high enough quasiparticle concentration the gap decreases so fast that a smaller chemical potential is needed to maintain the excess quasiparticle concentration. As we already mentionned, the results of the μ^* model in general are not on a rigorous theoretical basis and they should be considered with caution.

At low temperatures, it is possible to bypass the quasi-thermal models by using rate equations. This was done by Welte (1978). He performed a local stability analysis of the homogeneous superconducting state for the case where the perturbation is a fixed injection current density. He found that the gap is everywhere a double-valued function of the current density, the higher gap state being stable, and the lower gap state being always unstable, except for very small values of Δ (where the theory is not expected to be reliable). The analysis is interesting. It reveals the strong influence of phonons on the instability. (We saw a similar influence of phonons on μ^* instabilities). Unfortunately, the local stability analysis cannot tell us if a first order transition will occur before the system becomes locally unstable (See Chapter 10). Hence, it is hard to compare this theory with experiments. At best the theory gives upper limits on the value of current density for which the homogeneous superconductor is expected to be stable.

Finally, we briefly discuss Elesin's theory, which is on a much stronger theoretical basis and leads to predictions different from the μ^* and T^* models. It predicts a "coherent instability", whose name has been chosen for purposes of discussion by Elesin (The motivation for the name should become clear shortly). Elesin, Kondrashov and Sukhikh (1979) have solved the kinetic equations for the quasiparticles and phonons on a computer for a source of quasiparticles which injects uniformly at energies between nine and ten times the gap. The results are in agreement with various limiting cases which had been analyzed analytically by Elesin (1976, 1977a, 1977b) and Eckern et al. (1979). Within 5 to 10% the calculated quasiparticle distribution functions can be approximated by a Fermi function with a negative μ^* and an effective temperature larger than the bath temperature. A plot of μ^* and T^* as a function of the strength of the external source g is given in Fig. 4 for various values of the ambient temperature and phonon trapping factor $\gamma = \tau_{es}/\tau_B$. The inset shows the results extracted by Willemsen and Gray (1978) from experiments on tunnel junctions. At high voltages μ^* is indeed negative[1] and T^* increases in qualitative accord with the theory (The theory is valid only for high

[1] Aronov and Spivak (1979) say they can explain a negative μ^* by including electron-electron interactions in the collision operator. Aronov and Gurevich (1972) have shown that $\mu^* < 0$ implies that $\Delta > 0$ even if the quasiparticle temperature is larger than T_c.

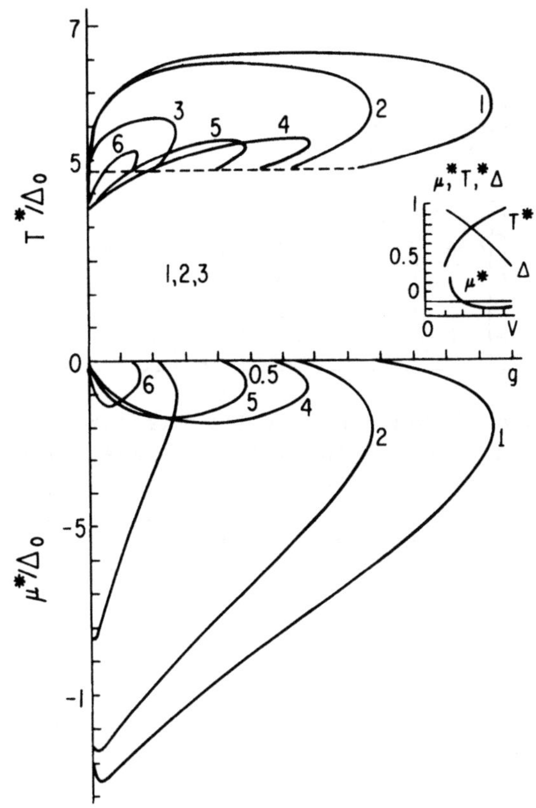

Fig. 4. Dependence of the effective temperature T* and effective chemical potential μ* on the quasiparticle generation rate g (in units of τ_R^{-1}) calculated for the following external temperature T and phonon trapping factor $\gamma = \tau_{es}/\tau_B$. (Δ_o is the equilibrium zero temperature gap) 1) $\gamma = 0$, $T = 0$; 2) $\gamma = 0.02$, $T = 0$; 3) $\gamma = 0.15$, $T = 0$; 4) $\gamma = 0$, $T/\Delta_o = 0.44$; 5) $\gamma = 0.02$, $T/\Delta_o = 0.44$; 6) $\gamma = 0.15$, $T/\Delta_o = 0.44$. (From Elesin et al. 1979).

voltages). Note that a negative μ* means the distribution function at the gap is always smaller than 1/2 as was proven by Elesin (1974) for optical illumination. These results suggest that theoretically the predictions of the positive μ* model (Hida 1978a) are not expected to be valid for laser illumination.

The numerical results of Elesin et al. (1979) for Δ as a function of g (Fig. 5) can be approximated by the analytical form which we looked at as an example in Chapter 10,

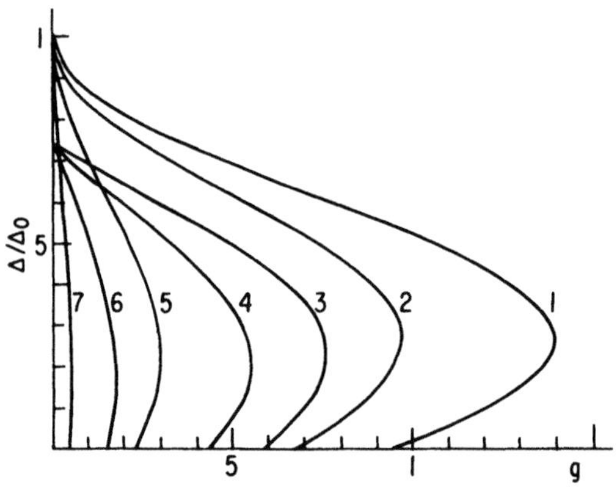

Fig. 5. Dependence of the gap Δ on the quasiparticle generation rate g (in units of τ_R^{-1}) for various values of the parameters $\gamma = \tau_{es}/\tau_B$ and T: 1) $\gamma = 0$, $T = 0$; 2) $\gamma = 0.02$, $T = 0$; 3) $\gamma = 0$, $T/\Delta_o = 0.44$; 4) $\gamma = 0.02$, $T/\Delta_o = 0.44$; 5) $\gamma = 0.15$, $T = 0$; 6) $\gamma = 0.15$, $T/\Delta_o = 0.44$; 7) $\gamma = 1.0$, $T = 0$. (From Elesin et al.1979).

$$\Delta\left(\frac{g - g_c}{g_c} + \bar{\alpha}\Delta^2 - b\Delta\right) = 0 .$$

Elesin et al. (1979) have given a table of values for g_c, $\bar{\alpha}$ and b for various values of ambient temperature and phonon trapping factors. There exists a region where Δ increases with g because the recombination rate can increase with Δ (hence more quasiparticles can be pumped) due to the presence of the term Δ^2/EE' in the coherence factors. This however immediately leads to the fact that this branch of the curve $\Delta(g)$ is locally unstable because a spontaneous increase in Δ leads to an enhanced recombination and hence to a decrease in the number of quasiparticles and (through the gap equation) to an accompanying further increase in Δ.

According to Elesin the superconductor should undergo a first order transition to the normal state at a critical value of the strength of the external source within the region where μ^*, T^* and Δ are double-valued functions of g. The argument is essentially as described in Chapter 10. For a certain value of g within the region where the gap is double-valued, a marginally stable wall can exist between the normal state and the largest of the two gaps. For a larger value of g, droplets of normal state grow until the superconducting region disappears completely. By contrast with the results of the μ^* model, in all limiting cases studied by Elesin (1977a) he finds no diffusive instability.

To close this Section, we should stress that even though Elesin et al. (1979) parametrized their results for the distribution function using μ^* and T^*, the fact that both parameters are needed simultaneously means that the predictions of their theory differs considerably from either the T^* or the μ^* model. This should be clear from what has been said above. Fig. 5 for the gap in Elesin's theory is also revealing. In the μ^* model, the gap is double-valued for the whole range of possible external source strengths while in the T^* model the gap is always single-valued.

Experiments with Lasers and Tunnel Junctions

Lasers. Let us first discuss time scales (Chang and Scalapino 1977). The energy of the incident radiation is much larger than either the Debye or gap energies. Excited electrons first decay by electron-electron interactions until their energies is about $(E_F\hbar\omega_D)^{\frac{1}{2}}$; then electron-phonon interactions become dominant. In about 10^{-14} second, the electrons have energies of the order of the gap energy where their decay then becomes much slower. They take of the order of 10^{-9} second to adjust to the local order parameter. Close to T_c the latter takes an even longer time to relax. Typically the light is applied in the form of a pulse of duration 10^{-6} second. During that time the quasi-particle and phonon distributions can relax to their local steady-

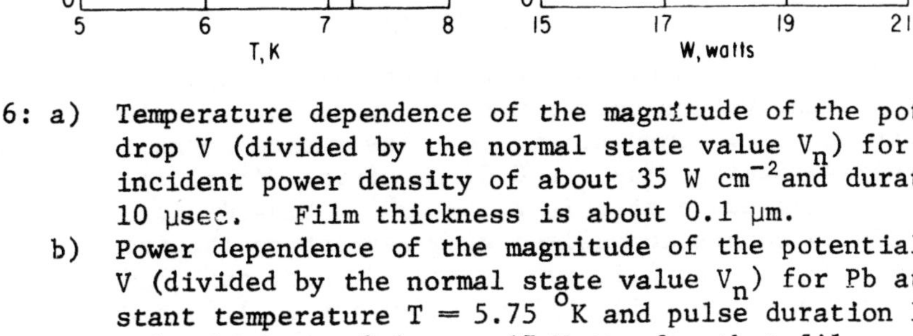

Fig. 6: a) Temperature dependence of the magnitude of the potential drop V (divided by the normal state value V_n) for Pb at incident power density of about 35 W cm^{-2} and duration of 10 μsec. Film thickness is about 0.1 μm.

b) Power dependence of the magnitude of the potential drop V (divided by the normal state value V_n) for Pb at constant temperature $T = 5.75\ ^oK$ and pulse duration 10 μsec. Film thickness 0.1 μm. 17 Watts for that film corresponds to a power density of 30 W cm^{-2}. Both figures are taken from Golovashkin et al. (1975).

state values. It is not clear that the gap and temperature do have time to relax to their true steady state value. As far as length scales are concerned typical films have widths of the order of millimeters (10^3 μm), lengths that can go up to centimeters (10^4 μm) and thicknesses of the order of 1000 Å (0.1 μm). This should be compared with the zero temperature coherence length $\xi(0)$ which ranges between 10^{-2} μm and 1 μm and with the quasi-particle diffusion length $(D\tau_E)^{\frac{1}{2}}$ which ranges between 1 μm and 10 μm. The radiation is usually incident on one of the two large sides of the films ("top" or "bottom").

Early experiments go back to Testardi (1971). In more recent experiments, Golovashkin et al. (1975) measured the potential drop which appears when a laser is shined on a superconductor in which a supercurrent is flowing (in a direction perpendicular to the incident light). This is taken as a measure of the resistance in the nonequilibrium state. Figure 6 contains the two main experimental phenomena which are consistently observed by this group as

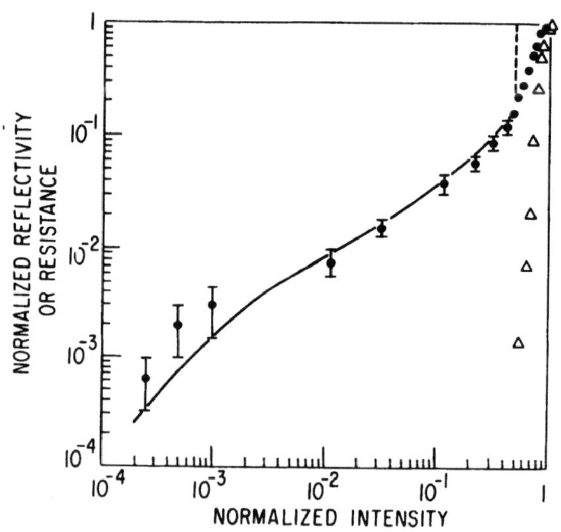

Fig. 7. Resistance and microwave reflectivity change as a function of relative incident power for Sn at 1.2 °K. Pulse duration, 60-90 nsec. Open triangles: resistance normalized to the normal state value for 0.05 μm strip. Absolute light power for this curve at 1 is 7.6 W cm^{-2}. Closed circles: reflectivity change normalized with respect to the full change between the equilibrium superconducting and normal states for a 0.04 μm thick film; intensity at 1 is 8 W cm^{-2}. Solid line: theoretical curve of Rothwarf et al. (1974). Dashed line: where the Owen-Scalapino transition should have occured. (From Sai-Halasz et al. 1974).

well as others. Figure 6a shows that as a function of temperature at fixed incident power, the measured signal rises gradually from zero at a threshold temperature up to its normal state value at T_c . The "width" of this resistive transition is much larger than what can be attributed to intrinsic inhomogeneities or to heating. Figure 6b shows the second result: the potential drop as a function of incident power at fixed temperature also has a threshold character. Figure 7 taken from Sai-Halasz et al. (1974) essentially shows the same result as Figure 6b: the open triangles on the right of the Figure represent measurements of resistance. The solid line and the dots on Figure 7 are a comparison of theory and experiment for the microwave reflectivity of the illuminated film. The theory (Rothwarf et al. 1974) is based on rate equations. The vertical dotted line represents the point where the reflectivity should have jumped to its normal state value according to the Owen-Scalapino theory (μ^* model). The current-voltage (I-V) curves for an illuminated film should give us additional information. Such curves are shown in Figure 8 (Hu et al. 1974). One sees a gradual broadening of the gap as the laser power is increased. The

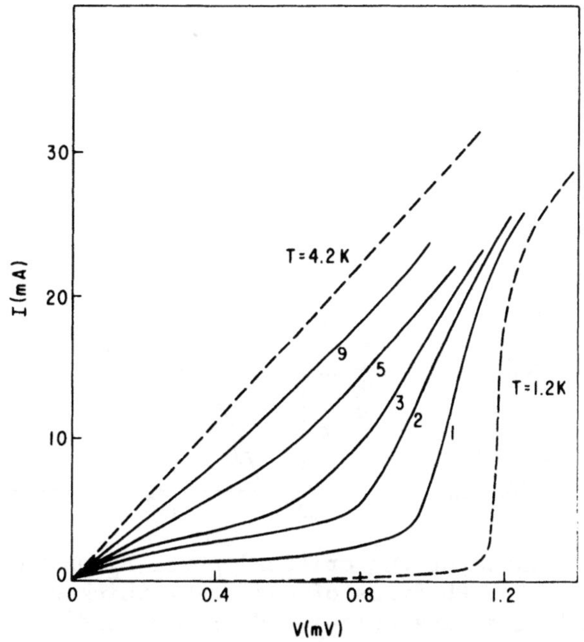

Fig. 8. The two dashed curves are taken in thermal equilibrium. The solid lines are I-V charactivistics under pulsed-light conditions (30 nsec). The number on each curve is the relative light intensity for measurements on 0.32 μm thick Sn films at T = 1.2 ^oK. (From Hu et al. 1974).

broadening is not inconsistent with either (Hu et al. 1974) heating or with a gap which varies smoothly between normal and superconducting regions in the junction. Hence these I-V measurements are not as conclusive as one may have hoped.

Which theory explains these experimental results? The T* model is consistent with a gradual decrease in the gap but not with a gradual increase in resistance. Even though the μ* model is not expected, theoretically, to be valid for laser illumination, how does it stand a comparison with experiment? Clearly, the gradual rise in resistance is inconsistent with the predictions: if we were in region II of Figure 1, the gap would be slightly modulated in space at the onset but it could still most probably carry a supercurrent. In region I it is not known exactly what type of inhomogeneous state is reached. Whatever it is, the predicted first order jump should be small for the theory to be consistent with experiment. The only way the μ* model could be contrived to give the correct answer would be to assume that the transition is the Owen-Scalapino first order transition but that, as in Elesin's theory (see below), the normal state does not have time to fill the whole sample during the laser pulse duration. Sai-Halasz et al. (1974) are to my knowledge the only ones who correlated the onset

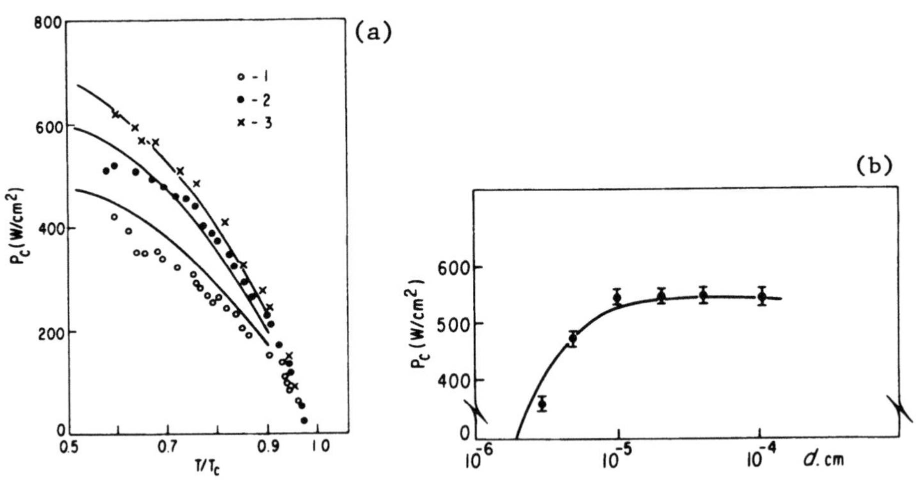

Fig. 9: a) Dependences of the critical power P_c on reduced temperature for Pb films of different thicknesses d. The solid lines are calculated from Elesin's theory. Thicknesses in microns: (1) Open circles 0.03 μm, (2) filled circles 0.05 μm and (3) crosses 0.1 μm. (From Golovashkin et al. 1980).

b) Dependence of the critical power P_c on film thickness for $T/T_c = .7$. The solid line is calculated from Elesin's theory (From Golovashkin et al. 1980).

of the resistive state with the Owen-Scalapino prediction. That prediction is not in very good agreement with the experimental results of Golovashkin et al. (1978).

That leaves Elesin's explanation. Theoretically it is certainly expected to be better. It does predict a threshold power. The experiments of Golovashkin et al. (1980) are in substantial agreement with Elesin's theory both for the temperature (Fig. 9a) and film thickness (Fig. 9b) dependences of the threshold power. The explanation for the gradual rise in resistance given by Elesin has been outlined in Chapter 10: beyond the critical power, droplets of normal state start to grow[1]. As we have seen however, very close to the critical power the nucleation fronts move very slowly and the normal state may not have time to fill up the whole sample during the length of the laser pulse. At higher incident power, the walls move faster and the measured resistance should be larger. The only qualitative verification of these nonstationary effects is the 1975 experiment of Golovashkin et al. where they checked

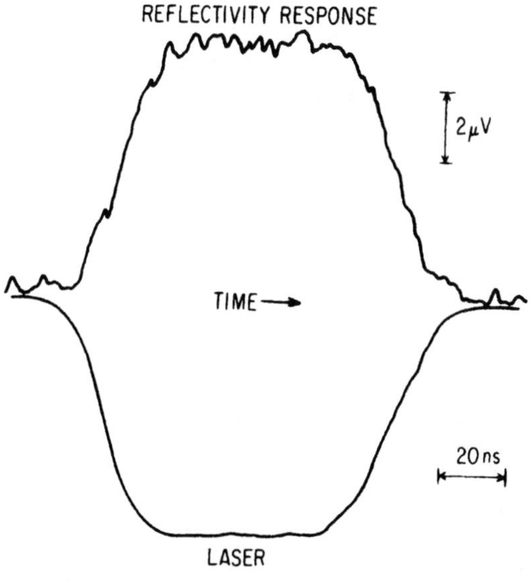

Fig. 10. Typical microwave reflectivity signal and laser pulse as a function of time. The signal amplitude is proportional to the difference in reflectivity with and without the laser pulse on. The voltage scale refers to the microwave detector output. The scale for the laser signal is arbitrary. (From Sai-Halasz et al. 1974).

[1] It may be helpful to have in mind the very rough analogy with vapor drops in boiling water.

that at fixed incident power the measured resistance did increase with the length of the laser pulse. No other check has ever been mentioned in the later papers of that group.

The problem with Elesin's explanation of the gradual rise in resistance is that the experiments do look stationary, as can be seen for example from Figure 10 (Sai-Halasz et al. 1974) which shows on the same plot the time dependence of the laser pulse and of the superconductor's response. The superconductor adjusts to the laser perturbation within a few nanoseconds (the quasiparticle relaxation time) and then seems to reach a steady state. A few nanoseconds is the time scale expected for relaxation in the superconducting or the normal state. However, Elesin's theory says that in the region where the nonequilibrium state is resistive, there is an additional much longer time scale associated with the growth time of normal state droplets. Has this longer time scale been overlooked in experiments?

If Elesin's theory is correct, there should also be an appreciable time delay between the onset of the laser pulse and the onset of resistance because the film should show up resistivity only when the normal state has somewhere filled the whole width of the strip. Such a delay was observed by Sai-Halasz et al. (1974) but it was independent of laser power while Elesin's theory would predict shorter delays for larger powers.

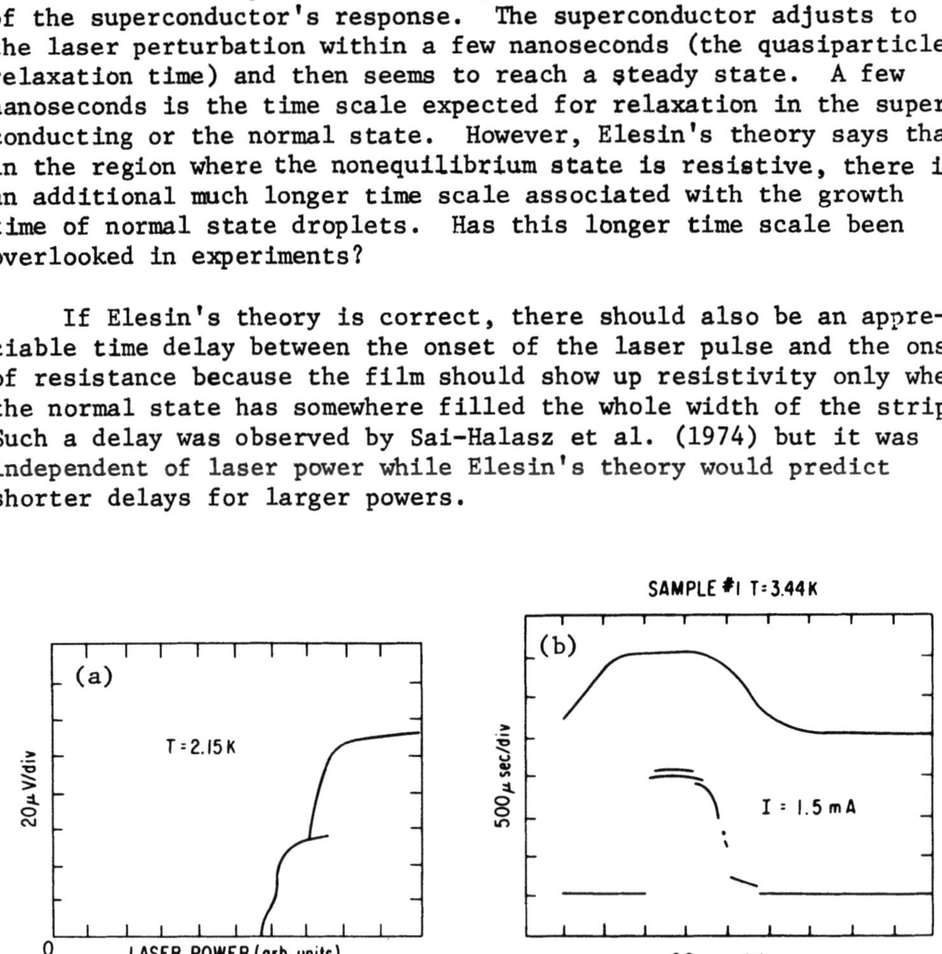

Fig. 11: a) Resistive transition for pulsed laser illumination measured on a one-dimensional Sn microstrip (0.068 μm x 4.4 μm x 51 μm). The horizontal axis is the laser power in arbitrary units. The vertical axis is the measured voltage for a current of 50 μA. (From Smith 1980)

b) Representative Sn microstrip response to a laser pulse. The laser pulse is the top curve; sample response is the bottom curve (From Smith 1980).

The two dimensional geometry clearly seems to complicate the problem. Smith (1980) has studied "one-dimensional" strips. Like those of Golovashkin et al. (1975, 1980) these experiments measure the voltage drop which appears when the laser is shined on a superconductor which carries a small supercurrent. A gradual rise in resistance (or voltage) as the laser power exceeds a threshold has also been observed in that experiment as illustrated in Figure 11a, but clearly the rise in resistance is not very "smooth". The time dependence of the superconductor's response also shows some "bumpiness" (compare Fig. 11b and Fig. 10). Smith has associated these features with phase slip centers. Instead of being a simplification of the two-dimensional case, the one-dimensional case seems to involve some rather different physical phenomena which deserve separate attention.

High Voltage Tunnel Injection. Typical junction dimensions are 10^3 μm x 10^3 μm x 0.1 μm. (Recall $\xi(0) \sim 10^{-2}$ to 1 μm, $(D\tau_E)^{\frac{1}{2}} \sim 1$ to 10 μm). Specific resistances between 10^{-2} and 10^{-4} Ohm - cm^2 and between 10^{-4} and 10^{-6} Ohm - cm^2 will be called respectively high and low specific resistance junctions. Experiments are usually performed in the double-junction geometry (e.g. Al - I - Al - I - Al) (See Chapter 5 for more details). The middle metal film which is common to both junctions is the film under study. One of the junctions usually has very high specific resistance and acts as a "detector" which probes the properties of the middle ("sample") film which is perturbed by the other "injector" junction.

For injection at energies much larger than twice the gap ($|eV| - \Delta_{injector} >> 2\Delta$) one expects that a tunnel junction will behave in the steady state in a manner very similar to films under laser irradiation. One rather obvious difference between the laser and injection cases however is that in the latter case, a current

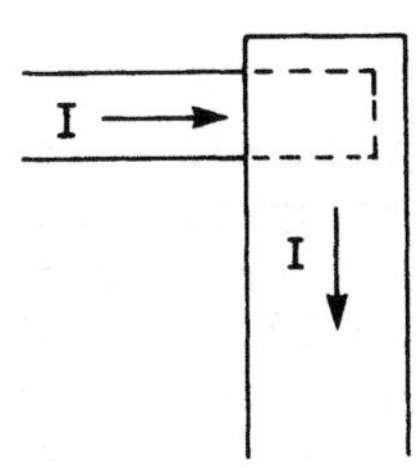

Fig. 12. Schematic view of two films forming a junction. The junction area is denoted by the dotted line. The current that flows through the junction must also flow in the films!

must also flow in and out of the film (Fig. 12). None of the theories we have mentioned accounts for that current. The belief is that if the current is much smaller than the critical current then its influence is minimal: the theories developed for the optical illumination case should apply (at least for the part of the distribution function which does not describe current flow). One should however be aware of the problem. Experiments such as those of Fuchs et al. (1977) which claimed to detect the first order phase transition predicted by the μ^* model (N_{os} line in Fig. 1) are rather ambiguous because the current at the instability point is about half the critical current. (Recall that when the critical current is reached in equilibrium, the value of the gap also jumps discontinuously to zero).

There is another difference between the laser and tunnel injection cases which cannot be overlooked. This is again related to the presence of an external circuit in the case of injection. If the experiment is done at fixed current the sample cannot jump discontinuously to the normal state even if the theory for the "currentless" sample says a first order transition to the normal state should occur. Since the normal state draws more current, in a constant current experiment the normal state could only grow, at the transition point, in proportion with the amount of current allowed in the system. Unless, for example, the superconducting part of the system simultaneously jumps to a higher gap value. In such a case one could observe jumps at the first order phase transition even in a constant current experiment. Clearly, the situation is different when the experiment is done at constant voltage. There, nothing forbids the system from jumping discontinuously to the normal state. In any case, constant voltage and constant current experiments are related. We will come back on these points in the next Section. In the experiments which will be mentioned in the following it is not always clear whether they were done in one of the two limiting cases (constant voltage or constant current) or in some intermediate regime. In most situations however it is difficult to have voltage-based injector junctions because their resistances are usually much smaller than that of the external circuit[1] .

One should also remark that all experiments we will shortly describe are steady state, not pulsed experiments, in contrast with the laser case. The experiment of Akoh et al. (1980) is an exception worth mentioning. Figure 13 shows the gap (in units of the zero temperature gap) as a function of time after the onset of a 1 msec injection pulse. The various curves correspond to different temperatures. The injection current was fixed. When the helium

[1] K. Gray, private communication

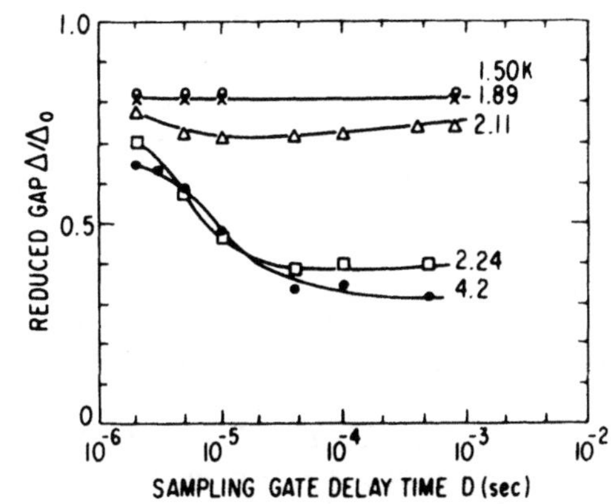

Fig. 13. Δ/Δ_0 for Pb-In (In: 10 wt%) versus sampling-gate delay time D at various temperatures for a fixed injection current from a low specific resistance injector (3×10^{-6} Ωcm^2). Junction size 10 μm x 10 μm x 0.12 μm. (From Akoh et al. 1980).

bath temperature is above the λ transition, clearly the system has response times of the order of 10^{-5} - 10^{-4} sec. These can be associated with thermal relaxation times. Akoh et al. (1980) further observed that as a function of injection current, a junction could exhibit a first order transition when one compared gaps measured 2 μsec after the pulse onset, while the same junction exhibited a second order transition when the experiment was done in true steady state conditions. Which experiment is closer to what the theory would predict? If the thermal relaxation time is associated with the sample and the phonon escape time τ_{es} then the steady state experiments are appropriate since these effects are accounted for in most theories. It is more likely however that these long thermal relaxation times (10 - 100 μsec) are associated with the whole system (including cooling) and not only with the sample and its immediate surroundings[1]. This means that these are additional heating effects which should be included in the theory if we want to compare with steady state experiments. It is plausible but not certain that such effects can be simply accounted for by an effectively higher bath temperature. We now proceed to discuss various steady state experiments.

In low specific resistance junctions, Rajeevakumar et al. (1979) have observed a gap decrease whose power dependence was consistent with a T* model. At temperatures above the λ transition of the helium bath, the jump of the order parameter to zero was associated with a critical current instability.

---

[1] K. Gray, private communication

Another group has not found any instabilities due to the nonequilibrium distribution in high voltage tunnel injection. This is illustrated in Figure 16 of Chapter 5. Using the method of Willemsen and Gray (1978) to extract an effective μ^* and T^* from detector I-V curves (see Chapter 5) Gray (unpublished) has plotted the gap as a function of T^* for many measurements on different junctions. The various symbols (triangles, open circles etc...) correspond to different junctions. In all cases, he has observed a smooth decrease of the gap. Note that all junctions tend to stay away from the Owen-Scalapino instability (bottom solid line). However, as we have seen previously, other μ^* model instabilities could in principle occur for smaller values of the excess quasi-particle concentration (and hence larger values of the gap) than predicted by Owen and Scalapino. Gray has not seen any of these yet. I do not have enough data to say if they should have been seen or not.

Iguchi (1978) on the other hand has done numerous experiments which have shown a variety of instabilities in (mainly) low specific resistance junctions. Some results are summarized in Figure 14.

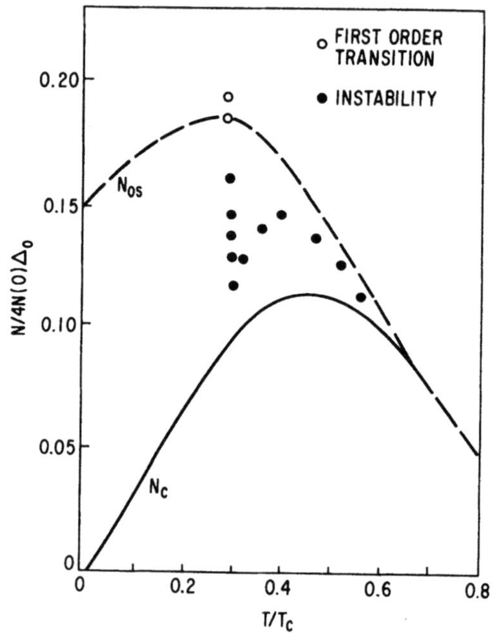

Fig. 14. The solid and dashed lines correspond to those of Fig. 1. The solid circles represent second order transitions and the open circles first order transitions. See text for determination of the vertical scale.(From Iguchi 1978).

The black dots represent the injection current (transformed to excess quasiparticle density) at which a change in slope of the resistance of the normal state detector was observed as a function of temperature. These points are associated with second order transitions. The open circles correspond to the sample jumping discontinuously (first order transition) to the normal state. The solid and dashed lines on that graph are those of Figure 1. While there seems to be a confirmation of the Owen-Scalapino instability line, the phonon trapping factor necessary to establish the vertical scale was chosen so that the experimental first order transition fit the theoretical one. Note that there are no points which verify the existence of the $3N_c/2$ line. According to Iguchi (private communication) the experiment was done in almost constant voltage conditions so that it is not too likely that first order transitions were mistaken for second order transitions as could have occurred in a constant current experiment. Note that the points on the Figure which are all vertically lined up were taken around the λ temperature where the phonon trapping factor changes drastically.

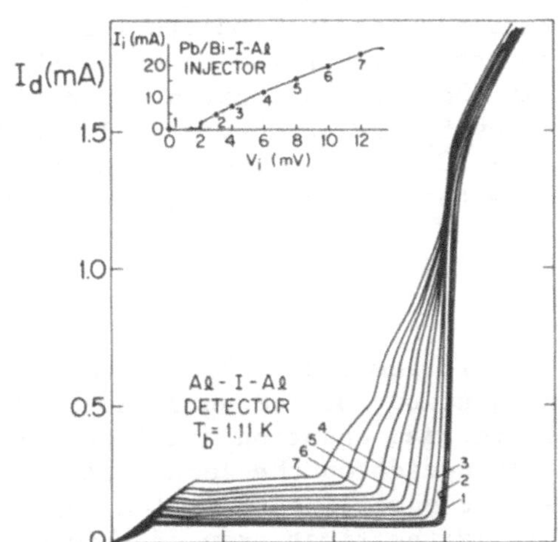

Fig. 15. Detector and injector (inset) current-voltage characteristics. The numbers on detector curves correspond to the value of the injector current and voltage (inset) at which the curves were measured. Junction size, 380 μm x 380 μm x 0.05 μm. High specific resistance Pb/Bi - Al injector. (From Iguchi and Langenberg 1980).

A most puzzling and fascinating result is that of Iguchi and Langenberg (1980) on high specific resistance junctions. The curves of Figure 15 represent detector I-V curves. The numbers on each curve correspond to the value of the injector current and voltage (inset) at which the detector curve was measured. The various steps on that curve indicate that many gaps may simultaneously exist in the perturbed film. Up to four coexisting gaps have been observed. Iguchi and Langenberg have proposed an extension of the Scalapino-Huberman (1977) theory to explain their experiment. The mathematical analysis is however very approximate. Apparently some additional experimental work would also be needed[1] to check whether effects due to the large inhomogeneities at the film edges were not inadvertently seen in the detector I-V curves. This experiment certainly deserves the attention of both theoreticians and experimenters.

The various experiments we have just described tend to indicate that the behavior of tunnel junctions is probably more complicated than just having a "first order transition" to the normal state as the theory of Elesin et al. (1979) would predict. It would be valuable to compare the experiments on junctions which do exhibit first order transitions with Elesin's theory. Such a comparison was done by Mitsen (1979). A first order transition to the normal state was observed in Sn for an excess quasiparticle density N of $.4 \times 10^{19}$ cm^{-3} . Elesin's theory predicts 1.3×10^{19} cm^{-3} . I could not verify from the data in the paper whether the current was much smaller than the critical current or not. Furthermore for that junction, the Owen-Scalapino theory would predict a transition when N is around $.3 \times 10^{19}$ cm^{-3} , much closer to the experimental value. Hence Mitsen's test of Elesin's theory is not really satisfactory!

NARROW SOURCES

In principle the problem should be approached as follows: solve the Boltzmann equation for the quasiparticle distribution function, then solve the gap equation and look for the various stationary states. To decide on the local stability of the solutions one should use the equations of motion, as outlined in the previous Chapter. It is the first step of this program, "solve the Boltzmann equation", which is avoided by the quasi-thermal models. For narrow sources that step can certainly not be avoided since for reasons mentioned earlier quasi-thermal models will be bad approximations to the true solution of the Boltzmann equation.

To decide on "global stability" one should use either the deterministic or, preferably, the probabilistic approach of the

---

[1] K. Gray, private communication

preceding Chapter. In the Section on Langevin and Fokker Planck equations of that Chapter, we described how to obtain a generalized free energy with the simplest probabilistic approach. The explicit form for the generalized free energy of nonequilibrium superconductors (Schmid 1977) can be found in Eq. (20) of Chapter 7. To date, such a generalized free energy has been found only for $T \sim T_c$. Furthermore, even at $T \sim T_c$, the explicit form of Eq. (20) is not always rigorously valid. (In this context, the discussion on "Unusual Features of Phase Transitions in Dissipative Systems" in the preceeding Chapter, should be read.). Specific results for the various sources can be obtained by substituting in Eq. (20) of Chapter 7 the appropriate value for δ_E : the latter can be found in Chapter 7, Eq. (14) for the microwaves, Eq. (31) for the phonons and Eq. (34) for tunneling. These results are due mainly to Eckern, Schmid, Schmutz and Schön (1979). They assumed that the phonons

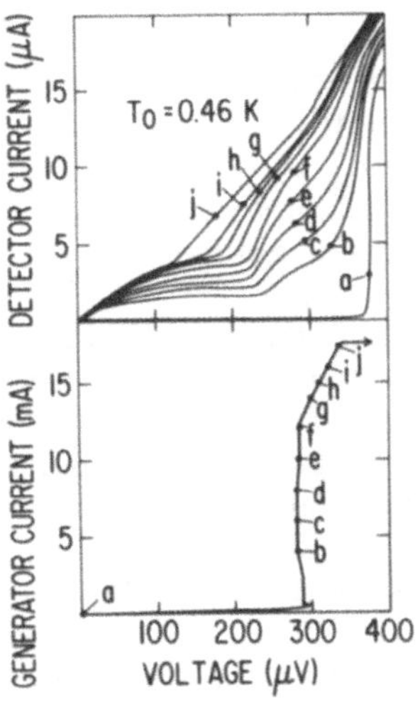

Fig. 16

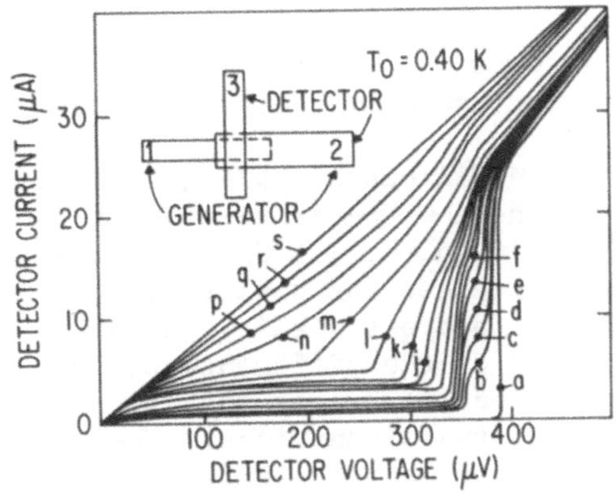

Fig. 17

Fig. 16. The top figure is a set of detector I-V curves. The labels identify for which value of the injector (also called generator) current and voltage each detector curve was measured. Film thickness 0.03 μm, specific resistance of the injector is about 10^{-5} Ωcm^2 . (From Gray and Willemsen 1978).

Fig. 17. Current-voltage characteristics of a particular detector junction for various injected currents. The inset shows the geometry. Curve a represents the I-V characteristic at $T = .4$ °K with no injection of quasiparticles. For the other curves, the injection rate I_o (in 10^{24} quasi-particles/sec/cm^3) is (b) 2.4; (c) 3.6; (d) 4.8; (e) 6.0; (f) 7.2; (g) 8.4; (h) 9.6; (i) 10.8; (j) 16.8; (k) 19.2; (ℓ) 24; (m) 30; (n) 33; (p) 36; (q) 42; (r) 48; (s) 60. (From Gray and Willemsen, 1978).

remain in equilibrium. This is never exactly the case in experiments but sometimes there are numerical calculations (Chang and Scalapino 1978) one can rely on to make sure no essential physics has been lost.

Theoretical and Experimental Results for Low Voltage Tunnel Injection

For voltages near the injection threshold ($|eV| \sim \Delta_1 + \Delta_2$) of a junction between two superconductors, Dynes et al. (1977) and Gray and Willemsen (1978) have observed a transition to a state where two superconducting phases can coexist in a single superconducting film. The experimental evidence is presented in Figure 16. Once again the labels on the various detector curves refer to various points on the injector I-V curve. The two "steps" on detector curves reflect the coexistence of two "phases" having different gap values. It can be seen that the growth of the lower gap phase is controled by the amount of current fed into the junction. This is clearly seen also from the detector curves of Figure 17 which were taken for a different junction. Curve a corresponds to the lowest injection current and s to the highest. Once the lowest gap state has filled the whole sample its gap decreases continuously as the injector current is increased.

Gray and Willemsen (1978), Hida (1978b) and Elesin (1979) working in the low temperature limit and Schön and Tremblay (1979) working mainly in the $T \sim T_c$ limit (See Chapter 7, Eqs. (20) and (34))have explained the qualitative features of the experiment including the proportionality between the junction conductance and the magnitude of the difference between the two observed gaps. This proportionality means that the effect is easiest to see in low specific resistance junctions. The theory for $T \sim T_c$ uses generalized free energy methods while the $T \sim 0$ theories rely on deterministic methods.

As usual a transition occurs because there exists two stable solutions to the gap equation. The physics of this phenomenon being here particularly simple, we explain it in more detail. At $T \sim 0$ (analogous arguments apply for $T \sim T_c$) this double-valuedness of the order parameter (or equivalently of the I-V characteristic) can roughly be understood from the following approximate form of the gap equation: $\Delta = \Delta_o(1 - 2N(\Delta))$ where Δ_o is the zero temperature gap and N is the excess quasiparticle concentration which self-consistently depends on the gap. When the voltage is very close to, but smaller than, the injection threshold ($|eV| < \Delta_{injector} + \Delta$), then, for $T = 0$, $\Delta = \Delta_o$. On the other hand, at that same voltage, the gap could be self-consistently smaller since then the junction could be above the injection threshold in which case we would have $\Delta = \Delta_o(1 - 2N)$ i.e. a smaller gap due to the injection produced excess quasiparticle number. There is a range of voltage close to the injection threshold for which the two stable solutions mentioned above (and a third unstable solution) are self-consistent.

Elesin (1979) has done the best solution of the transport and gap equations for T = 0. He predicts a similar instability for optical illumination near 2Δ[1]. He also shows how the phenomenon we are discussing now ("threshold instability") and the coherent instability we talked about earlier become intermixed for very low conductance junctions. The theories of Gray and Willemsen (1978) and Hida (1978b) are based on the double-valuedness of the I-V characteristic and on rate equations for the excess quasiparticle number. Hida used a disputable method however to determine at which of the possible values of voltage in the double-valued region the transition actually occurs. Gray and Willemsen have elucidated the role of inhomogeneities in nucleating the new phase. Note that close to T_c the two coexisting gaps (Schön et al. 1979) are an enhanced gap and an "unperturbed" gap (Chapter 7). More qualitative as well as quantitative predictions (including the effect of weak phonon trapping) have been made (Tremblay and Schön 1980) but no experiment has been done yet to check them.

This is an appropriate point to make a general remark concerning the analogy between equilibrium phase transitions and instabilities in superconductors. If we take as an analogy for first order phase transitions the liquid-gas transition, then all nonequilibrium phase transitions we have discussed up to now, excluding the case of tunnel junctions, are analogous to transitions where temperature and pressure are the controled variables: when the first order line in the phase diagram is crossed because one or the other of the variables is changed, then the system jumps "abruptly" from one phase to the other (given enough time for the system to reach equilibrium. In this context, see the laser case). In the tunnel injection case we are discussing now, close to T_c the generalized free energy reveals that voltage and current are like pressure and volume in the liquid-gas analogy[2]. Hence current controled experiments with the junctions are like volume controled experiments with the fluid: on the first order line, the relative amount of each phase is determined by the total current in the superconductor and by the total volume in the liquid-gas system. The importance of the external circuit (voltage vs current controled) in tunnel injection is thus easily understood by analogy with the equilibrium liquid-gas case. There as well, the transition occurs in a different way depending on the external world (pressure or volume controled). As we already said, in all

---

1 This instability should be in the microwave range. I do not think it has been observed yet.

2 Temperature in the liquid-gas case is analogous to conductance.

other cases of nonequilibrium superconductors (laser, microwaves, phonons) the external parameter we happen to be able to control is analogous to the pressure (or the temperature) of the liquid-gas system so first order transitions occur "abruptly" (Eckern et al. 1979). In tunnel junctions, current controled experiments are the easiest to achieve experimentally so up to now, at least for "low" voltage injection, only "smooth" transitions of the type described at the beginning of this section have been observed.

The liquid-gas analogy also illuminates the role of inhomogeneities. If there is a small amount of imperfections or impurities, the latter will act as nucleation centers but the properties of either phase will be very close to those of the homogeneous (pure) system. In particular, the possibility of coexistence of two phases will be determined by the free energy of the two phases of the pure system, not by the presence of inhomogeneities. Similarly, in the tunnel injected superconductor, Gray and Willemsen (1978) have discussed how the circuit may allow the low gap phase to nucleate at film inhomogeneities. Whether the small gap region will grow or not after it has nucleated and covered the inhomogeneous region then depends on the equations for the superconductor in the homogeneous region.

Large inhomogeneities however are a different question. In the nonequilibrium case this is further complicated by the fact that even if the perturbed film is homogeneous, the external source may be inhomogeneous. A laser with very non-uniform illumination is an example of an undesirable source. Finally, note that there is a difference between an inhomogeneous injector which has a large current at weak spots and an homogeneous injector which self-adjusts to the perturbed film, giving more current to the lower gap region than to the higher gap when there is coexistence. Experimentally however, it may be hard to distinguish between effects which are due to instabilities of the uniform system and those which are due to inhomogeneous manufacture of the film and/or non-uniform external perturbation.

Theoretical and Experimental Results for Monochromatic Microwaves and Phonons

For these two cases, there is very little to add to what has already been said in Chapter 7. We will basically extract a few essential points from that discussion and indicate where the results can be found.

For microwaves, the theoretical results have been discussed with Figure 5 of Chapter 7. Basically, as a function of temperature a first order transition is predicted. The relevant experimental results are in Figure 12 of Chapter 7. A first order transition has been observed essentially where it should be but a detailed comparison with the theory is rendered difficult not only because of the usual heating effects but also because there are probably hysteretic effects which are not included in a calculation which assumes equilibrium phonons but instead are related to the difference in dissipation between the normal and superconducting state (see Chapter 7). They could be accounted for theoretically by taking into account the different heating effects in the various states. In general, surface effects which influence the coupling of the external world to the bulk superconductor should also be included. A similar remark applies to all cases where a first order transition has been predicted on the basis of calculations which take into account only the nonequilibrium of the bulk electronic system.

The case of phonons is discussed in the context of Figure 13, Chapter 7. Again these results are due to Eckern et al. (1979). Some of the calculations were performed with a model which assumes that the superconductor is irradiated by phonons with a spectral distribution linear up to frequencies of order T. Nevertheless Eckern et al. (1979) have shown that most of the results will apply to monochromatic phonons as well. Other limitations of the theory are discussed in Chapter 7. The results are that, when $\xi(T) >> (D\tau_E)^{\frac{1}{2}}$, as the temperature is changed there is a first order transition to a normal or another superconducting state, depending on the strength of the external perturbation. When $(D\tau_E)^{\frac{1}{2}} >> \xi(T)$ and $2\Delta/\hbar$ is smaller than the phonon frequency, Eiler (1980) has found that the diffusive instability found by Eckern et al. (1979) develops as a second order transition when the temperature is reduced or the irradiation increased. Unfortunately the amplitude of the modulation is expected to be at most a few percent of the average gap value. This may make experimental confirmation difficult. At present, there are very few experiments. Those discussed in the context of Figure 14 Chapter 7 did not concentrate on the verification of the above predictions. Dharmadurai and Ratnam (1979) have performed experiments which exhibited a first order transition (as measured by the behavior of the critical current) but the spectral characteristics of their phonon source are not well defined.

CONCLUSION

Models for Wide Sources

The most "rigorous" theory to date for optical illumination and tunnel injection is that of Elesin et al. (1979). (For the laser case see also Elesin, 1976, 1977a, 1977b and Eckern et al., 1979). That theory leads to the so called "coherent instability" which is a "first order transition" to the normal state at a critical value of the incident power. The main limitation of the numerical part of the calculation (Elesin et al. 1979) is the assumption that both optical illumination and high voltage tunnel injection can be modeled by a term in the collision operator which creates quasiparticles between energies of 9Δ and 10Δ . Since the similarity between the laser and injection cases does not seem entirely verified by experiments it should not be overemphasized. It would be preferable to use more realistic drive terms in the collision operators.

As far as the μ^* model is concerned, its application to optical illumination is not supported by the calculations of Elesin et al. (1979). Furthermore it predicts (when $\mu^* > 0$) that the distribution function can be larger than 1/2, in contradiction with both analytical (Elesin, 1974) and numerical calculations (Elesin et al., 1979). For high voltage injection as well, the calculations of Elesin et al. (1979) do not support the μ^* model. Even if we suppose that there are certain intermediate regimes (neither "high" nor "low" voltage) of tunnel injection where the μ^* model is a good approximation, one should still beware of some of the conclusions which have been drawn from that model. For example, I do not think that the instability predicted by Scalapino and Huberman (1977) is on firm grounds. It is a consequence of the phenomenological assumption that the quasiparticle current is proportional to $-\nabla\mu^*$ (compare with Smith (1977)). A somewhat more "rigorous" calculation was performed by Baru and Sukhanov (1975). They did not rely on the phenomenological relation for the quasiparticle current but used a μ^* distribution in the complete set of transport equations for the superconductor, assuming equilibrium phonons. They did find a diffusive instability but unfortunately they did not include the Landau-Ginzburg deformation energy for the gap so they do not have a precise condition for the instability which can be compared directly with the appropriate limit of the Scalapino-Huberman result. It would be instructive to include explicitely the deformation energy in the Baru and Sukhanov (1975) calculation.

Experiments on Wide Sources

The comparison between theory and experiment for laser illumination is not completely satisfactory. The various experiments performed in the last six years all agree that with pulsed lasers one observes a smooth rise of the resistance as a function of temperature for fixed incident power and as a function of power at fixed temperature.

The only plausible theory to explain these experiments is Elesin's. It agrees well with the temperature and thickness dependence of the threshold power but the explanation for the gradual rise in resistance has not been explicitely checked by experiment yet. Note in passing that Elesin's explanation for the gradual rise in resistance applies to any theory which predicts a first-order transition. However, the threshold obtained from the μ^* model is not in very good agreement with experimental results (Golovashkin et al., 1978) in the narrow temperature range where the comparison can be legitimately made. What is clearly needed to check Elesin's nonstationarity hypothesis are experiments with variable pulse width or C.W. ("continuous wave" laser) experiments. In the latter case, Elesin's theory predicts that if there are no experimental complications due to substrate heating and the like, a first order transition to the normal state should be observed. If Elesin's theory turns out to be correct, then the detailed comparison between theory and pulsed experiments may turn out to be a quite formidable task in the time dependent regime. The reader should perhaps be reminded of the difficulties encountered in calculating time dependent phenomena when one equilibrium phase is rapidly quenched to another equilibrium phase. Here we would be interested in the transition from one nonequilibrium phase to another. If Elesin's theory turns out to be wrong and the nonequilibrium superconducting state is really stationary in the resistive state, then one should think about other theories... vortices or whatever.

On the experimental side, one should also be careful about laser speckles, inhomogeneities etc... which could all introduce artefacts which have not been accounted for in the theories. Smith (1980) for example gives a detailed account of the care necessary to make sure the laser illumination is uniform. Concerning his experiments on "one-dimensional" strips, if it was not for the considerable difficulty of the problem without a laser (cf. phase slip centers) it would be tempting to suggest that a theoretical treatment of that case should be done since it might lead to some insight.

With tunnel junctions the difference between constant current and constant voltage experiments should be kept in mind. Also, one should remember that critical current effects can be a problem since none of the theories have accounted for them yet. However, it is usually possible with a little experimental care to have currents much smaller than critical currents and still have appreciable nonequilibrium effects. Even when one is aware of that difficulty, the experimental results are rather confusing. Iguchi (1978) has observed instabilities at high injection voltage. These results and those of Iguchi and Langenberg (1980) are somewhat in conflict with those of Rajeevakumar et al. (1979) and Gray (Chapter 5) who did not observe any instabilities at high injection voltages. It should however be established whether phonon trapping factors, edge masking, etc. were comparable before concluding that there is a contradiction. More experiments are needed. Experiments which claim to verify Elsin's coherent instability theory (Mitsen 1979) are not really convincing and in general a more systematic comparison between experiments and

Elesin's theory should be made. The problem with tunnel injection could be that many of the observed instabilities are neither in the "high" nor "low" (threshold) voltage limit but rather in some intermediate regime where theoretical work is still lacking. The problem could also be with experiments.

In both the laser and tunnel injection cases, detailed measurements of the distribution function as we go towards the instability regime would be extremely valuable. This would require an extension of, for example, the work of Smith, Skocpol and Tinkham (1980) in the laser case and of Willemsen and Gray (1978) in the tunnel injection case. Also, for both laser and junction experiments, spatially resolved probing of the nonequilibrium state would be extremely valuable, if it can be done without introducing unwanted inhomogeneities.

Low Voltage Tunnel Injection

There exists viable theories for low voltage tunnel injection (injection at the gap edge). The comparison between theory and experiment is however only qualitative. Experimental verification of the many qualitative and quantitative predictions of the theory would be extremely valuable. This case and the case of phonons (following Section) represent in my (biased) opinion the best opportunities we have to really confront theory and experiment. In the junction case the already available results are encouraging. This is generally the case for "narrow sources". One can say with confidence that the case of "narrow sources" will soon be settled while that of "wide sources" is on more uncertain grounds.

We should also mention other theoretical predictions for instabilities in tunnel junctions which have not been experimentally observed yet. Ivlev (1978) has predicted an instability towards an inhomogeneous state for junctions at bias voltages around the injector and sample gap difference. Also, Shenoy and Agarwal (1980) have predicted an instability for Josephson junctions irradiated by coherent radiation.

Microwave and Phonon Irradiation

The effect of microwaves has been extensively studied, both experimentally and theoretically. The predicted first order transition to the normal state has been observed but the idealizations of the theory and the unknowns of the experiment forbid a more quantitative comparison. For phonon irradiation on the other hand, theory (Eckern et al., 1979; also Eiler, 1980) predicts a wide range of qualitatively different behaviors. Hence, experiments which would confirm even only the existence of the various regimes would be extremely convincing evidence for the theory. Unfortunately, no such experiment has been done yet.

In all cases discussed in this Chapter, phonon trapping is a main obstacle to a quantitative comparison between theory and experiment. The problem of instabilities in nonequilibrium superconductors is far from being completely solved but results are encouraging.

ACKNOWLEDGEMENTS

I should like to thank K. Gray whose advice on the interpretation of experiments and continuous support have been essential for the writing of this chapter. The responsibility for any mistakes or injustices is however entirely mine. I should like to thank W. Eiler for giving permission to quote his results before publication and H. Mooij for sending a preprint of Chapter 7. Innumerable illuminating discussions with G. Schön are also gratefully acknowledged.

This work was supported by NSERC (Canada) through a grant and through the program "Attaché de recherche CRSNG".

REFERENCES

Akoh, H., Kosaka, S., and Kajimura, K., 1980, Pulsed quasiparticle injection into superconducting lead-indium alloys via tunneling, Phys. Lett., 77 A:269.

Aronov, A.G., and Gurevich, V.L., 1972, Quasifield in superconductors and increase of T_c under nonequilibrium conditions. ZhETF Pis. Red., 15:564; Engl. transl. JETP Lett., 15:400.

Aronov, A.G., Spivak, B.Z., 1979, On the role of electron-electron collisions in the formation of nonequilibrium distributions in superconductors, Zh. Eksp. Teor. Fiz., 77:652; Engl. transl. Sov. Phys. JETP 50:328.

Baru, V.G., and Sukhanov, A.A., 1975, New types of instabilities in nonequilibrium excitation of a superconductor, ZhETF Pis. Red., 21:209; Engl. transl. JETP Lett., 21:93.

Chang, J.J., and Scalapino, D.J., 1974, New instability in superconductors under external dynamic pair breaking, Phys. Rev. B, 10:4047.

Chang, J.J., and Scalapino, D.J., 1977, Kinetic equation approach to nonequilibrium superconductivity, Phys. Rev. B, 15:2651.

Chang, J.J., and Scalapino, D.J., 1978, Nonequilibrium superconductivity, J. Low Temp. Phys., 31:1.

Chang, J.J., Lai, W.Y., and Scalapino, D.J., 1979, T* model of the energy distribution of excitations in nonequilibrium superconductors, Phys. Rev. B, 20:2739.

Dharmadurai, G., and Ratnam, B.A., 1979, Observation of a phonon-injection-induced first-order transition in Ag-Sn proximity-effect bridges, Phys. Rev. B, 20:4633.

Dynes, R.C., Narayanamurti, V., and Garno, J.P., 1977, Observation of a new superconducting state at high quasiparticle injection, Phys. Rev. Lett. , 39:229.

Eckern, U., Schmid, A., Schmutz, M., and Schön, G., 1979, Stability of superconducting states out of thermal equilibrium, J. Low Temp. Phys., 36:643.

Eiler, W., 1980, preprint.

Elesin, V.F., 1974, Nonequilibrium state of superconductors with optical excitation of quasiparticles, Zh. Eksp. Teor. Fiz., 66:1755; Engl. transl. Sov. Phys. JETP, 39:862.

Elesin, V.F., 1976, Features of the phase transition in nonequilibrium superconductors with optical pumping, Zh. Eksp. Teor. Fiz., 71:1490; Engl. transl. Sov. Phys. JETP, 44:780.

Elesin, V.F., 1977a, Nonstationary "intermediate" state of nonequilibrium superconductors, Zh. Eksp. Teor. Fiz., 73:355; Engl. transl. Sov. Phys. JETP, 46:185.

Elesin, V.F., 1977b, Theory of phase transitions in nonequilibrium superconductors with allowance for phonon reabsorption, Fiz. Tverd. Tela (Leningrad), 19:2977; Engl. transl. Sov. Phys. Solid State, 19:1744.

Elesin, V.F., 1979, Threshold instability and inhomogeneous states in nonequilibrium superconductors with optical and tunnel quasiparticle pumping, Zh. Eksp. Teor. Fiz., 76:2218; Engl. transl. Sov. Phys. JETP, 49:1121.

Elesin, V.F., Kondrashov, V.E., and Sukhikh, A.S., 1979, Kinetic theory of nonequilibrium superconductors with a wide-band source of quasiparticles, Fiz. Tverd Tela (Leningrad), 21:3225; Engl. transl. Sov. Phys. Solid State, 21:1861.

Fuchs, J., Epperlein, P.W., Welte, M., and Eisenmenger, W., 1977, Energy gap reduction in superconducting tin films by quasiparticle injection, Phys. Rev. Lett., 38:919.

Golovashkin, A.I., Mitsen, K.V., and Motulevich, G.P., 1975, Experimental investigation of the nonequilibrium state of superconductors excited by lasers, Zh. Eksp. Teor. Fiz., 68:1408; Engl. transl. 1976, Sov. Phys. JETP, 41:701.

Golovashkin, A.I., Ivanenko, O.M.,Mitsen, K.V., Motulevich, G.P., and Shubin, A.A., 1978, Temperature dependence of the critical power of optical pumping of nonequilibrium superconductors, Zh. Eksp. Teor. Fiz., 75:1517; Engl. transl. Sov. Phys. JETP, 48:766.

Golovashkin, A.I., Elesin, V.F., Ivanenko, O.M., Mitsen, K.V., and Motulevich, G.P., 1980, Conditions for the appearance of spatially inhomogeneous states in superconductors under laser pumping conditions, Fiz. Tverd. Tela (Leningrad), 22:105; Engl. transl. Sov. Phys. Solid State, 22:60.

Gray, K.E., and Willemsen, H.W., 1978, Inhomogeneous state of superconductors by intense tunnel injection of quasiparticles, J. Low Temp. Phys., 31:911.

Hida, K., 1978a, Phase transitions in nonequilibrium superconductors induced by quasiparticle injection, J. Low Temp. Phys., 32:881.

Hida, K., 1978b, The two-gap state induced by the singular I-V characteristics of Josephson tunnel junctions, Phys. Lett., 68A:71.

Hu, P., Dynes, R.C., and Narayanamurti, V., 1974, Dynamics of quasiparticles in superconductors, Phys. Rev. B, 10:2786.

Iguchi, I., 1978, Nonequilibrium gap instability and phase transition under intense tunnel injection in superconducting lead films, J. Low Temp. Phys., 33:439.

Iguchi, I., and Langenberg, D.N., 1980, Diffusive quasiparticle instability toward multiple-gap states in a tunnel-injected nonequilibrium superconductor, Phys. Rev. Lett., 44:486.

Ivlev, V.I., 1978, Resistive state in superconductor junctions, Zh. Eksp. Teor. Fiz., 75:1771; Engl. transl. Sov. Phys. JETP, 48:893.

Mitsen, K.V., 1979, Spatially inhomogeneous state of superconducting tin films under tunnel injection conditions, Fiz. Tverd. Tela (Leningrad), 21:3475; Engl. transl. Sov. Phys. Solid State, 21:2007.

Owen, C.S., and Scalapino, D.J., 1972, Superconducting state under the influence of external dynamic pair breaking, Phys. Rev. Lett., 28:1559.

Parker, W.H., 1975, Modified heating theory of nonequilibrium superconductors, Phys. Rev. B, 12:3667.

Rajeevakumar, T.V., Chang, J.J., and Chen, J.T., 1979, Energy gap suppression and instability in superconducting tin films under strong quasiparticle injection, J. Low Temp. Phys., 37:77.

Rothwarf, A., Sai-Halasz, G.A., and Langenberg, D.N., 1974, Quasiparticle lifetimes and microwave power response in nonequilibrium superconductors, Phys. Rev. Lett., 33:212.

Sai-Halasz, G.A., Chi, C.C., Denenstein, A., and Langenberg, D.N., 1974, Effect of dynamic external pair breaking in superconducting films, Phys. Rev. Lett., 33:215.

Scalapino, D.J., and Huberman, B.A., 1977, Onset of an inhomogeneous state in a nonequilibrium superconducting film, Phys. Rev. Lett., 39:1365.

Schmid, A., 1977, Stability of radiation-stimulated superconductivity, Phys. Rev. Lett., 38:922.

Schön, G., and Tremblay, A.-M., 1979, Nonequilibrium superconducting states with two coexisting energy gaps, Phys. Rev. Lett., 42:1086.

Shenoy, S.R., and Agarwal, G.S., 1980, Radiation induced bistability in Josephson junctions, Phys. Rev. Lett., 44:1524; erratum, 45:401.

Smith, A.D., Skocpol, W.J., and Tinkham, M., 1980, Quasiparticle-energy distribution in optically illuminated Al-PbBi superconducting tunnel junctions, Phys. Rev. B, 21:3879.

Smith, L.N., 1977, Diffusive instability in optically excited superconducting films, J. Low Temp. Phys., 28:519.

Smith, L.N., 1980, The resistive transition and superconducting properties of optically illuminated tin microstrips, J. Low Temp. Phys., 38:553.

Testardi, L.R., 1971, Destruction of superconductivity by laser light, Phys. Rev. B, 4:2189.

Tremblay, A.-M., and Schön, G., 1980, Effect of nonequilibrium phonons on superconducting states with two coexisting energy gaps, in: "Inhomogeneous superconductors-1979," D.U. Gubser, T.L. Francavilla, S.A. Wolf and J.R. Leibowitz, ed., American Institute of Physics, New York.

Welte, M., 1978, Instability of the energy gap in superconductors with nonthermal quasiparticle distributions, Z. Physik B, 29:107.

Willemsen, H.W., and Gray, K.E., 1978, Distribution function in nonequilibrium states of a superconductor, Phys. Rev. Lett., 41:812.

NORMAL-SUPERCONDUCTING BOUNDARIES

A.B. Pippard

Cavendish Laboratory
Cambridge
England

Normal-superconducting (NS) boundaries may be fabricated directly, e.g. by evaporation of lead onto a copper foil, or as a stack of alternating N and S laminae in the intermediate state of a type I superconductor. The technique for achieving regularly stacked laminae, transverse to a long rod, is described by Walton (1965) and a typical curve for the thermal resistivity of tin obtained in this work is shown in Fig. 1. The fraction of normal material rises linearly with magnetic field, and so does the electrical resistance, indicating that N and S laminae behave as a chain of resistors in series, with no extra contribution from the boundaries. By contrast the thermal resistivity is dominated by the boundaries. As the temperature approaches T_c the thermal boundary resistance becomes less marked and at the same time a boundary contribution to the electrical resistance becomes apparent, as illustrated in Fig. 2 (Landau, 1970; Pippard et al., 1971).

The electrical boundary resistance shows up very clearly in fabricated SNS samples, the current passing across a thin foil of pure copper between evaporated layers of lead (which wets, but does not alloy with, or appreciably penetrate, the copper). As Fig. 3 shows, the effect is prominent only in a narrow temperature range just below T_c for lead. It is considerably stronger when a lead-bismuth alloy is used instead of pure lead, but although the magnitude, normalized to the resistance at 0 K, initially rises sharply with Bi concentration, it hardly changes between 3% and 20%, the limit of substitutional solution (Harding et al., 1974).

Andreev's (1964) explanation of the difference between thermal and electrical boundary effects can be visualized in terms of the behaviour of an excitation arriving at the interface from the

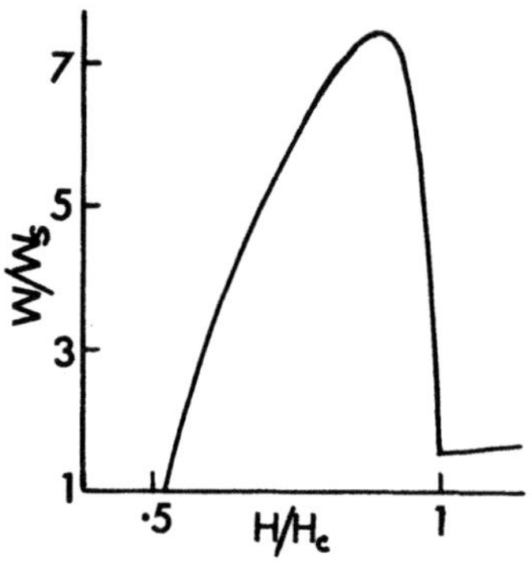

Fig. 1 Typical thermal resistance peak in the intermediate state (Walton, 1965); tin at 0.49 T_c. Magnetic field normalized to critical field, thermal resistance to resistance in superconducting state.

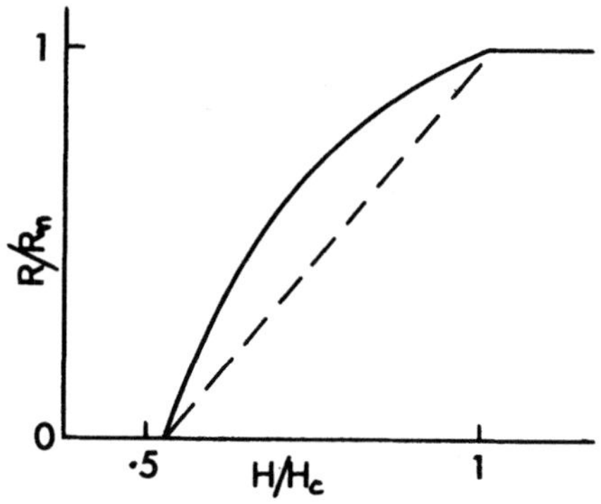

Fig. 2 Electrical resistance in the intermediate state (Pippard et al., 1971); tin at 0.994 T_c. Magnetic field normalized to critical field, resistance to resistance in normal state. The broken line shows the fraction of normal material.

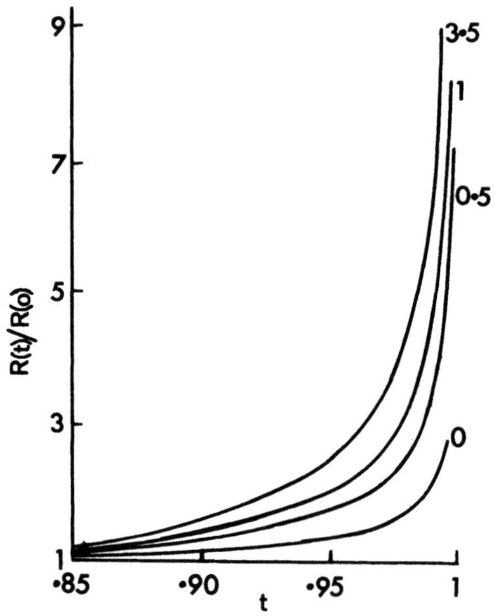

Fig. 3 Resistance of Pb-Cu-Pb sandwiches against temperature (Harding et al., 1974), normalized to resistance at 0 K and T_c for lead. The figures beside the curves show the percentage of bismuth in the lead.

normal metal. For convenience we start with a one-dimensional model of the metal, whose ground state is electrically neutral and has all electron states up to the Fermi energy, ε_F, occupied. Excited states of the metal are produced by injecting extra electrons from a reservoir whose chemical potential, μ, is the same as ε_F, or transferring electrons from the metal to the reservoir. If the metal is normal, an extra electron taken from the level μ in the reservoir and placed in a previously empty state of wavenumber k increases the total wavenumber of the metal by k, and energy $\xi = \varepsilon_k - \varepsilon_F$ must be provided; $\xi(k)$ defines the relation between the energy needed to create the electron-like excitation and the change in the total wavenumber. To create a hole-like excitation, an electron must be taken from a filled state k, below the Fermi level, given energy $\xi = \varepsilon_F - \varepsilon_k$, and transferred to the reservoir; the total wavenumber of the metal is reduced by k, and the wavenumber of the excitation is therefore -k, though the energy is still positive. The $\xi(k)$ relation for this one-dimensional normal metal thus takes the form of Fig. 4(a). There is no continuity across the cusp - an excitation on branch A can in principle be scattered to any point on A or D but not on to B or C. In the superconducting state, however, where the excitation energy E_k is $(\xi_k^2 + \Delta^2)^{\frac{1}{2}}$, it is possible for inelastic processes to transform an electron-like excitation on A into a hole-like excitation on B.

Consider now a graded junction between a normal and a superconducting metal, that is to say, one where Δ varies slowly in relation to the coherence length. An excitation entering from N suffers an adiabatic change in which its momentum k may be altered

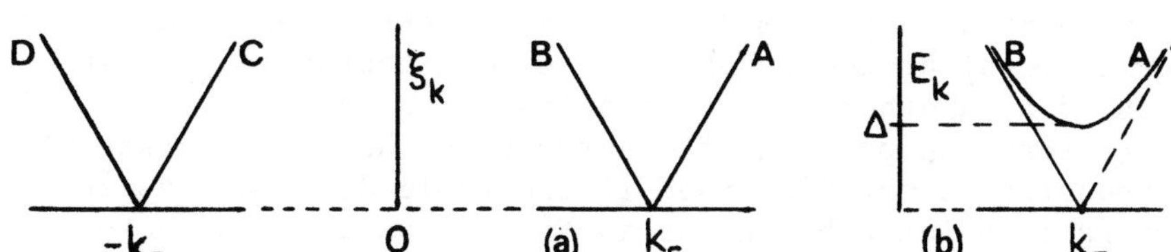

Fig. 4 Excitation representation of (a) normal metal, (b) superconductor with energy gap Δ; only states close to $\pm k_F$ are shown in (a), and close to $+k_F$ in (b). A and D are electron-like branches, B and C hole-like.

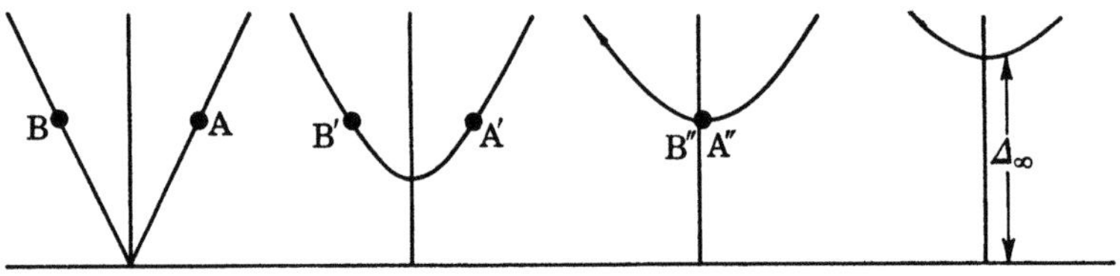

Fig. 5 Variation of local energy-level diagram at a graded interface between N at the left and S at the right; Δ_∞ is the energy gap deep in S. An excitation A entering from N penetrates as far as A", where it becomes identical with B" and emerges into N as B.

by interaction with the ground state, but the total energy E_k is conserved. As Fig. 5 makes clear, an electron entering is reflected as a hole; similarly, a hole is reflected as an electron. In the course of this change N has lost or gained two electrons, which means that a current is flowing across the interface, and the ground state of S has either gained or lost a pair and has taken over the task of carrying the current. It is important for what follows to remember that this process of Andreev reflection occurs in the same way at every point on the Fermi surface of a three-dimensional metal; an electron just outside $\underline{k}_F$ is reflected as a hole just inside $\underline{k}_F$. The wavenumber is only slightly affected, but the velocity of the excitation is reversed. This is quite unlike the usual process of wave reflection, where the normal component of $\underline{k}$ is reversed but the parallel component is unchanged. It has the consequence that the configuration of the interface is of minor importance - there is no randomness about Andreev reflection, even from a rough interface.

The classical visualization of the process which a graded junction allows is hardly permissible when the junction is narrower than the coherence length. It is not, however, necessary since one may rephrase the description in terms of the boundary condition that the superconductor imposes on the distribution function in the normal metal at the interface; any excitation at $\underline{k}_F$ must be accompanied by another of the opposite character. As a special case note that the pairing of electrons in the ground state of a superconductor is not disturbed if the whole distribution in k-space is translated by $\delta\underline{k}$, as in Fig. 6, which reverts to the one-dimensional model. In the normal state this displaced distribution can be described in terms of excitations as follows: the extra electrons on the right, at $k > k_F$, are electron-like excitations on branch A of Fig. 4, but those at $k < k_F$ are represented as a deficit of hole-like excitations on branch C;

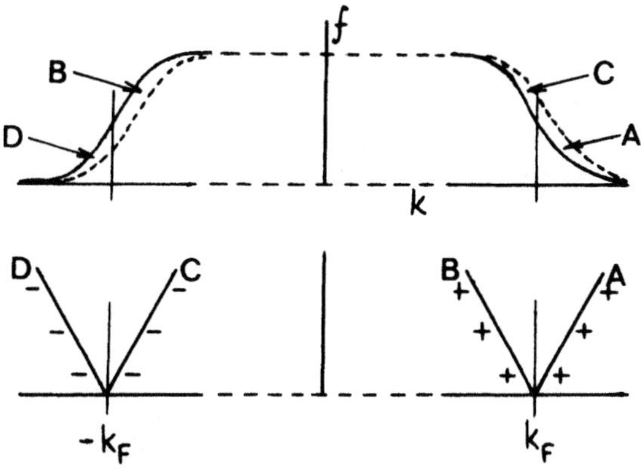

Fig. 6 The upper diagram shows the equilibrium (full line) and current-carrying (broken line) distribution functions; the lower diagram shows the displaced distribution in terms of excess and deficit of excitations.

similarly the left-hand side is represented as a deficit of electrons on D and an excess of holes on B. It is clear that the state of branch-balance, having equal populations of excitations on A and B, and equal deficits on C and D, satisfies Andreev's boundary condition and describes a current-carrying normal state which can condense into the superconducting state without rearrangement. Hence current-carrying N and S can be juxtaposed without modification and no extra resistance need be expected.

It is far otherwise with a thermal current. If the normal metal is hotter on the left than on the right, the Fermi tail is broader on the right than on the left and the excitation representation, as shown in Fig. 7, has precisely the opposite symmetry to what the superconductor demands. There is no temperature distribution in the vicinity of the interface that can reverse the symmetry, and consequently the Andreev boundary condition precludes heat flow. In practice, as Fig. 1 shows, it is difficult but not impossible to pass a heat current across the interface. This is because at non-zero T some of the excitations have energy greater than Δ and are therefore totally reflected with reversal of character*. The consequent loosening of the strict Andreev

*If the junction is smoothly graded, all excitations with $E<\Delta$ are reflected and all with $E>\Delta$ transmitted. With a more abrupt junction total Andreev reflection must still occur for $E<\Delta$, but excitations with $E>\Delta$ may be partially reflected. Similarly an excitation already in S may be partially reflected at the interface and return into S with its character reversed. This process is responsible for the Tomasch (1965) effect, according to McMillan and Anderson (1966).

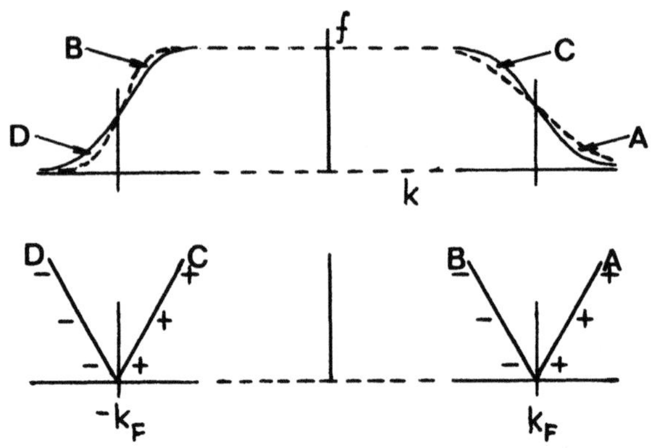

Fig. 7 The same as Fig. 6, except that the distribution function is disturbed by a temperature gradient; electrons moving to the right are hotter than those moving to the left.

condition is at the same time responsible for allowing heat flow and creating a boundary resistance for electric currents. We shall direct our attention to the latter problem only.

Let us ignore for the moment the confinement of supercurrent to the penetration layer, supposing the penetration depth to be very large. When a current flows across the interface an excess of electrons or holes, having energy greater than Δ, is injected from N into S, according to the sign of the current, let the injected excitations have an excess of electrons (conventional current S → N). Far away from the interface the current carried by these excitations has been taken over by the condensed state, a number of different processes being involved:

1) Elastic scattering by impurities etc. Normally such scattering does not change the character of the excitation, though in an anisotropic superconductor oblique branch-crossing elastic scattering is possible (Phillips, 1969); electron ⇌ hole transitions at constant E are not absolutely forbidden if the excitation moves to a new point on the Fermi surface with a different Δ. But if this possibility is ignored we may picture the injected electrons diffusing into S from the interface, with diffusion constant D. The distance travelled increases as $(Dt)^{\frac{1}{2}}$, so that as time goes on the excess concentration of electron-like excitations rises steadily as $(t/D)^{\frac{1}{2}}$. However, there can be no accumulation of space charge, so that as the excitations increase in density the electrons in the condensed state move out, and push a superflow ahead of them. If this were the only process, no steady state would be attained. However,

2) Oblique or inelastic phonon scattering with branch crossing converts the excess of electrons into holes. There is a characteristic temperature-dependent time constant, τ_{Q^*}, for the ironing-out of branch imbalance, and after this time has elapsed a steady-state distribution is well on the way to being established, with the difference between the rates of injection into the electron and hole branches matched by the rate of branch-crossing by inelastic processes. For a given injection rate the excess concentration of excitations at the interface $\propto (\tau_{Q^*}/\Delta)^{\frac{1}{2}}$.

3) While all this is going on, electrons and holes may recombine and new electron-hole pairs may be thermally excited, but this does not affect the argument.

To a good approximation* processes (1) and (2) may be summed up in a diffusion equation for the excess charge density, q, carried by the excitations:

$$D\partial^2 q/\partial x^2 - q/\tau_{Q^*} = \partial q/\partial t, \qquad (1)$$

in which D is the diffusion constant, $v\ell$, v being some average excitation velocity and ℓ the mean free path for randomization of direction (usually by impurity scattering). In the steady state q decays exponentially away from the interface, $q = q_o e^{-x/\lambda^S}$, where $\lambda^S = (D\tau_{Q^*})^{\frac{1}{2}} = (\ell\ell_{Q^*})^{\frac{1}{2}}$ and $\ell_{Q^*} = v\tau_{Q^*}$, the typical mean free path for interbranch scattering. At the interface the charge density is q_o and the diffusion current is Dq_o/λ^S, i.e. $v\ell q_o/\lambda^S$.

We must now relate q_o and the diffusion current to the total current density and the extra potential drop attributable to the interface. When the current is I, each branch in the normal metal carries $\frac{1}{4}I$, which is thus the flux of electrons on branch A entering the superconductor. Only the supergap fraction f actually enter as excitations, while each of the others contributes two electrons to give a supercurrent $\frac{1}{2}(1-f)I$. So too do the holes on branch C, and consequently the total supercurrent is $(1-f)I$ and the excitation current, i.e. the diffusion current at the interface, is fI. Hence

$$fI = Dq_o/\ell^S = v\ell q_o/\lambda^S. \qquad (2)$$

Now the excitations at the interface contribute a charge density q_o, and for neutrality the condensed state must be reduced in charge density by the same amount, so that the number density must be

*The following argument ignores many points of detail, but the physical principles are, I hope, sound and the result expressed by (4) is hardly improved, so far as the simplified one-dimensional model is concerned, by more scrupulous attention.

reduced by q_o/e. If the density of states in the normal state of the superconductor is $\mathcal{N}$, the Fermi level must be lowered by $q_o/\mathcal{N}e$. It was pointed out by Waldram (1975), however, that the chemical potential must be constant throughout the superconductor, and that the electrical potential V at the interface must be raised relative to the distant superconducting regions so that $q_o/\mathcal{N}e + eV = 0$. It is this potential difference that shows as the interface resistance:

$$R_i = |V/J| = f\lambda^s/\mathcal{N}e^2 v\ell = f\rho^s\lambda^s, \tag{3}$$

where ρ^s is the resistivity of S in its normal state: $\rho^s = 1/\mathcal{N}e^2 v\ell$ for a one-dimensional conductor. The same resistance would be presented by a length R_i/ρ^n of N, or r_i free paths in N, where $r_i = R_i/\rho^n\ell$:

$$r_i = f\lambda^s/\ell = f(\ell_{Q*}/\ell)^{\frac{1}{2}}. \tag{4}$$

As T_c is approached f rises from 0 to 1, but the major contributor to the rise of r_i is the temperature variation of ℓ_{Q*}, the branch-crossing mean free path, which must increase without limit since branch-crossing is absolutely forbidden in the normal state.

One may visualize the extra resistance as arising from the scattering of excitations in the diffusion layer of thickness λ^s. It does not matter that the scattering may be taking place deep in S; so long as there is branch imbalance there is a diffusion current of excitations, and so long as there is a diffusion current the distribution is asymmetrical, so that scattering can cause randomization and generation of entropy. The resistance may be interpreted as the concomitant of entropy production. It is not necessary for the appearance of resistance that there should be an electric current deep in S to interact with the electric field accompanying the charge imbalance, and in fact there will normally be no current. The ground state will react to the excitation current by developing a counterflow, and at the same time a surface current will be set up in the penetration layer to carry off the injected charge. There is no dissipation when a supercurrent flows in a region of constant chemical potential, even though an electric field is present to keep μ constant; but energy transfer is involved when the excitation current interacts with the electric field.

According to (4) the interface resistance increases with impurity concentration, c, in the superconductor as $c^{\frac{1}{2}}$, since $\ell \propto 1/c$. The saturation noted in Fig. 3 comes about because there is an interface resistance even at 0 K, not predicted by the above treatment, but itself rising as something like $c^{\frac{1}{2}}$, so that the ratio $R_i(T)/R_i(o)$ tends to a constant. In fact $R_i(o)$ probably rises a little faster than $c^{\frac{1}{2}}$, though the suggestion of $c^{2/3}$ by Harding et al. (1974) looks like an error. Fig. 8 compares the measured

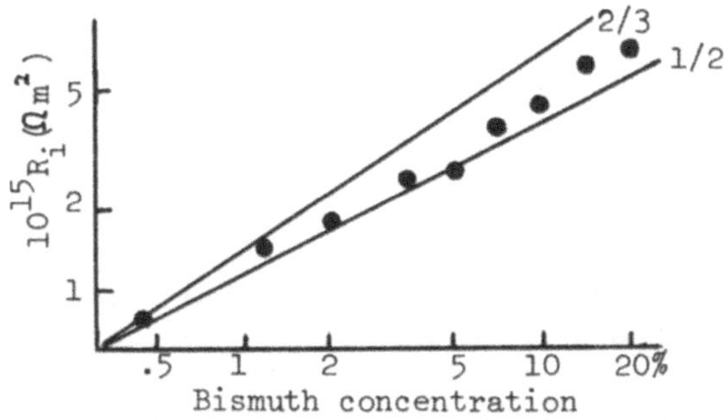

Fig. 8 Interface resistance in a Pb-Cu-Pb sandwich at 0 K, as a function of bismuth concentration in the lead (Harding et al., 1974). The straight lines are drawn for $R_i \propto c^{1/2}$ and $c^{2/3}$.

points with lines for $c^{\frac{1}{2}}$ and $c^{2/3}$, both of which lie rather outside the likely range of uncertainty. Whatever the precise form, however, the considerable rise in $R_i(o)$ makes clear that in dirty samples perfect Andreev reflection breaks down even though no excitation can surmount the energy gap.

To appreciate the problem which this raises consider the arrival from N of an electron-like excitation at a sharp (not graded) interface. It is now important to bear in mind the wave character of the excitations and, as in all cases of total reflection, the incident electron and reflected hole waves are accompanied by an evanescent wave which, if S were pure, would penetrate a distance of the order of the coherence length, ξ.

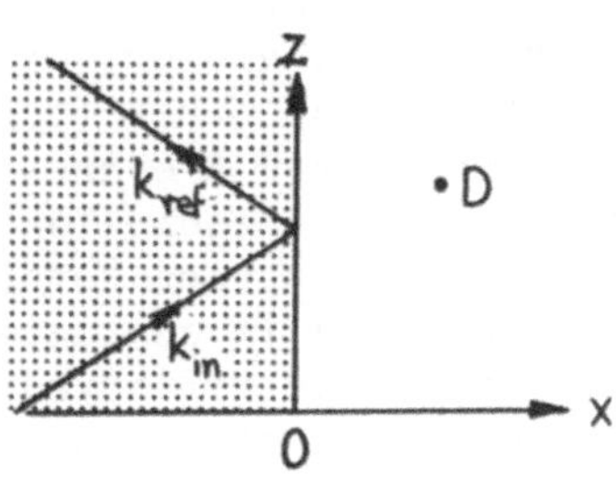

Fig. 9 Total reflection of a light wave at a dielectric (shaded) - free space boundary.

Impurities can scatter the evanescent wave so that wavelets return to the surface and generate disturbances in N which combine with and modify the reflected excitations. Let us now note the contrast between total reflection of a light wave and of an excitation. Fig. 9 shows the propagation vectors of a plane light wave in a dielectric, totally reflected at the interface. The evanescent wave in the region $x > 0$ has real k_z but imaginary k_x. If there is a scattering centre D the disturbance at D has its phase determined by z but independent of x, and the scattered evanescent wave has this same phase everywhere in the region $x > 0$, and an amplitude that decays exponentially with distance from D. What returns to the surface, then, is phase-coherent with the incident and reflected waves, and it can (indeed, must) generate a scattered component together with a coherent plane wave component that combines with the original reflected wave so as to reduce its amplitude. This process is exactly analogous to the usual scattering of a wave by a point defect, the amplitude and phase of the scattered wavelet being uniquely related so that the forward intensity is reduced in compensation for the scattered intensity. In a superconductor, however, the evanescent wave generated by an incident excitation of wavenumber $\underline{k_F}$ retains this wavenumber but is gradually attenuated over a distance of the order ξ, $k_F\xi$ being much larger than unity. As a result the scattered wavelet that returns to the surface from D has travelled many wavelengths and is effectively incoherent in phase, so that it cannot generate an outgoing wave to reduce the amplitude of the original reflection. Worse than this, an incident electron is totally reflected as a hole, but the incoherent scattered wavelet generates an incoherent electron-like component emerging into N. One must then ask how can the number of excitations be conserved, as it must be, and the extraordinary answer is that when multiple reflection occurs between D and the surface alternate reflections are phase-coherent.

A fairly detailed account of this matter is given by Harding et al. (1974) and will not be repeated here. The essential point is that at each point on the Fermi surface there are two propagating waves, electrons and holes, each having energy greater than Δ, and two evanescent waves with energy less than Δ. Of the latter, both of which have $\underline{k_F}$ as their real wave vector, one (type A) decays in the direction of increasing $\underline{k_F}.\underline{r}$ and the other (type B) in the direction of decreasing $\underline{k_F}.\underline{r}$. When a type A evanescent wave in S reaches the interface it suffers the equivalent of Andreev reflection, conserving all components of $\underline{k_F}$ and turning into a type B evanescent wave. The incident wave is characterized by decay of amplitude and increase of phase, the reflected wave by further decay of amplitude but decrease of phase. A spherical evanescent A-wavelet scattered from D bounces off the interface but does not continue to expand; rather, it reconverges as a B-wavelet on to D where it is once more scattered, reaches the interface, bounces and reconverges as an A-wavelet, and so on indefinitely, except that at every stage it decays exponentially until it is finally lost. Every time an A-wavelet reaches the interface its reflection is accompanied by an electron-like excitation emerging into N, while a B-wavelet generates a hole-like excitation in N. Now the phase change as the evanescent wave proceeds from the interface to D is perfectly undone on its return to the interface*. Consequently, wherever D may be, there is the same complete phase coherence between the originally reflected hole and the subsequent hole-like wavelets emerging after alternate scatterings; and the alternate electron-like wavelets are also phase coherent. It is therefore possible, when there are many scattering centres in the interface layer, for a strong scattered electron component to emerge while the initial hole-like reflected excitation is correspondingly reduced, and excitation number is thereby conserved.

It is one thing to give a qualitative account of the process but quite another to work it out quantitatively. The one-dimensional case, to be sure, presents no great difficulty since the distinction between an emergent plane wave and an omnidirectional scattered wave does not arise. But, even in one dimension , it is absolutely essential not to replace an assumed distribution of scattering centres by a random average before taking all the phase-coherent processes into account. Krähenbühl and Watts-Tobin (1979) have given a lengthy and difficult analysis which they believe solves the problem; insofar as I understand their treatment, I think it fails at this very point of phase coherence. The starting point, Schmid and Schön's (1975) Boltzmann equation for the excitations, gives the impression that the evanescent waves are regarded as probability waves for real excitations, which then diffuse under the influence of scattering by impurities. If this is the correct interpretation there is certainly no account taken of phase coherence. Their result, that the extra resistance at 0 K is proportional to $c^{\frac{1}{2}}$, is what one would expect for incoherent diffusion, as in (4); and it is worth noting that their argument would not be significantly affected by the dimensionality of the problem. The one-dimensional case, however, is so strongly influenced by phase coherence that the resistance is

*In the closely related Tomasch (1965) effect the excitations in S are real, not the evanescent waves considered here; electron-like excitations lie just above $\underline{k_F}$, hole-like just below. In the multiple reflections across a thick superconducting layer there is not exact phase cancellation, but constructive or destructive interference depending on the thickness of the layer and the energy of the excitation. This gives rise to oscillations in tunnelling analogous to the interference colours of oil films.

proportional to c, rather than $c^{\frac{1}{2}}$. In view of this, and the formidable difficulty of taking account of phase coherence in the 3-dimensional case, the problem must be regarded as so far unsolved.

REFERENCES

Andreev, A.F., 1964, The thermal conductivity of the intermediate state in superconductors, Sov. Phys. J.E.T.P., 19:1228.

Harding, G.L., Pippard, A.B. and Tomlinson, J.R., 1974, Resistance of superconducting-normal interfaces, Proc. R. Soc. Lond. A., 340:1.

Krähenbühl, Y. and Watts-Tobin, R.J., 1979, Microscopic theory of the current-voltage relationship across a normal-superconducting interface, J. Low Temp. Phys., 35:569.

Landau, I.L., 1970, Temperature dependence of the reflection coefficient of electrons on the interface between superconducting and normal phases, J.E.T.P. Lett., 11:437.

McMillan, W.L. and Anderson, P.W., 1966, Theory of geometrical resonances in the tunnelling characteristics of thick films of superconductors, Phys. Rev. Lett., 16:85.

Phillips, W.A., 1969, Ultrasonic attenuation at 500 kHz in superconducting tin, Proc. R. Soc. Lond. A., 309:259.

Pippard, A.B., Shepherd, J.G. and Tindall, D.A., 1971, Resistance of superconducting-normal interfaces, Proc. R. Soc. Lond. A., 324:17.

Schmid, A. and Schön, G., 1975, Linearized kinetic equations and relaxation processes of a superconductor near T_c, J. Low Temp. Phys., 20:207.

Tomasch, W.J., 1965, Geometrical resonance in the tunnelling characteristics of superconducting Pb, Phys. Rev. Lett., 15:672.

Waldram, J.R., 1975, Chemical potential and boundary resistance at normal-superconducting interfaces, Proc. R. Soc. Lond. A., 345:231.

Walton, A.J., 1965, The thermal and electrical resistance of tin in the intermediate state, Proc. R. Soc. Lond. A., 289:377.

CHARGE IMBALANCE

John Clarke

Department of Physics, University of California and Materials and Molecular Research Division, Lawrence Berkeley Laboratory, Berkeley, California 94720

I. INTRODUCTION

If an electron-like ($k>k_F$) or a hole-like ($k<k_F$) excitation is added to a superconductor, a charge imbalance is created. Charge imbalance is of practical importance first, because it gives rise to a measureable steady-state voltage in a superconductor, and second, because it provides a tool for measuring various electron-relaxation rates. The importance of the disequilibrium of the electron- and hole-like branches was first pointed out by Pippard et al. (1971) who measured the electrical resistance of superconductor-normal metal-superconductor (SNS) sandwiches as a function of temperature. Near the superconducting transition temperature, T_c, they found that the resistance increased with increasing temperature. They ascribed this increase to the propagation of quasiparticles having a branch imbalance distribution into or out of the superconductor, a process that, at the time, they believed generated a potential step at each NS interface, and thus produced an additional boundary resistance. However, the first quantitative understanding of charge imbalance arose from experiments in which electrons were injected via a tunnel junction into a superconducting film (Clarke 1972); at the time this experiment was performed, its connection with the NS interface resistance was not apparent.

To motivate the long theoretical discussion that follows, we will begin by briefly reviewing the essential ideas of the tunneling experiment. The configuration used is shown in Fig. 1. First, a film of Al (XX') was deposited on a glass substrate and thermally oxidized, and a strip of Sn (YY') was evaporated at right angles to the first film to form a tunnel junction. The volume of Sn

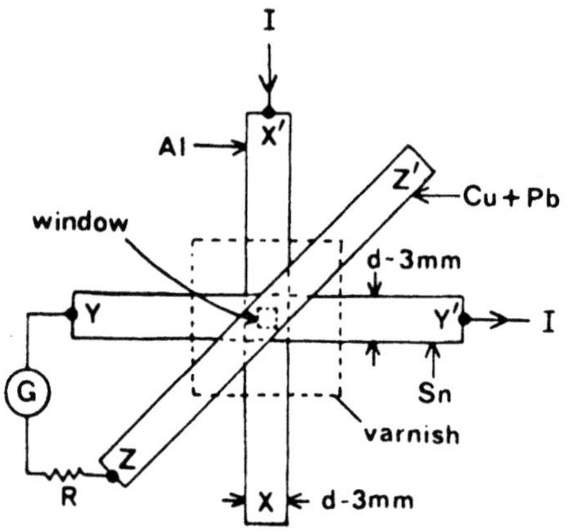

Fig. 1. Configuration of tunneling experiment to generate and detect charge imbalance in a Sn film. Galvanometer G (a SQUID) and resistor R are used to measure the voltage between the Sn and Cu films generated by the current passed between the Al and Sn films.

overlaying the Al was the region in which the charge imbalance was generated. The sample was removed from the evaporator, so that the Sn oxidized slightly, and a layer of varnish was applied leaving a window in the middle of the junction. A diagonal strip of Cu (ZZ') was evaporated to form the SIN junction that detects the charge imbalance. To reduce the resistance of the Cu strip, a film of Pb was deposited over it. With the sample immersed in liquid He^4, and with the Sn superconducting and the Al in the normal state, a current was passed between the Al and Sn films, and any voltage generated between the Sn and the Cu was measured with a SQUID voltmeter. A small voltage was detected that reversed sign when the sign of the injection current was reversed, and that was proportional to the injection current for injection voltages much higher than Δ/e, where $\Delta(T)$ is the energy gap in the Sn. The initial interpretation of this experiment by Tinkham and Clarke (1972) was later extended by Tinkham (1972) in a paper that laid the foundations of most of the important ideas of charge imbalance. Subsequently, there were two other theoretical papers of major importance, the first by Schmid and Schön (1975), who used a Green's function approach, and the second by Pethick and Smith (1979), who used a two-fluid approach involving the Boltzmann equation.

Charge imbalance has also been found to be central to several other experimental situations. In addition to the flow of current across an NS interface, these include the flow of a thermal current across an NS interface (Van Harlingen 1980), the resistance generated by phase slip centers in superconducting films (Skocpol et al.,

1974), and the flow of supercurrent along a superconductor in which a thermal gradient exists (Clarke et al., 1979). The purpose of this article is to review the theory of charge imbalance, and to discuss its relevance to a number of experimental situations. In Sec. II, we introduce the concepts of quasiparticle charge and charge imbalance, and discuss the generation and detection of charge imbalance by tunneling. In Sec. III we describe the relaxation of the injected charge imbalance by inelastic scattering processes, and show how the Boltzmann equation can be solved to obtain the steady state quasiparticle distribution and the charge relaxation rate. Section IV contains details of experiments to measure charge imbalance and the charge relaxation rate when inelastic scattering is the predominant relaxation mechanism. Section V is concerned with experiments on and theories of other charge relaxation mechanisms, namely relaxation via elastic scattering in the presence of energy gap anisotropy, or in the presence of a pair breaking mechanism such as magnetic impurities or an applied supercurrent or magnetic field. In Sec. VI we describe three other situations in which charge imbalance occurs, namely the resistance of the NS interface, phase slip centers, and the flow of a supercurrent in the presence of a temperature gradient. Finally, Sec. VII contains a concluding summary.

II. CHARGE IMBALANCE

In this section we introduce the concepts of quasiparticle charge and charge imbalance, and discuss the generation and detection of charge imbalance by tunneling. This discussion is based on the work of Tinkham and Clarke (1972), Tinkham (1972), Pethick and Smith (1979), Waldram (1975), and Kadin et al. (1980). These topics are also dealt with in detail by H. Smith in Chap. 15.

A. Quasiparticle Charge and Charge Imbalance

We begin by considering the total electronic charge, Q_{tot}, in a region of a superconductor, the overall charge at frequencies far below the plasma frequency being maintained at zero by the equal and opposite charge of the ion cores. From the BCS theory we have

$$Q_{tot} = \frac{2}{\Omega}\sum_{k} [u_k^2 f_k + v_k^2 (1-f_{-k})], \tag{2.1}$$

where the sum is over k states above and below k_F, and the factor of 2 arises from the sum over spin. In Eq. (2.1), Ω is the volume of the superconductor, and f_k is the occupancy of the state $\vec{k}$, which, in thermal equilibrium, is the Fermi function. The u_k and v_k are the usual BCS factors given by

$$u_k^2 = \tfrac{1}{2}(1 + \xi_k/E_k) \tag{2.2}$$

and

$$v_k^2 = \tfrac{1}{2}(1 - \xi_k/E_k), \tag{2.3}$$

where

$$\xi_k = \hbar^2 k^2/2m - \mu_s \tag{2.4}$$

is the kinetic energy of the electron of mass m in the state $\vec{k}$ referred to the chemical potential of the superfluid, μ_s, and

$$E_k = (\Delta^2 + \xi_k^2)^{\frac{1}{2}} \tag{2.5}$$

is the quasiparticle excitation energy. In Eq. (2.5), $\Delta(T)$ is the energy gap of the superconductor. The energy E_k is positive for all values of k, while ξ_k is positive for $k>k_F$ and negative for $k<k_F$. Figure 2(a) shows u_k^2 and v_k^2 vs. ξ_k/Δ. The quantity v_k^2 is the probability that the state $\vec{k}$ is occupied by a Cooper pair, while $u_k^2 = 1 - v_k^2$ is the probability that the state $\vec{k}$ is not occupied by a Cooper pair. In thermal equilibrium we can readily interpret Eq. (2.1). The quantity in square brackets is the probability that the electron state $\vec{k}$ is occupied. The first term is the product of the probability that the pair state $(\vec{k},-\vec{k})$ be empty, u_k^2, and the probability that the quasiparticle state be occupied, f_k. The second term is the product of the probability that the pair state be filled, v_k^2, and the quasiparticle state $-\vec{k}$ be empty, $(1-f_{-k})$.

We can regard Eq. (2.1) as the sum of a superfluid contribution

$$Q_s = \frac{2}{\Omega}\sum_k v_k^2 \tag{2.6}$$

and a quasiparticle contribution

$$Q^* = \frac{2}{\Omega}\sum_k (u_k^2 - v_k^2)f_k \equiv \frac{2}{\Omega}\sum_k q_k f_k, \tag{2.7}$$

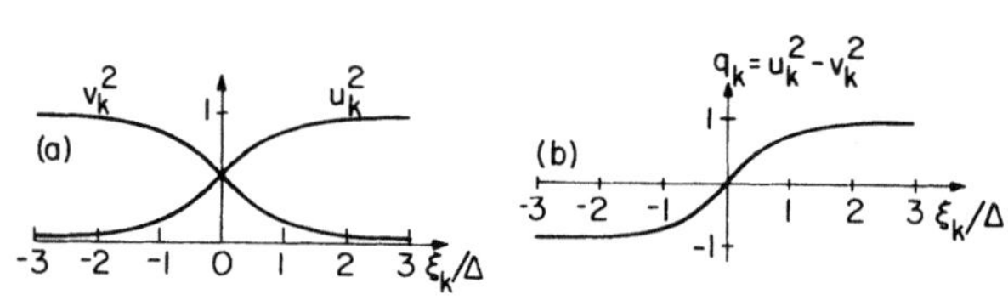

Fig. 2. (a) v_k^2 and u_k^2 vs. ξ_k/Δ and (b) q_k vs. ξ_k/Δ.

where we have set $f_{-k}=f_k$ in Eq. (2.1). In Eq. (2.7) we have introduced the very important concept of the effective quasiparticle charge

$$q_k \equiv u_k^2 - v_k^2 = \xi_k/E_k, \tag{2.8}$$

and defined the charge imbalance per unit volume, Q^*, which is the central subject of this paper. Figure 2(b) shows q_k vs. ξ_k; $q_k \to +1$ for $\xi_k \gg \Delta$ $(k>k_F)$, and $q_k \to -1$ for $\xi_k \ll -\Delta$ $(k<k_F)$. It is evident in Eq. (2.7) that if $f_k(\xi_k)$ is the Fermi function, which is even in ξ_k about μ_s, the quasiparticle charge, Q^*, is zero since q_k is odd in ξ_k. Thus, the total electronic charge is $\frac{2}{\Omega}\Sigma v_k^2$ at all temperatures, as we expect.

We now turn to a non-equilibrium situation in which f_k is no longer the Fermi function. In this case, Q^* may be non-zero, and μ_s may be shifted from its equilibrium value, the Fermi energy, by an amount $\delta\mu_s$. As we shall see, Q^* and $\delta\mu_s$ are intimately related. As an example, suppose that we add excitations to the $k>k_F$ quasiparticle branch in a particular region of the superconductor. Overall charge neutrality will ensure that pairs with an equivalent charge are removed from the region. Thus, the chemical potential of the pairs must decrease by an amount $\delta\mu_s$, as indicated in Fig. 3(a). At the same time, the excitation spectrum shifts because its minimum still occurs at the pair chemical potential, as shown in Fig. 3(b). The effective quasiparticle charge is still $q_k = \xi_k/E_k$, but ξ_k is now referred to the shifted chemical potential, $\mu_s - \delta\mu_s$. Thus, the addition of a single electron-like excitation induces some quasiparticles near the bottom of the excitation spectrum to cross branches, and produces a small increase in the charge of all the electron-like quasiparticles, and a small decrease in the charge of all the hole-like quasiparticles.

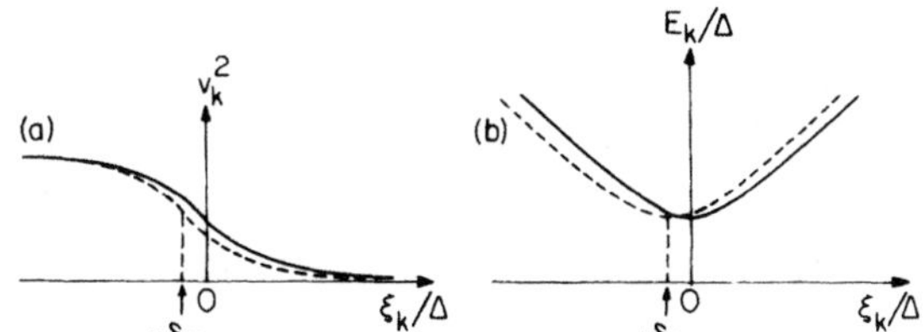

Fig. 3. The effect on (a) condensate density, and (b) the quasiparticle excitation spectrum when electron-like excitations are added to the superconductor. The solid and dashed lines represent equilibrium and the displacement from equilibrium, respectively.

The quantities Q^* and $\delta\mu_s$ are related as follows. By charge neutrality

$$Q^* + Q_s|_{\delta\mu_s \neq 0} = \frac{2}{\Omega}\sum_k v_k^2|_{\delta\mu_s=0} \quad , \tag{2.9}$$

or

$$Q^* = \frac{2}{\Omega}\sum_k (v_k^2|_{\delta\mu_s=0} - v_k^2|_{\delta\mu_s\neq 0}) \ . \tag{2.10}$$

Now the reduction in the number of pairs represented by the right hand side of Eq. (2.10) is just $2N(0)\delta\mu_s$, where $N(0)$ is the single-spin density of states per unit volume at the Fermi energy. Thus

$$Q^* = -2N(0)\delta\mu_s . \tag{2.11}$$

We see that Eq. (2.1) is really a self-consistent equation for μ_s: Given any distribution, f_k, the constancy of Q_{tot} determines μ_s. This is somewhat analagous to the BCS gap equation, in which f_k determines $\Delta(T)$.

In steady state with Q^* continuously injected at a rate $\dot{Q}_i^*$, the excess distribution relaxes at a rate τ_{Q*}^{-1} to produce a steady state charge imbalance Q^*. Thus

$$\tau_{Q*}^{-1} = \dot{Q}_i^*/Q^* . \tag{2.12}$$

For the moment we shall regard τ_{Q*}^{-1} as an empirical quantity, to be measured experimentally; we shall discuss its behavior in the presence of phonon scattering only in Sec. III, and contributions by other relaxation mechanisms in Sec. V.

To conclude our discussion on charge imbalance, it is instructive to consider qualitatively several disequilibrium situations. Figure 4(a) represents the addition of one excitation to the equilibrium distribution at some temperature. The quasiparticle charge, Q^*, is non-zero, as is the branch imbalance

$$Q = \frac{2}{\Omega}\left(\sum_{k>k_F} f_k - \sum_{k<k_F} f_k\right) \ . \tag{2.13}$$

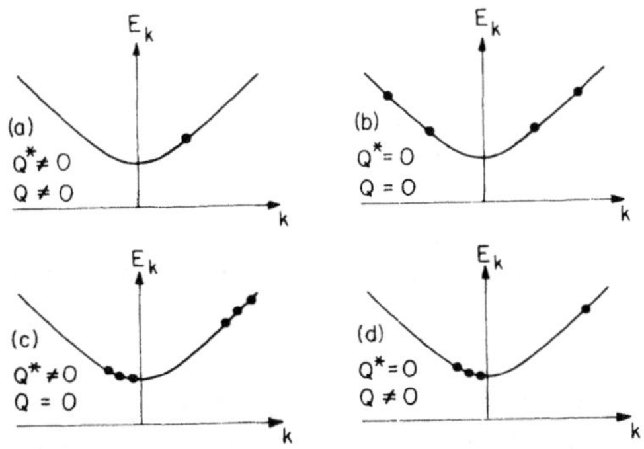

Fig. 4. Each dot represents a quasiparticle in excess of the equilibrium distribution: (a) $Q^*\neq 0$, $Q\neq 0$, (b) $Q^*=Q=0$, (c) $Q^*\neq 0$, $Q=0$, and (d) $Q^*=0$, $Q\neq 0$. In each case, the minimum value of E_k occurs at the pair chemical potential.

The quantity Q is the difference in the populations of the two branches, regardless of the effective charge, q_k. In Fig. 4(b), two excitations have been created on each branch in such a way that the net excitation population has been increased, but $Q^*=Q=0$. This is the so-called longitudinal mode of Schmid and Schön (1975). In Fig. 4(c), three quasiparticles have been added to each branch; thus $Q^*\neq 0$, because the excitations have different q_k, but $Q=0$. We note that this configuration of Q^* could relax if the quasiparticles on the $k>k_F$ branch scattered inelastically to lower energies; no interbranch scattering is required. Finally, Fig. 4(d) shows a branch imbalance, $Q\neq 0$, with $Q^*=0$: The charge of the single $k>k_F$ quasiparticle (nearly unity) is sufficient to cancel exactly the charges of the three low-lying $k<k_F$ quasiparticles that are each less than unity.

B. Generation and Detection of Charge Imbalance by Tunneling

We develop results for electron tunneling between a normal metal and a superconductor (NIS junction) that we can apply to both injection and detection processes. Single-particle tunneling is discussed at length by K. E. Gray in Chap. 5. We shall follow rather closely the approaches taken by Tinkham (1972) and Pethick and Smith (1979).

When a NIS junction is biased at a voltage V, the rate at which quasiparticles tunnel into the state $\vec{k}$ of the superconductor is given by a Golden Rule calculation:

$$\dot{f}_k\Big|_i = \frac{2\pi}{\hbar}|T|^2\aleph_n(0)\{u_k^2(1-f_k)f^o(E_k-eV) - v_k^2 f_k[1-f^o(E_k+eV)]$$
$$-u_k^2 f_k[1-f^o(E_k-eV)] + v_k^2(1-f_k)f^o(E_k+eV)\}. \tag{2.14}$$

Here, $|T|^2$ is the square of the tunneling matrix element, assumed to be independent of energy, and $\aleph_n(0)$ is the single-spin density of states in the volume of the metal. The Fermi functions $f^o(E_k \pm eV)$ refer to the normal metal, while f_k is the actual distribution in the superconductor, which is not, in general, the Fermi distribution. The four terms in braces in Eq. (2.14) correspond to the following processes respectively:

(i) An electron tunnels into the superconductor, creating an electron-like excitation with charge q_k; the remaining charge $(1-q_k)$ enters the condensate.

(ii) An electron tunnels into the superconductor, destroying a hole-like excitation, thus removing the charge q_k; the remaining charge $(1+q_k)$ enters the condensate.

(iii) An electron tunnels out of the superconductor, destroying an electron-like excitation and removing the charge q_k; the remaining charge $(-1+q_k)$ is supplied by the condensate [reverse of (i)].

(iv) An electron tunnels out of the superconductor, creating a hole-like excitation with charge q_k; the remaining charge $(-1-q_k)$ is supplied by the condensate [reverse of (ii)].

Cancelling some terms and using $u_k^2+v_k^2 = 1$, $u_k^2-v_k^2 = q_k$, we can rewrite Eq. (2.14) as

$$\dot{f}_k\Big|_i = \frac{2\pi}{\hbar}|T|^2\aleph_n(0)[u_k^2 f^o(E_k-eV) + v_k^2 f^o(E_k+eV) - f_k]$$
$$= \frac{2\pi}{\hbar}|T|^2\aleph_n(0)\{\tfrac{1}{2}[f^o(E_k+eV) + f^o(E_k-eV)] - f_k$$
$$+ \tfrac{1}{2}q_k[f^o(E_k-eV) - f^o(E_k+eV)]\}\ . \tag{2.15}$$

In Eq. (2.15), we emphasize that q_k is given by ξ_k/E_k where ξ_k is defined relative to the actual pair chemical potential.

From Eq. (2.15) we can derive an expression for the rate of injection of charge. Charge is injected into the state $\vec{k}$ at the rate of $q_k \dot{f}_k|_i$. If we multiply Eq. (2.15) by q_k and sum over $\vec{k}$, noting that only the even terms contribute, we find

$$\dot{Q}_i^* = \frac{2\pi}{\hbar\Omega} |T|^2 \aleph_n(0) \sum_k q_k^2 [f^o(E_k - eV) - f^o(E_k + eV)]. \quad (2.16)$$

The sum over spin produces a factor of 2 that cancels the factor ½ in Eq. (2.15). We have neglected the small contribution of the even part of $q_k f_k$ (i.e. we have neglected the odd part of f_k) on the grounds that the voltages generated by the non-equilibrium distribution are very small compared with the injection voltage. It is convenient to write Eq. (2.16) in integral form:

$$\begin{aligned}\dot{Q}_i^* &= \frac{4\pi}{\hbar\Omega} |T|^2 \aleph_n(0)\aleph(0) \int_0^\infty \frac{\xi^2}{E^2} [f^o(E-eV) - f^o(E+eV)] d\xi \\ &= \frac{G_{NN}}{e^2\Omega} \int_\Delta^\infty \rho^{-1}(E)\ [f^o(E-eV) - f^o(E+eV)] dE, \end{aligned} \quad (2.17)$$

where we have used $d\xi = (E/\xi)dE$. In Eq. (2.17), $\aleph(0)$ is the single-spin density of states in the volume of the superconductor when it is in the normal state, and $\rho(E) = E/(E^2-\Delta^2)^{\frac{1}{2}} = E/|\xi| = |q_k^{-1}|$ $(E>\Delta)$ is the normalized BCS density of states in the superconductor. We have introduced the conductance of the tunnel junction when the superconductor is in the normal state, $G_{NN} = 4\pi e^2|T|^2\aleph_n(0)\aleph(0)/\hbar$.

To obtain an expression for the net current injected into a superconductor from a normal metal, it is simplest to return to Eq. (2.14) and use the fact that the first two terms in braces add electrons to the superconductor, and make a positive contribution, while the last two terms remove electrons, and make a negative contribution. Summing over $\vec{k}$, after some simplification we find

$$I = \frac{4\pi e}{\hbar} |T|^2 \aleph_n(0) \sum_k [u_k^2 f^o(E_k - eV) - v_k^2 f^o(E_k + eV) - q_k f_k]. \quad (2.18)$$

We now write $u_k^2 = \frac{1}{2}(1+q_k)$, $v_k^2 = \frac{1}{2}(1-q_k)$, and $f_k = f^o + \delta f_k$ to obtain*

*We note that the E_k in f^o are referred to the modified chemical potential. PS denote this f^o by $f^{o(l.e.)}$, where "l.e." means "local equilibrium;" our δf_k is precisely the same as their $\delta f_k^{l.e.}$.

$$I = \frac{2\pi e}{\hbar} |T|^2 \aleph_n(0) \sum_k [f^o(E_k - eV) - f^o(E_k + eV) - 2q_k \delta f_k]. \tag{2.19}$$

In deriving Eq. (2.19), we used the fact that $\Sigma[\frac{1}{2}f^o(E_k+eV) + \frac{1}{2}f^o(E_k-eV) - f^o(E_k)] = 0$. Note that $2\,\Sigma\, q_k\delta f_k = 2\,\Sigma\, q_k f_k = \Omega Q^*$, since the sum over the even part of f_k vanishes. For any injection voltage of practical interest the last term is negligible, and we can write I in the integral form:

$$I = \frac{4\pi e}{\hbar} |T|^2 \aleph_n(0)\aleph(0) \int_0^\infty [f^o(E-eV) - f^o(E+eV)]d\xi$$

$$= \frac{G_{NN}}{e} \int_\Delta^\infty \rho(E)\,[f^o(E-eV) - f^o(E+eV)]dE\,. \tag{2.20}$$

We will find it useful to obtain an expression for Eq. (2.20) in the limit $eV \ll k_BT$ in which we can expand the Fermi functions to give $f^o(E\pm eV) = f^o(E) \pm (\partial f^o(E)/\partial E)eV$. This approximation leads to

$$I = G_{NN} \int_\Delta^\infty 2\rho(E) \left(-\frac{\partial f^o(E)}{\partial E}\right) VdE. \quad (eV \ll k_BT) \tag{2.21}$$

Since the conductance, G_{NS} is just I/V, we obtain the useful result for the reduced tunneling conductance for $eV/k_BT \to 0$, $g_{NS}(0) = G_{NS}(0)/G_{NN}$:

$$g_{NS}(0) \equiv Y(T) = 2\int_\Delta^\infty \rho(E) \left(-\frac{\partial f^o(E)}{\partial E}\right) dE, \tag{2.22}$$

where Y(T) is the Yoshida function.

We are now in a position to calculate the voltage generated by the steady state charge imbalance as measured by a second tunnel junction with a normal metal. This voltage, V_d, referred to the chemical potential of the superconductor far away from the injected region where equilibrium obtains, is measured with zero current flowing through the detector junction. Since Eq. (2.19) is equally true for the injection and detection junctions, we can set I=0, and use the low-voltage expansion of the Fermi functions to obtain

$$V_d = \frac{2\Sigma q_k f_k}{2e\Sigma(-\partial f^o(E)/\partial E)} = \frac{Q^*}{2N(0)e g_{NS}(0)}\,. \tag{2.23}$$

We have thus found a simple relation between Q^* or, equivalently, $-2N(0)\delta\mu_s$, and the voltage measured by the detector junction. It can be shown that if, as is often the case, the detector junction does not show the ideal BCS behavior, one should use the measured normalized conductance in Eq. (2.23), rather than the BCS value.

To complete our description of tunneling generation and detection of charge imbalance we need to combine Eqs. (2.12), (2.17), (2.20), and (2.23) so that we can deduce $\tau_{Q^*}^{-1}$ in terms of measureable or precisely calculable quantities. To do this, it is convenient to introduce the parameter F^* that enables us to express $\dot{Q}_i^*$ in terms of the injection current* I_i:

$$F^* = e\,\Omega\frac{\dot{Q}_i^*}{I_i} = \frac{\int_\Delta^\infty \rho^{-1}(E)[f^o(E-eV_i) - f^o(E+eV_i)]dE}{\int_\Delta^\infty \rho(E)[f^o(E-eV_i) - f^o(E+eV_i)]dE} \,. \qquad (2.24)$$

We have used V_i to denote the injection voltage. From Eqs. (2.12), (2.23) and (2.24) we find

$$\tau_{Q^*}^{-1} = \frac{\dot{Q}_i^*}{Q^*} = \frac{F^* I_i}{2N(0)e^2\Omega g_{NS}(0)V_d} \,. \qquad (2.25)$$

The quantity F^* is readily calculable, while the quantities I_i, Ω, $g_{NS}(0)$, and V_d are the quantities measured in the experiment.

In the limit of low injection voltages, $eV_i \ll k_BT$, the Fermi functions in Eq. (2.24) may be expanded to yield

$$F^* = Z(T)/Y(T), \quad (eV_i \ll k_BT) \qquad (2.26)$$

---

*It is evident that the integral expression for $\dot{Q}_i^*$ is less than that for I_i by a factor of $(\xi/E)^2$. To understand this, consider an electron injected at an energy $\sim\Delta$. Since the electron has a nearly equal probability of producing a quasiparticle on the $k_>$ and $k_<$ branches, the distribution imbalance is reduced by one factor of ξ/E. The second factor of ξ/E arises from the reduction in quasiparticle charge at low energies.

where $Z(T) \equiv 2 \int_{\Delta}^{\infty} \rho^{-1}(E) \left(- \frac{\partial f^{o}(E)}{\partial E} \right) dE.$ (2.27)

However, this limit is not very useful for most practical experiments with tunneling injections which are performed at injection voltages large compared with both $\Delta(T)/e$ and $k_B T/e$ (see Sec. IV). In these limits, one obtains

$$F^* = 1-\pi\Delta/2\,|eV_i|. \quad (eV_i >> k_B T,\ eV_i >> \Delta(T)) \qquad (2.28)$$

For a typical injection voltage, $eV_i/\Delta = 30$, we find $F^* = 0.95$.

At this point we have developed the ideas and results necessary to interpret a tunneling experiment that measures τ_{Q*}^{-1}. However, we shall continue our theoretical development, and in Sec. III discuss the form of the injected distribution, and the dependence of τ_{Q*}^{-1} on injection voltage and temperature for the case in which relaxation is via phonon scattering only.

III. CHARGE RELAXATION AND QUASIPARTICLE DISTRIBUTION FUNCTION IN THE PRESENCE OF PHONON SCATTERING

A. Introduction

When an electron-like excitation (for example) is injected into the superconductor, in general it not only creates a charge imbalance but also carries an energy well above thermal energies. Thus the processes that restore equilibrium involve both cooling and charge relaxation. However, these are not necessarily distinct processes. For example, if the quasiparticle is initially at an energy considerably above both $\Delta(T)$ and $k_B T$, it may emit a phonon and scatter on the same branch to an energy not too much greater than $\Delta(T)$. Clearly, the quasiparticle has cooled; in addition, it has given up some of its charge because $q(\equiv q_k)$ has been reduced from an initial value close to unity to a lower value. In this introductory discussion we shall not concern ourselves with cooling processes, but rather confine our attention to a qualitative discussion of charge relaxation that hopefully will set the scene for the quantitative description that follows in Sect. III B. Initially, we restrict ourselves to the temperature range near T_c where $\Delta/k_B T << 1$, and to a consideration of inelastic processes only. There are several calculations (Tinkham 1972, Schmid and Schön 1975, Pethick and Smith 1979) of the charge relaxation rate under these conditions, all leading to the result*

$$(F^* \tau_{Q^*})^{-1} = (\pi\Delta/4k_BT)\ \tau_E^{-1} \quad . \tag{3.1}$$

In Eq. (3.1), τ_E^{-1} is the inelastic scattering rate at T_c for a quasiparticle at the Fermi energy; the rate actually increases with the energy of the quasiparticle, but we shall neglect this dependence for the moment. A charge imbalance may relax through both quasiparticle scattering and recombination, and we now discuss these processes, giving qualitative arguments for Eq. (3.1).

We make the rather crude simplifying assumption that there are excess quasiparticles on the $k>k_F$ branch uniformly distributed in energy from Δ to k_BT. Consider first the scattering of a quasiparticle with energy $\sim k_BT$. This quasiparticle may scatter with the emission or absorption of a phonon with energy $\hbar\Omega$ to a new state E' on either the $k>k_F$ or the $k<k_F$ branch. The maximum phonon energy available for absorption is $\sim k_BT$. The coherence factor that governs these processes (Tinkham, 1975) is

$$(uu'-vv')^2 = \tfrac{1}{2}\left(1 + \frac{\xi\xi'-\Delta^2}{EE'}\right) \quad . \tag{3.2}$$

Suppose E' is on the $k>k_F$ branch [Fig. 5(a)]. Since ξ, ξ' are both positive and $E\sim k_BT_c>>\Delta$, the coherence factor is substantial for any value of E'. However, upward transitions, with phonon absorption, do not change q significantly, while downward transitions, with phonon emission, change q significantly only when $\Delta \leqslant E' \lesssim 2\Delta$ (say). Thus, of all possible transitions of this nature, only a fraction $\sim\Delta/2k_BT$ can relax Q^* significantly. Next, consider scattering onto the $k<k_F$ branch [Fig. 5(b)]. Given that $E\approx\xi>>\Delta$, we see that if $E'>>\Delta$ so that $\xi'\approx -E'$, the coherence factor is essentially zero. Thus, scattering to energies E' higher than E is rare. Moreover, scattering to energies lower than E is likely only if E' is reasonably close to Δ, say, between Δ and 2Δ; at $E'=2\Delta$, the coherence factor is $\approx \tfrac{1}{2}(1-\sqrt{3}/2) \approx 0.07$. Thus, again, only a fraction $\sim\Delta/2k_BT$ of transitions relax Q^*. We can extend our arguments to an injected $k>k_F$ electron in the range Δ to $\sim 2\Delta$, to find that it can scatter (upwards) into an energy range $\sim k_BT$ with significant change

*The paper by Tinkham (1972) contains a numerical error of order unity that was corrected in Ref. 10 of Clarke et al. (1979).

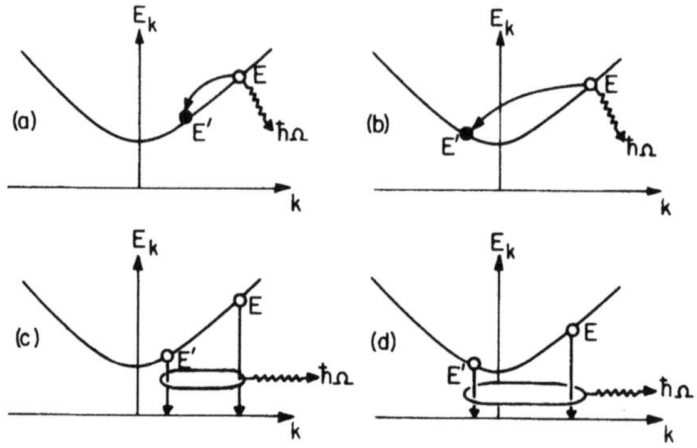

Fig. 5. Four inelastic processes that contribute to the relaxation of the charge of a quasiparticle with energy E. (a) scattering to a state of lower charge on the $k>k_F$ branch; (b) scattering to a state of lower charge of opposite sign on the $k<k_F$ branch; (c) recombination with a quasiparticle near the gap edge on the $k>k_F$ branch; (d) recombination with a quasiparticle near the gap edge on the $k<k_F$ branch. Processes (a) and (b) are governed by the coherence factor $(uu'-vv')^2$, while (c) and (d) are governed by the coherence factor $(vu'+uv')^2$.

of q. Clearly, a fraction $\sim\Delta/k_BT$ of the excess quasiparticles relax their charge in this way. We conclude that the overall charge relaxation rate for phonon scattering is $\tau_{Q*}^{-1} \sim (\Delta/k_BT)\,\tau_E^{-1}$.

We note in passing that elastic scattering from non-magnetic impurities can only relax Q* if the quasiparticle changes branches. In an isotropic superconductor, this process is strictly forbidden by the coherence factor Eq. (3.2), as can be seen immediately by realizing that if E and E' refer to different branches, u'=v, and v'=u. Consequently, it is perfectly reasonable to consider a theory in which only inelastic processes relax Q*. Later, in Sec. V, we discuss elastic scattering in an anisotropic superconductor, and also elastic scattering from magnetic impurities. For the moment, elastic scattering merely serves to scatter injected quasiparticles uniformly around the Fermi surface on the same branch without affecting the value of Q*.

We now turn our attention to Q^* relaxation by recombination, a process that is governed by the coherence factor (Tinkham 1975)

$$(vu'+uv')^2 = \tfrac{1}{2}\left(1 + \frac{\Delta^2-\xi\xi'}{EE'}\right), \qquad (3.3)$$

where E is the energy of the quasiparticle under consideration, and E' is the energy of the quasiparticle with which it recombines. First, consider recombination when E and E' refer to the same branch [Fig. 5(c)]. If E, E' are both large compared with Δ, the coherence factor becomes very small. However, if one energy is not too great, no greater than 2Δ, say, the coherence factor is relatively large, and recombination can make a substantial contribution to charge relaxation. A fraction $\sim\Delta/k_BT$ of the quasiparticles can relax Q^* in this way. Now consider recombination with E and E' on different branches [Fig. 5(d)]. Since the product $-\xi\xi'$ is now positive, the coherence factor is always at least ½. However, recombination processes have little effect on Q^* for energies where $|q|$ is reasonably close to unity. When one quasiparticle lies between Δ and $\sim 2\Delta$, and the other is at higher energies, Q^* does relax significantly, such processes representing a fraction $\sim\Delta/k_BT$ of the total. We conclude that recombination events again lead to an overall charge relaxation rate $\sim(\Delta/k_BT)\ \tau_E^{-1}$.

These qualitative arguments give some insight into the temperature dependence of $\tau_{Q^*}^{-1}$ near T_c. However, one is usually interested in a much wider temperature range, and, further, one would like to know the form of the steady state non-equilibrium distribution. To extend the theory one must resort to a numerical solution of the Boltzmann equation, which is the subject of Section III B.

B. The Boltzmann Equation for Charge Relaxation

This section follows closely the work of Chi and Clarke (1979, 1980) (CC). The Boltzmann equation simply expresses the total rate at which quasiparticles are injected into and scatter into and out of the state ξ_k, and can be written in the form

$$\dot{f}_\xi = G_\xi - G_{in\xi} - G_{el\xi} \ . \tag{3.4}$$

The first term G_ξ is the rate of injection into ξ_k, while $G_{in\xi}$ and $G_{el\xi}$ are the inelastic and elastic rates at which quasiparticles scatter out of ξ_k; for the moment we set $G_{el\xi}=0$. In a steady state situation $\dot{f}_\xi=0$. For tunneling from a normal metal into a superconductor the injection rate is given by the first line of Eq. (2.15) with the terms labeled with ξ rather than $\vec{k}$:

$$G_\xi = R_o\{\tfrac{1}{2}(1+\xi/E)[f^o(E-eV_i)-f^o(E)]-\tfrac{1}{2}(1-\xi/E)[f^o(E)-f^o(E+eV_i)]\}, \tag{3.5}$$

where

$$R_o \equiv G_{NN}/2N(0)e^2\Omega \; . \tag{3.6}$$

We have used the Fermi distribution for the distribution functions, thereby assuming the departure from equilibrium is small.

The inelastic scattering rate is proportional to $\alpha^2F(\Omega)$, the average of the product of the square of the matrix element for the electron-phonon interaction and the phonon density of states. We assume that $F(\Omega)$ is quadratic in Ω, and can be written in the form $F(k_BT_c/\hbar)(\Omega/k_BT_c)^2$. With this assumption, $G_{in\xi}$ can be written down using the Golden Rule (Kaplan et al. 1976, Chang and Scalapino 1977, Chang 1979, CC):

$$G_{in\xi} = \frac{2\pi}{\hbar}\frac{\alpha^2F(k_BT_c/\hbar)}{Z_1(0)(k_BT_c)^2}\int_{-\infty}^{\infty}d\xi' \;\times$$

$$\Big[\tfrac{1}{2}\left(1-\frac{\xi\xi'}{EE'}+\frac{\Delta^2}{EE'}\right)(E+E')^2\{[1+n^o(E+E')]f_\xi f_{\xi'}-n^o(E+E')(1-f_\xi)(1-f_{\xi'})\}$$

$$+\tfrac{1}{2}\left(1+\frac{\xi\xi'}{EE'}-\frac{\Delta^2}{EE'}\right)(E-E')^2\Theta(E-E')\{[1+n^o(E-E')]f_\xi(1-f_{\xi'})-n^o(E-E')f_{\xi'}(1-f_\xi)]\}$$

$$+\tfrac{1}{2}\left(1+\frac{\xi\xi'}{EE'}-\frac{\Delta^2}{EE'}\right)(E-E')^2\Theta(E'-E)\{n^o(E'-E)f_\xi(1-f_{\xi'})-[1+n^o(E'-E)]f_{\xi'}(1-f_\xi)\}\Big] . \tag{3.7}$$

In Eq. (3.7), $Z_1(0)$ is the electron-phonon renormalization factor, typically about 2 for most metals, Θ is the Heaviside function, which is zero for negative arguments and unity for positive arguments, and $n^o(\Omega)$ is the Bose-Einstein distribution function for the phonons. The first and second lines represent recombination, and contain a matrix element squared, a phonon density of states proportional to Ω^2, a coherence factor [Eq. (3.3)] , and various quasiparticle and phonon occupation numbers. The first term in braces represents recombination with phonon emission, while the second represents the reverse process, pair breaking, with phonon absorption. Since E and E' cannot be less than Δ, the phonon energies involved are all $>2\Delta$. The first and third lines represent phonon scattering between E and states $E'<E$, and again contain a matrix element squared, a phonon density of states, a coherence

factor [Eq. (3.2)], and various quasiparticle and phonon occupation numbers. The first term in braces represents scattering to lower energy states with phonon emission, while the second represents the reverse process, with phonon absorption. The first and fourth lines represent scattering between state E and states E'>E.

In Eq. (3.7), we have assumed that the phonons are in thermal equilibrium. Because of phonon trapping (Rothwarf and Taylor 1967), this is not a good approximation if one is interested in the total excess quasiparticle density. However, it is a valid approximation for the calculation of Q^* because the trapped excess phonons break pairs that on the average populate the two branches equally, and do not therefore affect the value of Q^*. We have also used the thermal equilibrium value of the energy gap, thereby assuming that the deviation from thermal equilibrium is very small.

The inelastic scattering rate, τ_E^{-1}, can be written in the form (Kaplan et al. 1976)

$$\tau_E^{-1} = \frac{14\zeta(3)\pi k_B T_c \alpha^2 F(k_B T_c/\hbar)}{\hbar Z_1(0)} , \qquad (3.8)$$

so that the prefactor of the integral in Eq. (3.7) is just $1/7\zeta(3)(k_B T_c)^3 \tau_E$.

C. Computer Solution of the Boltzmann Equation

It is convenient to write the quasiparticle distribution function as $f_\xi = f^o + \delta f_\xi$ (remembering that ξ is referred to the shifted chemical potential), where f^o is the Fermi function at energy $(\xi^2+\Delta^2)^{1/2}$. We then rewrite the Boltzmann equation in terms of the even (longitudinal) and odd (transverse) components of δf_ξ, $\delta f_\xi^L = \frac{1}{2}(\delta f_\xi + \delta f_{-\xi})$ and $\delta f_\xi^T = \frac{1}{2}(\delta f_\xi - \delta f_{-\xi})$. We make an initial guess at the values of δf_ξ^L and δf_ξ^T for values of ξ from zero to some energy much larger than the largest of Δ, $k_B T$, and $|eV_i|$, and use the Boltzmann equation to generate new values of δf_ξ^L and δf_ξ^T. This procedure is iterated until we find a consistent solution. We calculate Q^* from $4N(0)\Sigma q_\xi \delta f_\xi^T$, and find τ_{Q*}^{-1} from $\dot{Q}_i^*/Q^*$, using Eq. (2.17).

D. Computed Distribution Function

In Fig. 6 (from CC 1980) we plot $\delta f_>(E) - \delta f_<(E)$ vs. $\Delta(T)/k_B T$ for V_i=0.01 $\Delta(T)/e$ and T/T_c=0.99 and 0.9, where $\delta f_{\gtrless}(E)$ represents

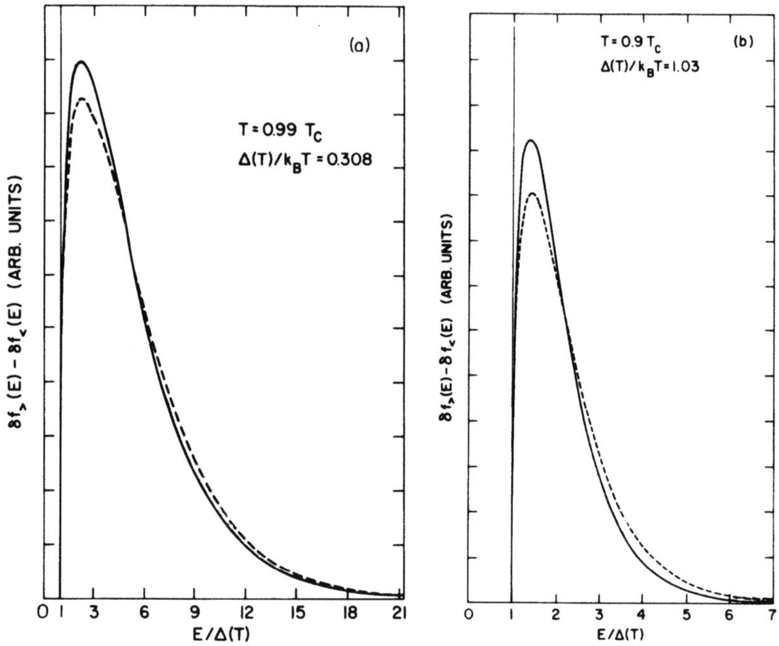

Fig. 6. $\delta f_>(E) - \delta f_<(E)$ vs. $E/\Delta(T)$ at (a) $T=0.99T_c$, and (b) $T=0.9T_c$:
——— computer calculation with $V_i=0.01\Delta(T)/e$, and
- - - - Pethick-Smith chemical potential model. Both curves are normalized to give the same value of Q*.

the deviation from equilibrium for $k \gtrless k_F$. The rapid decrease in the distribution below $\sim 2\Delta(T)$ as $E/\Delta(T) \to 1$ is due to the reduced generation rate at low energies described in Sec. IIB. The roll-off at high energies reflects the decrease in the injected distribution with increasing energy. At this very low injection voltage, electrons are injected almost equally into the $\vec{k}_>$ and $\vec{k}_<$ branches, so that the increase in the number of $\vec{k}_>$ quasiparticles is very nearly equal to the decrease in the number of $\vec{k}_<$ quasiparticles. Thus, the net number of quasiparticles is essentially unchanged (i.e. the longitudinal mode is not excited). In this limit, the chemical potential model of Pethick and Smith (1979) is expected to be a good description of the distribution. In their model, remembering that $E = [\Delta^2 + (\xi - \mu_s)^2]^{\frac{1}{2}}$, we can write the change in the distribution as

$$\delta f = \frac{\partial f^o}{\partial E} \frac{\partial E}{\partial \mu_s} \delta\mu_s = - \frac{\partial f^o}{\partial E} \frac{\xi}{E} \delta\mu_s, \tag{3.9}$$

so that

$$(\delta f_{>}-\delta f_{<}) = 2\left(-\frac{\partial f^{o}}{\partial E}\right)\frac{|\xi|}{E}\,\delta\mu_{s}. \tag{3.10}$$

This result is also shown in Figs. 6(a) and (b), normalized to represent the same value of Q^* as the computed curve. The chemcial potential curve is slightly lower than the computed distribution near the peak, and slightly higher at energies above about $6\Delta(T)$ and $2\Delta(T)$ for T/T_c=0.99 and 0.9 respectively.

In Fig. 7 (CC 1980) we plot the computed curves at V_i=0.01x $\Delta(T)/e$ and $20\Delta(T)/e$ for T/T_c=0.9. Both curves have been normalized to give the same value of Q^*. Injection at the higher voltages leaves the position of the peak virtually unchanged, but shifts part of the distribution from under the peak into the high energy tail. This shift indicates that some of the initial quasiparticles cross to the other branch as they cool from their high initial energy.

E. Voltage and Temperature Dependence of $\tau_{Q^*}^{-1}$

In the limit $T \to T_c$, theory predicts that to first order in $\Delta(T)/k_BT$, $\tau_{Q^*}^{-1} = (\pi\Delta(T)/4k_BT)\tau_E^{-1}$. To test the accuracy of this

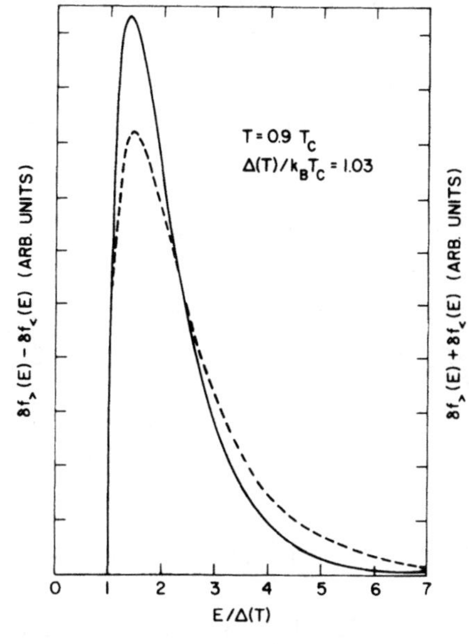

Fig. 7. $\delta f_>(E) - \delta f_<(E)$ vs. E/k_BT at T=0.9T_c: ——— V_i=0.01$\Delta(T)/e$, and - - - V_i=20$\Delta(T)/e$. Both curves are normalized to give the same value of Q*.

result at temperatures below T_c, CC computed τ_{Q*}^{-1} for $eV_i=0.01\Delta(T)$, $10\Delta(T)$, and $10k_BT_c$; the results are plotted in Fig. 8 (from CC 1980). At T_c, the computed values agree precisely with the theoretical values for all three injection voltages. However, as the temperature is lowered from T_c, $\tau_{Q*}^{-1}[4k_BT\tau_E/\pi\Delta(T)]$ initially increases and then decreases, the exact form of the curve depending on the injection voltage. At all temperatures below T_c, τ_{Q*}^{-1} increases with increasing injection voltage. This behavior reflects the increase in the inelastic scattering rate with increasing energy that is evident in the distribution of Fig. 7; inspection of Eq. (3.7) shows that the inelastic scattering rate out of a given state of energy E increases as E^3. In the low voltage limit and for $\Delta(T)/k_BT\leqslant 0.6$, CC found that the computed curve could be fitted to within ±1% by the empirical formula

$$\frac{\tau_{Q*}^{-1}4k_BT\tau_E}{\pi\Delta} = 1 + \tfrac{1}{2}\left(\frac{\Delta}{k_BT}\right) - \frac{3}{4}\left(\frac{\Delta(T)}{k_BT}\right)^2 . \quad (eV_i \ll \Delta(T), \quad \Delta(T)/k_BT\leqslant 0.6) \tag{3.11}$$

CC also computed the temperature dependence of τ_{Q*}^{-1} predicted by the Pethick-Smith distribution, Eq. (3.9), using the following procedure: (i) δf_ξ from Eq. (3.8) is used in Eq. (3.7) to calculate $G_{in\xi}$; (ii) the values of $G_{in\xi}$ are used in Eq. (3.4) to calculate G_ξ; (iii) the values of G_ξ are used to calculate $\dot{Q}_i^*$ from

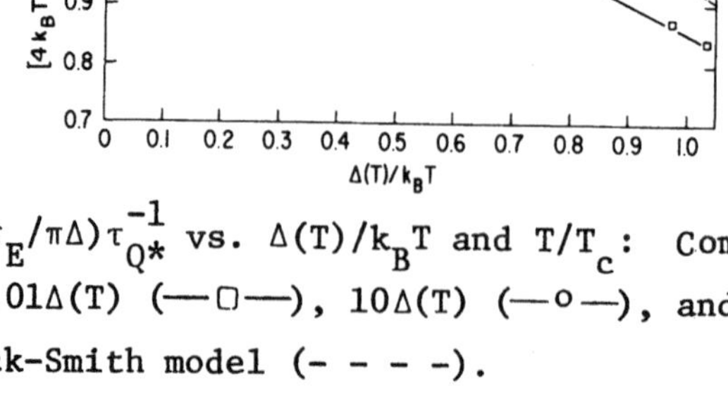

Fig. 8. $(4k_BT\tau_E/\pi\Delta)\tau_{Q*}^{-1}$ vs. $\Delta(T)/k_BT$ and T/T_c: Computed values for $eV_i=0.01\Delta(T)$ (—□—), $10\Delta(T)$ (—○—), and $10k_BT_c$ (—Δ—); Pethick-Smith model (- - - -).

$\dot{Q}_i^* = \int_0^\infty d\xi(\xi/E)(G_\xi - G_{-\xi})$; (iv) $\tau_{Q^*}^{-1}$ is found from $\dot{Q}_i^*/Q^*$. The result is shown as the dashed line in Fig. 8. At temperatures below T_c, the PS value of $\tau_{Q^*}^{-1}$ lies roughly 10% above the computed value for low injection voltages. This result can be understood by inspection of the distribution in Fig. 6. Since the PS distribution contains less quasiparticle charge near the peak and more at high energies than the computed distribution, it necessarily relaxes more rapidly because the inelastic scattering rate increases with energy.

The computed temperature dependence of $\tau_{Q^*}^{-1}$ is shown in Fig. 9 at temperatures down to $0.3T_c$ for $eV_i = 0.01\Delta(T)$ and $10\Delta(T)$, together with the PS prediction. As the temperature is lowered from T_c, the rate goes through a peak at a temperature that depends on the injection voltage, and then decreases steadily. The PS value always lies slightly above the low-injection voltage curve. It should be emphasized that these curves are computed only for inelastic-scattering processes, and that in real superconductors

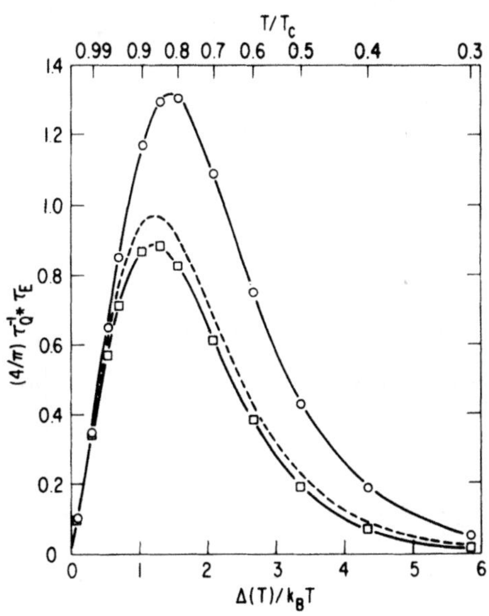

Fig. 9. Computed values of $(4/\pi)\tau_{Q^*}^{-1}\tau_E$ vs. $\Delta(T)/k_BT$ and T/T_c for $eV_i=0.01\Delta(T)$ (—□—), and $10\Delta(T)$ (—o—); dashed line is PS model.

elastic scattering in the presence of an anisotropic energy gap is likely to increase the low-temperature value of τ_{Q*}^{-1} very substantially (see Sec. V).

Finally, Fig. 10 shows τ_{Q*}^{-1} and $(F^* \tau_Q^*)^{-1}$, normalized to $\pi\Delta(T)/4k_B T\tau_E$, vs. $eV_i/\Delta(T)$ at $T=0.9T_c$. The experimentally measured voltage V_d is directly proportional to $(F^* \tau_Q^*)I_i$. We see that for $V_i \gtrsim 6\Delta(T)/e$ the voltage dependences of τ_Q^* and F^* cancel, so that $F^*\tau_Q^*$ becomes independent of voltage, and V_d becomes linear in I_i, as is observed experimentally (see Sec. IV).

This concludes our discussion of the theory of the relaxation of charge imbalance via inelastic scattering. We now move on to describe experimental measurements of Q^* and τ_{Q*}^{-1}.

IV. EXPERIMENTAL MEASUREMENTS OF Q* AND τ_{Q*}^{-1} IN TIN

We begin our description of the experimental work with measurements on Sn (Clarke 1972, Clarke and Paterson 1974, Moody and Paterson 1979) because in this material it is possible to obtain an identifiable temperature range near T_c in which inelastic charge relaxation dominates the elastic charge relaxation that is allowed in the presence of an anisotropic energy gap. A discussion of the latter process is deferred until Section VA. In this section we confine our discussion to the measurements by Clarke and Paterson (1974).

The basic configuration used in all tunneling injection experiments is shown in Fig. 1. Usually, four junctions were made simul-

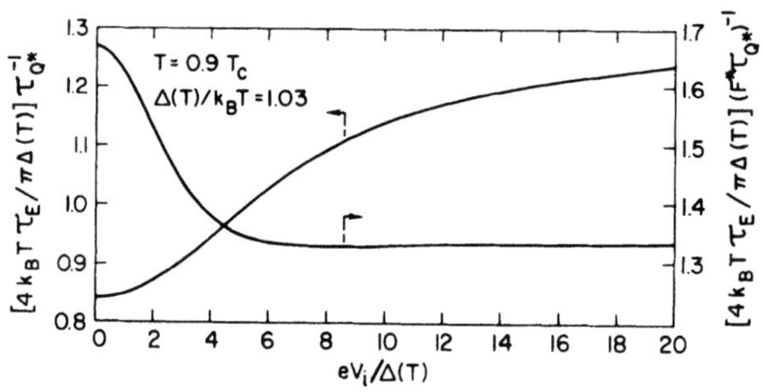

Fig. 10. $[4k_B T\tau_E/\pi\Delta(T)]\tau_{Q*}^{-1}$ and $[4k_B T\tau_E/\pi\Delta(T)](F^*\tau_{Q*})^{-1}$ vs. $eV_i/\Delta(T)$ for $T=0.9T_c$.

taneously. First, an Al film (XX') 120 to 200 nm thick and 3mm wide was evaporated through a metal mask onto a glass substrate. The Al was thermally oxidized in the evaporator, and a 3-mm wide cross strip of Sn(YY') 200 to 400 nm thick was deposited to complete the injector junction. The Sn film was then exposed to the atmosphere for 20 minutes to 2½h to produce a thin oxide barrier. The substrate was returned to the evaporator, and SiO deposited to mask off a $1 \times 1 mm^2$ window near the middle of the injector junction (SiO replaced varnish in all but the earliest measurements). A diagonal strip (ZZ') of CuAl (3 wt.%Al) about 2 μm thick was evaporated to make contact with the Sn oxide through the window in the SiO: This forms the detector junction. Because the rather high series resistance of the CuAl strip ($\gtrsim 1\Omega$) would have greatly reduced the voltage resolution, a Pb film was deposited over the CuAl. The Al in the Cu reduced the mean free path, ℓ, to $\sim$10nm, and thereby reduced the pair decay length, $(\hbar v_F \ell / 6\pi k_B T)^{\frac{1}{2}}$, sufficiently to eliminate Josephson tunneling between the Sn and Pb films (de Gennes 1964). The CuAl films were deposited by flash evaporation of pellets from a tungsten boat. The film thicknesses were monitored with a quartz crystal microbalance during evaporation; the thickness of each Sn strip was checked with an optical interferometer after completion of the electrical measurements.

The Sn-SnOx-CuAl+Pb junctions of the four samples on each substrate were connected in series with a calibrated resistor, R, and a superconducting coil coupled to a SQUID (see, for example, Clarke 1976). The output from the SQUID, suitably amplified, was fed back across the resistor R to produce negative feedback. Thus, any voltage developed across one of the detector junctions was measured at zero current, as required by Eq. (2.23). The circuit was carefully shielded to minimize the effects of fluctuating external fields, and of vibrations in static fields. The earth's magnetic field was reduced by about two orders of magnitude by means of two conductive μ-metal shields around the cryostat. A superconducting shield surrounded the low-temperature circuit, which was immersed directly in liquid helium. At temperatures below the λ-point the limiting noise in the measurement was determined by the Johnson noise developed in the probe junctions and the series resistor. Above the λ-point, the rms noise was several times greater, and probably arose from thermoelectric voltages developed by fluctuating temperature gradients in the bath.

We investigated each sample independently by connecting a current to the appropriate terminals. First, we obtained the $I_i - V_i$ characteristic of the injector junction by applying a current to X'Y' and measuring the voltage across XY. Next, we measured the resistance of the probe junction at low voltages by applying a current to Y'Z' and measuring the voltage across YZ with the SQUID voltmeter. Last, we measured the non-equilibrium voltage across YZ with the SQUID as a function of the injection current I_i.

The Al-AlOx-Sn injection junctions were of high-quality with a typical resistance of 1Ω at voltages above the energy gap. The Sn-SnOx-CuAl detector junctions had a resistance of typically 10^{-5}Ω at the transition temperature of the Sn. From the variation of resistance with temperature, one can obtain measured values of $g_{NS}(0)$. Figure 11 shows typical measured values of $g_{NS}(0)$ for 4 samples together with the behavior predicted by Eq. (2.22). In general the conductances lie above the theoretical value, often by a substantial amount. However, it was found that if one used the <u>measured</u> value of $g_{NS}(0)$ to compute τ_{Q*}^{-1} using Eq. (2.25), the values of τ_{Q*}^{-1} obtained from different samples at a given temperature and injection voltage agreed well with each other. Thus, it appears that the quality of the detector junction is not a vital factor in the determination of τ_{Q*}^{-1}; however, the barrier between the Sn and the CuAl must be sufficient to quench the proximity effect (DeGennes 1964) that would otherwise depress the order parameter in the superconductor. Very near T_c the conductance often dropped by a few percent with increasing temperature. This decrease in conductance is believed to arise from the resistance experienced by the quasiparticles injected into the superconductor, just as in the case of the

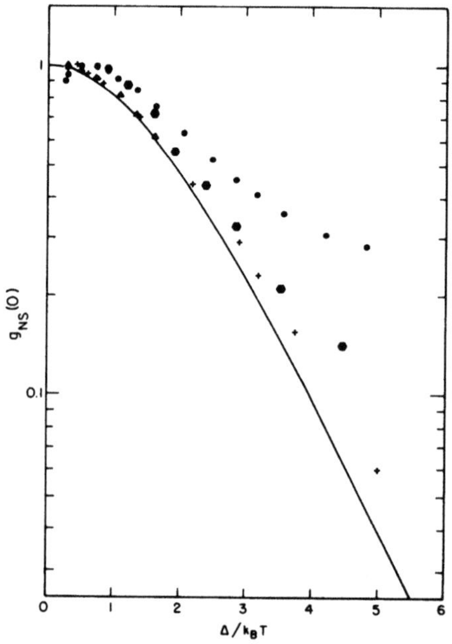

Fig. 11. Measured values of $g_{NS}(0)$ vs. Δ/k_BT for three Sn-SnOx-Cu (● ▲ +) and one SnIn-SnOx-Cu (●) detector junctions compared with theory (solid line).

excess resistance observed at the NS interface (see Sec. VA). The conductance dip is observed in these junctions because the resistance of the barrier is so low; for junctions with resistances of (say) 1mΩ or higher, the effect would be unobservably small. Since the excess resistance occurs only very close to T_c, the values of the conductance well below T_c were normalized to the maximum value of the conductance, rather than to the value at T_c.

At each temperature V_d was plotted vs. I_i for both polarities of the injection current. Near T_c, the values of V_d/I_i for both polarities were almost equal, but at lower temperatures there was often a marked asymmetry. This asymmetry probably arose from thermoelectric voltages generated by thermal gradients induced by heating in the injection junction [see Moody and Paterson (1979) for a discussion]. Provided the asymmetry is not too large, one can eliminate it in the data analysis by averaging together the two values of V_d/I_i. In the present analysis we confine our attention to high injection voltages, where $F^* \simeq 1$.

For each sample one computes $g_{NS}(0)\Omega V_d/I_i = F^*\tau_{Q^*}/2N(0)e^2$. The results for $V_i = 10\Delta(T)/e$ are plotted in Fig. 12 for three sets

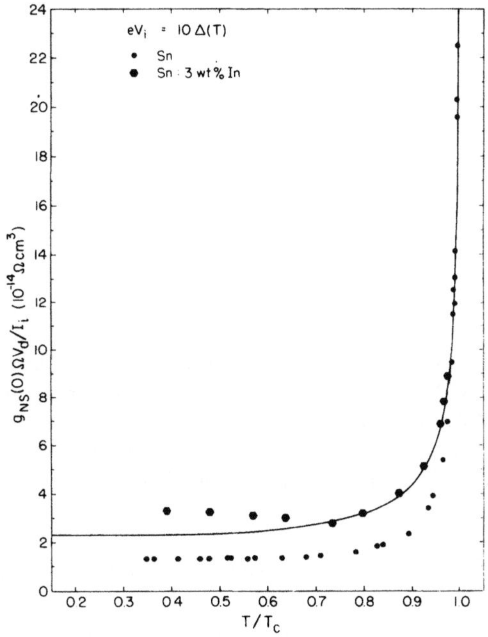

Fig. 12. $g_{NS}(0)\Omega V_d/I_i$ vs. T/T_c for pure Sn (•) and SnIn (●). The solid line is a fit of $k_BT_c/\Delta(T)$ to the data for SnIn for $T/T_c \gtrsim 0.7$.

of samples of pure Sn, and for one set of samples of Sn with 3wt. %In. The mean free path in the clean Sn samples was always boundary limited, while that in the alloy was estimated to be 42nm. The fact that all the data for the clean Sn samples lie on a smooth curve demonstrates that the predicted dependence on volume is correct. Near T_c, for both clean and dirty samples V_d/I_i, which is proportional to $F^*\tau_{Q^*}$, diverges with increasing temperature. Below $T/T_c \approx 0.95$, the value of $F^*\tau_{Q^*}$ for the dirty sample is significantly larger than for the clean sample. We postpone a discussion of this difference to Sec. VA, and focus our attention on the behavior near T_c, where inelastic scattering makes the only significant contribution to charge relaxation.

The solid line is a fit to the data of the curve $g_{NS}(0)\Omega V_d/I_i = 4.0\times10^{-14}\ k_BT_c/\Delta(T)\ \Omega cm^3$ in the limit $T\to T_c$. Using the value* $N(0) = 1.39 \times 10^{22} eV^{-1}cm^{-3}$ in Eq. (2.25) we calculate $(F^*\tau_{Q^*})^{-1} = 5.6 \times 10^9\ (\Delta/k_BT_c)\ s^{-1}$. Equating this value with the prediction of Eq. (3.1), we find $\tau_E = 1.4 \times 10^{-10}$ s. By way of comparison, Kaplan et al. (1976) have computed $\tau_E = 2.6 \times 10^{-10}$ s using measured values of α^2F in a collision integral. We regard the agreement as satisfactory.

In the case of the Sn alloy, the fit of $(F^*\tau Q^*)^{-1}$ to the predicted temperature dependence is excellent down to $T/T_c \approx 0.7$, while the pure Sn data diverge from the prediction below $T/T_c \approx 0.98$. To understand the differences between the clean and dirty samples, we must investigate the effects of gap anisotropy on charge relaxation. This is the subject of Sec. VA.

V. CHARGE RELAXATION VIA ELASTIC SCATTERING PROCESSES

We now turn to a discussion of charge relaxation by elastic scattering, beginning with scattering from non-magnetic impurities in the presence of an anistropic energy gap. We then consider elastic scattering in the presence of a pair-breaking mechanism such as magnetic impurities or an applied supercurrent or magnetic field.

A. Elastic Scattering with an Anisotropic Energy Gap

We have already seen from Eq. (3.2) that the coherence factor

*This value was computed from $N(0) = 3\gamma/2\pi^2k_B^2$, where γ is the coefficient of the electronic heat capacity taken from Kittel (1976).

$(uu'-vv')^2$ forbids charge relaxation via elastic scattering when the energy gap is isotropic. However, when the gap is anisotropic the coherence factor no longer vanishes, and elastic scattering can transfer quasiparticles between the $k_>$ and $k_<$ branches. Figure 13 shows the excitation spectra at two regions of the Fermi surface at which the energy gap is different. Since the values of $|\xi|$ and $|\xi'|$ are different for the initial and final states, the coherence factor is non-zero even though E and E' are identical. This situation was first discussed by Tinkham (1972), and subsequently, more quantitatively, by Chi and Clarke (1979). We shall follow the latter treatment.

The elastic scattering rate $G_{el\xi}$ that appears in the Boltzmann Equation [Eq. (3.4)] is given by

$$G_{el\xi} = \frac{\tau_1^{-1}}{2} \left\langle \rho_{\Delta'}(E) \left[1 - \frac{\Delta\Delta'}{E^2} - \left(1 - \frac{\Delta^2}{E^2}\right)^{\frac{1}{2}} \left(1 - \frac{\Delta'^2}{E^2}\right)^{\frac{1}{2}} \right] \right\rangle_{\Delta,\Delta'} \times$$

$$[f_\xi (1-f_{\xi'}) - f_{\xi'}(1-f_\xi)]. \tag{5.1}$$

Here, τ_1^{-1} is the elastic scattering rate of an electron when the metal is in the normal state, $\rho_{\Delta'}(E)$ is the final density of states, one-half times the first square bracket is the coherence factor, and the two terms in the second square bracket are the occupation factors for elastic scattering from ξ to ξ' and the reverse process, respectively. The symbol $\langle\ \rangle_{\Delta,\Delta'}$ indicates an angular average over the gap-anisotropy distribution. To average the coherence factor we set $\Delta = \bar{\Delta} + \delta$ and $\Delta' = \bar{\Delta} + \delta'$, where $\bar{\Delta}$ is the average energy gap. Our goal is to express the terms in $\langle\ \rangle$ in terms of the mean square gap anisotropy

$$\langle\delta^2\rangle = \langle(\Delta-\bar{\Delta})^2\rangle. \tag{5.2}$$

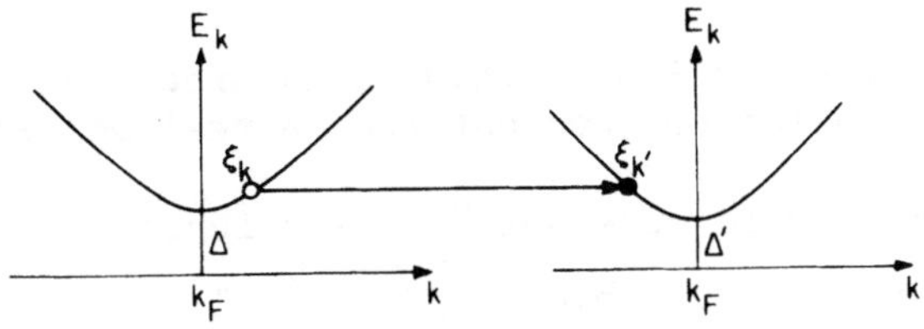

Fig. 13. Principle of charge relaxation by elastic scattering in the presence of gap anisotropy. The two excitation spectrum refer to two different regions of the Fermi surface where the energy gaps are different.

We first note that if one expands $\rho_{\Delta'}(E) = \rho_{\bar{\Delta}}(E)+(\partial\rho(E)/\partial\Delta')\delta\Delta'$, it can be shown that the second term is a higher order correction. Thus we set $\rho_{\Delta'}(E) = \rho_{\bar{\Delta}}(E) = E/(E^2-\bar{\Delta}^2)^{\frac{1}{2}}$. One substitutes for Δ and Δ' in the coherence factor, and expands with $E \gg \bar{\Delta}$. Using the fact that $\langle\delta\rangle_{\Delta} = \langle\delta'\rangle_{\Delta'} = 0$, we find $\langle (uu'-vv')^2\rangle_{\Delta,\Delta'} = \langle\delta^2\rangle/2(E^2-\bar{\Delta}^2)$. Clearly, this expression is meaningless for $E \le \bar{\Delta}$. However, for the purpose of the present calculation it is an adequate approximation to use this expression for $E > \bar{\Delta} + \langle\delta^2\rangle^{\frac{1}{2}}/2$, and to take the value at $E = \bar{\Delta} + \langle\delta^2\rangle^{\frac{1}{2}}/2$ for lower energies; the error introduced is of the order of $\langle\delta^2\rangle^{\frac{1}{2}}/\bar{\Delta}$. Thus, we find

$$\langle\rho_{\Delta}'(E)\ (uu'-vv')^2\rangle_{\Delta,\Delta'} = \langle\delta^2\rangle\, E/2(E^2-\Delta^2)^{\frac{3}{2}}\quad (E \geq \bar{\Delta}+\langle\delta^2\rangle^{\frac{1}{2}}/2) \tag{5.3}$$

We define the normalized mean-square gap anisotropy (Markowitz and Kadanoff 1963) by

$$\langle a^2\rangle = \langle\delta^2\rangle/\bar{\Delta}^2 . \tag{5.4}$$

For clean bulk superconductors ($\ell \gg \hbar v_F/\Delta$), $\langle a^2\rangle$ is a constant, independent of temperature. However, if the mean free path is shortened by impurity or surface scattering, $\langle a^2\rangle$ is reduced from its clean limit value, $\langle a^2\rangle_o$ (Anderson 1959), to a value given approximately by

$$\langle a^2\rangle = \langle a^2\rangle_o/[1 + (\hbar/2\tau_1\bar{\Delta})^2]. \tag{5.5}$$

Combining Eqs. (5.1), (5.3), (5.4) and (5.5) for $E \geq \bar{\Delta}+\langle\delta^2\rangle^{\frac{1}{2}}/2$ we arrive at our final expression for $G_{el\xi}$:

$$G_{el\xi} = \tau_1^{-1}\,\frac{\langle a^2\rangle_o}{1+(\hbar/2\tau_1\bar{\Delta})^2}\,\frac{\bar{\Delta}^2 E}{2(E^2-\bar{\Delta}^2)^{3/2}}\,(f_\xi - f_{-\xi}).\quad (E \geq \bar{\Delta}+\langle\delta^2\rangle^{\frac{1}{2}}/2) \tag{5.6}$$

In Eq. (5.6), we have set $\xi' = -\xi$; this is an excellent approximation for the values of anisotropy encountered in real metals.

It is instructive to consider $G_{el\xi}$ as a function of τ_1^{-1}. First, consider the clean limit in which $(\hbar/2\tau_1\bar{\Delta})^2 \ll 1$, i.e. $\ell \gg \hbar v_F/2\bar{\Delta}$. As we progressively increase τ_1^{-1}, for example, by adding impurities, $G_{el\xi}$ increases proportionately; in this limit, shortening the mean free path increases the scattering rate much more than it decreases the mean square anisotropy. On the other hand, in the dirty limit $(\hbar/2\tau_1\bar{\Delta})^2 >> 1$, $G_{el\xi}$ is proportional to τ_1. Thus, as we increase τ_Q^{-1}, $G_{el\xi}$ decreases; in this limit the mean square anisotropy is reduced more rapidly than the scattering rate is increased. From Eq. (5.6) it is evident that at a given temperature the scattering rate is a maximum when $\tau_1^{-1} = \hbar/2\bar{\Delta}$ or $\ell = \hbar v_F/2\bar{\Delta}$. Thus, sufficiently close to T_c, the sample must be in the limit in which $\hbar/2\tau_1\bar{\Delta} >> 1$, and $G_{el\xi}$ will be proportional to $\tau_1\bar{\Delta}^{-4}/(k_BT_c)^2$ for typical quasiparticle energies $\sim k_BT_c$. Thus, we expect the elastic charge relaxation rate to fall off very much more rapidly than the inelastic rate as $T \to T_c$, so that the inelastic rate eventually dominates above a temperature that depends on the relative magnitudes of τ_1^{-1} and τ_E^{-1}.

We are now in a position to understand qualitatively the behavior of the Sn and SnIn films in Fig. 12. For the clean Sn films, the mean free path is roughly the film thickness, say 300nm on the average. Consider first temperatures below about $0.8T_c$. In Table I we list values of τ_1^{-1}, $(\hbar/2\bar{\Delta}\tau_1)^2$ and $\tau_1^{-1}[1 + (\hbar/2\bar{\Delta}\tau_1)^2]^{-1}$ evaluated at low temperatures using $\bar{\Delta} = 1.76\ k_BT_c$, with $T_c = 3.8K$. We see that the elastic charge relaxation rate is expected to be

Table I. Parameters for the Sn and SnIn Films in Fig. 12[a].

	Sn	SnIn
ℓ (nm)	≈300	42
τ_1^{-1} (ps^{-1})	2.2	15.5
$(\hbar/3.52k_BT_c\tau_1)^2$	1.60	79
$\tau_1^{-1}[1 + (\hbar/3.52k_BT_c\tau_1)^2]^{-1}$ (ps^{-1})	0.85	0.19

a. We used $v_F = 0.65\times10^6 ms^{-1}$, calculated from $v_F = (\pi^2k_B^2/e^2\gamma)\sigma/\ell$ (Pippard 1965), where σ is the conductivity and $\sigma/\ell = 9.5 \times 10^{10}\Omega^{-1}cm^{-1}$ for Sn (Chambers 1952).

about 5 times faster in Sn than in SnIn. Since the total measured rate for both inelastic and elastic scattering is roughly two times higher in Sn than in SnIn, we conclude that the elastic contribution is relatively negligible for SnIn, but comparable with the inelastic rate for Sn. Now let us increase the temperature from $0.8T_c$. In the SnIn, the elastic charge relaxation rate becomes even more negligible compared with the inelastic rate. In the Sn, the elastic charge relaxation rate becomes progressively smaller than the inelastic rate, eventually becoming negligible when $\bar{\Delta}$ falls below a certain value. Thus, in the limit $T \rightarrow T_c$, the charge relaxation is due to inelastic scattering only, and the data for the Sn and SnIn samples merge together. Thus, sufficiently close to T_c, we are justified in using the data to extract values of τ_E, as we did in Sec. IV.

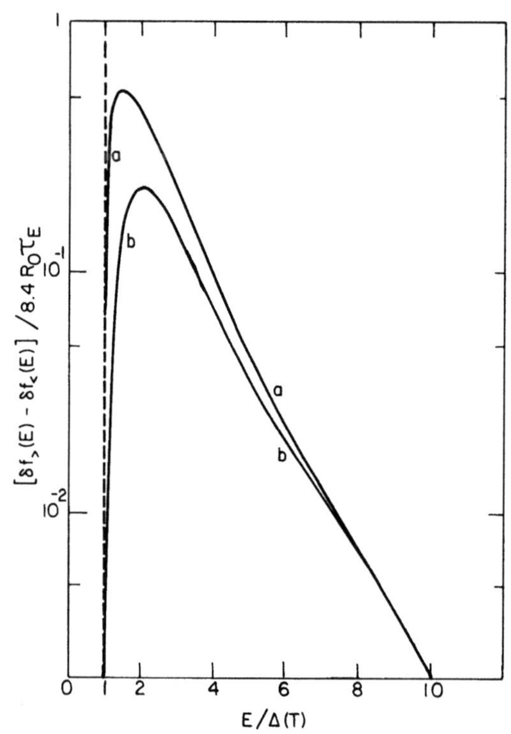

Fig. 14. Computed distribution for a 100nm-thick film at $0.9T_c$ with Δ=100meV for electrons injected from a normal film through a 1Ω 3x3mm tunnel junction biased at $10\Delta/e$. Curves a and b are for $\tau^{-1}_{Q*el}(0)\tau_E$=0 and 0.93, respectively.

To account quantitatively for the charge relaxation by elastic scattering one must insert $G_{el\xi}$ into the Boltzmann equation, Eq. (3.4), and solve the equation numerically, retaining the injection and inelastic collision terms in their previous forms. This was undertaken by CC, who calculated curves appropriate for their results on Al films. These films were always in the limit $(\hbar/2\bar{\Delta}\tau_1)^2 >> 1$ so that we can write Eq. (5.6) in the form

$$G_{el\xi} = \tau^{-1}_{Q^*el}(0)\left(\frac{\bar{\Delta}}{k_BT_c}\right)^2 \frac{E\bar{\Delta}^{-2}}{(E^2-\bar{\Delta})^{3/2}}(f_\xi - f_{-\xi}),\quad [(\hbar/2\bar{\Delta}\tau_1)^2 >> 1)] \tag{5.7}$$

where

$$\tau^{-1}_{Q^*el}(0) = 2\tau_1(k_BT_c)^2\langle a^2\rangle_o/\hbar^2 \tag{5.8}$$

is a characteristic elastic charge relaxation rate. Figure 14 shows the results of CC for $\tau^{-1}_{Q^*el}(0)\tau_E=0$, and $\tau^{-1}_{Q^*el}(0)\tau_E=0.93$. We see that the addition of the elastic relaxation process reduces the charge imbalance, and shifts the peak in the distribution to higher energies. This shift illustrates the fact that the elastic relaxation is much faster at low energies than at high energies,

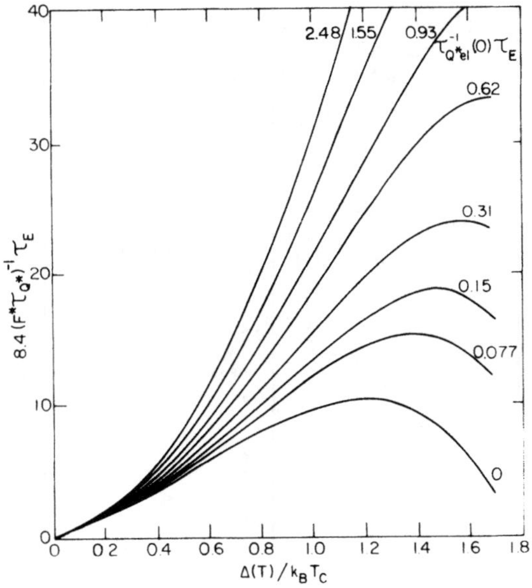

Fig. 15. Calculated values of $(F^*\tau_{Q^*}/8.4\tau_E)^{-1}$ vs. $\Delta(T)/k_BT_c$ for 8 values of $\tau^{-1}_{Q^*el}(0)\tau_E$ with $V_i=10\Delta(T)/e$.

because of the term $(E^2-\bar{\Delta}^2)^{-3/2}$ in Eq. (5.7). Figure 15 shows the computed curves of $(F^*\tau_{Q^*}/\tau_E)^{-1}$ vs. $\Delta(T)/k_BT_c$ for 8 values of $\tau^{-1}_{Q^*el}(0)\tau_E$. As $\Delta/k_BT_c \rightarrow 0$, the elastic rate becomes negligible, and all the curves tend to the same limit, which is just the inelastic charge relaxation rate. At lower temperatures, the charge relaxation rates increase monotonically with $\tau^{-1}_{Q^*el}(0)\tau_E$, as one would expect.

The sample configuration used by CC for the experiments was similar to that in Fig. 1, and is shown in the inset of Fig. 16. The injector film was in the superconducting state for most of the measurements. The mean free path of some of the samples was reduced, and, at the same time, the transition temperature increased by evaporating the Al in a partial pressure of oxygen. The transition temperatures and mean free paths of the cleanest and dirtiest samples were 1.22K and 409nm and 2.11K and 1nm, respectively. Since the detector junction resistances were relatively large ($\sim$1mΩ), and the values of V_d were also quite large because of the

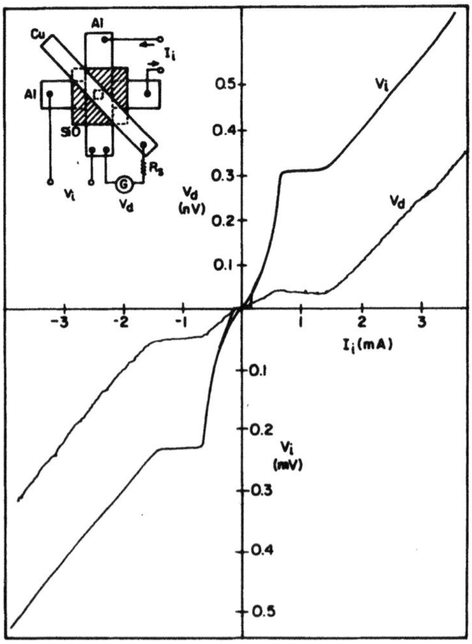

Fig. 16. Measured V_i and V_d vs. I_i for a representative sample at T/T_c=0.87. Note the two different voltage scales. Inset shows sample configuration.

long electron-phonon scattering time in Al, it was not necessary to deposit a Pb film over the Cu to reduce the series resistance. Consequently, pure Cu was used to complete the detector junction.

Figure 16 shows plots of V_i and V_d vs. I_i for a representative sample. It is noteworthy that when I_i increases at the sum of the gaps of the injector and injected films (approximately 0.23mV), V_d remains essentially constant. This is because additional electrons injected by the current increase have an energy Δ, and equally populate the $k_>$ and $k_<$ branches, thereby creating zero charge imbalance. This result is a direct proof that V_d is proportional to Q*.

The dots in Fig. 17 show $(F^*\tau_{Q^*})^{-1}$ vs. $\Delta(T)/k_BT_c$ for 5 representative samples with parameters listed in Table II; N(0) =

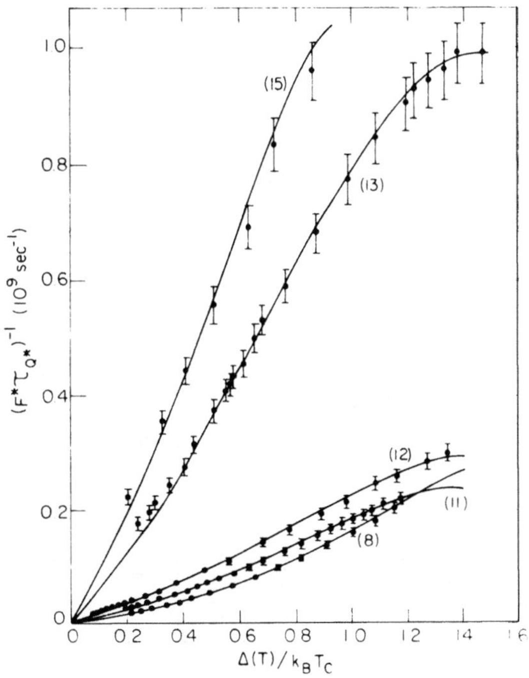

Fig. 17. Measured values (•) of $(F^*\tau_{Q^*})^{-1}$ vs. Δ/k_BT_c for 5 samples listed in Table II (sample numbers shown in parenthesis). The solid lines are a best fit to the solution of the Boltzmann equation.

Table II. Measured and Calculated Quantities for Five Al Samples[a]

Sample	T_c(K)	ℓ(nm)	τ_E(ns)	$\tau^{-1}_{Q*el}(0)\tau_E$	$\langle a^2 \rangle_o$
8	1.267	78	13	0.62	0.022
11	1.306	23	8	0.093	0.017
12	1.411	12.5	5.8	0.062	0.025
13	1.573	6.0	1.5	0.031	0.076
15	2.113	1.0	1.0	b	b

a. We used $v_F = 1.36 \times 10^6 ms^{-1}$, calculated from the free electron value divided by the thermal effective electron mass (Kittel 1976), $\rho\ell = 9 \times 10^{-16} \Omega m^2$ (Fickett 1971), and $N(0) = 1.74 \times 10^{28} eV^{-1} m^{-3}$ (Gschneider 1964).

b. The upward curvature was too small to enable an estimate of $\tau^{-1}_{Q*el}(0)$ to be made.

1.74×10^{28} $eV^{-1}m^{-3}$ (Gschneider 1964). The solid lines are the best fits to the computed solution of the Boltzmann equation using the fitting parameters τ_E^{-1} and $\tau_{Q*el}^{-1}(0)\tau_E$ shown in Table II. The fit to the data (and to the data of the 10 other samples measured) is excellent. For each sample we determined τ_E^{-1} from the slope in the limit $\Delta(T)/k_BT_c \to 0$, while we found $\tau_{Q*el}^{-1}(0)\tau_E$ from the upward curvature needed to fit the data at lower temperatures. The values of $\langle a^2 \rangle_o$ deduced from these data were usually 2 or more times larger than the value of 0.01 reported elsewhere (Leavens and Carbotte 1971, 1972, Blackford 1976). However, it should be realized that the values of $\langle a^2 \rangle$ in these thin Al films are very small: For the cleanest film measured, with ℓ=317nm, $\langle a^2 \rangle \approx 2 \times 10^{-4} \langle a^2 \rangle_o$ at $0.9T_c$. The remarkable sensitivity of τ_{Q*}^{-1} to small gap anisotropies reflects the weak electron-phonon interaction in Al. A similar degree of anisotropy in Sn would not have an observable effect on τ_{Q*}^{-1}.

For the cleanest Al films studied (not listed in Table II) the average value of τ_E was about 12ns. This value is in good agreement with that obtained from measurements of the 2Δ-phonon mean free path (Long 1973) and of gap relaxation (Schuller and Gray 1977), but about a factor of 4 shorter than the values obtained from recombination time measurements (Gray et al. 1969, Smith and Mochel 1976, Chi and Langenberg 1976) and from theoretical estimates (Kaplan et al. 1976, Lawrence and Meador 1978). The fact that the values of τ_E seem to fall into two groups, one clustered around 12ns and the other around 48ns is most peculiar, and at present remains an unsolved mystery. The dependence of τ_E^{-1} on T_c was also unexpected: One would expect τ_E^{-1} to be proportional to T_c^3 (Kaplan et al. 1976), but a much stronger dependence was observed, perhaps T_c^5. This observation could possibly be explained by an increase in $\alpha^2F(\omega)$ with increasing disorder (Bergmann 1971, Zavaritski 1969, Knorr and Barth 1970). Another suggestion made by CC is that oxygen atoms on the surface and/or at grain boundaries have dangling bands with magnetic moments due to the localized unpaired electrons. These magnetic moments could increase the charge relaxation rate through elastic spin-flip scattering (see Sec. VB). Since the density of magnetic moments is proportional to the oxygen doping level, and increases with T_c, this mechanism could explain the apparent rapid increase of τ_E^{-1} with T_c. In fact, in a quantitative analysis, CC showed that this hypothesis was consistent with the observations. At present, this matter remains unresolved. However, despite the questions remaining over the values of τ_E^{-1} and its dependence on T_c, the theory accounts satisfactority for the relaxations of Q* by elastic scattering in the presence of gap anisotropy.

B. Elastic Scattering from Magnetic Impurities

The addition of magnetic impurities to a superconductor destroys the degeneracy between time-reversed electron states through the exchange interaction between the conduction electrons and the impurities, and thereby gives the Cooper pairs a finite lifetime, the inverse of which is called the pair-breaking rate (Abrikosov and Gor'kov 1960). The pair breaking rate is the elastic spin-flip scattering rate, τ_S^{-1}, for electrons in the normal metal. A small concentration of impurities also smears out the peak in the BCS density of states over an energy range $\hbar/\tau_S$ (so that the energy gap and the order parameters are no longer equal), depresses the transition temperature by $\pi\hbar/4k_B\tau_S$, and alters the temperature dependence of the order parameter from the usual BCS form. In addition, the impurities can have a dramatic effect on the charge relaxation rate, as was first pointed out by Schmid and Schön (SS, 1975) and later studied by Pethick and Smith (1980), Entin-Wohlman and Orbach (1979), and Lemberger and Clarke (LC 1980). The essential reason can be understood by realizing that the coherence factor for elastic spin-flip scattering from one quasiparticle branch to the other is in an isotropic superconductor is not zero, but of the form

$$(uu'+vv')^2 = 4u^2v^2 = \Delta^2/E^2. \tag{5.9}$$

Since this factor approaches unity as $E\to\Delta$, we expect the spin-flip scattering to have an appreciable effect on τ_{Q*}^{-1} when $\tau_S^{-1} \gtrsim \tau_E^{-1}$.

SS found

$$(F^*\tau_{Q*})^{-1} = \frac{\pi\Delta}{4k_BT_c\tau_E}\left(1 + \frac{2\tau_E}{\tau_S}\right)^{\frac{1}{2}}. \quad (\tau_{Q*}^{-1} << \tau_E^{-1}) \tag{5.10}$$

A factor $(1 + \hbar^2\Gamma/\Delta^2\tau_E)^{\frac{1}{2}}$ has been omitted in Eq. (5.10), where $\Gamma = (2\tau_E)^{-1} + \tau_s^{-1}$, since it is very close to unity for all values of τ_E, τ_S, and Δ used in the experiments to be described later. This factor accounts for the effect of the smearing in the density of states on τ_{Q*}^{-1}. We will now interpret Eq. (5.10) physically. First, consider the limit $\tau_S^{-1} << \tau_E^{-1}$ in which the spin-flip scattering is a weak perturbation on the inelastic scattering, and to a first approximation does not affect the quasiparticle distribution created by the inelastic processes. Equation (5.10) can be expanded to give

$$\frac{1}{F^*\tau_{Q*}} = \frac{\pi\Delta}{4k_BT_c}\left(\frac{1}{\tau_E} + \frac{1}{\tau_S}\right). \quad (\tau_S^{-1} << \tau_E^{-1},\ \tau_{Q*}^{-1} << \tau_E^{-1}) \qquad (5.11)$$

The spin-flip term, $\pi\Delta/4k_BT_c\tau_S$, has this form because the coherence factor is substantial only for quasiparticles in the range from Δ to $\sim 2\Delta$, so that only a fraction $\sim\Delta/k_BT_c$ of the excess quasiparticles can relax through this channel.

In the limit $\tau_S^{-1} >> \tau_E^{-1}$, Eq. (5.10) reduces to

$$\frac{1}{F^*\tau_{Q*}} = \frac{\pi\Delta}{4k_BT_c}\left(\frac{2}{\tau_E\tau_S}\right)^{\frac{1}{2}}. \quad (\tau_S^{-1} >> \tau_E^{-1},\ \tau_{Q*}^{-1} << \tau_E^{-1}) \qquad (5.12)$$

In this limit, the spin-flip scattering modifies the quasiparticle distribution substantially because the lower energy excess quasiparticles undergo spin-flip scattering to the other branch more rapidly than higher energy quasiparticles can cool to replace them. As a result, the energy below which spin-flip charge relaxation is important is increased from $\sim 2\Delta$ to an energy E^*. We estimate this energy by equating the cooling rate, $\sim\tau_E^{-1}$, with the spin-flip branch crossing rate, $\sim\Delta^2/E^{*2}\tau_S$, to find $E^* \sim \Delta(\tau_E/\tau_S)^{\frac{1}{2}} >> \Delta$. Thus, at temperatures near T_c, of the quasiparticles scattered downwards by the cooling process, a fraction $\sim E^*/k_BT_c$ contributes significantly to charge relaxation, so that the rate is $\sim E^*/k_BT_c\tau_E \sim (\Delta/k_BT_c)(\tau_E\tau_S)^{-\frac{1}{2}}$, in essential agreement with Eq. (5.12). We emphasize that τ_E^{-1} enters the result not because it contributes to the charge relaxation *per se*, but because it determines the rate at which quasiparticles scatter downwards into the region from which they spin-flip scatter to the other branch.

LC extended CC's computer solution of the Boltzmann equation to include the spin-flip term. The purpose of this investigation was both to check the validity of Eq. (5.10) and to extend the theory to lower temperatures. The spin-flip scattering rate for a quasiparticle at energy ξ is

$$G_{sf\xi} = \frac{1}{\tau_S}\,\frac{\Delta^2}{E^2}\,\frac{E}{|\xi|}\,(f_\xi - f_{-\xi}). \tag{5.13}$$

The term τ_S^{-1} is just the elastic spin-flip scattering rate, Δ^2/E^2 is the coherence factor [Eq. (5.9)], $E/|\xi|$ is the final density of states, and $(f_\xi - f_{-\xi})$ is the usual occupation factor. LC added this term to the Boltzmann equation, Eq. (3.4), deleted the term $G_{el\xi}$, since this is negligible compared with $G_{sf\xi}$ for magnetic impurity concentrations of experimental interest, and solved the equation on a computer. Figure 18 shows the distribution for several values of τ_E/τ_S. As expected $(\delta f_\xi - \delta f_{-\xi})$ decreases at all energies as τ_E/τ_S increases. Furthermore, $(\delta f_\xi - \delta f_{-\xi})$ decreases more at low energies than at high energies, and the peak therefore moves to higher

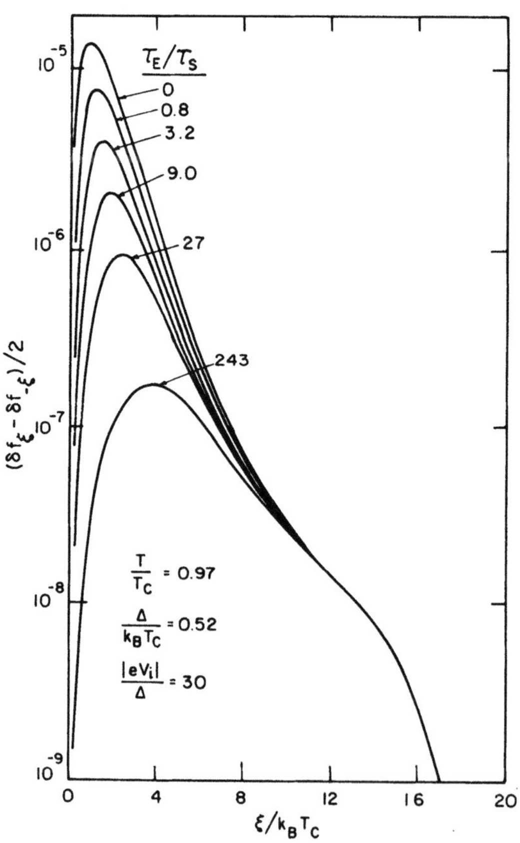

Fig. 18. $(\delta f_\xi - \delta f_{-\xi})/2$ vs. E/Δ for several values of τ_E/τ_S with $R_o = 200 s^{-1}$ and $\tau_E^{-1} = 10^8 s^{-1}$.

energies, because the spin-flip collision operator [Eq. (5.3)] is proportional to $\Delta^2/E|\xi|$. This is a graphic demonstration of the picture in which the low-lying excitations are rapidly relaxed by spin-flip scattering, thus depleting the charge-imbalance at low energies because of the relatively slow rate at which high energy injected quasiparticles can be cooled into this energy range by phonon scattering.

Figure 19 shows $(F^*\tau_{Q^*})^{-1}$, normalized to the SS factor $(\pi/4\tau_E)(1 + 2\tau_E/\tau_S)^{\frac{1}{2}}$ vs. Δ/k_BT_c for several values of τ_E/τ_S. For $\Delta/k_BT<1$ the computed values of $(F^*\tau_{Q^*})^{-1}$ lie above the SS result (except at $\Delta/k_BT_c=0$), by an amount that increases with τ_E/τ_S. This discrepancy persists to $\tau_E/\tau_S=0$, as is also shown in Fig. 8. Furthermore, for all values of Δ/k_BT_c, the calculated values of $[(\pi F^*\tau_{Q^*}/4\tau_E)(1 + 2\tau_E/\tau_S)^{\frac{1}{2}}]^{-1}$ increase monotonically with τ_E/τ_S.

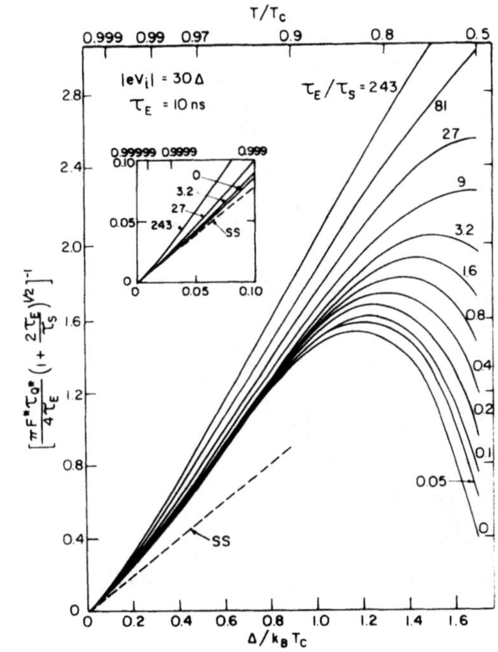

Fig. 19. $(F^*\tau_{Q^*})^{-1}$ normalized to $(\pi/4\tau_E)(1 + 2\tau_E/\tau_S)^{1/2}$ vs. Δ/k_BT_c. The curves approach the origin with unity slope. The dashed line is an extrapolation of the SS theory to low temperatures. The inset shows the region near the origin.

LC measured τ_{Q*}^{-1} in Al films doped with Er, a magnetic impurity. The samples had the configuration shown in Fig. 16, except in that the edges of both junctions were masked with SiO. The AlEr films were deposited by evaporating small pellets to completion. Figure 20 is a plot of $(F^*\tau_{Q*})^{-1}$ (obtained for $V_i >> \Delta/e$) vs. Δ/k_BT_c for three samples with different impurity concentrations. The relevant parameters are listed in Table III, with $\tau_E = 12\text{ns}(1.2/T_c)^3$ [Clarke and Chi 1979, Eq. (1.11)], and the spin-flip rate estimated from the data of Craven et al. (1971), $\tau_S^{-1} = (1.9\pm0.4) \times 10^{13} n_{Er} s^{-1}$, where n_{Er} is the concentration of Er atoms. For each sample plotted, $(F^*\tau_{Q*})^{-1}$ is linear in Δ/k_BT_c up to a value of about 0.8 ($T\approx 0.92T_c$), tending to flatten off as Δ/k_BT_c increases further. The initial linear behavior is consistent with the SS result Eq. (5.10) provided one extrapolates the SS result to substantially higher values of of Δ/k_BT_c than is justified *a priori*. The linear regions were used

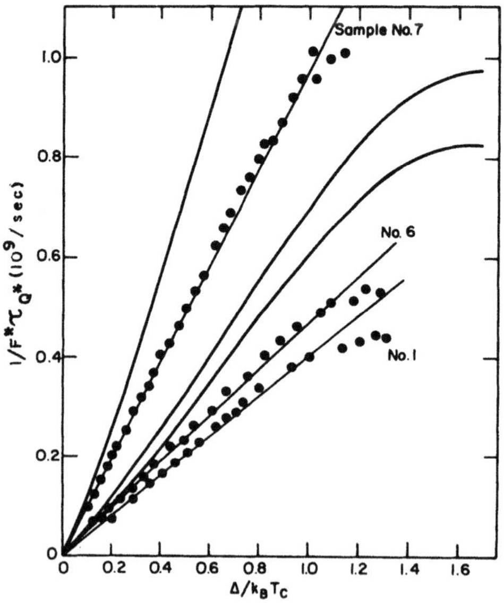

Fig. 20. Experimental values of $(F^*\tau_{Q*})^{-1}$ vs. Δ/k_BT_c for samples with Er concentrations of 21(#1), 81(#6), and 220(#7) at. ppm, with straight lines drawn through the data by eye. The other curves, which represent computer solutions to the Boltzmann equation, have the same slope as the data in the limit $\Delta/k_BT_c \to 0$.

Table III. Measured and Calculated Parameters for AlEr Films[a]

Sample	Er conc (ppm)	T_c (K)	ℓ (nm)	τ_E (ns)	τ_S (est.) (ps)	τ_S (meas.) (ps)
1	21	1.338	38	8.6	2500	1100
6	81	1.350	32	8.3	650	860
7	220	1.410	22	7.4	240	210

a. ℓ is obtained from $\rho\ell = 9 \times 10^{-16}\,\Omega m^2$ (Fickett 1971), and $N(0) = 1.74 \times 10^{28}\,eV^{-1}m^{-3}$ (Gschneider 1964). The value of τ_S(est.) is taken from Craven et al. (1971).

to obtain the values of τ_S (measured) listed in Table III. These values are all within a factor of 2 or better of the estimated values. As a further test of the applicability of the SS theory, in Fig. 21 we plot the measured values of the slope S of the linear region for all samples vs. the SS expression $(\pi/4\tau_E)(1 + 2\tau_E/\tau_S)^{\frac{1}{2}}$, using the values of τ_E and τ_S (estimated) from Table III. The general agreement is good, implying that the SS result accurately describes the observed dependence of $1/F^*\tau_{Q^*}$ on Δ/k_BT_c, and τ_E/τ_S for $\Delta/k_BT_c \lesssim 0.8$.

Figure 20 also shows curves computed from the Boltzmann equation using the values at τ_E and τ_S (measured) from Table III. It is immediately clear that, rather than providing a better fit to the data than the extrapolated SS result, the computed curves give a much worse fit. LC tried various other fitting procedures, but were not able to produce an acceptable fit. The discrepancy between the computed and experimental curves is further emphasized in Fig. 22, where $(SF^*\tau_{Q^*})^{-1}$ is plotted vs. Δ/k_BT_c for one sample of each Er concentration. Although there is a spread in the results for the larger values of Δ/k_BT_c, it is apparent that the data, normalized in this way, lie on a universal curve. This is in complete contrast to the computed curves shown in Fig. 19, in which the same quantity increases markedly with τ_E/τ_S at a given temperature.

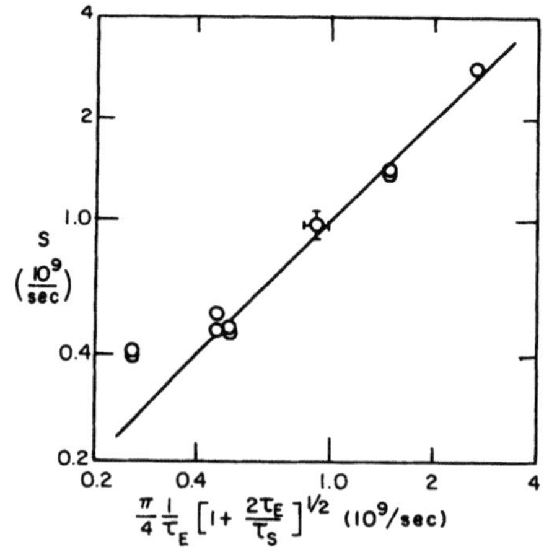

Fig. 21. S (slope of $(F^*\tau_{Q^*})^{-1}$ vs. Δ/k_BT_c for small Δ/k_BT_c) vs. $(\pi/4\tau_E)(1 + 2\tau_E/\tau_S)^{1/2}$ for all samples. The solid line through the origin has unity slope.

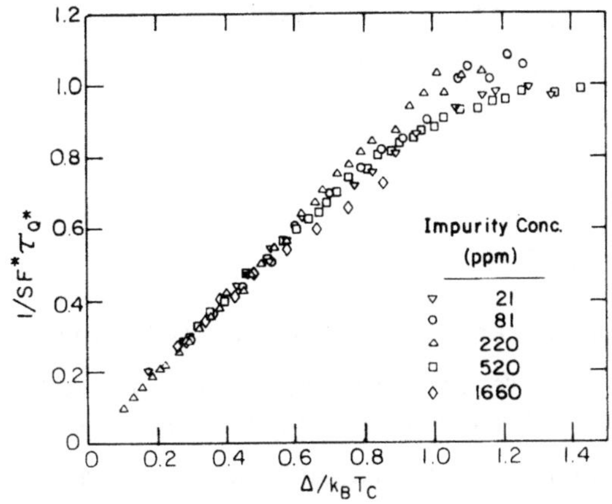

Fig. 22. $(SF^*\tau_{Q^*})^{-1}$ vs. Δ/k_BT_c for one representative sample of each impurity concentration.

This marked discrepancy between the data and the computed curves is extremely puzzling. It cannot be explained by invoking an additional charge-relaxation mechanism in the experiments, for example, Andreév (1964) reflection at a non-uniform gap or at the surface of the film, because the experimental rate $(F^*\tau_{Q^*})^{-1}$ lies substantially below, rather than above the computed rate. One can also rule out the possibility of impurity-impurity interactions, and of Kondo anomalies. It appears that either the collision operator for spin-flip scattering, Eq. (5.13), is inappropriate for charge relaxation or that the Boltzmann equation is not an adequate description of charge relaxation in the presence of magnetic impurities. On the other hand, the Schmid-Schön result, Eq. (5.10), provides a very satisfactory fit to the data over a much wider temperature range than can reasonably be expected.

C. Elastic Scattering in the Presence of a Supercurrent

When a supercurrent flows in a superconductor, the quasiparticle energies are raised by an amount $\vec{p}_k \cdot \vec{v}_s$, where $\vec{p}_k$ is the momentum of a quasiparticle in state $\vec{k}$, and $\vec{v}_s$ is the superfluid velocity (Aronov 1974, Galperin et al. 1974). Thus, quasiparticles at the Fermi energy where $\vec{v}_s$ and $\vec{p}_k$ are in the same direction are raised by p_Fv_s, while those where $\vec{v}_s$ and $\vec{p}_k$ are in opposite directions are lowered by p_Fv_S. This current-induced anisotropy allows elastic scattering between $\vec{k}_<$ and $\vec{k}_>$ states in much the same way as an intrinsic gap anisotropy (Fig. 13). Maki (1969) has shown that an appropriate electron relaxation rate for thin films $[d \ll (\ell\xi_o)^{1/2}]$ in the dirty limit ($\ell \ll \xi_o$) is

$$\frac{1}{\tau_S} = \frac{\ell v_F}{6}\left(\frac{p_S}{\hbar}\right)^2, \tag{5.14}$$

where d is the film thickness, $\xi_o = \hbar v_F/\pi\Delta(0)$, and $\vec{p}_S$ is the momentum of a Cooper pair. We can use Eq. (5.14) in Eq. (5.10) or (5.13) to predict the charge relaxation rate as a function of p_S. For the case of magnetic impurities LC found that at a given temperature $(F^*\tau_{Q^*})^{-1}$ was proportional to $(1 + 2\tau_E/\tau_S)^{1/2}$, as predicted by SS. Thus, we can immediately write down an expression for the detector voltage at a given temperature and injection current in the presence of a supercurrent, I_S:

$$V_d(I_S) = \frac{V_d(0)}{(1 + 2\tau_E/\tau_S)^{1/2}} = \frac{V_d(0)}{(1 + b^{SS}I_S^2)^{1/2}}, \tag{5.15}$$

where from Eq. (5.14),

$$b^{SS} = 2\tau_E/\tau_S I_S^2 = \tau_E \ell v_F p_S^2/3\hbar^2 I_S^2. \tag{5.16}$$

For uniform currents much less than the critical current we can use the relations (Tinkham 1975) $\vec{p}_S = 2m\vec{v}_S$, $\vec{j}_S = n_S e v_S$, $n_S = mc^2/4\pi e^2\lambda^2$ and, in the dirty limit, $\lambda(0) \approx \lambda_L(0)(\xi_o/\ell)^{1/2}$ to find

$$b^{SS} = \left(\frac{8\pi e}{\hbar c^2}\right)^2 \frac{\tau_E v_F \xi_o^2 \lambda_L^4(0)}{3\ell d^2 w^2} \frac{\lambda^4(T/T_c)}{\lambda^4(0)}. \tag{5.17}$$

Here, w is the film width, $\lambda_L(0)$ is the London penetration depth, and $\lambda(T/T_c)/\lambda(0)$ is a well-known function (Tinkham 1975). Equations (5.15) and (5.17) thus represent the SS prediction for the effect of a supercurrent on the value of Q* at fixed temperature and injection current.

Lemberger and Clarke (1980a) solved the Boltzmann equation in the presence of a supercurrent by replacing $1/\tau_S$ in Eq. (5.13) with Eq. (5.14) to obtain

$$G_{S\xi} = \frac{\ell v_F}{6}\left(\frac{p_S}{\hbar}\right)^2 \frac{\Delta^2}{E|\xi|}(f_\xi - f_{-\xi}). \tag{5.18}$$

With this substitution, all of the results computed in Sec. VB can be carried over to the present case. Thus, for fixed temperature and injection voltage,

$$V_d(I_S) = \frac{V_d(0)}{(1 + b^{num} I_S^2)^{1/2}}, \tag{5.19}$$

where $b^{num}(T)$ is the numerical value produced by the computer. In general, $b^{num}(T)$ will differ from $b^{SS}(T)$, as is evident in Sec. VB, and we can compare both values with the measured value, $b^{meas}(T)$.

Lemberger and Clarke (1980a) performed experiments using the usual geometry with a supercurrent introduced along the Al film in which the charge imbalance is generated. To increase the uniformity of the current distribution, a Nb groundplane was sputtered onto the substrate and covered with an insulating layer before the sample was deposited. A representative plot of $V_d(I_S)$ vs. I_S is shown in Fig. 23. As predicted by Eqs. (5.15) or (5.19), the value of $|V_d|$ decreased quadratically with increasing $|I_S|$ at low supercurrents, becoming linear in I_S at higher supercurrents. The slight asymmetry in the curves about $I_S=0$ was due to the non-negligible value of I_i. After shifting the origin appropriately, curves of the form $(1 + bI_S^2)^{1/2}$ were fitted; the quality of the fit is excellent. Experimentally determined values of b are shown in Fig. 24 for two samples with parameters listed in Table IV. The fitted curves b^{SS} and b^{num} are also shown. The temperature dependence of b^{SS} is in excellent agreement with the data, while b^{num} is in substantial disagreement. Furthermore, the ratio $\langle b^{SS}(T)/b^{meas}(T)\rangle$ varies between $\frac{1}{3}$ and $\frac{1}{2}$, an agreement which, given the uncertainties in the values of v_F, ℓ, $\lambda_L(0)$, and τ_E, is considered to be quite acceptable.

We therefore draw the same conclusions as for the case of magnetic impurities: The Schmid-Schön theory fits the measured data very accurately, even at values of Δ/k_BT_c much larger than one could reasonably expect, while the computed solution to the Boltzmann equation does not fit the data. Thus, the problems with the Boltzmann equation approach to charge relaxation appear to apply to pair breaking mechanisms in general.

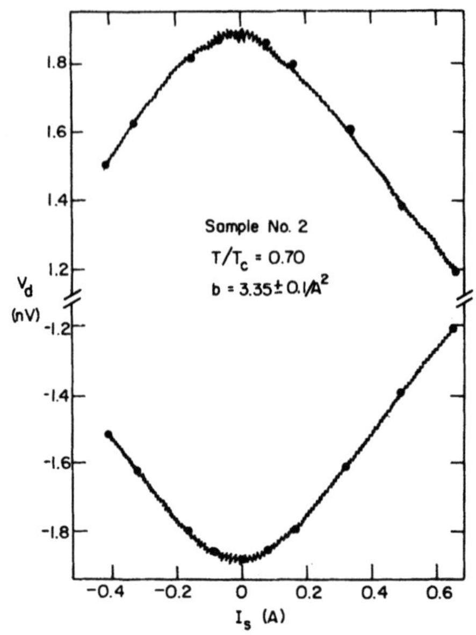

Fig. 23. Typical experimental plot of V_d vs. I_s for fixed I_i. The points are a fit to a function of the form $V_d(I_s) = V_d(0)/(1 + b\ I_s^2)^{1/2}$, with b as the fitting parameter.

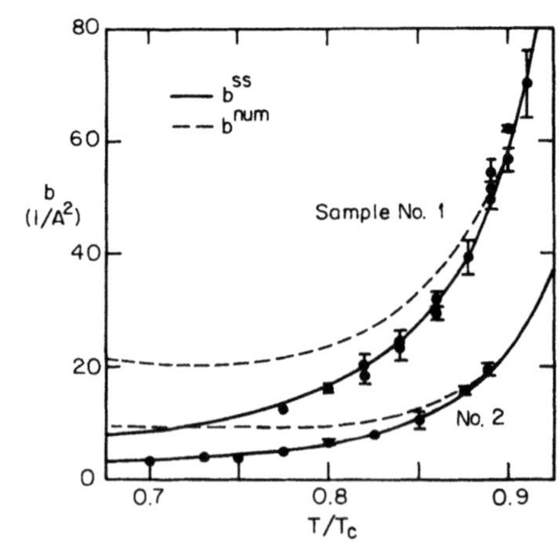

Fig. 24. Data points are measured values of b, solid and dashed lines are fits of b^{SS} and b^{num} to the data.

Table IV. Measured and Calculated Parameters for Samples 1 and 2[a]

Sample	T_c (K)	$\frac{R_{300}}{R_{4.2}}$	$\rho_{4.2}$ (nΩm)	ℓ (nm)	τ_E (ns)	$\left\langle \frac{b^{SS}}{b^{meas}} \right\rangle$
1	1.370	1.89	35.6	25	5.3	0.36
2	1.514	1.43	81.3	11	2.7	0.51

a. $R_{300}/R_{4.2}$ is the residual resistivity, $\rho_{4.2}$ is the resistivity of the Al film at 4.2K, ℓ is obtained from $\rho_{4.2}\ell = 9\text{x}10^{-16}\,\Omega\text{m}^2$ (Fickett 1971). The values $N(0) = 1.74\text{x}10^{28}\,\text{eV}^{-1}\text{m}^{-3}$ (Gschneidner 1964), $v_F = 1.36 \times 10^6\,\text{ms}^{-1}$ (Kittel 1976), and $\lambda_L(0) = 160$ nm (Meservey and Schwartz 1969) were used.

D. Elastic Scattering in the Presence of a Magnetic Field

An applied magnetic field is also a pair-breaking mechanism, and therefore creates a relaxation rate τ_S^{-1}. The exact form of this rate depends, for example, on whether the superconductor is a bulk sample or a thin film, and, in the latter case, whether the field is parallel or perpendicular to the plane of the film. Tinkham (1975) has given a useful summary of the results for τ_S^{-1}. The first demonstration of the essential correctness of the SS result Eq. (5.10), in the presence of a pair-breaking mechanism was by Kadin et al. in their study of phase slip centers. Subsequently, Hsiang (1980) measured the resistance of the SN interface in the presence of an applied magnetic field, and also found reasonable agreement with Eq. (5.10). For convenience, we will defer our discussion of these two experiments to Sec. VI.

VI. OTHER CHARGE IMBALANCE PHENOMENA

The final series of topics I am going to discuss concerns situations in which charge imbalance is generated by processes other than tunnel injection: The resistance of the normal metal-superconductor interface, phase slip centers, and the flow of a supercurrent in the presence of a temperature gradient.

A. The Resistance of the Normal Metal-Superconductor Interface

Pippard et al. (1971) measured the electrical resistance of superconductor-normal-metal-superconductor (SNS) sandwiches in which the normal metal was too thick and/or too dirty to sustain a Josephson supercurrent. They observed that, near the transition temperature of the superconductor, T_c, the resistance increased rapidly with increasing temperature. They ascribed this rise to the penetration of quasiparticles with energies greater than $\Delta_\infty(T)$ into the superconductor, where $\Delta_\infty(T)$ is the energy gap in S far from the interface. They also proposed that the additional boundary resistance was associated with a discontinuous jump in the electric potential at the NS interface, the electric field being zero throughout the superconductor. However, Yu and Mercereau (1972) showed that the potential did not fall abruptly to zero at the interface, but rather decayed exponentially in the superconductor. Subsequently, Harding et al. (1974) studied the resistance of SNS sandwiches in which the mean free path of the superconductor was shortened by alloying, and found an additional boundary resistance at low temperatures as well as a greatly enhanced rise in resistance near T_c.

The theory of the NS interface resistance has been widely investigated. The work of Rieger et al. (1971), who used a time-dependent Ginzburg-Landau theory, contained some essentially correct ideas, but did not produce the correct quasiparticle propagation length in the superconductor. Pippard et al. (1971) and Harding et al. (1974) used a Boltzmann equation approach that was later extended by Waldram (1975). The microscopic theory was developed by Schmid and Schön (1975), and has been extended by Ovchinnikov (1977, 1978), Artemenko and co-workers (1977, 1978) and Krähenbühl and Watts-Tobin (1978, 1979). Most recently, Hsiang and Clarke (1980) gave a simple description by adapting the tunnel injection theory that accounted quantitatively for their experimental data. In chapter 12, Pippard describes the Boltzmann equation approach for both the clean and dirty limits of the superconductor. I will confine myself to the clean limit ($\ell \gg \xi(T)$), and follow closely the work of Hsiang and Clarke (HC).

When a charge imbalance is created in one region of a superconductor, it relaxes over a length

$$\lambda_{Q^*} = (\ell v_F \tau_{Q^*}/3)^{\frac{1}{2}}, \qquad (6.1)$$

where we assume $\ell \ll v_F\tau_{Q^*}$. Very close to T_c $(\Delta \ll k_BT)$ almost all of the excitations incident from N propagate into S, so that, in the presence of an external current, a quasiparticle current flows in the superconductor. In the usual situation where the transverse dimensions of the interface are much larger than the London penetration depth, there is no net current in the interior of the superconductor. The internal quasiparticle current is cancelled by a pair current, with a corresponding flow of supercurrent on the surface. The electric field is continuous at the interface, and the electric field, the electric potential, Q*, and the quasiparticle current all decay exponentially into S with a characteristic length λ_{Q*} (see Fig. 25). The boundary resistance is of order $\lambda_{Q*}\rho_s/A$, where ρ_s is the normal-state resistivity of S, and A is the cross-sectional area of the interface. When the temperature is lowered somewhat, a substantial fraction of quasiparticles have energies $\lesssim\Delta_\infty(T)$, and are Andreév (1964) reflected at or near the NS interface. In this process, a $k_>(k_<)$ quasiparticle incident from N is scattered onto the $k_<(k_>)$ branch, and the current carried by these two excitations continues in the superconductor as a supercurrent. Thus, there is no boundary resistance associated with these quasiparticles, and there is a discontinuity in the electric field at the interface (Fig. 25). The potential is continuous at the interface, but its spatial derivative is not. In the presence of a current, the boundary scattering processes introduce disequilibrium in the quasiparticle distributions within an inelastic scattering length on either side of the interface.

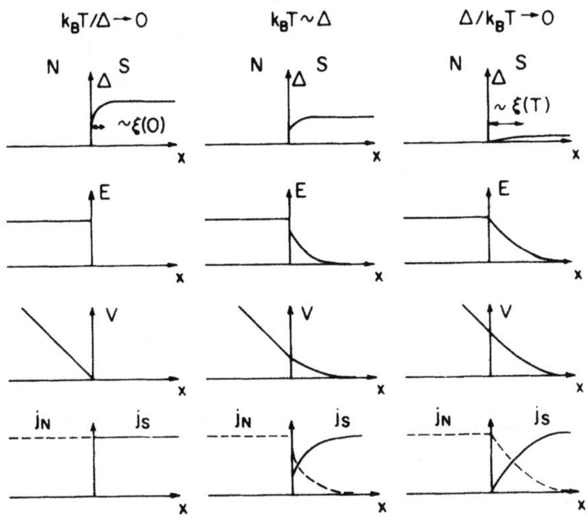

Fig. 25. Variation of energy gap, Δ, electric field, E, electric potential V, normal current, j_N, and supercurrent, j_S, across an NS interface for $k_BT/\Delta\to0$, $k_BT\sim\Delta$, and $\Delta/k_BT\to0$. The gap is taken to be zero in N, and the normal state properties are assumed to be the same in the two metals.

As the temperature is lowered still further ($\Delta_\infty >> k_BT$), essentially all of the quasiparticles are Andreév (1964) reflected at the interface, and there is no quasiparticle current in S. Correspondingly, the electric field and potential are zero in S (Fig. 25) and there is no boundary resistance. (In fact, the Andreév scattering process occurs over a distance $\sim\xi_0$, so that the discontinuities in the electric field and the derivative of the potential extend over this region. Furthermore, there will be a small boundary resistance of order $\xi\rho_S/A$ that is negligible in most practical situations.)

We now describe the simple model of HC that is valid in the limit $\Delta_\infty(T) << k_BT$. The model agrees well with the experimental data, and enables one to deduce values of τ_E. We assume that the transition temperature of the normal metal is much less than T, so that we can set $\Delta=0$ for $x<0$ (Fig. 25). In the superconductor, Δ rises from its value at the boundary, $\Delta_o(T)$, to its full value, $\Delta_\infty(T)$, over a distance of roughly the Ginzburg-Landau (1950) coherence length, $\xi(T)$, that is always much less than λ_Q^* in the temperature range investigated experimentally. However, we note that quasiparticles with energies greater than $\Delta_\infty(T)$ may undergo some charge relaxation in the region where Δ varies spatially. At least in the limit $\Delta<<k_BT$, this contribution to the overall relaxation rate is likely to be small, and we shall neglect it. We assume that the current densities are sufficiently low that they do not perturb Δ. We further assume that quasiparticles with energies greater than $\Delta_\infty(T)$ are transmitted into S with probability unity; this is a reasonable approximation because of the relatively slow change of Δ with x. Quasiparticles with energies $<\Delta_\infty(T)$ are Andreév (1964) reflected at a plane taken as $x = 0$ [since $\xi(T)<<\lambda_{Q^*}$]. Finally, we assume that the quasiparticles are close to thermal equilibrium even in the vicinity of the interface; we emphasize that this is a reasonable approximation only for $\Delta_\infty<<k_BT$.

The charge imbalance generated by the uniform injection of a current I_i into volume Ω of a superconductor in the limit $eV_i<<k_BT$ is given from Eqs. (2.24) to (2.26) by

$$Q^* = \frac{Z(T)}{Y(T)} \frac{I_i}{e\Omega} \tau_{Q^*} . \quad (eV_i<<k_BT) \tag{6.2}$$

In the case of the NS interface, I_i is just the quasiparticle current injected into S, and is related to the total current, I, by

$$Y(T) = I_i/I. \tag{6.3}$$

Equation (6.3) follows from the realization that in a SIN tunnel junction at low voltages a fraction $[1-Y(T)]$ of the current that flows at T_c cannot flow at a temperature $T < T_c$ because there are no states available in S at energies $<\Delta(T)$, whereas at the NS interface, in our approximation, this same fraction $[1-Y(T)]$ of the total current is transmitted into S as a pair current. Combining Eqs. (6.1) to (6.3), and replacing the exponentially decaying Q^* with a value that is constant at the value $Q^*(0)$ for $x \leqslant \lambda_{Q^*}$ and 0 for $x > \lambda_{Q^*}$, we find

$$Q^*(0) = \frac{Z(T) I \tau_{Q^*}}{eA\lambda_{Q^*}} . \tag{6.4}$$

The excess voltage, V_b, at each interface of the SNS sandwich adds to the voltage developed across the normal metal, and the total potential across the sandwich is measured with superconducting leads making metallic contact with the superconducting films. This is in contrast to the usual tunneling measurement of Q^*, where the potential is measured by a tunneling contact to a normal metal. Thus, we set $g_{NS}(0) = 1$ in Eq. (2.23), and combine the result with Eq. (6.4) to obtain

$$R_b = \frac{V_b}{I} = \frac{Z(T)\lambda_{Q^*}\rho_s}{A}. \qquad (\Delta \ll k_B T) \tag{6.5}$$

In Eq. (6.5) we have set $\tau_{Q^*} = 3\lambda_{Q^*}^2/\ell v_F$ and used the free electron model (Kittel 1976) to calculate $\rho_s = 3/2e^2 N(0)\ell v_F$. As $T \to T_c$, $Z(T) \to 1$, and the boundary resistance is just the resistance of a length λ_{Q^*} of the superconductor in the normal state. As the temperature is lowered, $Z(T)$ decreases, reflecting the fact that fewer quasiparticles are able to propagate into the superconductor. At low temperatures (Clarke et al. 1979) $Z(T) \approx (k_B T/\Delta)^{\frac{1}{2}} \exp(-\Delta/k_B T)$, so that R_b vanishes exponentially as $T \to 0$, as we expect. However, despite the fact that Eq. (6.5) is a good approximation both near T_c and in the limit $T \to 0$, we repeat our caution that it is not expected to be valid at intermediate temperatures where $\Delta_\infty \sim k_B T$.

If one measures R_b near T_c one can, at least in principle, deduce values of τ_{Q^*}. As was discussed in Secs. III and V, in general, both inelastic and elastic scattering contribute to

τ_{Q*}^{-1}. We can make a crude estimate of the elastic relaxation rate in the presence of gap anisotropy for the materials used in the experiments to be described by assuming $E \sim k_B T_c$ in Eq. (5.6). This leads us to the conclusion that inelastic scattering should dominate at temperatures above about $0.9T_c$. Since $\Delta/k_B T \simeq 1$ at $T/T_c = 0.9$, it appears that in the range of validity of Eq. (6.5), $\Delta/k_B T \ll 1$, the charge relaxation should be dominated by inelastic scattering, and we assume that $\tau_{Q*} = 4k_B T\tau_E/\pi\Delta_\infty(T)$.

Hsiang and Clarke (1980) measured the resistances of the SNS junctions listed in Table V. Each pair of materials chosen had a low mutual solubility and did not form intermetallic compounds. The PbBi - CuAl samples were made by evaporating the materials onto glass substrates; the CuAl was ~2μm thick, while the PbBi was ~20μm thick, considerably greater than λ_{Q*} over the experimental temperature range. In the remaining samples, the superconductor (up to 80 μm thick) was evaporated onto the two sides of Ir foils about 70 μm thick. The foils were cleaned by sputter etching in argon, the argon was pumped out of the system, and the superconducting material was evaporated onto each side of the foil. Two or three samples were connected in series and their resistances measured with a SQUID voltmeter. The samples were mounted in a vacuum can so that their temperature could be raised above 4.2K. The variation of resistance with temperature is shown in Fig. 26

Table V. Properties of SNS Junctions. The Measured Values of τ_E are Averaged Over 2 or 3 Samples of Each Type.

Superconductor/ normal metal	Type	Measured τ_E (10^{-10}s)	Calculated[a] τ_E (10^{-10}s)
$Pb_{0.99}Bi_{0.01}$- $Cu_{0.97}Al_{0.03}$	film	0.25	0.23(Pb)
Sn-Ir	foil	2.6	2.7
$Sn_{0.99}In_{0.01}$-Ir	foil	1.1	2.7(Sn)
In-Ir	foil	1.1	1.0

a. Kaplan et al. (1976)

for one representative sample of each type. Near T_c, the resistance rises rapidly with increasing temperature, while at low temperatures the resistance is nearly independent of temperature. To within the experimental accuracy, the low temperature resistance

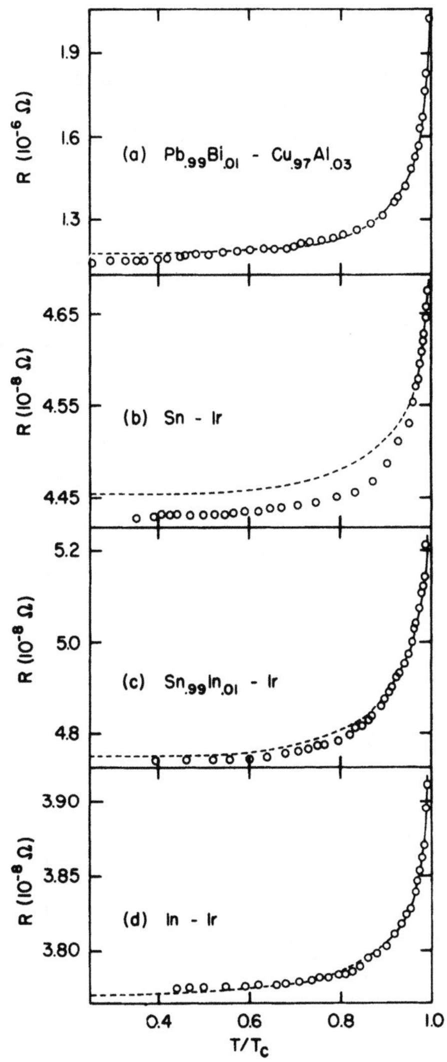

Fig. 26. Circles are measured total resistance vs. T/T_c for SNS junctions listed in Table V. The solid lines above 0.9 T_c [$0.96T_c$ in (b)] are the fit to Eq. (6.5), while the dashed lines show the extrapolation of the theory to lower temperatures.

was equal to the estimated resistance of the normal metal. In fact, an excess resistance equivalent to several mean free paths of the normal metal could not be detected because of the uncertainties in estimating the resistance of the normal layer.

To compare the data with Eq. (6.5), in Fig. 27 we plot the measured resistance vs. $Z(T)(k_BT/\Delta)^{\frac{1}{2}}$. The solid lines are a least squares fit to the data for $T>0.9T_c$ ($0.96T_c$ for Sn). The fit is good -- in fact, for (a), (c) and (d) it is surprising that such good agreement extends down to temperatures as low as $0.9\ T_c$ where $\Delta \sim k_BT$. The slope of the lines in Fig. 26 is $4(\ell v_F\tau_E/3\pi)^{\frac{1}{2}}\rho_s/A$, and

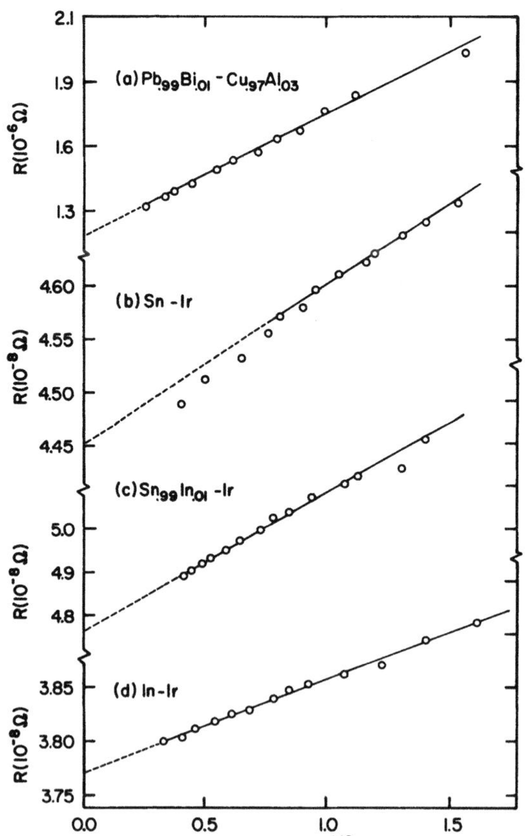

Fig. 27. Total measured resistance vs. $Z(T)(k_BT/\Delta)^{1/2}$ for the samples shown in Fig. 26. The solid lines are a least-squares fit to the data for $T>0.9T_c$ [$0.96T_c$ for (b)].

yields the average values of τ_E listed in Table V. The values of τ_E calculated by Kaplan et al. (1976) from $\alpha^2 F$ are also listed for comparison. The agreement is generally quite good. Using the fitted values of τ_E, we have plotted the resistance predicted by Eq. (6.5) in Fig. 26. Except for Sn, the extrapolated low-temperature resistance is in good agreement with the measured resistance, thus providing a good check on the consistency of our results. Particularly in the cases of PbBi and In, the fit is remarkably good even at intermediate temperatures, a result that is probably coincidental, particularly since the expression used for τ_{Q*} is quite inappropriate in this range. However, it may be that the increase in the elastic charge relaxation rate as the temperature is lowered tends to compensate for the decrease in the inelastic rate, thus keeping τ_{Q*} roughly constant at temperatures below about 0.9 T_c. One could, in principle, attempt to fit the data to more detailed theories, for example, that of Waldram (1975), at intermediate temperatures, but the difficulty of accounting for both elastic and inelastic scattering processes and their strong energy dependence makes this a formidable undertaking.

Hsiang (1980) studied the effect of a magnetic field applied parallel to the plane of SNS sandwiches in which the superconductor was $Pb_{0.98}Bi_{0.02}$, and the normal metal was Cd. The PbBi films are bulk, type-II superconductors, and the spin relaxation rate is given by (Maki 1969, de Gennes 1966, Tinkham 1975)

$$\frac{1}{\tau_s} = \frac{\ell v_F eH}{3c} \; . \tag{6.6}$$

Hsiang measured the boundary resistance as a function of applied magnetic field and temperature, and extracted values of τ_{Q*}^{-1} (T,H). Using Eq. (6.6) in Eq. (5.10) he was able to estimate τ_s^{-1} as a function of H. Although there was a good deal of spread in the results, he was able to establish reasonable agreement between the measured value of τ_s^{-1} and the value predicted by Eq. (6.6).

B. Phase-slip Centers

The second topic is the phase-slip center (PSC) in a superconducting filament (Skocpol et al. 1974). Since this topic is covered in detail by Skocpol in Chapter 18, we will discuss it only briefly, emphasizing those aspects involving charge imbalance.

PSC's manifest themselves as steps in the current-voltage characteristics of superconducting whiskers and bridges, usually rather close to T_c, as illustrated in Fig. 28(a). The model of the PSC is shown in Fig. 28(b). As the current, I, is increased eventually it exceeds the critical current at the weakest point in the microbridge. The order parameter in this region collapses to zero and the current is forced to flow as a normal current, thereby allowing the superfluid to build up again and to resume carrying the current. The cycle then repeats at a frequency 2eV/h, where V is the average voltage across the region. Each time the order parameter falls to zero, the phase difference between the two ends of the bridge slips by 2π: Hence the name "phase slip center." The width of the region over which the superfluid oscillates is roughly $2\xi(T)$. Now since the superfluid current oscillates between zero and its critical value, on the average approximately one-half of the current in this region flows as a supercurrent, and the other half as a normal current. One can then regard the PSC as an SNS junction. A fraction of the normal current (depending on temperature) that flows in the phase-slip region will propagate into the superconducting regions on either side, and decay into a supercurrent in a length λ_{Q^*}. Thus, from Eq. (6.5), the apparent resistance generated will be $\sim\frac{1}{2}Z(T)(2\lambda_{Q^*})\rho_s/A$.

Skocpol et al. (1974) made samples with many voltage leads. In the example shown in Fig. 28(a), they showed that the first four steps were produced by one PSC forming successively in each of the regions II, I, IV and III, while the fifth step was produced by a second PSC appearing in region IV. The slope, that is, the resistance, of each successive step increases by an approximately constant increment, each increment representing the resistance of one PSC. The only puzzling feature of this original work on PSC's was that the resistance did not appear to diverge as $(T_c-T)^{-1/4}$ as one would expect from the temperature dependence of τ_{Q^*}. However, Dolan and

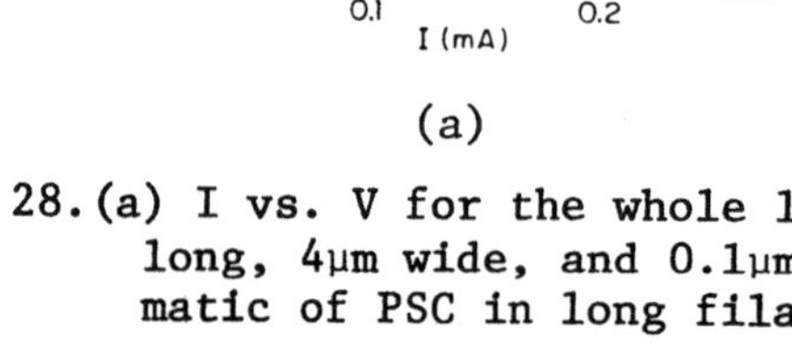

(a)

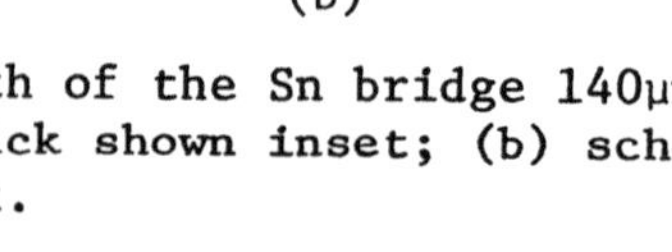

(b)

Fig. 28. (a) I vs. V for the whole length of the Sn bridge 140μm long, 4μm wide, and 0.1μm thick shown inset; (b) schematic of PSC in long filament.

Jackel (1977) resolved this difficulty in an elegant experiment in which they prepared a microbridge with a series of superconducting voltage leads along one side of the bridge and a series of normal voltage leads along the other. The probes were typically 2μm apart, and formed tunnel junctions with the microbridge. In this way, they were able to measure the average superfluid and normal potentials, $V_s = \bar{\mu}_s/e$ and $V_N = \bar{\mu}_N/e$, across a PSC, the position of which was defined by a small notch in the microbridge. Figure 29 shows their measured values of V_s and V_N. As expected, V_s changes abruptly, since $\xi(T)$ is less than the probe spacing, while V_N changes over a much greater length. Values of τ_{Q*} deduced from the spatial variation of V_N showed the predicted $(T_c-T)^{-\frac{1}{4}}$ temperature dependence, and produced the value $\tau_{Q*} = 6.3\times10^{9}\ (\Delta/k_BT_c)s^{-1}$, in excellent agreement with the results of Clarke and Paterson (1974) discussed in Sec. IV. Since this work on PSC was performed at $T \geq 0.98T_c$, inelastic scattering should completely dominate the charge relaxation. Thus, the model of the PSC proposed by Skocpol et al. is a complete and satisfying explanation of a phenomenon that had been a puzzle for many years.

More recently, Kadin et al. (1978) studied PSC's in the presence of a magnetic field, H, applied parallel to the Sn strip. By correcting for heating effects, they were able to obtain the expected $(T_c-T)^{-\frac{1}{4}}$ temperature dependence of the resistance near T_c They also found that the resistance associated with a PSC and hence the inferred value of τ_{Q*} was reduced by the magnetic field. The magnetic field dependence of τ_{Q*}^{-1} was consistent with the prediction of Schmid and Schön [Eq. (5.10)], with (Maki 1969, de Gennes 1966, Tinkham 1975)

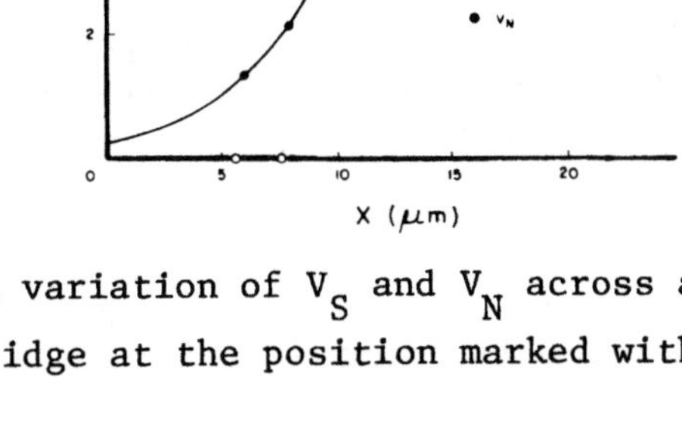

Fig. 29. Spatial variation of V_S and V_N across a PSC in a Sn microbridge at the position marked with the arrow.

$$\frac{1}{\tau_s} = \frac{\Delta(0,0)}{\hbar} \frac{H^2}{H_c^{\,2}||(0)} \quad . \qquad (6.7)$$

In Eq. (6.7), $\Delta(0,0)$ is the energy gap in zero field at zero temperature, and $H_c||(0)$ is the parallel critical field at zero temperature. This work was the first experimental demonstration of the essential correctness of the SS theory of charge relaxation in a magnetic field.

C. The Flow of Supercurrent in the Presence of a Temperature Gradient

Pethick and Smith (1979a) predicted that when a supercurrent, $\vec{I}$, flows in a superconductor along which there exists a thermal gradient, $\vec{\nabla}T$, a charge imbalance should be created that is proportional to $\vec{I}.\vec{\nabla}T$. Clarke et al. (1979a) observed this effect in Sn films, and established that the measured voltage was proportional to $\vec{I}.\vec{\nabla}T$, as predicted, but two to three orders of magnitude smaller than the predicted value. This discrepancy arose because the theory assumed that the elastic scattering rate was negligible compared with the inelastic rate, whereas the reverse was true in the experiment. Three further theories then appeared (Schmid and Schön 1979, Clarke and Tinkham 1979, and Beyer Nielsen et al. 1980) in attempts to account quantitatively for the temperature dependence and magnitude of the data. More recently, Heidel and Garland (1980) observed similar effects in Al films. I will begin this section by describing the experiment and the results of Clarke et al. (1979a), and then briefly compare the results with the theories. H. Smith and C. M. Falco will also discuss this topic in chapters 15 and 16.

The experimental configuration is shown in Fig. 30. The Sn (or Sn + 3wt.%In) film is typically 300nm thick and 0.1mm wide in the narrow region. After the Sn was oxidized, three Cu(+3%Al) disks $\sim$1μm thick were deposited, followed by three Pb strips. In a given experiment, one of the three Sn-SnOx-Cu tunnel junctions was used to measure the quasiparticle potential relative to the pair potential with a SQUID voltmeter. The substrate was mounted in a vacuum can, with a heater at each end to generate a temperature gradient. The temperature of the junction under investigation and the temperature gradient were estimated from two Allen-Bradley carbon resistors attached to the rear of the substrate.

In the measurements, the creation of a thermal gradient invariably produced a small voltage ($\lesssim$ 1pV). This offset, believed to arise from thermoelectric effects in non-superconducting components in the circuit, was eliminated by defining the voltage to be

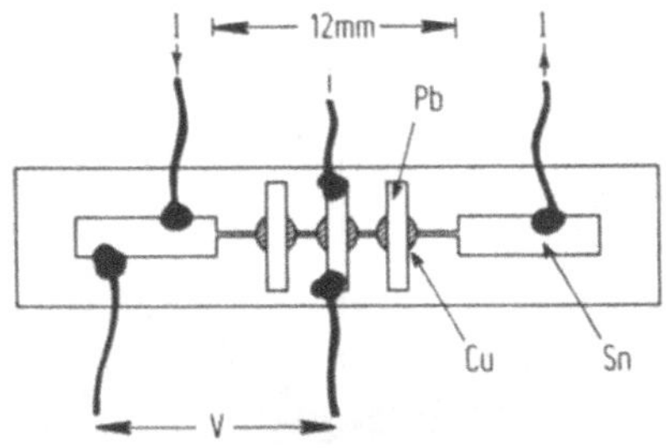

Fig. 30. Sample configuration.

zero when I=0. The current was then increased in steps, and the voltage measured at each step. In Figs. 31 and 32 we plot V vs. I for five values of ∇T, and V vs. ∇T for 10 values of I for a representative sample. Within the experimental accuracy, the voltage is bilinear in I and ∇T. In Fig. 33 we plot $Vg_{NS}(0)/I\nabla T$ vs. $t=T/T_c$. The charge imbalance per unit current and per unit temperature gradient diverges as $T/T_c \to 1$, and falls off steadily with decreasing temperature at low temperatures. To within the experimental resolution, the temperature dependence was the same for all 8 samples investigated. In Fig. 34 we plot $Vg_{NS}(0)/I\nabla T$ vs. (1-t) for the same sample as in Fig. 33; the divergence at temperatures above about $0.8T_c$ is close to 1/(1-t). In Table VI, we list the thickness and mean free path of the Sn or SnIn films and the quantity $Vg_{NS}(0)T(1-t) \times A/I\nabla T$ for four samples, two with $\ell > \xi_o$ and two with $\ell < \xi_o$. A is the

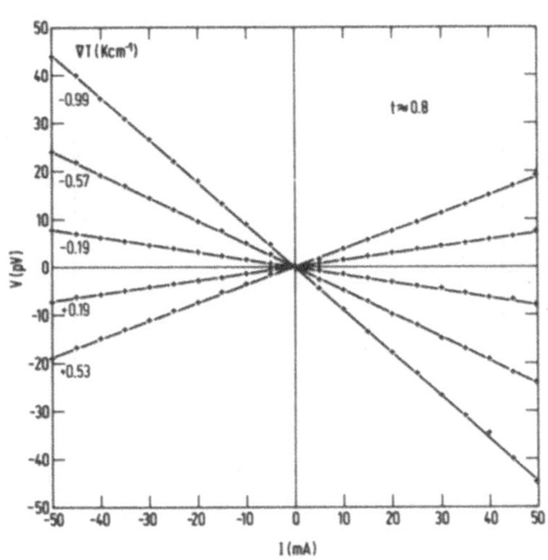

Fig. 31. V vs. I for 5 values of ∇T for sample 4.

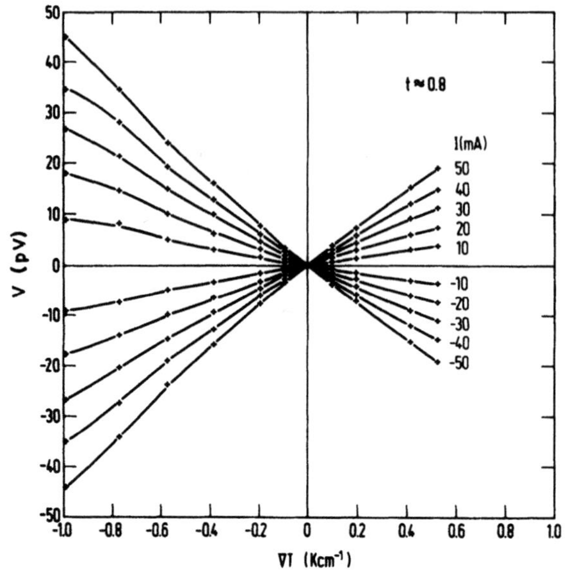

Fig. 32. V vs. ∇T for 10 values of I for sample 4. At each value of ∇T, the voltage is defined to be zero at I=0.

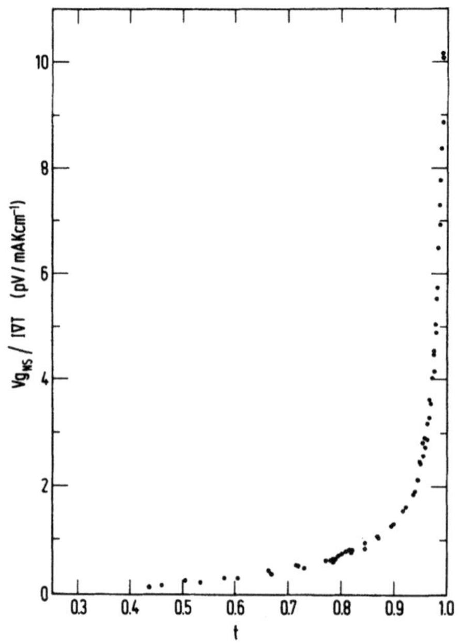

Fig. 33. $Vg_{NS}(0)/I\nabla T$ vs. reduced temperature, t, for sample 4.

cross-sectional area of the films. The results for the two clean samples agree well with each other, as do those for the two dirty samples.

The physical origin of the effect is indicated in Fig. 35. In Fig. 35(a), there is a thermal gradient, but no applied supercurrent. Quasiparticles moving from the left are at an effective temperature $T-\delta T$, while those from the right are at $T+\delta T$, where T is the local temperature. Thus, there is an imbalance in the populations of the $\vec{k}>\vec{k}_F$ and $\vec{k}<\vec{k}_F$ branches on the righthand side of the Fermi surface, but an equal and opposite imbalance on the lefthand side: As a

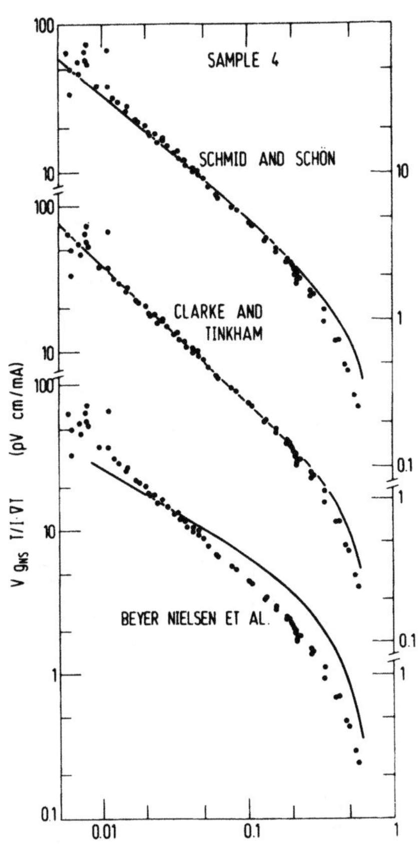

Fig. 34. $Vg_{NS}(0)/I\nabla T$ vs. (1-t) for sample 4. The three theoretical formulas have been fitted to the experimental data by scaling them appropriately.

Table VI. Properties of 4 Samples

Sample	Material	Thickness (nm)	ℓ (nm)	$Vg_{NS}(0)T(1-t)A/I\nabla T$ ($10^{-16}\Omega cm^3$)
4	Sn	320	428	1.2
6	Sn	430	294	1.0
7	Sn (+3% In)	190	61	0.8
8	Sn (+3% In)	190	61	0.8

result, Q*=0. If we now impose a superfluid velocity $\vec{v}_S$ [Fig. 35(b)] the excitation energies are raised and lowered on opposite sides of the Fermi surface by an amount $\vec{p}_k \cdot \vec{v}_S$ [see Sec. VC]. The induced asymmetry ensures that the population imbalances on opposite sides of the Fermi surface no longer cancel, and the resulting charge imbalance is the origin of the observed voltage.

Since the first paper by Pethick and Smith (1979a), there have been three separate approaches to the calculation of Q*. Schmid and Schön (1979) gave a result [Eq.(9)] that is valid for both $\ell > \xi_o$ and $\ell < \xi_o$ in the limit $\Delta/k_BT_c \to 0$. However, their formalism is actually valid at all temperatures, and, retroactively, one can readily modify

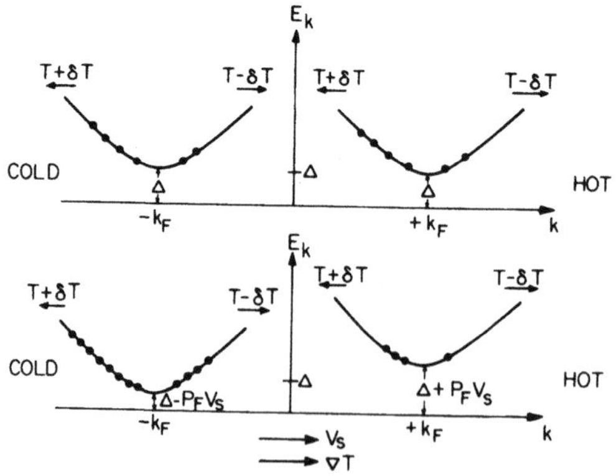

Fig. 35. Schematic representation of quasiparticle excitations in presence of (a) temperature gradient and (b) temperature gradient and applied supercurrent.

their result*. If one uses $\partial f^o/\partial E = -1/4\, k_BT ch^2(\Delta/2k_BT)$ in Eq. (7) in their paper, and inserts the result into Eq. (6), one finds

$$V = \frac{p_F \ell}{6eg_{NS}(0)} \frac{\vec{v}_s \cdot \vec{\nabla} T}{T} \frac{\Delta}{k_BT(1-Z)ch^2(\Delta/2k_BT)} \ln(8\Delta\tau_E/\hbar). \quad (6.8)$$

Here, p_F is the Fermi momentum, and Z is given by Eq. (2.27). Clarke and Tinkham (1980) used a simple kinetic approach to obtain

$$V = \frac{p_F \ell}{6eg_{NS}(0)} \frac{\vec{v}_s \cdot \vec{\nabla} T}{T} \frac{\Delta}{k_BT(1-Z)ch^2(\Delta/2k_BT)} \quad (6.9)$$

Equation (6.8) exceeds Eq. (6.9) by a factor of $\ln(8\Delta\tau_E/\hbar)$, which varies relatively slowly with temperature over the experimental range studied, with an average value of about 6. This factor is somewhat model dependent. Thus, apart from this factor, the Schmid-Schön and Clarke-Tinkham approaches yield essentially the same result. One can convert v_s into a current density by writing $v_s = j_s/n_se = \mu_o j_s \lambda^2(T)e/m$, and using the result $\lambda^2(T) = \lambda_L^2(0) \times (1 + \xi_o/\ell)(1-t^4)^{-1}$ (Tinkham 1979) to find

$$v_s = \frac{\mu_o j_s e\, \lambda_L^2(0)(1 + \xi_o/\ell)}{m(1-t^4)} \quad (6.10)$$

Near T_c, $1-Z \to \pi\Delta/4k_BT$, $ch^2(\Delta/2k_BT) \to 1$, and $1-t^4 \to 1-t$. Thus, near T_c, the temperature dependence of the voltage at fixed current is dominated by the temperature dependence of v_s, $(1-t)^{-1}$, in good agreement with the experimental results. The curves calculated for the two theories with $v_F = 6.5\times10^5 ms^{-1}$, $\lambda_L(0) = 5\times10^{-8}$m, and $\xi_o = 2.3\times 10^{-7}$m are plotted for sample 4 in Fig. 34. The curves were fitted at $T/T_c = 0.99$ by multiplying Eq. (6.9) by 2.4, and Eq. (6.8) by 0.4. It is evident that both theories predict the observed temperature dependence rather accurately.

Beyer Nielson et al. (1980) and, subsequently, Pethick and Smith (1980) have solved the Boltzmann equation for the case where $\tau_E^{-1} \to 0$ and for an isotropic energy gap to obtain

---

*G. Schön, private communication

$$V = \frac{p_F \ell}{e g_{NS}(0)} \frac{\vec{v}_s \cdot \vec{\nabla} T}{T} \left[\frac{1.93\Delta/k_B T}{ch^2(\Delta/2k_B T)} + \frac{8/15}{\exp(\Delta/k_B T) + 1} \right]. \quad (6.11)$$

Equation (6.11) is also plotted in Fig. 34, and evidently has a different temperature dependence than the experimental data. In Eq. (6.11), the first term in square brackets dominates the second for $T/T_c \lesssim 0.998$, that is, throughout the experimentally realizable range. The first term arises from quasiparticles in the energy range $\Delta - p_F v_s$ to $\Delta + p_F v_s$ in which the available phase space for charge relaxation via elastic scattering is reduced. This effect is particularly pronounced for quasiparticles with energies near $\Delta - p_F v_s$. However, Schön (1980) has recently suggested that elastic scattering may smear out the energy levels sufficiently to eliminate reduced relaxation rates at low energies, provided that $(p_F v_s/\Delta)^{1/2} << (\hbar/\tau_1)/\Delta$, a restriction that is strongly satisfied experimentally. If this suggestion is correct, the first term in square brackets in Eq. (6.11) will be greatly reduced, and may have a different temperature dependence. The second term is strictly valid only near T_c, and needs to be recalculated to be valid over a wide temperature range.

At the time of writing, these problems are not resolved, but one might reasonably hope for clarification in the very near future. However, it seems clear that an expression of the form of Eq. (6.8) or (6.9) is essentially correct in view of the excellent fit to the data on Sn, as well as to the data on Al obtained by Heidel and Garland.

VII. CONCLUDING SUMMARY

In this chapter I have attempted to develop the main ideas of charge imbalance in superconductors, and its application to various physical situations. We began by introducing the charge imbalance per unit volume, $Q^* = (2/\Omega)\sum_k q_k f_k$, where $q_k = \xi_k/E_k$ is the effective quasiparticle charge. Charge imbalance is tunnel injected at the rate $\dot{Q}_i^* = F^* I_i/e\Omega$, where F^* is a known function of V_i and T. The steady state charge imbalance, $Q^* = \dot{Q}_i^* \tau_{Q^*}$, where $\tau_{Q^*}^{-1}$ is the charge relaxation rate, produces a voltage $V = Q^*/2N(0)e\, g_{NS}(0)$ measured at zero detector current by a tunneling contact to a normal metal. In the absence of magnetic scattering or gap anisotropy, charge relaxation occurs via inelastic scattering at the rate $(F^* \tau_{Q^*})^{-1} = \pi\Delta(T)/4k_B T_c$ in the limit $\Delta/k_B T \to 0$. All of these results may be taken as exact.

Elastic scattering can relax charge imbalance under several different circumstances. If the gap anisotropy is non-zero, the coherence factor $(uu'-vv')^2$ for branch crossing is no longer zero. A computer solution of the appropriate Boltzmann equation yields results in good agreement with experiments on Al, using the normalized mean-square gap anisotropy as a fitting parameter. The presence of magnetic impurities can greatly enhance τ_{Q*}^{-1} because the coherence factor for branch crossing $(uu'+vv')^2$, approaches unity at low energies. Experimental data obtained on Al doped with Er is in excellent agreement with the Schmid-Schön prediction $(F^*\tau_{Q*}^{-1}) = [\pi\Delta(T)/4k_BT_c\tau_E](1 + 2\tau_E/\tau_S)^{1/2}$ over a wide range of values of τ_E/τ_S for $\Delta/k_BT_c \lesssim 0.8$. This is a very surprising result, since the theory is valid only for $\Delta/k_BT_c \ll 1$. On the other hand, however, a computer solution of the Boltzmann equation with the appropriate magnetic scattering term yields results that differ strikingly from the data. Charge relaxation can also be induced by a supercurrent in the film: The current induces an asymmetry across the Fermi surface that again allows branch relaxation via elastic scattering. The results from experiments on dirty Al are again in excellent agreement with the Schmid-Schön result, but in disagreement with the results of the computer calculation. Thus, charge relaxation in the presence of a pair breaking mechanism is in urgent need of attention, both because the Boltzmann equation produces results that are in distinct disagreement with the data, and because the Schmid-Schön theory is valid over a much wider temperature range than can reasonably be expected.

Finally, we discussed three other types of experiments involvcharge imbalance. The first was the resistance of the NS interface in the clean limit. A simple modification of the tunneling theory for the injection of Q* yields a result that is in excellent agreement with experiments on Pb, Sn, and In, and that produces very reasonable values of τ_E. The next example was the phase-slip center in a narrow superconducting strip. The magnitude of the resistance increment generated by each PSC is well explained by the theory of the NS interface. Finally, a charge imbalance is generated by the flow of a supercurrent in the presence of a temperature gradient. Theories by Schmid and Schön (1979) and by Clarke and Tinkham (1980) are in good agreement with data obtained from Sn and Al, while a theory by Beyer Nielsen et al. (1980) is in disagreement. Certain aspects of the theory of the effect, notably the relaxation rate of quasiparticles at low energies, remain to be explained.

In conclusion, I should like to point out that there are a number of other situations involving charge imbalance that I have not described in this chapter. These include branch imbalance waves (see chaps. 18 and 19), the existence of a temperature gradient in a superconductor (chap. 16) or across a NS interface, and the prediction that the charge relaxation rate near T_c in a pulsed measurement should be essentially τ_E^{-1} rather than $\tau_E^{-1}(\Delta/k_BT_c)$ (Kadin et al. 1979). These topics seem likely to receive considerable attention in the near future.

ACKNOWLEDGEMENTS

Much of the work described here was carried out in collaboration with others, namely C. C. Chi, U. Eckern, B. R. Fjordbøge, T-Y Hsiang, T. R. Lemberger, P. E. Lindelof, J. L. Paterson, A. Schmid, G. Schön, and M. Tinkham, to all of whom I express my grateful thanks. In addition I should like to thank J-J Chang, O. Entin-Wohlman, R. Orbach, A. B. Pippard, C. J. Pethick, D. J. Scalapino, H. Smith, D. J. Van Harlingen, and J. R. Waldram for helpful discussions. T. R. Lemberger and D. J. Van Harlingen made innumerable constructive suggestions during the writing of the manuscript, and H. J. Mamin carefully read the final version and eliminated a number of errors.

This work was supported by the Division of Materials Sciences, Office of Basic Energy Sciences, U.S. Department of Energy under Contract Number W-7405-ENG-48.

REFERENCES

Abrikosov, A. A., and Gor'kov, L. P., 1960, Contribution to the theory of superconducting alloys with paramagnetic impurities, Zh. Eksp. Teor. Fiz., 39:1781 [Sov. Phys. JETP, 12:1243].

Anderson, P. W., 1959, Theory of dirty superconductors, J. Phys. Chem. Solid., 11:26.

Andréev, A. F., 1964, The thermal conductivity of the intermediate state in superconductors, Zh. Eksp. Teor. Fiz., 46:182 [Sov. Phys. JETP, 19:1228].

Aronov, A. G., 1974, The influence of condensate motion on the thermoelectric effects in superconductors, Zh. Eksp. Teor. Fiz., 67:178 [Sov. Phys. JETP, 40:90].

Artemenko, N., and Volkov, A. F., 1977, Andréev reflection and the electric resistance of superconductors in the intermediate state, Zh. Eksp. Teor. Fiz., 72:1018 [Sov. Phys. JETP, 45:533].

Artemenko, N., Volkov, A. F., and Zaiteev, A. V., 1978, On the contribution of the superconductor resistance of a superconductor-normal metal system, J. Low Temp. Phys., 30:487.

Bergmann, G., 1971, Eliashberg function $\alpha^2(E)F(E)$ and the strong-coupling behavior of a disordered superconductor, Phys. Rev., B3:3797.

Beyer Nielsen, J., Ono, Y. A., Pethick, C. J., and Smith, H., 1980, The effect of impurity scattering on the thermally induced charge imbalance in a clean superconductor, Solid State Comm., 33:925.

Blackford, B. L., 1976, A tunneling investigation of energy-gap anisotropy in superconducting bulk aluminum crystals, J. Low Temp. Phys., 23:43.

Chambers, R. G., 1952, The anomalous skin effect, Proc. Roy. Soc. Lond., 215:481.

Chang, J-J, 1979, Relaxation of the quasiparticle charge imbalance in superconductors, Phys. Rev., B19:1420.

Chang, J-J, and Scalapino, D. J., 1977, Kinetic-equation approach to nonequilibrium superconductivity, Phys. Rev., B15:2651.

Chi, C. C., and Clarke, J., 1979, Quasiparticle branch mixing rates in superconducting aluminum, Phys. Rev., B19:4495.

Chi, C. C., and Clarke, J., 1980, Addendum to "Quasiparticle branch mixing rates in superconducting aluminum," Phys. Rev., B21:333.

Chi, C. C., and Langenberg, D. N., 1976, Microwave response of non-equilibrium superconductors, Bull. Am. Phys. Soc., 21:403.

Clarke, J., 1972, Experimental observation of pair-quasiparticle potential difference in nonequilibrium superconductors, Phys. Rev. Lett., 28:1363.

Clarke, J., 1976, Superconducting quantum interference devices for low frequency measurements in: "Superconductor Applications: SQUIDS and Machines," B. B. Schwartz and S. Foner, eds., Plenum, New York.

Clarke, J., Eckern, U., Schmid, A., Schön, G., and Tinkham, M., 1979, Branch-imbalance relaxation times in superconductors, Phys. Rev., B20:3933.

Clarke, J., Fjordbøge, B. R., and Lindelof, P. E., 1979a, Supercurrent induced charge imbalance measured in a superconductor in the presence of a thermal gradient, Phys. Rev. Lett., 43:642.

Clarke, J., and Paterson, J. L., 1974, Measurements of the relaxation of quasiparticle branch imbalance in superconductors, J. Low Temp. Phys., 15:491.

Clarke, J., and Tinkham, M., 1980, Theory of quasiparticle charge imbalance induced in a superconductor by a supercurrent in the presence of a thermal gradient, Phys. Rev. Lett., 44:106.

Craven, R. A., Thomas, G. A., and Parks, R. D., 1971, Sharpening of the resistive transition of a superconductor with the addition of paramagnetic impurities, Phys. Rev., B4:2185.

DeGennes, P. G., 1964, Boundary effects in superconductors, Rev. Mod. Phys., 36:225.

DeGennes, P. G., 1966, "Superconductivity of metals and alloys," Benjamin, New York.

Dolan, G. J., and Jackel, L. D., 1977, Voltage measurements within the non-equilibrium region near phase slip centers, Phys. Rev. Lett., 39:1628.

Entin-Wohlman, O., and Orbach, R., 1979, Effect of pair breaking on branch relaxation in nonequilibrium superconductors, Phys. Rev., B19:4510.

Fickett, F. R., 1971, Cryogenics, 11:349.

Galperin, Yu. M., Gurevich, V. L., and Kozub, V. I., 1974, Thermoelectric effects in superconductors, Zh. Eksp. Teor. Fiz., 66:1387 [Sov. Phys. JETP, 39:680].

Ginzburg, V. L., and Landau, L. D., 1950, Zh. Eksp. Teor. Fiz., 20:1064.

Gray, K. E., Long, A. R., and Adkins, C. J., 1969, Measurements of the lifetime of excitations in superconducting aluminum, Phil. Mag., 20:273.

Gschneidner, K. A., 1964, Physical properties and interrelationships of metallic and semi-metallic elements, Solid State Phys., 16:275.

Harding, G. L., Pippard, A. B., and Tomlinson, J. R., 1974, Resistance of superconducting-normal interfaces, Proc. Roy. Soc. Lond., A340:1.

Heidel, D. F., and Garland, J. C., 1980, Thermoelectric potentials in superconducting aluminum films, Bull. Am. Phys. Soc., 25:411.

Hsiang, T-Y, 1980, Magnetic field dependence of the superconductor-normal metal boundary resistance, Phys. Rev., B21:956.

Hsiang, T-Y, and Clarke, J., 1980, Boundary resistance of the superconducting-normal interface, Phys. Rev., B21:945.

Kadin, A. M., Skocpol, W. J., and Tinkham, M., 1978, J. Low Temp. Phys., 33:481.

Kadin, A. M., Smith, L. N., and Skocpol, W. J., 1980, Charge imbalance waves and nonequilibrium dynamics near a superconducting phase-slip center, J. Low Temp. Phys., 38:497.

Kaplan, S. B., Chi, C. C., Langenberg, D. N., Chang, J. J., Jafarey, S., and Scalapino, D. J., 1976, Quasiparticle and phonon lifetimes in superconductors, Phys. Rev., B14:4854.

Kittel, C., 1976, "Introduction to Solid State Physics," 5th edition, Wiley, New York.

Knorr, K., and Barth, N., 1970, Superconductivity and phonon spectra of disordered thin films, Solid State Comm., 8:1085.

Krähenbühl, Y., and Watts-Tobin, R. J., 1979, Microscopic theory of the current-voltage relationship across a normal-superconducting interface, J. Low. Temp. Phys., 35:569.

Lawrence, W. E., and Meador, A. N., 1978, Calculation of the order-parameter relaxation times in superconducting aluminum, Phys. Rev., B18:1154.

Leavens, C. R., and Carbotte, J. P., 1971, Contribution to the theory of weak coupling superconductors, Can J. Phys., 49:724; 1972, Gap anisotropy in a weak coupling superconductor, Annals of Physics, 70:338.

Lemberger, T. R., and Clarke, J., 1980, Charge-imbalance relaxation in the presence of a pair-breaking interaction in superconducting AlEr films, submitted to Phys. Rev. B.

Lemberger, T. R., and Clarke, J., 1980a, Charge imbalance relaxation in the presence of a pair-breaking supercurrent in dirty, superconducting films, submitted to Phys. Rev. B.

Long, A. R., 1973, The attenuation of high frequency phonons in metals, J. Phys., F3:2023.

Maki, K., 1969, Gapless superconductivity, in: "Superconductivity," R. D. Parks, ed., Marcel Dekker, New York.

Markowitz, D., and Kadanoff, L. P., 1963, Effect of impurities upon critical temperature of anisotropic superconductors, Phys. Rev., 131:563.

Meservey, R., and Schwartz, B. B., 1969, Equilibrium properties: comparison of experimental results with predictions of the BCS theory, in: "Superconductivity," R. D. Parks, ed., Marcel Dekker, New York.

Moody, M. V., and Paterson, J. L., 1979, Relaxation of the distribution function branch imbalance Q* in Sn and Sn-In alloys, J. Low Temp. Phys., 34:83.

Ovchinnikov, Yu. N., 1977, Penetration of an electrical field into a superconductor, J. Low Temp. Phys., 28:43.

Pethick, C. J., and Smith, H., 1979, Relaxation and collective motion in superconductors: a two-fluid description, Annals of Physics, 119:133.

Pethick, C. J., and Smith, H., 1979a, Generation of charge imbalance in a superconductor by a temperature gradient, Phys. Rev. Lett., 43:640.

Pethick, C. J., and Smith, H., 1980, Charge imbalance in nonequilibrium superconductors, to be published in J. Phys. C.

Pippard, A. B., 1965, "The Dynamics of Conduction Electrons," Gordon and Breach, New York.

Pippard, A. B., Shephard, J. G., and Tindall, D. A., 1971, Resistance of superconducting-normal interfaces, Proc. Roy. Soc. Lond., A324:17.

Rieger, T. J., Scalapino, D. J., and Mercereau, J. E., 1971, Charge conservation and chemical potentials in time-dependent Ginzburg-Landau theory, Phys. Rev. Lett., 27:1787.

Rothwarf, A., and Taylor, B. N., 1967, Measurement of recombination lifetimes in superconductors, Phys. Rev. Lett., 19:27.

Schmid, A., and Schön, G., 1975, Linearized kinetic equations and relaxation processes of a superconductor near T_c, J. Low Temp. Phys., 20:207.

Schmid, A., and Schön, G., 1979, Generation of branch imbalance by the interaction between supercurrent and thermal gradient, Phys. Rev. Lett., 43:793.

Schön, G., 1980, unpublished.

Schuller, I., and Gray, K. E., 1977, Temperature dependence of the relaxation time of the superconducting order parameter, Solid State Comm., 23:337.

Skocpol, W. J., Beasley, M. R., and Tinkham, M., 1974, Phase-slip centers and nonequilibrium processes in superconducting tin microbridges, J. Low Temp. Phys., 16:145.

Smith, L. N., and Mochel, J. M., 1976, Phonon and quasiparticle dynamics in superconducting aluminum tunnel junctions, Phys. Rev. Lett., 35:1597.

Tinkham, M., 1972, Tunneling generation, relaxation, and tunneling detection of hole-electron imbalance in superconductors, Phys. Rev., B6:1747.

Tinkham, M., 1975, "Introduction to Superconductivity," McGraw-Hill, Inc., New York.

Tinkham, M., and Clarke, J., 1972, Theory of pair-quasiparticle potential difference in non-equilibrium superconductors, Phys. Rev. Lett., 28:1366.

Van Harlingen, D. J., 1980, Thermoelectric generation of charge imbalance at superconductor interfaces, Bull. Am. Phys. Soc., 25:411.

Waldram, J. R., 1975, Chemical potential and boundary resistance at normal-superconducting interfaces, Proc. Roy. Soc. Lond., A345:231.

Yu, M. L., and Mercereau, J. E., 1972, Electric potentials near a superconducting-normal bondary, Phys. Rev. Lett., 28:1117.

Zavaritzkii, N. V., 1969, Increase of the critical temperature of superconductors condensed at low temperatures, Zh. Eksp. Teor. Fiz., 57:752 [Sov. Phys. JETP, 30:412].

KINETIC EQUATIONS FOR DIRTY SUPERCONDUCTORS

Albert Schmid

Institut für Theorie der Kondensierten Materie
Universität Karlsruhe, Karlsruhe, West Germany

I. INTRODUCTION

In this lecture, I will first derive from the Gorkov-Eliashberg theory a system of kinetic equations and then, I will show that these equations allow an adequate and convenient calculation of nonequilibrium phenomena in superconductors.

Let me first explain in a few words what I think is the new idea in the subject of nonequilibrium superconductivity. Recalling the history, one remembers that, since Bardeen, Cooper and Schrieffer (1957) discovered the theory of superconductivity, there have been many investigations on the response of a superconductor to perturbations, that is, there have been investigations on the behavior of a superconductor when driven out of thermal equilibrium. I shall mention only a few things such as electromagnetic response, attenuation of ultrasound and nuclear spin relaxation. Indeed, the quantitative explanation of such phenomena has been a major success of the BCS theory. A short time later, a great interest in the Ginzburg-Landau theory arose since this theory was able to explain the rather involved magnetic structure of superconductors of the second kind. At that time, Gorkov (1959) was able to show that the Ginzburg-Landau equations were a consequence of the BCS theory. Almost everybody seemed to be convinced then, that the theory of superconductivity could be expressed in relations involving only the order parameter.

However, the numerous investigations which have been carried through in recent times, and of which you have heard in previous lectures, have clearly shown that the distribution of the quasiparticles plays a most important role in the phenomena of nonequilibrium superconductivity.*

Therefore, an adequate theory of nonequilibrium superconductivity has, first of all, to establish a kinetic equation which allows the calculation of the quasiparticle distribution function f. Because of its similarity with the well-known prototype, I will call this a Boltzmann equation. Furthermore, self-consistency relations for the complex order parameter $\underset{\sim}{\Delta} = \Delta e^{-i\theta}$ and for the electromagnetic potentials $(\varphi, \vec{A})$ have to be found. As it turns out, it is more effective in the nonequilibrium theory, to work with the gauge invariant quantities Δ (magnitude of the order parameter); $\vec{v}_s$ (superfluid velocity); and Φ (chemical potential of the Cooper pairs).

In the derivation of these equations, the concept of quasiclassical Green's functions plays an important role. It has been introduced by Eilenberger (1968) and by Larkin and Ovchinnikov (1968) for equilibrium and nonequilibrium theory, respectively. Eliashberg (1971) has generalized this method for retarded interaction such that inelastic electron-phonon collisions can be included. Later Schmid and Schön (1975) have shown how to define a quasiparticle distribution in this theory.**

I wish to emphasize that the quasiparticle picture which emerges from this theory resembles very closely to what is known as the semiconductor model. There are quasiparticle states above as well as below the gap (such that the "valence" and the "conduction" band are filled and empty, respectively, at T=0), and these quasiparticles carry full electronic charge.

Although this theory can be formulated quite generally, I will restrict the following considerations to dirty superconductors, that is to superconductors where the mean free path is shorter than the coherence length. This is an important case in practice, and also, it allows us to demonstrate the flexibility of this theory.

Most of the equations are written down in a form which is valid at all temperatures. However, there appear expressions which are generally very difficult to calculate except in the so-called Ginzburg-Landau regime $\Delta \ll T_C$.+ Therefore we will restrict explicit

*The lack of emphasis of the equilibrium theory in the distribution function is understandable: there, the distribution function is the rather dull object known as the Fermi function.

**In this context, the very compact formulation of the problem by Larkin and Ovchinnikov (1977) should be mentioned.

+This is a problem already well known in the equilibrium theory where one has found it extremely hard to establish equations beyond this regime.

calculations to this limit. In most cases, we will also require that the characteristic frequencies of the nonequilibrium process are much smaller than the gap frequency.

The organisation of this lecture is displayed in the table of contents. In Sec. II, we establish the basic equations, and in Secs. III and IV, we apply these equations to some specific problems. Appendices are added for clarification of some special points.

II. KINETIC EQUATIONS

1. Basic Theory

The BCS pairing state consists of linear superpositions of particle and hole states. Therefore, we find it convenient to follow Gorkov (1958) who introduced one-particle correlation functions which are matrices in particle-hole space. In a first step, we define

$$\hat{G}^{>}(x_1,x_2) = -i\left\langle\begin{pmatrix} \psi_\uparrow(x_1)\psi_\uparrow^+(x_2) & \psi_\uparrow(x_1)\psi_\downarrow(x_2) \\ -\psi_\downarrow^+(x_1)\psi_\uparrow^+(x_2) & -\psi_\downarrow^+(x_1)\psi_\uparrow^+(x_2) \end{pmatrix}\right\rangle \qquad \text{(II.1a)}$$

where $\psi_\uparrow(x_1)$ etc. are the usual electron field operators, $x = (\vec{r},t)$; and where $\langle \ldots \rangle$ denotes an appropriate ensemble average. Similarly,

$$\hat{G}^{<}(x_1,x_2) = i\left\langle\begin{pmatrix} \psi_\uparrow^+(x_2)\psi_\uparrow(x_1) & \psi_\downarrow(x_2)\psi_\uparrow(x_1) \\ -\psi_\uparrow^+(x_2)\psi_\downarrow^+(x_1) & -\psi_\uparrow^+(x_2)\psi_\downarrow^+(x_1) \end{pmatrix}\right\rangle \qquad \text{)II.1b)}$$

It has been shown by Keldysh (1964), that in case of non-equilibrium processes it is advantageous to introduce the following combinations of $\hat{G}^{>}$ and $\hat{G}^{<}$, namely

$$\begin{aligned} \hat{G}^R(x_1,x_2) &= \{\hat{G}^{>}(x_1,x_2) - \hat{G}^{<}(x_1,x_2)\}\theta(t_1-t_2) \;; \\ \hat{G}^A(x_1,x_2) &= -\{\hat{G}^{>}(x_1,x_2) - \hat{G}^{<}(x_1,x_2)\}\theta(t_2-t_1) \;; \qquad \text{(II.1c)} \\ \hat{G}^K(x_1,x_2) &= \{\hat{G}^{>}(x_1,x_2) + \hat{G}^{<}(x_1,x_2)\} \;. \end{aligned}$$

Conventionally, $\hat{G}^{R(A)}$ are called the retarded (advanced) Green's function and I will call the combination in the last line Keldysh's function $\hat{G}^K$ for sake of definitiveness.

A most compact notation is obtained if one arranges these Green's functions as well as the corresponding self-energies in form of matrices

$$\check{G} = \begin{pmatrix} \hat{G}^R & \hat{G}^K \\ 0 & \hat{G}^A \end{pmatrix} ; \quad \check{\Sigma} = \begin{pmatrix} \hat{\Sigma}^R & \hat{\Sigma}^K \\ 0 & \hat{\Sigma}^A \end{pmatrix} , \tag{II.2}$$

and I will call them matrices in Keldysh space. With this notation, Dyson's equation assumes the conventional form

$$[\check{G}_o^{-1} - \check{\Sigma}]\ \check{G} = \check{1} , \tag{II.3a}$$

which is a short-hand notation of

$$\int dr_3 \int_{-\infty}^{+\infty} dt_3\ [\check{G}_o^{-1} - \check{\Sigma}]\ (x_1,x_3)\ \check{G}(x_3,x_2) = \check{1}\delta(x_1-x_2). \tag{II.3b}$$

It is interesting to note that the usual types of Feynman diagrams are valid for $\check{G}$ and $\check{\Sigma}$. For illustration, some comments pertaining to Keldysh's theory are given in Appendix A.

I will now discuss some properties of G for a translationally invariant state in thermal equilibrium which have been first derived by Eliashberg (1960). Let us consider first $\hat{G}^R$. Obviously, Eq. (II.3) connects only $\hat{\Sigma}^R$ with $\hat{G}^R$. After Fourier transforming the quantities with respect to the coordinate difference $x_1 - x_2 \equiv (\vec{r}_1-\vec{r}_2,\ t_1-t_2)$, we obtain*

$$\hat{G}^R(\vec{p},E) = [\hat{\tau}_3E - \hat{1}\xi_p - \hat{\Sigma}^R(E)]^{-1} , \tag{II.4}$$

where $\xi_p = p^2/2m - E_F$, and where $\hat{\tau}_3$ (similarly: $\hat{\tau}_1$ and $\hat{\tau}_2$) is the Pauli matrix. In general, the self-energy of a degenerate Fermi system depends weakly on the magnitude of the momentum, and therefore, this dependence has been neglected.

Consider the case where only the interaction with phonons contributes to the self-energy. Neglecting an irrelevant term proportional to the unit matrix and choosing the phase of the order parameter in a convenient way, we may write**

$$\hat{\Sigma}^R_{ph} = \hat{\tau}_3[1 - Z^R(E)]E - i\hat{\tau}_1\phi^R(E). \tag{II.5a}$$

---

* Throughout we will use units such that $\hbar = c = k_B = 1$.

** We follow here the notation used by Scalapino (1969).

In the weak coupling limit,+

$$\mathrm{Re}Z^R = 1+\lambda \; ; \; \mathrm{Im}(Z^R E) = (1+\lambda)/2\tau(E) \; ;$$
$$\phi^R = (1+\lambda)\Delta. \tag{II.5b}$$

In this relation, $\lambda(\equiv\lambda_{ep})$ is the dimensionless electron-phonon coupling constant; $1/\tau(E)$ ($\sim T^3/\theta^2$, if $E, \Delta \ll T$) is the inelastic collision rate, and Δ is the energy gap. There are similar relations for $\hat{G}^A$ with $Z^A(E) = \{Z^R(E)\}^*$; $\phi^A(E) = \{\phi^A(E)\}^*$.

An important information is contained in a quantity which is obtained by integrating $\hat{G}^R$ with respect to ξ_p, and which will be called $\hat{g}^R$,

$$\hat{g}^R(E) = \frac{i}{\pi}\int d\xi_p \; \hat{G}^R(\vec{p},E)$$

$$= \frac{\hat{\tau}_3 Z^R(E)E + i\hat{\tau}_1\phi^R(E)}{[(Z^R E)^2 - (\phi^R)^2]^{1/2}} \quad . \tag{II.6a}$$

In this relation, the square root has a positive imaginary part. Since it will be useful later, I introduce the notation

$$\hat{g}^{R(A)}(E) = \hat{\tau}_3[{}_{(-)}^{+}N_1 + iR_1] + \hat{\tau}_1[N_2 {}_{(-)}^{+} iR_2] \; , \tag{II.6b}$$

where the relation valid for the advanced quantity is given in parenthesis. Note that $N_j(E)$ and $R_j(E)$ are even and odd functions of E. The reduced density of states ($\equiv\rho(E)$) appears here as $N_1(E)$ and I will refer quite generally to N_j and R_j as generalized densities of states or spectral quantities. In the BCS limit (see Eq. (II.5b); $1/\tau(E)\to 0+$),

$$N_1 = |E|/\sqrt{E^2-\Delta^2} \; ; \qquad R_1 = -E/\sqrt{\Delta^2 - E^2} \; ;$$
$$N_2 = \Delta/\sqrt{\Delta^2 - E^2} \; ; \qquad R_2 = \Delta/\mathrm{sgn}E\sqrt{E^2-\Delta^2} \; . \tag{II.6c}$$

In particular, the spectral quantities vanish either inside or outside the gap region $|E|<\Delta$. As far as their physical significance is concerned, I remark that N_1 and R_2 occur in familiar combinations, that N_2 appears only if there is a disequilibrium between pairs and quasiparticles, and that R_1 plays an insignificant role. In the more general case, to which Eq. (II.5b) applies, the spectral quantities assume a form (for

\+ The term $\mathrm{Im}\phi^R$ has been neglected since it vanishes faster than Δ for $\Delta\to 0$. It has to be included at lower temperatures.

$\Delta\tau_E = 5$),as shown graphically in Fig. 1. Note that N_1 and R_2 though being small, do not vanish in the gap region and we may attribute this fact to the collision broadening $\propto 1/\tau(E)$ of the electronic levels. For the same reason, N_2 has a tail extending beyond the gap region.

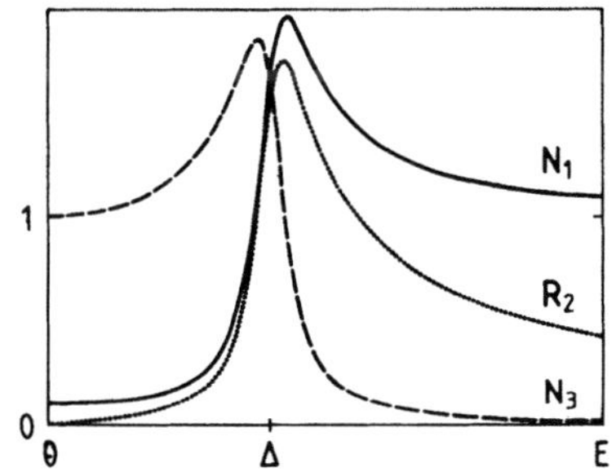

Fig. 1. The generalized densities of states $N_1(E)$ (solid line); $N_2(E)$ (dashed line); and $R_2(E)$ (dotted line), as defined by Eqs. (II.6) and (II.5) for $\Delta\tau_E = 5$.

Concerning the form of Keldysh's function, we should recall that in thermal equilibrium there exists a definite connection between commutator and anticommutator of a pair of operators*. This implies that

$$\hat{g}^{K}(E) = [\hat{g}^{R}(E) - \hat{g}^{A}(E)] \cdot \mathrm{th}\frac{E}{2T} . \tag{II.7}$$

Green's functions for the phonons can be defined similar to Eq. (II.1) where now the correlation functions contain the phonon field operators (see, for instance, Eq. (A.3)). Let us call B(E) the appropriate momentum and polarisation average of these Green's function, weighted with the electron-phonon coupling constant. Then

$$B^{R(A)}(E) = \int_{-\infty}^{+\infty} dE' \, \frac{\alpha^2 F(E')}{E-E' \underset{(-)}{+} i0} , \tag{II.8a}$$

where the Eliashberg function $\alpha^2 F(E)$ which is the effective phonon density of states is defined to be an odd function of its argument. Note that ** $B^{R(A)}(0) = -\lambda$. In thermal equilibrium, Keldysh's function for the phonons is given by

$$B^{K}(E) = [B^{R}(E) - B^{A}(E)] \cdot \mathrm{cth}\frac{E}{2T} . \tag{II.8b}$$

Using these definitions, we can write Eliashberg's equation for the electron-phonon self-energy in the following form (App.A)

$$\hat{\Sigma}^{R(A)}_{ph} = \frac{1}{4} \int_{-\infty}^{+\infty} dE' \, \{B^{A(R)}(E')\hat{g}^{K}(E+E') + B^{K}(E')\hat{g}^{R(A)}(E+E')\}. \tag{II.9a}$$

There is a similar relation for Keldysh's component of the self-energy,

$$\hat{\Sigma}^{K}_{ph} = \frac{1}{4} \int dE' \, \{B^{K} \hat{g}^{K} - (B^{R} - B^{A})(\hat{g}^{R} - \hat{g}^{A})\} , \tag{II.9b}$$

where the energy arguments are the same as above.

Concerning the inelastic collision time, we have in the normal state and at the Fermi level,

$$\frac{1}{\tau_E} = \frac{2}{1+\lambda} \int_{-\infty}^{+\infty} dE' \, \frac{\alpha^2 F(E')}{\mathrm{sh}E'/T} . \tag{II.9c}$$

*The fluctuation-dissipation theorem is a particular example of such a connection.

** In the paper of Schmid and Schön (1975), the normalisation has been chosen such that $B^{R(A)}(0) = -1$.

2. Quasiclassical Green's Function

In the case of space and time dependent states, the Green's function $\hat{G}(x_1,x_2)$ depends on all the coordinates in a non-trivial way. In this case, it will be convenient for us to use a mixed representation which displays the center of mass coordinates $\vec{r} = (\vec{r}_1 + \vec{r}_2)/2$; $t = (t_1 + t_2)/2$ together with the momentum $\vec{p}$ and the energy E which are the Fourier conjugate variables to the relative coordinates $(\vec{r}_1 - \vec{r}_2)$ and $(t_1 - t_2)$. Therefore,

$$\check{G}(\vec{p},E;\vec{r},t) = \int dr'dt' e^{-i\vec{p}\vec{r}' + iEt'} \check{G}(\vec{r}_1,t_1;\vec{r}_2,t_2), \qquad (II.10)$$

where $\vec{r}_{1(2)} = \vec{r} +(-)\vec{r}'/2$; $t_{1(2)} = t+(-)t'/2$. In the following, we will use this representation exclusively.

Let us consider the noninteracting part of the inverse Green's function which occurs in Dyson's equation. For the moment, we return to the pure coordinate representation where

$$\check{G}_o^{-1}(x_1,x_2) = \delta(x_1-x_2)\left[i\check{\tau}_3 \frac{\partial}{\partial t_2} - \check{1}e\varphi(x_2) + \frac{1}{2m}\left(\check{1}\frac{\partial}{\partial \vec{r}_2} - i\check{\tau}_3 e\vec{A}(x_2)\right)^2\right]. \qquad (II.11)$$

In this relation, $(\varphi,\vec{A})$ are the electromagnetic potentials and $\check{\tau}_3$ is $\hat{\tau}$ times the unit matrix in Keldysh's space.

Expression (II.11) can be simplified considerably by using the fact that in a superconductor the range of spatial variation of the physical quantities is much larger than the Fermi wave length p_F^{-1}. This means that, as a function of the space coordinates, $\check{G}(\vec{r}_1,\vec{r}_2)$ essentially vanishes, for $|\vec{r}_1-\vec{r}_2| \gg p_F^{-1}$. Consequently, we may approximate

$$\delta(x_1-x_2)\frac{1}{2m}\left(\check{1}\frac{\partial}{\partial \vec{r}_2} - i\check{\tau}_3 eA(x_2)\right)^2 \sim\Rightarrow -\check{1}\frac{p^2}{2m} + i\frac{\vec{p}}{m}\vec{\check{\partial}}, \qquad (II.12a)$$

where

$$\vec{\check{\partial}} = \check{1}\frac{\partial}{\partial \vec{r}} - i\check{\tau}_3 e\vec{A}(\vec{r},t) \qquad (II.12b)$$

is a gauge invariant derivative. Note that on the right hand side of Eq. (II.12a), we have changed to the mixed representation. (Here and below, this change is indicated by $\Rightarrow$).

The fact that the Green's function is a rather compact object as a function of its relative coordinates allows us to simplify also the spatial part of the convolution type of integral in Dyson's equation. Suppressing for simplicity time arguments and matrix indices, we may write

$$(\Sigma G)(\vec{r}_1,\vec{r}_2) \equiv \int d\vec{r}_3 \; \Sigma(\vec{r}_1,\vec{r}_3)G(\vec{r}_3,\vec{r}_2) \sim\Rightarrow \Sigma(\vec{p};\vec{r})G(\vec{p};\vec{r}) \; . \qquad \text{(II.12c)}$$

The approximations defined by Eq. (II.12) will be called the quasiclassical approximation.

In general, no simple approximation is possible with respect to the temporal part of the convolution integral. Thus, we will write

$$\int dt_3 \Sigma(t_1,t_3)G(t_3,t_2) = \Rightarrow \Sigma(E;t) \circ G(E;t), \qquad \text{(II.13a)}$$

where the associative dot-product can be expressed as an infinite series of differentations

$$A \circ B = AB + \frac{i}{2}\left\{\frac{\partial A}{\partial E}\frac{\partial B}{\partial t} - \frac{\partial A}{\partial t}\frac{\partial B}{\partial E}\right\} + \ldots\ldots\ldots\ldots \; . \qquad \text{(II.13b)}$$

In conventional units, this series means an expansion in powers of $\hbar$; and what is written down explicitly in above relation means specifically an expansion of the dot-product through first order in $\hbar$. Occasionally, we will make use of the notation

$$\{\{A,B\}\} = \left\{\frac{\partial A}{\partial E}\frac{\partial B}{\partial t} - \frac{\partial A}{\partial t}\frac{\partial B}{\partial E}\right\}, \qquad \text{(II.13c)}$$

which bears some resemblance with the classical Poisson bracket.

It is a consequence of the quasiclassical approximation that in Dyson's equation, the momentum appears only as a parameter. Therefore, we generalize Eq. (6a) in a straightforward way,

$$\overset{\frown}{g}(\vec{p},E;\vec{r},t) = \frac{i}{\pi}\int d\xi_p \; \overset{\frown}{G}(\vec{p},E;\vec{r},t) \; , \qquad \text{(II.14)}$$

and call $\check{g}$ the quasiclassical Green's function. The momentum variable on the left side of Eq. (II.14) means only a dependence on the direction of the momentum; its magnitude is understood to be fixed at $|\vec{p}| = p_F$. There is no loss of information since it is known that transitions in a Fermi system occur always close to the Fermi surface which means that is it necessary to specifiy only the direction of the momentum.

The equation of motion for the quasiclassical Green's function can be obtained as follows. We form the difference between Dyson's equation and its adjoint equation,

$$[\check{G}_o^{-1} - \check{\Sigma}]\,\check{G} - \check{G}\,[\check{G}_o^{-1} - \check{\Sigma}] = 0\ , \tag{II.15}$$

and in the quasiclassical approximation, the strongly momentum dependent terms $(p^2/2m)$ G cancel. This allows us to integrate Eq. (II.15) with respect to ξ_p, and we arrive at the desired equation

$$[\check{\tau}_3 E - \check{1} e\varphi + i\frac{\vec{p}}{m}\check{\vec{\partial}} - \check{\Sigma} \overset{\circ}{,} \check{g}]_- = 0\ , \tag{II.16}$$

where the commutator involves the dot-product*. Recall that we are using a mixed representation which means that the quantities in Eq. (II.16) are labeled by $(\vec{r},t)$ or by $(\vec{p},E;\vec{r},t)$, such that $|\vec{p}\,| \simeq p_F$.

The following remark is most important. Since Eq. (II.16) is homogeneous in $\check{g}$, it cannot determine the quasiclassical Green's function completely. Additional information is contained in the normalization condition

$$\check{g} \circ \check{g} = \check{1}\ , \tag{II.17}$$

which is valid for the following reason. In thermal equilibrium, and in a spatially homogeneous state, it follows from Eq. (II.6) that $(\hat{g}^R)^2 = (\hat{g}^A)^2 = \hat{1}$. This relation, together with Eq. (II.7) implies that $\hat{g}^R\hat{g}^K + \hat{g}^K\hat{g}^A = 0$, and we conclude, that Eq. (II.17) is correct. In the general case, I will argue as follows. Since the dot-product is associative, one can derive from Eq. (II.16) an equation of the same form where, however, $\check{g} \circ \check{g}$ replaces $\check{g}$. Evidently, ansatz (II.17) solves this equation; and since this solution joins up smoothly to the equilibrium solution, we conclude that this is the only possible solution.

We turn now our attention to the properties of the self-energy. The phonon part is still of the general form of Eq. (II.9) with a

*As an example, note that

$[\check{\tau}_3 E \overset{\circ}{,} \check{g}]_- = E[\check{\tau}_3,\check{g}]_- + (i/2)\,[\check{\tau}_3,\dot{\check{g}}]_+\ ;\ [\check{\vec{\partial}} \overset{\circ}{,} \check{g}]_- = \partial\check{g}/\partial\vec{r} - ie[\check{\tau}_3\vec{A} \overset{\circ}{,} \check{g}]_-$.

difference only as far as the angular dependence of $\hat{\bar{g}}(\vec{p},\ldots)$ is concerned. At one side, there are contributions to $\hat{\bar{\Sigma}}_{ph}(\vec{p},\ldots)$ which result from inelastic collisions. Since the change of the electronic momentum is insignificent, there we may calculate these inelastic contributions by inserting $\hat{\bar{g}}(\vec{p},\ldots)$ in the appropriate places. On the other hand, the contributions to $\hat{\bar{\Sigma}}_{ph}$ which lead to renormalisation and pair correlations originate from virtual transitions involving high momentum phonons. This fact suggests that the angular average $<\hat{\bar{g}}(\vec{p},\ldots)>$ enters into the corresponding expression. However, we will not need to go into these details since we will consider only cases where the angular dependence of $\hat{\bar{g}}$ is small. Thus, we use in our case Eq. (II.9) in a form where on the right hand side the space-time dependent angular average $<\hat{\bar{g}}(\vec{p},E;\vec{r},t)>$ is inserted everywhere.

Scattering at impurities leads to a self-energy which consists of two parts.

$$\hat{\bar{\Sigma}}_{imp}(E;\vec{r},t) = \hat{\bar{\Sigma}}_{NF} + \hat{\bar{\Sigma}}_{SF} \; ;$$

$$\hat{\bar{\Sigma}}_{NF} = -i\frac{1+\lambda}{2\tau} < \hat{\bar{g}}(\vec{p},E;\vec{r},t)> \; ; \tag{II.18}$$

$$\hat{\bar{\Sigma}}_{SF} = -i\frac{1+\lambda}{2\tau_s}\, \hat{\bar{\tau}}_3 < \hat{\bar{g}}(\vec{p},E;\vec{r},t)> \hat{\bar{\tau}}_3$$

where $1/\tau_s$ and $1/\tau$ denote the scattering rates with and without spin flip. For simplicity, the scattering has been assumed to be isotropic.

In the following, we will consider only strong impurity scattering $\tau T_c \ll 1$ (dirty limit) since this is the simplest and also the most important case. Then $\hat{\bar{g}}$ is predominatly isotropic with only a small angular-dependent part,

$$\hat{\bar{g}}(\vec{p},E;\vec{r},t) = \hat{\bar{g}}_s(E;r,t) + \frac{\vec{p}}{m}\cdot\hat{\vec{\bar{g}}}_p(E;rt). \tag{II.19a}$$

Within the same accuracy, it follows from Eq. (II.17) that

$$\hat{\bar{g}}_s \circ \hat{\bar{g}}_s = \hat{\bar{1}} \; ; \; \hat{\bar{g}}_s \circ \hat{\vec{\bar{g}}}_p + \hat{\vec{\bar{g}}}_p \circ \hat{\bar{g}}_s = 0. \tag{II.19b}$$

In that part of the equation of motion (II.16) which is odd in $\vec{p}$, we may neglect all terms which act on $(\vec{p}/m)\hat{\vec{\bar{g}}}_p$ except the term $\hat{\bar{\Sigma}}_{NF}$. Thus, we obtain

$$\hat{\vec{\bar{g}}}_p = -\frac{\tau}{2(1+\lambda)}[\hat{\bar{g}}_s \overset{\circ}{,} [\hat{\vec{\bar{\partial}}} \overset{\circ}{,} \hat{\bar{g}}_s]_-]_- \quad , \tag{II.20a}$$

where use has been made of the normalisation (II.19b). Inserting this result in Eq. (II.16), we average with respect to the directions of $\vec{p}$. Thus, we obtain

$$[\bar{\hat{\tau}}_3 E - \bar{\hat{1}} e\varphi - \bar{\hat{\Sigma}}_{ph} - \bar{\hat{\Sigma}}_{SF} \overset{\circ}{,} \bar{\hat{g}}_s]_- \\ - \frac{i}{2}(1+\lambda)D\, [\bar{\hat{\vec{\partial}}} \overset{\circ}{,} [\bar{\hat{g}}_s \overset{\circ}{,} [\bar{\hat{\vec{\partial}}} \overset{\circ}{,} \bar{\hat{g}}_s]_-]_-]_- = 0 \qquad \text{(II.30b)}$$

where $D = \tau v_F^2/3$ is the diffusion constant and where $v_F = (1+\lambda)^{-1} \times (p_F/m)$ is the renormalized Fermi velocity. Now the momentum variable has completely disappeared and the quantities depend only on $(\vec{r},t)$ or on $(E;\vec{r},t)$.

For sake of completeness, the expressions for charge density and charge current density are given as follows

$$\rho(\vec{r},t) = -2eN(o)\,\{\frac{1}{8}\int dE \;\mathrm{Sp}\; \hat{g}^K_s(E;\vec{r},t) + e\varphi\};$$
$$\vec{j}(\vec{r},t) = -\frac{e p_F^2 N(o)}{12 m^2}\int dE \;\mathrm{Sp}\; \hat{\tau}_3\, \hat{\vec{g}}^K_p\,(E;\vec{r},t)\;, \qquad \text{(II.21)}$$

where N(o) is the usual free particle density of states. The term porportional to the electric potential represents contributions to ρ which come from regions far off the Fermi surface and which the quasiclassical Green's function fails to include.

3. Boltzmann Equation

First, I will show how Keldysh's function can be used to define the distribution function of the electrons. Let us write the normalisation (II.17) in the detailed form

$$\hat{g}^R \circ \hat{g}^R = \hat{g}^A \circ \hat{g}^A = \hat{1}\;;$$
$$\hat{g}^R \circ \hat{g}^K + \hat{g}^K \circ \hat{g}^A = 0\;. \qquad \text{(II.22)}$$

One recognizes that the last relation can be solved by the ansatz

$$\hat{g}^K = \hat{g}^R \circ \hat{h} - \hat{h} \circ \hat{g}^A\;. \qquad \text{(II.23a)}$$

It is easily found that a substitution $\hat{h} \rightarrow \hat{h} + \hat{G}^R \circ \hat{B} + \hat{B} \circ \hat{G}^A$ leads to the same $\hat{g}^K$. There is strong evidence that this freedom of choice allows us to impose the condition that $\hat{h}$ be diagonal in particle-hole-space. Hence*

$$\hat{h} = \hat{1}\, h^{(L)} + \hat{\tau}_3 h^{(T)} , \tag{II.23b}$$

and $\hat{h}$ can be connected to a distribution function as follows

$$f = f^{(L)} + f^{(T)} ; \qquad f^{(L)} = (1-h^{(L)})/2 \; ; \; f^{(T)} = -h^{(T)}/2 . \tag{II.23c}$$

Comparing this ansatz with Eq. (II.7), we recognize that in thermal equilibrium $f^{(L)}$ is equal to the Fermi function

$$f_{th} = \frac{1}{e^{E/T}+1} \quad ; \quad (h_{th} = \mathrm{th}\frac{E}{2T}) , \tag{II.23c}$$

and that $f^{(T)}$ is zero. The properties of f which will emerge from the following discussion allow us to conclude that $f=f(\vec{p},E;\vec{r},t)$ is the number of electrons at $(\vec{r},t)$ in a state** labeled by the momentum direction $\vec{p}$ and the energy E.

In a dirty superconductor, the above relations are also valid for the isotropic part $f_s = f_s(E;\vec{r},t)$. A kinetic equation for f_s, which we will call a Boltzmann equation, can be derived from Eq. (II.20) by the following series of arguments.

In a preliminary step, it is useful to recall that physically, $\hat{\Sigma}_{ph}$ plays a twofold role. Inelastic collisions are contained in the components $\hat{\Sigma}^K_{ph}$ and $\hat{\Sigma}^R_{ph} - \hat{\Sigma}^A_{ph}$ which will be represented by a collision integral in the Boltzmann equation. Renormalisation and pair correlations are contained in $(\hat{\Sigma}^R_{ph} + \hat{\Sigma}^A_{ph})/2$. For sake of simplicity, we will discuss only the weak coupling limit $T_c \ll \omega_D$. Therefore, we put in correspondence with Eq. (II.5b),

---

*(L) and (T) are abbreviations of longitudinal and transverse, respectively. This terminology (Schmid and Schön, 1975) stems from the interpretation of the complex order parameter Δ as a vector in the complex plane. The changes $\delta\Delta$ are then parallel and perpendicular to Δ in the (L)- and (T)- mode, respectively. Physically, more appropriate terms would be energy mode (L) and charge mode (T) since they are characterized by changes in energy and charge respectively.

**This state is not a quasiparticle state of Bogoliubov's type, but rather a weighted sum of such states. See App. D.

$$\frac{1}{2}(\hat{\Sigma}^R_{ph} + \hat{\Sigma}^A_{ph}) = \lambda(\hat{\tau}_3 E - \hat{1} e\varphi) - i(1+\lambda)\hat{\Delta}(\vec{r},t) ;$$

$$\hat{\Delta} = \begin{pmatrix} 0 & \Delta \\ \Delta^* & 0 \end{pmatrix} . \qquad \text{(II.24a)}$$

There are, however, two changes which are noteworthy. Since in a gauge transformation, $\hat{\tau}_3 E$ changes as $\hat{1}e\varphi$, we have written the diagonal part leading to renormalisation* in a gauge invariant form. Furthermore, we have allowed for the moment that the order parameter $\Delta = \Delta e^{-i\theta}$ may become a complex quantity. In the weak coupling limit, it follows from Eq. (II.9a) that

$$\Delta(\vec{r},t) = \frac{1}{4i} \frac{\lambda}{1+\lambda} \int dE \, \{\hat{g}^K_s(E;\vec{r},t)\}_{12}, \qquad \text{(II.24b)}$$

where the subscript (12) specifies the matrix element in the particle-hole space.

In order to illustrate the principle, I will put apart for a moment the terms in Eq. (II.20) which lead to diffusion and collisions. Using the relation $[A \overset{o}{,} BoC]_- = Bo[A \overset{o}{,} C]_- + [A^o, B]_- o\, C$ together with Eq. (II.23a) we may write

$$0 = \frac{i}{4(1+\lambda)} [\hat{\tau}_3 E - \hat{1}e\varphi - \frac{1}{2}(\hat{\Sigma}^R_{ph} + \hat{\Sigma}^A_{ph}) \overset{o}{,} \hat{g}^K_s]_-$$

$$= \frac{i}{4} \hat{g}^R_s o[\hat{\tau}_3 E - \hat{1}e\varphi + i\hat{\Delta} \overset{o}{,} \hat{h}_s]_- \qquad \text{(II.25a)}$$

$$- \frac{i}{4} [\hat{\tau}_3 E - \hat{1}e\varphi + i\hat{\Delta} \overset{o}{,} \hat{h}_s]_- o\, \hat{g}^A_s ,$$

where we have exploited the fact that the equation of motion for $\hat{g}^R_s$ or $\hat{g}^A_s$ is identical to the first line of Eq. (II.25a) except for an obvious replacement. With regard to a convenient notation later, an extra factor $i/4(1+\lambda)$ has been inserted in the first line.

There are few cases only where a compact and explicit form of the dot-product exists. A linear perturbation above a time independent reference state is such a case; and an example will be presented later in Sec. (IV.1) where the Carlson-Goldman collective mode will be discussed.

---

*Note that λ is a constant independent of pair correlations.

In general, we have to resort to the expansion (II.13) of the dot-product in terms of a few time derivatives. Clearly such an expansion is the better, the slower the change in time of the quantities. Fortunately, there is at our disposal the gauge which we choose such that $\underset{\sim}{\Delta}$ is of least variation. This is obviously the case in a gauge where the phase $\Theta = 0$ (real Δ gauge). In other words, we select new electromagnetic potentials such that

$$- e\varphi \rightarrow - e\varphi' = \Phi = \frac{1}{2}\dot{\Theta} - e\varphi \; ;$$
$$- \frac{e}{m}\vec{A} - \frac{e}{m}\vec{A}' = \vec{v}_s = - \frac{1}{2m}(\nabla\Theta + 2e\vec{A}). \qquad \text{(II.26a)}$$

The quantities introduced above allow straightforward physical interpretation. Since $\dot{\Theta}/2$ can be considered as the electro-chemical potential of the Cooper pairs (that is, per particle), $\underset{\sim}{\Phi}$ has the meaning of a pair chemical potential. Furthermore, $\vec{v}_s$ is the superfluid velocity. Note that the gauge invariant derivative now assumes the form

$$\bar{\bar{\vec{\partial}}} = \bar{\bar{1}}\; \partial/\partial\vec{r} + i\hat{\tau}_3 m\vec{v}_s \; . \qquad \text{(II.26b)}$$

It has already been pointed out previously that an expansion of the dot-product in time derivatives is equivalent to an expansion in powers of $\hbar$. There are, however, in addition other quantities which involve powers in $\hbar$ when written in dimensional form. It is not difficult to convince oneself that one obtains a systematic expansion of the equation of motion (II.20) if the following rules are observed:

i) Time and space derivatives carry a factor $\hbar$;

ii) Collision rates (e.g. collision integral) are proportional to $\hbar$;

iii) Diffusion constant D carries a factor $\hbar^{-1}$.

All other quantities are of order $\hbar^o$.* There is strong evidence that an expansion of the equation of motion through first order $\hbar$ leads to a consistent theory. This theory is also quantitatively correct for processes which evolve sufficiently slowly in time such that the relevant frequencies are much smaller than the gap frequency i.e. if $\omega \ll \Delta$. If there is not an expressive statement to the

---

*Sometimes, it appears that the "strange" densities of states as N_2, R_1, etc. carry a factor $\hbar$ in the accessible energy range $|E| > \Delta$. For instance, it follows from Eq. (II.6) that in leading order, $N_2 = N_1\Delta/2\tau_E(E^2 - \Delta^2)$. However, we do not need to introduce such an additional rule.

contrary, we will in the following, strictly adhere to this principle.

As far as the structure of $g^{R(A)}$ is concerned, we will assume that it has only components proportional to $\hat{\tau}_3$ and $\hat{\tau}_1$, which means, that it is of the type (II.6b) though the detailed form of the density of states N_1 and R_1 may be different from Eq. (II.6a). It should be stressed that this assumption has to be checked for consistency in each case, since generally it may not be true. Later, there is a discussion on this assumption following Eq. (II.30b) of Sec. (II.4); but presently, we will use this simplified form since it is more convenient for demonstration. Furthermore, it has proven to be a correct or sufficiently accurate form in a number of important cases.

Let us now return to a further discussion of Eq. (II.25a). At this point, I wish to stress that a derivation of the kinetic equation requires only the diagonal component of the equation of motion.* Thus, we obtain from Eq. (II.25a) through first order $\not{\partial}$

$$0 = \hat{1}[N_1 \dot{f}^{(L)} - N_1 \dot{\Phi} \frac{\partial f^{(T)}}{\partial E} + R_2 \dot{\Delta} \frac{\partial f^{(L)}}{\partial E}] + \hat{\tau}_3[N_1 \dot{f}^{(T)} - N_1 \dot{\Phi} \frac{\partial f^{(L)}}{\partial E} + 2\Delta N_2 f^{(T)} - \{\{R_2, f^{(T)}\}\}] . \tag{II.25b}$$

Here and below, we omit the suffix s since no confusion may arise.

Postponing the discussion of the diffusion term for a moment, we turn our attention to the inelastic collisions. Clearly, we may replace the dot-product by simple multiplication. Furthermore, we will assume that the deviations of the distribution functions from local thermal equilibrium, namely $\delta f^{(L)} = f^{(L)} - f_{th}$ and $\delta f^{(T)}$, are small such that we may linearize the collision integrals. This assumption may not always be satisfied in a real situation. However, the full nonlinear form has a rather complicated structure and furthermore, we need to simplify later even the linear form in order to allow for analytical calculations. For sake of simplicity, we will assume that the phonons are in equilibrium**. Then, we obtain

---

* Due to this fact, spin-flip scattering does not appear explicitly in the Boltzmann-equation.

** In the general case, we have to replace $\mathrm{cth}\frac{E}{2T}$ of Eq. (II.8b) by $1+2n(E)$, where $n(E)$ is the distribution function of the phonons. For details we refer to the paper of Eckern and Schön (1978).

$$- \frac{i}{4(1+\lambda)} \{\frac{1}{2} [\hat{\Sigma}^R_{ph} - \hat{\Sigma}^A_{ph} , \hat{g}^K]_+ + \hat{\Sigma}^K_{ph} \hat{g}^A - \hat{g}^R \hat{\Sigma}^K_{ph}\}^{diag}$$

$$= - \hat{1}K^{(L)}(\delta f^{(L)}) - \hat{\tau}_3 K^{(T)}(f^{(T)}) , \qquad (II.25c)$$

where the collision integrals $K^{(L,T)}$ are given by

$$K^{(L,T)}(\eta) = \frac{-\pi}{1+\lambda} \int_{-\infty}^{+\infty} dE' \frac{\alpha^2 F(E - E')}{sh\frac{E-E'}{2T} ch \frac{E}{2T} ch \frac{E'}{2T}} \qquad (II.27a)$$

$$\times M^{(L,T)}(E,E') [ch^2 \frac{E}{2T} \eta(E) - ch^2 \frac{E'}{2T} \eta(E')],$$

and where $M^{(L,T)}$ is defined by*

$$M^{(L)}(E,E') = N_1(E)N_1(E') - R_2(E)R_2(E') ;$$
$$M^{(T)}(E,E') = N_1(E)N_1(E') + N_2(E)N_2(E') . \qquad (II.27b)$$

The contributions of the diffusion term are rather numerous and their derivation is explained in App. B . Repeatedly, there occurs the expression $[\hat{\vec{\partial}} \overset{\circ}{,} \hat{h}]_- = \partial\hat{h}/\partial\vec{r} + \hat{\tau}_3 m\vec{v}_s \partial\hat{h}/\partial E = 2[\hat{1}\vec{q}^{(L)} + \hat{\tau}_3\vec{q}^{(T)}]$, where we have introduced the abbreviations

$$\vec{q}^{(L)} = - \frac{\partial f^{(L)}}{\partial\vec{r}} - m\vec{v}_s \frac{\partial f^{(T)}}{\partial E} ;$$
$$\vec{q}^{(T)} = - \frac{\partial f^{(T)}}{\partial\vec{r}} - m\vec{v}_s \frac{\partial f^{(L)}}{\partial E} , \qquad (II.27c)$$

which have the meaning of quasiparticle currents pertaining to the (L) - and (T)- mode, respectively.

Collecting now the results on the terms containing time derivatives (Eq. (II.25b)), on the diffusion terms (Eq.(B.3)) and on the collision integrals (Eq. (II.27a)), we obtain a system of two Boltzmann equations which control the temporal evolution of the distribution function $f = f^{(L)} + f^{(T)}$. This set of equations is written in explicit form below:

*In the BCS limit $M^{(L)}/N_1N_1'$ is just the coherence factor $[1 - \Delta^2/EE']$.

$$N_1\dot{f}^{(L)} - N_1\dot{\Phi}\frac{\partial f^{(T)}}{\partial E} + R_2\dot{\Delta}\frac{\partial f^{(L)}}{\partial E}$$
$$+ \operatorname{div} DM^{(L)}(E,E)\vec{q}^{(L)} + 4Dm\vec{v}_s N_2R_2\vec{q}^{(T)}$$
$$+ D\{\{m\vec{v}_s(N_2^2 - R_2^2), \vec{q}^{(T)}\}\} + Dm\dot{\vec{v}}_s\frac{\partial}{\partial E}M^{(T)}(E,E)\vec{q}^{(T)} - K^{(L)}(\delta f^{(L)}) = 0 \ ;$$

$$N_1\dot{f}^{(T)} - N_1\dot{\Phi}\frac{\partial f^{(L)}}{\partial E} + 2\Delta N_2 f^{(T)} - \{\{R_2, \Delta f^{(T)}\}\}$$
$$+ \operatorname{div} DM^{(T)}(E,E)\vec{q}^{(T)} + 4Dm\vec{v}_s N_2R_2\vec{q}^{(L)}$$
$$+ D\{\{m\vec{v}_s(N_2^2 - R_2^2), \vec{q}^{(L)}\}\} + Dm\dot{\vec{v}}_s\frac{\partial}{\partial E}M^{(L)}(E,E)\vec{q}^{(L)} - K^{(T)}(f^{(T)}) = 0. \qquad \text{(II.27d)}$$

Quite generally, the distribution functions $\delta f^{(L)}$ and $f^{(T)}$ are even and odd functions of the energy, respectively. In some cases, for instance in the linear regime and for $\vec{v}_s = 0$, the equations for the (L)- and (T)- mode decouple.

Of the various terms in the Boltzmann equation, each of them may become important in a specific case, and some of them are unique to the superconducting state. The cases studied in Sections III and IV are meant to illustrate this point. Further information is contained in App. C, where a theorem on energy conservation is derived, and in App. D, where the transformation of the Boltzmann equation from the present E-representation to a generalised ξ-representation is studied.

However, inspite of the peculiarities which Eq. (II.27d) owes to the superconducting state, it is instructive to study the transition to the normal state, where $N_1 = 1$ and where all other spectral densities vanish. Then, the equations for the (L)- and (T)- mode become formally indentical; and one can add both to obtain one equation for $f = f^{(L)} + f^{(T)}$, namely

$$\dot{f} - \dot{\Phi}\frac{\partial f}{\partial E} - D\left[\frac{\partial}{\partial\vec{r}} - m\dot{\vec{v}}_s\frac{\partial}{\partial E}\right]^2 f - K(f) = 0 \ . \qquad \text{(II.28a)}$$

At this point, one should recall that the method of quasiclassical Green's function always leads to a kinetic equation where the quantities depend on the local energy E. In the case of a normal metal , this means that $E = \xi_p - \Phi$, and it is advantageous to introduce a distribution function

$$\tilde{f}(\xi;\vec{r},t) = f(\xi-\Phi;\vec{r},t) \ . \qquad \text{(II.28b)}$$

Evidently, we have $\dot{f} = \dot{f} - \dot{\Phi}\partial f/\partial E$, and a similar relation holds where the time derivative is replaced by the space derivative. It follows from Eq. (II.26) that $e\vec{E} = m\vec{v}_s + \nabla\Phi$. Therefore, we may write Eq. (II.28a) in the form

$$\dot{\tilde{f}} - D[\frac{\partial}{\partial\vec{r}} + e\vec{E}\frac{\partial}{\partial\xi}]^2 \tilde{f} - K(\tilde{f}) = 0 \ . \qquad \text{(II.28c)}$$

This is the usual form of the Boltzmann equation in the diffusive limit.

The collision integrals satisfy the relation

$$\int dE \ K^{(L,T)}(\eta) = 0 \ . \qquad \text{(II.27e)}$$

This is an important property in the (T)- mode which guarantees, as one can show, particle conservation*. This conservation law has to be preserved in any meaningful approximation. Therefore, we will frequently make use of the reduced collision operator.

$$K^{(T)}_{red} = -\frac{N_1}{\tau_E}\{f^{(T)}(E) - \frac{1}{4T\mathrm{ch}^2 E/2T}\frac{1}{Y(T)}\int dE' N_1(E')f^{(T)}(E')\}, \qquad \text{(II.27f)}$$

where $Y(T) = \int dE N_1/4T\mathrm{ch}^2 E/2T$ is the Yoshida function. Evidently $K^{(T)}_{red}$ has the property implied by Eq. (II.27e); it allows us to solve many problems by straightforward integration.

Since $\delta f^{(L)}$ is an odd function, Eq. (II.27e) is meaningless in the (L)-mode. Energy conservation would require that the integral of $\{EK^{(L)}\}$ should vanish. Presently, this is not the case since the phonons act as a heat bath. On the other hand, if $\Delta \ll T_c$, this circumstance allows us to make use of the relaxation ansatz

$$K^{(L)}_{red} = -\frac{N_1}{\tau_E}[f^{(L)} - f_{th}] \ , \qquad \text{(II.27g)}$$

which means a considerable simplification.

4. Self-Consistency Relations

Before we proceed to establish conditions which allows us to determine the order parameter and the fields $(\Phi, \vec{v}_s)$, we will

*For the normal state, this statement is evident.

investigate some properties of the spectral densities in non-equilibrium situations. From Eq. (II.20), it follows that the equation of motion for $\hat{g}^R$ is given by

$$[\hat{\tau}_3(E + \frac{i}{2\tau_E}) + \hat{1}\Phi + i\hat{\Delta} + \frac{i}{\tau_s}\hat{\tau}_3\hat{g}^R\hat{\tau}_3 \overset{\circ}{,} \hat{g}^R]_- \\ -\frac{i}{2}D\,[\hat{\vec{\partial}} \overset{\circ}{,} [\hat{g}^R \overset{\circ}{,} [\hat{\vec{\partial}} \overset{\circ}{,} \hat{g}^R]_-]_-]_- = 0\ , \tag{II.29a}$$

where the weak coupling expressions $(\tau(E) \underset{\sim}{\sim} \tau_E)$ have been inserted for $\hat{\Sigma}^R_{ph}$. We solve this equation using the ansatz

$$\hat{g}^R = \hat{\tau}_3\alpha + \hat{\tau}_1\beta + \hat{\tau}_2\gamma, \tag{II.30a}$$

which features an additional component* proportional to $\hat{\tau}_2$ in comparison with the equilibrium form (II.6b). I will argue later that this component is less important if the self-consistency conditions are formulated in a suitable way.

As previously, we will expand the dot-product through first order in $\hbar$. Fortunately, the first order time derivatives cancel, except for a term $\propto\dot{\Phi}$, in the two off-diagonal components of Eq. (II.29a) which are the only ones required here. The two resulting equations can be combined in the following form

$$(E + \frac{i}{2\tau_E})\,(\beta \pm i\gamma) \pm i\dot{\Phi}(\beta \pm i\gamma) - i\Delta\alpha + \frac{i}{\tau_s}\alpha(\beta \pm i\gamma) \\ -\frac{i}{2}D\,[\alpha(\frac{\partial}{\partial\vec{r}} \pm 2im\vec{v}_s)^2(\beta \pm i\gamma) - (\beta \pm i\gamma)\frac{\partial^2}{\partial\vec{r}^2}\alpha] = 0. \tag{II.29b}$$

To the same degree of accuracy, the normalisation (II.17) requires

$$\alpha^2 + \beta^2 + \gamma^2 = 1\ . \tag{II.29c}$$

For illustration, let us write down an approximate solution of Eq. (II.29b) in powers of E^{-1}. One obtains

---

*This component has been introduced explicitly first by Hu (1980). Actually, $\hat{g}^R$ contains also a component $\propto\hat{1}$. We have found presently no situation where it has been needed.

$$\alpha = 1 + \frac{1}{2}\Delta^2 E^{-2} + \ldots ;$$

$$\beta = i\Delta E^{-1} + [\frac{1}{\tau_s} + \frac{1}{2\tau_E} + \frac{D}{2}(4m^2 v_s^2 - \frac{\partial^2}{\partial \vec{r}^2})\Delta]\, E^{-2} - \frac{i}{2}\Delta^3 E^{-3} + \ldots ;$$

$$\gamma = -[\frac{mD}{\Delta}\frac{\partial}{\partial \vec{r}}\vec{v}_s\Delta^2 + i\dot{\Phi}]E^{-2} + \ldots\ldots . \qquad \text{(II.31a)}$$

The expansion (II.31a) is the basis for the derivation of the Ginzburg-Landau equation in the equilibrium theory. Obviously, γ is connected with the divergence of the supercurrent.

One may infer from Eq. (II.31a) that

$$\Gamma = \frac{1}{\tau_s} + \frac{1}{2\tau_E} + \frac{D}{2}(4m^2 v_s^2 - \Delta^{-1}\frac{\partial^2}{\partial \vec{r}^2}) \qquad \text{(II.31b)}$$

acts as a pair breaking energy. In the limit of large pair breaking $\Gamma >> \Delta$, one finds the following result

$$\alpha = 1 ; \quad \beta = \frac{i\Delta}{E+i\Gamma} . \qquad \text{(II.31c)}$$

In intermediate cases we have frequently used the approximation

$$\alpha^{R(A)} = \frac{E_{(\pm)} i\Gamma}{[(E_{(\pm)} i\Gamma)^2 - \Delta^2]^{1/2}} ;$$

$$\beta^{R(A)} = \frac{i\Delta}{[(E_{(\pm)} i\Gamma)^2 - \Delta^2]^{1/2}} , \qquad \text{(II.31d)}$$

$(\mathrm{Im}[..]^{1/2} > 0)$, which agrees with Eqs. (II.6a) and (II.5b) if $\Gamma = 1/2\tau_E$.

One obtains similar results for the advanced Green's function $\hat{g}^A$. Quite generally, one finds that

$$\alpha^{R(A)} = {}_{(\pm)}N_1 + iR_1 ;$$

$$\beta^{R(A)} = N_{2(\pm)} iR_2 ; \qquad \text{(II.30b)}$$

$$\gamma^{R(A)} = N_{3(\pm)} iR_3 ,$$

where N_j and R_j are even and odd functions of E as before.

A Boltzmann equation which includes the spectral densities N_3 and R_3 (Hu, 1980) differs from form (II.27) derived in the previous section in two respects. Firstly, some terms acquire corrections of an obvious type; in the definition of Eq. (II.27b) for instance, one now obtains $M^{(L)} = N_1N_1' - R_2R_2' - R_3R_3'$, etc. Considering Eq. (II.31a), we estimate that γ leads to contributions smaller than those of β by at least a factor $(\Delta/T)^4$. Secondly, new terms arise which are formally different from previous ones. However, in the problems which we have studied so far these terms did not lead to new physical effects and numerically, their contribution has been found negligible.*

Let us now turn our attention to the self-consistency conditions (II.24b) for the order parameter. Neglecting for a moment terms containing time derivatives, we have

$$\frac{1}{4i}\{\hat{g}^K\}_{12} = \frac{1}{2}[R_2(1-2f^{(L)}) - 2N_3f^{(T)}] - \frac{i}{2}[N_2f^{(T)} + R_3(1-2f^{(L)})]. \tag{II.32}$$

Accordingly, there follow from Eq. (II.24b) two separate conditions one for the real and one for the imaginary part. We will see later that the condition on the imaginary part guarantees conservation of charge: indeed, the term $R_3(1-2f^{(L)})$ does lead to the expression div $\vec{J}_s$ in the equilibrium theory. On the other hand, the term $N_3f^{(T)}$ leads only to minor corrections in the condition for the real part. These two facts suggest the following procedure. Charge conservation will be enforced directly through the continuity equation, which allows us to abandon the condition on the imaginary part. Thus, (in most cases), we do not need* to calculate the spectral quantities N_3 and R_3.

Of the contributions to $\{\hat{g}^K\}_{12}/4i$ linear in the time derivatives, only the term $-(1/2)\dot{N}_2\partial f_{eq}/\partial E$ is important. Note that, in connection with f_{eq} (and $\partial f_{eq}/\partial E$), we may use the expanded form (II.31a) for the spectral quantities from which we obtain the equilibrium part (and the "simple" time derivative) of the following Ginzburg-Landau equation,

$$\frac{\pi}{8T_c}\dot{\Delta} = -\{\alpha + \beta\frac{\Delta^2}{T_c^2} + \xi(o)^2[4m^2\vec{v}_s^{\,2} - \frac{\partial^2}{\partial r^2}] - \chi\}\Delta. \tag{II.33a}$$

---

* In general, one should check whether γ is negligible, indeed.

In this equation, $\alpha = (T-T_c)/T_c$; $\beta = 7\zeta(3)/8\pi^2$; $\xi^2(o) = \pi D/8T_c$, and

$$\chi = -\frac{1}{\Delta}\int dE\ R_2 \delta f^{(L)} . \tag{II.33b}$$

Since χ controls the magnitude of the order parameter, we will call it the control function.

In order to illustrate the role of the self-consistency condition on the charge conservation, we examine a simple type of equation for Keldysh's function. Taking an appropriate trace with respect to particle-hole space, we obtain

$$\begin{aligned} 0 &= \mathrm{Sp}\ \hat{\tau}_3\ [\hat{\tau}_3 E + i\hat{\Delta} \overset{\circ}{,} \hat{g}^K]_- \\ &= i\mathrm{Sp}\ \dot{\hat{g}}^K + i\mathrm{Sp}\ \hat{\tau}_3\ [\hat{\Delta}, \hat{g}^K]_- + \mathrm{Sp}\ \hat{\tau}_3\{\dot{\hat{\Delta}}\ \frac{\partial \hat{g}^K}{\partial E} + \ldots\ldots\} . \end{aligned} \tag{II.34}$$

If we integrate this equation with respect to E, all the contributions to the third term in the second line vanish. It follows from Eq. (II.21), that the first term is essentially the time derivative of the charge density. We conclude that this time derivative vanishes if $\int dE\ \hat{g}^K$ has only an off-diagonal component proportional to $\hat{\Delta}$. This is, however, equivalent to the condition that the imaginary part of $\int dE\{\hat{g}^K\}_{12}/4i$ should vanish.

Therefore, we introduce explicitly the continuity equation $\dot{\rho} + \mathrm{div}\vec{j} = 0$. A considerable simplification is possible if one exploits the fact that in a metal, space charges are suppressed almost entirely by Coulomb forces. Thus, we may introduce

$$\rho = 0 \ ; \qquad \mathrm{div}\ \vec{j} = 0 \ , \tag{II.35a}$$

as an extra condition. Subsequently, Maxwell's equation concerning the currents can be written in a simplified form

$$\mathrm{curl}\ \mathrm{curl}\ \vec{v}_s = -\frac{4\pi e}{mc}\vec{j} \ , \tag{II.35b}$$

where Eq. (II.26) has been used to substitute the vector potential.* Concerning the expression for the charge density, we obtain from Eq. (II.21) within our standard accuracy (zero order in $\hbar$)

*If necessary, the curl of two dimensional δ- functions centered at the position of vortices have to be added to the r.h.s. of Eq. (II.35b).

$$\rho = 2eN(o)\{\int dE\ N_1 f^{(T)} + \Phi\}, \qquad (II.36a)$$

In a first step, we obtain from Eq. (II.21) for the current density the following expression

$$\vec{j} = \frac{\sigma}{8e}\int dE\ Sp\ \hat{\tau}_3\{[\hat{\vec{\partial}} \overset{\circ}{,} \hat{h}]_- - \hat{g}^R \circ [\hat{\vec{\partial}} \overset{\circ}{,} \hat{h}]_- \circ \hat{g}^A + \frac{1}{2}[\hat{g}^R \overset{\circ}{,} [\hat{\vec{\partial}} \overset{\circ}{,} \hat{g}^R]_-]_- \circ \hat{h} - \frac{1}{2}\hat{h} \circ [\hat{g}^A , [\hat{\vec{\partial}} \overset{\circ}{,} \hat{g}^A]_-]_-\} , \qquad (II.36b)$$

where $\sigma = 2(1+\lambda)e^2N(o)D$. Again, the expression is evaluated to zero order in $\not{M}$. From a formal point of view at least, it seems natural to separate the current density in a normal and in a superfluid part as follows

$$\vec{j}_n = \frac{\sigma}{e}\int dE\ [M^{(T)}(E,E)\vec{q}^{(T)} - \{\{m\vec{v}_s(N_2^2 - R_2^2), f^{(L)}\}\}] \ ;$$

$$\vec{j}_s = \frac{2\sigma}{e}\ m\vec{v}_s\int dE\ N_2R_2\ [1-2f^{(L)}] \ . \qquad (II.36c)$$

Recall that $\vec{q}^{(T)}$ is defined in Eq. (II.27c). In writing down this expression, it has been assumed additionally that $\hat{g}^{R(A)}$ acquires no $\gamma\hat{\tau}_2$ which would appear as an extra contribution to the supercurrent in form of a space derivative.

It is interesting to note that in cases where the pair breaking is small, N_2R_2 acts almost like a δ-function at $|E|=\Delta$, that is $N_2R_2 \simeq (\pi/2)\Delta E\delta(E^2-\Delta^2)$. In this case, one finds

$$\vec{j}_s = \frac{\pi\sigma}{e}\ m\vec{v}_s\Delta[1-2f^{(L)}(E=\Delta)] \ . \qquad (II.36d)$$

Another problem frequently encountered is the linear response in a spatially homogeneous electric field. There, we have $m\vec{v}_s = e\vec{E}$, and $\partial f^{(T)}/\partial\vec{r} = 0$. Then, we obtain

$$\vec{j}_n = \vec{E}\ \sigma\int dE\ [N_1^2 + R_2^2\ ](-\frac{\partial f_{th}}{\partial E}) \ . \qquad (II.36e)$$

In conclusion we may say that a time dependent Ginzburg-Landau equation (III.33) has been established which allows us to calculate the real part Δ of the order parameter. The phase of the order parameter is contained together with the electromagnetic potentials in the gauge invariant combinations Φ and

$\vec{v}_s$ which are the pair chemical potential and the superfluid velocity. Both quantities can be determined from Eq. (II.35). Note that from Eq. (II.26), there follows a connection between Φ and $\vec{v}_s$,

$$m\vec{v}_s = - \nabla\Phi + e\vec{E} , \qquad \text{(II.37)}$$

which has the meaning of an acceleration equation for the superfluid According to Eq. (II.37), an electric field may exist; this field, however, plays no role in the present theory. In particular, $\vec{E} \neq 0$ does not contradict the condition $\rho = 0$ since nowhere do we rely on Maxwell's equation connecting the electric field with space charges.

III. COLLECTIVE RELAXATION PROCESSES

1. Relaxation of the Magnitude of the Order Parameter

We will study here the temporal behaviour of small changes $\delta\Delta$ of the magnitude of the order parameter from its equilibrium value which will be denoted by Δ.

The simplest case is when $\Delta = 0$, that is, for temperatures above the transition temperature. Obviously, we also have $\chi = v_s = 0$. Therefore, if $\delta\Delta$ is constant in space, it follows from Eq. (II.33) that

$$\frac{\pi}{8T}\, \delta\dot{\Delta} = - \alpha\delta\Delta , \qquad \text{(III.1a)}$$

from where we deduce a relaxation time

$$\tau_R^{(L)} = \frac{\pi}{8} \cdot \frac{1}{|T-T_c|} . \qquad \text{(III.2a)}$$

This relaxation time can be seen experimentally by observing the decay of thermal fluctuations of the order parameter. Specifically, Anderson, Carlson, and Goldman (1972) have analysed the fluctuation enhanced Josephson current of a junction with an Aluminum film as one electrode. One can derive from a more detailed theory that this current has a peak at a bias voltage $2eV_p = \hbar/\tau_R(L)$. Except for a region very close to T_c, excellent agreement is found if the relaxation time given by Eq. (III.2a) is used for comparison.

In the superconducting state, the relaxation of the order parameter is controlled by rearrangements in the quasiparticle population of the electronic levels, which are induced by the change in time of the order parameter. Again, we consider a spatially homogeneous situation in which case $f^{(T)} = 0$ in the linearized form of the theory. Then, the (L)- mode Boltzmann equation is given by

$$N_1 \delta \dot{f}^{(L)} + R_2 \delta \dot{\Delta} \frac{\partial f_{th}}{\partial E} + \frac{N_1}{\tau_E} \delta f^{(L)} = 0 \, , \tag{III.3}$$

where the relaxation ansatz (II.27g) has been used for the collision integral and where we have put $\vec{v}_s = 0$ for simplicity. We assume that the time dependence of $\delta f^{(L)}$ is controlled by $\delta\dot{\Delta}$, and that the relaxation time $\tau_R^{(L)} \gg \tau_E$. Then, the first term of Eq. (III.3) can be neglected and we may solve this equation for $\delta f^{(L)}$. In the next step, we calculate the control function

$$\chi = \tau_E \delta\dot{\Delta} \frac{1}{\Delta} \int dE \frac{R_2^2}{N_1} \frac{\partial f_{th}}{\partial E} = - \frac{\pi}{4T_c} \tau_E \delta\dot{\Delta} \, . \tag{III.4}$$

We observe that R_2^2/N_1 is sharply peaked at $|E| = \Delta$ and that it decreases rapidly $\propto E^{-2}$ for large energies. Therefore, we may put $\partial f_{th}/\partial E = -1/4T_c$ in the above integral provided that $\Delta \ll T_c$. The precise numerical coefficient in the last line of Eq. (III.4) has been obtained in the limit of vanishing pair-breaking Γ.

In most cases, $\tau_E\Delta \gg 1$. Then, the "simple" time derivative in the Ginzburg-Landau equation (II.33) can be neglected and we may write down the following linearized form

$$\frac{\pi}{4T_c} \tau_E \Delta \delta\dot{\Delta} = - 2\beta \frac{\Delta^2}{T_c^2} \delta\Delta \, , \tag{III.1b}$$

where the equilibrium order parameter is determined by $\beta\Delta^2/T_c^2 = -\alpha$. This equation implies a relaxation time

$$\tau_R^{(L)} = \frac{\pi^3}{7\zeta(3)} \frac{T_c}{\Delta} \tau_E \, , \tag{III.2b}$$

with the property that $\tau_R^{(L)} \propto \Delta^{-1} \propto (T_c - T)^{-1/2}$ for $\Delta \to 0$. Very close to T_c the assumption $\Delta\tau_E \gg 1$ becomes invalid. Eckern and Schön (1978) have calculated $\tau_R^{(L)}$ for arbitrary pair breaking Γ (c.f. Eq. (II.31d)), and for arbitrary values of the order parameter Δ, and they have obtained the result

$$\tau_R^{(L)} = \frac{\pi^3}{7\zeta(3)} \frac{T_c}{\Delta} \tau_E \left[\left(1 + \frac{\Gamma^2}{\Delta^2}\right)^{1/2} - \frac{\Gamma}{\Delta} + \frac{1}{2\Delta\tau_E} \right] \, , \tag{III.2c}$$

provided that $\Delta \ll T_c$. In the limit $\Delta \to 0$, it follows that $\tau_R^{(L)} = (\pi^3/14\zeta(3))T_c/\Delta^2 = \pi/16(T-T_c)$, which is just half* the value given by Eq. (III.2a). Clearly, the assumption $\tau_R^{(L)} \gg \tau_E$ made at the beginning is satisfied if $\Delta \ll T_c$.

Schuller and Gray (1977) were able to observe the relaxation of the magnitude of the order parameter in Aluminium after having it driven out of thermal equilibrium by a short laser pulse. The measured relaxation time was indeed found to diverge $\propto (T_c-T)^{-1/2}$ in the vicinity of the transition temperature. Earlier, Peters and Meissner (1973) observed a similar behaviour for this relaxation time by measuring the high frequency impedance of superconducting tin. In their experiments, the pair breaking effect of a supercurrent was used as a tool to manipulate the magnitude of the order parameter. In both experiments, the inferred values of τ_E were in reasonable agreement with those obtained from independent sources.

A few remarks are in order concerning the question how the relaxation of the order parameter is related to the recombination of quasiparticles.

It is commonly said that in the latter process, a pair of quasiparticles recombines to form a Cooper pair. There are some problems which arise from this interpretation. It is known that the recombination phonon carries away all the momentum so that the quasiparticles recombined have zero momentum. This result leads however, to a conflict in a situation where the common momentum of the pairs is non-zero. Furthermore, if we consider these processes from a purely formal point of view, we realize that these recombination processes continue to exist even after the transition to the normal state.

In the picture which emerges from the present theory, the order parameter is a different word for pair correlations and a change in the order parameter is equivalent to a change in pair correlations. A recombination process means that a particle recombines with a hole. In a superconductor, such a process** clears some part of the phase space to pair correlations. In a second step the pairs adjust their correlations to this new situation at an intrinsic rate $v_F/\xi_o \sim T_c$ which leads to an effective rate $\sim |T-T_c|$ near the transition temperature on account of competing processes. Except for temperatures very close

* The stiffness against order parameter fluctuations is twice as large below T_c than above T_c at the same value of $|\alpha|$.

** There are other processes which lead to the same end. Microwave stimulated superconductivity may serve as an example.

to T_c (or above T_c), this rate is the largest and therefore, it is the slow dynamics of the quasiparticle which controls the rate of changes of the order parameter. On the other hand, the dynamics of the quasiparticle has to be found under condition of a changing energy gap and thus, it may be different from the dynamics at constant gap.

2. Conversion of Currents

This phenomenon has been discovered as well as explained by Clarke (1972) and Tinkham (1972). Commonly, it is called the relaxation of branch imbalance; in the framework of the present theory, however, it seems to be more appropriate if I refer to it as the conversion of currents, specifically the conversion of normal current into supercurrent and vice versa.

From a phenomenological point of view, a finite rate of conversion produces a finite difference in the electochemical potential $\mu_P(=\dot{\theta}/2)$ and μ_Q of the Cooper pairs and quasiparticles, respectively. Therefore, we put

$$\mu_Q - \mu_P = - \frac{J_s}{2(1+\lambda)eN(o)} \tau_R^{(T)} , \qquad \text{(III.5a)}$$

where $\tau_R^{(T)}$ is the characteristic time for conversion*, and where J_s is the supercurrent converted per unit volume.

As far as the definition of chemical potentials is concerned we consider Eq. (II.37) and conclude that supercurrents, if stationary, connect regions having the same μ_P. Usually, diffusive equilibrium in the quasiparticle system is thought to indicate a constant μ_Q. However, one has to keep in mind that for non-thermal states, this definition depends on the nature of the diffusion process. In the case of superconductors, one is led by experience to define

$$\mu_Q - \mu_P = \int dE\, N_1 f^{(T)} . \qquad \text{(III.5b)}$$

In equilibrium, $f^{(T)}=0$, and hence $\mu_Q=\mu_P$, as it should be.

In the experiment of Clarke, a quasiparticle current J_n (per unit volume) is injected via a tunnel junction from a normal metal (N) into the superconductor. There it is converted into a supercurrent J_s (hence $J_n=J_s$) which flows out of the active volume. The dimension of this volume has been chosen such that the non-equilibrium distribution $f^{(T)}$ is constant in space. A second tunnel junction couples a detector (D, which is a normal metal) to the

*In the hydrodynamic theory, $\zeta = \tau_R^{(T)}/2(1+\lambda)N(o)$ is the coefficient of second viscosity (Schmid, 1980).

superconductor, and one measures the voltage V_D that is required to reduce the tunnel current to the detector to zero.*

We model our calculation of $\tau_R^{(T)}$ according to this experimental arrangement. In order to describe mathematically the injection of electrons from the normal metal (N) at voltage V_N to the super-conductor, we start from the tunneling Hamiltonian $\bar{\hat{H}}_{NS} + \bar{\hat{H}}_{SN}$, where

$$\bar{H}_{NS} = \bar{H}^{+}_{SN} = T_{NS} e^{i\bar{\hat{\tau}}_3 eV_N t} \tag{III.6a}$$

In second order perturbation theory, we obtain the following contribution to the self-energy of the superconductor

$$\bar{\Sigma}_T = - i\pi N_N(o)\ \bar{H}_{SN} \circ \bar{g}_N \circ \bar{H}_{NS} \tag{III.6b}$$

The normal metal (of which the density of states is $N_N(o)$) may be assumed to be thermal equilibrium;hence

$$\hat{g}_N^K = 2\hat{\tau}_3 \text{th}\,\frac{E}{2T} \quad ; \qquad \hat{g}_N^{R(A)} = \binom{+}{-}\hat{\tau}_3\ , \tag{III.6c}$$

and one finds that

$$\hat{\Sigma}_T^K = - i\pi N_N(o)|T_{NS}|^2 \cdot 2\hat{\tau}_3 \text{th}\,\frac{E+\hat{\tau}_3 eV}{2T} \quad ; \qquad \hat{\Sigma}_T^{R(A)} = \binom{-}{+} i\pi N_N(o)|T_{NS}|^2 \hat{\tau}_3\ . \tag{III.6d}$$

The contribution of this self-energy part to the Boltzmann equation can be determined with the dot-product in zero order of time derivatives. Thus, we obtain

$$Q_T^{(L)} = \frac{2\pi}{1+\lambda} N_N(o)|T_{NS}|^2 N_1(E)\{f^{(L)}(E) - \tfrac{1}{2}[f_{th}(E+eV_N)+f_{th}(E-eV_N)]\};$$

$$Q_T^{(T)} = \frac{2\pi}{1+\lambda} N_N(o)|T_{NS}|^2 N_1(E)\{f^{(T)}(E) + \tfrac{1}{2}[f_{th}(E+eV_N)-f_{th}(E-eV_N)]\}. \tag{III.7a}$$

In the stationary and spatially homogeneous case which is considered here, the (L)- and (T)- mode Boltzmann equations decouple. Specifically, for $f^{(T)}$ we obtain the equation

$$2\Delta N_2 f^{(T)} + Q_T^{(T)} - K^{(T)} = 0. \tag{III.7b}$$

In order to obtain analytical solutions, we replace the collision integral by its reduced form (II.27f). One can show that this approximation leads to exact results in the limit

---

* An analysis of the tunnel current reveals that $\mu_Q-\mu_P=g_{NS}\cdot eV_D$, where g_{NS} is the normalised conductance of the detector junction in the limit $eV_D \ll T$.

$\Delta\to 0$. Additional simplifications are possible if the tunneling probability is small and if $eV_N \ll T$. Then,

$$Q_T^{(T)} = - \frac{J_N}{2e(1+\lambda)N(o)} \cdot \frac{1}{Y(T)} \cdot \frac{N_1}{4T\mathrm{ch}^2 \frac{E}{2T}} , \qquad \text{(III,7c)}$$

where $J_N = - 2e(1+\lambda)N(o) \int dE\, Q_T^{(T)}$ is the tunnel current per unit volume from the injector metal and where Y(T) is the Yoshida function. Then, the distribution function is of the form

$$f^{(T)} = \frac{N_1}{N_1 + 2\Delta\tau_E N_2} \frac{1}{4T\mathrm{ch}^2 \frac{E}{2T}} \eta , \qquad \text{(III.8a)}$$

and we find that $\tau_R^{(T)} = J_1/2\Delta J_2$, where

$$J_1 = \frac{1}{\eta} \int dE\, N_1 f^{(T)} = Y(T) - 2\Delta\tau_E J_2 ;$$

$$J_2 = \frac{1}{\eta} \int dE\, N_2 f^{(T)} . \qquad \text{(III.8b)}$$

In the evaluation of J_2, we have to include the pair breaking energy Γ in the definition of N_2. If

$$T_c \gg \Delta(\Gamma\tau_E)^{1/2} , \qquad \text{(III.9a)}$$

then both expressions (II.31c,d) lead essentially to the same result which is

$$\tau_R^{(T)} = \frac{4T_c}{\pi\Delta} \left(\frac{\tau_E}{2\Gamma}\right)^{1/2} \left(1 + \frac{\Gamma}{2\Delta^2\tau_E}\right)^{1/2} . \qquad \text{(III.9b)}$$

Consider the case $\Gamma = 1/2\tau_E$ where only inelastic phonon collisions contribute to the pair breaking. If $\Delta\tau_E \gg 1$, we obtain $\tau_R^{(T)} = (4T_c/\pi\Delta)\tau_E$ which is, apart from a numerical coefficient, of the same form as $\tau_R^{(L)}$ given by Eq. (III.2b). However, the dependence of the relaxation times on Γ is different in the intermediate range, and only in the limit $\Delta\to 0$, both are similar again.

The connection between the branch imbalance relaxation time τ_Q^* and $\tau_R^{(T)}$ has been discussed by Clarke et al. (1979). It has been found that in the limit $\Delta \ll T_c$, both definitions coincide.

Concerning the appearance of the renormalisation constant $(1+\lambda)$ in Eq. (III.5a), one should recall that N(o) is the bare density of states. This means for instance that it is $(1+\lambda)N(o)$ which appears as density of states in the specific heat expression. In this context, I wish to stress that τ_E (c.f. Eq. (II.9c) is defined such that it has the meaning of an (observable) decay time of an excess quasiparticle population.

3. Decay of Normal Current

It is known from experimental and theoretical investigations of Pippard and coworkers (1971) and from the work of Hsiang and Clarke (1979), that an extra resistance arises when a current passes from a normal metal into a superconductor. This extra resistance originates from the finite rate of conversion and it is accompanied by a penetration of the normal current into the superconductor.

From a theoretical point of view, the problem of the extra resistivity of an N-S interface is very complex and the proximity effect plays an important role.* In order to avoid this difficulty, I consider the penetration of a normal current into an homogeneous superconductor. Such a model may apply quite well to the situation some distance away from the N-S boundary.

In the stationary case, and in the linear regime, the (T)-mode Boltzmann equation then assumes the form

$$2\Delta N_2 f^{(T)} - DM^{(T)}(E,E)\,\frac{\partial^2}{\partial x^2} f^{(T)} - K^{(T)} = 0\ . \qquad \text{(III.10)}$$

This is a homogeneous equation in $f^{(T)}$ which has infinitely many solutions of the form $f^{(T)} \propto e^{-\kappa x}$. We will determine only the solution of smallest value of κ. Note that this solution will dominate at larger distances, particularly, if the other κ values are distinctively larger which is the case for $\Delta \to 0$.

As done previously, we replace the collision integral by its reduced form. Then, the distribution function is of the same form as given by Eq. (III.8a) where, however, N_2 is replaced by $N_2 - D\kappa^2 M^{(T)}/2\Delta$. The same substitution should be carried through in the definitions (III.8b) of the relevant integrals, and the proper value for κ can be found from the condition

$$J_2 = \int dE\, \frac{N_1}{4T\mathrm{ch}^2 \frac{E}{2T}} \cdot F(E, D\kappa^2) = 0\ , \qquad \text{(III.11a)}$$

where

$$F(E,D\kappa^2) = \frac{N_2 - D\kappa^2 M^{(T)}/2\Delta}{N_1 + 2\Delta\tau_E N_2 - D\kappa^2 \tau_E M^{(T)}}\ . \qquad \text{(III.11b)}$$

This is a rather complicated integral for which we will find a suitable approximation as follows. Considering a form for the densities of states as given by Eq. (II.31d), we observe that the denominator of F is always positive (if $D\kappa^2\tau_E < 1$)

---

* See for instance, Schmid and Schön (1975)

and decreases as E increases. On the other hand, the numerator changes sign from plus to minus as E increases at some distance above the gap. A close inspection shows that we may use quite well the following approximations

$$N_1 \simeq M^{(T)} \simeq 1 \; ; \quad N_2 \simeq \frac{\Gamma\Delta}{E^2} \quad . \tag{III.12}$$

Then, the integration can be done explicitly and we obtain

$$J_2 = \frac{1}{\tau_E} \cdot \frac{1}{1-b} \{1 - y\psi^{(1)}(\tfrac{1}{2} + y)\}, \tag{III.13a}$$

where $y = \pi^{-1}(a/(1-b))^{1/2}$; $a = \Delta^2 \tau_E \Gamma / 2T^2$; $b = D\kappa^2 \tau_E$; and where $\psi^{(1)}$ is the derivative of the Digamma function. If $y \ll 1$, there is a solution of $J_2 = 0$ if

$$\Lambda^{(T)} = \kappa^{-1} = (D\tau_R^{(T)})^{1/2} \quad ;$$
$$\tau_R^{(T)} = \frac{2\tau_E}{\pi a^{1/2}} = \frac{4T}{\pi\Delta}\left(\frac{\tau_E}{2\Gamma}\right)^{1/2} \quad , \tag{III.14}$$

where $\Lambda^{(T)}$ is the current penetration length.* This relation is valid if $a \ll 1$ which is equivalent to condition (III.9a). On the other hand, the form (III.13a) allows a solution $J_2 = 0$ only if $a < \pi^2/12$, and then the maximal value of $\Lambda^{(T)}$ is $\Lambda = (D\tau_E)^{1/2}$. This failure to have solutions for larger values of a is a result of the approximations made previously which do not allow the integral (III.11a) to acquire sufficient negative contributions at larger values of the energy E. In essence, this failure does not result from approximation (III.12), but it is rather due to the neglect of the energy dependence of the inelastic collision rate.

In order to improve the calculations in this respect, we put

$$\tau(E) = \tau_E\left[1 + c\left|\frac{E}{2T}\right|^3\right]^{-1} \tag{III.15}$$

where $c = 16/21\zeta(3)$. Strictly, this expression is valid only if $\Delta \ll T$, and if either $|E| \ll T$ or $|E| \gg T$; for the present purpose, however, it is a sufficient approximation. Thus, we arrive at the integral

$$J_2 = \frac{1}{\tau_E} \int_0^\infty dx \, \frac{1}{ch^2 x} \, \frac{(1-bx^2)(1-cx^3)}{a + (1-b)x^2 + cx^5} \quad , \tag{III.13b}$$

---

*Essentially, $\Lambda^{(T)}$ agrees with Λ_{Q^*}.

which has to be evaluated numerically. For various values of the parameter $a = \Delta^2 \tau_E \Gamma / 2T^2$, the result is shown in the following table

a	0.01	0.1	1.0	5.0	10.0
$(\Lambda/\Lambda^{(T)})^2$	0.14	0.5	1.7	4.2	6.2
$\tau_E/\tau_R^{(T)}$	o.16	0.5	1.6	3.5	5.0

The second line lists the values of the current penetration length $\Lambda^{(T)}$, whereas the last line shows $\tau_E/\tau_R^{(T)} = \pi a^{1/2}/2$ for comparison.

We conclude that, due to some cancelations, the expression (III.14) for the current penetration has a larger range of applicability than expected. Experimental evidence for the dependence of the penetration length on Γ and τ_E has been found recently by Kadin et al. (1978).

The penetration of a normal current into a superconductor is accompanied by an electric field $\vec{E}$. Equating the expression (II.36a) for the charge density ρ to zero, we find that the pair chemical potential is equal to $\Phi = -\int dE\, N_1 f^{(T)}$. On the other hand, $\vec{v}_s = 0$ in the stationary case we are considering, and it follows from Eq. (II.37) that $e\vec{E} = \nabla\Phi$. Thus, $\vec{E}$ and Φ penetrate into the superconductor with the same characteristic length $\Lambda^{(T)}$.

4. Time and Space Dependent Conversion

Again, we consider the linearized form of the (T)- mode equations where the deviations $f^{(T)}$; Φ ; and $\vec{v}_s$ from zero are the relevant first order quantities. In the appropriate Boltzmann equation,

$$N_1 \dot{f}^{(T)} - N_2 \dot{\Phi} \frac{\partial f_{th}}{\partial E} + 2\Delta N_2 f^{(T)} - \frac{\partial R_2}{\partial E} \Delta \dot{f}^{(T)}$$
$$- D \frac{\partial}{\partial \vec{r}} M^{(T)} \left[\frac{\partial f^{(T)}}{\partial \vec{r}} + m\dot{\vec{v}}_s \frac{\partial f_{th}}{\partial E}\right] - K^{(T)} = 0 , \qquad \text{(III.16)}$$

the time derivatives of the distribution function can be incorporated by the following substitutions. Firstly, the term $N_1 \dot{f}^{(T)}$ can be combined with the scattering-out term $N_1 f^{(T)}/\tau_E$ of the collision integral by the following replacement

$$\frac{1}{\tau_E} \to \frac{1}{\tau_E} + \frac{\partial}{\partial t} . \qquad \text{(III.17a)}$$

Secondly, the term $(\partial R_2/\partial E)\Delta \dot{f}_2^{(T)}$ can be included in $2\Delta N_2 f_2^{(T)}$, if we substitute

$$\Gamma \to \Gamma + \frac{1}{2}\frac{\partial}{\partial t} . \qquad \text{(III.17b)}$$

This procedure can be justified as follows. According to Eq. (III.12), we may approximate $N_2 \simeq \Delta\Gamma/E^2$. Within the same degree of accuracy (c.f. Eq. (II.31c)), we are allowed to put $R_2 \simeq \Delta/E$. This concludes the argument.

With this reinterpretation of Γ and $1/\tau_E$, the distribution function is of the same general type as found previously. In particular, we obtain

$$\frac{2\Delta \int dE\ N_2 f_2^{(T)}}{\int dE\ N_1 f^{(T)}} = \frac{\pi\Delta}{4T_c}\left[(2\Gamma + \frac{\partial}{\partial t})(\frac{1}{\tau_E} + \frac{\partial}{\partial t})\right]^{1/2} . \qquad \text{(III.17c)}$$

Taking into account the condition $\int dE\ N_1 f^{(T)} + \Phi = 0$ for charge neutrality, and using the expression (II.36c) for the normal current, we can write the energy-integrated form of Eq. (III.16) as follows

$$-\frac{\pi\Delta}{4T_c}\left[(2\Gamma + \frac{\partial}{\partial t})(\frac{1}{\tau_E} + \frac{\partial}{\partial t})\right]^{1/2}\Phi + \frac{eD}{\sigma}\operatorname{div}\vec{j}_n = 0, \qquad \text{(III.18)}$$

provided that we omit the negligible contribution $-(1-Y(T))\dot{\Phi}$.

For sake of simplicity, we calculate the normal current in the approximation (III.12) where $M^{(T)} \simeq 1$. Taking into account charge neutrality, we obtain

$$\vec{j}_n = \sigma E , \qquad \text{(III.19)}$$

since $\nabla\Phi + m\dot{\vec{v}}_s = e\vec{E}$.

The consequences which follow from Eq. (III.18) have been discussed in detail by Kadin et al. (1980) for the case $\Gamma = 1/2\tau_E$. Two simple cases are noteworthy. In the stationary case where the time derivatives vanish and where $\nabla\Phi = e\vec{E}$, we are able to reproduce relation (III.14) for the current penetration length $\Lambda^{(T)}$. On the other hand, in the case of spatially homogeneous state, we find that the decay time of the pair chemical potential is just τ_E.

IV. OSCILLATORY PHENOMENA

1. The Carlson-Goldman Collective Mode

This mode considered here is characterized by oscillations in space and time of the superfluid velocity with almost sound-like dispersion. It was discovered by Carlson and Goldman (1975, 1976) who observed a new structure in the fluctuations of the Josephson current between two superconductors. An increase in this current, that is an increase in the dissipation indicates the excitation of this mode.

The frequency ω of this mode is rather large; and in the limit

$$\Gamma \ll \omega \ll \Delta \ll T_c \quad , \tag{IV.1}$$

and one can find simple equations which describe this mode. Then, we can neglect 2Γ and $1/\tau_E$ in comparison with $\partial/\partial t$ and Eq. (III.18) assumes the form

$$-\frac{\pi\Delta}{4T_c}\dot{\Phi} + D \operatorname{div} e\vec{E} = 0 \ . \tag{IV.2a}$$

In addition, we have to require

$$\begin{aligned} m\dot{\vec{v}}_s &= -\nabla\Phi + e\vec{E} \ ; \\ \operatorname{div}(\vec{j}_s + \vec{j}_n) &= 0 \ , \end{aligned} \tag{IV.2b}$$

where $\vec{j}_s = \pi\sigma m\Delta^2\vec{v}_s/2eT_c$ and $\vec{j}_n = \sigma\vec{E}$.

From these equations, a damped oscillating solution is obtained with a dispersion relation

$$\omega = -i\frac{\pi\Delta^2}{4T_c} \pm \sqrt{2D\Delta q^2 - \left(\frac{\pi\Delta^2}{4T_c}\right)^2} \ . \tag{IV.3}$$

Since $\operatorname{div}\vec{j} = 0$, one may say that the superfluid and the normal fluid move in opposite directions in order to avoid space charges. The ohmic losses of the normal component are responsible for the dissipation. Close to T_c, where the superfluid density is small, the superfluid oscillations can be compensated by small oscillations of the normal component and hence there is little damping, and the mode is propagating with almost the bare wave velocity $\sqrt{2D\Delta} = v_F\sqrt{2\tau\Delta/3}$. At lower temperatures, the superfluid density and the damping increase rapidly.

An accurate description can be obtained on the basis of the theory developed in Sec. II. A considerable simplification is possible due to the fact that the frequencies are large such that $\omega\tau_E \gg 1$ which allows us to neglect inelastic collisions. In fact, the mode in consideration is a collisionless mode, and in most cases, the frequency is even so large that we may neglect pairbreaking due to magnetic fields, etc.* On the other hand, the frequencies may become comparable with the gap which requires an exact evaluation of the dot-product. This is possible if we restrict ourselves to the linear regime.

Consider for instance, Eq. (II.25a) in which the contribution

$$\hat{g}^R \circ [i\hat{\Delta} \overset{\circ}{,} \hat{h}]_- - [i\hat{\Delta} \overset{\circ}{,} \hat{h}]_- \circ \hat{g}^A \tag{IV.4a}$$

occurs. Since we are concerned with a (T)-mode, we have $\delta f^{(L)} = \delta\Delta = 0$. Thus, in the real Δ gauge only $f^{(T)} = -h^{(T)}/2$ appears in first order whereas the other quantities are those of the equilibrium state. Thus, we obtain

$$\hat{g}^R(E + \tfrac{\omega}{2})\ [i\hat{\Delta}, \hat{\tau}_3 h^{(T)}(E)]_- - [i\hat{\Delta}, \hat{\tau}_3 h^{(T)}(E)]_- \hat{g}^A(E - \tfrac{\omega}{2}), \tag{IV.4b}$$

if $h^{(T)}$ oscillates $\propto \exp[-i\omega t]$. Repeating these arguments step by step, one can derive rather easily the following linearized, collisionless (T) - mode Boltzmann equation ,

$$\begin{aligned} &N_1 \dot{f}^{(T)} - N_1 \dot{\Phi} \frac{\partial f_{th}}{\partial E} + 2\Delta N_2 f^{(T)} \\ &- \frac{\partial}{\partial \vec{r}} DM^{(T)}(E,E) \left[\frac{\partial f^{(T)}}{\partial \vec{r}} + m\dot{\vec{v}}_s \frac{\partial f_{th}}{\partial E}\right] = 0 , \end{aligned} \tag{IV.5}$$

where the densities of state are defined by

$$\begin{aligned} N_1(E) &= \frac{1}{2}\left\{ \frac{E^+}{[(E^+ + i\Gamma)^2 - \Delta^2]^{1/2}} - \frac{E^-}{[(E^- - i\Gamma)^2 - \Delta^2]^{1/2}} \right\} ; \\ N_2(E) &= \frac{1}{2}\left\{ \frac{i\Delta}{[(E^+ + i\Gamma)^2 - \Delta^2]^{1/2}} + \frac{i\Delta}{[(E^- - i\Gamma)^2 - \Delta^2]^{1/2}} \right\} . \end{aligned} \tag{IV.6}$$

In this relation, $\mathrm{Im}[\cdots\cdot]^{1/2} > 0$; and $E^{\pm} = E \pm \omega/2$.** Furthermore, $M^{(T)}(E,E) = N_1^2 + N_2^2$. In the following, we consider the limit $\Gamma \to 0$.

---

*The effect of pairbreaking due to paramagnetic impurities has been studied by Entin-Wohlman and Orbach (1978).

**In the Boltzmann equation, $i[f_{th}(E^+) - f_{th}(E^-)]$ has been replaced by $-(\partial f_{th}/\partial E)(\partial/\partial t)$ with negligible error.

The condition for charge neutrality is satisfied by

$$\int dE\ N_1 f^{(T)} + \Phi = 0 \ , \qquad \text{(IV.7a)}$$

where, as before, and as in the following, the definition (IV.6) applies. In the continuity equation div $(\vec{j}_n+\vec{j}_s)$ = (source), we have to introduce explicitly a source term which represents in some standard form the drive due to the Josephson current. As far as the normal current is concerned, it is given formally by the expression of Eq. (II.36c) and div $\vec{j}_n$ is obtained directly by integrating Eq. (IV.5) with respect to the energy. Some extra care is needed in the calculation of the supercurrent. It follows from Eq. (II.36b), that in linear approximation only the thermal distribution $h_{th}=1-2f_{th}$ enters in this expression. This fact allows us to determine $\hat{g}^{R(A)}$, and particularly, its nonequilibrium part $\partial\hat{g}^{R(A)}/\partial\vec{r}$ (proportional to $\hat{\tau}_2$) by an expansion in powers of E^{-1} which in turn means an expansion of $\vec{j}_s$ in powers of T_c^{-1}. In the present case, an expansion through order $E^{-3}(T_c^{-2})$ is sufficient.

Thus, the continuity equation assumes the form

$$iq\frac{\pi D m v_s \Delta^2}{2T_c}\left[1 + \frac{7\zeta(3)}{2\pi^3 T_c}(Dq^2+i\omega)\right]$$
$$= 2\Delta\int dE\ N_2 f^{(T)} + \frac{7\zeta(3)}{4\pi^2}\frac{\Delta^2}{T_c}(-i\omega)\,\Phi - \frac{\Delta T_c}{N(o)}, \qquad \text{(IV.7b)}$$

where it has been assumed that the deviations from equilibrium oscillate $\propto \exp\,[i\vec{q}\vec{r} - i\omega t]$ with $\vec{q}\,||\,\vec{v}_s$. Except for a factor $2(1+\lambda)eN(o)$, the left side of Eq. (IV.7b) is equal to div $\vec{j}_s$. On the right side, the term $-\Delta T_c/N(o)$ represents the above mentioned source term.

It is easy to solve Eq. (IV.5) for $f^{(T)}$ in terms of Φ and $\vec{v}_s$. Since $\Delta \ll T_c$, we may replace in the integral $\int dE\ N_2 f^{(T)}$ of Eq. (IV.7b), the derivative $\partial f_{th}/\partial E$ by $-1/4T_c$. After suitable subtractions, the same procedure may be applied to the integral $\int dE\ N_1 f^{(T)}$ of Eq. (IV.7a). Then the integrals can be evaluated exactly in terms of complete elliptic integrals of all three kinds.* Eventually, we obtain $v_s = v_s(q,\omega)$, etc.

*For details see G. Schön (1976) ; G. Schön and V. Ambegaokar (1979). One should observe that the gauge used there is different from the present one, which accounts for some formal differences. The results are, of course, the same.

Customarily, the result is expressed in a different form. In a gauge where the vector potential is zero one has to interpret $\vec{v}_s$ as oscillations in the phase, and hence in the transverse part $\delta\underset{\sim}{\Delta}^{(T)}$ of the complex order parameter $\underset{\sim}{\Delta}$. These quantities are connected as follows:

$$\delta\Delta^{(T)} = 2m\Delta\, v_s/q \ . \tag{IV.8a}$$

Furthermore $\delta\Delta^{(T)}$ is the response of the order parameter to a normalized force and hence, it has the meaning of a pair field susceptibility. Thus, the dissipation-fluctuation theorem requires

$$S^{(T)}(q,\omega) = \frac{2T}{\omega}\ \mathrm{Im}\ \delta\Delta^{(T)}(q,\omega) \ , \tag{IV.8b}$$

where $S^{(T)}$ is the (T)-mode part of the order parameter structure factor which is defined in the usual way

$$\begin{aligned} S(q,\omega) &= \langle\langle \underset{\sim}{\Delta}(\vec{r},t)\underset{\sim}{\Delta}^{*}(o,o)\rangle\rangle_{q\omega} \\ &= S^{(T)} + S^{(L)} \ . \end{aligned} \tag{IV.9}$$

The contribution $S^{(L)}$ from the (L)-mode part of the order parameter fluctuations can be calculated directly from the linearized Ginzburg-Landau equation (II.33) where in addition a normalized drive term $1/2N(o)$ has to be added to the right hand side. Since the (L)-mode contributes to the structure factor particularly at low frequencies, the Boltzmann equation (II.27) (but without collision integral) can be used after linearizing it properly. Eventually, one calculates $S^{(L)}$ from $\delta\Delta^{(L)}$ by a relation similar to Eq. (IV.8b).

Graphs of $S(q,\omega)$ as a function of ω for various temperatures at a fixed value of q are shown in Fig. 2 as broken lines. The (L)- mode is responsible for the central peak whereas the (T)-mode oscillations appear as resonances at finite ω-values. For comparison, the experimental results of Carlson and Goldman (1975,76) are shown as solid lines. Essentially, the experimental and the theoretical curves agree rather well within the limits of the experimental accuracy. No adjustable parameters have been used.

There is also good agreement between experiment and theory in the temperature and q-dependence of the resonant frequency. Compared with the result (IV.3) of the simple model defined by Eq.(IV.2), the resonant frequency is depressed and the damping is increased due to quasiparticle diffusion which has been neglected in the derivation of Eq. (IV.2).

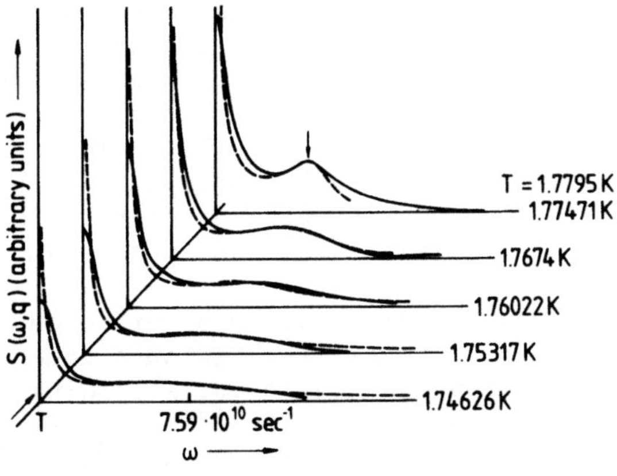

Fig. 2. The order parameter structure factor in arbitrary units plotted as a function of frequency and temperature. The solid lines show the experimental results of Carlson and Goldman (1975) for Aluminium films, and the broken lines represent the theory. The wave vector $q=4.10^{-4}cm^{-1}$ and the diffusion coefficient $D=18.6sec^{-1}cm^{2}$ are obtained from independent measurements. The unknown absolute scale of the measured structure factor has been chosen such that it agrees at the point marked by the arrow.

2. Dynamic Properties of Weak Links

We consider in the following a model where a narrow link ("bridge") connects two bulk superconductors ("banks") which are of the same material. The local state of the bridge will depend only on the distance from the banks since its cross-section is uniform and small. An important feature is that at the end points, this local state attains the equilibrium values characteristic for each of the banks because the three dimensional diffusion in the banks restores equilibrium effectively.

For the sake of simplicity, we restrict our attention to the case of short bridges where the length (2a) is small compared to the coherence length $\xi(T)$. In such a case, the quasiparticle distribution in the bridge is determined by very efficient diffusion processes with the banks acting as reservoirs. There is however, the following important exception: quasiparticles with energy less than the gaps of the banks, will be confined to the bridge region, and their number only changes by inelastic electron-phonon scattering. This peculiar fact has been stressed by Aslamasov and Larkin (1976) as well as by Tinkham (1979).

In the present theory, which has been worked out by Schmid, Schön and Tinkham (1980), emphasis is laid on a nonequilibrium contribution to the supercurrent which is, according to Eq. (II.36d) proportional to the (L)-mode change $\delta f^{(L)}(E=\Delta)$ of the quasiparticle distribution at the gap edge. In short bridges (which means $2a \ll \xi(t)$), this contribution produces the largest nonequilibrium effect.

Furthermore, if $2a \ll \xi(T)$, the terms containing spatial derivatives play the most important role in the Ginzburg-Landau equation. Thus, in a gauge where the vector potential vanishes,* $\underset{\sim}{\Delta}$ is determined in leading order by the equation $\nabla^2 \underset{\sim}{\Delta} = 0$. This means that

$$\underset{\sim}{\Delta}(x) = \frac{1}{2}\, \Delta_b [(1 - \frac{x}{a}) e^{-i\varphi/2} + (1 + \frac{x}{a}) e^{i\varphi/2}] \quad , \qquad \text{(IV.10a)}$$

where Δ_b is the gap of the banks at $x = \pm a$. The phase difference φ obeys the Josephson relation $\dot{\varphi} = 2eV$, where $V = V(t)$ is the voltage difference between the banks. It follows from Eq. (IV.10a) that

$$\Delta^2(x) = \Delta_b^2 [\cos^2 \varphi/2 + \frac{x^2}{a^2} \sin^2 \varphi/2] \quad , \qquad \text{(IV.10b)}$$

*We will return to the customary real Δ gauge in a short while.

and furthermore, that the superfluid velocity $2m\vec{v}_s = -\nabla\Theta$ is given by

$$2mv_s = \frac{1}{2a} \frac{\sin^2\varphi/2}{\cos^2\varphi/2 + \frac{x^2}{a^2}\sin^2\varphi/2} \quad . \tag{IV.10c}$$

In the static (equilibrium) case, we find from Eqs. (II.36d) and (IV.10) the supercurrent density in the bridge equal to

$$j_{eq} = j_o \sin\varphi \; ; \qquad j_o = \frac{\sigma}{2a} \frac{\pi\Delta_b^2}{4eT_c} \quad . \tag{IV.11}$$

In order to determine the quasiparticle distribution in the dynamic (nonequilibrium) case, we assume $eV \ll \Delta_b$, which means that the frequencies are still small compared with the gap. This allows us to use the Boltzmann equation (II.27) provided that we return to the real Δ gauge, which means that we chose the vector potential $A = -mv_s/e$.

For sake of simplicity, the following arguments are based on the BCS densities of states, i.e. on the form (II.31d) in the limit $\Gamma \to 0$. It is important to realize the local nature of these densities because $\Delta = \Delta(x,t)$ is a function of space and time as given by Eq. (IV.10b). Therefore, $N_1(E)$ is finite even for energies $|E|$ below the gap Δ_b of the banks provided that $|E| > \Delta(x,t)$. This feature is also found for $M^{(L)}(E,E)$ which is equal to $\Theta(E^2 - \Delta^2(x,t))$ in the BCS limit. Thus, we conclude that the diffusion of quasiparticles in the (L)- mode is confined to the region $x^2 < a^2[(E/\Delta_b)^2 - \cos^2\varphi/2]/\sin^2\varphi/2$, provided that $|\cos\varphi/2| < |E/\Delta_b| < 1$, whereas quasiparticles of higher energies can diffuse freely. In cases where diffusion is unrestricted, this process relaxes the nonequilibrium quasiparticles at a fast rate of the order $D/a^2 \gg \Delta_b^2/T_c$ (where the strong inequality follows from the condition $2a \ll \xi(T)$). This rate is much faster than any competing generation process. This means, that we have

$$f^{(L)} = f_{th} \; ; \text{ for all } x, \text{ if } |E| > \Delta_b \quad , \tag{IV.12a}$$

since then, the banks can act as reservoirs.

By contrast, in the (T)-mode, $M^{(T)}$ is non-zero and large everywhere, and there is no restriction imposed on the diffusion of quasiparticles in this mode. Again, diffusion is the fastest process and therefore, we conclude that

$$f^{(T)} = 0 \; ; \text{ for all } x \text{ and } E \; , \tag{IV.12b}$$

which is the same value as in the banks.

Thus, the problem is reduced to solving for energies $|E|<\Delta_b$ the (L)-mode Boltzmann equation *

$$N_1 \dot{f}^{(L)} + R_2 \dot{\Delta} \frac{\partial f^{(L)}}{\partial E} - D \frac{\partial}{\partial x} M^{(L)} \frac{\partial}{\partial x} f^{(L)} + \frac{N_1}{\tau_E} [f^{(L)} - f_{th}] = 0 \; , \tag{IV.13}$$

where we have replaced the collision integral by the relaxation ansatz (II.27g).

This equation can be solved in a closed form as follows. Taking into account the step like dependence of $M^{(L)}$ on x, we conclude that $\partial f^{(L)}/\partial x = 0$ at the boundaries of the accessible region $\Delta(x,t)>|E|$. Thus, the diffusion operator has one eigenvalue zero, whereas the other ones are at least of the order D/a^2. This means that the spatial average $<f^{(L)}>$ of $f^{(L)}$ is much larger than any of its Fourier components of finite wave vector. Consequently, in the spatial average of Eq. (IV.13), we may replace $<N_1 f^{(L)}>$ by $<N_1><f^{(L)}>$ with very good accuracy. Similar substitution can be made with other terms of Eq. (IV.13).

A further point deserves attention. If we define

$$\xi = \int_0^E dE' \; N_1(E') \; , \tag{IV.14}$$

we may convince ourselves that $N_1 = (\partial\xi/\partial E)_t$ and $R_2\dot{\Delta} = -(\partial\xi/\partial t)_E$ (the subscript denotes the variable which is kept constant). Since the spatial averaging is interchangeable with integration and differentation, we may write the space average of Eq. (IV.13) in the following form

$$\left(\frac{\partial<\xi>}{\partial E}\right)_t \left(\frac{\partial<f^{(L)}>}{\partial t}\right) - \left(\frac{\partial<\xi>}{\partial t}\right)_E \left(\frac{\partial<f^{(L)}>}{\partial E}\right)_t + \frac{1}{\tau_E} \left(\frac{\partial<\xi>}{\partial E}\right)_t [<f^{(L)}> - f_{th}] = 0. \tag{IV.15a}$$

---

* Terms quadratic in v_s have been omitted. One can show that in the present case, this is a consistent procedure. (See, for instance App. D.).

This equation is equivalent to

$$\left(\frac{\partial\langle\delta f^{(L)}\rangle}{\partial t}\right)_{\langle\xi\rangle} + \frac{1}{\tau_E}\langle\delta f^{(L)}\rangle = -\frac{\partial f_{th}}{\partial E}\left(\frac{\partial E}{\partial t}\right)_{\langle\xi\rangle} \quad ; \tag{IV.15b}$$

where we have assumed that the space average of Eq. (IV.14) has been inverted to yield $E = E(\langle\xi\rangle;t)$. Eq. (IV.15b) can be integrated to yield

$$\langle\delta f^{(L)}(t)\rangle = \frac{1}{4T_c}\int_{t_o}^{t} dt' e^{-(t-t')/\tau_E}\left(\frac{\partial E}{\partial t'}\right)_{\langle\xi\rangle} , \tag{IV.16a}$$

where we have put $\partial f_{th}/\partial E = -1/4T_c$. Condition (IV.12a) implies, that at a given $\langle\xi\rangle$, t_o is the latest time where $|E(\langle\xi\rangle;t_o)| = \Delta_b$. Clearly, if $|E(\langle\xi\rangle;t)| \gtrless \Delta_b$, we have $\delta f^{(L)} = 0$.

In the case of slow motion where the Josephson frequency is small such that $eV\tau_E \ll 1$, we obtain

$$\langle\delta f^{(L)}\rangle = \frac{\tau_E}{4T_c}\left(\frac{\partial E}{\partial t}\right)_{\langle\xi\rangle} , \tag{IV.16b}$$

provided that $|E(\langle\xi\rangle;t)| < \Delta_b$, and zero otherwise. In the opposite case of fast motion where $eV\tau_E \gg 1$, the limiting form of the distribution function is

$$\langle\delta f^{(L)}\rangle = \frac{1}{4T_c}\left[E(\langle\xi\rangle;t) - \Delta_b \operatorname{sgn} E\right]. \tag{IV.16c}$$

The last relation means simply that $|f^{(L)}(E)| = f_{th}(\Delta_b)$ for $|E| < \Delta_b$.

We will now look for a more explicit form of the results obtained so far. For the BCS density of states $N_1 = |E/\sqrt{E^2-\Delta^2}|$, we obtain from Eq. (IV.14) the result $\xi = \sqrt{E^2-\Delta^2}$ where $\Delta = \Delta(x,t)$ is given by Eq. (IV.10b). In the energy range $|\cos\gamma| < |E/\Delta_b| < 1$, the spatial averaging leads to the simple result

$$\langle\xi\rangle = \frac{\pi}{4}\,\frac{E^2 - \Delta_b^2\cos^2\varphi/2}{\Delta_b|\sin\varphi/2|} . \tag{IV.17}$$

According to Eq. (II.36d), the nonequilibrium contribution to the supercurrent is

$$\delta j_s = -\frac{\pi\sigma}{e}\,2mv_s\Delta\langle\delta f^{(L)}(E=\Delta)\rangle . \tag{IV.18}$$

Strictly, the spatial variation of δj_s requires a correction δv_s in the leading term of the supercurrent. However, in an adequate approximation, we may neglect this correction if we calculate the spatial average $<\delta j_s>$.

In the case of slow motion, we obtain

$$<\delta f^{(L)}> = -\frac{\tau_E \Delta_b^2}{8T_c E}\left[1+\sin^2\varphi/2-\frac{E^2}{\Delta_b^2}\right]\frac{\cos\varphi/2}{\sin\varphi/2}, \tag{IV.19a}$$

from which we calculate

$$<\delta j_s> = \tau_E \dot{\varphi} p_1(\varphi/2)$$
$$p_1(\varphi/2) = \{(\varphi/2)(1+\sin^2\varphi/2)\mathrm{ctg}\varphi/2 - \cos^2\varphi/2\}/2. \tag{IV.19b}$$

The function $p_1(\varphi/2)$ is even in φ; it is of period 2π; and the expression given above is valid in the range $|\varphi| \leq \pi$. Furthermore, $p_1 \geq 0$; it vanishes at $\varphi = 0, \pm\pi$ and it has a maximum of 0.40 at $\varphi = \pm 2.1$. We emphasize the fact that $<\delta j_s>$, though being a pair current, is dissipative since it carries the factor $\dot{\varphi} = eV$.

In the case of fast motion, the excess current is given by

$$<\delta j_s> = [r_1(\varphi/2) - \sin\varphi]j_o \quad ; \tag{IV.20}$$

$$r_1(\varphi/2) = \cos\varphi/2 \, \ln\frac{1+\sin\varphi/2}{1-\sin\varphi/2} \quad .$$

This contribution to the pair current is nondissipative, and hence we have written it in a form which allows us to add the equilibrium contribution (IV.11) easily. Thus, the total supercurrent is equal to $r_1(\varphi/2)j_o$. The function $r_1(\varphi/2)$ is odd in φ; it is of period 2π; and it is non-negative in the range $0 \leq \varphi \leq \pi$. Furthermore, it vanishes at $\varphi = 0$ and $\varphi = \pm\pi$, and it has a maximum value of 1.325 at $\varphi = 1.98$.

In the expression (II.36b) for the normal current we have $f^{(T)} = 0$, and in the most important case of fast motion, $\partial f^{(L)}/\partial E = \partial f_{th}/\partial E$ for $|E| < \Delta_b$; and zero elsewhere. One can argue that in sufficient approximation, this expression can be evaluated in the limit $\Delta_b \to 0$. Thus, we obtain $j_n = \sigma V/2a$, since $<m\dot{v}_s> = eV/2a$.

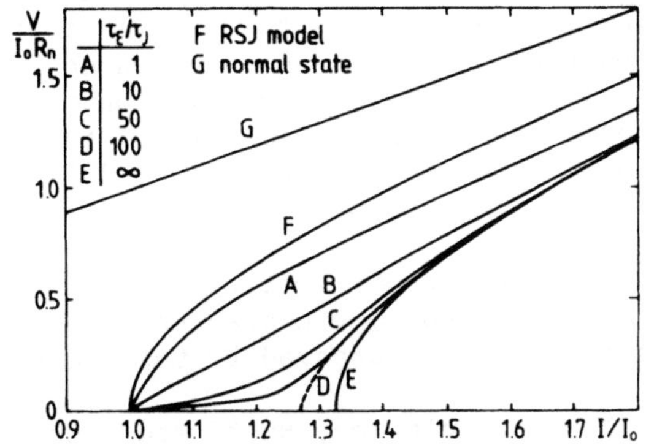

Fig. 3. Theoretical current-voltage characteristic of a short weak link. The various curves labelled by A through E are obtained by numerical integration for various values of the parameter τ_E/τ_J. Curve F is obtained from the RSJ model with normal resistivity and with I_o as the threshold current. Curve G is the CVC in the normal state.

Concerning the current-voltage characteristic (CVC) of a weak link, one has to keep in mind that the voltage V=V(t) is determined by the condition that the total current $j_{eq} + \langle\delta j_s\rangle + j_n$ is constant in time. Nevertheless, we can immediately draw qualitatively the following picture. In the case of slow motion, to which Eq. (IV.19b) applies, there is a dissipative pair current $\langle\delta j_s\rangle = \sigma_{eff} V/2a$ which is larger than the ohmic current by a factor

$$\frac{\sigma_{eff}}{\sigma} = \frac{1}{2}\,\overline{p_1(\gamma)}\cdot\frac{\tau_E}{\tau_J} \quad ; \qquad \text{(IV.21)}$$

where $\tau_J = 2T_c/\pi\Delta_b^2$ and where $\overline{p_1}$ is some suitable time average. This means that for currents above the threshold current j_o, the CVC is similiar to the one of a resistively shunted junction (RSJ-model) with however, an enhanced conductance σ_{eff}. On the other hand, in the limiting case of fast motion we expect a RSJ-type behaviour where, according to Eq. (IV.20) the threshold current appears to be enhanced to the value $1.325 j_o$, and where the conductivity has the normal state value.

If we join the two types of limiting behaviour with a smooth curve, we obtain a current-voltage characteristic with a foot-like structure as it has been observed by Octavio et al (1977).

In a quantitative calculation, Rueff and Schön (1980) have integrated the Boltzmann equation (IV.15b) numerically. In Fig. 3, the CVC thus obtained is shown graphically for various values of τ_E/τ_J.

I wish to thank Yuri Ovchinnikov and Gerd Schön for valuable comments and useful suggestions.

V. APPENDICES

A. Some Comments on Keldysh's Theory

The following considerations are not specific to the theory of superconductivity and therefore, we may restrict our attention to the normal state where $G^{>}(x_1,x_2) = - i\langle\psi_{\uparrow}(x_1)\psi^{+}_{\uparrow}(x_2)\rangle$, etc. Concerning Dyson's equation (II.3) where the product of self-energy $\bar{\Sigma}$ and the Green's function $\bar{G}$ (now matrices in Keldysh's space only) occur, we form the combination

$$\begin{aligned}(\Sigma G)^{>} &= \frac{1}{2}\{(\Sigma G)^{R} - (\Sigma G)^{A} + (\Sigma G)^{K}\} \\ &= \frac{1}{2}\{\Sigma^{R}G^{R} - \Sigma^{A}G^{A} + \Sigma^{K}G^{A} + \Sigma^{R}G^{K}\} \\ &= \frac{1}{2}\{\Sigma^{R}[G^{R} - G^{A} + G^{K}] + [\Sigma^{R} - \Sigma^{A} - \Sigma^{K}]\, G^{A}\}\ .\end{aligned} \tag{A.1}$$

The first line follows from an obvious generalisation of Eq. (II.1c) and in the second line, use has been made of the matrix representation (II.2). In all the above expressions, integration with respect to the internal coordinate is implied as it is shown, for instance, in Eq. (II.3b). Obviously, the expressions in the square brackets of the last line in Eq. (A.1) are $G^{>}$ and $\Sigma^{>}$, respectively. Using the representation (II.1c) for Σ^{R} and G^{A}, we find

$$\begin{aligned}(\Sigma G)^{>}(t_1,t_2) &= \int_{-\infty}^{t_1} dt_3\,[\Sigma^{>}(t_1,t_3) - \Sigma^{<}(t_1,t_3)]G^{>}(t_3,t_2) \\ &- \int_{-\infty}^{t_2} dt_3\,\Sigma^{>}(t_1,t_3)\,[G^{>}(t_3,t_2) - G^{<}(t_3,t_2)]\ .\end{aligned} \tag{A.2}$$

In this relation, we have written down only the dependence on the time variable since this is of interest here. It is worthy of note that relation (A.2) can be recovered from equation (8-27) which is found in the book of Kadanoff and Baym (1962) and which has been derived by different methods (and also under limiting assumptions).

The self-energy of the electrons due to their interaction with the phonons can be calculated in Migdal's approximation. This is shown diagrammatically in Fig. 4, where the straight

Fig. 4

Self-energy of the electrons.

and the wavylines may be considered to represent the correlation functions of electrons and phonons, respectively. In the case of phonons, we have

$$D^{>}(x_1,x_2) = D^{<}(x_2,x_1) = -i\,\langle\phi(x_1)\phi(x_2)\rangle\,, \tag{A.3}$$

where $\phi(x)$ are the real phonon field operators.* Specifically, the self-energy diagram is meant to imply that

$$\begin{aligned} \Sigma^{>}(x_1,x_2) &= G^{>}(x_1,x_2)\,D^{<}(x_2,x_1)\;; \\ \Sigma^{<}(x_1,x_2) &= G^{<}(x_1,x_2)\,D^{>}(x_2,x_1)\;, \end{aligned} \tag{A.4}$$

and it follows from this relation that

$$\begin{aligned} \Sigma^{R}(x_1,x_2) &= \{\Sigma^{>}(x_1,x_2) - \Sigma^{<}(x_1,x_2)\}\theta\,(t_1-t_2) \\ &= \tfrac{1}{2}\{\,[G^{>}(x_1,x_2) - G^{<}(x_1,x_2)][\,D^{>}(x_2,x_1) + D^{<}(x_2,x_1)] \\ &\quad - [G^{>}(x_1,x_2) + G^{<}(x_1,x_2)][D^{>}(x_2,x_1)-D^{<}(x_2,x_1)]\}\phi(t_1-t_2) \\ &= \tfrac{1}{2}\,\{G^{R}(x_1,x_2)D^{K}(x_2,x_1) + G^{K}(x_1,x_2)D^{A}(x_2,x_1)\}\;. \end{aligned} \tag{A.5a}$$

Similarly, we obtain

$$\begin{aligned} \Sigma^{A} &= \tfrac{1}{2}\,\{G^{A}\cdot D^{K} + G^{K}\cdot D^{R}\}\;; \\ \Sigma^{K} &= \tfrac{1}{2}\,\{G^{K}\cdot D^{K} - [G^{R}-G^{A}][\,D^{R}-D^{A}]\}\;, \end{aligned} \tag{A.5b}$$

*In comparison with the case of electrons, there is an obvious change in sign which takes into account the difference in the commutation relation between Fermions and Bosons. Otherwise, the definition of the Green's function follows exactly Eq. (II.1c) where the arguments are the same as above. One clearly recognizes the connection of these relations with Eq. (II.9).

B. Diffusion Terms in the Boltzmann Equation

We start from the diagonal part of the Keldysh component of the diffusion term,

$$\begin{aligned}\check{T} &= \frac{D}{8}\{[\check{\vec{\partial}} \overset{\circ}{,} [\check{g} \overset{\circ}{,} [\check{\vec{\partial}} \overset{\circ}{,} \check{g}]_-]_-]_-\}^{K,\ \mathrm{diag}} \\ &= \frac{D}{8}\{[\hat{\vec{\partial}} \overset{\circ}{,} \hat{g}^R \circ [\hat{\vec{\partial}} \overset{\circ}{,} \hat{g}^K]_- + \hat{g}^K \circ [\hat{\vec{\partial}} \overset{\circ}{,} \hat{g}^A]_- \\ &\quad - [\hat{\vec{\partial}} \overset{\circ}{,} \hat{g}^R]_- \circ \hat{g}^K - [\hat{\vec{\partial}} \overset{\circ}{,} \hat{g}^K]_- \circ \hat{g}^A]_-\}^{\mathrm{diag}} \quad .\end{aligned} \tag{B.1}$$

Next, we insert the ansatz (II.23a) for $\hat{g}^K$ and, using repeatedly the normalisation, we obtain

$$\check{T} = T_1 + T_2 + T' \quad , \tag{B.2a}$$

where

$$\begin{aligned}T_1 &= \frac{D}{4}\{ [\hat{\vec{\partial}} \overset{\circ}{,} [\hat{\vec{\partial}} \overset{\circ}{,} \hat{h}]_- - \hat{g}^R \circ [\hat{\vec{\partial}} \overset{\circ}{,} \hat{h}]_- \circ \hat{g}^A]_-\}^{\mathrm{diag}} \quad ; \\ T_2 &= \frac{D}{8}\{ [\hat{g}^R \overset{\circ}{,} [\hat{\vec{\partial}} \overset{\circ}{,} \hat{g}^R]_-]_- \circ [\hat{\vec{\partial}} \overset{\circ}{,} \hat{h}]_- \\ &\quad - [\hat{\vec{\partial}} \overset{\circ}{,} \hat{h}]_- \circ [\hat{g}^A \overset{\circ}{,} [\hat{\vec{\partial}} \overset{\circ}{,} \hat{g}^A]_-]_-\}^{\mathrm{diag}} \quad ; \\ T' &= \{[\hat{\vec{\partial}} \overset{\circ}{,} [\hat{g}^R \overset{\circ}{,} [\hat{\vec{\partial}} \overset{\circ}{,} \hat{g}^R]_-]_-]_- \circ \hat{h} \\ &\quad - \hat{h} \circ [\hat{\vec{\partial}} \overset{\circ}{,} [\hat{g}^A \overset{\circ}{,} [\hat{\vec{\partial}} \overset{\circ}{,} \hat{g}^A]_-]_-]_-\}^{\mathrm{diag}} .\end{aligned} \tag{B.2b}$$

Clearly, the contribution T' is canceled by other terms which add up to complete the equation of motion for $\hat{g}^R$ and $\hat{g}^A$.

Let us put $[\hat{\vec{\partial}} \overset{\circ}{,} \hat{h}] = 2[\hat{1}\vec{q}^{(L)} + \hat{\tau}_3\vec{q}^{(T)}]$, where $\vec{q}^{(L)}$ and $\vec{q}^{(T)}$ are given by Eq. (II.27c). This relation is correct to $O(\hbar^2)$. Thus, we find easily that

$$T_1 = D[\hat{1}\mathrm{div} + \hat{\tau}_3 m\vec{v}_s \frac{\partial}{\partial E}][\hat{1}M^{(L)}(E,E)\vec{q}^{(L)} + \hat{\tau}_3 M^{(T)}(E,E)\vec{q}^{(T)}] \ , \tag{B.3a}$$

within the limits of the required accuracy. As far as the contribution T_2 is concerned, we should observe that

$$\{[\hat{g}^R \overset{\circ}{,} [\hat{\vec{\partial}} \overset{\circ}{,} g^R]_-]_-\}^{diag} = -4im\vec{v}_s[N_2+iR_2]^2\,\hat{\tau}_3 + O(\hbar^2). \tag{B.4}$$

Therefore, through first order $\hbar$, we have

$$T_2 = 4Dm\vec{v}_s N_2 R_2 \hat{\tau}_3[\hat{1}\vec{q}^{(L)} + \hat{\tau}_3\vec{q}^{(T)}]$$
$$+ D\{\{m\vec{v}_s(N_2^2-R_2^2), \hat{\tau}_3[\hat{1}\vec{q}^{(L)} + \hat{\tau}_3\vec{q}^{(T)}]\}\}. \tag{B.3b}$$

This completes this Appendix.

C. Energy Theorem

It is comparatively easy to show that the equation of motion (II.20) satisfies the requirements of charge and energy conservation* if we keep it in its original form. Specifically, this means that we do not transform it into an equation for the distribution function. The difference namely is, that in the latter form, we have included implicitly the requirement that $\hat{g}^{R(A)}$ must satisfy its own equation of motion.

For simplicity, let us assume that $\hat{g}^R = \alpha^R \hat{\tau}_3 + \beta^R \hat{\tau}_1$. Then, this equation of motion requires that we have

$$\dot{\alpha}^R - i\dot{\Delta}\frac{\partial \beta^R}{\partial E} + 2iDm^2 \vec{v}_s \dot{\vec{v}}_s \frac{\partial {\beta^R}^2}{\partial E} = 0 \; ;$$
$$\dot{\Phi}\frac{\partial \alpha^R}{\partial E} - 2i\,\mathrm{div}\, Dm\vec{v}_s {\beta^R}^2 = 0. \qquad \text{(C.1)}$$

Most easily, these equations are obtained from the diagonal part of the equation of motion (II.29a) by an expansion through first order in $\hbar$**. There is a similar relation for the advanced quantities.

Let us call the left hand side of the first and second line of Eq. (C.1) Z_1^R and Z_2^R, respectively. Then, we form

$$-\frac{1}{4}\{Z_1^R \circ h^{(L)} - h^{(L)} \circ Z_1^A - Z_2^R \circ h^{(T)} + h^{(T)} \circ Z_2^A\}$$
$$= [\dot{N}_1 + \dot{\Delta}' \frac{\partial R_2}{\partial E} - 4Dm^2 \vec{v}_s \dot{\vec{v}}_s \frac{\partial}{\partial E} N_2 R_2](-\frac{1}{2}h^{(L)})$$
$$- D\{\{\frac{\partial m^2 \vec{v}_s \dot{\vec{v}}_s (N_2{}^2 - R_2{}^2)}{\partial E}, -\frac{1}{2}h^{(L)}\}\} \qquad \text{(C.2)}$$
$$- [\dot{\Phi}' \frac{\partial N_1}{\partial E} + 4D\,\mathrm{div}\, m\vec{v}_s N_2 R_2](-\frac{1}{2}h^{(T)})$$
$$- D\{\{\mathrm{div} m\vec{v}_s (N_2{}^2 - R_2{}^2), -\frac{1}{2}h^{(T)}\}\} = 0.$$

---

* Strictly speaking, energy is only conserved if we include the phonons in our considerations. This is, however, not advantageous at the present level of discussion. Thus, we allow a change in the electronic energy through inelastic electron-phonon collisions. A similar comment should be given concerning the coupling to the electromagnetic field.

** It is worthy to note, that Eq. (C.1) can also be derived from Eqs. (II.29b + c).

This expression will be added to the (L)- mode Boltzmann equation (II.27d) and, after some rearrangement of various terms, we obtain

$$\frac{\partial}{\partial t} N_1 [f^{(L)} - \frac{1}{2}] - \dot{\Phi} \frac{\partial}{\partial E} N_1 f^{(T)} + \dot{\Delta} \frac{\partial}{\partial E} R_2 [f^{(L)} - \frac{1}{2}]$$

$$+ \text{div } D[M^{(L)}\vec{q}^{(L)} - 4m\vec{v}_s N_2 R_2 f^{(T)} - \{\{m\vec{v}_s(N_2{}^2 - R_2{}^2), f^{(T)}\}\}]$$

$$+ Dm\dot{\vec{v}}_s \frac{\partial}{\partial E} [M^{(T)}\vec{q}^{(T)} - 4m\vec{v}_s N_2 R_2 [f^{(L)} - \frac{1}{2}] \tag{C.3}$$

$$- \{\{m\vec{v}_s(N_2{}^2 - R_2{}^2), f^{(L)}\}\}] - K^{(L)} = 0.$$

Next, we operate on this expression by $2N(o)\int dE\, E \ldots.$ and transform the appropriate terms by partial integration. At some places the expressions (II.36a) and (II.36c) for charge density ρ and current density $\vec{j}$ emerge* as well as the gap equation, in the form $\Delta = -\lambda\int dER_2[f^{(L)} - 1/2]$. Furthermore, we will make use of the continuity equation $\dot{\rho} + \text{div}\vec{j} = 0$ as well as of the relation $e\vec{E} = m\dot{\vec{v}}_s + \nabla\Phi$. Eventually, we obtain an equation for the energy balance in the form

$$\dot{\varepsilon} + \text{div}\vec{J}_\varepsilon - \vec{E}\vec{j} - 2N(o)\int dE\, EK^{(L)}(\delta f^{(L)}) = 0 \;, \tag{C.4}$$

where the energy density and the energy current are given by

$$\varepsilon = N(o)[2\int dEEN_1 f^{(L)} + \frac{1}{\lambda}\Delta^2 - \Phi^2] + \frac{1}{e}\Phi\rho \quad ;$$

$$\vec{J}_\varepsilon = \frac{\sigma}{e^2}\int dE\; E\; [M^{(L)}\vec{q}^{(L)} - 4m\vec{v}_s N_2 R_2 f^{(T)} \tag{C.5}$$

$$- \{\{m\vec{v}_s(N_2{}^2 - R_2{}^2), f^{(T)}\}\}] + \frac{1}{e}\Phi\vec{j} \quad .$$

Clearly, Eq. (C.4) tells us that there is an overall change in energy which is caused by the coupling of the electrons to the electromagnetic field ($\vec{E}\vec{j}$) and to the phonons ($K^{(L)}$).

* Apparently, an inconsistency arises in connection with the renormalisation factor $(1+\lambda)$. In the following, we therefore entirely neglect renormalisation.

D. Transformation from E- to ξ- Representation

The energy E which appears in the Boltzmann equation (II.27) has the meaning of a local energy. This has already been illustrated in the main text, and for a normal metal by the transformation of the Boltzmann equation into the form (II.28). We will call such a form a ξ- representation.

An analogous transformation of the Boltzmann equation for the superconducting state is much more involved and therefore, we will discuss this transformation step by step.

Consider first the case where $f^{(T)} = 0$. In addition, we will also neglect for the moment terms containing space derivatives. In a first step, we add the (L)- and (T)- mode Boltzmann equation (II.27) with a special weight, namely, such that the (T)- mode equation carries a factor N_1^{-1} in comparison with the (L)- mode equation. Thus, we obtain

$$N_1\dot{f}^{(L)} - \dot{\Phi}\,\frac{\partial f^{(L)}}{\partial E} + R_2\dot{\Delta}\,\frac{\partial f^{(L)}}{\partial E} - 4Dm^2\vec{v}_s\dot{\vec{v}}_s N_2R_2\,\frac{\partial f^{(L)}}{\partial E}$$

$$- D\,\{\{m\vec{v}_s(N_2{}^2-R_2{}^2),\, m\dot{\vec{v}}_s\,\frac{\partial f^{(L)}}{\partial E}\}\}_{(E,t)} - Dm^2\dot{\vec{v}}_s^{\,2}\,\frac{\partial}{\partial E}\,M^{(L)}\,\frac{\partial f^{(L)}}{\partial E}$$

$$- K^{(L)}(\delta f^{(L)}) = 0\ . \qquad \text{(D.1)}$$

Note that the Poisson bracket has been labeled by (E,t) for a clear distinction.

As it will become clear in a short while, the natural definition of the ξ- variable is

$$\xi = \int_0^E dE'N_1(E') + \Phi. \qquad \text{(D.2)}$$

We now observe that Eq. (C.1) implies

$$\dot{N}_1 + \dot{\Delta}\,\frac{\partial R_2}{\partial E} - 4Dm^2\vec{v}_s\dot{\vec{v}}_s\,\frac{\partial}{\partial E}\,N_2R_2 = 0\ , \qquad \text{(D.3a)}$$

and from this relation, we obtain the important result

$$\left(\frac{\partial \xi}{\partial t}\right)_E = \dot{\Phi} - \dot{\Delta}R_2 + 4Dm^2\vec{v}_s\dot{\vec{v}}_s N_2R_2. \qquad \text{(D.3b)}$$

According to the theory of Jacobian determinants, we have

$$\{\{A,B\}\}_{(E,t)} = N_1\{\{A,B\}\}_{(\xi,t)}, \tag{D.4a}$$

and in particular

$$\left(\frac{\partial A}{\partial E}\right)_t = N_1\left(\frac{\partial A}{\partial \xi}\right)_t . \tag{D.4b}$$

Hence, it follows that the first line of Eq. (D.1) is of the form $\{\{\xi,f^{(L)}\}\}_{(E,t)} = N_1(\partial f^{(L)}/\partial t)_\xi$,

and therefore, Eq. (D.1) can be written in the form

$$\left(\frac{\partial f^{(L)}}{\partial t}\right)_\xi - D\{\{m\vec{v}_s(N_2{}^2-R_2{}^2),\ m\dot{\vec{v}}_sN_1\ \frac{\partial f^{(L)}}{\partial \xi}\}\}_{(\xi,t)}$$
$$- Dm^2\vec{v}_s^{\,2}\ \frac{\partial}{\partial \xi}\ M^{(T)}N_1\ \frac{\partial f^{(L)}}{\partial \xi} - N_1^{-1}K^{(L)}(\delta f^{(L)}) = 0. \tag{D.5}$$

In principle, terms containing space derivatives can be included. In this case, one has to observe that

$$\left(\frac{\partial A}{\partial \vec{r}}\right)_E = N_1\{\{A,E\}\}_{(\vec{r},\xi)} . \tag{D.6}$$

At this point, the following conclusions can be drawn. The E-representation is characterized by simple forms in the spatial derivatives whereas the terms containing time derivatives are rather involved. The situation is reversed in the ξ-representation which features a simple time derivative and complicated spatial derivatives. It is worthy of note, that for the problem of a weak link presented in Sec. (IV.2), we have been able to take advantage of both representations.

It is important to realize , that the ξ- representation can be defined for arbitrary pairbreaking though it might be difficult to find an explicit form for $E = E(\xi)$. In the case of zero pair-breaking, we obtain from Eq. (D.2) immediately the BCS relation

$$E = (\mathrm{sgn}(\xi-\Phi))\ \sqrt{(\xi-\Phi)^2 + \Delta^2} . \tag{D.7}$$

Next, we turn our attention to the (T)- mode distribution function $f^{(T)}$. Again, we form the sum of the two Boltzmann equations weighted as described previously. Of course, all the terms containing $f^{(L)}$ can be disregarded since we have already taken care of them. To the resulting equation, we add

$$N_1^{-1}\, f^{(T)}[N_1 + \Delta \frac{\partial R_2}{\partial E} - 4Dm^2 \vec{v}_s \vec{v}_s \frac{\partial}{\partial E} N_2 R_2]$$
$$- f^{(T)}[\mathring{\Phi} \frac{\partial N_1}{\partial E} + 4D\, \mathrm{div} m\vec{v}_s N_2 R_2] = 0 \quad , \qquad (D.8)$$

which is, according to Eq. (C.1), zero. Omitting for a moment space derivatives, we obtain

$$N_1^{-1} \frac{\partial}{\partial t} N_1 f^{(T)} - \dot{\Phi}\frac{\partial}{\partial E} N_1 f^{(T)} + 2N_1^{-1} \Delta N_2 f^{(T)}$$
$$-\Delta N_1^{-1}\{\{R_2, f^{(T)}\}\}_{(E,t)} - 4N_1^{-1} Dm^2 \vec{v}_s \dot{\vec{v}}_s \frac{\partial}{\partial E} N_2 R_2 f^{(T)}$$
$$- N_1^{-1} D\{\{m\vec{v}_s (N_2{}^2 - R_2{}^2),\ m\dot{\vec{v}}_s \frac{\partial f^{(T)}}{\partial E}\}\}_{(E,t)} \qquad (D.9)$$
$$-N_1^{-1} Dm^2 \mathring{\vec{v}}_s^{\,2} \frac{\partial}{\partial E} M^{(L)} \frac{\partial f^{(T)}}{\partial E} - N_1^{-1} K^{(T)}(f^{(T)}) = 0.$$

We have been able to reduce this equation to a simple form only in the case of negligible pairbreaking where $N_2 = N_1{}^3\Gamma\Delta/E^2$. In this case, it is consistent to neglect the terms quadratic in $\vec{v}_s$. Then, we obtain

$$\left(\frac{\partial N_1 f^{(T)}}{\partial t}\right)_\xi + \frac{2\Delta^2\Gamma}{E^2}(N_1 f^{(T)}) - N_1^{-2} K^{(T)}(f^{(T)}) = 0, \qquad (D.10)$$

where $E = E(\xi)$ is given by Eq. (D.7). If necessary, the terms containing spatial derivatives have to be added.

Eventually, we mention that (in the case of vanishing pairbreaking) the ξ- representation is very close to a representation based on Bogolyubov quasiparticles (excitation representation). There, the distribution function $\tilde{f}$ is connected with the distribution function here through the relation

$$\tilde{f}(\xi) = \mathrm{sgn}E\ f^{(L)}(E) + \mathrm{sgn}\ (\xi - \Phi)\ N_1 f^{(T)}(E). \qquad (D.11)$$

The excitation representation has been laid out nicely by Aronov and Gurevich (1974) and it has been applied to nonequilibrium superconductivity by various people in particular, by Pethick and Smith (1979).

REFERENCES

Anderson, J. T., Carlson, R. V., and Goldman, A. M., 1972, Pair tunneling as a probe of order-parameter fluctuatins in superconductors: zero magnetic field effects, J. Low Temp. Phys., 8:29.

Aronov, A. G., and Gurevich, V. L., 1974, Response of a pure superconductor to a slowly varying perturbation, Fiz. Tverd. Tela, 16:2656; Sov. Phys. Solid State, 16:1722.

Aslamasov, L. G., and Larkin, A. I., 1976, Superconducting contacts with a nonequilibrium electron distribution function, Zh. Eksp. Theor. Fiz., 70:1340; Sov. Phys. JETP, 43:698.

Bardeen, J., Cooper, L. N., and Schrieffer, J. R., 1957, Theory of superconductivity, Phys. Rev., 108:1175.

Carlson, R. V., and Goldman, A. M., 1975, Propagating order-parameter collective modes in superconducting films, Phys. Rev. Letter, 34:11.

Carlson, R. V., and Goldman, A. M., 1976, Dynamics of the order parameter of superconducting aluminum films, J. Low Temp. Phys., 25:67.

Clarke, J., 1972, Experimental observation of pair-quasiparticle potential difference in nonequilibrium superconductors, Phys. Rev. Letters, 28:1363.

Clarke, J., Eckern, U., Schmid, A., Schön, G., and Tinkham, M., 1979, Branch-imbalance relaxation time in superconductors, Phys. Rev. B, 20:3933.

Eckern, U., and Schön, G., 1978, Relaxation processes in superconductors, J. Low Temp. Phys. 32:821.

Eilenberger, G., 1968, Transformation of Gorkov's equation for type II superconductors into transport-like equations, Z. Phys., 214:195.

Eliashberg, G. M., 1960, Interactions between electrons and lattice vibrations in a superconductor, Zh. Eksp. Teor. Fiz., 38:966; Sov. Phys. JETP, 11:696.

Eliashberg, G. M., 1971, Inelastic electron collisions and non-equilibrium stationary states in superconductors, Zh. Eksp. Theor. Fiz. 61:1254; Sov. Phys. JETP, 34:668.

Entin-Wohlman, O., and Orbach, R., 1978, Mode propagation in superconductors near T_c, Ann. Phys. 116:35.

Gorkov, L. P., 1958, On the energy spectrum of superconductors, Zh. Eksp. Theor. Fiz. 34:735; Sov. Phys. JETP, 7:505.

Gorkov, L. P., 1959, Microscopic derivation of the Ginzburg-Landau equations in the theory of superconductivity, Zh. Eksp. Theor. Fiz., 36:1918; Sov. Phys. JETP, 9:1364.

Gorkov, L. P., 1959, Theory of superconducting alloys in a strong magnetic field near the critical temperature, Zh. Eksp. Theor. Fiz. 37:1407; Sov. Phys. JETP, 10:998.

Hsiang, T. Y., and Clarke, J., 1980, Boundary resistance of the superconducting-normal interface, Phys. Rev. B 21:945.

Hu, C. R., 1980, New set of time-dependent Ginzburg-Landau equations for dirty superconductors near T_c, Phys. Rev. B, 21:2775.

Kadanoff, L. P., and Baym, G., 1962, Quantum statistical mechanics, (W. A. Benjamin, Inc., New York).

Kadin, A. M., Skocpol, W. J., and Tinkham, M., 1978, Magnetic field dependence of relaxation times in nonequilibrium superconductors, J. Low Temp. Phys., 33:481.

Kadin, A. M., Smith, L. N., and Skocpol, W. J., 1980, Charge imbalance waves and nonequilibrium dynamics near a superconducting phase-slip center, J. Low Temp. Phys., 38:497.

Keldysh, L. V., 1964, Diagram Technique for nonequilibrium processes, Zh. Eksp. Theor. Fiz., 47:1515; Sov. Phys. JETP, 20:1018.

Larkin, A. I., Ovchinnikov, Yu. N., 1968, Quasiclassical method in the theory of superconductivity, Zh. Eksp. Theor. Fiz., 55:2262; Sov. Phys. JETP, 28:1200.

Larkin, A. I., and Ovchinnikov, Yu. N., 1977, Nonlinear effects during the motion of vortices in superconductors, Zh. Eksp. Theor. Fiz., 73:299; Sov. Phys. JETP, 46:155.

Octavio, M., Skocpol, W. J., and Tinkham, M., 1977, Nonequilibrium-enhanced supercurrents in short suerconducting weak links, Phys. Rev. B, 17:159.

Pethick, C. J., and Smith, H., 1979, Relaxation and collective motion in superconductors: a two-fluid description, Ann. Phys., 119:133.

Peters, R., and Meissner, H., 1973, Relaxation time of the Cooper-pair density in tin, Phys. Rev. Letters, 30:965.

Pippard, A. B., Shepherd, J. G., and Tindall, D. A., 1971, Resistance of superconducting-normal interfaces, Proc. Roy. Soc. A (London), 324:17.

Rueff, M., and Schön, G., 1980, Quasiparticle dynamics in superconducting weak links, to be published in J. Low Temp. Phys.

Scalapino, D. J., 1979, The electron-phonon interaction and strong-coupling superconductors, Superconductivity I, Ch. 10, Ed. R. D. Parks (M. Dekker, N. Y.).

Schmid, A., and Schön, G., 1975, Linearized kinetic equations and relaxation processes of a superconductor near T_c, J. Low Temp. Phys., 20:207.

Schmid, A., Schön, G., and Tinkham, M., 1980, Dynamic properties of superconducting weak links, Phys. Rev. B, 21:5076.

Schmid, A., 1980, Phenomenological theory of dissipation in a superconductor, J. Low Temp. Phys., 41:37.

Schön, G., 1976, Propagating collective modes in superconductors, Thesis (unpublished), Universität Dortmund.

Schön, G., and Ambegaokar, V., 1979, Collective modes and non-equilibrium effects in current-carrying superconductors, Phys. Rev. B, 19:3515.

Schuller, I., and Gray, K. E., 1977, Temperature dependence of the relaxation time of the superconducting order parameter, Solid State Commun., 23:337.

Tinkham, M., 1972, Tunneling generation, relaxation and tunneling detection of hole-electron imbalance in superconductors, Phys. Rev. B, 6:1747.
Tinkham, M., 1979, Nonequilibrium superconductivity, Festkörperprobleme (Advances in Solid State Physics), Vol. XIX, 363, J. Treusch (ed.) Vieweg, Braunschweig.

CHARGE IMBALANCE: ITS RELAXATION, DIFFUSION AND OSCILLATION

C.J. Pethick

Department of Physics
University of Illinois at Urbana-Champaign
Urbana, Illinois 61801

and

Nordita, Blegdamsvej 17
DK-2100 Copenhagen Ø, DENMARK

and

H. Smith

Physics Laboratory I, H.C. Ørsted Institute
University of Copenhagen
DK-2100 Copenhagen Ø, DENMARK

I. INTRODUCTION

Broadly speaking, one can distinguish three different levels of description which have made decisive contributions to our understanding of condensed matter. On the coarsest time and length scales phenomena may be described in terms of a limited number of variables. This leads to descriptions such as hydrodynamics for liquids (two-fluid hydrodynamics for superfluids), elasticity theory for solids, and the Landau-Ginzburg equations for superconductors. At the other extreme one has fully microscopic descriptions, which are able to take into account rapid spatial and temporal variations. The price one pays for the increase in generality is the increased complexity of the formalism, and a large increase in the number of degrees of freedom that must be considered. For example, distribution functions must be used, rather than simply the total density. At this microscopic level the theory of superconductors, the BCS theory, has been spectacularly successful in making possible the understanding of a vast range of phenomena. By comparison, microscopic theories have had relatively little impact on work on the helium liquids.* In these areas a third level of description, intermediate between the macroscopic and fully microscopic ones mentioned above, has had a profound impact. The basic ingredients of the description are quasiparticles, and in the case of superfluid systems, a condensate. Physical quantities are expressed in terms of the distribution function for the quasiparticles, and a small number of variables specifying the state of the condensate. For the helium liquids, the properties of quasiparticles cannot be calculated accurately using microscopic methods, and therefore the intermediate level approach is a semi-phenomenological one, employing quasiparticle properties deduced from experiment. In the case of superconductors one is in a better position, since there quasiparticle properties may be evaluated using the microscopic BCS theory, and do not have to be treated as phenomenological parameters.

The distinction between the properties of the quasiparticles and those of the condensate cannot, however, be performed under the most general non-equilibrium conditions. When spatial and temporal variations occur rapidly (over scales comparable with the temperature dependent coherence length and the inverse gap frequency), or in the presence of strong depairing one must use the more general matrix Green's function approach described in the chapter by A. Schmid. It is still useful, however, to employ a quasiparticle formalism in situations where it can be justified, since the concept of quasiparticles is a familiar one. Also, the language of quasiparticles and their coupling to the condensate is a good one for exploring analogies between superconductors and the helium liquids where the lack of a detailed microscopic theory has resulted in much more emphasis being placed on the quasiparticle approach.

The simplicity of the quasiparticle approach suits it ideally for developing concepts and obtaining physical insight. In this chapter we shall show how it leads naturally to a two-component picture for the charge in a superconductor, some of the charge being associated with the condensate (the superfluid) and the remainder with the quasiparticles (the normal component). Under non-equilibrium conditions the charges associated with the normal and superfluid components may differ from their equilibrium values,

---

* The superfluid phases of ^3He are perhaps something of an exception, but here the successful microscopic calculations have for the most part been based on extensions of the BCS theory of superconductors.

and this leads to the concept of charge imbalance. Also the separation of the total charge into a normal and a superfluid component leads naturally to the introduction of currents of normal and superfluid charge, which differ somewhat from the normal current and the supercurrent occurring in the standard two-fluid model.

The basic ingredients of this two-component model for the charge are contained in the article by Betbeder-Matibet and Nozières (1969), but it was only after Leggett and Takagi's (1977) work on an analogous two-component model for the spin of superfluid ^{3}He that the physical content was brought out (Pethick and Smith, 1978, 1979a). In the limit when the normal component may be treated as a fluid, we are able to derive two-fluid equations. Since our two-fluid equations are derived directly from the kinetic equation for the quasiparticles they possess a firm microscopic basis and a well-defined region of validity. Phenomenological models had been proposed earlier by Bray and Schmidt (1975) and by Putterman (1977). Our equations are in many respects similar to theirs but with the important difference that we are able to obtain explicit expressions for the phenomenological coefficients from microscopic theory.

For clean superconductors, the quasiparticle formalism we discuss is valid provided that spatial variations are slow on the scale of the temperature dependent coherence length $\xi(T) \sim \hbar v_F/\Delta$, and that temporal variations are slow on a scale $\hbar/\Delta$. Here v_F is the Fermi velocity and Δ the energy gap. Under these circumstances the internal structure of quasiparticles responds essentially instantaneously to the local gap and the state of the normal component may be specified in terms of a scalar quasiparticle distribution function rather than, as under more general conditions, a matrix distribution function involving anomalous correlations. Likewise the gap responds essentially instantaneously to the local quasiparticle distribution, and is therefore given by the static gap equation, for weak coupling superconductors, the BCS-equation. The time development of the superconducting component is given by the usual superfluid acceleration equation, and the development of the quasiparticle distribution function is governed by the quasiparticle Boltzmann equation. In the case of dirty superconductors some of these statements must be modified, as we shall specify in more detail later.

The remainder of this chapter is organized as follows. In Section II we discuss the foundations of the two-component model for the charge, calculate susceptibilities for the normal and superfluid charge, and derive expressions for the corresponding currents. Generation of charge imbalance by tunnel injection and by the combined action of a superfluid velocity and a temperature gradient is considered in Sec. III. We consider the processes by which charge imbalance relaxes in Sec. IV, and in Sec. V describe applications of the concept of charge imbalance to spatially inhomogeneous situations, such as the collective mode and normal-superconducting boundaries. Section VI is a brief conclusion.

II. QUASIPARTICLE AND CONDENSATE CHARGE

Within the BCS framework the expectation value of the total charge per unit volume is given by

$$Q_{tot} = \sum_{i\sigma} < c_i^+ c_i > = \sum_{i\sigma} u_i^2 f_i + v_i^2 (1-f_i) \tag{1}$$

measured in units of the electronic charge -e. A quasiparticle in the superfluid is characterized by its spin σ and the quantum numbers i , which for a clean superconductor may be identified with the crystal momentum. In a dirty superconductor, the quantum numbers i are those necessary to specify the true eigenstates of the normal metal (Anderson, 1959). The distribution function for the quasiparticles is denoted by f_i , while u_i and v_i denote the BCS coherence factors

$$u_i^2 = \frac{1}{2}(1 + \frac{\xi_i}{E_i}) ,$$

and

$$v_i^2 = \frac{1}{2}(1 - \frac{\xi_i}{E_i}) . \tag{2}$$

In Eq.(2) the energy variable ξ denotes the normal state quasiparticle energy ε measured with respect to the chemical potential μ , or $\xi_i = \varepsilon_i - \mu$. As we shall explain later, the chemical potential μ is the energy required to add a particle to the condensate. The quasiparticle energy E_i is given by

$$E_i = \sqrt{\xi_i^2 + \Delta^2} . \tag{3}$$

From the expression (1) one sees that changes in the total charge may arise in two distinct ways, namely from changes in the quasiparticle distribution function f_i and from changes in the coherence factors u_i and v_i. We may write for first order changes

$$\delta Q_{tot} = \delta Q_n + \delta Q_s \tag{4}$$

where

$$\delta Q_n = \sum_{i\sigma} (u_i^2 - v_i^2)\delta f_i \tag{5}$$

and

$$\delta Q_s = \sum_{i\sigma} (1 - 2f_i)\delta v_i^2 \quad . \tag{6}$$

Here δQ_n arises from changes in the quasiparticle distribution, and we therefore refer to it as the quasiparticle (or normal) charge, whereas the pair charge δQ_s is seen to involve only changes in the coherence factors. This separation of changes in the density into two components exactly parallels Leggett and Takagi's (1977) discussion for the spin density in superfluid ^{3}He.

One can immediately see from Eq.(5) that the addition of a quasiparticle adds a charge $q = u^2 - v^2$ to the system. This effective charge q depends on the quasiparticle energy variable ξ as shown in Fig. 1 .

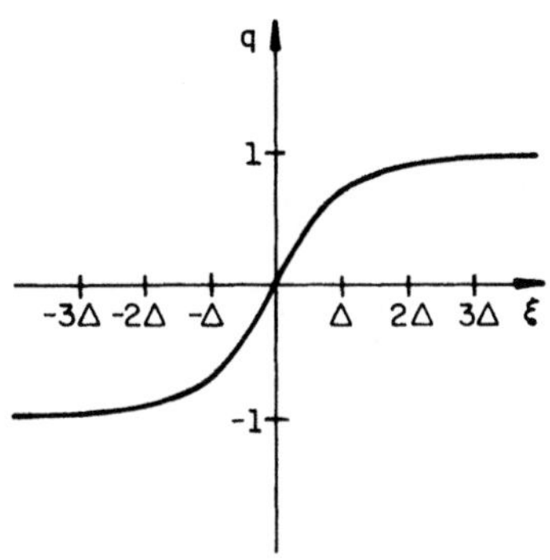

Fig. 1. Effective charge of a quasiparticle as a function of the normal state quasiparticle energy ξ .

A very useful concept is that of the susceptibility of the superfluid, which is the response of the superfluid part of the density δQ_s to a chemical potential change. If one regards the total energy E_{tot} of the superconductor as a functional of the total density $n(\equiv Q_{tot})$ and the quasiparticle distribution function f_i the chemical potential is given by

$$\mu = \frac{\partial E_{tot}}{\partial n}(n, f_i) \; . \tag{7}$$

Thus μ is the energy per particle required to add particles to the superconductor without changing the quasiparticle distribution, i.e. it is the energy required to add a particle at fixed entropy. From Eqs.(4-6) one can see immediately that since the quasiparticle distribution is not changed, the added particles must go into the superfluid part of the density. Thus we may write

$$\mu = \left(\frac{\partial E_{tot}}{\partial Q_s}\right)_{f_i} \tag{8}$$

which indicates that μ is the chemical potential associated with the superfluid (the pairs), and we shall henceforth denote it by μ_s. The response of Q_s is given by Eq.(6), and the changes in the coherence factor v_i^2 (Eq.(2)) are given by

$$\delta v_i^2 = \frac{\Delta^2}{2E_i^3}\,(\delta\mu_s - \delta\varepsilon_i) \; . \tag{9}$$

In the absence of Fermi liquid effects ε_i does not depend on μ_s and therefore one finds

$$\delta Q_s = \chi_s^o \delta\mu_s \tag{10}$$

where we have introduced the susceptibility χ_s^o given by

$$\chi_s^o = \sum_{i\sigma} (1 - 2f^o)\,\frac{\Delta^2}{2E_i^3} \; . \tag{11}$$

The superscript on χ indicates that Fermi liquid effects are neglected. The equilibrium Fermi function is denoted by f^o. In thermal equilibrium a change in μ_s alters the quasiparticle energies and hence the quasiparticle distribution. This changes the quasiparticle charge by an amount

$$\delta Q_n = \chi_n^o \delta\mu_s \tag{12}$$

where

$$\chi_n^o = \sum_{i\sigma} q_i \frac{\partial f^o}{\partial E_i} \frac{\partial E_i}{\partial \mu_s} = \sum_{i\sigma} - q_i^2 \frac{\partial f^o}{\partial E_i} \tag{13}$$

since $\partial E_i / \partial \mu_s = - q_i$. The total susceptibility of the superconductor, χ_{tot}^o , is equal to the normal-state value 2N(0):

$$\chi_{tot}^o = \chi_s^o + \chi_n^o = 2N(0) \tag{14}$$

where N(0) is the density of states per spin at the Fermi energy in the normal state. A plot of the reduced susceptibility of the superfluid component χ_s^o / χ_{tot}^o (= 1-Z(T)) is given in Fig. 2 ; in the calculation we assumed the gap to be given by the BCS expression. For $T \to 0$, Z(T) tends to 0 and for $T \to T_c$, Z(T) is approximately equal to $1 - \pi\Delta/4k_BT_c$. With an eye to later applications one should keep in mind the rapid variation of $\chi_n^o (= Z\chi_{tot}^o)$ near T_c . At $T = 0.9\, T_c$, χ_n^o has dropped to about 40% of its value at T_c . Note that the susceptibilities are unaffected by non-magnetic impurities.

We now consider how to calculate currents of the charge associated with the quasiparticles and with the pairs, and how to allow for spatial inhomogeneities. The separation of the total current into superfluid and normal parts is one of the basic steps in constructing two-fluid models for liquid helium and superconductors, and we briefly review the arguments in order to clarify the differences between these older two-fluid ideas and the one based on the two-component picture for the density. For this discussion we consider the case of a clean superconductor, and we shall therefore work with quasiparticle states having definite momenta $\underset{\sim}{p}$. The total current of electrons may be written as

$$\underset{\sim}{J} = n_e \underset{\sim}{v}_s + \sum_{\underset{\sim}{p}\sigma} \frac{\underset{\sim}{p}}{m} f_{\underset{\sim}{p}} \tag{15}$$

if one neglects electronic band structure effects that destroy the Galilean invariance of the electrons. Here n_e is the total electron number density, and $m \underset{\sim}{v}_s$ is the momentum of an electron in the condensate. The distribution function for quasiparticles in

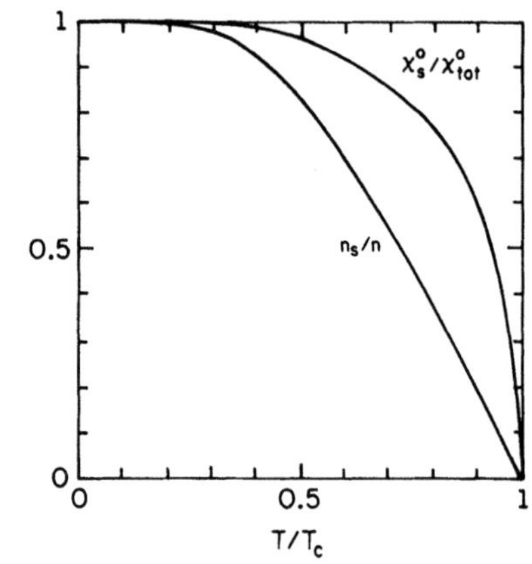

Fig. 2. The fractional density of superfluid electrons and the fraction of the susceptibility due to the superfluid electrons as a function of temperature.

equilibrium with the lattice in the presence of a superfluid flow is

$$f_{\tilde{p}}^{\ell.e.} = \frac{1}{e^{(E_{\tilde{p}}+\tilde{p}\cdot \tilde{v}_s)/k_BT}+1} \tag{16}$$

and therefore to first order in $\tilde{v}_s$ the current is

$$\tilde{J} = n_e\tilde{v}_s - n_n\tilde{v}_s = n_s\tilde{v}_s \tag{17}$$

where the excitation backflow $(-n_n\tilde{v}_s)$ is characterized by the normal density

$$n_n = \sum_{\underset{\sim}{p}\sigma} \frac{p^2}{3m} \left(- \frac{\partial f^o}{\partial E_{\underset{\sim}{p}}}\right) \tag{18}$$

and

$$n_s = n_e - n_n . \tag{19}$$

As this discussion shows, the superfluid density n_s and the normal density n_n are measures of the response of currents to vector potentials, and are quite distinct from the scalar densities δQ_s and δQ_n introduced earlier in this paper. The quantity n_s/n_e is plotted in Fig. 2. These results for n_s/n_e and χ_s^o / χ_{tot}^o apply provided the scale of spatial variations is large compared with the temperature dependent coherence length $\sim\hbar v_F/\Delta$. Note that the result for χ_s^o / χ_{tot}^o applies to both clean and dirty superconductors.

To derive the two-fluid equations we consider a space- and time-dependent situation. When spatial variations are absent Q_n is changed only by collisions,

$$\frac{\partial Q_n}{\partial t} = \left(\frac{dQ_n}{dt}\right)_{coll} \tag{20}$$

but when there are spatial inhomogeneities Q_n can also change locally because of the transport of normal charge from elsewhere. Thus Eq.(20) must be generalized to

$$\frac{\partial Q_n}{\partial t} + \underset{\sim}{\nabla} \cdot \underset{\sim}{J}_n^Q = \left(\frac{dQ_n}{dt}\right)_{coll} \tag{21}$$

where $\underset{\sim}{J}_n^Q$ is the current associated with Q_n, which we shall refer to as the current of normal charge.

To obtain some insight into the physical meaning of this current, let us consider the case of a clean superconductor, where the quantum numbers i characterizing the quasiparticle states may be identified with their momentum. The quasiparticle distribution is then determined by the Boltzmann equation, which when linearized has the form

$$\frac{\partial f_{\underset{\sim}{p}}}{\partial t} + \underset{\sim}{v}_{\underset{\sim}{p}} \cdot \underset{\sim}{\nabla}\, \delta f_{\underset{\sim}{p}} - \left(\frac{\partial f^{o}}{\partial \underset{\sim}{p}} \right) \cdot \underset{\sim}{\nabla}\, \delta E_{\underset{\sim}{p}} = \left(\frac{df_{\underset{\sim}{p}}}{dt} \right)_{coll} \tag{22}$$

where in the absence of Fermi liquid effects $\delta E_{\underset{\sim}{p}} = \underset{\sim}{p} \cdot \delta \underset{\sim}{v}_s - q\, \delta\mu_s$. The velocity $\underset{\sim}{v}_{\underset{\sim}{p}} = \underset{\sim}{\nabla}_{\underset{\sim}{p}} E_{\underset{\sim}{p}}$ is the group velocity of an excitation. Note that $-\nabla \delta E$ plays the role of an effective force on a quasiparticle. The gradient terms in Eq.(22) may be combined to give

$$\frac{\partial f_{\underset{\sim}{p}}}{\partial t} + \underset{\sim}{v}_{\underset{\sim}{p}} \cdot \underset{\sim}{\nabla}\, \delta f^{\ell.e.}_{\underset{\sim}{p}} = \left(\frac{df_{\underset{\sim}{p}}}{dt} \right)_{coll} \tag{23}$$

where $\delta f^{\ell.e.}$ is the deviation from local equilibrium, that is $\delta f^{\ell.e.} = \delta f - (\partial f^{o}(E)/\partial E)\delta E$.

Multiplying Eq.(23) by q and summing over $\underset{\sim}{p}$ and σ yields Eq. (21) with Q_n given by (5) and

$$\underset{\sim}{J}^{Q}_{n} = \sum_{\underset{\sim}{p}\sigma} q_{\underset{\sim}{p}} \underset{\sim}{v}_{\underset{\sim}{p}} \delta f^{\ell.e.}_{\underset{\sim}{p}} \quad . \tag{24}$$

This has the obvious physical interpretation that an excitation carries a current of normal charge given by its group velocity times its effective charge. We now compare this current with the normal current as usually defined. When the distribution function differs by an amount $\delta f^{\ell.e.}$ from the local equilibrium distribution function (16) the total current $\underset{\sim}{J}$ may be written as

$$\underset{\sim}{J} = \underset{\sim}{J}_s + \underset{\sim}{J}_n \tag{25}$$

where $\underset{\sim}{J}_s = n_s \underset{\sim}{v}_s$ is given by (17-19) and the normal current is

$$\underset{\sim}{J}_n = \sum_{\underset{\sim}{p}\sigma} \frac{\underset{\sim}{p}}{m}\, \delta f^{\ell.e.}_{\underset{\sim}{p}} \quad . \tag{26}$$

Since $\underset{\sim}{v}_p = q_{\underset{\sim}{p}} \underset{\sim}{p}/m$ we see that the current $\underset{\sim}{p}/m$ carried by a quasiparticle consists partly of a current of normal charge $q^2_{\underset{\sim}{p}} \underset{\sim}{p}/m$. The remainder, $(1-q^2_{\underset{\sim}{p}})\underset{\sim}{p}/m$, is a current of superfluid charge associated with a quasiparticle.

The continuity equation for the superfluid follows from the fact that the total particle number is conserved. It therefore has the form

$$\frac{\partial Q_s}{\partial t} + \underset{\sim}{\nabla} \cdot \underset{\sim}{J}_s^Q = \left(\frac{dQ_s}{dt}\right)_{coll} = -\left(\frac{dQ_n}{dt}\right)_{coll} \tag{27}$$

where the current of superfluid charge is $\underset{\sim}{J}_s^Q = \underset{\sim}{J} - \underset{\sim}{J}_n^Q$. The current of superfluid charge differs from the conventional supercurrent $\underset{\sim}{J}_s$ according to

$$\underset{\sim}{J}_s^Q = \underset{\sim}{J}_s + \sum_{\underset{\sim}{p}\sigma} \frac{\underset{\sim}{p}}{m} (1-q_{\underset{\sim}{p}}^2)\, \delta f_{\underset{\sim}{p}}^{\ell.e.} \quad . \tag{28}$$

Near T_c, $1-q_{\underset{\sim}{p}}^2$ is only significantly different from zero when the normal state quasiparticle energy is less than or comparable to Δ. For all but pathological forms of $\delta f^{\ell.e.}$ one then finds that close to T_c

$$\underset{\sim}{J}_s^Q = \underset{\sim}{J}_s + \beta \underset{\sim}{J}_n \tag{29}$$

where β is of order Δ/k_BT. In the applications we consider, $|\underset{\sim}{J}_n| \simeq |\underset{\sim}{J}_s|$ and therefore the second term in (29) can be neglected.

The above discussion has been for clean superconductors. In a dirty superconductor we expect Eq.(29) to still hold, with $\beta \sim \Delta/k_BT$ near T_c, but in evaluating $\underset{\sim}{J}_s$ one must use the expression for the superfluid density in a dirty superconductor rather than the result obtained from (18-19) for the clean limit.

To proceed further we need the acceleration equations for the normal fluid and the superfluid. The latter is the well-known equation for $\underset{\sim}{v}_s$,

$$m \frac{\partial \underset{\sim}{v}_s}{\partial t} = - \underset{\sim}{\nabla} (\mu_s - e\phi) \tag{30}$$

where ϕ is the electrostatic potential.

Near T_c the normal current $\underset{\sim}{J}_n$ and the current of normal charge $\underset{\sim}{J}_n^Q$ satisfy similar equations of motion. The normal currents relax due to scattering with ordinary impurities. The equation of motion for the normal current $\underset{\sim}{J}_n^Q$ may be obtained by multiplying the Boltzmann equation (23) by $q_{\underset{\sim}{p}} \underset{\sim}{v}_{\underset{\sim}{p}}$ and summing over all states $\underset{\sim}{p}\sigma$.

Apart from a term involving the gradient of the electrochemical potential $\mu_s - e\phi$ the current equation of motion includes a diffusive term associated with spatial variations of the quasiparticle charge. We find

$$\frac{\partial \underset{\sim}{J}_n^Q}{\partial t} + \underset{\sim}{\nabla} \left(\sum_{\underset{\sim}{p}\sigma} q_{\underset{\sim}{p}} \frac{1}{3} v_{\underset{\sim}{p}}^2 \, \delta f_{\underset{\sim}{p}}^{\ell.e.}\right) + \frac{1}{3} v_F^2 \, \chi_n^o \, \underset{\sim}{\nabla} (\mu_s - e\phi) = \left(\frac{d\underset{\sim}{J}_n^Q}{dt}\right)_{coll} \qquad (31)$$

where we neglect for simplicity the contributions to $\delta f^{\ell.e.}$ which behave as higher (even) spherical harmonics (the $\ell = 2$ contribution describes the effects of viscosity). Such terms are generally negligible in the situations of interest. Near T_c, by utilizing the fact that the current of normal charge relaxes at a rate $1/\tau_{tr}$, one obtains

$$\frac{\partial \underset{\sim}{J}_n^Q}{\partial t} + \frac{v_F^2}{3} \underset{\sim}{\nabla} \, \delta Q_n^{\ell.e.} + \frac{n}{m} \underset{\sim}{\nabla}(\mu_s - e\phi) = - \frac{\underset{\sim}{J}_n^Q}{\tau_{tr}} \quad . \qquad (32)$$

The equation of motion in this form is only valid near the transition temperature T_c and for spatial variations that occur over lengths much longer than the impurity mean free path $v_F \tau_{tr}$ (Artemenko and Volkov, 1975).

In the next two sections we discuss the generation and relaxation of quasiparticle charge. As we shall see, the collision term $\left(\frac{dQ_n}{dt}\right)_{coll}$ often assumes a simple form near the transition temperature. This allows us to combine the two-fluid equations into a single differential equation in space and time, which we use to discuss diffusive and oscillatory motion of quasiparticle charge.

III. GENERATION OF QUASIPARTICLE CHARGE

To demonstrate why the separation of the charge into normal and superfluid components is useful, let us consider the tunnelling injection experiment pioneered by J. Clarke (1972). This experiment and its interpretation in terms of the two-fluid picture of the charge has been discussed in the chapter by J. Clarke. Here we shall only recapitulate the main points. The experimental set-up, consisting of a N-S-N sandwich, is shown schematically in Fig. 3. A current is passed through the right-hand (injector) junction, thereby generating a non-equilibrium quasiparticle population in the superconductor. The non-equilibrium population leads to a

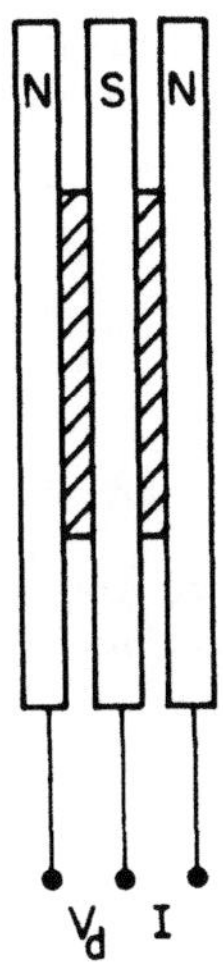

Fig. 3. Schematic diagram of the N-S-N sandwich used in charge imbalance experiments.

voltage across the left-hand (detector) junction in the absence of any net current flow across the detector junction. To analyse both the injector and detector junctions, consider the rate at which quasiparticles are added, by tunnelling, to the state i in the superconductor. From a simple perturbation calculation the injected current I may be shown to be given by (Tinkham, 1972; Pethick and Smith, 1979a)

$$\frac{I}{(-e)} = \frac{2\pi}{\hbar} |T|^2 N_n(0) [\sum_{i\sigma} \frac{1}{2}(f^o(E_i+eV) - f^o(E_i-eV)) - q_i \delta f_i^{\ell.e.}] \quad (33)$$

where T is the tunnelling matrix element, $N_n(0)$ the density of states per spin in the normal metal at the Fermi energy, $f^o(E_i)$ the Fermi function, and $\delta f_i^{\ell.e.}$ the deviation from local equilibrium, given by

$$\delta f_i^{\ell.e.} = f_i - f^o(E_i)$$

where E_i is the actual quasiparticle energy in the non-equilibrium situation, and not its equilibrium value. The energy E_i differs from its equilibrium value as a consequence of, for example, non-equilibrium contributions to the chemical potential.

The total injected current in (33) consists of two terms: the usual voltage-dependent tunnelling contribution and a non-equilibrium term which is simply proportional to the deviation of the quasiparticle charge from local equilibrium,

$$\delta Q_n^{\ell.e.} = \sum_{i\sigma} q_i \delta f_i^{\ell.e.} \quad (= Q^*) \quad . \tag{34}$$

The quantity $\delta Q_n^{\ell.e.}$ corresponds to what has often been referred to as Q^* in the literature. We have used the notation $\delta Q_n^{\ell.e.}$ to stress the fact that it is the deviation from local equilibrium that is important, not the deviation from equilibrium. The distinction between these two quantities is extremely important in many applications we shall consider, and has generally not received sufficient attention. In the remainder of this chapter we shall, however, denote $\delta Q_n^{\ell.e.}$ by Q^* .

The detection of a quasiparticle charge imbalance is achieved by measuring the voltage V_d across the detector junction when no current flows through it. According to (33) this voltage is given by

$$V_d = - \frac{Q^*}{2eN(0)g_{ns}} \tag{35}$$

where $g_{ns} = \sum_{i\sigma} (-\partial f^o / \partial E) / 2N(0)$ is the normalized tunnelling conductance and N(0) the density of states per spin at the Fermi energy in the normal state of the superconducting material.

We now determine the rate at which quasiparticle charge is added to the superconductor by tunnelling across the injector junction. By a perturbation calculation similar to the one above, one finds that the rate $(df_i/dt)_{inj}$ at which quasiparticles are added to the state i in the superconductor is given by

$$\left(\frac{df_i}{dt}\right)_{inj} = \frac{2\pi}{\hbar} |T|^2 N_n(0)\Big[\frac{1}{2} q_i (f^o(E_i + eV) - f^o(E_i - eV)) + \frac{1}{2} (f^o(E_i - eV) + f^o(E_i + eV) - 2f^o(E_i)) - \delta f_i^{\ell.e.}\Big]. \tag{36}$$

Since the non-equilibrium contribution to the injected quasiparticle charge is always small compared with the total, we may neglect it in (36). The rate, $(dQ_n/dt)_{inj}$, at which quasiparticle charge is injected is obtained by multiplying (36) by q_i and summing over all quasiparticle states i, σ :

$$\left(\frac{dQ_n}{dt}\right)_{inj} = \sum_{i\sigma} q_i \left(\frac{df_i}{dt}\right)_{inj}$$

$$= \frac{2\pi}{\hbar} |T|^2 N_n(0) \sum_{i\sigma} \frac{1}{2} q_i^2 \left(f^o(E_i + eV) - f^o(E_i - eV)\right) . \tag{37}$$

The second term in (36) is not small compared with the first one, but it does not contribute to the injected charge, since it is even in the variable ξ. It does, however, give rise to changes in the magnitude of the gap (see, e.g., Chang 1978) .

The steady-state value of the non-equilibrium quasiparticle distribution is determined by balancing the injection rate (36) against the rate $(df/dt)_{coll}$ at which the occupation of the quasiparticle state i is changed by collisions. The collision term of this Boltzmann equation involves the deviation $\delta f^{\ell.e.}$ from local equilibrium, which in turn determines Q^* according to (34). In its general form the Boltzmann equation is an inhomogeneous integral equation which requires use of numerical methods for its solution. From the solution of the Boltzmann equation one may determine the charge relaxation time, τ_{Q^*} , which is defined by

$$\left(\frac{dQ_n}{dt}\right)_{coll} = - \frac{Q^*}{\tau_{Q^*}} . \tag{38}$$

It is then evident from Eqs.(32), (35), (37) and (38) that in a steady state the detected voltage V_d can be written as

$$V_d = \frac{F^*}{2e^2 N(0) g_{ns}} \tau_{Q^*} I \tag{39}$$

where $F^* = F^*(\Delta/k_B T, eV/k_B T)$ is a known function of temperature and applied voltage which is given explicitly in the chapter by J. Clarke. Note that F^* is the function F' of Pethick and Smith (1979a). The result (39) was also derived independently by Moody and Paterson (1979). The function F^* tends to unity as T approaches T_c and remains near unity for arbitrary temperatures provided the injection voltage V is much larger than the energy gap.

Experimentally one measures V_d for a given total current I and extracts τ_{Q^*} from the relation (39). To compare theory with experiment one must determine τ_{Q^*} from the solution $\delta f^{\ell.e.}$ of the Boltzmann equation using the definition (38).

As another example of quasiparticle charge generation we consider next the simultaneous action of an imposed uniform supercurrent (velocity $\underset{\sim}{v}_s$) and a thermal gradient $\underset{\sim}{\nabla} T$. It was pointed out by Pethick and Smith (1979b) that under these circumstances one should detect a charge imbalance voltage proportional to $\underset{\sim}{v}_s \cdot \underset{\sim}{\nabla} T$. This prediction was subsequently established experimentally by Clarke, Fjordbøge and Lindelof (1979).

The amount of charge generated by the temperature gradient may be calculated from the kinetic equation for the distribution function of the BCS quasiparticles, just as for the case of tunnelling injection discussed above. The Boltzmann equation has the same driving term as that in the standard thermal conductivity calculation, except that both the quasiparticle energy $\tilde{E}_p = E_p + \underset{\sim}{p} \cdot \underset{\sim}{v}_s$ and velocity $\nabla_{\underset{\sim}{p}} \tilde{E}_{\underset{\sim}{p}} = \underset{\sim}{v}_{\underset{\sim}{p}} + \underset{\sim}{v}_s$ depend on $\underset{\sim}{v}_s$ (Aronov, 1974), and is

$$-(\underset{\sim}{v}_{\underset{\sim}{p}} + \underset{\sim}{v}_s) \cdot \frac{\underset{\sim}{\nabla} T}{T} \tilde{E} \frac{\partial f^o}{\partial \tilde{E}} = \left(\frac{df_{\underset{\sim}{p}}}{dt}\right)_{coll} , \qquad (40)$$

where $\left(\frac{df_{\underset{\sim}{p}}}{dt}\right)_{coll}$ is the collision term. The part of interest for charge imbalance problems is that odd in ξ, and the only part having this symmetry is the $\underset{\sim}{v}_{\underset{\sim}{p}}$ term. The $\underset{\sim}{v}_s$ term, which is even in ξ, may be dropped. The rate at which quasiparticle charge is generated is found by multiplying the driving term (40) by $\xi_{\underset{\sim}{p}}/E_{\underset{\sim}{p}}$ and summing over momentum space. We find

$$\left[\frac{dQ_n}{dt}\right]_{gen} = - n_e \, \alpha \, \underset{\sim}{v}_s \cdot \frac{\underset{\sim}{\nabla} T}{T} , \qquad (41)$$

where the dimensionless quantity α is given by

$$\alpha = 2 \int_0^\infty d\xi \, \frac{\xi^2}{E^2} \frac{\partial}{\partial E} \left[E \frac{\partial f^o}{\partial E} \right] , \qquad (42)$$

and n_e is the electron density. For $T \to T_c$, α tends to

$(\pi/4)\Delta/k_BT_c$. This shows that, as one might expect, for a normal metal no charge imbalance is generated, since all the charge is then in the normal component.

In the next section we discuss a variety of relaxation mechanisms, which determine the steady state value of Q^* and hence the observed voltage.

IV. CHARGE RELAXATION

We shall now discuss the different types of relaxation processes that may contribute to the decay of the quasiparticle charge.

a) Phonons

Quasiparticles in a superconductor may be scattered by emitting or absorbing phonons, just as in a normal metal. In addition, because quasiparticle number is not conserved in a superconductor, other processes can occur in which pairs of excitations are created by a phonon, or a phonon is created by annihilation of a pair. Since quasiparticles have effective charges that depend on the quasiparticle energy, these processes produce changes in the total quasiparticle charge. The total charge is conserved in these processes, and therefore the changes in quasiparticle charge are accompanied by equal and opposite changes in the pair charge. These processes may therefore be regarded as ones in which quasiparticle charge and pair charge are converted into each other.

An important case to consider is when the temperature is close to T_c , where the superfluid gap is small compared with the temperature. The quasiparticle spectrum may be divided into three regions, as shown schematically in Fig. 4 . The effective charge of a quasiparticle is essentially 1 in region 1 , varies between +1 and -1 in region 2, and is close to -1 in region 3. Scattering processes in which the initial and final quasiparticles are either both in region 1 or both in region 3 produce very little change in the quasiparticle charge, since the initial and final quasiparticles have essentially the same effective charge. On the other hand, processes in which the initial quasiparticle is in region 1 and the final one is in region 3, or vice versa, change the total quasiparticle charge by 2 units. However, the matrix elements for these processes involve factors such as $u^i u^f$ or $v^i v^f$, where i and f refer to the initial and final states, and are small since one of the coherence factors is always small. For example, if i is in 1 and f is in 3, $u^i \simeq 1$, u^f is small, v^i is small and $v^f \simeq 1$. The only scattering processes which lead to appreciable changes in the quasiparticle charge (~ 1) and have matrix elements comparable to those for a normal state scattering process are those in which either the initial or the final

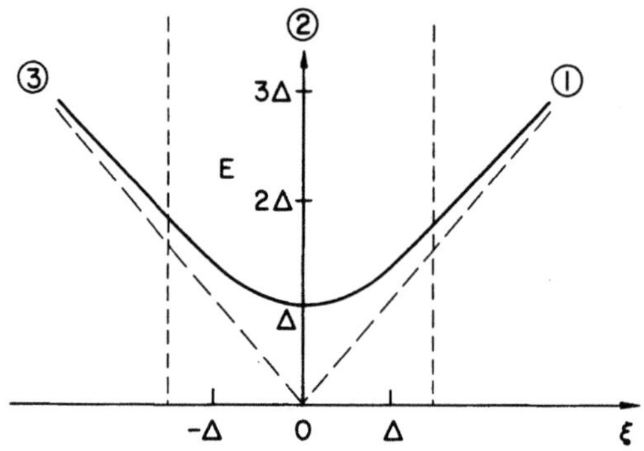

Fig. 4. Schematic representation of the various parts of the quasiparticle spectrum important in charge imbalance relaxation.

quasiparticle is in region 2. By similar arguments one can show that the only important pair creation and annihilation processes are those in which one of the quasiparticles is in region 2. We may estimate the rate at which quasiparticle charge decays as a typical normal state quasiparticle scattering rate, reduced by the number of quasiparticles in region 2 compared with the total number of thermally excited quasiparticles. Region 2 has a width $\sim \Delta$ in ξ, whereas thermally excited quasiparticles are spread over a width $\sim k_BT$. Thus we obtain the estimate

$$\frac{1}{\tau_{Q^*}} \simeq \frac{\Delta}{k_BT} \frac{1}{\tau_{in}(0)}$$

where we have used for the characteristic normal state scattering rate the inelastic relaxation rate $\tau_{in}^{-1}(\xi)$ of a normal state quasiparticle at the Fermi energy ($\xi = 0$) at the transition temperature. For metals with an effective phonon density of states varying as the square of the frequency, one obtains the estimate $\hbar/\tau_{in}(0) \sim k_BT_c^3/\theta_D^2$, where θ_D is the Debye temperature. The T^3

dependence is due to the fact that the rates of the various processes are proportional to the number of phonon states that can be thermally excited. Electron-electron scattering contributes to $\tau_{in}^{-1}(0)$ as well, but its effect is generally negligible, except at very low temperatures, due to the T^2 dependence of the electron-electron scattering rates.

The vanishing of the decay rate for quasiparticle charge as $\Delta \to 0$ reflects the fact that in the normal state there can be no conversion of quasiparticle charge into pair charge, since all the charge then resides in the normal component. One can exploit this property to solve the Boltzmann equation exactly to first order in the parameter $\Delta/(k_B T_c)$. One finds

$$\frac{1}{\tau_{Q^*}} = \frac{\pi}{4} \frac{\Delta}{k_B T_c} \tau_{in}^{-1}(0) , \tag{43}$$

a result first derived by Schmid and Schön (1975a). It differs from the approximate answer obtained by Tinkham (1972) only by the numerical factor $\pi/4$. An important point is that this result holds for almost all possible source terms in the Boltzmann equation, and does not depend on any assumptions about the electron-phonon matrix elements or the phonon spectrum. We shall briefly indicate how this result is obtained from the Boltzmann equation by expanding the solution in terms of the normal state eigenfunctions of the collision operator (Pethick and Smith 1979a). It is convenient to write the deviation from local equilibrium $\delta f^{\ell.e.}$ in the form

$$\delta f^{\ell.e.} = f^o(E)\,(1 - f^o(E))\; q\,\Psi \; . \tag{44}$$

If Ψ is independent of energy, this non-equilibrium distribution function corresponds to a Fermi function $f^o(E)$ with $E = \sqrt{(\varepsilon-\mu_n)^2 + \Delta^2}$, where the chemical potential μ_n is shifted relative to μ_s by an amount $\Psi k_B T$. As we shall now explain, the solution of the Boltzmann equation to leading order in $\Delta/k_B T_c$ has precisely the form Ψ = const. Near T_c the Boltzmann equation has the following structure:

$$X_1 = \sum_2 I^n(1,2) \left\{ (\Psi_1 - \Psi_2) + (1 - q_2^{\,2})\, \Psi_2 \right\} , \tag{45}$$

where $q_i X_i = \frac{\pi}{\hbar} |T|^2 N_n(0)\; q_i (f^o(E_i + eV) - f^o(E_i - eV))$ is the part of the driving term (36) which is odd in the energy variable ξ_i . The integral kernel $I^n(1,2)$ is that of the normal state

Boltzmann equation and we have made the appropriate approximations in the coherence factors corresponding to the limit $\Delta \ll k_BT$. By expanding Ψ in a complete set of eigenfunctions ψ_ν for the normal state collision operator I^n, we see that to leading order only the eigenfunction ψ_0 = const. contributes, since this eigenfunction has eigenvalue zero associated with particle conservation in the normal state. Inserting $\Psi = c$ in Eq.(45) we may therefore determine c by summing over all states 1

$$\sum_1 X_1 = c \sum_{12} I^n(1,2)(1-q_2^2) \tag{46}$$

and using the property

$$1 - q^2 = \frac{\Delta^2}{\Delta^2 + \xi^2} \simeq \pi \Delta \delta(\xi) \tag{47}$$

Since the normal state inelastic relaxation rate $\frac{1}{\tau_{in}(\xi)}$ is given by

$$\frac{1}{\tau_{in}(\xi_1)} = \frac{\sum_2 I^n(1,2)}{f^o(\xi_1)(1-f^o(\xi_1))} \tag{48}$$

we therefore obtain Eq.(43) on using $f^o(0)(1 - f^o(0)) = \frac{1}{4}$.

Since Δ/k_BT_c is a small parameter only very near T_c one must in general solve the Boltzmann equation numerically in order to determine the charge relaxation rate τ_{Q*}^{-1}. A detailed comparison between theory and experiment for Aℓ has been made by Chi and Clarke (1979,1980) and is discussed in detail in the chapter by J. Clarke. It is also possible to use a variational approach (Pethick and Smith 1980) to calculate the charge relaxation rate at arbitrary temperatures. For the standard variational method (Ziman, 1960) to be applicable, the quantity on which a bound is required must be proportional to $\sum_i Y_i \Phi_i$, where Y_i is the driving term in the Boltzmann equation and $\Phi_i = \delta f_i^{\ell.e.}/[f_i^o(1-f_i^o)]$.

Here we derive a bound on $Q^* = \sum_i q_i f^o(1-f^o)\Phi_i$, and therefore a bound can be obtained only if $Y_i \propto q_i f^o(1-f^o)$. Since the driving term in this problem is $Y_i = q_i X_i$, which is defined immediately following Eq.(45), one easily sees that this condition implies that the variational method may be applied for small injection voltages ($|eV| << k_BT$) . As a trial function we have used a deviation function corresponding to $\Phi_i \propto q_i$, or Ψ_i = const. in Eq.(44). The resulting charge relaxation rate $1/\tau_{Q*}$ exhibited in Fig. 5 is an upper bound to the actual one, but it differs by no more than 10% from that obtained from the numerical solution of the Boltzmann equation at not too low temperatures ($T > 0.6\ T_c$) . While the variational calculations are strictly applicable only for low injection voltages, they give good estimates of τ_{Q*} at the higher injection voltages typically employed in experiments. This may easily be seen by referring to the numerical solutions of the Boltzmann equation described in John Clarke's chapter in this volume.

From the work by Chi and Clarke it is evident that electron-phonon scattering by itself is not sufficient to explain the observed temperature dependence. The phonons dominate the scattering only near T_c . At lower temperatures elastic collisions of quasiparticles with impurities play an important role due to the anisotropy of the energy gap.

b) Impurity Scattering

Let us consider first a clean superconductor in which the concentration of non-magnetic impurities is low enough that the mean free path is much longer than the zero temperature coherence length. A quasiparticle with momentum $\underset{\sim}{p}$ may be scattered to states in some other direction $\hat{\underset{\sim}{p}}'$ on both sides of the Fermi surface, with the energy being conserved. The processes are illustrated schematically in Fig. 6. The matrix elements for processes 1 and 2 are proportional to $uu' - vv'$, where the coherence factors u and v are defined in Eq.(2), and the prime indicates the appropriate final states allowed by energy conservation. For the case of an isotropic gap $|\xi|$ is equal to $|\xi'|$, by energy conservation. The matrix element for 1 is ξ/E , since $\xi' = \xi$ and therefore $u = u'$ and $v = v'$. For process 2 the matrix element vanishes since $\xi' = -\xi$ and therefore $u = v'$ and $u' = v$. Process 1 conserves quasiparticle charge, since for the case of an isotropic gap quasiparticles with the same ξ have the same effective charge. Thus impurity scattering leads to no change in the quasiparticle charge if the energy gap is isotropic.

The situation is different, however, if the energy gap is a function of direction, as pointed out by Tinkham (1972). Both

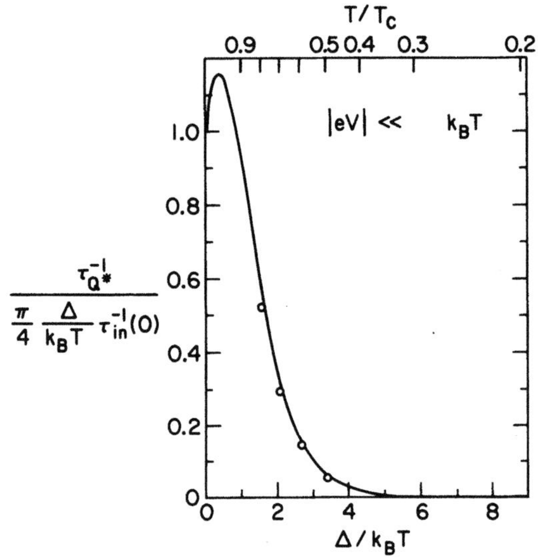

Fig. 5. The charge relaxation rate $1/\tau_{Q^*}$ for low injection voltages as a function of Δ/k_BT. Results obtained by Chi and Clarke (1979, 1980) are shown as points.

processes then have finite matrix elements, and also give rise to changes in the quasiparticle charge. When the anisotropy of the gap is small the branch mixing process 2 provides the dominant contribution to the changes in quasiparticle charge.

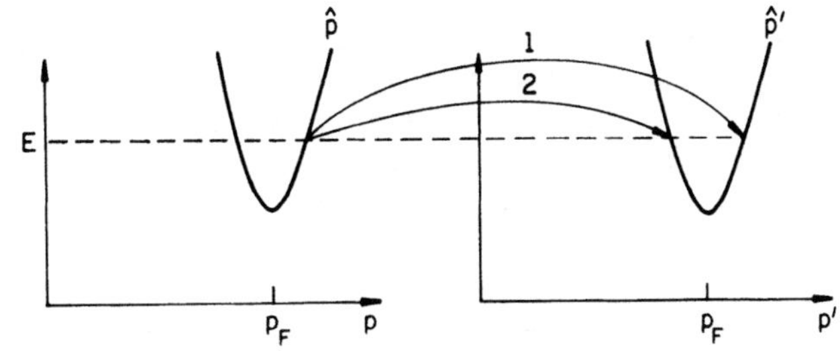

Fig. 6. Schematic representation of elastic scattering between quasiparticle states in two different directions $\hat{p}$ and $\hat{p}'$. In process 1 the quasiparticle remains on the same branch, while in process 2 it changes branches.

When the gap anisotropy is small it is a simple matter to include the effect of branch mixing processes in the Boltzmann equation discussed earlier. For branch mixing processes like 2 in Fig. 6 the angle-averaged collision term is

$$\left(\frac{df}{dt}\right)_{coll} = -\frac{1}{\tau_{av}}\left(\delta f(\xi) - \delta f(-\xi)\right) = -\frac{2}{\tau_{av}}\,\delta f(\xi) \qquad (49)$$

with

$$\frac{1}{\tau_{av}} = \frac{\overline{\delta\Delta^2}}{2\xi^2}\,\frac{E}{|\xi|}\,\frac{1}{\tau_{tr}}\,. \qquad (50)$$

The first term in the relaxation rate (50) comes from the coherence factors, the second from the density of states in the superconductor, compared with that in the normal metal, and τ_{tr} is the impurity scattering time in the normal metal. $\overline{\delta\Delta^2} = \langle(\Delta - \langle\Delta\rangle)^2\rangle$ is the mean square deviation of the gap from the average value and $\langle\ldots\rangle$ denotes an angular average.

c) Magnetic Impurities

Magnetic impurities, unlike non-magnetic impurities, relax the quasiparticle charge even if the energy gap is isotropic. The process represented by 2 in Fig. 6 has a matrix element proportional to $uu' + vv' = \Delta/E$ for magnetic scattering, compared with $uu' - vv'$ for non-magnetic scattering. The characteristic time for scattering is

$$\frac{1}{\tau_{mag}} = \left(\frac{\Delta}{E}\right)^2 \frac{E}{|\xi|} \frac{1}{\tau_s} , \tag{51}$$

where $(\Delta/E)^2$ comes from the square of the matrix element, and $E/|\xi|$ from the density of states. The time τ_s is the relaxation time for magnetic scattering in the normal metal. The collision term may be written in a form similar to (49) for non-magnetic impurities:

$$\left(\frac{df}{dt}\right)_{coll} = - \frac{1}{\tau_{mag}} (\delta f(\xi) - \delta f(-\xi)) = - \frac{2\delta f(\xi)}{\tau_{mag}} . \tag{52}$$

Despite the simplicity of (51-52) it is in practice quite complicated to account theoretically for the role of magnetic impurity scattering. This is because the scattering by magnetic impurities is never the only important mechanism, so it is necessary to consider the combined effects of magnetic impurities and phonons. Analytical results have been obtained only for $\Delta \ll k_B T$. The simplest case to consider is for $\tau_s^{-1} \ll \tau_{in}^{-1}(0)$, which can be treated by perturbation theory starting from the solution to the problem with only phonon scattering. We find

$$\frac{1}{\tau_{Q*}} = \frac{\pi}{4} \frac{\Delta}{k_B T} \frac{1}{\tau_{in}(0)} + \frac{\Delta}{k_B T} \frac{1}{\tau_s} . \tag{53}$$

For $\tau_s^{-1} \ll (k_B T/\Delta)^2 \tau_{in}^{-1}(0)$ Schmid and Schön (1975a) obtained the interpolation formula

$$\frac{1}{\tau_{Q*}} = \frac{\pi}{4} \frac{\Delta}{k_B T} \sqrt{\frac{1}{\tau_{in}(0)} \left[\frac{1}{\tau_{in}(0)} + \frac{2}{\tau_s}\right]} . \tag{54}$$

For $\tau_s^{-1} = 0$ this reduces to (43). For $\tau_s^{-1} \ll \tau_{in}^{-1}(0)$ it yields $1/\tau_{Q*} = (\pi\Delta/4k_B T)(\tau_{in}^{-1}(0) + \tau_s^{-1})$ instead of (53).

Let us now exhibit the general solution of the Boltzmann equation valid near T_c, when $\Delta \ll k_B T$ and in addition $\tau_{in}^{-1}(0) \gg (\Delta/k_B T)^2 \tau_s^{-1}$. Under these conditions the magnetic impurity scattering dominates at energies $\xi \sim \Delta$, but is negligible at energies $\xi \sim k_B T$, and the deviation function $\delta f^{\ell.e.}$ in this case assumes the form (Entin-Wohlman and Orbach, 1979)

$$\delta f^{\ell.e.} = Aq(\partial f^o/\partial E)(\tau_{in}^{-1}(0) + 2\tau_{mag}^{-1}(\xi))^{-1} \tag{55}$$

where A is a normalization constant that can be obtained by inserting (55) into the Boltzmann equation. One finds as a result

$$\tau_{Q*}^{-1} = \tau_{in}^{-1}(0)\,\frac{\pi}{4}\,\frac{\Delta}{k_B T_c} + \frac{1}{4}\int_{-\infty}^{\infty} d(\xi/k_B T_c)\,\frac{q^2}{\tau_{in}(0) + \tau_{mag}(\xi)/2} \tag{56}$$

$$= \tau_{in}^{-1}(0)\,\frac{\Delta}{k_B T_c}\left[\frac{\pi}{4} + \frac{\beta}{2}\int_1^{\infty}\frac{dy\sqrt{y^2-1}}{y}\,\frac{1}{(y\sqrt{y^2-1}+\beta)}\right]$$

where $\beta = 2\tau_{in}(0)/\tau_s$. For $\beta \ll 1$ one recovers (53) while for $\beta \gg 1$ the result is the same as for (54), $\tau_{Q*}^{-1} = (\pi\Delta/4k_B T_c)\sqrt{2/\tau_s\tau_{in}(0)}$. The relative difference between the Schmid-Schön result (54) and the complete expression (56) is 3% at the most (for $\beta \simeq 1$) and vanishes in the limits $\beta \to 0$ and $\beta \to \infty$.

d) Impurity Scattering in the Presence of Supercurrents

As a final example we discuss the role of impurity scattering in relaxing the charge which is generated by the simultaneous presence of a supercurrent and a thermal gradient. In the absence of a supercurrent, impurity scattering does not affect charge relaxation for a superconductor with an isotropic gap, since the matrix element for process 2 in Fig. 6 vanishes, and process 1 does not change the effective charge. When a superflow is imposed the quasiparticle spectrum becomes anisotropic. The matrix element for scattering is $uu' - vv' = \frac{1}{2}[(1+\frac{\xi}{E})^{\frac{1}{2}}(1+\frac{\xi'}{E'})^{\frac{1}{2}} - (1-\frac{\xi}{E})^{\frac{1}{2}}(1-\frac{\xi'}{E'})^{\frac{1}{2}}]$, where u and v are the coherence factors for the initial state and u' and v' are those for the final state, and the change in quasiparticle charge in the process is $\frac{\xi}{E} - \frac{\xi'}{E'}$. Scattering occurs on surfaces of constant energy, but since $|\xi|$ is not a constant on such surfaces, a process like 1 in Fig. 6 has a finite matrix element and process 2 does change the effective charge of a quasiparticle. Several calculations of this have appeared recently (Schmid and Schön, 1979; Clarke and Tinkham, 1980; Beyer Nielsen, Ono, Pethick and Smith, 1980). We shall here discuss the solution obtained by Beyer Nielsen et al. (1980) and compare to the other results available.

The collision term in the Boltzmann equation for the case of non-magnetic impurities is

$$\left(\frac{df_{\underset{\sim}{p}}}{dt}\right)_{coll} = -\sum_{\underset{\sim}{p}'} (uu' - vv')^2\, w(\underset{\sim}{p},\underset{\sim}{p}')(f_{\underset{\sim}{p}} - f_{\underset{\sim}{p}'})\delta(\tilde{E}_{\underset{\sim}{p}} - \tilde{E}_{\underset{\sim}{p}'}) , \quad (57)$$

where w is the normal state collision probability and $\tilde{E}_{\underset{\sim}{p}} = E_{\underset{\sim}{p}} + \underset{\sim}{p}\cdot\underset{\sim}{v}_s$.

Since in the situations of interest here the distribution function differs little from the equilibrium one, we may linearize the collision term. An important point to notice is that the energy that occurs in the energy-conserving delta function is the actual quasiparticle energy in the non-equilibrium situation $\tilde{E}_{\underset{\sim}{p}}$, not the one for equilibrium, $\tilde{E}^o_{\underset{\sim}{p}}$. To make this point explicit, we recall that in a situation where there is charge imbalance the chemical potential of the pairs differs from its equilibrium value, and this alters the dispersion relation of the BCS quasiparticles. For such a case the conserved energy is the one calculated using the shifted chemical potential, not the equilibrium one. Quite generally, the collision term vanishes if $f_{\underset{\sim}{p}}$ is the local equilibrium distribution function $f^o(\tilde{E}_{\underset{\sim}{p}})$, where f^o is the Fermi function. The Boltzmann equation becomes

$$-\underset{\sim}{v}_{\underset{\sim}{p}} \cdot \underset{\sim}{\nabla} T \frac{\tilde{E}}{T}\frac{\partial f^o}{\partial \tilde{E}} = -\sum_{\underset{\sim}{p}'} (uu' - vv')^2 w(\underset{\sim}{p},\underset{\sim}{p}')(\delta f^{\ell.e.}_{\underset{\sim}{p}} - \delta f^{\ell.e.}_{\underset{\sim}{p}'})\delta(\tilde{E}_{\underset{\sim}{p}} - \tilde{E}_{\underset{\sim}{p}'}) , \quad (58)$$

where we have defined the deviation from local equilibrium as

$$\delta f_{\underset{\sim}{p}}^{\ell.e.} = f_{\underset{\sim}{p}} - f^{o}(\tilde{E}_{\underset{\sim}{p}}) \; , \qquad (59)$$

$$= \delta f_{\underset{\sim}{p}} - \delta E_{\underset{\sim}{p}} \frac{\partial f^{o}}{\partial \tilde{E}_{\underset{\sim}{p}}} \; ,$$

with $\delta f_{\underset{\sim}{p}} = f_{\underset{\sim}{p}} - f^{o}(\tilde{E}^{o}_{\underset{\sim}{p}})$ the deviation from equilibrium, and $\delta E_{\underset{\sim}{p}} = \tilde{E}_{\underset{\sim}{p}} - \tilde{E}^{o}_{\underset{\sim}{p}}$. The distinction between $\delta f^{\ell.e.}$ and δf is very important in the present problem since close to T_c the $\delta E_{\underset{\sim}{p}}$ term in (59) completely dominates $\delta f_{\underset{\sim}{p}}$. In view of the fact that the collision term (58) is explicitly linear in $\delta f^{\ell.e.}$, we can linearize it immediately simply by replacing all quantities except $\delta f^{\ell.e.}$ by their equilibrium values (i.e. $\tilde{E}_{\underset{\sim}{p}} \to \tilde{E}^{o}_{\underset{\sim}{p}}$, etc.).

When the scattering in the normal state is s-wave, the collision probability $w(\underset{\sim}{p},\underset{\sim}{p}')$ is a constant. The integral kernel in Eq.(58) is therefore separable in the variables $\underset{\sim}{p}$ and $\underset{\sim}{p}'$, and the integral equation itself is trivially soluble (Schmid and Schön 1979, Beyer Nielsen et al. 1980).

The charge imbalance was obtained by Beyer Nielsen et al. for all temperatures and superfluid velocities $\underset{\sim}{v}_s$. In general this requires numerical integration, but when $p_F v_s \ll \Delta$ a good interpolation formula for the voltage is

$$(-e)\, V_d g_{ns} = \frac{Q^*}{2N(0)} \simeq p_F v_F \tau_{tr} \frac{(\underset{\sim}{v}_s \cdot \underset{\sim}{\nabla} T)}{T} \left\{ 1.93 \frac{\Delta/k_B T}{\cosh^2(\Delta/2k_B T)} + \frac{8}{15} \frac{1}{e^{\Delta/k_B T}+1} \right\} , \qquad (60)$$

where the first term comes from $\tilde{E} < \Delta + p_F v_s$ and the second from $\tilde{E} > \Delta + p_F v_s$. Although the energy range for $\tilde{E} < \Delta + p_F v_s$ is only $2p_F v_s$, $Q(\tilde{E})$ there is independent of v_s and it therefore gives a contribution of order v_s to Q^* . From Eq.(60) one can see that, while in the limit $T \to T_c$ the contribution from $\tilde{E} > \Delta + p_F v_s$ dominates, the contribution from the lower energy region is the more important one for $\Delta/k_BT_c \gtrsim 1/7$ or $|T-T_c| \gtrsim 0.002\ T_c$. Very near T_c the right-hand side of Eq.(60) becomes $(4/15)\ p_Fv_F\tau_{tr}\ (\underset{\sim}{v}_s \cdot \underset{\sim}{\nabla}T)/T$ in agreement with the result of Schmid and Schön (1979) for the case of clean superconductors with only impurity scattering present. We stress that although (60) is a good overall interpolation formula for the total voltage, the separate contribution of the energy region $\tilde{E} > \Delta + p_Fv_s$ is not well represented by the second term except very near T_c. Since the contribution from this energy region is generally small compared with that of the lower energy region except very near T_c, one can therefore use (60) as an approximate formula for the total voltage.

The result of Tinkham and Clarke (1980) was also obtained for the clean case with only impurity scattering present. These authors employed a relaxation time approximation for the non-equilibrium distribution function. Their result near T_c differs only by the numerical coefficient $5/2\pi$ from the $\Delta/T \to 0$ limit of (60). At lower temperatures, however, the two results differ significantly. In part this is due to the significant contribution of the low energy region $\tilde{E} < \Delta + p_Fv_s$ obtained from the solution by Beyer Nielsen et al. The experimental observations are in fair agreement with the result of Tinkham and Clarke, as discussed in the chapter by J. Clarke. Recently G. Schön (1980) has suggested that this may be due to the fact that smearing of the density of states by impurity scattering, an effect not included in the simple Boltzmann equation approach, is important, and that the condition for being able to neglect states in the pocket where $\tilde{E} < \Delta + p_Fv_s$ is $v_s/v_c \gg (\xi/\ell)^2$. Here $v_c \sim \Delta/p_F$ is the critical velocity and ξ the temperature dependent coherence length. Within the framework of the Boltzmann equation calculation by Beyer Nielsen et al. (1980) this condition amounts to the identification of a characteristic broadening of the quasiparticle states with the rate of branchcrossing, which in that calculation was found to be $\sim (1/\tau_{tr})(v_s/v_c)^{\frac{1}{2}}$ for states within the pocket. This equivalent broadening is only small compared with the width of the pocket, if the condition $v_s/v_c \gg (\xi/\ell)^2$ is satisfied, where $\ell = v_F\tau_{tr}$.

When smearing is important one would expect the contribution of the pocket to be greatly reduced. In even the cleanest experimental specimens ξ/ℓ is at least of order unity, which means that smearing is expected to be important for superfluid velocities right up to the critical one. More work is needed to obtain detailed quantitative estimates of the effects of smearing.

V. DIFFUSION AND OSCILLATORY MOTION

The two-component picture discussed in Section II can be used in situations when the quasiparticles and superfluid are not in equilibrium, and it is indeed in such contexts that it is most useful. One particularly instructive case is provided by a charge imbalance in a superconductor close to T_c, and with phonons as the dominant source of relaxation. For slow time variations, we saw in Section IV by solving the Boltzmann equation that the quasiparticle distribution function may be written as

$$f_i = f_i^{\ell.e.} - q_i \frac{\partial f^o}{\partial E_i} \delta\mu \tag{61}$$

where $\delta\mu$ is independent of the state i. Here $f_i^{\ell.e.}$ is the local equilibrium distribution, equal to the Fermi function $f^o(E_i)$ evaluated with the quasiparticle energy equal to its local equilibrium value $E_i(\mu_s)$. Since $q_i = -\partial E_i/\partial\mu_s$, the distribution is, to first order in $\delta\mu$, the same as

$$f_i = f^o(E_i(\mu_s + \delta\mu)) \tag{62}$$

which is a local equilibrium distribution function with the quasiparticle energy E_i shifted by an amount corresponding to a chemical potential change $\delta\mu$. It is therefore natural to regard the quantity $\mu_s + \delta\mu$ as a chemical potential for the normal component, and we shall denote it in future by μ_n.

The susceptibilities which were introduced in Section II enable us to rewrite the collision term in some useful forms. The relaxation of the quasiparticle charge is described phenomenologically by

$$\left(\frac{dQ_n}{dt}\right)_{coll} = - \frac{Q^*}{\tau_{Q^*}} \tag{63}$$

Q^* is the deviation of Q_n from its local equilibrium value, that is from its value when the normal component is in equilibrium with the superfluid at the actual value of μ_s. If we denote by $\delta\mu_s$ the deviation of μ_s from its value in global equilibrium, we may write $Q^* = \delta Q_n - (\partial Q_n/\partial\mu_s)\delta\mu_s = \delta Q_n - (\chi_n^o/\chi_s^o)\delta Q_s$ where δQ_n is the deviation of Q_n from its global equilibrium value, and we have used Eqs.(10) and (12). Since $\delta Q_s = \delta Q_{tot} - \delta Q_n$ we can finally write

$$\left(\frac{dQ_n}{dt}\right)_{coll} = - \frac{1}{(1-Z)\,\tau_{Q*}} \left(\delta Q_n - \delta Q_n{}^{eq}\right) \qquad (64)$$

where $\delta Q_n{}^{eq} = (\chi_n^o / \chi_{tot}^o)\delta Q_{tot}$ is the value of δQ_n for equilibrium in the presence of a density fluctuation δQ_{tot}. Therefore Q_n relaxes towards the local equilibrium value at a rate $1/\tau_{Q*}$ according to (63) but towards the global equilibrium value at a rate $1/(1-Z)\tau_{Q*}$ according to (64). For the case of phonon scattering near T_c, $\tau_{Q*} = (\pi\Delta/4k_BT_c)^{-1}\,\tau_{in}(0)$, and $1-Z = (\pi/4)(\Delta/k_BT_c)$. Therefore the rate for relaxation towards global equilibrium is just $1/\tau_{in}(0)$ which remains finite in the limit $T \to T_c$. In Fig. 7 the differences between the various density deviations are shown schematically for a temperature close

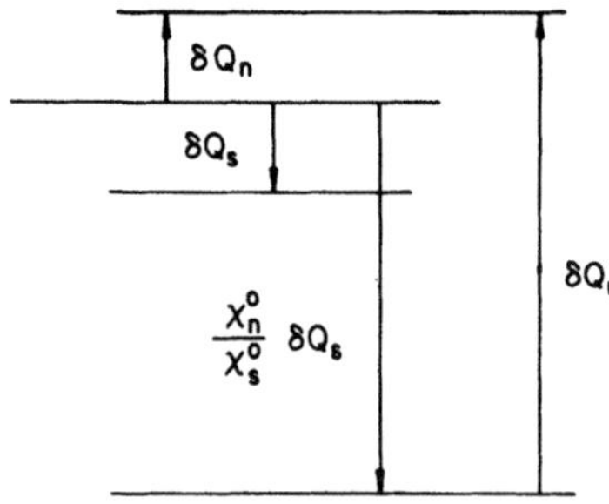

Fig. 7. Schematic representation of the relationship between the deviation of the quasiparticle charge from equilibrium, δQ_n, the deviation of the superfluid charge from equilibrium, δQ_s, and the deviation of the quasiparticle charge from local equilibrium, $\delta Q_n^{\ell.e.} = Q^*$. We assume the total charge is unaltered, and therefore $\delta Q_n = -\delta Q_s$.

to T_c, when $\chi_n^o / \chi_s^o = Z/(1-Z)$ is large. The positive δQ_n is accompanied by an equal and opposite δQ_s, since we assume here that δQ_{tot} vanishes. Because the susceptibility of the superfluid is small near T_c, the change in Q_s leads to a large change in μ_s which in turn shifts the local equilibrium value of δQ_n by a large amount.

Let us now consider the conversion term that enters the continuity equations for quasiparticle and pair charge. We have argued above that $\left(\frac{dQ_n}{dt}\right)_{coll}$ may be written in the two equivalent ways (63) and (64). In the situations we consider, the change, δQ_{tot}, in the total charge density is negligible and therefore

$$\left(\frac{dQ_n}{dt}\right)_{coll} = - \frac{\delta Q_n}{\tau_{in}(0)} \tag{65}$$

at temperatures near the transition temperature T_c.

We may combine the equation (65) together with (21),(27) and (28-31) to get a single differential equation in space and time for the quasiparticle charge δQ_n : (Kadin, Smith and Skocpol, 1980)

$$\left(\frac{n_s}{(1-Z)\,n_e}\,\frac{1}{3}\,v_F^2\,\nabla^2 - \frac{\partial^2}{\partial t^2} - \left(\frac{n_s}{n_e}\,\frac{1}{\tau_{tr}} + \frac{1}{\tau_{in}(0)}\right)\frac{\partial}{\partial t} - \frac{n_s}{n_e}\,\frac{1}{\tau_{tr}}\,\frac{1}{\tau_{in}(0)}\right)\delta Q_n = 0. \tag{66}$$

We shall now use this equation to discuss a number of physically interesting situations involving spatial and temporal variations.

a) Carlson-Goldman Mode

The collective mode observed by Carlson and Goldman (1976) is basically an oscillation of the charge associated with the superfluid. The normal fluid plays little role, except in so far as it moves in such a way as to preserve overall charge neutrality. To derive the expression for the mode frequency we combine the superfluid equation of motion (30) $m\,\partial \underset{\sim}{v}_s/\partial t = - \underset{\sim}{\nabla}(\delta\mu_s - e\phi)$, the continuity equation for charge associated with the superfluid component (27) $\frac{\partial \delta Q_s}{\partial t} + \underset{\sim}{\nabla}\cdot(n_s \underset{\sim}{v}_s) = -\left(\frac{dQ_n}{dt}\right)_{coll}$, and the susceptibility relation (10) $\delta Q_s = \chi_s^o\,\delta\mu_s$. To close the set of equations one needs an equation for ϕ. This is determined by Eq.(32) for the motion of the normal component. Close to T_c the normal density n_n is large compared with n_s, and therefore the velocity of the normal component required to ensure charge neutrality ($\underset{\sim}{v}_n = -(n_s/n_n)\underset{\sim}{v}_s$) is small, consequently the electrostatic potential necessary to drive it is small, and ϕ may be neglected in the superfluid equation of motion. For frequencies high compared with $1/\tau_{in}$, the relaxation term in the continuity equation may be neglected, and therefore one finds a sound-like mode with a velocity c given by

$$c^2 = \frac{n_s}{m\chi_s^o} \tag{67}$$

or

$$c = \sqrt{\frac{1}{3}\frac{n_s}{(1-Z)n_e}}\, v_F \tag{68}$$

This result (67) may also be obtained from the solution of Eq.(66), which includes the damping term. One easily sees from that equation that damping is negligible if the frequency is large compared with $n_s/n_e\,\tau_{tr}$ and $1/\tau_{in}(0)$. In the dirty limit the expression for n_s is

$$n_s = \frac{\pi}{2}\frac{\Delta^2\tau_{tr}}{k_BT_c\hbar}\, n_e \quad . \tag{69}$$

Using $1-Z \simeq \frac{\pi}{4}\frac{\Delta}{k_BT}$, we then find for the velocity of the mode

$$c = \sqrt{\frac{2D\Delta}{\hbar}} \tag{70}$$

when written in terms of the diffusion coefficient $D = v_F^2\tau_{tr}/3$. The result (70) was first obtained by Schmid and Schön (1975b).

In the clean limit for which $n_s \simeq 2n_e(1-T/T_c)$ the mode velocity agrees with that derived earlier by Artemenko and Volkov (1975). This result will, however, be modified when the non-locality of the electrodynamic response, which alters the effective value of n_s, is taken into account.

The propagating mode has been observed by Carlson and Goldman (1975,1976) in a narrow temperature interval below T_c, and is discussed in detail in Goldman's chapter. In dirty superconductors the main cause of damping is associated with ohmic dissipation resulting from the motion of normal component which gives rise to the damping term involving $(n_s/n_e)\ 1/\tau_{tr} \simeq \Delta^2/\hbar(k_BT_c)$ in Eq.(66). To observe the mode one therefore needs frequencies which are comparable to or greater than $\Delta^2/(\hbar k_BT_c)$. On the other hand the frequency must be less than $\Delta/\hbar$, since the two-fluid description is limited to long wavelengths and low frequencies. These conditions restrict the observation of the collective mode to a narrow range below T_c . The observed value of the velocity agrees well with the result (70).

The collective mode in superconductors is similar to a number of sound waves in liquids, and we shall here make comparison with first, second and fourth sound. A first sound mode involves a fluctuation of the total density n , and its velocity may be written

$$c_1^2 = \frac{\gamma}{m}\frac{n}{\chi} , \tag{71}$$

where γ is the ratio of adiabatic and isothermal bulk moduli, and $\chi = (\partial n/\partial\mu)_T$ is the isothermal susceptibility. At low temperatures γ tends to 1, and therefore one has

$$c_1^2 = \frac{n}{m\chi} , \tag{72}$$

which has precisely the same form as Eq.(67) for the collective mode in superconductors, but with a different density and susceptibility.

In a superfluid contained in a porous medium the normal component can be locked and a density oscillation mode can exist in which only the superfluid flows. This is fourth sound, and its velocity is given by

$$c_4^2 = \frac{n_s}{m\,\chi} = \frac{n_s}{n}\, c_1^2 \tag{73}$$

if one neglects a small contribution due to temperature fluctuations. In this mode the restoring force comes from the variations of the chemical potential with density, as it does in first sound. However, the density that enters the dispersion relation is the superfluid density, rather than the total density, since it is only the superfluid component that moves.

A bulk superfluid is also able to support coupled counterflows of the normal and superfluid components, so-called second sound. In practical cases the mode is basically a temperature fluctuation. Its velocity is much less than that of first sound and is given to a good approximation by

$$c_2^2 = \frac{1}{m}\frac{n_s}{n_n}\frac{T s^2}{c_V} \tag{74}$$

where s is the entropy per particle and c_V is the specific heat per particle. The basic restoring force comes from the temperature changes in the normal component. The fluctuations of the total density are small in a second sound wave, but the chemical potential gradients resulting from them are sufficient to drive the superfluid component in such a way that the total current remains close to zero. This mode is similar to the collective mode in superconductors in two respects, i) they both involve counterflows of the normal and superfluid components and ii) fluctuations in the total density are small. In a superconductor density fluctuations are strongly inhibited by the long-range Coulomb interaction, and in second sound by the relatively large amount of energy required to create density fluctuations compared with that to create temperature fluctuations. An important difference between all the above types of sound and the collective mode is that in the sound modes the matter is locally in thermodynamic equilibrium, whereas departures from equilibrium are essential in the collective mode, which is driven by differences between the chemical potentials of the normal and superfluid components. Second sound and the collective mode differ in two important respects. The basic restoring force is due to temperature changes in the first case, and due to chemical potential difference (charge imbalance) in the second case. Since restoring forces associated with temperature changes are smaller than those associated with density or charge imbalance changes by a factor $\sim(T/T_F)^2$, where T_F is the Fermi temperature, the velocity of second sound is much smaller than that of the other sorts of sound, all other things being equal. Also in the case of the superconductor the motion of the normal component is dissipative, due to scattering from impurities. As we have seen, this leads to a damping rate $(n_s/n_e)(1/\tau_{tr})$ for the collective mode, which results in the mode being well defined only at temperatures close to T_c. In bulk liquids motion of the normal component is not damped by any process analogous to impurity scattering in a superconductor, and therefore second sound can exist over a wider temperature range than can the collective mode in a superconductor.

The expression for the velocity of the collective mode is similar to that for first sound. However the density that enters is the superfluid density, since essentially all the kinetic energy in the mode is in the superfluid component, and the susceptibility that enters is the one associated with charge imbalance, rather than with changes in the total density. For small changes in the charge imbalance in a superconductor (at constant total density) the change in the energy per unit volume is $\frac{1}{2}(\delta Q_s - \delta Q_n)^2 / \chi_s^o$, while for changes in the total density in a liquid the analogous result is $\frac{1}{2}\,\delta n^2/\chi$.

b) Diffusion of Quasiparticle Charge

Under stationary but spatially inhomogeneous conditions the normal charge may diffuse over a characteristic length Λ_{Q*} which from Eq.(66) is seen to be given by

$$\Lambda_{Q*}^2 = \frac{1}{3} v_F^2 \tau_{in}(0) \tau_{tr} \frac{1}{1-Z} \quad . \tag{75}$$

The concept of a characteristic diffusion length for quasiparticle charge which may be much larger than the mean free path in the normal state of the metal was first introduced by Pippard, Shepherd and Tindall (1971) as a means of explaining the observed N-S boundary resistance. In such experiments one passes a current across a normal-superconducting interface and measures the resistance as a function of the temperature and the mean free path of the normal state of the superconducting material. In their classic paper Pippard and his co-workers argued that near T_c the quasiparticles may diffuse into the superconducting material over a characteristic length which is of order $\sqrt{\tau_{in}\tau_{tr}}\, v_F$, thus increasing the effective length of the normal region where dissipation occurs. Near T_c , where Andreev reflection of the quasiparticles entering the superconductor is unimportant, the extra resistance ΔR per unit area may be expected to be given by this diffusion length times the normal state resistivity of the superconducting material, leading to an overall dependence $\Delta R \sim \tau_{tr}^{-\frac{1}{2}}$ for the excess boundary resistance in good agreement with observation.

Waldram (1975) pointed out that the appropriate diffusion length involves the charge relaxation time τ_{Q*} instead of τ_{in} , with a resulting diffusion length given by (75). The excess boundary resistance per unit area is then given by $\Delta R = \Lambda_{Q*}\rho_n$ where ρ_n is the normal state resistivity of the superconducting material. This result is a simple consequence of the exponential decay of the quasiparticle charge over the region inside the superconductor where the normal current is converted into supercurrent. The result $\Delta R = \Lambda_{Q*}\,\rho_n$ was derived microscopically by Artemenko, Volkov and Zaitsev (1978) and was also discussed by Waldram (1975). Recent experiments by Hsiang and Clarke (1980), who also presented a simple derivation of this result, are in very good agreement with this prediction, both qualitatively and quantitatively. The experiments have been used to extract the inelastic electron-phonon scattering times $\tau_{in}(0)(=\tau_E)$ at T_c for various metals.

Within the two-fluid picture we have developed it is easy to derive the simple expression for the N-S boundary resistance near T_c. Consider the variation in Q^* or equivalently the variation in quasiparticle chemical potential difference $\delta\mu = \delta(\mu_n - \mu_s) = Q^*/\chi_n^o$ away from an N-S boundary at x = 0. Inside the superconductor (x > 0) $\delta\mu$ varies exponentially,

$$\delta\mu = \delta\mu(0)e^{-x/\Lambda_{Q^*}} \tag{76}$$

in consequence of (66). The variation of the chemical potential drives a normal current. Close to T_c the normal current $\underset{\sim}{J}_n$ is essentially the same as the current of normal charge $\underset{\sim}{J}_n^Q$, which is given by Eq.(32). One therefore finds

$$J_n = -D\frac{d}{dx}\delta Q_n = -e\,\frac{1}{3}\,v_F^2\tau_{tr}\Lambda_{Q^*}^{-1}\,e^{-x/\Lambda_{Q^*}}\,\delta\mu(0)\chi_n^o \tag{77}$$

which similarly vanishes exponentially inside the superconductor. At the boundary (x = 0) the normal current density times the cross sectional area A must equal the total applied current I. This condition determines the boundary value $\delta\mu(0)$ of the chemical potential difference:

$$\delta\mu(0) = -\frac{I}{eA}\,\frac{3}{v_F^2}\,\Lambda_{Q^*}\,\frac{1}{\tau_{tr}}\,\frac{1}{\chi_n^o}\quad . \tag{78}$$

The extra voltage drop associated with the quasiparticle diffusion into the superconductor is therefore given by

$eV = \delta\mu_s(0) = -\dfrac{\chi_n^o}{\chi_{tot}^o}\,\delta\mu(0)$ yielding the excess resistance per unit

area

$$\Delta R = \frac{V}{(I/A)} = \Lambda_{Q^*}\,\frac{m}{n_e e^2}\,\frac{1}{\tau_{tr}} = \Lambda_{Q^*}\rho_n\ , \tag{79}$$

where use has been made of the relations $\chi_{tot}^o = 2N(0) = 3n_e/mv_F^2$, and $\rho_n^{-1} = n_e e^2 \tau_{tr}/m$. We emphasize that (79) may be expected to hold only near T_c where Andreev reflection can be ignored.

A direct demonstration of the importance of the diffusion process characterized by the length Λ_{Q*} has been furnished by the beautiful experiments of Dolan and Jackel (1977), who used superconducting and normal probes to measure the spatial dependence of the chemical potentials μ_n and $\mu_s - e\,\phi$ near a phase slip center. They showed that the spatial dependence of μ_n was given accurately by the exponential $e^{-x/\Lambda_{Q*}}$ and demonstrated the weakly divergent behaviour of Λ_{Q*} near the transition temperature, where $\Lambda_{Q*} \propto (T_c - T)^{-\frac{1}{4}}$. The magnitude of Λ_{Q*} agreed well with (75) when the charge relaxation time τ_{Q*} was identified with the one measured in the tunnelling injection experiment.

Diffusion of quasiparticle charge near phase-slip centres has also been studied by Skocpol, Beasley and Tinkham (1974) from measurements of differential resistance. These experiments did not reveal the diverging temperature dependence of the diffusion length but later ones by Kadin, Skocpol and Tinkham (1978) gave clear evidence of the divergence. These authors studied also the effects of pair breaking on the charge relaxation time τ_{Q*} which determines the diffusion length according to $\Lambda_{Q*} = (\frac{1}{3} v_F^2 \tau_{Q*} \tau_{tr})^{\frac{1}{2}}$.

The observations were consistent with the result (54) derived by Schmid and Schön (1975a) for an arbitrary (weak) pair-breaking mechanism near T_c , although a detailed test requires that the theoretical considerations be extended to lower temperatures where $\Delta/(k_B T_c)$ is not a small parameter (within weak-coupling BCS theory one finds $\Delta/k_B T_c = 0.5$ when $T \simeq 0.975\ T_c$).

As a final remark we note that the equation (66) shows that δQ_n relaxes to zero on a time scale $\tau_{in}(0)$ rather than τ_{Q*} (Leggett and Takagi, 1977; Pethick and Smith, 1979a; Kadin, Smith and Skocpol, 1980). Under spatially homogeneous conditions a solution of the equation is $\delta Q_n = \delta Q_n^o e^{-t/\tau_{in}(0)}$. This exponential decay in time has not yet been directly observed in an experiment.

VI. CONCLUSION

In this chapter we have illustrated how the two-component model for the charge density, and the concept of charge imbalance provide a natural framework for describing a variety of phenomena in non-equilibrium superconductors. Our discussion has been based on the quasiparticle Boltzmann equation, which allows us to exhibit the fundamental ideas with a minimum of mathematical machinery. The Boltzmann equation approach is justified in a number of realistic situations, and can take into account many aspects of the physics of real metals, for example some of the most important strong-coupling effects. However, there are situations where the simple Boltzmann equation approach fails, such as for example in the presence of strong pair breaking, or in considering some phenomena in dirty superconductors. Then one must use the more general approach described in the chapter by A. Schmid.

The merit of the Boltzmann equation approach is its conceptual simplicity and wide applicability. To properly identify its limits of validity in a given situation of physical interest it is, however, very important that one be guided by a microscopic theory, in which the Boltzmann equation description is contained as a well-defined limit.

REFERENCES

Anderson, P.W., 1959, Theory of dirty superconductors, J. Phys. Chem. Solids, 11:26.

Aronov, A.G., 1974, The influence of condensate motion on the thermoelectric effects in superconductors, Zh. Eksp. Teor. Fiz., 67:178 [1975, Sov. Phys. JETP, 40:90].

Artemenko, S.N. and Volkov, A.F., 1975, Collective excitations with a sound spectrum in superconductors, Zh. Eksp. Teor. Fiz., 69:1764 [1976, Sov. Phys. JETP, 42:896].

Artemenko, S.N., Volkov, A.F., and Zaitsev, A.V., 1978, On the contribution of the superconductor to the resistance of a superconductor-normal metal system, J. Low Temp. Phys., 30:487.

Betbeder-Matibet, O. and Nozières, P., 1969, Transport equations in clean superconductors, Ann. Phys. (N.Y.), 51:392.

Beyer Nielsen, J., Ono, Y.A., Pethick, C.J., and Smith, H., 1980, The effect of impurity scattering on the thermally induced charge imbalance in a clean superconductor, Solid State Commun., 33:925.

Bray, A.J. and Schmidt, H., 1975, Collective modes in charged superconductors near T_c, Solid State Commun., 17:1175.

Carlson, R.V. and Goldman, A.M., 1975, Propagating order-parameter collective modes in superconducting films, Phys. Rev. Lett., 34:11.

Carlson, R.V. and Goldman, A.M., 1976, Dynamics of the order parameter of superconducting aluminum films, J. Low Temp. Phys., 25:67.

Chang, J.J., 1978, Gap enhancement in superconducting thin films due to quasiparticle tunnel injection, Phys. Rev., B17:2137.

Chi, C.C. and Clarke, J., 1979, Quasiparticle branch mixing rates in superconducting aluminum, Phys. Rev., B19:4495.

Chi, C.C. and Clarke, J., 1980, Addendum to "Quasiparticle branch mixing rates in superconducting aluminum", Phys. Rev., B21:333.

Clarke, J., 1972, Experimental observation of pair-quasiparticle potential difference in nonequilibrium superconductors, Phys. Rev. Lett., 28:1363.

Clarke, J., Fjordbøge, B., and Lindelof, P.E., 1979, Supercurrent-induced charge imbalance measured in a superconductor in the presence of a thermal gradient, Phys. Rev. Lett., 43:642.

Clarke, J. and Tinkham, M., 1980, Theory of quasiparticle charge imbalance induced in a superconductor by a supercurrent in the presence of a thermal gradient, Phys. Rev. Lett., 44:106.

Dolan, G.J. and Jackel, L.D., 1977, Voltage measurements within the nonequilibrium region near phase-slip centers, Phys. Rev. Lett., 39:1628.

Entin-Wohlman, O. and Orbach, R., 1979, Effect of pair breaking on branch relaxation in nonequilibrium superconductors, Phys. Rev., B19:4510.

Hsiang, T.Y. and Clarke, J., 1980, Boundary resistance of the superconducting-normal interface, Phys. Rev., B21:945.

Kadin, A.M., Skocpol, W.J., and Tinkham, M., 1978, Magnetic field dependence of relaxation times in nonequilibrium superconductors, J. Low Temp. Phys., 33:481.

Kadin, A.M., Smith, L.N., and Skocpol, W.J., 1980, Charge imbalance waves and nonequilibrium dynamics near a superconducting phase-slip center, J. Low Temp. Phys., 38:497.

Leggett, A.J. and Takagi, S., 1977, Orientational dynamics of superfluid ^{3}He: A "two-fluid" model. I. Spin dynamics with relaxation, Ann. Phys. (N.Y.), 106:79.

Moody, M.V. and Paterson, J.L., 1979, Relaxation of the distribution function branch imbalance Q^* in Sn and Sn-In alloys, J. Low Temp. Phys., 34:83.

Pethick, C.J. and Smith, H., 1978, Relaxation and collective motion in superconductors, a two-fluid description, J. Physique Supplement, 39:C6-488.

Pethick, C.J. and Smith, H., 1979a, Relaxation and collective motion in superconductors: a two-fluid description, Ann. Phys. (N.Y.), 119:133.

Pethick, C.J. and Smith, H., 1979b, Generation of charge imbalance in a superconductor by a temperature gradient, Phys. Rev. Lett., 43:640.

Pethick, C.J. and Smith, H., 1980, Charge imbalance in nonequilibrium superconductors, J. Phys. C., to be published.

Pippard, A.B., Shepherd, J.G., and Tindall, B.A., 1971, Resistance of superconducting-normal interfaces, Proc. Roy. Soc. Lond., A324:17.

Putterman, S.J., 1977, Phenomenological theory of collective modes and relaxation effects in type II superconductors, J. Low Temp. Phys., 28:339.

Schmid, A. and Schön, G., 1975a, Linearized kinetic equations and relaxation processes of a superconductor near T_c, J. Low Temp. Phys., 20:207.

Schmid, A. and Schön, G., 1975b, Collective oscillations in a dirty superconductor, Phys. Rev. Lett., 34:941.

Schmid, A. and Schön, G., 1979, Generation of branch imbalance by the interaction between supercurrent and thermal gradient, Phys. Rev. Lett., 43:793.

Schön, G., 1980, private communication, this summer school.

Skocpol, W.J., Beasley, M.R., and Tinkham, M., 1974, Phase-slip centers and non-equilibrium processes in superconducting tin microbridges, J. Low Temp. Phys., 16:145.

Tinkham, M., 1972, Tunneling generation, relaxation, and tunneling detection of hole-electron imbalance in superconductors, Phys. Rev., B6:1747.

Tinkham, M. and Clarke, J., 1972, Theory of pair-quasiparticle potential difference in non-equilibrium superconductors, Phys. Rev. Lett., 28:1366.

Waldram, J.R., 1975, Chemical potential and boundary resistance at normal-superconducting interfaces, Proc. Roy. Soc. Lond., A 345:231.

Ziman, J.M., 1960, "Electrons and Phonons," Oxford University Press, London.

THERMOELECTRIC EFFECTS IN SUPERCONDUCTORS

Charles M. Falco

Solid State Science Division
Argonne National Laboratory
Argonne, IL 60439

James C. Garland

Department of Physics
Ohio State University
Columbus, OH 43210

1. INTRODUCTION

The first measurements of thermoelectricity in superconductors were made by Meissner (1927) who found the Seebeck coefficient S (thermopower) in a superconducting thermocouple to be at least an order of magnitude below the equivalent normal state value. This result was subsequently verified by Borelius et al., (1931); Burton et al., (1935); Keesom and Matthijs (1938); and Casimir and Rademakers (1947). In 1953, Pullan showed unequivocally that the thermopower of a super-conductor decreased at T_c by at least four decades.

In addition to the thermopower, other thermoelectric coefficients were found to vanish in superconductors. Steiner and Grassman (1935) showed that the Peltier coefficient of a bi-metallic Pb-Sn loop was at least five orders of magnitude lower in the superconducting state than in the normal state. A similiar decrease in the Thomson coefficent at the transition temperature was observed by Daunt and Mendelssohn (1945) for a Pb loop. Despite these early findings, all thermoelectric effects do no vanish in superconductors. Ginzburg (1944) was the first to point out the possible existence of superconducting thermoelectric effects. A great deal of recent work has now established that there are thermoelectric effects which are actually unique to the superconducting state. Furthermore, as we shall show in this review, the study of these effects provides an interesting and fruitful way to investigate the non-equilibrium dynamics of the superconducting state.

Section 2 of this paper introduces the two fluid model of superconductivity which is analogous to that commonly used to describe super-fluid helium. The impossibility of observing "normal" thermoelectric coefficients in superconductors is shown to follow from the spatial rigidity of the electrochemical potential. Recent electrostatic measurements of the chemical potential in superconductors are also discussed. In Section 3, we discuss attempts to study the "fountain effect" in superconductors; in principle, these measurements, which make use of the asymmetry in the current-voltage characteristic of Josephson junctions, should allow determination of the thermopower of a superconductor. Section 4 summarizes recent measurements of the thermoelectrically generated flux in bimetallic superconducting rings. This effect, which is based upon the concept of fluxoid quantization, yields information about the non-equilibrium quasiparticle distribution of a superconductor. The accousto-electric effect, which is conceptually similiar to the thermoelectric flux effect, is described in Section 5.

Very recently, there has been a flurry of theoretical and experimental research on thermoelectric charge imbalance effects in superconducting thin films. This work is reviewed in Section 6.

2. THERMOELECTROSTATIC EFFECTS

Thermoelectrostatic effects are most easily described by invoking a two-fluid model of a superconductor in which the normal fluid, carried by the quasiparticle excitations, behaves similarly to the ordinary conduction electron gas of a normal metal, while the superconducting properties are exhibited by a condensate of Cooper pairs. The relationship of the two fluids to the microscopic parameters of the superconductor is actually quite subtle and has only recently been put on a solid theoretical foundation by Pethick and Smith (1977). However, the original idea of a two-fluid model can be traced to Landau (1941), who postulated a set of phenomenological equations of motion for liquid helium. The extension of this idea to superconductivity was obvious, of course, although it was recognized very early that there would be profound differences between a charged and uncharged superfluid.

One difference, suggested by Putterman and deBruyn Ouboter (1970), pertains to the strong coupling between the electronic excitations of a superconductor and the phonon spectrum. This coupling, they reasoned, should effectively clamp the normal fluid, thus leading to the following equation of motion:

$$m \frac{\partial v_s}{\partial t} = -\nabla(\mu + e\phi + (1/2)\, mv_s^2 + mH) \quad . \tag{1}$$

In the above expression, v_s is the superfluid velocity, m is the electronic mass, μ is the chemical potential, and ϕ is the electrostatic potential. Dissipation in the superconductor is contained in the ∇H term, introduced originally by Khalatnikov (1965) for superfluid helium. For a superconductor, we can associate this dissipation with the divergence of the supercurrent:

$$mH = -\ \xi \nabla \cdot \vec{j}_s \tag{2}$$

The term in parenthesis on the right of Eq. (1) is the total electrochemical potential for the superconductor and can be seen to be the total driving force for the superfluid.

Equation (1) was derived assuming the normal fluid to be clamped. For stationary flow ($\partial\vec{v}_s/\partial t = 0$) at constant temperature ($\nabla\mu, \nabla H = 0$), Eq. (1) reduces to

$$\Phi_2 - \Phi_1 = \frac{m\rho_s}{2e\rho_n} (v_{s,2}^2 - v_{s,1}^2) \quad , \tag{3}$$

which shows that an electrostatic potential difference should be established between two points in a superconductor having different superfluid velocities. This potential difference must be measured by a noncontacting device, such as a vibrating reed electrometer; an ordinary voltmeter measures difference in electrochemical potential difference and would read zero in this case because $\nabla\bar{\mu} = 0$.

The Bernoulli potential of Eq. (3) was observed in the experiments of Bok and Klein (1968) and of Morris and Brown (1971). In the latter experiment, a 9 kHz ac "Bernoulli current" was driven through a 0.1 mm diameter Pb wire and the resulting potential was capacitively detected using an electrostatic voltmeter. Their results, shown in Fig. 1 illustrate the quadratic dependence of the Bernoulli potential on superfluid velocity predicted by Eq. (3). From the magnitude of the measured potential at I_c, it is possible to obtain a value for the penetration depth. Morris and Brown found 394 ± 20 Å for their Pb sample. These results provide a nice confirmation of this particular prediction of the two fluid theory.

In the presence of a temperature gradient, the electrochemical potential becomes

$$\nabla\bar{\mu} = \nabla\mu + e\nabla\phi = 0 \tag{4}$$

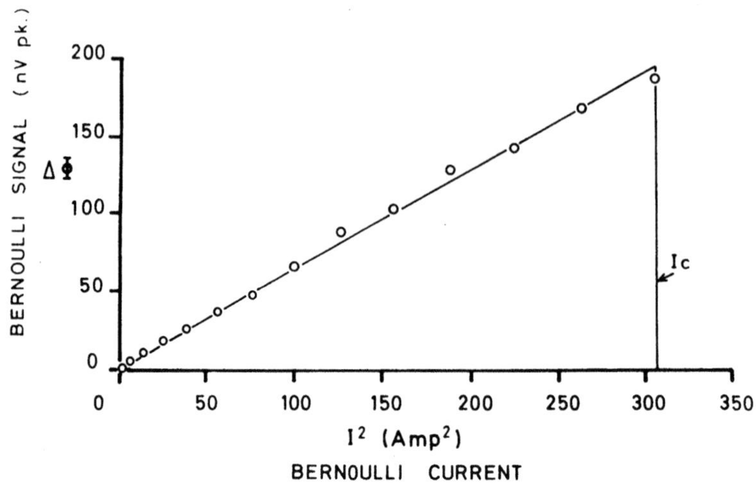

Fig. 1. Bernoulli potential measured in superconducting Pb at 4.2 K. Morris and Brown (1971).

or

$$\phi_2 - \phi_1 = \frac{1}{e}[\mu(T_2) - \mu(T_1)] \quad . \tag{5}$$

Equation (5) suggests that an electrostatic potential difference should be induced between the ends of a superconductor if the ends are maintained at different temperatures. An attempt to measure this electrostatic potential was first made by deWaele, deBruyn Ouboter and Pipes (1973). One of their results is shown in Fig. 2, where the work function, W, of Pb is plotted as a function of temperature. (The electrostatic potential is given by dW/dT.) It is clear from the figure that no discontinuity could be observed at the transition temperature T_c. The absence of any discontinuity in Nb was also noted by Pipes and Darling (1975). Subsequently, however, Shott and Walton (1977) repeated these experiments using a Sn sample and found

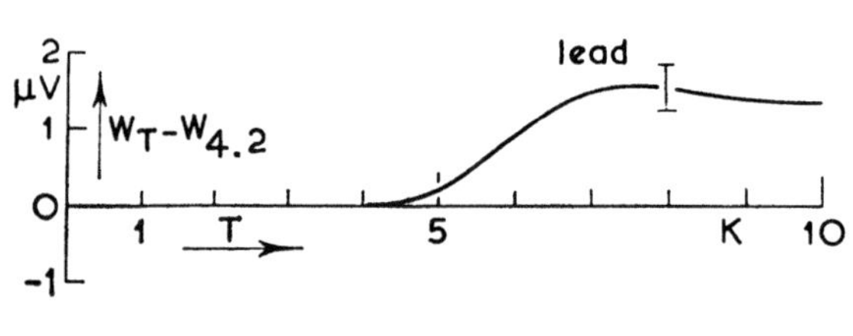

Fig. 2. Work function of Pb as a function of temperature measured with respect to its value at 4.2 K. deWaele, deBruyn Ouboter, and Pipes (1973).

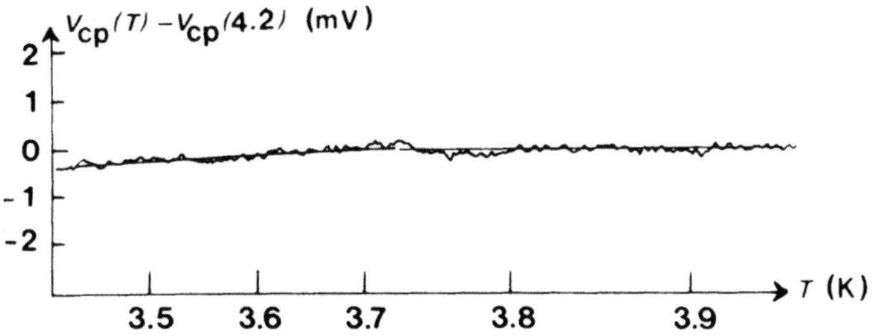

Fig. 3. Contact potential difference between Sn and Cu as a function of temperature measured with respect to their value at 4.2 K. Shott and Walton (1977).

a small change in the temperature dependence of the electrostatic potential at T_c. Below T_c, the work function was found to decrease at a rate of approximately 2mV/K. This is the same rate at which the energy gap changes in Sn near T_c. Their results, expressed in terms of the work function, are shown in Fig. 3. This is in apparent disagreement with the work of deWaele et al. (1973) and Pipes and Darling (1975) which reported sufficient resolution to see such an effect. However, whichever case turns out to be correct, such experiments seem to measure properties unrelated to thermoelectric transport coefficients in superconductors.

3. SUPERCONDUCTING FOUNTAIN EFFECT

The electrostatic potential predicted by Eq. (5) leads to an effect which parallels the fountain effect observed in liquid helium. In the ordinary fountain effect, liquid helium flows through a thin capillary tube joining two reservoirs, one of which contains a heater. The superfluid component flows without friction through the capillary, so that there is no difference in chemical potential between the reservoirs. The normal fluid flows with difficulty, however, resulting in both a pressure and temperature gradient across the capillary. Although the fountain effect derives its name from the accumulation and eventual overflow of liquid in the warmer reservoir, the essential physical idea is the establishment of a pressure difference proportional to the difference in temperature.

In a superconductor, the supercurrent always maintains the electrochemical potential at a constant value. Because the normal current is "clamped" by impurity and phonon scattering, however, there can be a difference in electrostatic potential. The relationship between the electrostatic potential and the temperature difference described by Eq. (5) is thus analogous to the pressure/temperature relationship in liquid helium.

There is obviously no "fountain" of charge in the superconducting effect, although there would be a slight variation in charge density along the length of the sample. This would give rise to an electrostatic potential as discussed in Section 2. By analogy to normal metal transport theory, we can write

$$\vec{J}_n = L_T(-\nabla T) = -\vec{J}_s \quad , \tag{6}$$

where $\vec{J}_n$ is the normal current, $\vec{J}_s$ is the supercurrent, and L_T is a thermoelectric transport coefficent which, in a normal metal, would be given by $L_T = \sigma S$ (σ is the conductivity and S is the thermoelectric power). Because the total current density in the wire is zero, and because there is no electrochemical voltage gradient along the wire, there is no obvious way to verify Eq. (6) directly.

Clarke and Freake (1972), however, noted that if the superconducting wire contained a Josephson weak link, then it would be possible to "unlock" the normal and superfluid currents. Their experiment arrangement is shown in Fig. 4. A sharpened Pb screw in contact with a Pb foil was used as a point contact Josephson junction. When a temperature difference was impressed across this junction, a thermally generated normal current and counterflow of supercurrent flowed through the junction, in accordance with Eq. (6). An additional

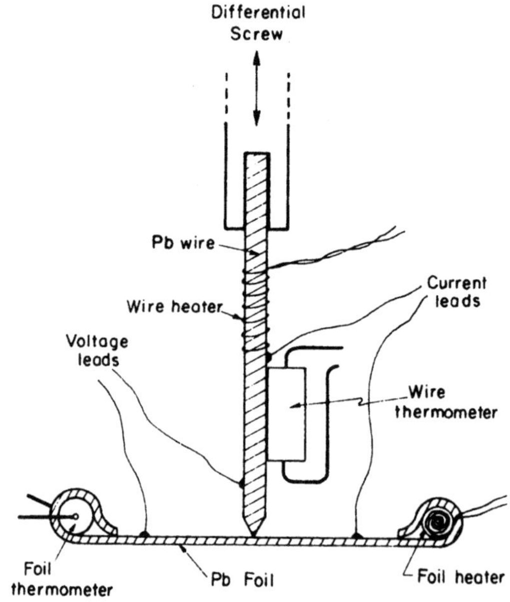

Fig. 4. Experimental arrangement used for measuring asymmetry in I-V characteristics induced by temperature gradient. Clarke and Freake (1972).

supercurrent, injected from an external power supply through wires connected across the junction, was used to drive the junction into a dissipative state. The magnitude of the additional current depended in its direction, being greatest when it opposed the thermally induced current. By measuring the asymmetry in the external current needed to produce a voltage across the junction, Clarke and Freake were thus able to estimate the thermoelectric transport coefficient L_T. Although the scatter in their data was large, they concluded that the sign of L_T was the same as for Pb in the normal state.

Similar measurements have recently been performed on Nb by Welker and Bedard (1977), and on both Pb and Nb by Matsinger et. al., (1978). In the Welker and Bedard experiment, a thin Nb wire was broken in vacuum at 4.2 K, and a point contact weak link was formed by touching the ends of the wire together without warming the sample or breaking the vacuum. No asymmetry could be detected in this exceptionally clean system, although an effect could sometimes be seen in junctions contaminated by an oxide. A null result was also found by Matsinger et al. (1978) on Pb and Nb junctions prepared at room temperature. Figure 5 shows the I-V characteristics of one of their Pb point contact junctions. The theory curves also shown in Fig. 5 illustrate the asymmetry expected (but not observed) from reversing the direction of the temperature gradient.

Very recently, Smith, Tinkham and Skocpol (1980) have proposed an alternate model to explain the results of the Clarke and Freake experiment. Their theory examines the properties of a superconducting tunnel junction in the presence of a temperature gradient and concludes that an asymmetry in the tunneling probability for electrons and holes should result in a tunneling current which is linear in ∇T. Unlike the thermoelectrically induced current, this tunneling current does not depend on the thermoelectric transport coefficient L_T in

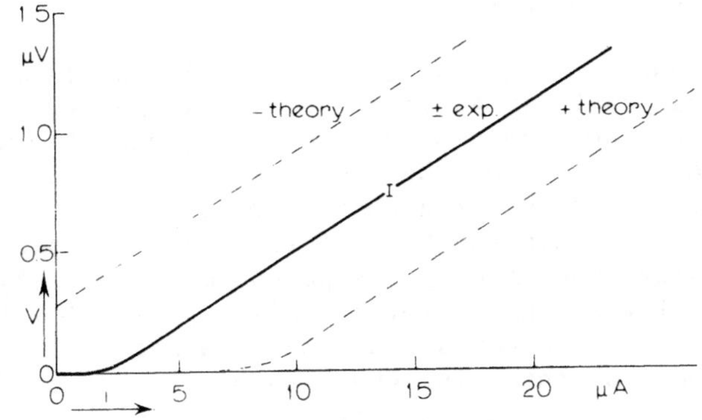

Fig. 5. Measured positive and negative branches of the I-V characteristic of a Pb point contact held in a temperature gradient. Theory curves are expected asymmetry based on Eq. (13). Matsinger, de Bruyn Ouboter and van Beelen (1978).

the superconductor; it is solely due to the tunneling barrier. This theory appears to be compatible with the Clarke and Freake (1972) data in sign. It is also speculated that the one positive result by Welker and Bedard on a "dirty" contact could be due to the presence of an oxide. However, the model is not supported by the null results of Matsinger et al. (1978). The magnitude of the predicted effect is also smaller than that which is measured. Given these several sets of results, the correct interpretation will probably not be settled until additional measurements are performed.

4. THERMOELECTRICITY IN SUPERCONDUCTING RINGS

In a normal metal, the electric current responds to an electric field and temperature gradient. Similiarly, quasiparticle excitations in superconductors are governed by a transport equation which resembles that for conduction electrons in normal metals:

$$\vec{J}_n = \frac{\sigma}{e} \nabla(\mu_s - \mu_n) + L_T(-\nabla T) \quad . \tag{7}$$

In Eq. (7) μ_n is the chemical potential of the quasiparticles and μ_S is the chemical potential of the superfluid condensate. It is clear from the equation that the difference in these two chemical potentials may be regarded as the electrical driving force for the quasiparticles. As discussed recently by Pethick and Smith (1977), this difference is zero in the interior of a bulk homogeneous superconductor, so that the only driving force is the thermoelectric term $L_T(-\nabla T)$. In ordinary geometries, the thermoelectric current driven by a temperature gradient cannot be measured because it produces neither a voltage gradient nor a magnetic field. This point was first recognized by Ginzburg (1944), who reasoned that any experimental manifestation of the normal current would be neutralized by the counterflow of supercurrent J_S.

Recently, however, Garland and Van Harlingen (1974) and Gal'perin, Gurevich and Kozub (1974) independently suggested that thermoelectric currents could be observed in a ring geometry where there is a constraint on the phase of the order parameter. Figure 6 shows a bimetallic superconducting ring whose junctions are fixed at different temperatures. Within the body of the ring the thermoelectric current is locally cancelled by a supercurrent, i.e., $\vec{J}_n + \vec{J}_s = 0$. On the surface of the ring, however, this cancellation is not exact; the resulting surface currents inducing a magnetic flux through the ring. Measurements of this flux then can be related to the thermoelectric response of the quasiparticles. An insight into the origin of this flux can be obtained by the following simple argument, based on the two fluid model.

We let the superfluid be described by a wave function whose amplitude corresponds to the superfluid density n_s, and whose phase is related to the momentum of the superfluid, i.e.,

$$\psi_s = (n_s/2)^{1/2}\, e^{i\theta} \quad . \tag{8}$$

The phase is constant in a singly-connected geometry, but in a ring geometry the phase integral can be integer multiples of 2π:

$$\begin{aligned} \nabla\theta\cdot d\vec{l} &= \hbar^{-1} \quad \rho_s\cdot d\vec{l} \\ &= \hbar^{-1}(2m\vec{v}_s + \frac{2e}{c}\vec{A})\cdot d\vec{l} \\ &= 2\pi n \quad . \end{aligned} \tag{9}$$

Pethick and Smith (1978) showed that the potential difference $\mu_s-\mu_n$ decays exponentially into the bulk of a superconductor. Thus if we now let the path of integration be the interior of the bimetallic ring of Fig. 6, and we make the substitution (using Eq. (7) with the first term zero)

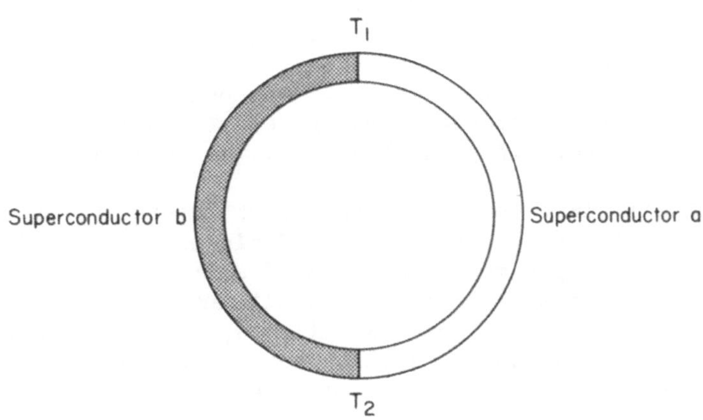

Fig. 6. Bimetallic loop made up of two superconductors with different transport coefficients L_T.

$$\vec{v}_s = \frac{\vec{J}_s}{en_s} = \frac{(-\vec{J}_n)}{en_s} = \frac{L_T \nabla T}{en_s} \quad , \tag{10}$$

then Eq. (9) reduces to the following expression:

$$\Phi = n\Phi_o + \frac{mc}{e^2}\left(\frac{L_T^a}{n_s^a} - \frac{L_T^b}{n_s^b}\right)(T_1 - T_2) \quad . \tag{11}$$

In Eq. (11), Φ is the magnetic flux through the hole in the ring (obtained by integrating the vector potential), Φ_o is the flux quantum hc/2e, L_T^a and L_T^b are the thermoelectric coefficients for metals A and B, and n_s^a and n_s^b are the superfluid densities for metals A and B.

Equation (11) shows that flux is not quantized inside a bimetallic superconducting ring which is out of thermal equilibrium. The non-quantized flux is linear in the applied temperature gradient and proportional to the difference in thermoelectric coefficients of each component of the ring. If we make the assumption that L_T for a superconductor is comparable to L_T for a normal metal then, as pointed out by Garland and Van Harlingen (1974), the non-quantized component of flux would typically be a fraction of a flux quantum; small yet easily resolvable with a SQUID magnetometer. The argument leading to Eq. (11) may be put on a more sophisticated footing (see Van Harlingen, Heidel and Garland (1980)) and may be easily generalized to other geometries.

The first attempt to measure the flux predicted by Eq. (11) was made by Selzer and Fairbank (1974). In this experiment, the anisotropy of the thermoelectric properties of single crystalline tin was used to provide the required inhomogeneity (instead of two superconductors with different values of L_T). The only observed flux was attributed to spurious causes, leading Selzer and Fairbank to conclude that the effect, if present at all, was smaller than predicted.

Zavaritskii (1974) performed the first experiment on a bimetallic superconducting ring, using lead and tin as the constituents of his sample. This experiment was followed shortly by measurements on tantalum-niobium rings by Pegrum, Guenault and Pickett (1975), and by Falco (1976). In each of these experiments, a flux was observed which was proportional to the applied temperature difference and which diverged as the transition temperature of the ring was approached from below. Such a divergence arises from the factor of n_s which appears in the denominator of the right side of Eq. (11) and is expected to have a $(T_c-T)^{-1}$ temperature dependence. Surprisingly, the magnitude of the flux measured in each experiment was much larger than a flux quantum for $T \to T_c$. A typical result from the Falco (1976) experiment is shown in Fig. 7. Although these bimetallic ring experiments were initially interpreted as supporting--at least qualitatively--the thermoelectric flux theory, this interpretation was challanged in a subsequent paper by Pegrum and Guenault (1976). On the basis of further measurements on their tantalum-niobium samples, Pegrum and Guenault concluded that their original data had a spurious origin arising from the temperature dependence of the superconducting penetration depth.

Partially to resolve these doubts, Van Harlingen, Heidel and Garland (1978, 1980) subsequently performed thermoelectric flux measurements on a series of lead-indium samples. This later experiment made use of self-shielding toroidal samples, illustrated in Fig. 8, with which it was possible to circumvent some of the experimental problems associated with the simple ring geometry. Measurements of the thermoelectrically induced flux from one of the toroidal lead-indium samples is shown in Fig. 9. To obtain the data in the figure, the bottom face of the toroid was fixed at a temperature 7 mK below the transition temperature of indium, while the top face of the toroid was heated in 10μK steps. The flux generated within the toroidal cavity of the sample was coupled, via a superconducting flux transformer, into a SQUID detector. The data in the figure clearly demonstrates the divergence observed in the other experiments, although

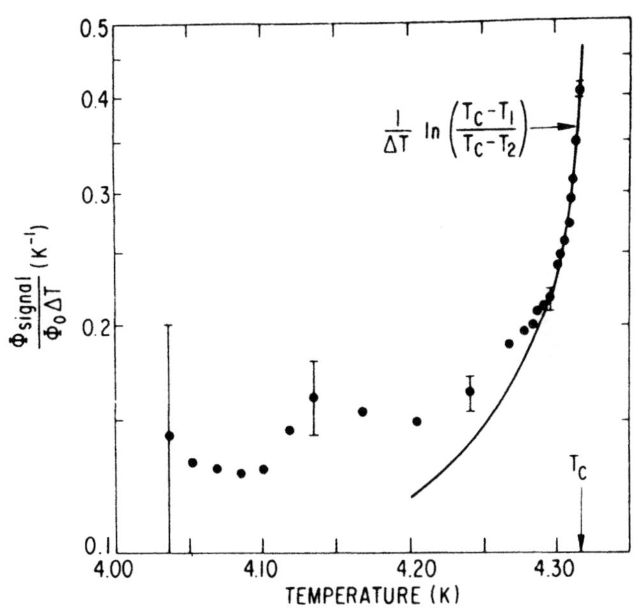

Fig. 7. Magnetic flux in a bimetallic loop near T_c. Falco (1976).

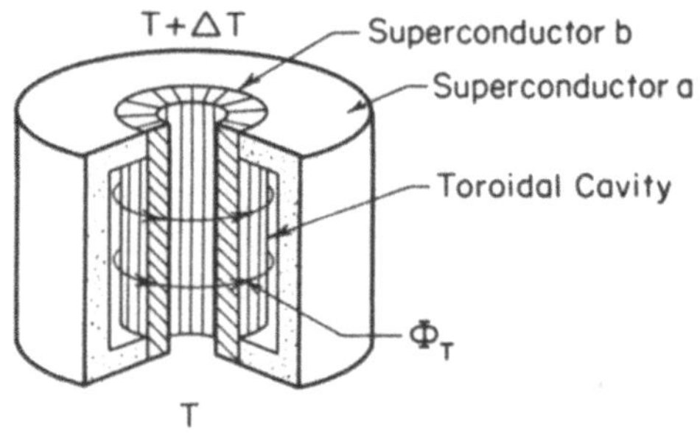

Fig. 8. Toroidal bimetallic loop geometry. Van Harlingen, Heidel and Garland (1980).

the power law was found to have a $(T_c-T)^{-3/2}$ dependence rather than the $(T_c-T)^{-1}$ dependence which was expected. In addition, the magnitude of the observed effect was very large, nearly five orders of magnitude greater than predicted by Eq. (11). By performing tests on a homogeneous all-indium toroid, and by specifically looking for the spurious problems suggested by Pegrum and Guenault, Van Harlingen et al., concluded that their measurements were correct. This leaves a serious unresolved problem. Current theories of thermoelectricity in superconductors appear unable to explain the large effects observed in bimetallic loops.

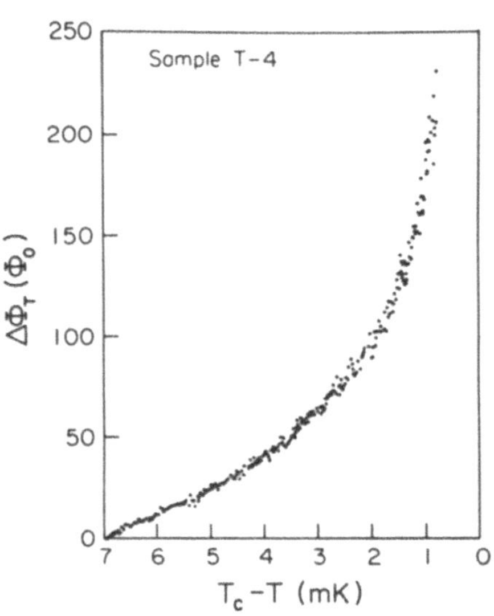

Fig. 9. Magnetic flux in a toroidal bimetallic loop near T_c. Van Harlingen, Heidel and Garland (1980).

5. ACOUSTOELECTRICITY IN SUPERCONDUCTING RINGS

Though technically not thermoelectricity, the acoustoelectric effect is included in this review because of its conceptual similiarity to the thermoelectric flux effect discussed in the previous section. In the acoustoelectric effect, a normal current in a superconductor is driven by a sound wave rather than by a temperature gradient. The resulting counterflow of supercurrent then leads to a gradient of the phase which, in a ring geometry, produces a magnetic flux. The influence of a sound wave on the phase of a superconducting wavefunction was first discussed by Gal'perin, Gurevich and Kozub (1974), who speculated that the phase gradient could be made large enough to produce a measurable magnetic flux in a superconducting ring. Falco (1976) subsequently showed that the effect could be enhanced in a bimetallic ring geometry.

The basic result of the Gal'perin et al. (1974) theory is that an acoustic wave propagating through a superconductor produces a normal current J_n given by

$$\vec{J}_n = (\sigma\Gamma\vec{S}/en_o s)F(e\Delta/kT) \quad . \tag{13}$$

In the above expression, σ is the normal state conductivity, Γ is the absorption coefficient for sound, $\vec{S}$ is the acoustic energy flux density, n_o is the total electron density, and s is the velocity of sound. The function $F(e\Delta/kT)$ varies slowly with temperature and is of order unity near T_c (the region of interest). The mechanism which produces a magnetic flux then exactly parallels that leading to Eq. (11) for the thermoelectric flux effect. For the bimetallic ring geometry illustrated in Fig. 10, the magnetic flux generated by a sound wave is given by

$$\Phi = n\Phi_o + \frac{mc}{e^2}\left(\frac{\sigma^a F^a}{n_o^a n_s^a s^a} - \frac{\sigma^b F^b}{n_o^b n_s^b s^b}\right)(S_a - S_b) \quad . \tag{14}$$

The superscripts a and b in the above equation identify parameters associated with the A and B superconductors in the ring. In order to estimate the magnitude of the accoustoelectric effect, it is useful to simplify Eq. (14) by choosing metals having $\sigma^a \gg \sigma^b$ and by restricting the temperature range to the region just below T_c where $n_s \cong n_o(1-T/T_c)$. With these assumptions, Eq. (14) reduces to

$$\Phi/\Phi_o \cong n + \frac{m\sigma F}{he^2 n_o^2 s}\,\frac{(S_a - S_b)}{(1-T/T_c)} \tag{15}$$

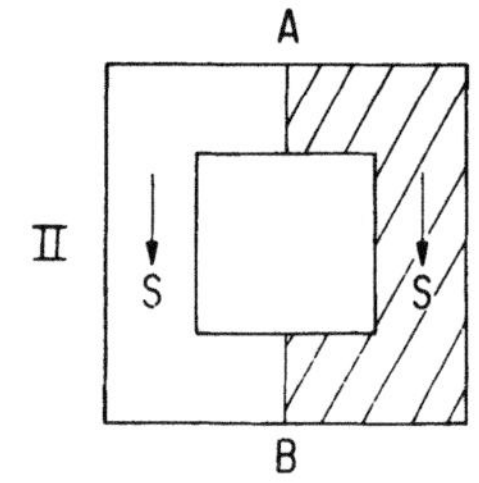

Fig. 10. Geometry used to generate and acoustically induced magnetic flux. Falco (1976).

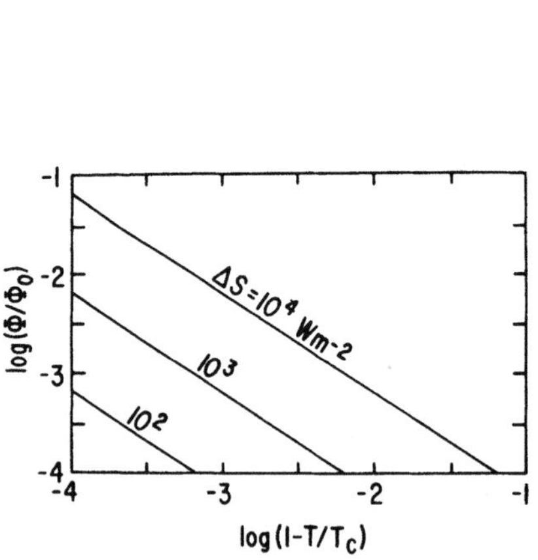

Fig. 11. Calculated acoustically generated magnetic flux in a bimetallic loop as a function of reduced temperature for several values of acoustic energy density. Falco (1976).

Figure 11 illustrates he temperature dependence of the magnetic flux predicted by Eq. (15) for various levels of sound intensity in superconducting tin. The flux is smaller than a flux quantum but still within the resolution of a SQUID magnetometer for temperatures no more than a few mK below T_c. To date, there have been no attempts to measure the acoustically generated flux predicted by Eq. (14). It would be very interesting to learn whether the effect is consistent with the analysis presented here or whether it displays the same anomalous behavior as the thermoelectric flux effect.

6. THERMOELECTRIC POTENTIALS IN SUPERCONDUCTORS

As previously discussed, there are no electrochemical potential differences in superconductors which result from the application of a temperature gradient. In 1977, however, Falco suggested that a temperature gradient could lead to a difference between the chemical potentials of the superconducting condensate and of the quasiparticles, and that this difference could be detected with a tunnel junction probe. Tunnel junctions had been used for several years to study non-equilibrium quasiparticle distributions in superconductors; as explained in detail by Tinkham (1972), a tunnel voltage is produced whenever there are an unequal number of electron-like and hole-like quasiparticles.

The general idea of a branch imbalance, or charge imbalance as it is now called, was considered in detail by Pethick and Smith (1979) who showed that the concept arises naturally out of the symmetry of the quasiparticle distribution function. Their method was to classify the non-equilibrium distribution arising from any source into a component which has even symmetry about the Fermi surface and a component which has odd symmetry. In 1978, Heidel and Garland concluded that a temperature gradient would lead to a charge imbalance by calculating the quasiparticle distribution function directly from the Boltzmann equation. The term in their solution which produced a charge imbalance was shown to be linear in the superfluid velocity v_s. In the particular situation studied by Heidel and Garland (1978) and by Falco (1977), v_s arose self-consistently out of the supercurrent response to the normal thermoelectric current. The value of v_s was quite small and, not surprisingly, the voltage measured by Falco (1977) turned out also to be quite small.

Subsequently, Pethick and Smith (1979) proposed that the charge imbalance potential could be greatly enhanced by treating v_s as an independent parameter. One could do this by simultaneously injecting a heat current and an electric curent into a thin film superconductor. Pethick and Smith also derived an explicit expression for the potential in terms of experimentally accessible quantities. Their result was

$$V = \frac{\pi}{6} \frac{E_f \Delta}{ekT} \frac{\tau}{g_{ns}} \vec{v}_s \cdot \frac{\nabla T}{T} . \qquad (16)$$

In this equation, g_{ns} is the normalized conductance between the superconductor and the normal probe and τ is a relaxation time for the charge imbalance (the other symbols have their customary meanings).

Clarke, Fjordbøge and Lindelof (1979) measured the potential predicted by Eq. (16) in thin films of superconducting tin and found that V was indeed proportional to $\vec{v}_s \cdot \nabla T$ and that $Vg_{ns}T/J_s \nabla T$ diverged as $(1-T/T_c)^{-1}$. Although this temperature dependence was consistent with Pethick and Smith's interpretation of the time τ as that associated with inelastic charge relaxation, the size of the effect was approximately two orders of magnitude smaller than predicted. Figure 12 shows the dependence of their measured potentials on an externally driven current.

Subsequently, Schmid and Schon (1979), Clarke and Tinkham (1980), Nielsen, Ono, Pethick and Smith (1980) and Entin-Wohlman and Orbach (1980) have independently examined the thermoelectric charge imbalance problem and have each derived expressions for the voltage which are analogous to Eq. (16). All authors now seem to agree that the original association of the time τ with inelastic scattering was incorrect and that it is elastic impurity scattering which dominates

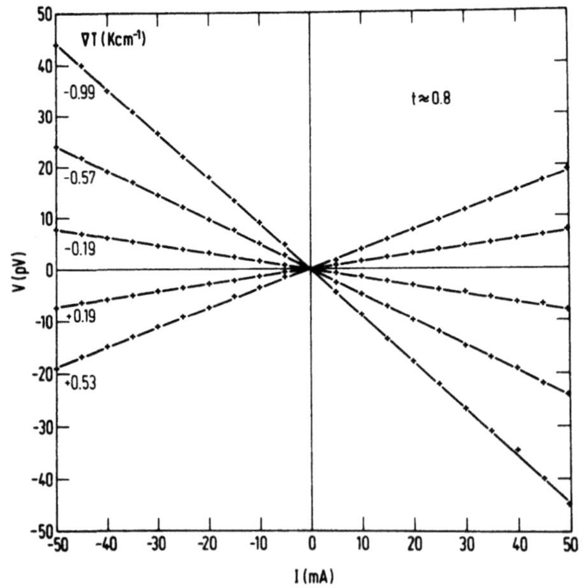

Fig. 12. Dependence of branch imbalance potential in Sn on externally driven supercurrent for several values of temperature gradient. Clarke, Fjordbøge and Lindelof (1979).

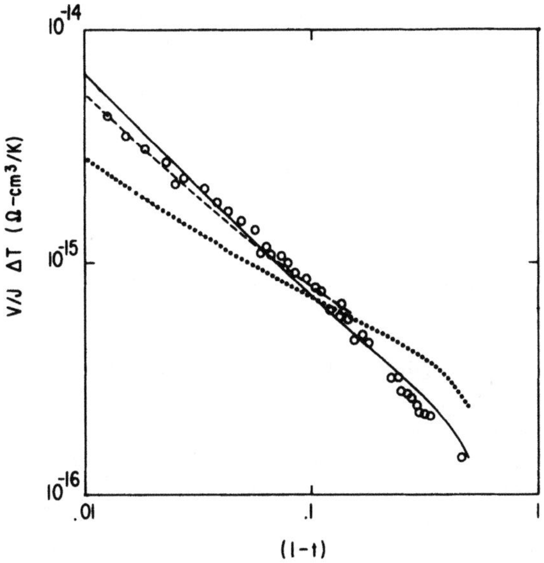

Fig. 13. Comparison of measured branch imbalance potential in an aluminum sample with the various theories of the effect. Schmid and Schön (1979) and Entin-Wohlman and Orbach (1980) (dashed line); Beyer Nielsen, Ono, Pethick and Smith (1980) (dotted line); and Clarke and Tinkham (1980) (solid line). The magnitudes of each theory have been adjusted for a "best fit" to the data. Heidel and Garland (1980).

the thermoelectric relaxation process. Figure 13 shows a comparison of the different theoretical expressions for the charge imbalance voltage as a function of reduced temperature. In the figure, the magnitude of each theoretical expression was scaled to provide a best fit to some very recent data on aluminum by Heidel and Garland (1980). It is clear from the quite different temperature dependences predicted by each calculation that there are still unresolved theoretical issues associated with this problem. In particular, the various theoretical results shown in Fig. 13 have been calculated from either a Boltzman equation approach or using Green's functions. These calculations have been carefully examined by workers in the field who have at present been unable to locate errors in any calculation which would alter Fig. 13.

ACKNOWLEDGMENTS

We would each like to thank our collaborators, Thomas R. Werner (CMF) and Dale Van Harlingen and David F. Heidel (JCG) for essential contributions to portions of the research reported here. In addition, one of us (CMF) would like to acknowledge many useful discussions on this subject with Ivan K. Schuller. This research partially supported under the auspices of the Solid State Physics and Materials Branch of the U.S. Department of Energy and by the U.S. National Science Foundation Grant No. DMR 78-09428.

REFERENCES

Aronov, A. G., 1975, The influence of condensate motion on the thermo-electric effects in superconductors, Sov. Phys.-JETP, 40:90.

Bok, J., and Klein, J., 1968, "Electric fields" in superconductors, Phys. Rev. Letters, 20:660.

Borelius, G., Keesom, W. H., Johansson, C. H., and Linde, J. O., 1931, Measurements of thermo-electric forces of lead and tin down to the temperature of liquid helium, Proc. Koninhl. Ned. Akad. Wetenschap, 34:1365.

Burton, E. F., Tarr, F.G.A., and Wilhelm, J. O., 1935, Nature, 136:141.

Casimir, H.B.G., and Rademakers, A., 1947, The thermo-electric behaviour of a superconductor in the neighborhood of the transition point, Physica, 13:33.

Clarke, J., and Freake, S. M., 1972, Superconducting fountain effect, Phys. Rev. Letters, 29:588.

Clarke, J., Fjordbøge, B. R., and Lindelof, P. E., 1979, Supercurrent induced charge imbalance measured in a superconductor in the presence of a thermal gradient, Phys. Rev. Letters, 43:642.

Clarke, J., and Tinkham, M., 1980, Theory of quasiparticle charge imbalance induced in a superconductor by a supercurrent in the presence of a thermal gradient, Phys. Rev. Letters, 44:106.

Daunt, J. G., and Mendelssohn, K., 1945, An experiment on the mechanism of superconductivity, Proc. Roy. Soc. A, 185:225.

de Waele, A.Th.A.M., de Bruyn Ouboter, R., and Pipes, P. B., 1973, Thermo-electrostatic effects in superconductors, Physica, 65:587.

Entin-Wohlman, O., and Orbach, R., 1980, Generation of charge imbalance in a superconductor by a temperature gradient, Phys. Rev., to be published.

Falco, C. M., 1976, Acoustically generated magnetic flux in a bi-metallic superconducting loop, Phys. Rev., B14:3853.

Falco, C. M., 1976, Thermally induced magnetic flux in a superconducting ring, Solid State Commun., 19:623.

Falco, C. M., 1977, Observation of thermally induced potential in a superconductor, Phys. Rev. Letters, 39:660.

Gal'perin, Yu. M., Gurevich, V. L., and Kozub, V. I., 1974, Thermoelectric effects in superconductors, Sov. Phys.-JETP, 39:680.

Gal'perin, Yu. M., Gurevich, V. L., and Kozub, V. I., 1974, Nonlinear acoustic effects in superconductors, Sov. Phys.-JETP, 38:517.

Gal'perin, Yu. M., Gurevich, V. L., and Kozub, V. I., 1974, Acousto-electric effect in superconductors, Sov. Phys. Solid State, 16:783.

Gal'perin, Yu. M., Gurevich, V. L., and Kozub, V. I., 1976, Non-stationary acoustoelectric effects in propagation of transverse sound in superconductors, Sov. Phys.-JETP, 42:296.

Garland, J. C. and Van Harlingen, D. J., 1974, Thermoelectric generation of flux in a bimetallic superconducting ring, Phys. Letters, 47A:423.

Ginzburg, V. L., 1944, On the thermoelectric phenomena in superconductors, J. Phys. USSR, 8:148.

Heidel, D. F. and Garland, J. C., 1978, Thermoelectric branch imbalance currents in superconductors, J. de Phys. 39:C6-492.

Heidel, D. F., and Garland, J. C., 1980, to be published.

Keesom, W. H. and Matthijs, C. J., 1938, Physica, 5:437.

Khalatnikov, I. M., 1965, "An introduction to the theory of super-fluidity," W. A. Benjamin, New York.

Lindelof, P. E., Fjordbøge, B. R., and Hansen, J. B., 1979, A thermally induced charge imbalance and other non-equilibrium effects in a superconductor with a moving condensate, preprint.

Matsinger, A. A. J., de Bruyn Ouboter, R., and van Beelen, H., 1978, On the possible existence of a thermopower in the superconducting state, Physica, 93B:63.

Morris, T. D. and Brown, J. P., 1971, The temperature dependence of the Bernoulli effect in a type-I superconductor, Physica, 55:760.

Pegrum, C. M., Guénault, A. M., and Pickett, G. R., An experi-mental investigation of the thermoelectric generation of flux in a superconducting bimetallic loop, in: "Low Temperature Physics-LT14," M. Krusius and M. Vuorio, eds., North Holland, Amsterdam (1975).

Pegrum, C. M. and Guénault, A. M., 1976, Thermoelectric flux effects in superconducting bimetallic loops, Phys. Letters, 59A:393.

Pethick, C. J. and Smith, H., 1978, Relaxation and collective motion in superconductors, a two-fluid description, J. de Phys. 39:C6-488.

Pethick, C. J. and Smith, H., 1979, Generation of charge imbalance by a temperature gradient, Phys. Rev. Letters, 43:640.

Pipes, P. B. and Darling, D. H., Temperature dependence of contact potential in superconducting niobium, in: "Low Temperature Physics-LT14," M. Krusius and M. Vuorio, eds., North Holland, Amsterdam (1975).

Pullan, G. T., 1953, The thermo-electric behaviour of tin and silver at liquid-helium temperatures. Proc. Roy. Soc. A, 217:280.

Putterman, S. and de Bruyn Ouboter, R., 1970, Are there thermo-electric effects in superconductors?, Phys. Rev. Letters, 24:50.

Putterman, S. J., 1974, "Superfluid Hydrodynamics," North Holland, Amsterdam.

Schmid, A. and Schön, G., 1979, Generation of branch imbalance by the interaction between supercurrent and thermal gradient, Phys. Rev. Letters, 43:793.
Schuller, I. K. and Falco, C. M., 1979, Optically induced non-equilibrium and heating effects in superconductors, Sov. J. Low Temp. Physics, 4:446.
Selzer, P. M. and Fairbank, W. M., 1974, Thermally generated magnetic fields in an anisotropic metallic crystal at low temperatures, Phys. Letters, 48A:279.
Shott, M. and Walton, A. J., 1977, Changes in the chemical potential of tin on becoming superconducting, Phys. Letters, 60A:53.
Smith, A. D., Tinkham, M., and Skocpol, W. J., 1980, A new thermoelectric effect in tunnel junctions, preprint.
Steiner, V. K., and Grassmann, P., 1935, Eine obere grenze der thermokraft zwischen supraleitern, Phys. Z., 36:527.
Tinkham, M., 1972, Tunneling generation, relaxation, and tunneling detection of hole-electron imbalance in superconductors, Phys. Rev., B6:1747.
Van Harlingen, D. J., Heidel, D. F., and Garland, J. C., 1980, Experimental study of thermoelectricity in superconducting indium, Phys. Rev., B21:1842.
Welker, N. K., and Bedard, F. D., Anomalous thermoelectric effects in point contact tunnel junctions, in: "SQUID Superconducting Quantum Interference Devices and their Applications," H. D. Hahlbohm and H. Lübbig, eds., Walter de Gruyter, Berlin (1977).
Zavaritskii, N. V., 1974, Observation of superconducting current excited in a superconductor by heat flow, JETP Letters, 19:126.

COLLECTIVE MODES OF THE SUPERCONDUCTING ORDER PARAMETER

A. M. Goldman

School of Physics and Astronomy
University of Minnesota
Minneapolis, MN 55455, USA

INTRODUCTION

A new propagating collective mode with an acoustic dispersion relation was detected in dirty superconductors near T_c by Carlson and Goldman (1973,1975,1976). Such a mode was later obtained theoretically in the dirty-limit by Schmid and Schön (1975) and in the clean-limit by Artemenko and Volkov (1975). The possibility of a collective mode in superconductors was first considered shortly after the BCS theory was developed. Bogoliubov (1959) and Anderson (1958) found weakly damped oscillations of the order-parameter phase with an acoustic dispersion relation in the case of an uncharged superconductor, a result consistent with nonrelativistic generalizations of the Goldstone theorem which imply the existence of such a mode as a consequence of the gauge symmetry breaking at the superconducting phase transition (Wagner, 1966).

However, in a charged superconductor the long-range coulomb force renders the Goldstone theorem inconclusive and drives the Bogoliubov-Anderson mode up to the plasma frequency. Although such a mode is still necessary to maintain gauge invariance, it plays no role in the superconducting transition which involves much lower energies. A review of this subject covering work through the late 1960's has been given by Martin (1969).

In order to understand the physical character of the new mode in superconductors, it is useful to consider the several types of propagating excitations with acoustic spectra known in superfluid ^{4}He. Of these, first, second and fourth sound are bulk oscillations which can be described in terms of two-fluid hydrodynamics with first and fourth sound involving density oscillations. In the case of first sound there are oscillations of both the normal and superfluid components. In fourth sound the superfluid component alone oscillates. For a charged system these modes would all be converted into plasma oscillations (Enz, 1974).

On the other hand, second sound involves temperature or entropy oscillations. Because such oscillations are not in the liquid density but in the quasiparticle density, a mode of this sort might exist in superconductors. However, it is expected to be highly damped (Bardeen, 1958; Ginzburg, 1961) except under stringent conditions which cannot easily be met in real superconductors.

The new collective mode in superconductors found in pair-field susceptibility measurements is different from any of the acoustic modes of superfluid ^{4}He that might be relevant to a charged system and is not associated with second sound. Furthermore, the new mode cannot be obtained from hydrodynamic equations which are derived assuming local thermodynamic equilibrium.

The new mode exists at frequencies in excess of the reciprocals of the energy and branch-imbalance relaxation times and consists of oscillations of the branch imbalance Q which in turn are associated with oscillations of the phase of the order parameter and the condensate velocity. The quasiparticle and condensate currents are oppositely directed so that the total current vanishes. Since there is no net current, there is no magnetic field and the oscillations can penetrate bulk superconductors as they are not hindered by the Meissner effect. In addition, as there are no charge density oscillations, the mode is not driven up to the plasma frequency.

The detection of collective oscillations in superconductors was accomplished in tunneling experiments being used to study the pair-field susceptibility and structure factor of the superconducting order parameter. The relationship between $\chi''(\omega,q)$, the imaginary part of the frequency and wave-vector dependent pair-field susceptibility of the superconducting order parameter, and pair tunneling in Josephson junction structures was first pointed out by Ferrell (1969 and Scalapino (1970). The determination of $\chi''(\omega,q)$ has been established as a broad and sensitive, although not universal, probe of order parameter dynamics (Anderson et al., 1972). The range of frequencies, wave vectors and temperatures accessible in such studies are subject to certain limitations. Furthermore, data analysis is model dependent in that one measures a sum of contributions to χ'' from the space and time variation of the amplitude and phase of the order parameter. Nevertheless, the important new physical result found in these experiments, the propagating mode of the order parameter in the superconducting state is a qualitative feature of the data.

Here a discussion of measurements of the pair-field susceptibility of superconductors which reveal the existence of a propagating mode will be presented. For completeness a discussion of the regime above T_c will be included. In this regime the experiments test the time-dependent Ginzburg-Landau equations (Cyrot, 1973). In the regime below T_c the complexity of the theory of Eliashberg (1968, 1971) is required to understand the results, in particular the propagating mode.

THE EXCESS CURRENT DUE TO PAIR TUNNELING AND THE GENERALIZED SUSCEPTIBILITY

Detailed microscopic treatments of the connection between the excess current due to pair tunneling and the generalized susceptibility have been given by Scalapino (1970), Takayama (1971), Shenoy and Lee (1974), and Dinter (1978). The imaginary part of the wave-vector and frequency dependent order-parameter susceptibility $\chi''(\omega,q)$ is proportional to an excess current in the dc I-V characteristic of a tunneling junction in which the superconductor under study is incorporated as one electrode of a junction with the second electrode a metal with a much higher transition temperature. The arguments ω and k, the frequency and wave vector, respectively, are related to the dc voltage across the junction and the magnetic field H applied in the junction plane through the relations $\omega = 2eV/h$ and $q = \frac{2e}{\hbar}[d/2 + \lambda']B$, where d is the thickness of the film whose susceptibility is being determined and λ' is the magnetic penetration depth of the electrode with the higher transition temperature. The simple form of the theory requires that for the electrode of interest $d \ll \xi(T)$ and for the reference electrode, $\lambda' \ll d'$.

The simplest way to see the connection between the generalized susceptibility and the excess current due to pair tunneling is to examine the linear response model of Scalapino (1970). In this model the coupling between the two superconductors is described by a Hamiltonian of the form

$$H_I = \overline{C}e^{-i\omega t}\int d^2x e^{iq\cdot \underset{\sim}{x}}\hat{\Delta}(\underset{\sim}{x},t) + \text{h.c} \tag{1}$$

where $\hat{\Delta}(x,t)$ is the fluctuation pair-field operator

$$\hat{\Delta}(\underset{\sim}{x},t) = g\hat{\psi}_\downarrow(\underset{\sim}{x},t)\hat{\psi}_\uparrow(\underset{\sim}{x},t) \tag{2}$$

Here the quantities $\hat{\psi}_\uparrow$ and $\hat{\psi}_\downarrow$ are field operators for spin-up and spin-down quasiparticles, g is a coupling constant of the fluctuating electrode and q is the wave vector of the coupling which is given by $\overline{C}$. This effective Hamiltonian implies an average pair transfer current

$$\langle I_1 \rangle = \frac{4e\bar{C}}{\hbar d} \operatorname{Im} \left\{ e^{-i\omega t} \langle \hat{\Delta}(-\underset{\sim}{q}, t) \rangle \right\} \tag{3}$$

The commonly understood situation when $T > T_c$ is for $\langle \hat{\Delta}(-q,t) \rangle_o = 0$ which results in vanishing pair-transfer or Josephson current. Here $\langle .... \rangle_o$ denotes an average with respect to an isolated superconductor. However, because Eq. (3) involves an average with respect to a superconductor coupled to a reference electrode, there is actually a contribution to the pair current from an induced order parameter which is the linear response to the coupling $H_I(t)$ given by

$$\hat{\Delta}(-\underset{\sim}{q}, t) = \frac{1}{i\hbar} \int_{-\infty}^{t} dt' \langle [\hat{\Delta}(-q,t), H_I(t')] \rangle_o \tag{4}$$

This induced order parameter can be shown to be given by

$$\hat{\Delta}(-q,t) = \bar{C} \int d^2\chi' e^{-i\underset{\sim}{q} \cdot \underset{\sim}{x}'} \int_{-\infty}^{\infty} dt' e^{i\omega t'} \chi(\underset{\sim}{x} - \underset{\sim}{x}', t-t') \tag{5}$$

where $\chi(x-x', t-t')$ is the response function

$$\chi(\underset{\sim}{x} - \underset{\sim}{x}', t-t') = \frac{i}{\hbar} \theta(t-t') \langle [\hat{\Delta}(\underset{\sim}{x}', t), \hat{\Delta}(x,t)] \rangle_o \tag{6}$$

where $\theta(x)$ is the unit step function. Carrying out the various integrals, one obtains for the pair transfer current the expression

$$\langle I_1 \rangle = \frac{4e|\bar{C}|^2 A}{\hbar d} \chi''(\omega, q) \tag{7}$$

which is second order in the coupling. Here $\chi''(\omega,q)$ is the imaginary part of the susceptibility which is the space and time Fourier transform of the response function.

When $T < T_c$ there is a contribution to the pair-transfer current which is the usual Josephson tunneling current, first-order in $\bar{C}$. This must be augmented by an additional contribution to $\langle I_1 \rangle$ associated with $\chi''(\omega,q)$ which is second order $\bar{C}$. This is an unusual feature of pair tunneling evidently not anticipated by Josephson (1963).

The contribution to the tunneling current second order in $\bar{C}$

and proportional to $\chi''(\omega,q)$ is seen to be the result of a kind of proximity effect in which the tunneling interaction induces an order parameter in the normal state or increases the order parameter relative to its equilibrium value in the superconducting state. The pair transfer Hamiltonian characterizing the Josephson coupling of an asymmetric junction in effect provides an external field which couples to the order parameter. The response of the order parameter to this coupling determines the characteristic susceptibility associated with the phase transition and results in an excess current due to pair tunneling in the junction proportional to $\chi''(\omega,q)$.

EXPERIMENTAL

The crossed-film tunneling junctions required for these experiments were prepared in a conventional manner in an oil-free vacuum chamber capable of maintaining a pressure in the evaporation chamber of less than 5 x 10^{-8} Torr (Carlson and Goldman, 1976). Since important data is associated with the superconducting transition, care was taken to mask the edges of the low-T_c electrode. Quality checks on junctions included testing for zero-field steps in the I-V characteristic, which, if found, are evidence of weak shorts, study of the magnetic field dependence of the zero-voltage current which will exhibit a well-defined Fraunhofer pattern if the insulating layer is uniform, and selecting for study only junctions in which the product of the leakage conductance and normal-state tunneling resistance was 10^{-3} or lower.

Measurements were carried out in an apparatus designed to give substantial isolation from electromagnetic noise and at the same time allow for substantial temperature regulation. I-V characteristics were obtained using four-terminal ac techniques in which the junctions were usually current-biased.

The excess current due to a pair tunneling was determined by subtracting the quasiparticle current from the total measured tunneling current. The appropriate quasiparticle current in early work was initially determined by an extrapolation to low voltages of an expression derived from the BCS theory (Giaever, 1969) parameterized by fitting over a range of voltages for which the excess current due to pair tunneling was small. A typical excess-current voltage characteristic resulting from such an analysis is shown in Fig. 1. Actually, the excess current may be significant with the Al electrode in the normal state even for voltages greater than 200 μV (F. Aspen and A. M. Goldman, 1979). Thus very precise determinations of the excess-current voltage characteristic required a fit of the total measured current to a model containing a sum of contributions from the quasiparticle, excess and leakage currents, respectively. A set of parameters of the excess-current voltage characteristic, R_N, G_L (leakage conductance) and $\Delta_{Pb}(0)$

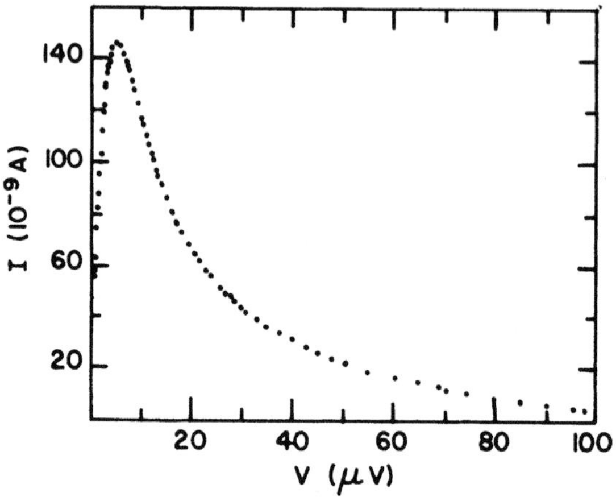

Fig. 1. Excess current-voltage characteristic of an $Al-Al_2O_3-Pb$ junction ($T = 1.847$, $T_c = 1.751$).

were thus determined for $T > T_c$. These same parameters were then used to compute the background quasiparticle and leakage current below T_c. Such a procedure was essential for the analysis of the regime below T_c for which the functional form for the excess current was in doubt.

THE PAIR-FIELD SUSCEPTIBILITY

The Normal State ($T > T_c$)

Measurements in this regime are a critical test of dynamical behavior of superconducting order parameter governed by a diffusive time-dependent Ginzburg-Landau equation. When $d < \xi(T)$, the imaginary part of the pair-field susceptibility is given by

$$\chi''(\omega,q) = \frac{2m\xi^2(0)}{\hbar\varepsilon} \frac{\omega/\Gamma_o}{[(1+q^2\xi^2(T))^2 + \omega^2/\Gamma_o^2]} \tag{8}$$

where $\Gamma_o = \frac{8}{\pi}\frac{k_B T_c}{\hbar}\varepsilon$ is the relaxation frequency of the order parameter, and $\varepsilon = \frac{T-T_c}{T_c}$ is a reduced temperature (Scalapino, 1970). More complicated expressions are relevant for the case of thick films (Shenoy and Lee, 1974). It should be emphasized, as was pointed out by Scalapino, that $1/\Gamma_o$ is not the pair lifetime but is the time for a perturbation of the order parameter to diffuse a coherence distance $\xi(T)$.

Figure 2 contains a plot of the peak voltage of the excess current-voltage characteristic as a function of temperature. This is in good agreement with Eq. (8) except near T_c. There have been a number of discussions of the physical origins of the disagreement of experiment and theory for this regime. Film inhomogeneity and disorder (Cremer and Simanek, 1976), finite size effects, critical behavior and circuit effects have all been treated or have been the subject of speculation (Anderson et al., 1972). The subject is not quantitatively resolved at this time.

Pair-breaking such as that which might result from magnetic impurities results in a significant modification of $\chi''(\omega,q)$ (Entin-Wohlman and Orbach, 1975). This inverse susceptibility can be written as

$$\chi^{-1}(\omega,q) = N(0)\left[\ln\frac{T}{T_c} + \psi\left(\frac{1}{2} + \rho + (Dq^2 - i\omega)/4\pi k_B T\right) - \psi\left(\frac{1}{2} + \rho_c\right)\right] \tag{9}$$

where the pair breaking parameter $\rho = \hbar\left(\frac{1}{2\tau_E} + \frac{1}{\tau_S}\right)/2\pi k_B T$ and $\rho_c = \rho(T_c)$. Here τ_E is the phonon-inelastic scattering time and τ_S is the spin-flip scattering time. In the regime where $T - T_c \ll T_c$, χ'' is of the same general form as Eq. (8). However, the peak voltage

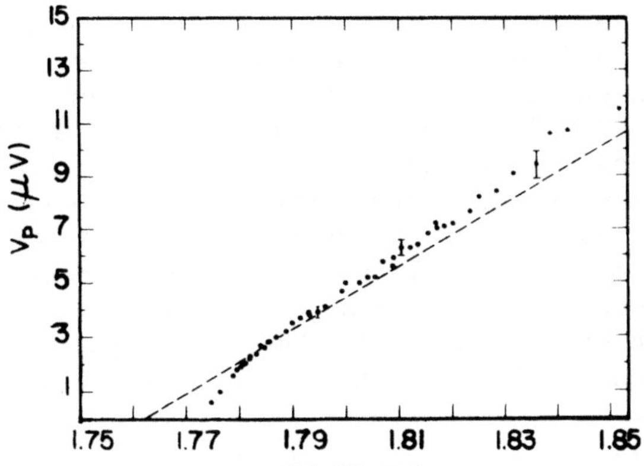

Fig. 2. Peak voltage $V_p = \frac{\hbar}{2e}\Gamma_o$ vs. T for an $Al-Al_2O_3-Pb$ junction.

of the excess current is reduced from its value in the absence of magnetic impurities and is given by

$$V_p = (\hbar/2e)[\Gamma_o f(\rho) + Dk^2] \tag{10}$$

where

$$f(\rho) = (1-\rho\psi'(\tfrac{1}{2} + \rho))/(\psi'(\rho + \tfrac{1}{2})\psi'(\tfrac{1}{2})) \tag{11}$$

Here $\psi(x)$ and $\psi'(x)$ are digamma and trigamma functions, respectively. In the first Born approximation in the Abrikosov-Gor'kov (1960) model

$$\frac{1}{\tau_s} = (\pi/2)C\, J_{sf}^{\;2}\, N(E_F)(g_j-1)^2 j(j+1) \tag{12}$$

where J_{sf} is the exchange integral, $N(E_F)$ is the single-spin density of states at the Fermi level, g_j is the Lande g-factor of the impurity, j is its total angular momentum, and C is the magnetic impurity concentration (C. Rettori et al., 1973).

In Fig. 3 relaxation frequencies $\omega_R = \frac{2eV_p}{\hbar}$ of the superconducting order parameter normalized to $8(T-T_c)/\pi$ are plotted as a function of temperature for samples with different values of ρ_c. The spin-flip scattering times determined from plots of V_p vs. T have been found to be consistent with Eq. (12) with the Er concentration determined by Auger analysis (Aspen and Goldman, 1979).

The Superconducting State ($T < T_c$)

Measurements of $\chi''(\omega,q)$ below T_c must be made at finite q (nonzero H-field) because of the necessity of suppressing the usual Josephson effects which are first order in the coupling constant of the effective Hamiltonian. Magnetic fields used in the measurements were never in excess of the critical field of the lead electrode. Below T_c three main features are found in the excess current-voltage characteristic. Just below T_c a shoulder develops on the low-voltage side of the main peak. As T is reduced further this moves to lower voltages. The second feature is a sharpening of the main peak and its movement to higher rather than lower voltages as T is reduced. At lower temperatures the peak again broadens. The third feature is the appearance of a temperature-dependent secondary peak, which occurs at a voltage given by $eV = \Delta(T)$ where $\Delta(T)$ is the energy gap of the fluctuating electrode (Al in this instance). The general behavior described above is documented in Fig. 4.

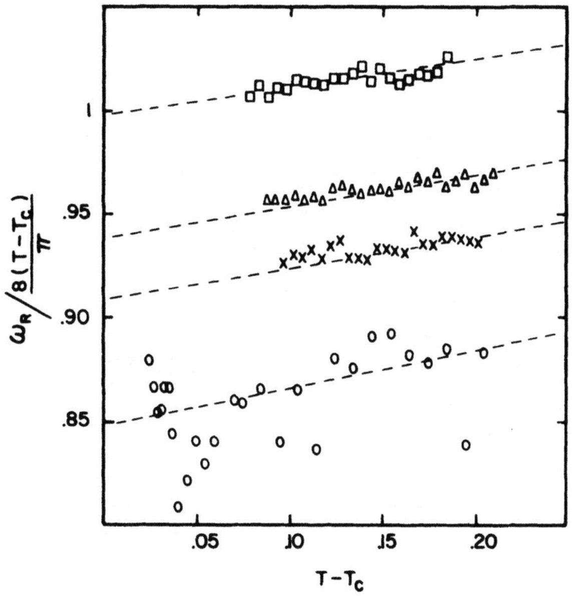

Fig. 3. Relaxation frequencies normalized to $8(T-T_c)/\pi$ with different values of ρ_c and T_c; □ , $\rho_c = 0$, $T_c = 1.221$ K; Δ, $\rho_c = 0.042$, $T_c = 1.153$; x, $\rho_c = 0.063$, $T_c = 1.196$ K; O, $\rho_c = 0.111$, $T_c = 1.079$. Dashed lines are from Entin-Wohlman and Orbach (1975). From Aspen (1980).

It is somewhat more revealing to study the structure factor $S(\omega,q)$ rather than $\chi''(\omega,q)$ since from the former one can ascertain the character of the dynamics without a detailed model. The structure factor is the space and time Fourier transform of the order parameter correlation function whereas the susceptibility is the transform of the response function. The two are related through the fluctuation-dissipation theorem

$$S(\omega,q) = \frac{\hbar}{\pi}\left[\exp\left(\frac{h\omega}{k_B T}\right) - 1\right]^{-1}\chi''(\omega,q) \tag{13}$$

In the limit $\frac{\hbar\omega}{k_B T} << 1$ one finds

$$S(\omega,q) = \frac{k_B T}{\pi}\frac{\chi''(\omega,q)}{\omega} \tag{14}$$

Since $\chi'' \sim$ excess current and $\omega = \frac{2eV}{\hbar}$, $S(\omega,q)$ is proportional to the quotient of the excess current and the voltage at which it is

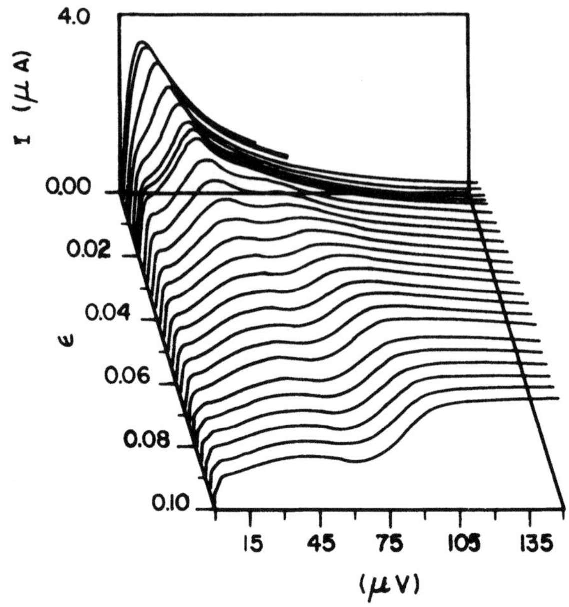

Fig. 4. Excess current-voltage characteristic vs. $\varepsilon = \frac{T_c - T}{T_c}$ with $2\pi\rho_c T_c$ = 26 μV, H = 31.2 Oe, Dk^2 = 20.2 μV, T_c = 1.155 K and R_N = 0.053 Ω. From Aspen (1980).

measured. Typical results are shown in Fig. 5. With the frequency scale used, the aluminum gap peak is not seen. Above T_c the structure factor is a Lorentzian centered at $\omega = 0$. Below and near T_c, $S(\omega,q)$ has a peak at a nonzero frequency in addition to the usual one at the origin. Usually a peak in the dynamical structure factor at nonzero frequency implies the existence of a propagating mode, which is the new feature below T_c in addition to the diffusive mode.

An approximate dispersion relation for the propagating mode can be determined by measuring the wave-vector dependence of the finite frequency peak of the structure factor where the wave-vector is set by the magnetic field. Figure 6 shows the result for an Al film. In Fig. 6a the dispersion relation is plotted at fixed T and in Fig. 6b at fixed reduced temperature $\varepsilon(H,T)$. The bending of the curves in Fig. 6a is a consequence of the approach to the superconductor-normal phase boundary with increasing H. Because an applied magnetic field shifts T_c as well as sets the wave vector, the physical character of the mode is best revealed by the plot at fixed ε. The curve in Fig. 6b corresponds to $\varepsilon(H,t) = 6.32 \times 10^{-3}$.

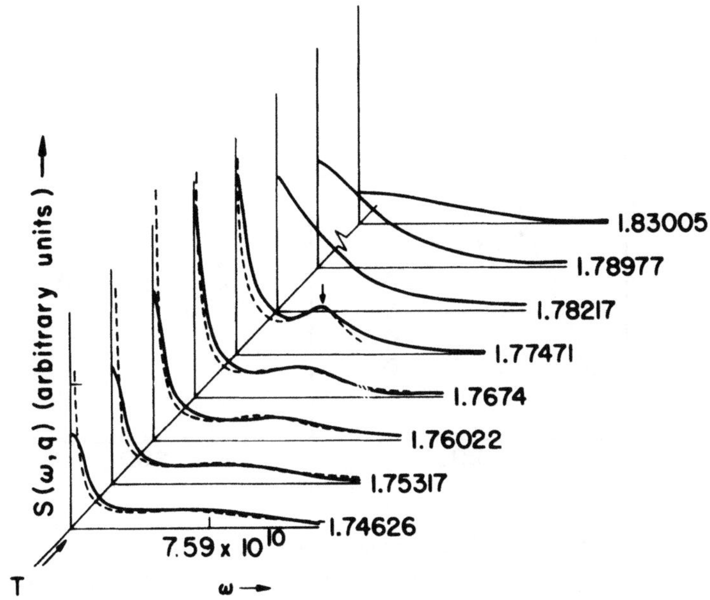

Fig. 5. Structure factor I_{ex}/V in arbitrary units vs. T and ω. The magnetic field is 125 Oe. Curves are shown only for $\omega > 0$ portion and are symmetric about $\omega = 0$. $T_c(H) = 1.780$ K. Dashed lines are from Schmid and Schön (1975).

For this sample $\xi(0) = 5.46 \times 10^{-6}$ cm and $T_c(H) = 1.780$ K. The apparent propagation velocity is 1.74×10^6 cm/sec. The dashed line in Fig. 6b was computed from the theory of Schmid and Schön (1975).

Another way to test the theory is to introduce magnetic impurities, thus exploiting the sensitivity of the propagating mode to pair-breaking effects (Entin-Wohlman and Orbach, 1978). The effect of impurities is to postpone to a temperature below T_c, the opening of the gap in the excitation spectrum and also the development of the propagating mode. If the pair-breaking parameter ρ is large enough, the propagating mode may never develop as it is possible for the gapless regime to end at a temperature below that at which the window of propagation of the mode has closed (Aspen and Goldman, 1979).

An analysis which gives physical insight into the character of the modes can be carried out by fitting a qualitatively valid

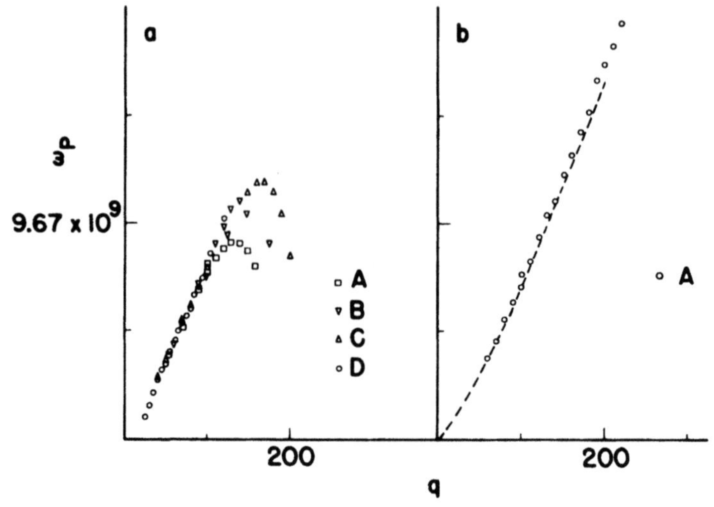

Fig. 6(a). Position of peak in $S(\omega,q)$ of Al plotted vs. q in units of magnetic field at several temperatures. The points A-D correspond to T = 1.771, 1.767, 1.764 and 1.753 K, respectively.

(b) Frequencies of peaks in $S(\omega,q)$ at fixed $\omega = 6.32 \times 10^{-3}$. Dashed line is from Schmid and Schon (1975).

expression for the pair-field susceptibility data, treating the resonant frequency and the damping of the propagating T-mode, the damping of the diffusive L-mode, and the voltage feature at the Al gap as adjustable parameters (F. Aspen and A. M. Goldman, 1979). This can be done taking

$$\chi''(\omega,q) = \frac{A\omega}{\omega^2+B^2} + \frac{C\omega}{(\omega-D)^2+E^2} + \frac{C\omega}{(\omega+D)^2+E^2} + \frac{1}{1+(\frac{G}{\omega})^4} \tag{15}$$

This form is suggested by both previous experiments and theory. Here A through I are adjustable parameters. The second and third terms are the positive and negative frequency contributions of the T-mode and the fourth term is a step taken to represent the feature in the excess current at 2eV = 2Δ(T). Originally it was suggested that this feature was a tunneling anomaly associated with the ac Josephson current (Simanek and Hayward, 1974). Recent theoretical work suggests it may be associated with the L-mode susceptibility (Baramidze and Cheishvilli, 1975; Dinter, 1978; Schön and

Ambegaokar, 1978). Both effects (anomaly and L-mode peak) may be present simultaneously. Four of the eight parameters used in the fit, B, D, E and G, have physical significance and are in reasonable agreement with theory which makes specific predictions. They are respectively the low-frequency limit of the L-mode relaxation frequency, the real and imaginary parts of the T-mode eigenfrequency and the high-frequency limit of the L-mode eigenfrequency. In Fig. 7 we show the real and imaginary parts of the T-mode eigenfrequency ω_T, (H and E) determined by a fit to Eq. (13). The results of the fit are superimposed against three different theories parameterized by values of ρ, T_c and D determined from data taken above T_c (Aspen, 1980).

DISCUSSION

The understanding of the above experimental results involves a theory in which there is a separation of the equation for the order parameter into an equation for the change in the magnitude of the order parameter (the longitudinal mode or L-mode) and an equation for the change in phase (the transverse mode or T-mode) (Schmid and Schön, 1975). A key feature of the theory is that these equations are coupled to the transport equation for the quasiparticle disequilibrium. The resultant equations are thus more complicated than the usual Ginzburg-Landau equation.

There is also a profound physical distinction between the

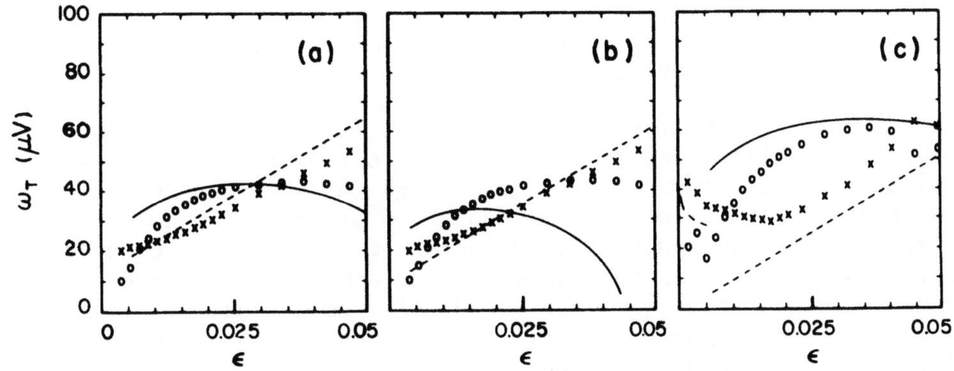

Fig. 7. Real (0) and imaginary parts (x) of the T-mode eigenfrequencies for Al(Er) compared with theory with the real part given by a solid line and the imaginary by a dashed line for (a) Dinter (1978), (b) Schön-Ambegoakar (1979) and (c) Entin-Wohlman and Orbach (1978). Here $2\pi\rho_c T_c$ = 26 µV, Dk^2 = 20.2 µV and T_c = 1.155 K. From Aspen (1980).

modes. In the L-mode the absolute value of the order parameter changes and the disequilibrium of the quasiparticle distribution function δF_E^L is an odd function of energy measured from the Fermi level. Since the number of Cooper pairs is proportional to $|\Delta|^2$, relaxation in the L-mode is a Cooper pair condensation. Here $\int dEN_1(E)\ \delta f_E^L = 0$ resulting in the number of quasiparticles remaining constant, but with their energy distribution changing.

In the T-mode the order parameter phase varies and the disequilibrium of the quasiparticle distribution function δf_E^T is an even function of energy as measured from the Fermi surface. Since spatial variations of the phase are supercurrents, relaxation processes in this mode relate to the interconversion between normal and supercurrents. As quasiparticle number changes, this mode can be excited by quasiparticle injection. The T-mode can also be described in the language applied to describe branch or charge imbalance (Pethick and Smith, 1979; A. M. Kadin et al., 1980). In particular, the propagating T-mode is the high-frequency limit of the charge imbalance dynamics. The low-frequency limit is relevant to the description of electric field penetration into a superconductor in the presence of current flow across a normal superconductor interface.

In the language of a two-fluid model (Bardeen, 1958) the propagating T-mode corresponds to a counterflow of normal and supercurrents. To order $\frac{\omega}{\omega_p^2 \tau}$, $\nabla \cdot \underset{\sim}{J}_s = -\nabla \cdot \underset{\sim}{J}_n$ (Entin-Wohlman and Orbach, 1978). The motion of the superfluid generates the dispersion law. The T-mode when it propagates is a collisionless mode in that its frequency $\omega > 1/\tau_E$. However, ω is still less than the reciprocal of the impurity scattering time. Examination of Fig. 5 indicates that even in the absence of magnetic impurities there is a temperature window of finite width below T_c in which the mode propagates. This window is a consequence of the fact that only near T_c is the superelectron density low enough for the required counterflow of normal fluid to be small enough to permit the mode not to be overdamped. The counterflowing normal fluid,as a consequence of impurity scattering, is the source of the damping of the mode. As T drops below T_c, the superelectron density grows and the damping increases. When magnetic impurities are present, the mode propagates only in the gap regime which develops at a temperature below T_c. The window in this instance is clearly displayed in Fig. 7 where the results of a fit to Eq. 15 are displayed against the predictions for the T-mode eigenfrequency of three different theories. In a gapless superconductor the T-mode would be expected to always be diffusive (Entin-Wohlman and Orbach, 1978).

The low-frequency and high-frequency limits of the L-mode eigenfrequency have been studied by Aspen (1980) and because of space limitations will not be discussed in detail here. The L-mode damping is found to be in qualitative agreement with theory. It is also important to note that the anomaly in the tunneling current at the gap voltage may be interpreted as a peak in the L-mode contribution to the pair-field susceptibility. This peak (Schön and Ambegaokar, 1979) is a resonance and not a propagating mode as it is independent of wave vector.

Detailed theories of the transverse collective mode have been presented by a number of authors. In addition to articles in this volume by H. Smith (1981), and Albert Schmid (1981), and those previously mentioned in the present discussion, there have been papers by Artemenko and Volkov (1975) who considered the clean limit, Bray and Schmidt (1975) and Putterman (1977) who developed phenomenological theories, Galaika et al. (1978), Cheishvilli (1977) and Ovchinikov (1977). A very complete microscopic theory has been given by Dinter (1978) and the subject has been reviewed by Artemenko and Volkov (1979). Giaquinta and Mancini (1978) have reviewed pair-field susceptibility experiments.

At the moment there is no data at all on clean-limit systems for which $\xi_o < \ell$, but in which impurity scattering is still the dominant relaxation mechanism. Calculations for this limit (Artemenko and Volkov, 1975) yield expressions identical to those of Schmid and Schön (1975) when the results of the latter workers are transformed from the dirty to the clean limit.

Numerical evaluations of the formulae suggest that in the clean limit the transverse mode may propagate with relatively low attenuation. However, this is not absolutely certain. Recently Kulik (1980), in considering the problem of the penetration of longitudinal electric fields into superconductors, has treated the interaction of the quasiparticles with the quanta of the transverse modes of the order parameter which he has called "boglons". Kulik finds that for nonmagnetic superconductors the "electron-boglon" interaction may be more efficient than the usual mechanisms of branch mixing. Although it is not certain that this proposal correctly accounts for the penetration of electric fields into superconductors, it contains an explicit prediction for the damping of the propagating mode which acquires a dependence on the "electron-boglon" scattering time, which is of the order of the electron-phonon time at low temperatures, but which is very small as $T \to T_c$. This effect may make a noticeable contribution to the temperature-dependent damping of the propagating mode near T_c, especially when the usual damping mechanism is reduced as would be the case in the extreme limit when the elastic (electron-impurity) scattering time is long. On the other hand, in what might be called the "superclean" limit, when the frequency is greater than the reciprocal of the electron-impurity scattering time, the T-mode should be nonpropagating as a result of a Landau damping mechanism (Artemenko and Volkov, 1979).

ACKNOWLEDGEMENTS

The development of this subject is the result of the labors of a number of graduate students including J. T. Anderson, R. V. Carlson and F. E. Aspen. Stimulating discussions have also been held with R. A. Ferrell, D. J. Scalapino, L. Nosanow, P. G. Hohenberg, J. W. Halley, Jr., C. E. Campbell, R. Orbach, O. Entin-Wohlman, P. Lee, S. R. Shenoy, G. Schön, A. Schmid, H. Schmidt and I. O. Kulik. The work was supported by the Department of Energy and by the National Science Foundation under Grant NSF/DMR80-06959.

REFERENCES

Abrikosov, A. A., and Gor'kov, L. P., 1960 Contribution to the theory of superconducting alloys with paramagnetic impurities, Zh. Eksp. Teor. Fiz. 39: 1781 [1961, Sov. Phys. JETP 12: 1243].

Anderson, J. T., Carlson, R. V., and Goldman, A. M., 1972, Pair tunneling as a probe of order-parameter fluctuations in superconductors: zero magnetic field effects, J. Low Temp. Phys. 8: 29.

Anderson, P. W., 1958, Coherent excited states in the theory of superconductivity: gauge invariance and the Meissner effect, Phys. Rev. 110: 827; Random-phase approximation in the theory of superconductivity, Phys. Rev. 112: 1900.

Artemenko, S. N., and Volkov, A. F., 1975, Collective excitations with a sound spectrum in superconductors, Zh. Eksp. Teor. Fiz. 69: 1764 [1976, Sov. Phys. JETP 42: 896].

Artemenko, S. N., and Volkov, A. F., 1979, Electric fields and collective oscillations in superconductors, Usp. Fiz. Nauk. 128: 3 [1980, Sov. Phys. Usp. 22: 295].

Aspen, F. E., 1980, Effects of spin-flip scattering on the dynamics of the superconducting order parameter, Doctoral Dissertation, U. of Minnesota, unpublished.

Aspen, F., and Goldman, A. M., 1979, Spin-flip scattering and the dynamics of the superconducting order parameter, Phys. Rev. Lett. 43: 307.

Baramidze, G. A., and Cheishville, O. D., 1975, Effects of fluctuational pairing on electron tunneling, Sov. J. of Low Temp. Phys. 1: 430.

Bardeen, J., 1958, Two-fluid model of superconductivity, Phys. Rev. Lett. 1: 399.

Bogoliubov, N. N., Tolmachev, V. V., and Shirkov, D. N., 1958, "New method in the theory of superconductivity," English translation: Consultants Bureau Enterprises, Inc., New York.

Bray, A. J., and Schmidt, H., 1975, Collective modes in charged, gapless superconductors, J. Low Temp. Phys. 21: 669.

Carlson, A. V., and Goldman, A. M., 1973, Superconducting order-parameter fluctuations below T_c, Phys. Rev. Lett. 31: 880.

Carlson, R. V. and Goldman, A. M., 1975, Propagating order-parameter collective modes in superconducting films, Phys. Rev. Lett. 34:11.

Carlson, R. V. and Goldman, A. M., 1976, Dynamics of the order parameter of superconducting aluminum films, J. Low Temp. Phys. 25: 67.

Cheishvilli, O. D., 1977, Effect of electron-phonon scattering on collective excitations in "dirty" superconductors, Fiz. Nizk. Temp. 3: 736 [1977, Sov. J. Low Temp. Physics. 3: 357].

Cremer, S., and Simanek, E., 1976, Critical dynamics of inhomogenous superconducting films, Phys. Rev. B 14: 1927.

Cyrot, M., 1973, Ginzburg-Landau theory for superconductors, Rep. Prog. Phys. 36: 103.

Dinter, M., 1978, Superconductor fluctuations in tunneling junctions below the transition temperature, J. Low Temp. Phys. 32: 529.

Eliashberg, G. M., 1968, Nonstationary equations for superconductors with low concentration of paramagnetic impurities, Zh. Eksp. Teor. Fiz. 55: 2443 [1969, Sov. Phys. JETP 28: 1298].

Eliashberg, G. M., 1971, Inelastic electron collisions and non-equilibrium stationary states in superconductors, Zh. Eksp. Teor. Fiz. 61: 1254 [1972, Sov. Phys. JETP 34: 668].

Entin-Wohlman, O., and Orbach, R., 1975, Effect of magnetic impurities on fluctuations in superconducting films above T_c, Phys. Rev. B 12: 4812.

Entin-Wohlman, O., and Orbach, R., 1978, Mode propagation in superconductors near T_c, Ann. Phys. (New York) 116: 35.

Enz, Charles P., 1974, Two-fluid hydrodynamic description of ordered systems, Rev. Mod. Phys. 46: 705.

Ferrell, R. A., 1969, Fluctuations and the superconducting phase transition: II Onset of Josephson tunneling and paraconductivity of a junction, J. Low Temp. Phys. 1: 423.

Galaiko. V. P., Glushchik, N. I., and Shumeiko, V. S., 1979, Oscillations in electric charge density in superconductors with paramagnetic impurities, Fiz. Nizk. Temp. 4: 289 [1978, Sov. J. Low Temp. Phys. 4: 139].

Giaever, I., 1969, Tunneling between superconductors, in:"Tunneling Phenomena in Solids," E. Burstein and S. Lundquist, eds, Plenum, New York.

Giaquinta, G., and Mancini, N. A., 1978, Superconducting order parameter fluctuations and pair tunneling, La Rivista del Nuovo Cimento 1: 1.

Ginzburg, V. L., 1961, Second sound, the convective heat transfer mechanism, and exciton excitations in superconductors, Zh. Eksp. Teor. Fiz. 41: 828 [1962, Sov. Phys. JETP 14: 594].

Kadin, A. M., Smith, L. N., and Skocpol, W. J., 1980, Charge imbalance and nonequilibrium dynamics near a superconducting phase-slip center, J. Low Temp. Phys. 38: 487.

Kulik, I. O., 1980, The penetration depth of the electric field in a superconductor, Sov. J. Low Temp. Phys., to be published.
Martin, P. C., 1969, Collective modes in superconductors, in: "Superconductivity," Vol. 1, R. D. Parks, ed., Marcel Dekker, New York.
Ovchinikov, Yu. N., 1977, Longitudinal oscillations in superconducting alloys, Zh. Eksp. Teor, Fiz. 72: 773 [1977, Sov. Phys. JETP 45: 404].
Pethick, C. J., and Smith, L. N., 1979, Relaxation and collective motion in superconductors: a two-fluid description, Ann. Phys. (New York) 119: 133.
Putterman, S., 1977, Phenomenological theory of collective modes and relaxation effects in Type-II superconductors, J. Low Temp. Phys. 28: 339.
Rettori, C., Davidov, D., Orbach, R., Chock, E. P., and Ricks, B., 1973, Electron-spin resonance of rare earths in aluminum, Phys. Rev. B 7: 1.
Scalapino, D. J., 1970, Pair tunneling as a probe of fluctuation in superconductors, Phys. Rev. Lett. 24: 1052.
Schmid A., and Schön, G., 1975, Collective oscillations in a dirty superconductor, Phys. Rev. Lett. 34: 941.
Schmid, Albert, 1981, Kinetic equations for dirty superconductors, to be published.
Schön, G., and Ambegaokar, B., 1979, Collective modes and nonequilibrium effects in current-carrying superconductors, Phys. Rev. B 19: 3515.
Shenoy, S. R., and Lee, P. A., 1974, Probe of superconductor fluctuations by a tunneling junction in a magnetic field, Phys. Rev. B 10: 2744.
Simanek, E., and Hayward, J. C., 1974, Superconducting fluctuations and tunneling anomalies below T_c, Physica 78: 199.
Smith, H., 1981, Charge imbalance in nonequilibrium superconductors, to be published.
Takayama, H., 1971, Superconducting fluctuation effects on the S-I-N junction current, Prog. Theoret. Phys. (Kyoto) 46: 1.
Wagner, H., 1966, Long-wavelength excitations and the Goldstone theorem in many-particle systems with "broken" symmetries, Z. Physik 195: 273.

NONEQUILIBRIUM EFFECTS IN 1-D SUPERCONDUCTORS

W. J. Skocpol

Department of Physics
Harvard University
Cambridge, MA 02138

INTRODUCTION

The nature of the current-induced transition from perfect superconductivity to full recovery of the normal state has been of recurrent interest since the earliest days of H. Kamerlingh Onnes, who reported (1913): "At the end of the superconductive wire there is no potential difference as long as the current remains below the threshhold value for this wire. As soon as this value is surpassed potential differences originate and the wire by a process that is not quite cleared up obtains ordinary resistance." Such a transition is inherently a nonequilibrium phenomenon, and given what we have learned from the preceding chapters it should be possible to see what aspects of this process have and have not yet been cleared up.

I shall restrict my discussion to the properties of thin wires, sufficiently narrow that their properties can be considered uniform across their cross-section, varying only along their length. (This omits a number of interesting phenomena covered in Chapter 20.) In principle this requires transverse dimensions smaller than the coherence length $\xi(T)$ and the penetration depth $\lambda(T)$ of the material, so that the superconducting order parameter and current density are uniform; this is usually sufficient to guarantee that other characteristic lengths such as the normal-state inelastic diffusion length also correspond to a one-dimensional situation. In practice, thin-film microstrips or whisker crystals with transverse dimensions somewhat larger (of the order of one micrometer for my favorite material, tin) are sufficient. In this case (Skocpol, 1976) the current flow in the superconductor is sufficiently uniform that the critical current I_c corresponds to the full Ginzburg-Landau critical current density

$$J_c^{GL}=(c/3\sqrt{6}\pi)\ H_c(T)/\lambda(T)\ \propto\ (1-t^2)^{3/2}(1+t^2)^{\frac{1}{2}} \qquad (1)$$

of order 10^7 A/cm^2 for tin at low temperatures. Beyond this current density, solutions of the static, equilibrium Ginzburg-Landau equations do not exist, and some sort of nonequilibrium, resistive phenomenon must occur.

SUPERCONDUCTING/NORMAL INTERFACES

A good starting point for discussion of these nonequilibrium resistive phenomena is the normal/superconducting interface discussed in Chapters 12, 13, and 15. If a segment of length L_n of a 1-D wire of cross-sectional area A is normal for some reason, such as being a different material with much lower T_c, the electrical resistance will exceed the resistance of the segment itself because of the boundary resistance associated with charge imbalance in the superconducting region at either end. According to the simple model presented by John Clarke in Chapter 13, the total resistance should be

$$R(T) = \rho_n L_n/A + 2Z(T)\ \rho_s \Lambda_{Q*}(T)/A \qquad (2)$$

where ρ_n and ρ_s are the normal state resistivities of the normal and superconducting regions, Λ_{Q*} is the charge imbalance diffusion length, and the factor $Z(T) \approx 1 - \pi\Delta/4k_BT$ (near T_c) accounts for the fraction of the total current that is injected as a quasiparticle current at the interface.

The new feature which is encountered in the narrow 1-D wire is that the current density can be very large, and proper account must be taken of the depairing effects of large supercurrents and of the heating-type disequilibrium associated with the localized power dissipation I^2R. According to Schön and Ambegaokar (1979), large supercurrents should reduce the order parameter $|\psi|$ (hence Δ) in accordance with the Ginzburg-Landau result

$$|\psi|^2 \sim (1 - \xi^2 p_s^2/\hbar^2) \qquad (3)$$

and also should provide a pair-breaking mechanism (rate $Dp_s^2/\hbar^2$) analogous to spin-flip scattering which increases the charge imbalance relaxation rate. Since $|\psi|^2$ is only reduced to 2/3 even at J_c^{GL}, the former effect should cause at most a 10% increase of the boundary resistance, while the latter mechanism should act to reduce it. Galaiko (1977) also emphasizes the importance of this elastic branch mixing in the presence of a supercurrent, except near T_c where the supercurrents are small and inelastic processes dominate. Whether these effects are observable depends on whether temperature rises due to dissipation can be minimized, since the boundary resistance increases with increasing temperature, and indeed the size of the normal region will grow if portions of the superconducting region are heated above T_c. In recent experiments that I have conducted the latter effects predominate.

To understand this, let us consider a simple-heating model, in which the electron and phonon systems are characterized by the same elevated temperature T*. As Chang has discussed in Chapter 9, such a description is most appropriate when phonon trapping is large because of poor acoustic matching between the thin film and the substrate (Chapter 1), as with tin on glass or sapphire. Joule heat is generated in the resistive regions of the filament and flows by electronic conduction along the filament and across Kapitza resistances R_K into both the substrate and the helium bath, if present. As a first approximation we will assume that the temperature throughout and nearby a normal region with fixed boundaries is given by

$$T^* - T_o = (I^2 R_n) R_K = I^2 \rho_n / Y_K w^2 d \equiv C_o I^2 \tag{4}$$

where R_n and R_K are determined by the dimensions of the normal segment and Y_K is the Kaptiza conductance per unit area. Combining (2) and (4), the voltage across the SNS region is simply I times R(T*(I)), and the differential resistance dV/dI can also be calculated explicitly. As the current increases, the temperature near the interface rises and the resistance increases, tending to diverge as T* approaches the T_c of the superconducting region. This defines a critical current

$$I_c^* = C_o^{-\frac{1}{2}} (T_c - T_o)^{\frac{1}{2}} = [Y_K w^2 d (T_c - T_o)/\rho_n]^{\frac{1}{2}} \tag{5}$$

at which both R and dV/dI diverge (in this uniform-temperature approximation).

This simple model gives a good account of the data that I have recently observed in tin microstrips on glass substrates with normal segments defined by a proximity-effect underlay of gold in that region (Skocpol, unpublished). Figure 1a shows experimental data for dV/dI on a microstrip with w=3 μm and d = 0.12 μm over a range of temperatures above the T_c' of a tin-gold segment about 6 μm long. Figure 1b shows a fit of the above theory for a single set of reasonable values of the somewhat adjustable parameters (T_c, C_o, ρ_n, τ_E/ρ_s). At low currents one sees Ohmic behavior corresponding to the usual temperature-dependent boundary resistance. At higher currents, the divergence of dV/dI is quite pronounced, and the T_o dependence of I_c^* is well accounted for by the theory. Except very near to T_c, I_c^* is much less than the I_c^{GL} calculated for the microstrip from Ginzburg-Landau theory.

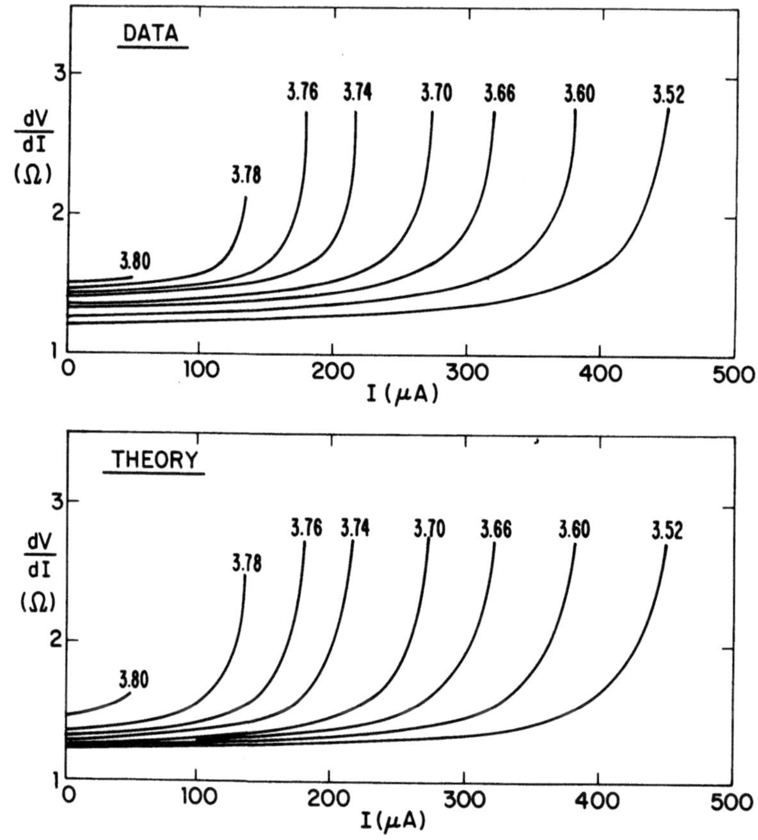

Fig. 1. Heating effects at superconducting/normal boundaries.

HOTSPOT MODEL

As I approaches or exceeds I_c^*, more careful account must be taken of the spatial profile of the temperature T*, which satisfies the heat flow equation

$$-K \frac{d^2T^*}{dx^2} + \frac{Y_K}{d} (T^*-T_o) = \frac{I^2\rho_n}{w^2d^2} \theta(x) \qquad (6)$$

where K is the local thermal conductivity and θ goes from 1 in the normal region to 0 in the superconducting region. The combination of conduction within the film and surface heat transfer leads to a characteristic thermal healing length given by $\eta_{th} = (Kd/Y_k)^{\frac{1}{2}}$, which typically exceeds Λ_{Q*} for the case of large phonon trapping in which the simple heating model is justified. For constant coefficients these equations involve appropriate combinations of functions of the form $\exp(\pm x/\eta_{th})$, chosen to satisfy the boundary conditions. As $T^* \to T_c$ at an interface defined by material inhomogeneity, the boundary resistance does not actually diverge. Because of the decrease of dissipation in the superconducting region, the temperature will fall as one moves away from the interface, so that even if $T^* = T_c$ at the interface, giving no charge imbalance relaxation locally, charge imbalance relaxation will still occur further away leading to a finite limit on the boundary resistance contribution. With further increase of the current, T^* will reach T_c at a position in the previously superconducting material, and the size of the normal region will spread beyond the area defined by the original material inhomogeneity. In this case it is referred to as a self-heating normal hotspot.

In earlier work (Skocpol, et al. 1974a,; also Volkov and Kogan, 1974), I have discussed in considerable detail the current-voltage characteristics that result from self-heating hotspots, and we have drawn attention to their importance in setting limits on the

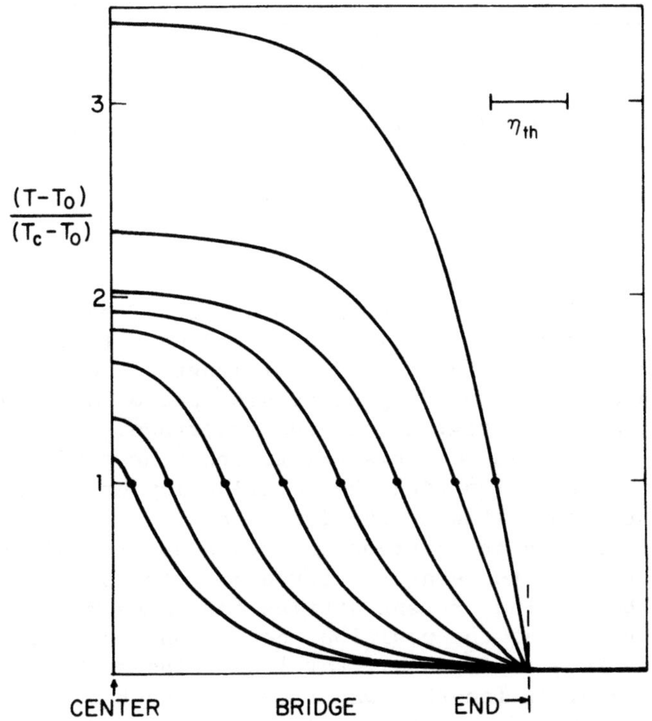

Fig. 2. Temperature distributions for hotspots of various lengths in long microstrips. (Skocpol et al., 1974c)

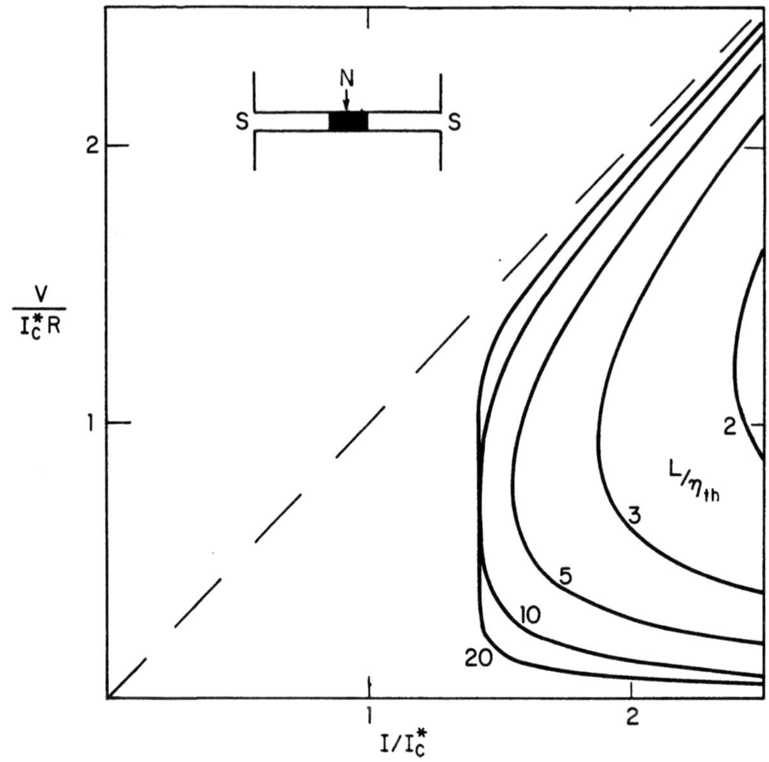

Fig. 3. Hotspot model I-V curves. (Skocpol et al., 1974c)

high-frequency performance of Josephson devices (see Chapter 8). Figure 2 shows temperature profiles corresponding to various positions of the interface for a bridge of length $L=10\eta_{th}$ with ideal cooling at its ends, computed ignoring dissipation associated with the boundary resistance region near the interface. Figure 3 shows the resulting I-V curves for this bridge and others of different relative lengths, also ignoring the contribution of the boundary resistance. At high voltages, the interface is pinned near the end of the bridge, and the current increases quasi-Ohmically leading to very large temperatures throughout the bridge. At intermediate voltages, the current attains the value which causes the interfaces to remain at the positions corresponding to the appropriate length, hence resistance. For long bridges when the interfaces are many healing lengths away from both each other and the ends, a single value of the current is required, and any deviation from it causes the normal domain to grow or shrink. This accounts for the vertically rising portion of the long-bridge I-V characteristics at a current I_h. Under these conditions the temperature rise at the center of the bridge is $2(T_c-T_o)$ (assuming h_K independent of T^*, and $K_s=K_n$). Thus the required current I_h is $\sqrt{2}$ times the previously defined I_c^*, which corresponded to a temperature rise just to T_c. As Yamasaki and Aomine (1979) have shown, when proper account is taken of the T^{*4} dependence of the heat transfer, the required I_h decreases from the above value as the temperature is lowered, being approximately I_c^* at $T_o=\frac{1}{2}T_c$ and $0.7I_c^*$ at $T_o=0$. [In this approach, I_c^* is defined in terms of $Y_K(T_c)$.] At low voltages a proper description of the I-V curves requires careful inclusion of the boundary resistance effects. If they are ignored as in Figure 3, then the normal dissipative region in a homogeneous microstrip can be allowed to shrink arbitrarily small while the effective cooling area is at least $2\eta_{th}$ in length. This requires more current to provide the necessary heat to maintain the hotspot above T_c, and therefore the I-V characteristics show negative differential resistance at low voltages. This description is appropriate in situations of large phonon trapping where η_{th} exceeds Λ_{Q^*}.

An example of how well this model can work is shown in Figure 4 for a tin bridge (L=42 μm, w=3 μm, d=0.11 μm) on a sapphire substrate. The fitted heat transfer coefficient Y_K(which is an average

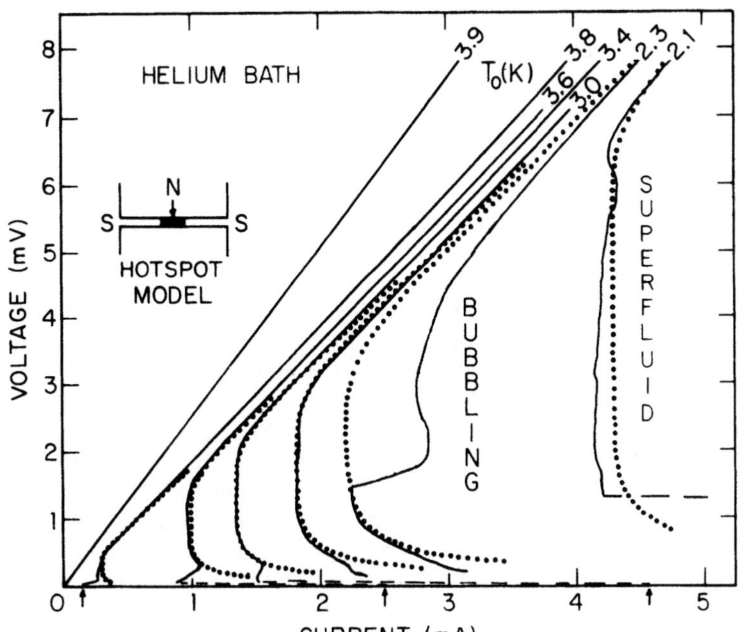

Fig. 4. Fit of hotspot theory (dotted curves) to data from a tin microstrip on a sapphire substrate. (Skocpol et al., 1974c)

constant value approximating the effects of the actual dependence on T^{*3}) declines slightly from 3.0 to 2.2 W/cm^{2}K as the bath temperature is reduced, then jumps to 7.3 W/cm^{2}K below the lambda point as the superfluid contributes strongly to the Kapitza conductance. On the curve for 2.3K, no attempt has been made to follow quantitatively the increase of Y_K above a critical power flux due to the onset of nucleate boiling, but the change is clearly in the proper direction. Note that the region of negative differential resistance is clearly visible in this data taken in a voltage-biased configuration, although some deviations from the simple theory are observed when the resistance is small enough to be comparable to the boundary resistance.

Although I shall never forget the personal thrill of working out this hotspot model after suddenly realizing that heating effects could be the primary explanation of this data, rather than merely a worrisome complication, the fact is that these ideas are so "obvious" that each generation and scientific tradition can have the thrill of discovering it for themselves. Kamerlingh-Onnes (1913) states, for example, "When the current surpasses the threshold value, a local heating of a very small portion of the superconductive wire takes place, so that the step-up point of resistance is reached; at greater densities this temperature spreads over the wire, till there is equilibrium between the heat given off and the Joule-heat generated. I cannot go into all the details of the phenomena, but the bearing upon the question of experiments with superconductors as soon as we go to great current densities is evident." Similar temperature domain structures and I-V curves were analyzed in considerable detail by Busch (1921) for the system consisting of a wire with temperature-dependent resistivity encapsulated in a gaseous environment, which was used as a current stabilization device called the ballast resistor. In the late 1950's, much work was done on the nucleation and propagation of resistive domains in superconducting thin films in connection with research on cryotron computer devices, which probably failed as a competitive technology precisely because of the relatively slow thermal time constants and long thermal length scales. Much of this work has been summarized by Newhouse (1964). The formation of hotspots protected by a vapor bubble even in superfluid helium above a critical power flux was literally seen (through a microscope) by Garbuny and Gottlieb (1964), and this tradition for investigating critical heat fluxes has been continued by Volotskaya et al. (1978). The boundaries of hotspots have also been "seen" recently using a scanning electron microscope (Stöhr and Huebner, private commun.) A qualitative description of the hotspot model applied to two-dimensional constriction-type microbridges was articulated by Wyatt (1969), and a semi-quantitative treatment was given by Nad' and Polyanskii (1973). Heating models appropriate to point contact geometries were worked out by Gubankov, et al. (1972) and Iwanyshyn and Smith (1972). Given the diffusive character of information flow and retention, it is not surprising that during the early 1970's a number of researchers were independently beginning to rediscover the importance of nonequilibrium heating phenomena as manifested in the new context of narrow microstrips and Josephson devices. But the long history of earlier, similar efforts should remind us all that we are but the latest generation to be fascinated by nonequilibrium phenomena. For a recent survey of applications and modification of the hotspot model, see Dharmadurai (1980).

Thus far we have considered only the simple heating model appropriate for conditions of poor phonon removal. If the thin films are made much thinner and narrower, and placed in intimate contact with single-crystal acoustically matched substrates of high thermal conductivity and low Kapitza resistance, then phonon removal can be greatly increased so that the phonon system remains near the ambient temperature T_o. Even under these conditions, it is possible for the electron system to be heated with respect to the lattice, and for phenomena similar to the thermal hotspot model to occur. A theory of this for spatially uniform conditions was worked out be Shklovskii (1975) and extended to a hotspot configuration by Kashchei (1976), based on the early calculations of Kaganov et al, (1957). The thermal impedance between electrons and phonons in copper has been measured by Anderson and Peterson (1972), as discussed in Chapter 1.

A simple description of the removal of heat from the electrons the phonons can be obtained by considering a transient situation, where an energy $C_V \Delta T$ per unit volume is removed in a time of order τ_E. Therefore the conductance per unit area Y_E is approximately $C_V d/\tau_E$. For an 0.1 μm tin film with τ_E=2 x 10^{-10} sec, this corresponds to a conductance of 18 W/cm^{2}K, which will be in series with the Kapitza conductance Y_K. Because the reflection coefficients from tin to sapphire are only 0.2 and 0.06 for the longitudinal and transverse polarizations respectively (Kaplan, 1979), the acoustic mismatch theory (Chapter 1) gives Y_K= 2.6 W/cm^{2}K at T_c. This clearly will be the bottleneck for 0.1 μm films, even if an additional Y_K = 6.6 W/cm^{2}K due to transfer to the helium bath is included. (This is calculated at T_c using the experimental reflection coefficient of 0.2; above the λ-point the thermal conductivity of the helium is so poor that the effective heat transfer coefficient is smaller.) For much thinner films such as those studied by Volotskaya et al (1976), Y_E will be more nearly comparable to Y_K.

These expectations give rather good agreement with the data from the bridge in Fig. 4. The appropriate series/parallel combinations of the expected values give conductances of 2.3 W/cm^{2}K and 6.1 W/cm^{2}K without and with helium respectively. The experiment gives 1.9 W/cm^{2}K without and helium, 3.0 W/cm^{2}K with normal helium, and 6.8 W/cm^{2}K with superfluid helium, in excellent agreement considering the limitations of using differential conductances evaluated at T_c. Further comparison can be made with the data for a bridge 0.02 μm thick on either a sapphire or a quartz substrate presented by Volotskaya et al. (1976). From their data I calculate an observed conductance of 3.8 W/cm^2K which is larger than Y_K for sapphire, comparable to the expected Y_E, and smaller than Y_K for quartz by about a factor of 2. Thus whether electron heating dominates in this sample is not clear.

In order to achieve the best phonon removal possible, I have undertaken a new series of experiments with tin bridges on Cd Te substrates, which is one of the best combinations surveyed by Kaplan (1979), with longitudinal and transverse reflection coefficients of 0.98 and 0.95, better even than the well-publicized matching of Al to BaF_2. According to the acoustic mismatch model, this yields Y_K = 36 W/cm^2K at T_c, sufficient to make the electron-phonon impedance in large part the bottleneck.

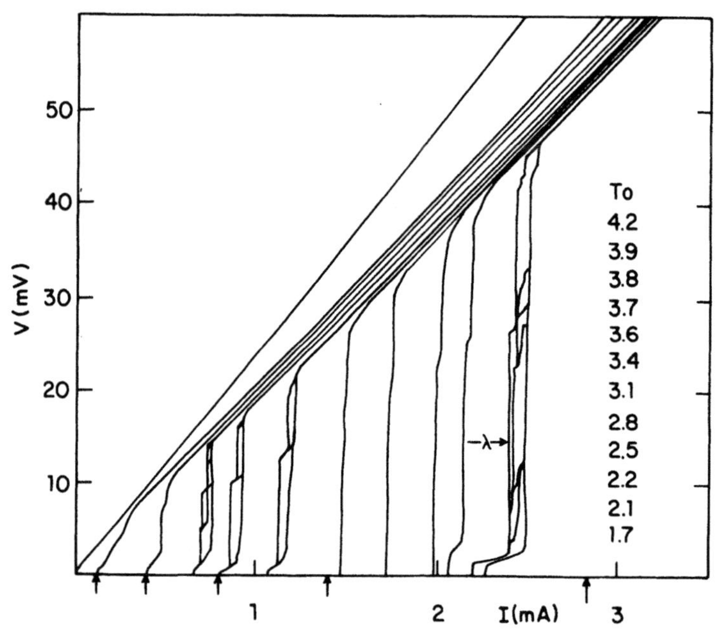

Fig. 5. I-V data showing improved heat transfer on a CdTe substrate.

Figure 5 shows the current-biased I-V curves of a tin microbridge with length 110 μm, width 2.2 μm, and thickness 100 nm cut with a diamond knife on CdTe. The basic shape of the curves is that of the hotspot model, with sharp corners corresponding to a very short thermal healing length. At low voltages, at least near T_c and at low temperatures, the curve has positive differential resistance, suggesting that the thermal healing length is smaller than the diffusion length. The curves are still substantially hysteretic with respect to the zero voltage state, and also show some hysteresis in the hotspot region, probably dependent on the precise pinning of the boundary of the hotspot at imperfections along the bridge. The current I_h at a given resistance level is approximately proportional to $(1-T_o/T_c)^{\frac{1}{2}}$ as expected from the hotspot model. There is a definite but small increase of I_h at the lambda point (compare the curves for 2.2 and 2.1 degrees) corresponding to an increase of the apparent heat transfer coefficient Y_K from 12 W/cm^{2}K to 15.5 W/cm^{2}K. For comparison, the appropriate series/parallel combinations of Y_E, Y_K(substrate), and Y_K(helium) correspond to expected values of 12 W/cm^{2}K and 12.6 W/cm^{2}K, respectively. The fact that the observed change at the lambda point is larger than expected may indicate that part of the series impedance comes not from Y_E, which is unaffected at the λ-point, but rather from the unknown thermal impedance of the substrate, which could be partly shorted out by improved heat removal at its top surface. In general, the magnitude of the observed conductance is sufficiently smaller than the theoretical value for Y_K that the importance of electron heating seems clearly established. In thinner films, as expected, the observed heat transfer coefficient is even smaller.

INJECTION-CONTROLLED JOSEPHSON DEVICES

Thus far we have explored some of the nonequilibrium processes whereby quasiparticle resistive effects may extend into the superconducting regions of a 1-D microstrip. By the proximity effect, however, the superconducting order parameter can penetrate on the scale of the coherence length into the normal region as well. If the normal region is short enough, small current densities can be carried as supercurrent through the region, up to a temperature-dependent critical current density which depends exponentially on $L_n/\xi(T)$. When I_c is exceeded and an average voltage drop $\overline{V}$ develops across the weak link, Josephson oscillation at the frequency $\omega=2e\overline{V}/\hbar$ can be observed and the resulting device can be used as a detector or generator of electromagnetic radiation.

Devices with controllable Josephson characteristics across a nonequilibrium region have been made by locally perturbing a short segment of a one-dimensional strip by quasiparticle injection, phonon absorption, or the influence of resistive behavior in a cross strip. Volkov (1971) was the first to discuss theoretically the properties of a short illuminated region, concluding that it could show Josephson behavior. Wong et al. (1976, 1977) were the first to observe these effects. Optical perturbation of a tin strip with all but a short segment shielded by a thick silver overlay yielded strong Josephson behavior only in a narrow range of thermal environments (Yeh and Langenberg, 1978), but Gilabert et al. (1977) had more success by crossing the microstrip with a narrow optical waveguide. Wong et al. achieved better control by using tunneling injection of quasiparticles from a small tunnel junction into a tin strip to form the nonequilibrium weak link. The required injection corresponded to a critical power level, of the order of 16 W/cm^2 at 2K (referred to the junction area). They explained their results for the reduction of the critical current in this region in terms of a model based on the number density of excited quasiparticles and the Rothwarf-Taylor equations modified to include diffusion. More recently Kaplan (1980) has measured similar devices on Pb-alloy and In_2Bi strips, for which a simple heating model seems adequate. Josephson effects were observed to as high as 450 μV. A related sort of device has also been studied by Bevza et al. (1978, 1979a,b) who passed a control current through a perpendicular cross-strip in electrical contact with the strip to be controlled. When both strips were superconducting they failed to observe the dc control effects predicted by Kulik et al. (1977), but once the cross-strip switched into a resistive state above its critical current, reductions of the critical current of the other strip, and Josephson effects across it, were observed. In this case the detailed nature of the nonequilibrium perturbation of the other strip is not clear.

TIME-DEPENDENT CHARGE IMBALANCE PHENOMENA

Since the Josephson devices described above are localized in regions of the order of the coherence length, typically much less than the diffusion length, the boundary resistances will be very important in determining their resistance unless the weak region is made of a radically more resistive material (such as an oxide barrier). Under usual experimental circumstances the total current is fixed, and therefore the oscillatory supercurrent in the Josephson devices requires that the quasiparticle currents also be time dependent. Therefore, let us see how the time-dependent aspects of the charge imbalance phenomena discussed in Chapters 12-17 manifest themselves in this case.

I will do so using the equations of the two-fluid charge imbalance theories in the simple linearized form valid near T_c ($\Delta << k_BT$), since this is the region where heating effects ($\sim I^2(T)R$) play the least role. The basic equation are (Kadin et al., 1980):

$$\tau_J \dot{J}_S = \sigma E + eD\nabla Q^* \tag{7}$$

$$J_n \cong \sigma E \tag{8}$$

$$(1-Z)\dot{Q}^* + \nabla\cdot J_n/e + Q^*/\tau_{Q^*} = 0 \tag{9}$$

where near T_c, $Z(T) = 1-\pi\Delta/4k_BT$, and $\tau_{Q^*}=(4k_BT/\pi\Delta)\tau_E$. The time τ_J which governs the acceleration of the supercurrent is given by

$$\tau_J = (n/n_s)\tau_{tr} = 2\hbar k_B T_c/\pi\Delta^2 = \tau_{GL}/5.79 \tag{10}$$

The quantity eQ* is the now-familiar charge density stored in the quasiparticle system, which is counterbalanced by an opposing change of the charge density in the condensate in order to maintain overall electroneutrality. As a result there is a difference of the chemical potential of the quasiparticles and superfluid such that

$$Z(\mu_n-\mu_s) = Q^*/2N(0) = -\mu_s \tag{11}$$

At the level of approximation inherent in Eq. (8), we can consider $Z \approx 1$ and therefore $\mu_n \approx 0$. Neglecting charge imbalance effects, equations (7) and (8) describe the usual two-fluid model, in which the normal channel is resistive and the superconducting channel is inductive in character, with time contant $L/R = \tau_J$. The charge imbalance theory establishes the proper connection between these channels, such that a shift of charge from one to the other is accompanied by a difference of potential (capacitance C) and by a tendency for the charge imbalance to relax toward equilibrium (leakage conductance G). This suggests that a 1-D filament of nonequilibrium superconductor can be modeled by a simple transmission line circuit (Figure 6) with R, L, C, and G (per unit length).

Indeed, the equations above (at least in the stated level of approximation) are isomorphic to Kirchoff's laws for the transmission line (e.g., Karakash 1950):

$$\frac{dI_n}{dx} + GV + C\dot{V} = 0$$

$$\frac{dV}{dx} + RI_n - L\dot{I}_s = 0 \tag{12}$$

where $I = JA$ is the current in the filament of cross-sectional area A, $V = (\mu_n - \mu_s)/e$ is the potential difference, and the circuit

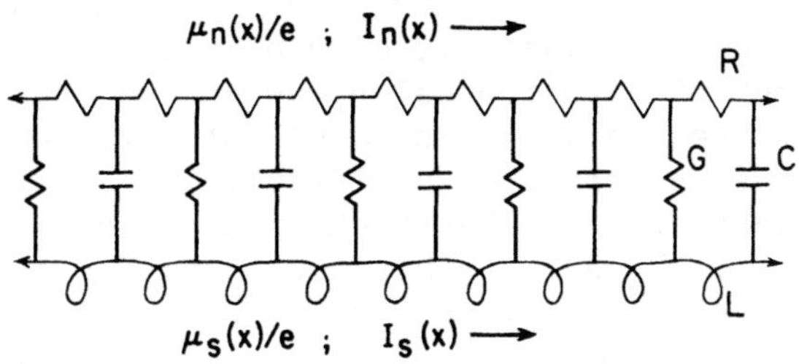

Fig. 6. Transmission-line equivalent of the charge imbalance theory.

parameters are given by

$$R = 1/\sigma A; \quad G = 2N(0)e^2A/\tau_{Q*} = \sigma A/\Lambda^2_{Q*}$$
$$L = \tau_J/\sigma A; \quad C = 2N(0)e^2A(1-Z) = (\sigma A/\Lambda^2_{Q*})\tau_E \tag{13}$$

The new time constant $C/G = \tau_E$ is just the relaxation time $(1-Z)\tau_{Q*} = \tau_E$ of the isolated system (see Eq. (9)). For a steady state situation, a potential difference at one end will decay exponentially with a characteristic length $(RG)^{-\frac{1}{2}} = \Lambda_{Q*}$, so that the injected quasiparticle current gradually converts to supercurrent, as at a normal/superconducting boundary. In a time-dependent situation the transmission line will propagate damped, dispersive charge imbalance waves. If the driving potential varies as $V_o \exp i\omega t$ the waves will vary as $\exp i(\omega t - Kx)$ where the propagation constant K is given by

$$-K^2 = (R + i\omega L)(G + i\omega C) = \Lambda_{Q*}^{-2}(1 + i\omega\tau_J)(1 + i\omega\tau_E) \tag{14}$$

This, of course, is just the dispersion relation associated with the second order partial differential equation for Q* obtained by eliminating J_n from equations (7)-(9). At high frequencies the phase velocity of the waves is $(LC)^{-\frac{1}{2}} = \Lambda_{Q*}/(\tau_J\tau_E)^{\frac{1}{2}} = (2D\Delta/\hbar)^{\frac{1}{2}}$, which is just the velocity of the Carlson-Goldman modes (Chapter 17). The ac decay length Λ_{ac} for the amplitude of the oscillations $(\mathrm{Im}K)^{-1}$ is given at high frequencies by

$$\Lambda_{ac} = 2(R/L + G/C)^{-1}(LC)^{-\frac{1}{2}} = 2\Lambda_{Q*}(\tau_J\tau_E)^{\frac{1}{2}}/(\tau_J + \tau_E) \tag{15}$$

while at zero frequency we recover the static decay length Λ_{Q*}. The characteristic impedance of the transmission line is given by

$$Z(\omega) = [(R + i\omega L)/(G + i\omega C)]^{\frac{1}{2}} =$$
$$= (\Lambda_{Q*}/\sigma A)[(1 + i\omega\tau_J)/(1 + i\omega\tau_E)]^{\frac{1}{2}} \tag{16}$$

which proves very useful in solving the electrical problem of a Josephson oscillator in a 1-D filament.

When connected between two such "transmission lines" in a 1-D geometry, a Josephson oscillator is driven by the difference of condensate electrochemical potential $\mu_s + e\varphi$ across it, where φ is the electrostatic potential. If the oscillator region is sufficiently short that it contributes negligible quasiparticle resistance, then the quasiparticle electrochemical potential should be continuous across it. Figure 7 shows a plot of the instantaneous electrochemical potentials near such a Josephson oscillator with a sawtooth-shaped current-phase relation, calculated from equations (7)-(9) for parameters corresponding to tin at $0.99T_c$ $(\tau_E/\tau_J = 11)$. The oscillator is running at 1.5 times its critical current,

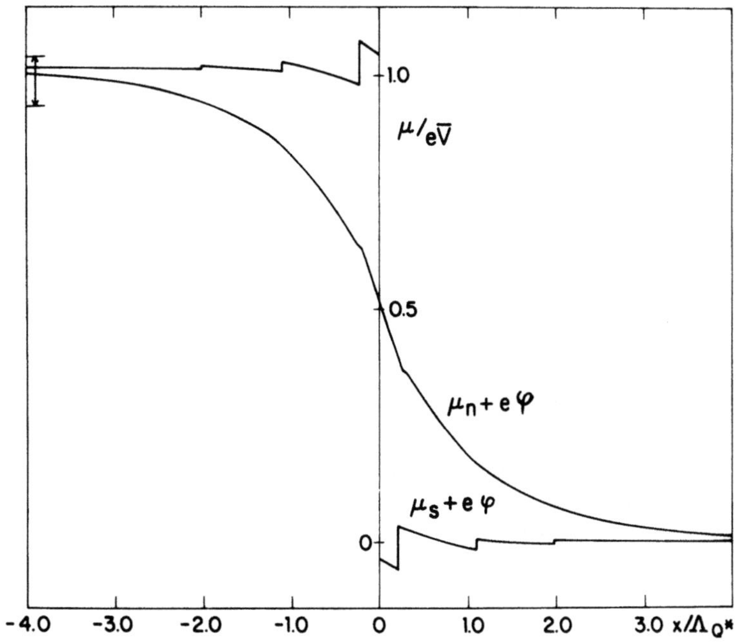

Fig. 7. Instantaneous potentials in the nonequilibrium regions near a Josephson junction in a 1-D filament. (Kadin et al., 1980)

corresponding to an average voltage $\overline{V}$ of $I_c(2\Lambda_{Q*}/\sigma A)$. The sawtooth-shaped waves that are launched travel outwards and decay in amplitude, with oscillations of $\mu_s+e\varphi$ much larger than oscillations of $\mu_n+e\varphi$. For $\tau_E=11\tau_J$, the high-frequency limit of Λ_{ac} should be about $\frac{1}{2}\Lambda_{Q*}$, so that the ac oscillations are not expected to die out too much more rapidly than the dc average of $\mu_n+e\varphi$, which varies as $\exp(-x/\Lambda_{Q*})$. In contrast, the dc average of $\mu_s+e\varphi$ is everywhere constant (otherwise steady acceleration of the supercurrents would occur), except for the sharp discontinuity at the center where the idealized Josephson oscillator has been placed. The arrow at the far left of the picture indicates the range of variation during the cycle of the asymptotic value of $e\varphi$ at large distances, where μ_s and μ_n have decayed to zero. This total voltage drop oscillates because we have assumed a constant total current.

With this picture in mind, we can now assess the relative importance of static and dynamic effects near such an oscillator. The dc decay length of $\mu_n+e\varphi$ is always Λ_{Q*}, whereas the ac decay length lies somewhere between its low frequency limit of Λ_{Q*} and its high frequency limit of $2\Lambda_{Q*}(\tau_J\tau_E)^{\frac{1}{2}}/(\tau_J + \tau_E)$. Unlike τ_E which is slowly varying near T_c, τ_J diverges at T_c. The two times for tin are equal at about 4 mk below T_c; far from T_c, τ_J is much shorter than τ_E. Where they are equal, the ac decay length is just Λ_{Q*}. It grows relatively shorter both nearer to and farther from T_c. Very near to T_c the minimum ac decay length is given by

$$\Lambda_{ac} = 2\Lambda_{Q*}(\tau_E/\tau_J)^{\frac{1}{2}} = \Lambda_{Q*}(8\tau_E\Delta/\hbar)^{\frac{1}{2}} \tag{17}$$

while farther from T_c it is given by

$$\Lambda_{ac} = 2\Lambda_{Q*}(\tau_J/\tau_E)^{\frac{1}{2}} \cong (k_BT/\Delta)^{\frac{1}{2}}\xi(T) \tag{18}$$

Since for fixed T_c-T, τ_J is independent of material while τ_E varies by orders of magnitude from material to material, the range for which the above regimes apply can vary considerably. For tin, one finds that the ac and dc decay lengths are comparable over much of the range of practical interest near T_c. For aluminum with its much longer values of τ_E, the dc decay length is considerably longer than the ac decay length, which is independent of τ_E over the range of practical interest. (Lindelof, 1978)

The above model can also be used to calculate I-V curves,

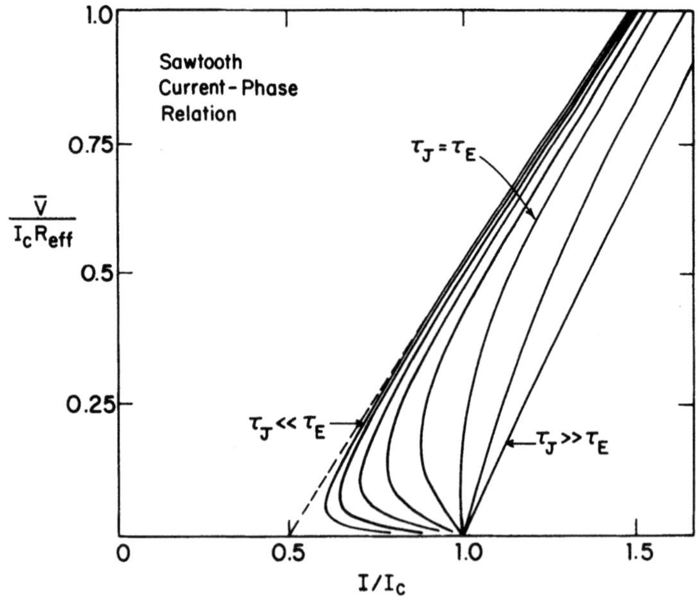

Fig. 8. Theoretical I-V curves for a Josephson junction shunted by the characteristic impedance of the transmission-line equivalent circuit. (Kadin et al., 1980)

shown in Figure 8. From the transmission line picture it is clear that the Josephson oscillator should be shunted by a complex impedance $R_J + 2Z(\omega)$; R_J is the ohmic resistance of the oscillator and $Z(\omega)$ is the complex impedance of the transmission line. For the usual situation of $\tau_E > \tau_J$, the impedance is smaller at high frequencies than at low frequencies and therefore, as in the capacitively shunted junction model is predicted to cause hysteresis of the I-V curves. The precise shape of the I-V curve is strongly influenced by the current-phase relation of the junction (Jackel, et al.,1977). For a sinusoidal current-phase relation the I-V curve at high voltages becomes ohmic with $R = R_J + 2\Lambda_{Q*}/\sigma A$, but for other current phase relations such as the sawtooth which have a time-averaged supercurrent during the cycle, it approaches a line of constant differential resistance parallel to, but offset from, the ohmic line by the amount of the "excess supercurrent", which would be a definite fraction of I_c.

I should mention here that there are at least three other mechanisms by which such an apparent "excess supercurrent" might also arise. For certain geometries, heating induced spreading of the resistive region leads to a nonlinear resistance resulting in an offset from the apparent ohmic line (Skocpol et al., 1974), but this bears no relation to I_c. Another approach, which has had success in explaining the excess supercurrent observed in point contacts as a function of temperature (Divin and Nad, 1979), relates the excess current at high voltages to the energy gap (Artemenko et al.,1978). A simple explanation of this result based on calculating the factor Z which describes the direct conversion of current to supercurrent at a normal/superconducting interface has been developed by Tinkham (Klapwijk et al., unpublished).

PROXIMITY EFFECT BRIDGES

One situation in which the above model is at least a good starting point for explaining the phenomena is the "proximity effect bridge" developed and extensively studied by Mercereau and coworkers. Mercereau made pioneering contributions both in the physics of nonequilibrium superconductors and in the technological aspects of Josephson devices. Yu and Mercereau (1972) were the first to demonstrate experimentally that μ_s and μ_n were different in the nonequilibrium region near a current-carrying interface. The proximity-effect bridge, a localized weak segment of a superconducting microstrip, was demonstrated to function as a Josephson device as early as 1968 (Notrays and Mercereau, 1971).

Over the years, this group worked with bridges of various sorts. One type consisted of a segment of a tin microstrip weakened by an underlayer of gold, a shorter version of the normal segments that I discussed at the beginning of this chapter.

Another type, which is quite rugged and can be fabricated with considerable precision, consisted of a bilayer microstrip with refractory materials such as tantalum on the bottom and niobium on the top, or else just thin tantalum with a thickness-dependent T_c. Where etched or ion milled, the strip has a considerably lowered T_c, allowing the formation of very short S/N/S structures with proximity-effect Josephson coupling across them (Notarys and Mercereau, 1973).

By using superconducting and normal probes near a current-carrying S/N interface of this sort, Yu and Mercereau (1972) demonstrated that the nonequilibrium potentials extended into the superconducting region much farther than the coherence length. They subsequently showed (1975) that the measured length scale of the nonequilibrium potential was Λ_{Q^*} as expected from the above arguments. The measured diffusion length was a few microns, corresponding to a value of τ_E of 0.65×10^{-10} seconds. They also showed that if the "normal" region is really a superconductor above its critical current, then only the resistive portion of the current across it contributes to the quasiparticle potentials in the superconductor. The I-V curves of short weak links which functioned as Josephson oscillators showed a definite excess supercurrent of $\sim\frac{1}{2}I_c$, which can be explained in terms of a time-averaged supercurrent through the device. The existence of time-dependent oscillations of the potentials near such a bridge was revealed most clearly by Palmer and Mercereau (1974, 1975, 1977) who fabricated highly uniform arrays of hundreds of proximity effect bridges spaced only a few micrometers apart and observed clearcut synchronization effects between the oscillators when the spacing was reduced to be comparable to the diffusion length.

Heating effects, analyzed by Decker and Palmer (1977), also proved to be quite important in these devices as one might expect from the relatively poor geometry for cooling by metallic conduction. Joule heating causes a local temperature rise which reduces the critical current in the nonzero-voltage state causing critical current hysteresis and an upper voltage limit on the Josephson rf step response, similar to that discussed in Chapter 8.

It should be noted here that the Mercereau group has also conducted a number of complex and innovative nonequilibrium experiments which have not yet been comprehensively explained. Yu and Mercereau (1976), for example, have measured rf-induced nonequilibrium potential differences <u>inside</u> single proximity effect bridges. Palmer (1975) reports extensive measurements of rf-induced step structures with fractional integer Josephson relations at single superconducting/weak/normal interfaces, and Franson (1977) observed critical current reductions of a distinctly nonthermal character in proximity effect bridges subjected to quasiparticle tunneling injection. These experiments show that much still remains to be learned about nonequilibrium phenomena in current-carrying superconducting weak links.

PHASE-SLIP CENTERS

Finally, we come to a remarkable phenomenon which occurs in principle even in perfectly homogeneous superconducting filaments. I refer to the spontaneous appearance, once the critical current is exceeded, of "phase-slip centers"-- a periodic array of spatially localized Josephson oscillators connected by nonequilibrium regions of just the sort that we have been discussing. One might think of this as an alternation of normal and superconducting regions reminiscent of the dynamic intermediate state proposed by London (1937, 1950) to explain the resistive transition of large-diameter Type-I superconducting wires. But actually the concept is a much more intrinsically quantum phenomenon, analogous to the motion of a regular vortex lattice in an ideal current-carrying Type-II superconductor.

Warburton and Webb (1968; also Warburton 1972) were the first to report experimental indications of unexplained phenomena in tin whisker crystals near T_c at large current densities. While studying fluctuation-induced resistance below T_c (caused by thermally acti-

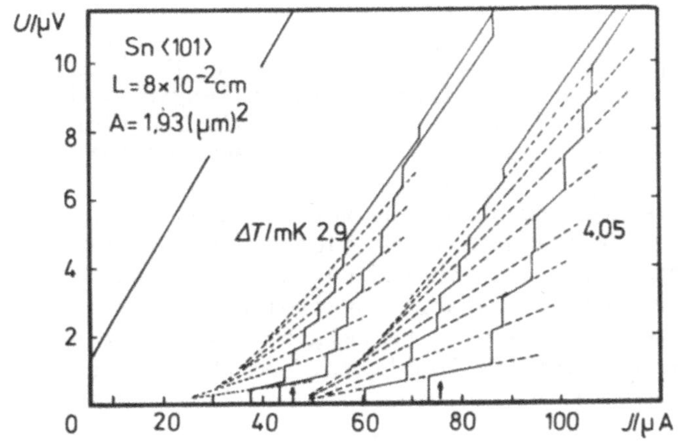

Fig. 9. Experimental I-V curves for phase-slip centers in a long tin whisker. (Meyer, 1973)

vated isolated phase-slip events), they also saw the appearance of a series of discrete voltage jumps in the resistive transition at current levels exceeding several microamperes. V. Ambegaokar (1969) quickly suggested the possibility that the jumps were related to the appearance of "a sort of resistive mixed state with normal regions of the wire alternating with superconducting ones," and called for further experiments.

Subsequent experiments on tin whiskers (Meyer and von Minnigerode, 1972; Meyer, 1973) and on tin microstrips (Skocpol et al., 1974a,b) helped to clarify the experimental regularities. Figure 9 shows the highly regular steplike structures measured in the I-V curves of tin whiskers. Similar structures spread out over a wider range of currents could also be seen in the less homogeneous microbridges. The curves appeared to be the sum of voltage increments in series, each characterized by a constant differential resistance and an apparent excess supercurrent of somewhat greater than $\frac{1}{2}I_c$. Comparison of microbridges of roughly equal mean free path but different dimensions established that the differential resistance equalled the normal resistance of a length of the filament of order 10 μm in moderately clean bridges but much longer in very pure whiskers.

With these experimental clues, Tinkham ventured a phenomenological model to explain them. Starting with ideas about quasiparticle diffusion introduced by Pippard et al. (1971) and about repetitive phase-slip processes and unequal condensate and quasiparticle chemical potentials introduced by Rieger et al. (1971,1972), he integrated these into a picture of the structure of phase-slip centers (Skocpol et al.,1974b) which differed in detail but not in spirit from the behavior pictured in Fig. 8. Josephson phenomena localized on the short coherence length scale were described in terms of an oscillatory acceleration and collapse of the supercurrent with phase change of 2π per cycle, corresponding to the same total drop of time-averaged electrochemical potential as occurred on the longer inelastic diffusion length scale for the quasiparticles. Qualitatively at least, this behavior is the 1-D limit of the regular passage of vortices (see Chapter 20.)

Experiments confirmed various key features of this model. Differential resistance data from bridges and whiskers agreed with the expected mean free path dependence and magnitude of the diffusion length for inelastic scattering from phonons. Bridges with multiple voltage probes along the length proved that the phase-slip centers were spatially localized, and the presence of Josephson phenomena was confirmed by direct observation of the ac Josephson effect (Skocpol et al.,1974b). Comparable experiments have also been conducted on tin whiskers (Meyer and Tidecks, 1977a; Tidecks and Meyer, 1977,1979 and Tidecks and von Minnigerode, 1979). As for other materials, the study of phase-slip centers in aluminum, where the inelastic diffusion length is very much longer because of the large values of τ_E, helped verify the generality of the phenomenon and the importance of inelastic phonon scattering processes (Klapwijk and Mooij, 1976; Klapwijk et al; 1977). Other observations of phase-slip centers have included those of Rachford et al. (1975) who detected changes in the microwave impedance of tin and lead bridges, Jillie (1976) who studied indium microbridges, Laibowitz et al. (1979) who measured them in extremely narrow niobium "nanostripes", Tidecks and Slama (1980) who observed them in indium whiskers, and Chaudari et al. (preprint) who measured very narrow amorphous W-Re strips. In the latter case, the ability to measure the unknown inelastic scattering time from the observed differential resistance of phase-slip centers played a key role in evaluating the agreement of observed normal-state behavior above T_c with theories of electronic localization (maximum metallic resistance) in which inelastic processes contribute to delocalization.

One aspect of the model that has received subsequent clarification is the question of which quasiparticle diffusion length is appropriate. In our original work we were unable to obtain convincing evidence for a divergence of the diffusion length at T_c and therefore suggested that $\Lambda_E=(D\tau_E)^{1/2}$ rather than Λ_{Q^*} might be the appropriate length. Subsequently, Dolan and Jackel (1977) performed the definite experiment of directly measuring the quasiparticle and condensate electrochemical potentials with a series of normal and superconducting probes closely spaced inside the nonequilibrium region of a phase-slip center in a tin microstrip. To their initial surprise, the results were in excellent agreement with the time-average of Fig. 7. Their direct measurements of the spatial falloff of the quasiparticle potential clearly showed the divergent behavior $\sim(1-T/T_c)^{-1/4}$ and magnitude expected for Λ_{Q^*}, and proved more reliable than estimates based on the total differential resistance, which near T_c must be determined over a very narrow range of currents and far from T_c is strongly affected by heating. Although resistance measurements on aluminum do not reveal a divergence (Klapwijk et al.,1976, and private communication) and results on tin whiskers are subject to differing interpretations (compare Tidecks and Meyer (1979) to Galaiko et al. (1976)), nevertheless divergent behavior has been seen clearly in indium microbridges (Jillie, 1976), indium whiskers (Tidecks and Slama, 1980), and tin microbridges with localized weak spots (Skocpol et al., 1978; Kadin et al.,1978). I believe that it is present in all cases, but can be obscured by heating effects and other experimental difficulties. In addition to the $(\Delta T)^{-1/4}$ divergence of dV/dI that they detect in Meyer's whisker data, Dmitriev et al. (1977,1978) measure a strong $(\Delta T)^{-1}$ dependence in tin microbridges farther from T_c, which they attribute to elastic scattering interacting with current-induced gap anisotropy, as urged by Galaiko (1975).

Another aspect of continued experimental attention is the magnitude of the observed excess supercurrent interpreted from the I-V characteristics. Tidecks has reported careful measurements in tin and indium whiskers which show that the fraction of I_c measured increases from about ½ towards 1 as T_c is approached and in very clean whiskers, but no explanation has been advanced. A related question is the remarkable hysteresis of the I-V curves, which in tin whiskers first develops a few millidegrees below T_c. This result appears to be in at least qualitative agreement with the intrinsic hysteresis predicted by the transmission line model described previously (Kadin et al., 1980).

Finally, an active area of investigation has been the influence of inhomogeneities and mutual interaction effects on the spatial location and critical current for the appearance of succesive phase-slip centers. In my original work with tin microbridges it was clear that the successive appearance of phase-slip centers was strongly influenced by material inhomogeneities, particularly of T_c, so that the first phase-slip center to appear did so at the location with the smallest critical current, and thus phase-slip centers tended to be "pinned" as are vortices in real Type-II materials. In the more nearly perfect single-crystal whiskers, this need not be the case. Clearly from the model in Fig. 7, normal currents proportional to $\nabla(\mu_n+e\varphi)$ flow within a few diffusion lengths of a phase-slip center, and therefore in a whisker, the first phase-slip center can appear at the middle and remain stable with increasing total current until the reduced supercurrent far away again reaches the critical current; then at least one more phase-slip center will appear, probably causing a symmetrical rearrangement. An explanation of this sort [briefly stated in the original Skocpol et al. (1974b) article and elaborated most recently by Tinkham (1979)] can account for the sort of mutual interaction effects between phase-slip centers observed by Meyer and Tidecks (1977b). In principle, this repeated formation of additional phase-slip centers continues with increasing current until the phase-slip centers are so closely packed together that they are indistinguishable from the normal state. In practice, heating effects soon set in and recover the normal state more rapidly.

The results of this process can be expressed more quantitatively (Galaiko 1977, Tinkham 1979). When appropriate boundary conditions are imposed on the exponential variations of $\mu_n+e\varphi$ the total voltage drop across n equally spaced phase-slip centers in a bridge of length L (taking into account a time-averaged supercurrent $\overline{I_s}=\beta I_c$ through each "Josephson oscillator") is

$$V_n=\rho_n L(J-\beta J_c)[(2n\Lambda_{Q*}/L)\tanh(L/2n\Lambda_{Q*})] \qquad (19)$$

which shows integer increments of differential resistance for widely separated phase-slip centers, but which saturates at the normal resistance of the filament for closely spaced ones. The critical current for the appearance of the (n+1)th center is reached when the current is

$$J_{max,n} = J_c \frac{\cosh(L/2n\Lambda_{Q*}) - \beta}{\cosh(L/2n\Lambda_{Q*}) - 1} \tag{20}$$

In the limit of a very long homogeneous filament, the voltage steps associated with the successive nucleations will be comparatively small, and the resulting smooth I-V characteristic can be calculated. Near J_c it takes the following form, based on the above model:

$$(J-J_c)/J_c=2(1-\beta)\exp[-\rho_n(J-\beta J_c)/\bar{E})] \tag{21}$$

where $\bar{E}$ is the average voltage drop per unit length. This describes the curve for increasing current only, because the criteria leading to hysteresis are not sufficiently well understood to be incorporated. The form of this result is typical of the expressions obtained when other theories of phase-slip centers are applied to very long homogeneous filaments (Galaiko, 1975,1976; Bezuglyi et al., 1977; Ivlev et al., 1980.)

THEORIES OF PHASE-SLIP CENTERS

The comprehensive theoretical explication of the nature and structure of phase-slip centers is one of the most important challenges in the field of nonequilibrium superconductivity. By their very nature, phase-slip centers involve time-dependent nonequilibrium phenomena occurring under conditions of large supercurrent density comparable to the critical current. Thus the full solution requires consideration of phenomena on several different length scales, including regions of rapid oscillatory phenomena and with full consideration of the nonlinear effects of large current densities. Although considerable progress has been made, this task is by no means yet complete. One line of development, culminating in the work of Kramer and Watts-Tobin (1978), has been to explore the time dependent Ginzburg-Landau (TDGL) equations, and nonequilibrium extensions thereof. Another line of work, summarized by Galaiko (1977), has emphasized static solutions of equations taking full account of the nonlinear effects of large supercurrent densities. By comparing some of the key results of these theoretical studies with the linearized two-fluid model presented previously we can begin to understand what the outlines of a synthetic approach might look alike.

The basic TDGL equations can be written (eg. Kramer and Baratoff, 1977)

$$u\left(\frac{\partial\psi}{\partial t} + i\mu\psi\right) - \frac{\partial^2\psi}{\partial x^2} - (1-|\psi|^2)\Psi = 0$$
$$j = \mathrm{Im}\psi^*\partial\psi/\partial x - \partial\mu/\partial x \tag{22}$$

where ψ is the normalized complex order parameter (magnitude 1 in spatially uniform equilibrium), lengths are in units of the coherence length $\xi(T)$, times are in units of τ_J, the current is normalized to $J_o = (3\sqrt{3}/2)J_c^{GL}$, and the electrochemical potential μ is $\mu_n + e\varphi$ normalized to $eJ_o\xi/\sigma$. The parameter u is equal to 12 in a gapless superconductor with magnetic impurities and would be 5.79 in a dirty superconductor if other nonequilibrium effects could be neglected. The first application of the TDGL equations to long filaments with spatially periodic resistive structures was carried out by Fink (1973a,b; Fink and Poulsen, 1974,1975). He assumed a static structure (since shown not to be a stable solution of the equations) and arrived at a picture similar to Fig. 7 but for which the only length scale was $\xi(T)$. The TDGL equations have been explored numerically for the case of a long filament by Likharev (1975) and by Kramer and Baratoff (1977). The latter workers found that if they started the calculations from a slightly perturbed version of the Langer-Ambegaokar (1967) solution for a single isolated phase-slip event, there existed a very narrow range of currents (j=0.326-0.335 for u=5.79; 0.284-0.291 for u = 12, where j_c^{GL} = 0.385) in which stable repetitive phase-slip events occurred regularly at the Josephson frequency.

In order to see the connection to our equations (7)-(9), it is useful to rewrite the TDGL equations in a form used by Ivlev et al. (1980), separating the complex order parameter into normalized magnitude f and phase ϕ, then subsuming the phase terms into a gauge-invariant form using the quantities

$$a = A - (c/2e)\nabla\phi = -(c/2e\xi)q$$
$$\Phi^* = \varphi + \frac{1}{2e}\frac{\partial\phi}{\partial t} = Q^*/2N(0)e \tag{23}$$

referred to in the Russian literature as the gauge invariant vector and scalar potentials, since

$$E = -\frac{1}{c}\frac{\partial a}{\partial t} - \nabla\Phi^* = -\frac{1}{c}\frac{\partial A}{\partial t} - \nabla\phi \tag{24}$$

which is a restatement of Eq. (7). The total current is given by

$$J = J_s + J_n = f^2 q J_o + \sigma E \tag{25}$$

which is just our equation (7) generalized to include the f^2 dependence of the supercurrent, where J_n is handled in the same level of approximation as (8). The complex TDGL equation can then be separated into two real equations:

$$u\ \partial f/\partial t - \partial^2 f/\partial x^2 + f(f^2 + q^2 - 1) = 0 \tag{26}$$

$$uf^2\Phi^* - \frac{\partial}{\partial x} f^2 q = 0 \tag{27}$$

The first of these equations deals with variations of f (n_s) in space and time, not considered in our transmission line model. The second has the same structure as our equation (9) [since $\Phi^* \propto Q^*$ and div J_s = - div J_n] except that the $\dot{Q}^*$ term which describes the "capacitance" in our model is missing. Ivlev et al. (1980) suggest that u should be treated as a phenomenological parameter adjusted to be equal to $(\xi^2/\Lambda^2_{Q^*})$ so that these equations correctly describe the static decay length for the charge imbalance. Unfortunately this adjustment of u also affects the time dependence of the first equations, with no obvious justification. Thus when Ivlev et al. examine these equations they conclude that ac effects are localized to a small length scale $(\Lambda_{Q^*}\xi)^{1/2}$, much less than Λ_{Q^*}. Probably the omission of the capacitive term is the most important factor in reaching this conclusion, since I note that putting C=0 in (15) reduces Λ_{ac} in our model to zero. Subsequent work by Ivlev and Kopnin (1980) has been much closer to that of Kramer and Watts-Tobin (1978) and Galaiko (1977).

Rather than making phenomenological adjustments of uncertain validity Kramer and Watts-Tobin (1978) adopted a different approach. Starting from the Schmid and Schön formalism, they derived a modified TDGL equation

$$\frac{u}{[1+\gamma^2|\psi|^2]^{1/2}}\left[\frac{\partial}{\partial t} + i\mu + \tfrac{1}{2}\gamma^2\frac{\partial|\psi|^2}{\partial t}\right]\psi = (\nabla - 2ieA/\hbar c)^2\psi + (1-|\psi|^2)\psi \tag{28}$$

where $\gamma = 2\tau_E\Delta/\hbar$. This equation builds in a longer decay length for charge imbalance and also increases the coefficient of the $\partial|\psi|^2/\partial t$ term so that it emphasized the dynamic gap enhancement effects discussed in Chapter 8. It does not, however, include the "capacitive" term discussed above.

Nevertheless, numerical computations with this theory show a promising progress in the treatment of phase-slip centers. For values of γ=10 and 20, the current range for phase-slip solutions is greatly expanded, and the resulting I-V curves (Fig. 10) show many realistic features, including a region of nearly constant differential resistance corresponding to a length of $\sim 7.5\xi(T)$ for each phase-slip center, an excess supercurrent of $\sim \tfrac{1}{2}I_c$, and considerable hysteresis of the I-V curves. Since the theory is restricted to a rather narrow temperature range of validity near T_c ($\sim$12 mK for tin and indium) thorough comparisons with data are not possible. However, the length scale and temperature dependence are in reasonable agreement with data for clean indium whiskers (Tidecks and Slama, 1980) which show a divergence stronger than Λ_{Q^*} very near

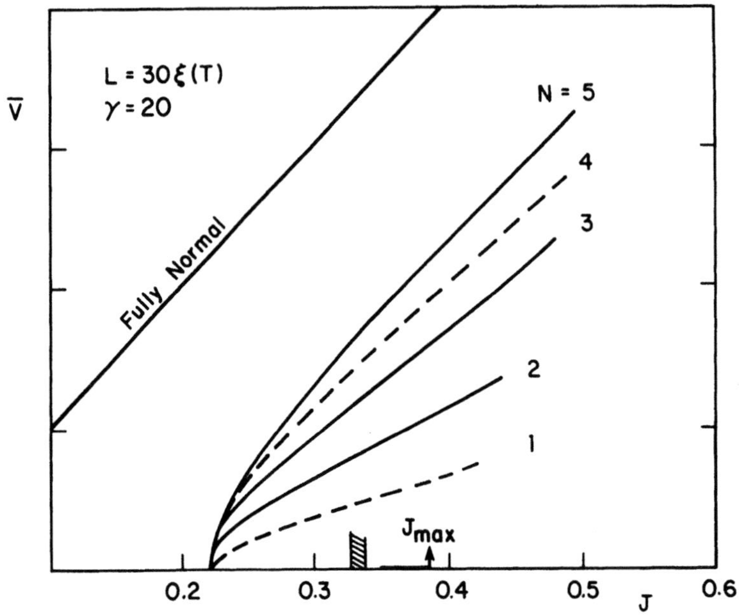

Fig. 10. Theoretical I-V curves for N phase slip centers in a long filament. (Kramer and Watts-Tobin, 1978; N=1 and N=4 added.)

to T_c. The measured critical currents are proportionately somewhat smaller than required to produce the theoretically expected hysteresis ratios and excess supercurrent ratios, but the J_c in the theory is actually the upper limit of the local metastability of the superconducting state, and thus the observed critical currents may be reduced as a result of premature switching. Very recently, Baratoff (1980) has investigated numerical solutions of a different (gapless) set of coupled equations which place more emphasis on charge imbalance and show overdamped ac oscillations near the center of the phase-slip center solutions.

Another line of approach has been pushed for a number of years by Galaiko, who has emphasized primarily static, nonlinear aspects of the problem in order to mediate between his high-powered theoretical formalisms and experimental results on tin microstrips obtained by several of his experimental coworkers (Churilov et al. 1969). The picture developed converges nicely to the same general ideas discussed above. A useful summary of the entire body of this work appeared in Galaiko (1977). Because of his desire to include the strongly nonlinear aspects of the problem, Galaiko calls attention to an additional term in the equation for the magnitude of the order parameter which is proportional to Φ^{*2}. He also points out that when large supercurrents are flowing, the resulting "gap anisotropy" allows elastic scattering to accomplish charge-imbalance relaxation, and that this mechanism should be more important than inelastic phonon scattering in dirty superconductors except very near to T_c. Because of the strong energy dependence of this relaxation process, approximations which involve high frequency limits in the equations prove necessary in order to simplify them. The resulting diffusion length is found to be of the order of magnitude of $(p_s\Delta/T_c)^{-1}$ which is approximately $\xi(T)k_BT_c/\Delta$. Perhaps it is only coincidence, but this is just the dependence for Λ_{ac} that we obtained in the high frequency limit away from T_c (Eq. (18)). Perhaps Galaiko's approach to simplifying his rather forbidding microscopic equations has succeeded in building in certain aspects of what I have referred to as the "capacitive" term, which is missing from the other approaches. Dmitriev and Khristenko (1977) show data from tin microstrips consistent with this "elastic branch mixing" mechanism.

Galaiko (1975, 1976) has emphasized primarily the very long ideally homogeneous filament, calculating expressions for the smooth variation of the I-V curves of very many phase-slip centers, although as mentioned previously, he has also addressed the case of a shorter filament where discrete voltage jumps for each phase-slip center are observed (Bezuglyi et al.,1977). Other topics considered by Galaiko are the regime of currents $J >> J_c$ where superconductivity first nucleates from the normal state (of little practical importance because of heating effects), and also the possibility of low-frequency positional motion of the phase-slip-center structure, which may explain low frequency electromagnetic oscillations observed experimentally (Churilov et al.,1969; Galaiko et al.,1973,1974). Circuit-controlled relaxation oscillations (Skocpol et al.,1974c) may be an alternative explanation of the observed phenomenon, however.

Finally, I would like to draw together a number of the various ideas noted above in order to suggest the form of a set of equations which should include as many correction factors as possible given our present knowledge. First, I would like to consider an issue raised by Pethick and Smith (1979), who make a distinction between the usual $J_n=J-J_s$ and the current associated with flow of quasiparticle charge, which they call J_n^Q. According to their theory $J_n^Q=Z(T)J_n$ (near T_c), and J_n^Q is the quantity which properly enters the divergence term in (9). (This automatically accounts for the factor of Z(T) entering Eq. (2)). If the factor of Z(T) in this divergence term is included, this allows the identification of the voltage V in the transmission line model with $(\mu_n-\mu_s)/e=\Phi^*/Z$ without the approximation $\mu_n\approx 0$, although it also requires modification of the definitions of L and R by a factor of $Z^{-1}(T)$. The change of L then shifts the velocity of the Carlson-Goldman modes in agreement with Pethick and Smith's equation (5.68). Thus this effect can be incorporated into these theories by a slight renormalization of coefficients. The term proportional to $(e\Phi^*/\Delta_o)^2$ noted by Galaiko can be incorporated explicitly into the Kramer and

Watts-Tobin equations, and any imbalance relaxation mechanisms other than phonon scattering can presumably be included in a redefinition of the parameter γ. The key feature highlighted by our transmission line model, its capacitive term which leads to wavelike behavior at high frequencies, can also be incorporated in a very direct manner by replacing the term $\mu=2e$ by $2e(+ \tau_E\dot{\Phi}^*)$. To see that this is the case, we can substitute the form $\psi=f\exp(i\phi)$ into the complex form of the proposed equation:

$$\frac{u\tau_J}{[1+\gamma^2 f^2]^{1/2}} \left(\frac{\partial}{\partial t} + i2e(+\tau_E\dot{\Phi}^*) + \tfrac{1}{2}\gamma^2 \frac{\partial^2|\psi|^2}{\partial t}\right)\psi = (1-|\psi|^2 + 2e\Phi^{*2}/\Delta_o^2)\psi+\xi^2 \frac{\partial^2\psi}{\partial x^2} \qquad (29)$$

and separate the result into two real equations after eliminating derivatives of ϕ in favor of gauge invariant quantities):

$$\frac{u\tau_J}{[1+\gamma^2 f^2]^{1/2}} \left[\frac{\partial f}{\partial t} + \frac{1}{2}\gamma^2 \frac{\partial f^2}{\partial t}\right] = [1-f^2-q^2+ 2(e\Phi^*/\Delta_o)^2]f+\xi^2\frac{\partial^2 f}{\partial x^2} \qquad (30)$$

and

$$\frac{u\tau_J}{[1+\gamma^2 f^2]^{1/2}} [2e(\Phi^*+\tau_E\dot{\Phi}^*)] = (\xi/f^2)\,\frac{\partial}{\partial x}\,(f^2q) \qquad (31)$$

As intended, the latter equation is equivalent to Eq. (9) with the identification $\tau_E=(1-Z)(1+\gamma^2f^2)^{1/2}\tau_J/Zf^2$, a correspondance which holds in the large γ limit near T_c. As far as I can tell, these changes should help to extend the Kramer and Watts-Tobin equations to be applicable over a wider range of temperatures and voltages. Whether a full derivation would also turn up other terms of which I am unaware remains to be seen. However, it would be interesting to obtain numerical solutions of such equations in the temperature range where $\Lambda_{Q^*} >> \xi(T)$ to see whether the results are qualitatively similar to the model in Fig. 7 of charge-imbalance waves near an ideal Josephson oscillator.

CONCLUSION

We have seen that a great variety of nonequilibrium phenomena can be studied in a 1-D geometry which helps to simplify the behavior and clarify the length scales. In its simplest form the 1-D nonequilibrium filament can be described in terms of the spatially varying quantities T*, Q* or Φ^*, and the quantities f or n_s and q or p_s which characterize the order parameter, governed by heat flow equations and a substantially modified form of the TDGL equations. The phenomena covered range from normal hotspots which Kamerlingh Onnes himself imagined, to phase-slip centers with intrinsic quantum mechanical (Josephson) properties, that are presently understood in general terms but not complete detail. But even in the latter case Kamerlingh Onnes (1913) perhaps pointed the way. While speculating on the resistive transition in metal filaments which are "homogeneous, unstrained, everywhere crystallized in the same manner" (whiskers) he turned toward explanations "on the lines of the explanation given for the mechanism of ordinary resistance" (scattering by "vibrators" with quantum properties) and drew speculative analogies to the generation of waves on water beyond a certain critical wind velocity. I'm sure that we all can concur with his statement that "the superconductive state raises questions of fundamental interest for the theory of quanta, which in the last three years has taken a prominent place in very different departments of physics."

ACKNOWLEDGEMENTS

The author appreciates the support given over the years to this research by the National Science Foundation, the Joint Services Electronics Program, and the Office of Naval Research. I also would like to thank M. R. Beasley and M. Tinkham for their guidance and help throughout the early stages of this research.

REFERENCES

Ambegaokar, V., 1969, Critical currents and the onset of resistance in superconductors, in:"Superconductivity," P. R. Wallace, ed., Gordon and Breach, New York.

Anderson, A. C., and Peterson, R. E., 1972, The thermal resistance between electrons and phonons in copper, Phys. Lett., 38A: 519.

Artemenko, S. N., Volkov, A. F., and Zaitsev, A. V., 1978, The excess current in superconducting point junctions, JETP Lett., 28:589.

Baratoff, A., 1980, Inhomogeneous non-equilibrium states in current-driven superconducting filaments, private commun.

Bevza, Yu. G., Karamushko, V. I., and Dmitrenko, I. M., 1978, Cross-like film structures, microwave irradiation stimulated superconductivity and Josephson effect, J. Physique,39:C6-537.

Bevza, Yu. G., Karamushko, V. I., and Dmitrenko, I. M., 1979a, Microwave irridation stimulated superconductivity and Josephson effect in cross-like film structures, J. Physique, 40: L-101.

Bevza, Yu. G., Karamushko, V. I., Kel'man, E. A., Tzach,G. G., and Dmitrenko, I. M., 1979b, Electrically controlled Josephson elements, Sov. J. Low Temp. Phys., 5:332.

Bezuglyi, E. V., Bratus, E. N., and Galaiko, V. P., 1977, Voltage jumps in the current-voltage characteristics of superconductive channels, Sov. J. Low Temp. Phys., 3:491.

Busch, H., 1921, Über die Erwärmung von Drähten in verdünnten Gasen durch elektrischen Strom, Ann. Phys., 64:26.

Chaudhari, P., Broers, A. N., Chi, C. C., Laibowitz, R., Spiller, E., and Viggiano, J., 1980, Phase-slip and localization diffusion lengths in amorphous W-Re alloys, unpublished.

Churilov, G. E., Dmitriev, V. M., and Beskorsyi, 1969, Generation of high frequency oscillations by thin superconducting films, JETP Lett., 10:146.

Decker, S. K., and Palmer, D. W., 1977, Thermal effects in superconducting proximity-effect bridges, J. Appl. Phys., 48:2043.

Dharmadurai, G., 1980, Self-heating induced phenomena in long thin-film superconductors and their applications, Phys. Stat. Sol. (a), to appear.

Divin, Yu. Ya.,and Nad', F. Ya., 1979, Dependence of the excess current in superconducting point contacts on temperature and voltage, JETP Lett., 29:516.

Dmitriev, V. M., and Khristenko, E. V., 1977, Temperature dependence of the nonequilibrium field penetration depth in a superconductor, Sov. J. Low Temp. Phys., 3:587.

Dmitriev, V. M., Khristenko, E. V., Churilov, G. E., and Svetlov, V. N., 1978, Structure of nonequilibrium resistive state of superconducting strips, J. Physique, 39:C6-507.

Dolan, G. J., and Jackel, L. D., 1977, Voltage measurements within the nonequilibrium region near phase-slip centers, Phys. Rev. Lett., 39:1628.

Eru, I. I., Peskovatskii, S. A., and Poladich, A. V., 1973, Non-thermal discontinuity in the current-voltage characteristic of a thin superconducting film, Sov. Phys. Solid State, 15: 1069.

Eru, I. I., Peskovatskii, S. A., and Poladich, A. V., 1979, Movement of resistive region boundaries in a current-carrying superconducting film, Sov. Phys. Solid State, 21:1149.

Fink, H. J., 1973, Current transitions of superconducting whiskers, Phys. Lett. A., 42:465.

Fink, H. J., 1973b, A new superconducting state, Phys. Stat. Sol. (b), 60:843.

Fink, H. J., and Poulsen, R. S., 1974, Exact solution of a new superconducting state, Phys. Rev. Lett., 32:762.

Fink, H. J., and Poulsen, R. S., 1975, Complete solution of a new superconducting current state above the critical current, Phys. Rev. B., 11:1870.

Franson, D. W., 1977, "Effects of Nonequilibrium Quasiparticle Distribution in Superconducting Weak Links," Cal. Tech. Ph.D.

Galaiko, V. P., 1975, Microscopic theory of resistive current states in superconducting channels, Sov. Phys. JETP, 41:108.

Galaiko, V. P., 1976, Features of the volt-ampere characteristics and oscillations of the electric potential in superconducting channels, Sov. Phys. JETP, 44:141.

Galaiko, V. P., 1977, Kinetic theory of resistive states and electric oscillations in superconducting channels, J. Low Temp. Phys., 26:483.

Galaiko, V. P., Dmitriev, V. M., and Churilov, G. E., 1973, Dynamic phase stratification and generation of electromagnetic waves in thin current-carrying superconducting films, JETP Lett., 18:213.

Galaiko, V. P., Dmitriev, V. M., Kristenko, E. V., and Churilov, G. E.,1974, Mechanism for nonstationary behavior of thin-film superconducting bridges in the resistive current state. Rev. Phys. Appl. 9: 161.

Galaiko, V. P., Dmitriev, V. M., and Churilov, G. E., 1976, Resistance current states and the nonstationary properties of superconducting channels, Sov. J. Low Temp. Phys., 2: 148.

Garbuny, M. and Gottlieb, M., 1966, Critical heat transfer in liquid helium II as controlling factor in superconducting phase transitions, in: "Heat Flow below 100°K and Its Technical Applications", Pergamon Press, New York.

Gilabert, A., Ostrowsky, D. B., Vanneste, C., Papuchon, M., and Puech, B., 1977, Superconducting weak links induced by optical guided waves, Appl. Phys. Lett., 31: 590.

Gubankov, V. N., Likharev, K. K., and Margolin, N. M., 1972, Properties of superconducting point contacts, Sov. Phys. in Solid State, 14: 819.

Ivlev, B. I., Kopnin, N, B., and Maslova, L. A., 1980, Dynamics of a resistive state in superconductors, submitted to Zh. ETF.

Iwanyshyn, O. and Smith, H. J. T., 1972, Thermal effects at superconducting point contacts, Phys. Rev. B., 6: 120.

Jackel, L. D., Henkels, W. H., Warlaumont, J. M., and Buhrman, R. A., 1976, Current-phase relations as determinants of superconducting thin-film weak-link I-V characteristics, Appl. Phys. Lett., 29: 214.

Jillie, D. W., 1976, "Interactions between coupled thin film microbridge Josephson junctions," Ph.D. Thesis, State University of New York at Stony Brook.

Kadin, A. M., Skocpol, W. J., and Tinkham, M., 1978, Magnetic field dependence of relaxation times in nonequilibrium superconductors, J. Low Temp. Phys., 33: 481.

Kadin, A. M., Smith, L. N., and Skocpol, W. J., 1980, Charge imbalance waves and nonequilibrium dynamics near a superconducting phase-slip center, J. Low Temp. Phys., 38: 497.

Kaganov, M.I., Lifshitz, I. M., and Tanatarov, L. V., 1957, Relaxation between electrons and the crystalline lattice, Sov. Phys. JETP, 4:173.

Kamerlingh Onnes, H., 1913, Superconductors, Commun. Kamerlingh Onnes Lab. Univ. Leiden, Suppl. 34b:67.

Kaplan, S. B., 1979, Acoustic matching of superconducting films to substrates, J. Low Temp. Phys., 37:343.

Kaplan, S. B., 1980, Simple-heating-induced Josephson effects in quasiparticle-injected superconducting weak links, J. Appl. Phys., 51:1682.

Karakash, John J., 1950, "Transmission Lines and Filter Networks", Macmillan, New York.

Kashchei, V. A., 1976, Resistive domain in long superconductor of small cross section at temperatures close to critical, Sov. Phys. Solid State, 19:934.

Klapwijk, T. M., and Mooij, J. E., 1976, Phase-slip centers in superconducting aluminum, Phys. Lett. A, 57:97.

Klapwijk, T. M., Sepers, M., and Mooij, J. E., 1977, Regimes in the behavior of superconducting microbridges, J. Low Temp. Phys., 27: 801.

Klapwijk, T. M., Blonder, G. E., and Tinkham, M., private commun.

Kramer, L., and Baratoff, A., 1977, Lossless and dissipative current-carrying states in quasi-one-dimensional superconductors, Phys. Rev. Lett. 38: 518.

Kramer, L., and Watts-Tobin, R. J., 1978, Theory of dissipative current-carrying states in superconducting filaments, Phys. Rev. Lett. 40: 1041.

Kulik, I. O., Omel'yanchuk, A. N., and Kel'man, E. A., 1977, Current-controlled superconductive bridge contacts, Sov. J. Low Temp. Phys., 3: 537.

Laibowitz, R. B., Broers, A. N., Yeh, J. T. C., and Viggiano, J. M., 1979, Josephson effect in Nb nanobridges, Appl. Phys. Lett., 35: 891.

Langer, J. S. and Ambegaokar, V., 1967, Intrinsic resistive transition in narrow superconducting channels, Phys. Rev., 164: 498.

Likharev, K. K., and Yakobson, L. A., 1975, Dynamical properties of superconducting filaments of finite length, Sov. Phys. JETP, 41: 570.

Little, W. A., 1959, The transport of heat between dissimilar solids at low temperatures, Can. J. Phys., 37: 334.

London, F., 1937, "Une Conception Nouvelle de la Supraconductibilite,", Hermann & Cie, Paris.

London, F., 1950, "Superfluids", Wiley, New York.

Meyer, J. D., 1973, Spannungsstufen in den U(T)-Übergangskurven and U(I)-kennlinien stromtragender Zinn-Wisker, Appl. Phys., 2: 303.

Meyer, J. D. and Tidecks, R., 1976, The influence of the mean free path on the current induced superconducting/normal conducting transition of tin whiskers with indium impurities, Solid State Commun., 18: 305.

Meyer, J. D. and Tidecks, R., 1977b, Mutual influence of voltage steps in the I-V characteristics of current carrying superconducting tin whiskers, Solid State Commun., 24:643

Meyer, J. D., and Tidecks, R., 1977a, Investigation of the current induced phase transition of superconducting tin whiskers with several potential probes, Solid State Commun., 24: 639.

Meyer, J. and von Minnigerode, G., 1972, Instabilities in the transition curve of current carrying one-dimensional superconductors, Phys. Lett. A., 38A: 529.

Nad', F. Ya and Polyanskiy, O. Yu., 1973, Hysteresis in narrow superconducting bridges, Radio Eng. and Electron Phys., 18: 1784.

Newhouse, V. L., 1964, "Applied Superconductivity", Wiley, New York. (Sec. 17).

Notarys, H. A. and Mercereau, J. E., 1971, Dynamics of small superconductors, Physics, 55: 424.

Notarys, H. A. and Mercereau, J. E., 1973, Proximity effect bridges and superconducting microcircuitry, J. Appl. Phys. 44: 1821.

Palmer, D. W. and Mercereau, J. E., 1975, Coherent effects in series arrays of proximity effect superconducting bridges, IEEE Trans. Magnetics. MAG-11: 667

Palmer, D. W. ánd Mercereau, J. E., 1974, Coherent effects in large arrays of superconducting bridges, Appl. Phys. Lett., 25: 467.

Palmer, D. W. and Mercereau, J. E., 1977, Mode locking in arrays of superconducting weak links, Phys. Lett. A., 61: 135.

Pippard, A. B., Shepherd, J. G., and Tindell, D. A., 1971, Resistance of superconducting-normal interfaces, Proc. Roy. Soc. Lond. A., 324: 17.

Rachford, F. J., Huang, C. Y., Wolf, S. A., and Nisenoff, M., 1975, Observation of steps in the microwave impedance of superconducting microbridges, Solid State Commun., 17: 1493.

Rieger, T. J., Scalapino, D. J., and Mercereau, J. E., 1971, Charge conservation and chemical potentials in time-dependent Ginzburg Landau theory, Phys. Rev. Lett., 27: 1787.

Rieger, T. J., Scalapino, D. J., and Mercereau, J. E., 1972, Time-dependent superconductivity and quantum dissipation, Phys. Rev. B., 6: 1734.

Schön, G. and Ambegaokar, V., 1979, Collective modes and nonequilibrium effects in current-carrying superconductors, Phys. Rev. B., 19: 3515.

Shklovskii, V. A., 1975, Heating of the electrons as the cause of hysteresis of the critical current in the resistive state of superconductors, Sov. Phys. Solid State, 17: 2040.

Skocpol, W. J., Beasley, M. R., and Tinkham, M., 1974a, The electrical behavior of thin-film microbridges: self-heating and superconducting quantum processes, Rev. Phys. Appl., 9: 19.

Skocpol, W. J.,Beasley, M. R., and Tinkham, M., 1974b, Phase-slip centers and nonequilibrium processes in superconducting tin microbridges, J. Low Temp. Phys. 16: 145.

Skocpol, W. J., Beasley, M. R., and Tinkham, M., 1974c, Self-heating hotspots in superconducting thin-film microbridges, J. Appl. Phys., 45: 4054.

Skocpol, W. J., 1976, Critical currents of superconducting microbridges, Phys. Rev. B., 14: 1045.

Skocpol, W. J., Kadin, A. M., and Tinkham, M., 1978, Measurement of potential differences in nonequilibrium superconductors, J. Physique, 39: C6-1421.

Tidecks, R. and Meyer, J. D., 1979, Current-induced resistive state of superconducting tin whiskers with indium impurities, Z. Phys. B., 32: 363.

Tidecks, R. and Slama, G., 1980, Breakdown of superconductivity in current-carrying indium whiskers, preprint.

Tidecks, R., and von Minnigerode, G., 1979, The influence of high-frequency radiation on the U-I characteristics of superconducting tin whiskers, Phys. Stat. Sol. (a) 52: 421.

Tinkham, M., 1979, The interaction of phase-slip centers in superconducting filaments, J. Low Temp. Phys., 35: 147.

Volkov, A. F., 1971, Weak coupling produced in superconductors by irradiation, Sov. Phys. JETP, 33: 811.

Volkov, A. F. and Kogan, Sh. M., 1974, Temperature domains in superconductor films, JETP Lett. 19: 4.

Volotskaya, V. G., Musienko, L. E., Dmitrenko, I, M., and Kalekin, Yu. V., 1976, Critical current hysteresis in thin superconducting films, Sov. J. Low Temp. Phys., 2: 248.

Volotskaya, V. G., Bogdzevich, A., Musienko, L. E., and Kalekin, Yu. V., 1978, Critical heat flux in He-II for narrow films, Cryogenics, 18: 557.

Warburton, R. J. and Webb, W. W., 1968, Voltage steps in the resistive transition of superconducting microcrystals, Bull. Am. Phys. Soc., 13: 379.

Webb, W. W. and Warburton, R. J., 1968, Intrinsic quantum fluctuations in uniform filamentary superconductors, Phys. Rev. Lett. 20: 461.

Wong, T.-W., Yeh, J. T. C., and Langenberg, D. N., 1976, Quasi-particle-injection-induced superconducting weak links, Phys. Rev. Lett., 37: 150.

Wong, T.-W., Yeh, J. T. C., and Langenberg, D. N., 1977, Controllable superconducting weak links, IEEE Trans. Magnetics, MAG-13: 743.

Wyatt, A. F. G., 1969, Weakly coupled superconductors in: "Tunneling Phenomena in Solids", E. Burstein and S. Lundquist, eds., Plenum, New York.

Yamasaki, S. and Aomine, T., 1979, Self-heating effects in long superconducting thin films over a wide temperature range, Jap. J. Appl. Phys., 18: 667.

Yeh, J. T. C. and Langenberg, D. N., 1978, Light-induced superconducting weak links, Appl. Phys. Lett., 32: 191.

Yu, M. L., and Mercereau, J. E., 1972, Electric potentials near a superconducting-normal boundary, Phys. Rev. Lett., 28: 1117.

Yu, M. L., and Mercereau, J. E., 1975, Nonequilibrium quasiparticle current at superconducting boundaries, Phys. Rev. B., 12: 4909.

Yu, M. L. and Mercereau, J. E., 1976, rf-induced potential inside a superconducting weak link, Phys. Rev. Lett., 37: 1148.

SHORT RANGE INTERACTION BETWEEN TWO SUPERCONDUCTING WEAK LINKS

P. E. Lindelof and J. Bindslev Hansen

Physics Laboratory I, H. C. Ørsted Institute
University of Copehagen
DK-2100 Copenhagen Ø, Denmark

INTRODUCTION

The study of interacting weak links has been stimulated by the interest in constructing coupled coherent arrays of Josephson elements for technical applications and by the desire to achieve a better understanding of the fundamental static and dynamic processes taking place in and around weak links. It should be noted that these phenomena are often most sensitively monitored by using another weak link or a tunnel junction (Giaever, 1965) as a detector.

In this chapter we shall describe short range static and dynamic interactions between weak links that are placed close together in the same superconducting medium, i.e., the spacing between the weak links is of the order of or smaller than at least one of the three characteristic lengths in the superconductor, ξ, λ and Λ_Q*. We shall not have space here for going into the related topic of long range electrodynamic interactions between two weak links or to review the present state of the art of constructing large coherent networks of Josephson junctions. These topics were treated in the summer school lectures.

Weak links

The properties of all weak links are basicly characterized by the two-fluid model in combination with the Josephson effect; this is the well known resistively shunted junction (RSJ) model. The quantum phase difference ϕ of the superconducting order parameter across the weak link determines the supercurrent:

$$I_s = I_c \sin \phi \quad , \tag{1}$$

and the voltage:

$$V = \frac{\hbar}{2e} \frac{d\phi}{dt} \quad . \tag{2}$$

The remaining normal electrons shunt the supercurrent by a resistive current

$$I_n = V/R \tag{3}$$

where R is the quasiparticle resistance of the junction. For the different types of weak links, R must be determined individually (see Waldram (1976) and Likharev (1979) for reviews of the Josephson effect).

The ac Josephson effect

If a steady voltage is maintained across a Josephson device, the supercurrent will oscillate at a frequency $2eV/\hbar$. This ac Josephson effect is easily seen by examining Eqns. 1 and 2. However, the complete solution of Eqns. 1, 2 and 3 depends on whether the current or the voltage is the controlled parameter. The plate capacitor-like shape of a tunnel junction (Fig. 1) has a relatively large capacitance in parallel with the RSJ model, so its external impedance is low over most of its operating frequency range. In this case the current oscillates sinusoidally. However, due to this low impedance capacitive shunt, the high frequency oscillation is not easily matched to external loads.

The other structures in Fig. 1 typically have intrinsically lower impedances in comparison to the film inductance and external impedances. The current is the controlled parameter in these cases. If the total current $I = I_s + I_n$ is controlled, we have to solve a first order differential equation. The result is a non-sinusoidal oscillation of the ac voltage for small dc voltages ($V_{dc} < I_c R$) consisting roughly of pulses occurring at the Josephson frequency $2eV_{dc}/h$, with a height $2RI_c$ and width $\hbar/2eRI_c$. This oscillation, shown in Fig. 2, generates electromagnetic radiation which can be detected. Normally the detector is sensitive only in a small bandwidth around the detector frequency ν_D and the voltage of the weak link is varied. Since the non-sinusoidal current contains a large number of higher harmonics of the Josephson frequency, such a measurement will result in a signal within the detector bandwidth every time $2neV_{dc}/h = \nu_D$ i.e., at a number of harmonics.

A characteristic property of weak links is their ability to pull their own frequency toward an applied signal frequency ω and to phase-lock with it. This injection locking happens for certain ranges

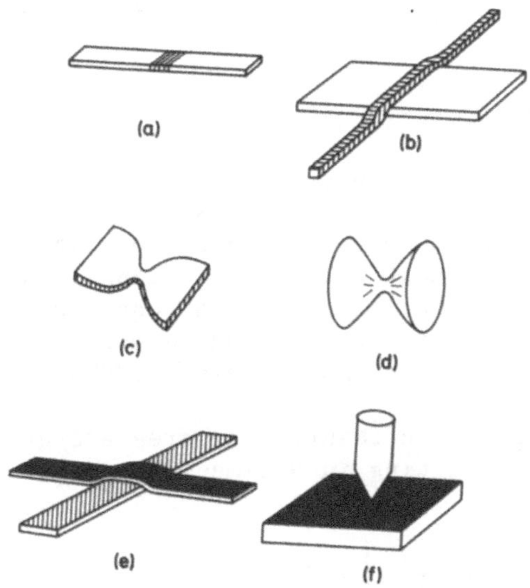

Fig. 1. Different types of Josephson weak links: (a) 1-dim. phase-slip center; (b) proximity effect bridge; (c) 2-dim. microbridge; (d) 3-dim. microbridge; (e) tunnel junction and (f) point contact.

of the applied dc current and discrete values of the voltage $V_{dc} = n\hbar\omega/2e$. These steps can be found as the solution to the equation:

$$I_{\omega} \sin\omega t + I = I_c \sin\phi + \frac{\hbar}{2eR}\frac{d\phi}{dt} \quad . \tag{4}$$

For $\hbar\omega/2eRI_c >> 1$ the locking range is $2I_c \cdot J_n\ (2eRI_{\omega}/\hbar\omega)$ where J_n

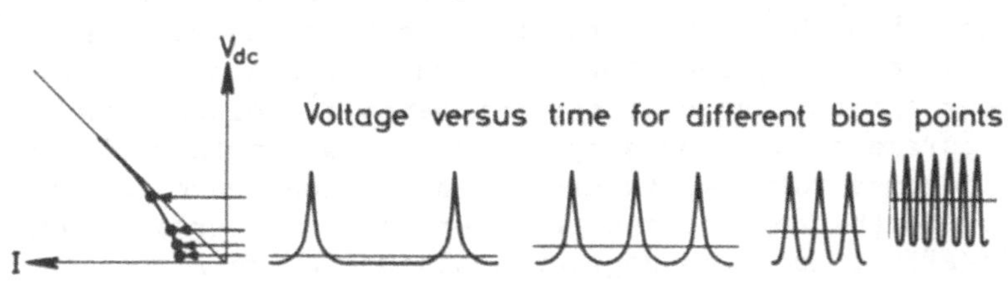

Fig. 2. The current controlled RSJ model. Results of an analog computer simulation showing the highly non-sinusoidal voltage waveform at a number of low voltage bias points on the hyperbolic dc I-V characteristic.

is the Bessel function of n'th order and n is the step number.

Characteristic lengths

While the tunnelling theory very accurately describes the properties of Josephson tunnel junctions, the theory for other types of weak links is not so complete. Predictions about the properties of the other weak links must rely on modelling. In particular, the theory for superconductors out of thermal equilibrium (Artemenko and Volkov, 1979) must be added to the usual Ginzburg-Landau theory (Lindelof, 1978; Bindslev Hansen and Lindelof, 1980a; Tinkham, Chap. 8).

In the Ginzburg-Landau theory, the free energy in the superconductor is minimized resulting in a time-independent complex equation for the order parameter ψ:

$$\xi^2(\vec{\nabla}-\vec{A}\ 2ie/\hbar)^2\ \psi + \left(1 - \left|\frac{\psi}{\psi_o}\right|^2\right)\psi = 0 \quad , \tag{5}$$

where ψ_o is the amplitude of the order parameter when the superconductor is unperturbed and ξ is the temperature dependent coherence length which is the characteristic length scale for variations of the order parameter. The quantum expression for the current density

$$\vec{J}_s = \frac{e\hbar}{2im}(\psi^*\vec{\nabla}\psi - \psi\vec{\nabla}\psi^*) - \frac{4e^2}{m}\psi^*\psi\,\vec{A} \quad , \tag{6}$$

along with Maxwell's equations, determines the characteristic length λ for the penetration of magnetic fields and currents into the superconductor.

For a one-dimensional phase-slip center (Chap. 18), it is particularly simple to discuss the various characteristic lengths as illustrated in Fig. 3. According to the Ginzburg-Landau theory a supercurrent through a long strip will depress the order parameter. This suppression will, even for very small irregularities in the strip, occur spontaneously at one (or several) points along the strip. The order parameter will be suppressed over a length scale which naturally becomes the temperature dependent coherence length. The penetration depth confines the supercurrent to a surface layer; however, this effect is often neglected since the cross section of the strip is usually smaller than λ^2.

If the critical current of the strip is exceeded, a voltage appears across the phase-slip center and both the amplitude and the

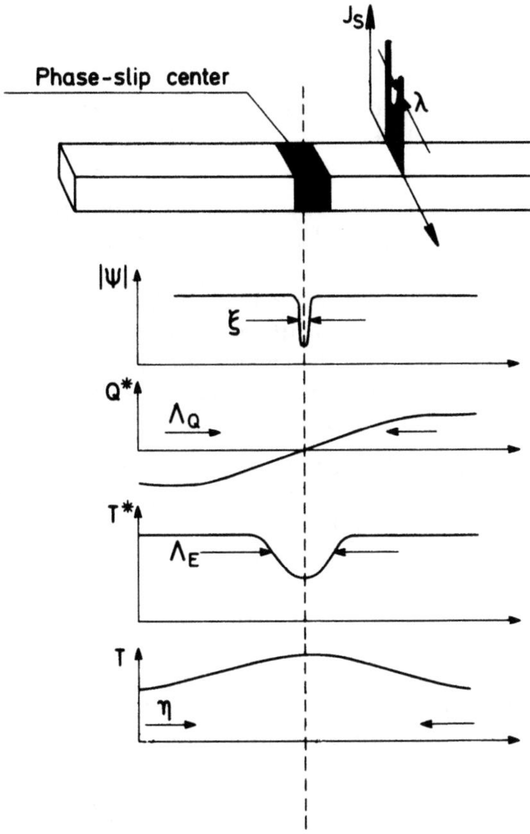

Fig. 3. Spatial variation of the relevant parameters around a one-dimensional phase slip center (PSC) illustrating the important characteristic lengths. From the top: variation of the supercurrent density, J_s, across the strip (λ); order parameter, ψ, variation (ξ); quasiparticle charge imbalance, Q^*, variation (Λ_{Q^*}); effective temperature, T^*, of the quasiparticle distribution (Λ_E); and the equilibrium temperature, T, of the lattice (η).

phase of the order parameter will oscillate with the Josephson frequency ω_J. As long as $\omega_J < \hbar/\Delta$ this will still take place over a length scale of the coherence length. The excitations will be brought out of equilibrium by the oscillating order parameter. However, since a redistribution of excitations in energy typically takes a characteristic time of the order of the inelastic (energy) relaxation time τ_E, the length scale for variation of properties pertinent to the excitations will be much longer than the coherence length. The voltage change in the phase-slip center means that the excitations must change their electrochemical potential μ_n which also means that the effective charge of the excitations Q^* changes. This can only happen over a characteristic charge imbalance diffusion length $\Lambda_{Q^*} = [v_F^2 \tau_E \tau_{tr} \Delta(0)/\Delta(T)]^{1/2}$. The quasiparticle resistance of a phase slip center is determined by this length.

The excitations can be out of equilibrium due to charge build-up, but they can also be redistributed from their thermal energy distribution. Such an energy imbalance decays with the inelastic relaxation time and over a characteristic length scale $\Lambda_E = (v_F^2 \tau_E \tau_{tr})^{1/2}$. The non-thermal energy distribution of the excitations does not necessarily mean that the energy gap is decreased (as if there is an increased effective temperature T^* in the equation for the energy gap). Redistribution of excitations can also lead to an increased energy gap, or a reduced T^*, as indicated in Fig. 3.

In a completely isolated superconducting strip, the Joule heat generated at the phase slip center will only escape through the ends, and the equilibrium temperature will therefore vary linearly from the phase slip center toward the two ends. However, if the phase slip center is in contact with a substrate (or the helium) then the equilibrium temperature will vary much faster from the value at the phase slip center toward the bath. If K is the thermal conductivity of the strip and if α is the heat transfer coefficient across the interface, then the equilibrium temperature will decay toward the bath temperature over a characteristic thermal healing length $\eta = (Kt/\alpha)^{1/2}$, where t is the film thickness. For an indium or tin film on a glass substrate and with a thickness of 100 nm, one finds roughly $\eta \approx 5\ \mu m$. If the dissipation in the phase slip center raises the temperature above the transition temperature a normal hot spot is created and, for instance, the ac Josephson effect is extinguished.

In proximity effect bridges (1-dim.) the characteristic lengths will correspond to those for the phase slip center. The relaxation around a two-dimensional bridge with smallest dimension ℓ will be somewhat faster, however, since a logarithmic factor $\ln(\ell/r + 1)$ will be multiplied on all the exponential functions of the distance r from the bridge middle. A three dimensional bridge will have the recovery speeded up by a factor ℓ/r, and, for instance, Λ_{Q^*} will not enter the expression for the electrical resistance of the three dimensional microbridge.

SHORT RANGE INTERACTION BETWEEN TWO WEAK LINKS

We shall deal with two closely spaced weak links located in the same superconducting medium and interacting through order parameter variations, and quasiparticle diffusion currents. The nonequilibrium situation in the immediate vicinity of a weak link in the non-stationary state must be taken into consideration, and we can not expect the simple RSJ model to give a fully adequate description of such proximity coupled weak links.

It is by no means trivial to place two microbridges within a distance of only a few μm. Different methods have a number of conflicting advantages. The use of electron beam lithography allows small two-dimensional bridges to be placed at almost any distance (Lukens et al., 1979; Jillie et al., 1977a; see Fig. 4). However, three-dimensional bridges (and sometimes variable thickness bridges) have important advantages, since they localize the order parameter better and are also more efficiently cooled. Another fabrication method is based on the cross scratch method, where a groove in the substrate will leave a bridge when a perpendicular scratch in a superconducting film is made. Diamond tools smash the film at a large distance from the bridges, because the edge of the tool has a wide angle (Octavio and Skocpol, 1979). Razor blades on the other hand are so sharp that even 1 μm thick films will be affected only a short distance from the bridges as shown in Fig 5. We may point out here that there is an argument which seems to favor two-dimensional bridges in spite of their inferior properties. The fact that the

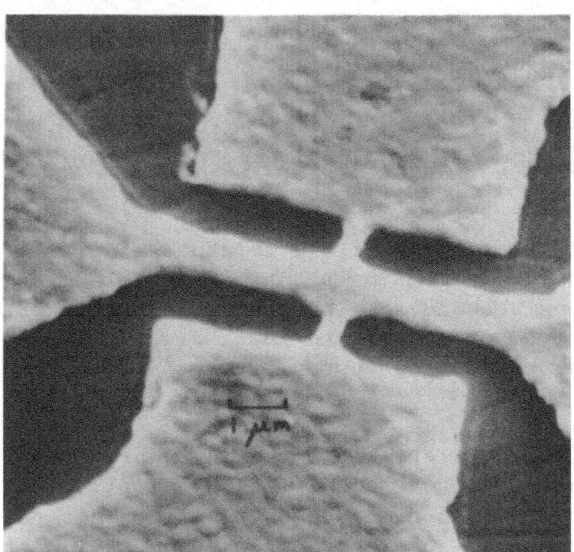

Fig. 4. SEM micrograph of one of the samples used by Jillie (1976). The light areas are a 100 nm indium film.

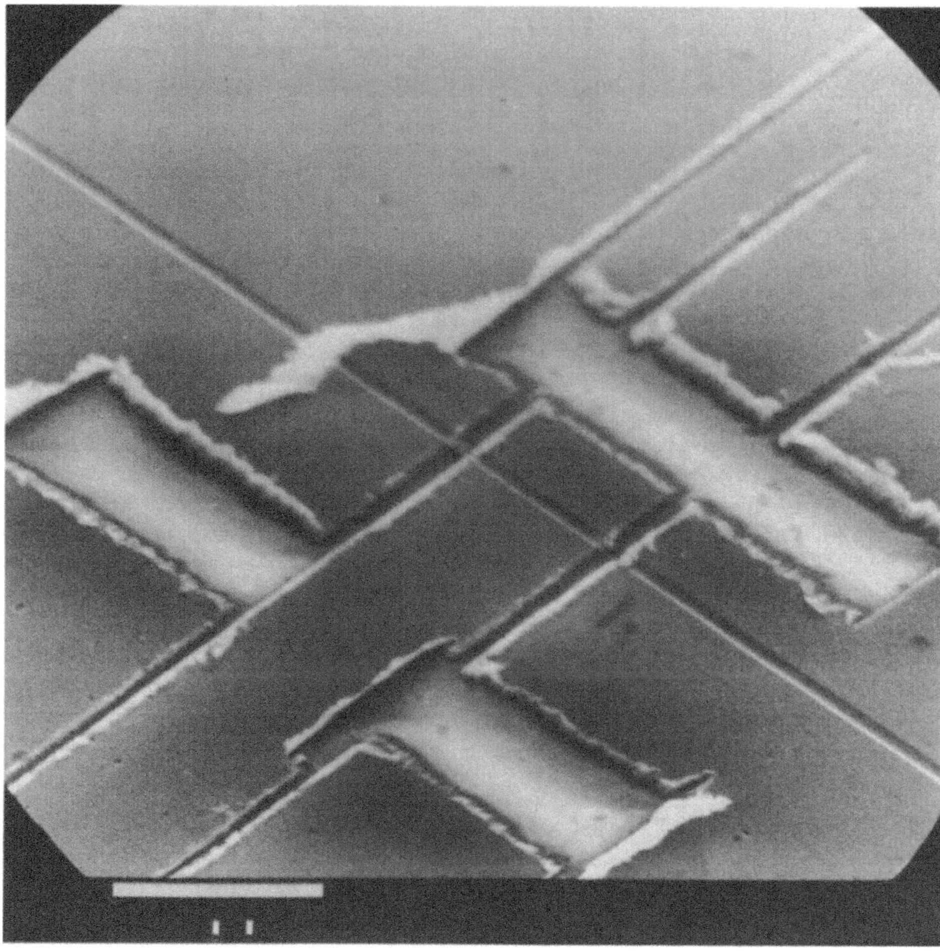

Fig. 5. SEM picture of two variable thickness indium bridges cut in a 1 μm thick film using a sharp razor blade by the cross-scratching technique. The groove in the substrate is seen running from upper left to lower right. The distance between the bridges is approximately 8 μm in this sample.

two-dimensional geometry lets the temperature drop only slowly from the bridge out into the background (Tinkham et al., 1977) also means that the quasiparticle potential drops slowly. This means that two-dimensional bridges will interact over a somewhat longer distance than three-dimensional bridges. In particular if the heat transfer coefficient across the interface between the superconducting film and the substrate can be improved, two-dimensional bridges should be superior for short range interaction experiments.

Order Parameter Interaction

When the distance between two weak links is of the order of the coherence length, ξ, one link can feel the order parameter variations taking place in the center of the adjacent junction. In equilibrium, i.e., in the stationary state where the order parameter only depends on position and not on time, this interaction is caused by the super-current-induced depression of the order parameter which reduces the critical current. Close to T_c this dc interaction can be fully described by the Ginzburg-Landau equations with appropriate boundary conditions. These non-linear differential equations were solved numerically by Blackburn et al. (1972 and 1975); by Howard et al. (1975) and by Way et al. (1977) and also in a perturbation calculation by Deminova et al. (1979). The results agree qualitatively with experimental observations on pairs of thin-film microbridges by Jillie et al. (1975), Jillie (1976) and Lindelof et al. (1977).

The change in the critical current of one bridge, ΔI_{c1}, was measured as a function of the current flowing through the other bridge. Jillie's results are shown in Fig. 6 for the sample geometry of Fig. 4. The depression of the order parameter induced by the supercurrent should be independent of the relative current direction, i.e., the depression should be symmetrical under a reversal of the direction of the current through one of the bridges. This symmetry is, however, masked by a dc effect which is asymmetrical, namely the effect of the current distribution in the region between the bridges which depends strongly on whether the currents flow in the same

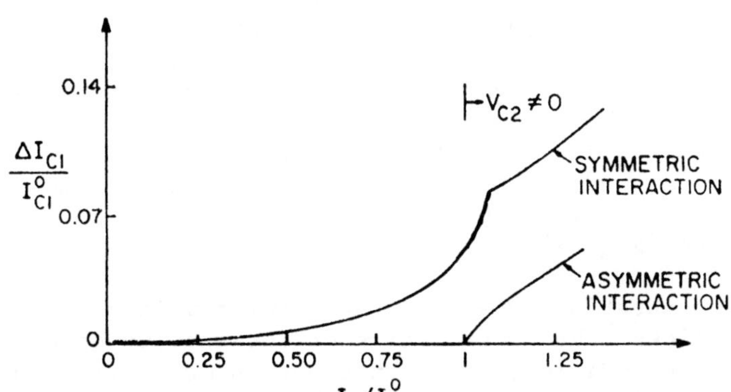

Fig. 6. The change of the critical current of bridge 1 of Fig. 4, normalized to its unperturbed value, plotted against the current through bridge 2 (Jillie, 1976). For $I_2 < I^o_{c2}$ both bridges are in the stationary state (only the series aiding case is plotted here, see Fig. 10 for both); for $I_2 > I^o_{c2}$ ($V_{c2} \neq 0$) the interaction measured by ΔI_{c2} is divided up into the parts symmetrical and asymmetrical with respect to dc bias.

direction through the bridges (series aiding) or in opposite directions (series opposed). Both experimentally (Jillie 1976) and theoretically (Way et al., 1977), the smallest depression of the order parameter is found in the series opposed configuration where the supercurrents tend to cancel each other in the inter-bridge region.

We will follow Jillie (1976) and give a simple estimate of the magnitude of the symmetrical dc interaction. We want to know how much I_{c1} is depressed when $I_2 = I_{c2}$. From one-dimensional Ginzburg-Landau theory it is known that at the critical current the order parameter in the middle of a phase-slip center is only 82% of its unperturbed value ($\psi^2 = 2\psi_o^2/3$, however for a small microbridge $\psi^2 = \psi_o^2/2$). Hence at $I_2 = I_{c2}$ we have $\psi(o) = 82\%\ \psi_o$ as in Fig. 7. The depression falls off exponentially when we go away from bridge 2. Therefore at the middle of bridge 1, we have $-\Delta\psi(d) = 0.18\ \psi_o \exp(-d/\xi)$. Now, since $I_c \propto \psi^2$, it follows that

$$I_{c1} = [1-0.18\ \exp(-d/\xi)]^2 I^o_{c1} \quad . \tag{7}$$

For one of Jillie's samples with two indium thin-film microbridges 1.6 μm apart, and at a temperature of 0.99 T_c, for which $\xi(T) = 1.2$ μm, the result becomes $I_{c1} = 90\%\ I^o_{c1}$ for $I_2 = I_{c2}$. The experimental value is 95% I^o_{c1}.

Other aspects of this static interaction are presented in Fig. 8 (Jillie, 1976 and Jillie et al., 1975) showing: 1) that the relative depression of the critical current, $\Delta I_{c1}/I^o_{c1}$, measured at $I_2 = I_{c2}$ increases as the temperature is raised toward T_c due to the increase of $\xi(T)$; and 2) that close to T_c the critical currents of the series connected pair tend to lock to a common value and at the same time the transition to the resistive non-stationary state becomes sharper. These two phenomena are only observed for T_c-T<20 mK. Both of these observations are readily understood on the basis of a mutually reinforced depression of ψ which follows naturally from the simple model presented above (Fig. 7).

In the non-stationary state, when finite voltages develop across the two weak links, the order parameter interaction becomes much more complex, and, in addition, other interaction mechanisms come into play (charge imbalance waves, lattice heating and/or an excess of "hot" quasiparticles). The order parameter in the middle of the bridges goes to zero periodically at the Josephson frequency. Therefore its time average, $\langle\psi\rangle$, in the center of a bridge will be much lower and the corresponding depression of $\langle\psi\rangle$ caused in the other bridge is greater than in the stationary state. Consequently, $|\Delta I_{c1}|$ will increase. This accounts for the rise of $|\Delta I_{c1}|$ when I_2 exceeds I_{c2} as shown in Fig. 6.

Naturally there also exists a high-frequency interaction between

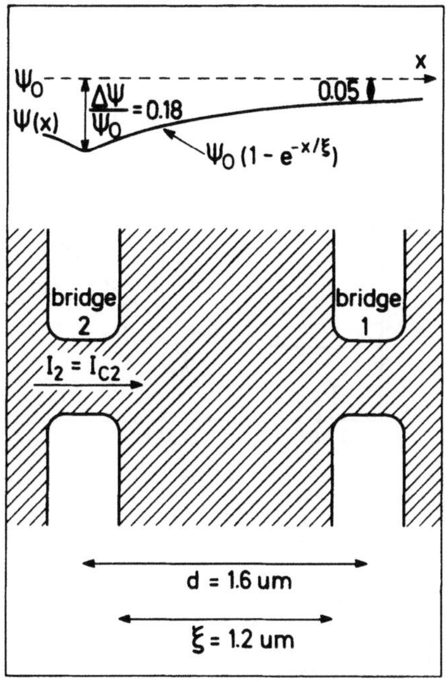

Fig. 7. This diagram illustrates the supercurrent-induced depression of the order parameter with both microbridges in the stationary state ($I_2 = I_{c2}$, $I_1 = 0$). The maximum supercurrent that flows through bridge 2 causes a depression of the order parameter, ψ, to 82% of its unperturbed value in the middle of the bridge. ψ then recovers exponentially as shown in the upper part of the figure reaching 95% of its full value at the middle of bridge 1 (Jillie, 1976).

two weak links less than ξ apart and both biased at finite voltages. The order parameter oscillation in one bridge modulates ψ in the other bridge directly. If the two voltages are nearly equal (or one is close to a simple fraction of the other) the ac Josephson oscillations in the two bridges will tend to phase-lock. This dynamic interaction was treated in detail by Deminova et al., (1979) who extended their perturbation calculation to the non-stationary state with current-control. In their solution to the Ginzburg-Landau equation the interaction took the form of a first-order correction to the zero-order approximation which was the well-known solution to Laplace's equation $\nabla^2\psi = 0$ (see Likharev, 1979) with the usual boundary condition $d\psi/dr_{\perp} = 0$. They sought solutions describing the phase-locked system with the two weak links running at a single frequency and obtained curves representing the boundaries of the phase-locked regions in currents and voltages for both current configurations and for various values of the coupling strength and the ratio between the two critical currents. They predicted that for strongly coupled

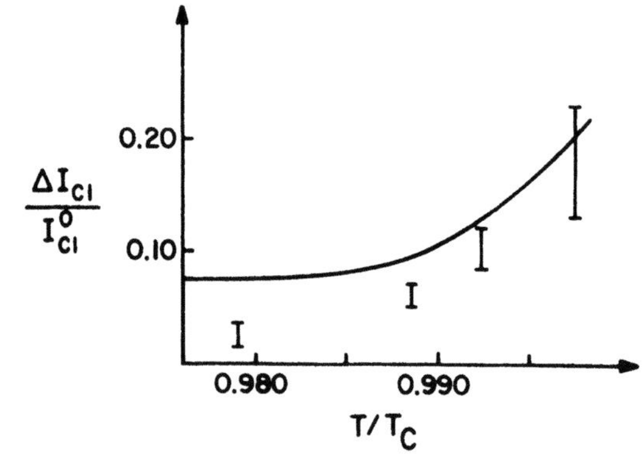

(a)

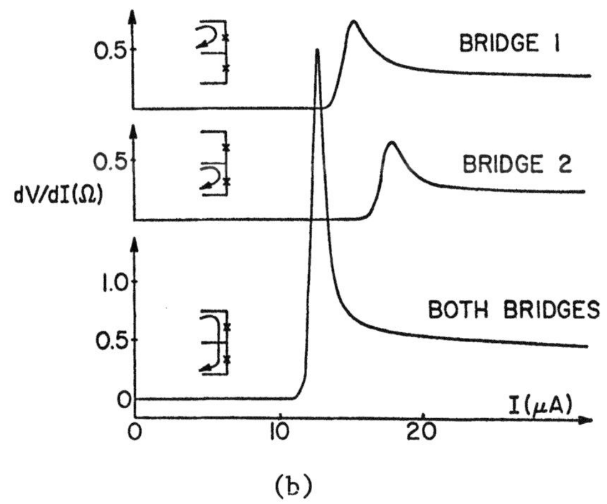

(b)

Fig. 8 (a) Temperature dependence of the stationary order parameter interaction (Jillie, 1976). The change in critical current is plotted as a function of temperature. ΔI_{c1} is taken at $I_2 = I^o_{c2}$, and the spread in ΔI_{c1} is due to the difference between ΔI_{c1} in the two current configurations. The solid curve is a theoretical estimate that follows from the simple picture given in Fig. 7.
(b) Critical current locking between two coupled planar microbridges biased in series ($d \approx 2\mu m$), Jillie, 1976. Note the reduction in the critical currents and the sharpening of the resistive transition when both bridges are biased as compared to one being biased alone.

planar thin-film microbridges (like Jillie's) the widest locking range, ΔI_{lock}, should be observed for the series opposed configuration. This agrees with observations which we shall discuss later. For pairs of point contacts or variable-thickness bridges (VTB), in which the coupling should be weaker, the opposite should be the case. So far, there are no reports on measurements that could test this prediction.

The upper cut-off frequency for this dynamic interaction will presumably be of the order of the gap frequency, $\nu_{2\Delta} = 2\Delta/\hbar$. Even close to T_c this limit is so high (e.g. $\nu_{2\Delta} \approx 90$ GHz in indium at 0.99 T_c) that very well-cooled weak links (VTB's or point contacts) would have to be used in order to reach it. In the samples of closely spaced planar microbridges studied so far, heating effects will destroy the dynamic direct order parameter interaction even at low voltages (frequencies). In order to distinguish between such an interaction and the quasiparticle charge imbalance interaction that we shall deal with subsequently, a very careful study of the frequency and spatial dependence of the coupling would be needed for a pair of well-cooled weak links. In this connection we note that the order parameter coupling is strongest precisely between planar weak links that do fulfill the rigid boundary conditions well, i.e., between structures without current concentrations (Likharev, 1979), and that these types of weak links are also the ones in which the heating problem is most severe (Tinkham et al., 1977, and Tinkham, Chap. 8).

Quasiparticle Interaction

We now turn to the other important type of proximity coupling between weak links, the quasiparticle charge imbalance coupling. We note that neither type of proximity coupling is contained within the RSJ model. The reason for this is that the RSJ model with current control does not allow any net ac current to be set up. The ac supercurrent and the ac normal current must cancel each other in the bridge region. But, since the quasiparticles diffuse out to a distance which at low frequencies is determined by Λ_Q^*, the quasiparticle charge imbalance diffusion length, the picture given by the RSJ model cannot be a correct description of the nonequilibrium situation around the weak link.

For two closely spaced weak links (now in the sense that $d \lesssim \Lambda_Q^*$, we may try to include the quasiparticle current injected from one weak link to the other in the RSJ model (Jillie et al., 1980). For the series opposed current configuration, we may write

$$I_1 + i_{q2} = I_{c1}\sin\phi_1 + V_1/R_1 \tag{8}$$

$$I_2 + i_{q1} = I_{c2}\sin\phi_2 + V_2/R_2 \quad , \tag{9}$$

where i_{q2} is the part of the quasiparticle current generated in bridge 2 that flows through bridge 1, and similarly for i_{q1}; also R_1 and R_2 are the normal state resistances of the two junctions. The total quasiparticle current through junction 2 is

$$I_{q2} = I_2 - I_{c2}\sin\phi_2 + i_{q1} \ , \tag{10}$$

of which a fraction α_2 flows through junction 1. Hence

$$i_{q2} = \alpha_2 I_{q2} = \alpha_2(I_2 - I_{c2}\sin\phi_2 + i_{q1}) \ , \tag{11}$$

and similarly for i_{q1}

$$i_{q1} \ \alpha_1 I_{q1} = \alpha_1(I_1 - I_{c1}\sin\phi_1 + i_{q2}) \ . \tag{12}$$

Combining these equations and assuming that the coupling parameters, α_1 and α_2, are small so we can neglect all terms in $\alpha_1\alpha_2$, we get

$$I_1 = I_{c1}\sin\phi_1 + V_1/R_1 - \alpha_2(I_2 - I_{c2}\sin\phi_2) \ , \tag{13}$$

$$I_2 = I_{c2}\sin\phi_2 + V_2/R_2 - \alpha_1(I_1 - I_{c1}\sin\phi_1) \ . \tag{14}$$

These two equations are identical to those describing a system of two weak links with a common resistive shunt. Therefore, to first order, the equations describing the quasiparticle coupling and resistive shunt coupling are the same. For larger values of α_1 and α_2, we should expect this analogy to break down since in that case there will be non-negligible feedback currents, which means that bridge 1 will be effected by the quasiparticle current it injects into bridge 2. In the non-stationary situation these feedback currents, which also contain information of the phases, could become important for the locking mechanism even for small α's.

A consequence of this model is that a finite voltage across bridge 2 will influence the critical current of bridge 1. Although this is a simple consequence of Eqns. 13 and 14, we shall present a qualitative argument as illustrated in Fig. 9. In Fig. 9, bridge 2 is biased at a voltage which drives a quasiparticle current $I_{qp2} = V_2/R_2$, of which a small fraction, $\alpha_2 I_{qp2}$, is injected through bridge 1. Bridge 2 is shown biased in the same direction as bridge 1 (series aiding) or in the opposite direction (series opposed). If bridge 1 carries only a supercurrent ($I_{qp1} = 0$), then the maximum zero voltage current, i.e., the critical current I_{c1}, will be the sum of the intrinsic critical current of bridge 1 and $\alpha_2 I_{qp2}$. Therefore I_{c1} will be $\alpha_2 I_{qp2}$ larger than its intrinsic value by $\alpha_2 I_{qp2}$ in the series aiding case and be smaller by the same amount in the series opposed case. The experimental results which demonstrate this (Jillie et al., 1977) are shown in Fig. 10. The inset

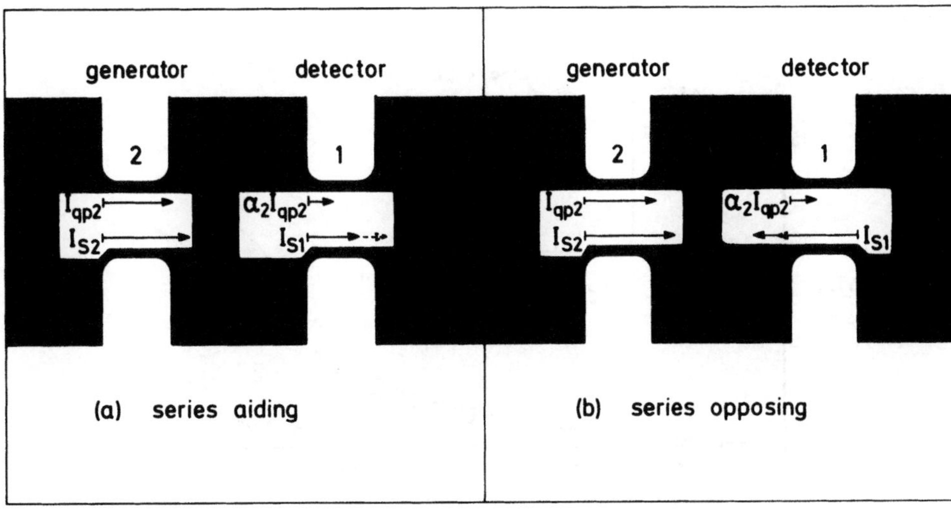

Fig. 9. Asymmetrical dc quasiparticle interaction between the closely spaced microbridges (after Jillie et al., 1977b).

shows the asymmetry (i.e., $\alpha_2 I_{qp2}$) as a function of I_{qp2}, so the slope of this plot is α_2. Jillie et al., (1977b) find that α_2 thus determined agrees very well with a quasiparticle current which decays over Λ_Q^*. They also find that α_2 vanishes for samples with a distance between their indium bridges which exceeds 4 μm, consistent with these findings.

A similar although less transparent stationary interaction is expected in the situation where both bridges are biased at a finite voltage ($I_{qp1} \neq 0$ and $I_{qp2} \neq 0$ in Fig. 9.) Again an asymmetric situation appears for the series aiding and series opposed cases, which helps the total current I_1 and I_2 in one case, and hampers them in the other. The series aiding and series opposed cases also give a quite different quasiparticle potential μ_n in the region between the bridges as illustrated in Fig. 11 (Lindelof and Bindslev Hansen, 1977). The pair potential μ_s has no anomalies in the region between the bridges until the coherence length becomes comparable to the bridge separation ($\xi \ll \Lambda_Q^*$).

Voltage Locking

One particularly interesting interaction between two bridges which are within a quasiparticle diffusion length of each other, occurs if the voltage bias of the bridges has the same absolute value. Figures 12 and 13 show how such an interaction influences the current-voltage characteristics of two microbridges for the series opposing bias and series aiding bias respectively (Jillie et al., 1980). For

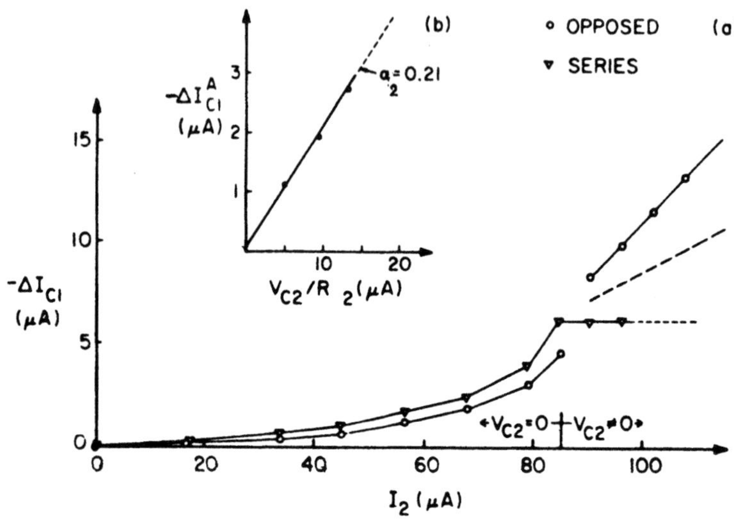

Fig. 10 Experimental results on two planar indium microbridges (d=2μm), showing the interaction sketched in Fig. 9 (Jillie et al., 1977b). (a) The change in the apparent critical current of bridge 1 for a range of fixed currents through bridge 2. The data include both directions of I_1 relative to I_2 (aiding and opposing). The voltage across bridge 2 when $I_1 = I_{c1}$ is denoted V_{c2}, which is zero for $I_2 < 85$ μA. The dashed line (long dash marks) is the symmetric or average component of ΔI_{c1} for the two directions of I_2. (b) inset: The antisymmetric component, ΔI^A_{c1}, of ΔI_{c1}, as a function of the quasiparticle part of I_2, i.e., $I_{qp2} = V_{c2}/R_2$ where $R_2 = 0.32\ \Omega$.

the series opposing case, the two voltages lock to each other. This is hardly surprising since a microbridge will try to lock to an external ac source (in this case the other bridge). This locking is also borne out by computer simulations based on Eqns. 13 and 14. The surprising observation of Jillie et al., (1977a) which is not borne out by their simulations is the difference between the series opposing and series aiding case. In the series aiding case, the two bridges hardly lock. This difference can presumably not be explained on the basis of an ac interaction alone, and we shall return to this problem. Another interesting observation of Jillie et al., (1977a) was that the dc locking was voltage dependent and went through several minima as a function of voltage. This is shown in Fig. 14 where the temperature dependence is also indicated.

A more direct way to demonstrate the locking of the ac Josephson oscillation in two bridges is by detecting the total radiated power, P. For two closely spaced bridges, this was investigated by Lindelof

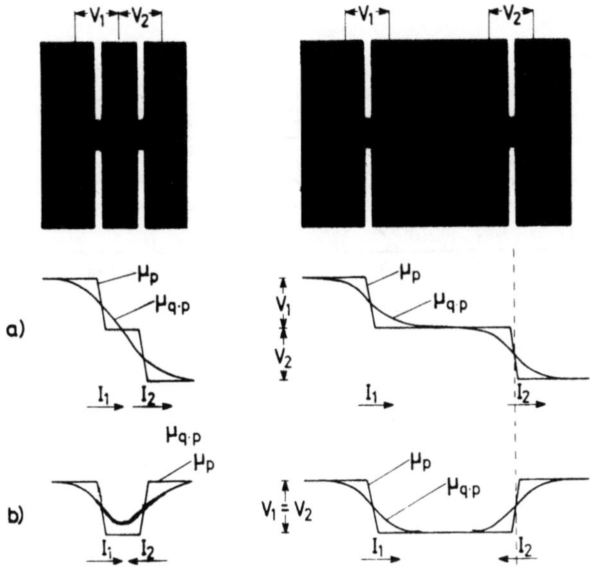

Fig. 11. Qualitative picture of the instantaneous variation of the pair potential μ_s and the quasiparticle potential $\mu_{qp} \equiv \mu_n$ in the region around two closely spaced thin-film micro-bridges (shown in the upper part of the figure). The variations are shown for two different inter-bridge spacings and for both current configurations, (a) series aiding and (b) series opposed.

and Bindslev Hansen (1977) and the result is shown in Fig. 15. They did not observe voltage asymmetry for the aiding and opposing situation. Another difference from the experiments of Jillie et al., (1979a) is the more 3-dimensional character of their bridges (variable thickness bridges). The degree of coherence C was measured by the degree to which amplitude (and not power) addition was observed. Assuming that the amplitude added in phase, the individually emitted powers P_1 and P_2 allowed the following ratio to be formed

$$C = (P - P_1 - P_2)/2 \sqrt{P_1 P_2} \quad . \tag{15}$$

This ratio is plotted versus temperature in Fig. 16 for the measurements of Fig. 15. Close to the critical temperature, where the charge imbalance diffusion length becomes long, C approaches one, corresponding to full coherence as expected. Observations were also made which showed coherent behavior of a pair of bridges where the signal from the bridges added out of phase and was a function of the detection frequency as well as temperature. This indicated a type of resonance for the coupling wave.

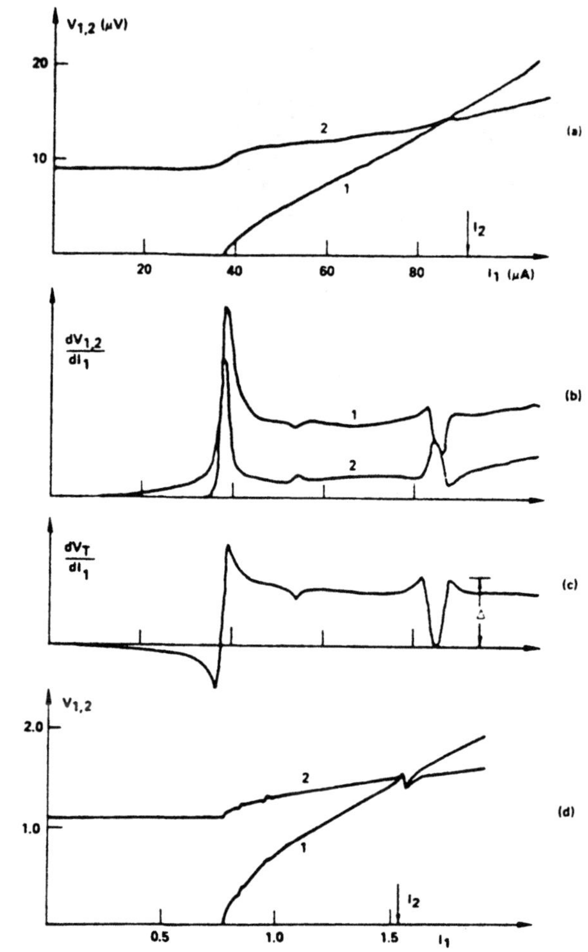

Fig. 12. Experimental results (a), (b), and (c) and numerical simulation (d) for a series opposed biased pair of indium microbridges 2 μm apart showing voltage locking. (a) I-V curves with I_2 held constant (91 μA) and I_1 swept; (b) the differential resistance of the two curves shown in (a); and (c) the total differential resistance $dV_T/dI_1 = dV_1/dI_1 - dV_2/dI_1$. During voltage locking $dV_T/dI_1 = 0$. In the locking region $\Delta(dV_T/dI_1)$ is a measure of the strength of the interaction; (d) Numerical simulation using the correct parameters for the coupled bridges ($\alpha_1 - \alpha_2 = 0.21$) and Eqns. 13 and 14. (Jillie et al., 1980).

Quasiparticle Voltage and Frequency Locking

The quasiparticle mediated voltage and frequency locking between two microbridges can have two distinctly different origins, as we shall now describe.

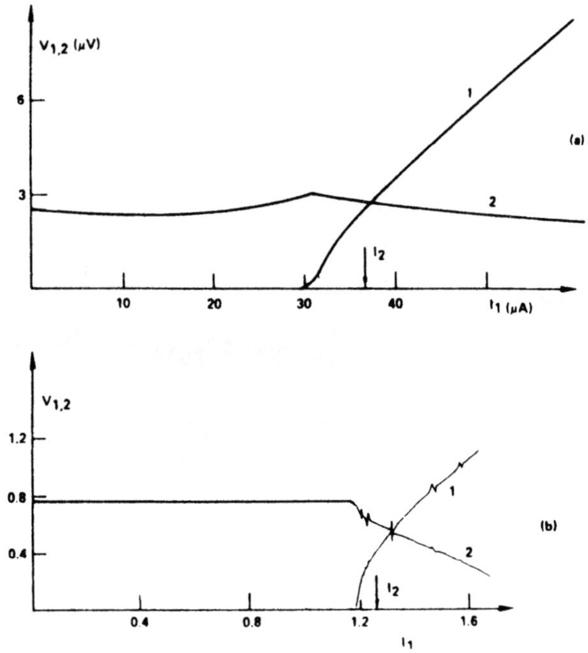

Fig. 13. Experimental results (a) and numerical simulation (b) for a series aiding biased pair of bridges, same sample as in Fig. 12. (a) I-V curves with I_2 held constant (37 μA) and I_1 swept; (b) numerical simulation, $\alpha_1 = \alpha_2 = 0.21$. Locking of the dc voltages is predicted but not observed. (From Jillie et al., 1980).

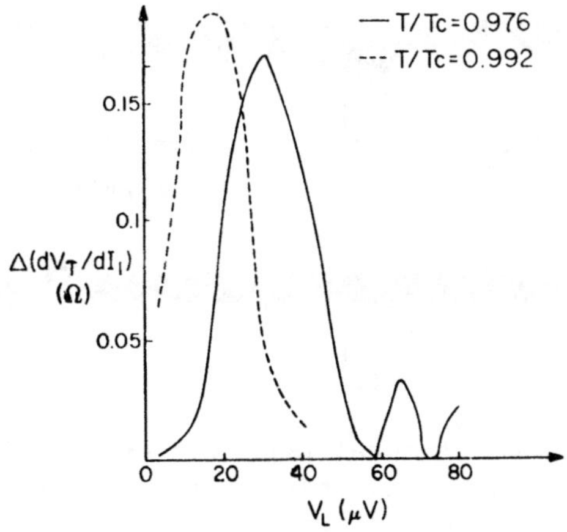

Fig. 14. Voltage and temperature dependence of voltage locking interaction strength, Δ, as defined in Fig. 12 for the series opposed case (Jillie et al., 1977a).

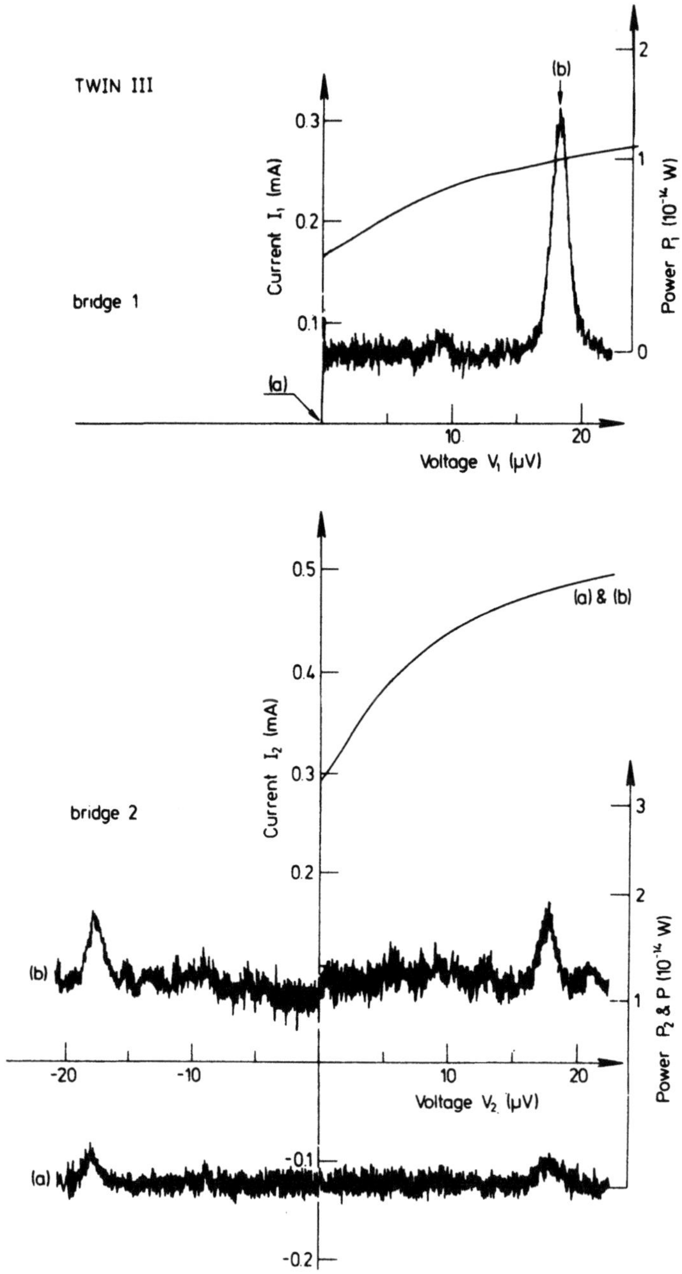

Fig. 15. Example of experimental observation of coherent in-phase radiation from two tin microbridges 12 μm apart, at $T/T_c = 0.966$. The interaction is symmetrical. Trace (a) is for bridge 2 alone ($I_1 = 0$), while trace (b) is taken with both bridges radiating simultaneously into the receiver at 8.7 GHz. (Lindelof and Bindslev, 1977).

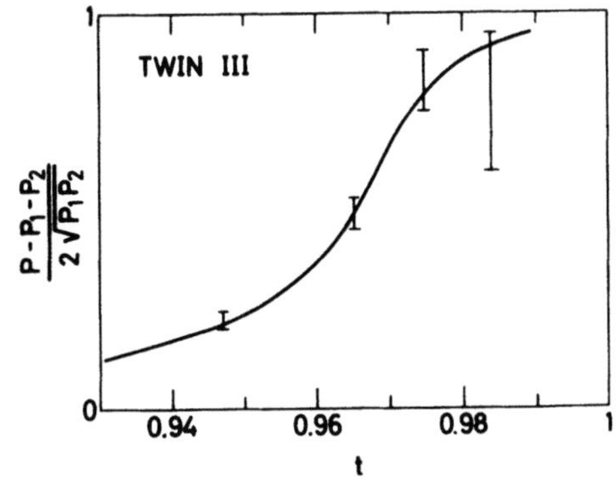

Fig. 16. Observed temperature dependence of the high frequency interaction in Fig. 15. The coherence ratio, C, as defined in the text, is plotted vs. the reduced temperature for the same pair of tin microbridges as in Fig. 15, both biased at the same voltage corresponding to the receiver frequency of 8.7 GHz (Lindelof and Bindslev Hansen, 1977).

ac injection locking. The quasiparticle potential decays over a length which at low frequencies is the charge imbalance diffusion length. Above a frequency of the order of the inverse inelastic relaxation time $\omega_E \sim 1/\tau_E$, this characteristic length becomes shorter due to a skin depth effect. However, at high frequencies a charge imbalance collective mode may also propagate through the superconductor (Artemenko and Volkov, 1978) and give rise to a phase of the oscillation which varies according to the same equation as a transmission line (Lindelof, 1978). This ac Josephson oscillation of one bridge will lock the other through this oscillating potential and give rise to mutually induced steps in their I-V characteristics and to coherent radiation. One bridge receives a reflection of its own signal back from the other bridge. Its phase depends on the frequency and the distance between the bridges while its amplitude gradually dies out over a quasiparticle diffusion length or skin depth depending on the frequency. Figure 17 illustrates the signal of one bridge received by itself after reflection in the other bridge. Constructive or destructive interference will be observed depending on the distance between the two bridges. The fact that the charge imbalance oscillation propagates in the superconductor allows an equivalent circuit, as shown in Fig. 18, to be set up for a series of two or more closely spaced microbridges (Kadin et al., 1980; Artemenko et al., 1979), and this is a matter of current investigation (see Chap. 18).

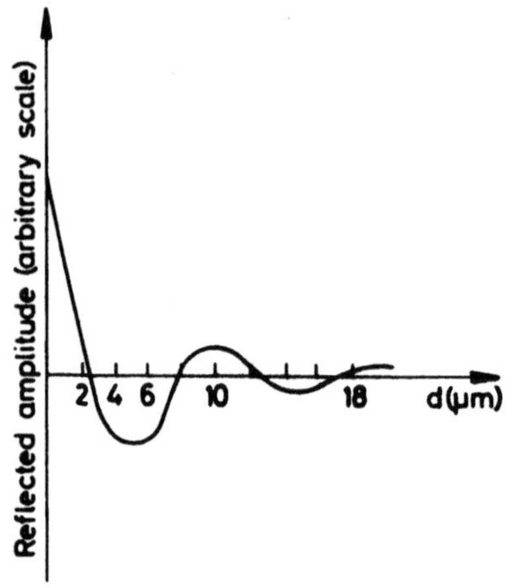

Fig. 17. Qualitative picture (for tin) of the reflected amplitude of the Josephson oscillation from bridge 1 via bridge 2 onto itself. Such a "Fabry-Perot resonance" is connected with a charge imbalance wave.

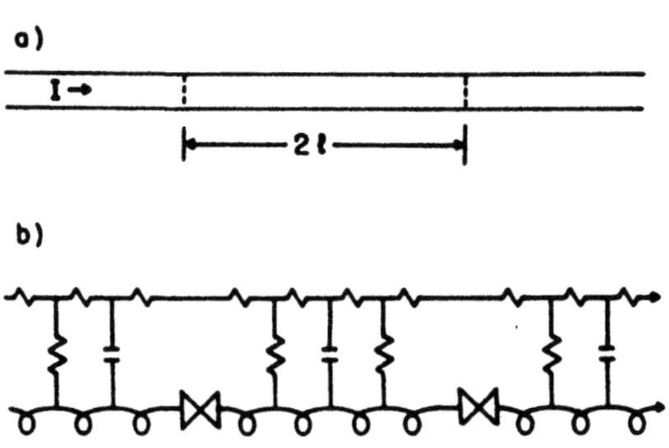

Fig. 18. (a) Two closely spaced phase-slip centers in series on a superconducting strip; (b) transmission line equivalent diagram of (a) (Kadin et al., 1980).

dc injection locking. The phenomenon of dc locking can also occur because the diffusion current can take advantae of states at the gap energy. It is based on the assumption that the energy distribution of the quasiparticles diffusing through the region between the bridges is not a smooth function but has a sharp maximum at the energy corresponding to the gap singularity in the superconductors beyond the bridges. Figure 19 illustrates this behavior. The voltage bias V_2 across the bridge shown to the right is constant. The three pictures display the quasiparticle distributions in the semiconductor picture for three different voltages V_1 across the left bridge. From top to bottom, this corresponds to the series aiding situation, $V_1 = 0$ and the series opposed situation. The small peaks in the distributions correspond to quasiparticle injection from the gap singularity through a bridge to another superconductor. Only the series opposed case will show a voltage locking. The locking comes about because the diffusion current from bridge 1 which is injected through bridge 2 (compare Fig. 9) will be particularly large when the singularity of the right superconductor is on the same level as the peak in the injected distribution of the middle superconductor. A similar phenomenon has been observed in experiments with two superimposed tunnel junctions (Kaplan et al., 1977 and Chap. 5).

Such a model can possibly explain the asymmetry in the locking as found by Jillie et al., (1977a). This model can also give rise to resonances. The quasiparticles in the injected peak can have a particularly long de Broglie wavelength $2\pi/q$ where q is given by:

$$h\, v_F\, q = [(eV+\Delta)^2 - \Delta^2)]^{1/2} \approx (2eV\, \Delta)^{1/2} \quad , \qquad (16)$$

(Lindelof and Hansen, 1977), where v_F is the Fermi velocity of the metal. This mode appears as a superposition of the excited states k_2 and k_4 in Fig. 20, situated at energy eV above the singularity. Equation 16 predicts a decreasing wavelength with increasing voltage and an increasing wavelength with increasing temperature. It is not clear how this can be understood in the light of experiments, and for indium, an estimate of the wavelength $2\pi/q$ at $V = 20\ \mu V$, $T/T_c = 0.976$ and $\Delta \approx 120\ \mu eV$ is $2\pi/q \approx 0.3$ mm (compare Fig. 14). Finally we might note, that when the voltages across the microbridges are multiples of Δ, a particularly strong interaction should be observed. Such a phenomenon is, we believe, the most likely explanation of the periodic variation of the voltage locking shown in Fig. 14.

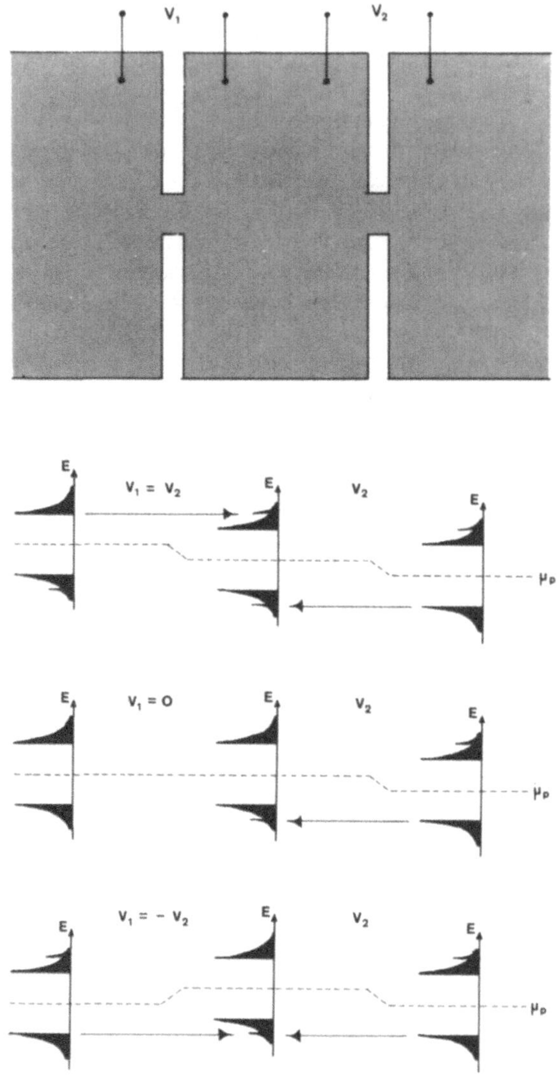

Fig. 19. This figure gives a qualitative illustration of the quasiparticle dc locking mechanism. The superconducting film (grey) with the two closely spaced microbridges is shown at the top. Below that and with the same x-axis, the variation of the pair potential through the system is shown for three different bias situations (V_2 is constant): $V_1 = V_2$ (aiding), $V_1 = 0$, and $V_1 = -V_2$ (opposing). The quasiparticle distribution in the three regions, with the injected peaks due to the diffusive currents, are also shown (semiconductor picture). The dc locking occurs only for the series opposed current configuration (lowest picture) as discussed in the text.

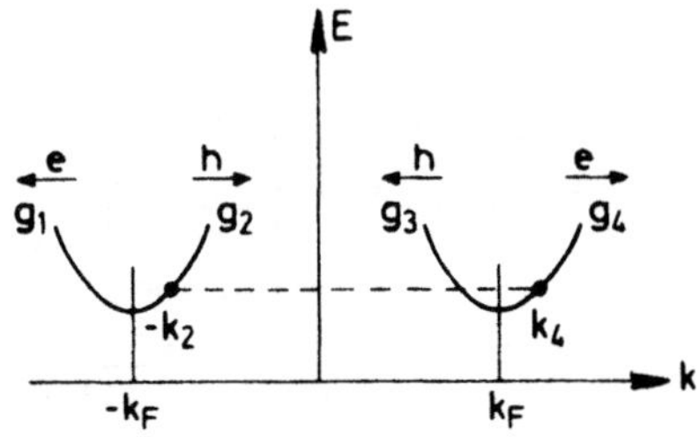

Fig. 20. Dispersion curves for the quasiparticles in a one-dimensional superconductor.

ACKNOWLEDGEMENT

For cooperation and advice we would like to thank O. Eg, B. R. Fjordbøge and H. Højgaard Jensen. We also acknowledge the continued interaction with the "Josephson group" at the Technical University of Denmark at Lyngby. We are supported by the Danish Natural Science Research Council (511-7048, 511-10160, and 511-15187).

REFERENCES

Artemenko, S. N., Volkov, A. F., and Zaitsev, A. V., 1978a, Collective modes in arrays of superconducting bridges, J. de Phys. (Paris) 39:C6-588 (Proc. of LT-15).

Artemenko, S. N., Volkov, A. F., and Zaitsev, A. V., 1978b, Josephson effect and collective modes in arrays of superconducting bridges, Pis'ma Zh. Eksp. Teor. Fiz., 27:122 [JETP Lett. 27:113, 1978]

Artemenko, S. N., and Volkov, A. F., 1979, Electric fields and collective oscillations in superconductors, Uspekhi Fiz. Nauk 128:3.

Bindslev Hansen, J., and Lindelof, P. E., 1980a, Non-equilibrium phenomena in Josephson junctions, Proc. of 2nd IC-SQUID Conf. Berlin 1980, to be published.

Blackburn, J. A., Schwartz, B. B., and Baratoff, A., 1972, One-dimensional periodic superconducting weak link systems, Phys. Lett., A42:31.

Blackburn, J. A., Schwartz, B. B., and Baratoff, A., 1975, Single and multiple superconducting weak-link systems, Low Temp. Phys., 20:523.

Deminova, G. F., and Kovalenko, P. S., 1979, A calculation of interaction between two Josephson junctions, IEEE Trans. Mag., MAG-15:291.

Giaever, I., 1965, Detection of the ac Josephson effect, Phys. Rev. Lett., 14:904.

Gray, K. E., 1975, Superconducting tunneling into films with voltage steps, Proc. 14th Conf. on Low Temp. Phys., vol. 2, ed. M. Krusius and M. Vuorio (Amsterdam:North-Holland), p. 207.

Howard, W., and Kao, Y. H., 1975, Current-phase relationship of coupled superconducting weak links, Proc. 14th Conf. on Low Temp. Physics, vol. 4, ed. M. Krusius and M. Vuorio (North-Holland, Amsterdam), p. 144.

Jillie, D. W., Lukens, J. E., and Kao, Y. H., 1975, Observation of microwave synchronization of thin film microbridge arrays, IEEE Trans Mag., Mag-11:671.

Jillie, D. W., Lukens, J. E., Kao, Y. H., and Dolan, G. J., 1976, Observation of voltage locking and other interactions in coupled microbridge Josephson junctions, Phys. Lett., 55A:381.

Jillie, D. W., 1976, Interactions between coupled thin-film microbridge Josephson junctions, Ph.D. Thesis, State University of New York, Stony Brook, unpublished.

Jillie, D. W., Lukens, J. E., and Kao, Y. H., 1977a, Voltage locking in two coupled microbridge Josephson junctions, IEEE Trans. Mag., MAG-13:578.

Jillie, D. W., Lukens, J. E., and Kao, Y. H., 1977b, Observation of interactions between two superconducting phase-slip centers, Phys. Rev. Lett., 38:915.

Jillie, D. W., Nerenberg, M. A. H., and Blackburn, J. A., 1980, Voltage locking and other interactions in coupled superconducting weak links. II Experiment., Phys. Rev., 21B:125.

Kadin, A. M., Smith, L. N., and Skocpol, W. J., 1980, Charge imbalance waves and nonequilibrium dynamics near a superconducting phase-slip center, J. Low Temp. Phys., 38:497.

Kaplan, S. B., Kirtley, J. R., and Langenberg, D. N., 1977, Experimental determination of the quasiparticle energy distribution in a non-equilibrium superconductor, Phys. Rev. Lett., 39:291.

Kulik, I. O., 1966, Motion of Abrikosov lines in an electric field and the energy dissipation mechanism in homogeneous superconductors, Zh. Eksp. Teor. Fiz. 50:1617 [Sov. Phys. JETP 23:1077, 1966].

Likharev, K. K., 1979, Superconducting weak links, Rev. Mod. Phys., 51:101.

Lindelof, P. E., Bindslev Hansen, J., Mygind, J., Pedersen, N. F., and Sørensen, O. H., 1977, Coherent Josephson radiation from an array of two Dayem bridges, Phys. Lett., 60A:451.

Lindelof, P. E., and Bindslev Hansen, J., 1977, Coherent behavior of two interacting Dayem bridges, J. Low Temp. Phys., 29:369.

Lindelof, P. E., 1978, Non-equilibrium effects in constricted superconductors, J. de Phys. (Paris), 39:C6-1411. (LT-15).

Lukens, J. E., Sandell, R. D., and Varmazis, C., 1979, Fabrication of microbridge Josephson junctions using electron beam lithography, "Future Trends in Superconductive Electronics," A.I.P. Conf. Proc. No. 44, ed. by B. S. Deaver et al., (AIP New York), p. 298.

Meyer, J. D., and Tidecks, R., 1977, Mutual influence of voltage steps in the V-I characteristics of current-carrying superconducting tin whiskers, Solid. State Comm., 24:643.

Nerenberg, M. A. H., Blackburn, J. A., and Jillie, D. W., 1980, Voltage locking and other interactions in coupled superconducting weak links. I. Theory, Phys. Rev., 21B:118.

Nerenberg, M. A. H., and Blackburn, J. A., "Mutual Voltage Locking in Linear Arrays of Josephson Weak Links," preprint, 1980.

Octavio, M., and Skocpol, W. J., 1979, Microwave response of superconducting variable-thickness microbridges, J. Appl. Phys., 50:3505.

Palmer, D. W., and Mercereau, J. E., 1974, Coherent effects in large arrays of superconducting bridges, Appl. Phys. Lett., 25:467.

Palmer, D. W., and Mercereau, J. E., 1975, Coherent effects in series arrays of proximity effect superconducting bridges, IEEE Trans. Mag. MAG-11:667.

Palmer, D. W., and Mercereau, J. E., 1977, Mode locking in arrays of superconducting weak links, Phys. Lett., 61A:135.

Tinkham, M., Octavio, M., and Skocpol, W. J., 1977, Heating effects in high-frequency metallic Josephson devices: voltage limit, bolometric mixing and noise, J. Appl. Phys., 48:1311.

Tinkham, M., The interaction of phase-slip centers in superconducting filaments, J. Low Temp. Phys., 35:147.

Waldram, J. R., 1976, The Josephson effects in weakly coupled superconductors, Rep. Prog. Phys., 39:751.

Way, Y. C., Hsu, K. C., and Kao, Y. H., 1977, Interactions between two superconducting weak links in the stationary (V=0) states, Phys. Rev. Lett., 39:1684.

MOTION OF MAGNETIC FLUX STRUCTURES

R. P. Huebener

Physikalisches Institut II
University of Tübingen
7400 Tübingen, FRG

1. INTRODUCTION

The Abrikosov lattice of flux lines in the mixed state of a type-II superconductor and the flux-containing domains in the intermediate state of a type-I superconductor represent spatially inhomogeneous states of the material. The more or less localized normal and superconducting regions may be looked at as "particles", with a complementary character (such as electrons and holes). These "particles" are subjected to the laws of non-equilibrium thermodynamics, i.e., to the concepts of flows and generalized forces, valid in the linear range of the thermodynamics of irreversible processes (Nicolis and Prigogine, 1977). The forces we have to deal with here are those arising from an applied electric current (note that an electric field can usually not be sustained in a superconductor) or temperature gradient, resulting in the flow of the magnetic flux structure (our "particles") and in the flow of heat.

Not far from equilibrium we expect a linear response of the superconductor to the external forces, the flow of "particles" and heat varying linearly with the applied electric current or temperature gradient, and the corresponding transport coefficients being truly constants. It is this linear regime of the non-equilibrium behavior with which we are dealing in the dominant part of my lectures. In the last Section treating the current-induced dissipative state we are confronted with non-linear phenomena. Here the superconductor is subjected to conditions far from equilibrium, and we must discuss the development of dissipative structures (Nicolis and Prigogine, 1977). This regime is characterized by the generation of new spatial and/or temporal order through the nonequilibrium conditions. We see a highly structured response of the superconductor, and the transport coefficients are not any more simple constants.

In addition to the nonlinear behavior observed sufficiently far from equilibrium resulting in the formation of dissipative structures, there is another mechanism leading to a nonlinear sample response: flux pinning. This effect is caused by deviations of the specimen material from perfect homogeneity. Sample inhomogeneities, which are always present, act as pinning sites for the magnetic flux structure and hinder the free motion of magnetic flux in response to an external force. Some comments on flux pinning will be presented in Section 5.

Since magnetic flux flow is associated with energy dissipation, flux flow and flux pinning are highly important subjects for many practical applications of superconductors.

More than 25 years ago, Shoenberg (1952) pointed out that in the presence of an electric current applied to a type-I superconductor residing in the intermediate state domain motion is expected to occur in a direction perpendicular to the current and to the magnetic field. He appears to have been the first to suggest the phenomenon of current-induced flux motion. Further details on subsequent historic developments of the flux-flow concept can be found in recent summaries (Kim and Stephen, 1969; Huebener, 1972, 1974, 1979).

As we will see, the motion of magnetic flux structures in superconductors results in various thermoelectric and thermomagnetic effects, which can be much larger than these effects in the normal metallic state. These phenomena must clearly be distinguished from the thermoelectric effects observed in a superconductor in the absence of magnetic flux structures and discussed in Chapter 16 of this volume.

2. LORENTZ FORCE

2.1 Phenomenological Force Equation

For a phenomenological discussion of the Lorentz force acting upon flux lines (containing one flux quantum) or flux tubes (containing more than one flux quantum) we consider a flat superconductor with its broad surfaces located in the x-y plane and a magnetic field applied in z-direction generating the magnetic flux structure. An electric current with density $\underline{J}_x$ applied in x-direction causes flux motion with velocity $\underline{v}_f$ in a direction given by the Hall angle Θ between the y-axis and $\underline{v}_f$ (see Fig. 1). The flux motion obeys the following force equation

$$(\underline{J}_x \times \underline{\Phi})/c - fn_s e(\underline{v}_f \times \underline{\Phi})/c - \eta \underline{v}_f - \underline{f}_p = 0. \qquad (1)$$

The Lorentz force $(\underline{J}_x \times \underline{\Phi})/c$ is compensated by the Magnus force $fn_s e(\underline{v}_f \times \underline{\Phi})/c$, the damping force $\eta \underline{v}_f$, and the pinning force $\underline{f}_p$. Here $\underline{\Phi}$ is the flux contained in a flux-line or flux tube and n_s the number density of superconducting electrons. In (1) the forces are given per unit length of flux line or flux tube.

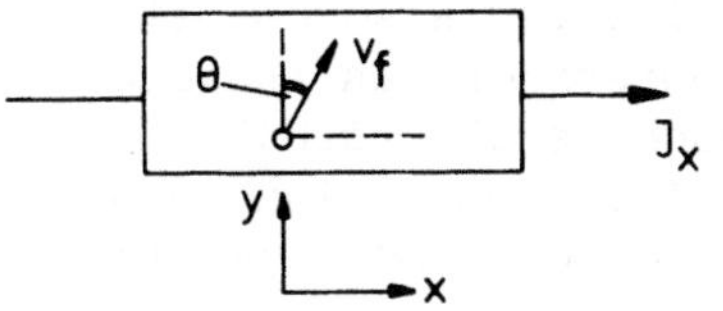

Fig. 1. Geometry for current-induced flux flow.

The Lorentz force can be understood in terms of the Maxwell stresses in the following way. Starting from Maxwell's equation

$$\text{curl } \underline{H} = \frac{4\pi}{c} \cdot \underline{J} \qquad (2)$$

and using the vector identity

$$\frac{1}{2} \text{ grad } (\underline{H} \cdot \underline{H}) = (\underline{H} \cdot \text{grad}) \underline{H} + \underline{H} \times \text{curl } \underline{H} \qquad (3)$$

we have

$$\frac{n_f}{c}(\underline{J} \times \underline{\Phi}) = \frac{\mu_m}{c}(\underline{J} \times \underline{H}) = -\frac{\mu_m}{8\pi}\text{grad}(H^2) + \frac{\mu_m}{4\pi}(\underline{H} \cdot \text{grad})\underline{H}. \qquad (4)$$

Here $\underline{H}$ is the spatially averaged magnetic field, μ_m the relative permeability, and n_f is the density of flux lines or flux tubes

$$n_f = B / \Phi \quad . \qquad (5)$$

In (4) the Lorentz force is given per unit volume of the superconductor. We see that this force arises from a gradient in the magnetic hydrostatic pressure $\mu_m H^2/8\pi$ and a term describing an additional tension along the field lines. The combination of both terms yields a tension $\mu_m H^2/8\pi$ along the field lines and an equal pressure transverse to them. The resulting forces may be illustrated in two limiting cases: an assembly of straight flux lines with a density gradient and an assembly of curved flux lines with constant density. In the first case the Lorentz force is directed down the density gradient and in the second case toward the center of curvature.

In the motion of flux lines the Magnus force plays a similar role as in the motion of quantized vortices in superfluid helium. Neglecting dissipation and pinning, the Magnus force results in flux-line motion perpendicular to the driving force, i.e., parallel to the current density $\underline{J}$. The Magnus force corresponds to the hydrodynamic lift force as expressed in the Kutta-Joukowski relation. The constant f in the Magnus term, introduced by Van Vijfeijken (1968), indicates the fraction of the Magnus force that is active. Except for extremely pure superconductors, f is much smaller than 1.

The damping force, directed opposite to $\underline{v}_f$, is described by the scalar damping factor η. It contains all dissipative processes, such as eddy current damping within the normal core of a flux line.

The pinning force $\underline{f}_p$ is directed also opposite to $\underline{v}_f$. Flux pinning results from local deppressions in the Gibbs free energy of the flux line arising through metallurgical inhomogeneities in the sample material. Clearly, for flux motion to occur, the Lorentz force must exceed the pinning force. Some comments on flux pinning will be given in Section 5.

We note that, so far, our discussion applies equally to flux lines and flux tubes, the number of flux quanta in the quantity $\underline{\Phi}$ of (1) being kept arbitrary.

2.2 Flux Flow Resistance

The steady-state motion of an assembly of flux lines or flux tubes induced by the Lorentz force causes the time-averaged macroscopic electric field $\underline{E}$ according to Faraday's law

$$\underline{E} = - \text{grad } V = -(\underline{v}_f \times \underline{B})/c \quad , \tag{6}$$

where V is the voltage. For the geometry of Fig. 1 the component v_{fy} causes the (longitudinal) resistive voltage, whereas the component v_{fx} causes the (transverse) Hall voltage. From (1) we see that the direction of current-induced flux flow depends upon the two parameters f and η. Taking the case f = 1 and neglecting for simplicity the pinning force, (1) can be written as

$$\frac{n_s e}{c} (\underline{v}_s - \underline{v}_f) \times \underline{\Phi} - \eta \, \underline{v}_f = 0 \quad , \tag{7}$$

if we substitute for the supercurrent density $\underline{J}_s$

$$\underline{J}_s = n_s \, e \, \underline{v}_s \tag{8}$$

$\underline{v}_s$ being the electron velocity. We see that $\underline{v}_f$ must be perpendicular to the difference $\underline{v}_s - \underline{v}_f$. As indicated in Fig. 2, this requires that the component v_{fy} cannot be larger than $v_s/2$, independent of the value of η. For a given value of v_s (or J_s) this would lead to a maximum value of the longitudinal electric field E_x.

For $\eta \to 0$ we find that $\underline{v}_f \to \underline{v}_s$, i.e., the flux line would move along with the supercurrent velocity. Then E_x would be zero, and there would only exist a transverse Hall field E_y. On the other hand, for the case $f = 0$ the velocity $\underline{v}_f$ is oriented perpendicular to $\underline{v}_s$. Then the Hall field would be zero and there would only exist a longitudinal field E_x. We shall return to the question of the Hall effect in Section 2.3.

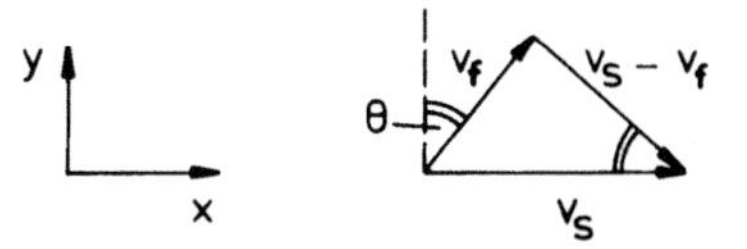

Fig. 2. Vectors $\underline{v}_s$, $\underline{v}_f$, and $\underline{v}_s - \underline{v}_f$ according to (7).

Neglecting the Magnus force and inserting (6) into (1) we obtain

$$\frac{\partial V}{\partial x} = \Phi\, B(J_x - J_c)/\eta\, c^2 \,, \tag{9}$$

where J_c is the critical current density

$$J_c = f_p \cdot c/\Phi \quad . \tag{10}$$

According to (9) the longitudinal electric field increases linearly with J_x. The slope of this dependence yields the flux-flow resistivity

$$\rho_f = \Phi\, B/\eta\, c^2 \,. \tag{11}$$

A typical plot of the flux-flow voltage versus current is shown in Fig. 3. Following the onset of a non-zero voltage at the critical current I_c, we note a regime where the V-I characteristic is curved upwards and, finally, the linear behavior expressed in (9). The curved region can easily be explained through a spatially inhomogeneous distribution of the magnitude of the pinning force $\underline{f}_p$. Such a distribution of the strength of the different pinning sites leads to an increase in the number of moving flux lines with increasing current, in addition to the increase in flux-flow velocity, thereby causing the non-linear behavior. In Section 5 we will return to this question of a distribution in the pinning force.

The slope of the V-I curve in the linear regime defines the flux-flow resistivity ρ_f. The variation of ρ_f in a type-II superconductor with magnetic field and temperature is shown schematically in Fig. 4.

Ideal flux-flow behavior can be studied only if the influence of flux pinning is reduced as much as possible. Such a reduction can be achieved by "shaking" the assembly of flux lines through the

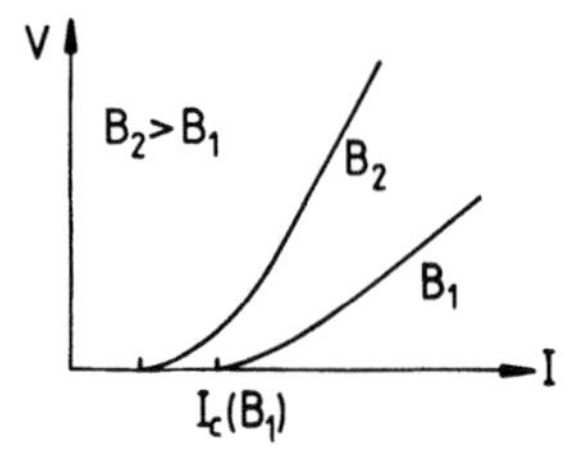

Fig. 3. Flux-flow voltage versus current.

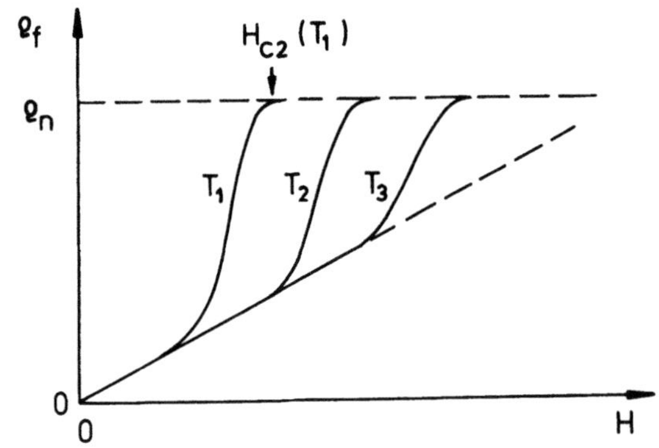

Fig. 4. Flux-flow resistivity ρ_f versus magnetic field for different temperatures ($T_1 > T_2 > T_3$).

application of an alternating magnetic field or an alternating current of small amplitude superimposed on the dc component. Extremely weak flux pinning has been found in certain type-II alloys, in particular niobium alloyed with Mo or Ta. Also oxygen-doped aluminum films have shown to be a useful material for studying near-ideal flux-flow behavior.

The early measurements of the flux-flow resistivity in type-II superconductors indicated the empirical relation in the low-temperature limit

$$\rho_f/\rho_n = H/H_{c2}(0) \, , \tag{12a}$$

where ρ_n is the normal-state resistivity. With (11) we obtain

$$\eta = \Phi_o \cdot H_{c2}(0)/\rho_n c^2 \, . \tag{12b}$$

The ratio $H/H_{c2}(0)$ is approximately the fraction of the volume occupied by the cores of the flux lines. Hence, relation (12a) indicates that the dissipation takes place predominantly in the core. At high magnetic fields an interesting deviation from (12a) is caused by the influence of paramagnetism. Because of the paramagnetic alignment of the electron spins, the normal-state energy is reduced, thus leading to additional pair breaking and an excess amount of electrical resistivity (Clogston, 1962; Chandrasekhar, 1962).

The concept of flux flow developed originally for explaining the resistive voltages in the mixed state of type-II superconductors has been verified in many experiments in which the motion of flux lines or flux tubes could be measured more or less directly. Some of these experiments will be discussed in Section 4. Further details can be found in a recent monograph (Huebener, 1979). The flux-flow concept holds equally well for an assembly of flux tubes in the intermediate state of a type-I superconductor. Here magneto-optical flux detection can serve for directly observing the flux-flow behavior with high time resolution.

During flux flow, energy is dissipated via eddy current damping in the normal core of a flux line or flux tube. We first address ourselves to the case of flux tubes with a diameter large compared to the coherence length. The electric field induced in a flux tube moving with velocity v_f is

$$E = v_f \cdot B_c/c \quad . \tag{13a}$$

Here we consider the absolute value only and assume $\underline{v}_f$ and $\underline{B}_c$ to be oriented perpendicular to each other. The power dissipated per unit length of flux tube is

$$\eta v_f^2 = \int \sigma_n \cdot E^2 \cdot dA = (\sigma_n \cdot v_f^2 \cdot B_c/c^2) \cdot \int B_c \cdot dA \ , \tag{13b}$$

where σ_n is the normal conductivity and dA the area element of integration. In (13b) we must integrate over the area of the normal core. With $\int B_c \cdot dA = \Phi$ we find

$$\eta = \Phi \cdot \sigma_n B_c / c^2 \ . \tag{13c}$$

Because of our assumption of a large flux-tube diameter we could ignore effects arising within the wall region of the tube. However, such effects become appreciable if the flux-tube diameter becomes comparable to the coherence length, and even more, if we are dealing with flux lines in a type-II superconductor. In the latter case the normal vortex core and the volume penetrated by magnetic flux are not identical any more. The straigthforward application of our arguments from above leading to (13c) to the case of a type-II superconductor does not yield the empirical relation (12b). Here the geometrical environment of the normal vortex core must be included in our discussion, as we will see in more detail in Section 6.

Recently, through magneto-optical experiments in Pb-films of type I it has been possible to verify directly the terms describing the Lorentz force and the damping force in (1) (Selig, 1980). The idea is as follows. The power dissipated by a flux tube moving with velocity v_f is per unit length ηv_f^2. If there are N_f flux tubes traveling through a superconducting film of thickness d, the total dissipation is $N_f \cdot d \cdot \eta \cdot v_f^2$. Equating this with the electric power dissipated we obtain

$$I \cdot V = N_f \, d \, \eta \, v_f^2 \ . \tag{14}$$

Following an independent magneto-optical measurement of v_f and N_f, we can calculate the damping coefficient η from (14). Neglecting the Magnus force (which is usually possible), (1) reduces to

$$(\underline{J}_x \times \underline{\Phi})/c = \eta \underline{v}_f \tag{15}$$

in the case when flux pinning becomes unimportant. From the direct determination of η and $\underline{v}_f$ the validity of (15) could be verified within an accuracy of 10 - 15 %. The experimental values of η were found to agree with the result in (13c) within 30 %.

In the intermediate state also current-induced motion of superconducting domains can be observed slightly below H_c where most of the sample is in the normal state. The superconducting domains move in the same direction as the normal domains at lower magnetic fields. Keeping in mind the complementary character of normal and superconducting domains discussed in Section 1, the motion of a superconducting region is equivalent to the motion of a corresponding flux tube in opposite direction. Hence the generated "flux flow voltage" is opposite to the regular resistive voltage, thus reducing the electric resistance below its normal-state value.

It is important to note that, so far, we have discussed the time-averaged voltage only. However, because of the discreet nature of the magnetic flux structure, during flux flow voltage pulses with a characteristic time dependence are generated by the individual flux lines. The total voltage is then the result of the linear superposition of all individual voltage pulses. This leads to important consequences for the electrical noise associated with the flux-flow voltage, as will be discussed in Section 4.3.

Aside from the effects of flux pinning, current-induced flux flow in an applied magnetic field can be considered a linear response (see Section 1) if we treat the assembled flux lines of flux tubes as existing "particles" and do not worry about their origin and decay at the sample edges during flux flow. However, the formation and the decay of the individual flux lines are, clearly, nonlinear phenomena. On the other hand, this nonlinearity is insignificant if the number of flux lines participating in the flux-flow process at any time is much larger than the number which is in the process of nucleation and decay.

2.3 Hall Effect

Flux flow induced by the Lorentz force results in a finite transverse voltage component (Hall voltage) if the Magnus force is active. Referring to the geometry of Fig. 1, we see that it is the velocity component v_{fx} parallel to the applied current which causes the transverse Hall field. In addition to the detection of a transverse Hall voltage, the velocity component v_{fx} in x-direction has also directly been observed in magneto-optical flux-flow experiments in type-I superconductors (König and Kirchner, 1975). As shown in a beautiful experiment by Haenssler and Rinderer (1967), the influence of the Magnus force can be seen during the penetration of magnetic flux into a circular disk of a type-I superconductor. Here it results in a motion of the flux toward the center in spirals, the azimuthal component of the flux motion being the consequence of the Magnus force.

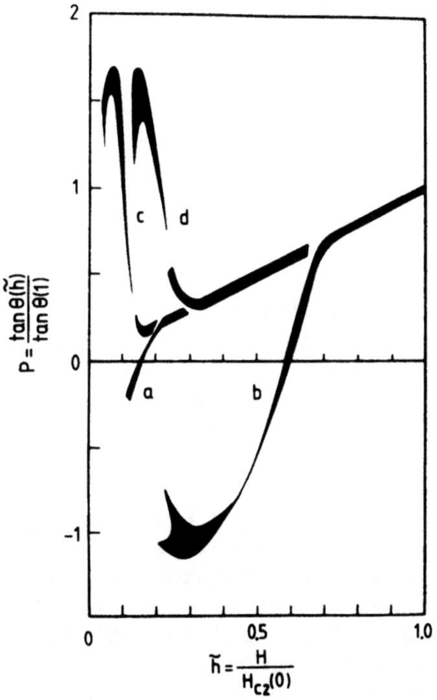

Fig. 5. Reduced Hall angle $P = \tan\Theta(\tilde{h})/\tan\Theta(1)$ versus reduced field $\tilde{h} = H/H_{c2}(0)$ for two vanadium samples of different purity. (a) 4.22 K and (b) 2.13 K for a sample with a residual resistance ratio RRR = 14.5; (c) 4.22 K and (d) 3.76 K for a sample with RRR = 7.2. (Usui et al., 1969).

The Hall effect has been investigated in the mixed state of various type-II superconductors. Extensive measurements were performed in particular in niobium and vanadium, where the Hall angle was found to depend sensitively on the purity of the metal. Figure 5 shows as an example the reduced Hall angle plotted versus the reduced magnetic field for two vanadium specimens of different purity. Further details may be obtained from a recent monograph (Huebener, 1979). A comparison of the experimental results with theoretical models will be given in Section 6.

2.4 Ettinghausen and Peltier Effect

In the normal core of a flux line or (multiquantum) flux tube the entropy density is larger than in the surrounding superconducting phase. Hence, during the current-induced motion of the magnetic flux structure entropy is transported across the superconductor. The process of flux nucleation at the sample edge takes up heat energy, which is dumped back into the superconductor during the flux annihilation process at the opposite edge. The heat-current density $\underline{U}$ associated with the moving magnetic flux-structure is

$$\underline{U} = n_f \; T \; S_\varphi \; \underline{v}_f \; , \tag{16}$$

where n_f is the number density of vortex lines or flux tubes per unit area and S_φ the transport entropy per unit length of flux line or flux tube. Under stationary conditions the heat-current density (16) is compensated by regular heat conduction, and we have for the component in y-direction

$$U_y = n_f \; T \; S_\varphi \; v_{fy} = -K \frac{\partial T}{\partial y} \quad . \tag{17}$$

A similar equation holds for the component in x-direction. K is the heat conductivity. With (16) we find

$$\left|\frac{\partial T}{\partial y}\right| = c \; \frac{T}{K} \; \frac{S_\varphi}{\Phi} \; \left|\frac{\partial V}{\partial x}\right| \quad . \tag{18}$$

The transverse temperature gradient (Ettinghausen effect) is proportional to the longitudinal electric field. The proportionality constant contains the quantities S_φ/Φ , K, and T. A relation similar to (18) holds for the longitudinal temperature gradient (Peltier effect), $|\partial T/\partial x|$ being proportional to the Hall field $|\partial V/\partial y|$.

Experimental observations of the Ettinghausen and Peltier effect have been reported for various type-I and type-II superconductors (Huebener, 1979). Because of the particular form of the Lorentz force, the (transverse) Ettinghausen effect is usually 2-3 orders of magnitude larger than the (longitudinal) Peltier effect, in contrast to the behavior in the normal metallic state. In the mixed or intermediate state the Ettinghausen effect is 2-3 orders of magnitude larger than in the normal state. The ratio of the Peltier and the Ettinghausen heat current is given by the Hall angle Θ:

$$\tan \Theta = \frac{v_{fx}}{f_{fy}} = \frac{U_x}{U_y} = \left| \frac{\partial T/\partial x}{\partial T/\partial y} \right| \quad . \tag{19}$$

From the Onsager reciprocity relations we conclude that the Nernst and Seebeck effect must also exist in the mixed and the intermediate state. These phenomena will be discussed in Section 3. The quantity regulating all these thermomagnetic effects is the transport entropy S_φ.

2.5 Josephson Relation

In (6) we have related the flux-flow voltage in a superconductor to Faraday's law. However, it is not obvious how an electric field can be sustained in a multiply connected superconductor containing a moving assembly of flux lines or flux tubes. An answer can be found, if we remember that a superconductor constitutes a macroscopic quantum system described by the complex wave function

$$\psi(\underline{r},t) = |\psi| \cdot e^{i\phi} \quad . \tag{20}$$

($\underline{r}$ = spatial coordinate; t = time). This wave function obeys the time-dependent Schrödinger equation

$$i\hbar \frac{\partial \psi}{\partial t} = E \cdot \psi \quad , \tag{21}$$

E being the energy. We consider two weakly coupled superconductors (1) and (2) described by the wave functions ψ_1 and ψ_2, respectively. If the voltage V exists between both superconductors, they differ in energy by the amount $\Delta E = 2eV$. Here the factor 2e results from the Cooper pairs constituting our macroscopic quantum system. From (20) and (21) we then obtain the Josephson relation (Josephson, 1962)

$$\frac{2e}{\hbar} V = \frac{\partial \phi}{\partial t} \quad . \tag{22}$$

Here

$$\phi = \phi_2 - \phi_1 \tag{23}$$

is the phase difference between the two superconductors (1) and (2), or between two points (1) and (2) in space located on both sides of the path along which a flux line or flux tube moves across the superconducting specimen.

From (22) we expect a time-dependent voltage between the two points (1) and (2) if a vortex line traverses the region between them. The voltage due to the uniform motion of an assembly of vortex lines consists of a linear superposition of many individual voltage pulses. This temporal structure of the individual contributions to the total time-averaged voltage has important consequences for the electrical noise power, as will be discussed in Section 4.3.

We can calculate the time-averaged voltage generated by a moving assembly of vortex lines by replacing the temporal structure of the derivative $\partial\phi/\partial t$ by a phase jump of 2π taking place when the vortex line crosses the straight line connecting point (1) and (2). For a vortex-line assembly of density n_f we obtain

$$\overline{\partial\phi/\partial t} = 2\pi \, s_{12} \, v_f \, n_f \, , \tag{24}$$

where s_{12} is the distance between (1) and (2). With (22) we find

$$|\overline{E}| = \frac{\overline{V}}{s_{12}} = \frac{1}{c} v_f \, B \quad , \tag{25}$$

i.e., the same result as in (6). Here we have utilized the expression for the fluxoid quantum

$$\Phi_o = \frac{hc}{2e} = 2.07 \times 10^{-7} \text{ G cm}^2 \quad . \tag{26}$$

For simplicity we assumed the orientation of $\underline{v}_f$, s_{12}, and the vortex lines to be perpendicular each other and omitted the vector notation.

3. THERMAL FORCE

In addition to the Lorentz force of an electric current, another force can act upon an assembly of flux lines or flux tubes: the thermal force. This force always acts upon a local region of matter in the presence of a temperature gradient, if the local entropy density differs from that of the environment. The thermal force is given by

$$f_{th} = - S_\varphi \cdot \text{grad } T \quad , \tag{27}$$

where S_φ is the transport entropy. It is fundamental to all thermal-diffusion phenomena. In our case the thermal force results from the fact that within the vortex core or within a flux tube the entropy density is higher than in the superconducting phase. It is directed from the hot end to the cold end of the sample. The quantity S_φ in (27) is identical to that appearing in the heat-current density (16), as can be shown from the Onsager reciprocity relations (Huebener, 1979).

3.1 Phenomenological Force Equation

For a phenomenological discussion of the force equation we assume again a flat superconductor with its broad surfaces located in the x-y plane and a magnetic field applied in z-direction such that the plate resides in the mixed or the intermediate state. A temperature gradient is applied in x-direction (Fig. 6). In analogy to (1) thermally induced flux motion obeys the following force equation

$$-S_\varphi \text{ grad } T - fn_s e(\underline{v}_f \times \underline{\Phi})/c - \eta\underline{v}_f - \underline{f}_p = 0. \qquad (28)$$

The thermal force $-S_\varphi$ grad T is compensated by the Magnus force $fn_s e(\underline{v}_f x \underline{\Phi})/c$, the damping force $\eta\underline{v}_f$, and the pinning force $\underline{f}_p$. Again, $\underline{\Phi}$ is the flux contained in a flux line or flux tube, and (28) applies to both magnetic objects. Again, the forces are given per unit length of flux line or flux tube.

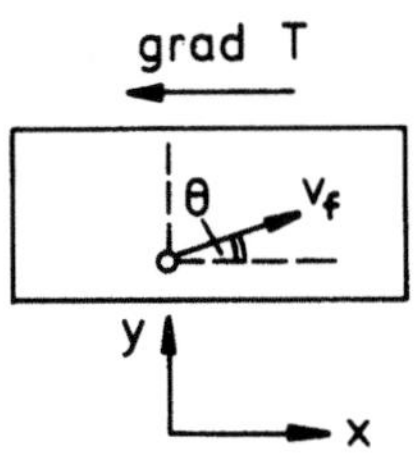

Fig. 6. Geometry for thermally induced flux flow.

3.2 Nernst and Seebeck Effect

According to (28) a temperature gradient in x-direction generally causes flux motion in the x- and y-direction. The velocity components v_{fx} and v_{fy}, together with (6), lead to a transverse and longitudinal electric field, corresponding to the Nernst and Seebeck effect, respectively. The voltages from thermally induced flux flow are expected to increase proportional to the temperature gradient, at least for small values of grad T and if flux pinning is relatively weak. If we neglect the Magnus force and insert (6) into (28) we obtain for the Nernst voltage

$$\frac{\partial V}{\partial y} = (-S_\varphi) \frac{B}{\eta c} (\text{grad}_x T - \text{grad}_c T) \quad . \qquad (29)$$

Here we have defined the critical temperature gradient $\text{grad}_c T$ by

$$\text{grad}_c T = - f_p / S_\varphi \quad . \qquad (30)$$

We note from (9) and (29) the similarity between flux flow induced by an electric current and by a temperature gradient. A linear flux-flow regime exists in both cases. Further, in both cases the onset of flux flow is determined by the pinning force which is related to the critical current density and the critical temperature gradient according to (10) and (30), respectively.

From (28) we see that without the Magnus force the component v_{fy} would be zero and the Seebeck effect would vanish. The Nernst effect arising from thermally induced flux flow has been experimentally studied in a number of type-I and type-II superconductors (Huebener, 1979). Again, because of the particular form of the flux flow voltage (Eq. (6)), in the mixed and the intermediate state the Nernst effect is 2-3 orders of magnitude larger than the Seebeck effect, in contrast to the behavior in the normal metallic state. In the mixed and the intermediate state the Nernst effect is 2-3 orders of magnitude larger than in the normal state and can be distinguished easily from the normal contribution. The Seebeck effect from thermally induced flux flow is expected to have similar magnitude as the normal contribution below the superconducting transition temperature. Again, the ratio of the electric fields from the Seebeck and Nernst effect is given by the Hall angle Θ:

$$\tan \Theta = \frac{v_{fy}}{v_{fx}} = \left|\frac{\partial V/\partial x}{\partial V/\partial y}\right| \quad . \tag{31}$$

3.3 Transport Entropy

The transport entropy S_φ represents the main quantity to be derived from the various thermomagnetic measurements (see (18) and (29)). We see from (18) that the Ettinghausen effect directly yields the quantity S_φ/Φ and that possible complications due to flux pinning are automatically eliminated. We note that from thermomagnetic measurements only the transport entropy <u>per unit flux</u> can be obtained.

For the temperature dependence of S_φ we expect the following behavior: an increase from zero with T at low temperature, a maximum at an intermediate temperature, and a decrease to zero at the critical temperature T_c. The vanishing of S_φ at $T = 0$ and $T = T_c$ results, of course, from the fact that here the difference between the entropy density in the normal and superconducting state reaches zero. With increasing magnetic field we expect a gradual decrease of S_φ from a maximum value at low fields reaching zero at the critical field H_c or H_{c2}. This behavior, shown schematically in Fig. 7, has generally been observed experimentally.

An interesting situation exists slightly below H_c or H_{c2}. Here the majority of the superconductor is in the normal state, and the material contains isolated superconducting domains or local regions with a finite value of the order parameter. The current-induced flow of these superconducting regions, discussed in Section 2.2, represents a heat current in opposite direction as that due to regular flux flow. From these considerations we expect a sign reversal of the Ettinghausen and the Peltier effect just below the critical field (which is as yet unobserved). The appearance of highly localized superconducting regions imbedded in the normal phase just below the critical field is expected to result in thermally induced motion of these superconducting domains in opposite direction as

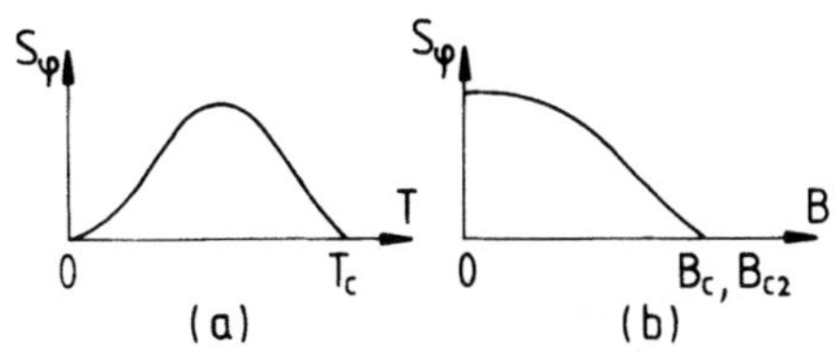

Fig. 7. Transport entropy S_φ versus temperature (a) and magnetic field (b), schematically.

that of thermally induced flux flow. It appears that a shoulder or a second peak observed in the Nernst voltage of type-I superconducting films just below H_c is caused by this effect (Huebener and Seher, 1969).

In a type-I superconductor the difference between the entropy density in the normal and superconducting state is

$$S_n - S_s = -\frac{H_c(T)}{4\pi}\frac{\partial H_c(T)}{\partial T} . \tag{32}$$

This difference vanishes at T = 0 and $T = T_c$, as expected. From (32) we find for the excess entropy per unit length of a flux tube containing the flux Φ

$$\frac{S_\varphi}{\Phi} = -\frac{1}{4\pi}\frac{\partial H_c(T)}{\partial T} . \tag{33}$$

Using the approximation

$$H_c(T) \approx H_c(0)\left(1 - \frac{T^2}{T_c^2}\right) \tag{34}$$

we obtain

$$\frac{S_\varphi}{\Phi} = \frac{H_c(0)}{2\pi}\frac{T}{T_c^2} . \tag{35}$$

At intermediate temperatures and magnetic fields the order of magnitude calculated from (35) agrees with the experimental results. Several corrections to be applied to the simple expressions (33) and (35) have been discussed (Huebener, 1979). We note that the contribution from the wall energy has been neglected so far. Such a contribution will become important for small flux tubes.

For a type-II superconductor, in analogy to (33), one finds for the transport entropy per unit length of vortex line

$$\frac{S_\varphi}{\Phi_o} = - \frac{1}{4\pi} \frac{\partial H_{c1}(T)}{\partial T} \quad . \tag{36}$$

This expression should be valid for small magnetic fields, at which the flux lines are well separated from each other. Well below T_c, (36) is in reasonable agreement with the experimental results. Close to T_c expression (36) cannot be valid any more, since here the vortices are not so well localized and their interaction cannot be neglected any more.

4. EXPERIMENTS

An experimental study of the phenomena discussed in Sections 2 and 3 essentially requires straightforward measurements of voltages or temperatures. In the following we briefly describe some experiments which have provided further insight into the dynamics of magnetic flux structures in superconductors. We do not present a complete list, and a more complete treatment can be found elsewhere (Huebener, 1979).

4.1 Magneto-optical Flux Detection

The inhomogeneous distribution of magnetic flux and its temporal changes in a superconductor can be studied using the Faraday rotation of the polarization vector of a beam of polarized light.

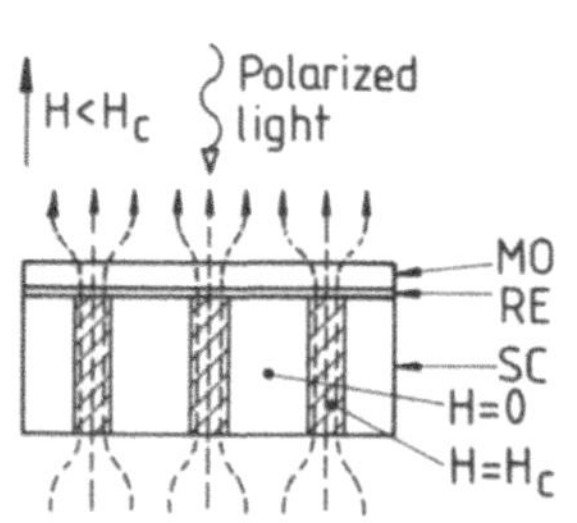

Fig. 8. Scheme for magneto-optical flux detection (SC: Superconductor; MO: magneto-optical material; RE: reflecting metal film).

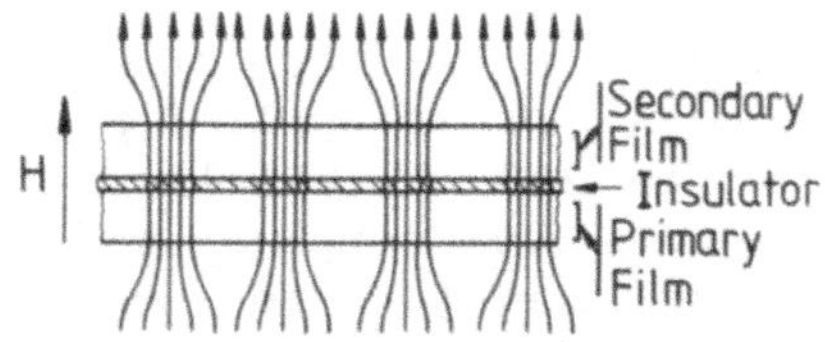

Fig. 9. Magnetic coupling between two superconducting films electrically insulated from each other.

The schematic of this magneto-optical flux detection is shown in Fig. 8. A superconducting plate, SC, containing normal domains because of the applied magnetic field H, is covered on top with a sheet of some magneto-optical material, MO. Upon this arrangement polarized light is directed from the top. It passes through the magneto-optical sheet before and after its reflection at the highly reflecting film RE. During both passages through MO the light experiences Faraday rotation, the rotation angle depending on the local magnetic field. By passing the outgoing light through an analyzer, a magneto-optical image of the inhomogeneous field distribution at the surface of the specimen can be obtained.

From Fig. 8 we see that the spatial resolution of this method cannot exceed roughly the thickness of the magneto-optical sheet. As first demonstrated by Kirchner (1971), films of the rare-earth compounds EuO, EuS, or EuSe of only 1000 - 2000 Å thickness can provide the necessary amount of Faraday rotation. Observation with a powerful microscope then yields a spatial resolution of about 0.5 µm. Therefore, applications of this technique are restricted to the intermediate state of type-I superconductors. Under optimum conditions, magnetic field modulations down to 20 Oe can be detected.

High temporal resolution of the magneto-optical flux detection can be achieved using the stroboscopic principle. In this way the flux-tube nucleation process in the current-induced dissipative state has been investigated (Chimenti and Huebener, 1977) with a time resolution of better than 0.1 µs (see Section 7).

4.2 Magnetic Coupling

An interesting configuration for flux flow is that of two superconducting films separated from each other only by a thin insulating layer (see Fig. 9). In an applied perpendicular magnetic field the flux structures in both films will perfectly register with each other, this magnetically coupled state representing the configuration of lowest free energy. Here we assume that the thickness of the insulating layer is smaller than the spacing of the vortex lines or flux tubes. If we induce flux flow in the primary film through the application of a primary current, flux motion will also take place in the secondary film because of the magnetic coupling effect. Therefore, in addition to the primary voltage, a voltage in the secondary film can also be observed, although no transport current has been applied to this film. Because of this, the configuration in Fig. 9 is sometimes referred to as superconducting dc transformer.

We simplify our discussion by neglecting flux pinning. Complete magnetic coupling between the moving flux structures in both films then prevails as long as the viscous drag force in the secondary film is smaller than the coupling force. Noting that the viscous drag force increases proportional to $\underline{v}_f$, it will exceed the coupling force at some stage. Here slippage between the flux structures in both films will set in, and the secondary voltage will decrease with increasing primary current. This behavior is summarized schematically in Fig. 10. We note that the details depend on the magnitudes of η_1 and η_2 and on the thickness of both films.

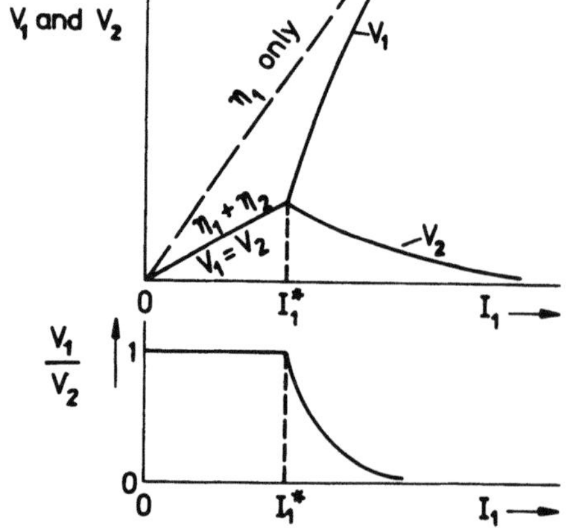

Fig. 10. Magnetic coupling between two superconductors: primary voltage V_1, secondary voltage V_2, and the ratio V_1/V_2 versus primary current I_1. Complete coupling exists up to $I_1 = I_1^*$.

Magnetic coupling experiments have first been performed by Giaever (1965, 1966). Ekin and coworkers (Ekin et al., 1974; Ekin and Clem, 1975) have studied these effects in granular aluminum films with extremely small flux pinning. A detailed theory of magnetic coupling including pinning effects has been developed by Clem (1974, 1975).

4.3 Electrical Noise Power

As we have pointed out in Section 2.5, the time-averaged dc voltage generated by a moving assembly of flux lines or flux tubes represents a linear superposition of individual voltage pulses. The shape of these pulses depends on the microscopic nature of the moving flux structure. A direct measurement of these individual voltage pulses is possible only if they are sufficiently far apart. In general such resolution of the individual pulses is impossible. However, the temporal pulse structure can be extracted from the electrical noise power spectrum.

We consider a statistically fluctuating voltage $V(t)$ with a constant time average $\overline{V}$ (statistically stationary). With $\delta V(t) = V(t) - \overline{V}$ the total mean square noise voltage is related to the spectral noise power $w(\omega)$ by

$$\overline{(\delta V)^2} = \int_0^\infty w(\omega)\, d\omega \quad . \tag{37}$$

The spectral noise power represents the Fourier transform of the autocorrelation function $\chi(\tau)$:

$$w(\omega) = 4 \int_0^\infty \chi(\tau) \cos\omega\tau \, d\tau \quad , \tag{38}$$

where $\chi(\tau)$ is

$$\chi(\tau) = \overline{V(t) \cdot V(t + \tau)} \quad . \tag{39}$$

We note that the spectral noise power is the temporal part of what is usually called dynamical structure factor (Van Hove, 1954a, 1954b). As an example we assume a sequence of statistically independent rectangular voltage pulses of constant amplitude ΔV and average rate ν_p. The lifetime τ_p of the pulses is distributed exponentially according to the autocorrelation function

$$\chi(\tau) = (\Delta V)^2 \cdot \nu_p \cdot \tau_p \cdot \exp(-\tau/\tau_p) \quad . \tag{40}$$

For this case the spectral power is

$$w(\omega) = 4\, \overline{V} \cdot \Delta V \cdot \tau_p / (1 + \omega^2 \tau_p^2) \quad . \tag{41}$$

Relations (40) and (41) are schematically shown in Fig. 11.

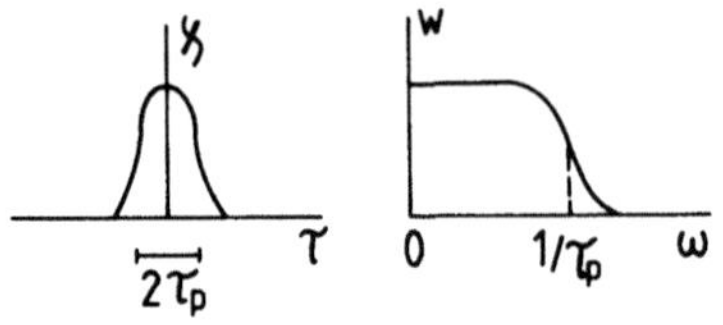

Fig. 11. Autocorrelation function $\chi(\tau)$ and power spectrum $w(\omega)$ from (40) and (41).

For an assembly of moving flux quanta the shape of the voltage pulses can be obtained from (22). If we approximate this shape by a rectangular pulse of amplitude ΔV and duration τ_p we find from (22)

$$\frac{\partial\phi}{\partial t} = \frac{2\pi}{\tau_p} = \frac{2e}{\hbar} \cdot \Delta V \quad , \tag{42}$$

or with (26)

$$\Delta V = \Phi_o / c\tau_p \quad . \tag{43}$$

In the low-frequency limit this yields for the spectral noise power (Huebener, 1979)

$$w(0) = 2 \cdot \overline{V} \cdot \Phi_o \,/\, c \quad . \tag{44}$$

This result is analogous to the Schottky relation

$$w(0) = 2 \cdot \overline{I} \cdot e \tag{45}$$

for the shot noise associated with the current I in an electron tube. If we are dealing with moving flux tubes containing the flux $\Phi = N_t \cdot \Phi_o$, the quantity Φ_o must be replaced by Φ in (44). Similarly, if short range correlation exists between flux lines, the total flux of the correlated flux lines appears in (44) instead of Φ_o. We note that generally the low-frequency limit $w(0)$ yields information on the amount of flux involved in the microscopic process during flux flow. On the other hand the break in the $w(\omega)$ curve indicates the average duration of this process (see Fig. 11).

Uniform motion of a perfect monocrystalline vortex lattice leads to long-range correlations between flux lines. In this case the statistical noise in the flux-flow voltage is absent. The autocorrelation function is then perfectly periodic, and the noise power spectrum consists of a sharp line at some distinct frequency.

The time-derivative of the phase-difference ϕ in (22) and, hence, the pulse shape of the flux-flow voltage depends sensitively on the geometry of the voltage probes and their location relative to the path of the moving flux line. A detailed discussion of this question has been given by Clem (1970). As an example we show in Fig. 12 the voltage pulse generated by a flux line moving at constant velocity along a straight line passing between the contact points (1) and (2). The voltage pulse consists of two superimposed Lorentzians which attain their maximum at the time of closest approach of the flux line to contact point (1) and (2), respectively. The time integral of each Lorentzian has the value $\Phi_o/2c = h/4e$.

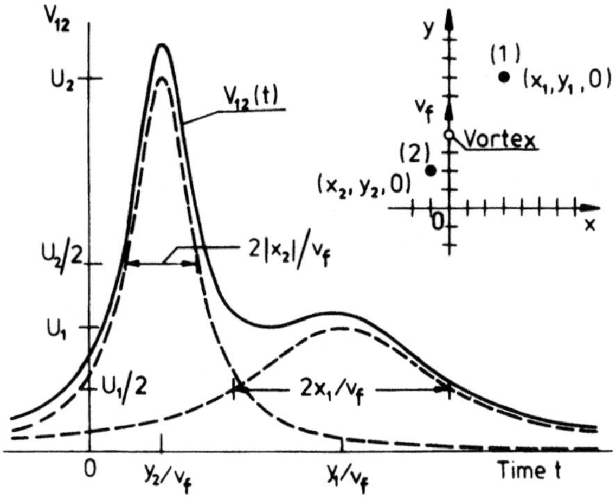

Fig. 12. Voltage V_{12} between contacts (1) and (2) versus time. The flux line moves along the y-axis. The coordinates of the contacts are shown in the inset. The quantities U_1 and U_2 are defined as $U_1 = \hbar v_f/2ex_1$ and $U_2 = \hbar v_f/2e|x_2|$. The total voltage pulse (solid line) is the sum of two Lorentzians (dashed lines). (Clem, 1970).

Experiments on flux-flow noise have been performed in type-II and type-I superconductors. They have addressed such questions as the bundle size of strongly correlated flux lines, the pinned fraction of all flux lines during flux flow, the existence of cross correlations between different pairs of voltage probes placed at different locations, and delay-time effects arising during flux flow. In some cases the noise data indicate that flux motion takes place in some form of a hopping process, the vortex lines being held up momentarily at pinning sites during their transit across the sample. A more detailed discussion of these experimental observations can be found elsewhere (Huebener, 1979).

5. COMMENTS ON FLUX PINNING

As expressed in the phenomenological force equations (1) and (28), flux flow induced by the Lorentz force or the thermal force can occur only if the driving force exceeds the pinning force present in the material. The pinning force determines the critical current and the critical temperature gradient according to (10) and

(30), respectively. In the following we discuss some fundamental concepts associated with flux pinning. Further details can be found in a series of recent monographs (Campbell and Evetts, 1972; Ullmaier, 1975; Huebener, 1979).

Flux pinning arises from inhomogeneities in the sample material causing a spatial variation of the Gibbs free energy of the system of flux lines or flux tubes. Obviously, the magnetic flux structure prefers a configuration representing a local minimum in the Gibbs free energy. The pinning force is then given by

$$\underline{f}_p = - \text{grad } G_f , \tag{46}$$

where G_f is the Gibbs free energy per unit length of flux line or flux tube. In general, for a lattice or an assembly of flux lines or flux tubes pinning represents a complex problem. In addition to dealing with the interaction between the individual flux line and the pinning site, the summation of the forces from all pinning sites must be carried out. Here we must take into account also the interaction between the flux lines themselves, i.e., the elastic properties of the flux-line lattice. Further, we usually encounter a distribution of pinning forces in the material.

As we have pointed out, sample inhomogeneity (geometrical, compositional, structural, etc.) always plays a central role in flux pinning. Here the sample edges represent an important case resulting in an energy barrier against entry and exit of magnetic flux. Ideal flux flow without pinning effects can be established only in a perfectly homogeneous sample. The influence of the sample edges can be eliminated in a geometry resulting in circular flux motion (Corbino disk).

As an example of a pinning center we consider a thickness modulation in a thin-film superconductor. We assume an indentation of depth δd and radius R_p at the surface of a film of thickness d (see Fig. 13a). In an applied perpendicular magnetic field a flux line or flux tube can lower its energy by residing within the indentation. If the radius r of the flux tube is less than R_p, this energy gain is per unit length of flux tube

$$\delta G_f = \frac{H_c^2}{8\pi} \cdot r^2 \pi \cdot \frac{(\delta d)}{d} . \tag{47}$$

Note that for a (single-quantum) flux line we have $r = \xi$ in (47). As we see from (46), for the pinning force the spatial structure of this energy gain becomes important (reflecting simply the mechanical principle of the inclined plane). Hence, we must take into account the width (δR_p) of the edge of the indentation. We consider two different cases: $(\delta R_p) \ll r$ and $(\delta R_p) \gg r$. In the first case (steep edge) we have

$$f_p = \frac{H_c^2}{8} \cdot r \cdot \frac{(\delta d)}{d} \; ; \qquad (\delta R_p) \ll r \quad , \tag{48}$$

and in the second case (flat edge)

$$f_p = \frac{H_c^2}{8} \cdot r \cdot \frac{r}{(\delta R_p)} \cdot \frac{(\delta d)}{d} \; ; \qquad (\delta R_p) \gg r \quad . \tag{49}$$

The factor $r/(\delta R_p) \ll 1$ in (49) again contains the principle of the inclined plane and indicates that an energy barrier with a small energy increment per distance only causes weak flux pinning.

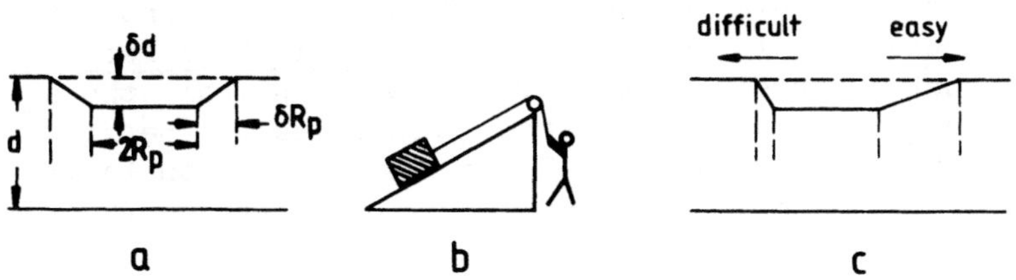

Fig. 13. Thickness modulation in a thin-film superconductor as pinning center for symmetric (a) and nonsymmetric (c) edge structure of the center. Note principle of the inclined plane (b).

An interesting situation arises if the edge structure on two opposite sides of the indentation is nonsymmetrical (Fig. 13c). The pinning force then differs in the two opposite directions, resulting in preferred flux flow in one direction and rectification effects for an alternating driving force. Such rectification can easily be observed experimentally (Swartz and Hart, 1965; Morrison and Rose, 1970).

Maximum flux pinning can be achieved if the indentation penetrates through the total thickness of the film, i.e., in the presence of a hole in the superconductor. If a lattice of holes exists in the film, the maximum pinning force acting upon a flux-line lattice can easily be calculated in the case of perfect spatial matching between the pinning holes and the vortex lattice. In this case elastic distortions of the vortex lattice need not be considered, and the summation of the pinning forces does not present any problem. From (48) the maximum pinning force per unit length is

$$f_p = H_c^2 \cdot \xi / 8 \quad . \tag{50}$$

The maximum pinning-force density is found by multiplication with the vortex-line density n_f from (5):

$$n_f \cdot f_p = B \cdot H_c^2 \cdot \xi / 8\Phi_o = J_c \cdot B/c \quad . \tag{51}$$

Using the relation

$$H_{c2} = \sqrt{2} \cdot H_c \cdot \lambda/\xi = \Phi_o/2\pi \, \xi^2 \tag{52}$$

we obtain from (51) for the maximum critical current density

$$J_c = c \cdot H_c/16\pi \sqrt{2} \, \lambda \quad . \tag{53}$$

This result is close to the value calculated from the Ginzburg-Landau theory for the case of constant order parameter.

So far we have treated only the interaction between the individual flux line and the pinning center, including the case of perfect geometrical matching between the flux structure and the pinning-center configuration. In the case of a statistical geometric distribution of pinning centers (which may all be identical) the summation of the pinning forces and the elastic properties of the flux-line lattice must be handled more carefully. This situation is illustrated in Fig. 14. In a perfectly rigid flux-line lattice the forces from the pinning centers act in all directions perpendicular to the line orientation, thereby cancelling each other and resulting in zero overall flux pinning. On the other hand, if elastic distortions of the flux-line lattice are possible, under the influence of an external driving force the vortex lines will be held locally to the various pinning centers such that the total pinning force in any particular direction is maximized (Alden and Livingston, 1966; Yamafuji and Irie, 1967). However, the consumption of elastic energy in the vortex lattice tends to reduce the overall pinning force. Again, we expect maximum flux pinning for perfect geometrical matching between the arrangement of the pinning sites and the undistorted flux-line lattice.

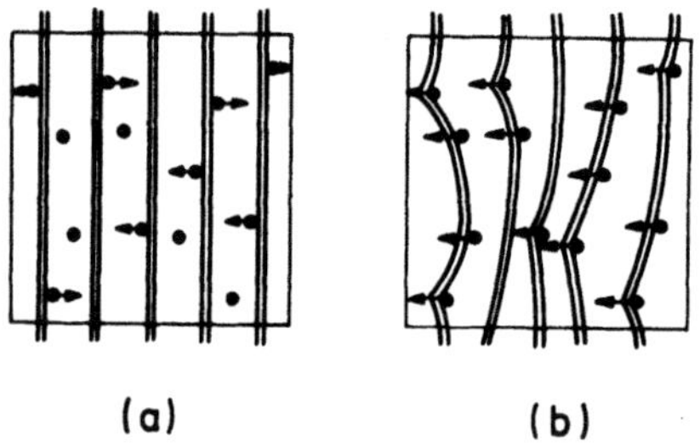

Fig. 14. Summation of pinning forces for an arbitrary distribution of pinning sites. (a) rigid flux-line lattice, (b) soft flux-line lattice. Arrows indicate the forces from the pinning interaction; driving force directed to the right.

Generally, in a material flux pinning cannot be described simply by a configuration of identical pinning sites. Instead we have to deal with a distribution of the local pinning force, i.e., a distribution of the values of f_p and J_c in (1) and (9), respectively. Such a distribution leads to a regime in the V-I characteristics with upward curvature before the linear flux-flow regime is attained at higher currents. Equation (9) can then be written as

$$V = \frac{\Phi^2 \cdot L_s}{\eta \cdot c^2} \cdot \sum_i n_i \cdot (J - J_{ci}) \quad , \tag{54}$$

where L_s is the sample length and n_i the fraction of the density n_f of the flux lines or flux tubes experiencing the critical current density J_{ci}. At a particular current density J only the fractions n_i with $J_{ci} < J$ participate in flux flow. For describing this fraction we define the distribution function P(J)

$$\sum_i n_i \; [J_{ci} < J] = P(J)\; n_f \quad , \tag{55}$$

where the sum is taken over all fractions i with $J_{ci} < J$. The behavior expressed in (54) is illustrated in Fig. 15a. At the different values of J_{ci} an additional term of the sum in (54) must be added. It is important to note that each term is linear in J and starts from zero at $J = J_{ci}$. Figure 15b shows the derivative $\partial V/\partial J$ for the given example. We see that this derivative is directly related to the distribution function P(J), and we can write

$$\frac{\partial V}{\partial J} = \frac{\Phi^2 \cdot L_s}{\eta \cdot c^2} n_f \cdot P(J) \quad . \tag{56}$$

The distribution function P(J) increases from zero at the smallest value of J_{ci} and reaches one at the maximum J_{ci} where all flux lines participate in flux flow (see Fig. 16a). The derivative

$$p(J) = \frac{\partial P(J)}{\partial J} \tag{57}$$

defines the distribution density and indicates the relative probability per current-density intervall for the flux lines to experience the critical current density J (Fig. 16b). The distribution density

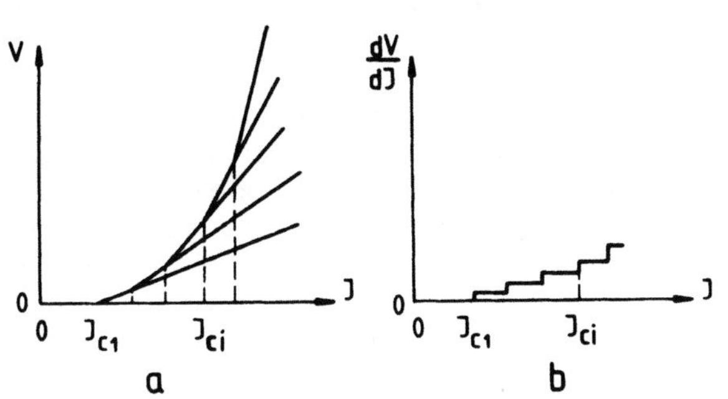

Fig. 15. Distributed strength of the local pinning force: voltage versus current density (a) and derivative ∂V/∂J versus J (b), as indicated in (54).

is related to the second derivative $\partial^2 V/\partial J^2$ according to (56):

$$\frac{\partial^2 V}{\partial J^2} = \frac{\Phi^2 \cdot L_s}{\eta c^2} \cdot n_f \cdot p(J) \quad . \tag{58}$$

We see that in principle the distribution density of the local pinning force can be obtained from the derivative $\partial^2 V/\partial J^2$. In such an analysis of experimental V(I) curves it is important to include the effects of both bulk pinning and edge pinning.

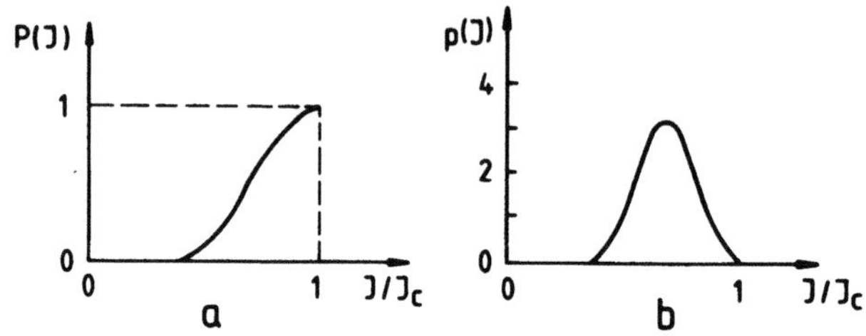

Fig. 16. Distributed pinning force: (a) distribution function P(J) defined in (55) and (b) derivative $p(J) = \partial P(J)/\partial J$ versus the normalized current density J/J_c. (At $J = J_c$ the function P(J) reaches the value 1.0).

Finally, we discuss the frictional force f_p acting upon the moving flux structure observed experimentally and contained in (1) and (9). This force results in a constant reduction of v_f and V for a given current density J compared with the case $f_p = 0$, shifting the V(I) curves to higher currents along the current axis. The appearance of this frictional force is not as obvious as the action of the pinning force against the onset of flux motion. At first sight, the total work done against the pinning force by a flux line moving across the potential well of a pinning center appears to be zero, However, from (1) and (6) together with (5) the power dissipation per unit length of vortex line is

$$\frac{J \cdot |E|}{n_f} = (\eta \cdot v_f + f_p) \cdot v_f \quad , \tag{59}$$

the second term on the right containing the contribution from the dynamic pinning force.

During flux flow the reduction of the Lorentz force due to the pinning force f_p can be seen from a simple argument. We note that the effective mass of a flux line or flux tube is extremely small and that the inertia force is at least several orders of magnitude smaller than the other forces acting upon the magnetic flux structure. (For this reason we have neglected the inertia force in the phenomenological force equations (1) and (28). The effective mass of a moving flux line can be estimated by considering the kinetic energy contained in the currents induced by the flux motion (Kim and

Stephen, 1969)). What happens when a moving flux line passes the potential well of a pinning center ? As a consequence of its small effective mass, the flux line cannot recover its previous potential energy in form of kinetic energy when reaching the bottom of the well. Instead this potential energy is totally dissipated. More quantitatively, we must look at the time-averaged force acting upon the flux line when passing the pinning site. Assuming a symmetric pinning-potential well with linear banks of the same slope f_p on opposite sides, the forces $(J\cdot\Phi/c)+f_p$ and $(J\cdot\Phi/c)-f_p$ will act upon the flux line when entering and leaving the well, respectively. Noting the different velocities on the two banks, the time-average of the two forces is

$$\overline{\text{force}} = \frac{J \cdot \Phi}{c} - \frac{f_p}{J \cdot \Phi/c} \cdot f_p \quad . \tag{60}$$

Equation (60) is restricted to the regime $(J\phi/c)\geq f_p$, the time-averaged force being zero for $(J\phi/c)\leq f_p$. (In our simple argument the Magnus force has been neglected). We see from (60) that the time-averaged force acting upon a single flux line or flux tube is reduced below the value of the Lorentz force because of flux pinning. However, in (60) this reduction is not independent of the current density J and vanishes for $J\Phi/c \gg f_p$.

For explaining the constant frictional force $\underline{f}_p$ in the flux-flow state one must consider the elastic distortions of the configuration of vortex lines taking place as it passes through a material with pinning sites. As shown by various authors (Yamafuji and Irie, 1967; Lowell, 1970; Good and Kramer, 1970; Larkin and Ovchinnikov, 1979; Brandt, 1980) the elastic distortions can lead to a frictional force which is independent of the external driving force and about equal to the force exerted by the pinning sites on the static vortex lattice.

6. TIME-DEPENDENT THEORIES

Flux flow has been treated in a number of phenomenological theories. Here the relatively elementary Bardeen-Stephen model provides good insight into the basic physics of flux flow and of the dissipative processes. It concentrates on the dynamics of a single flux line. A more general analysis has been given in terms of the time-dependent Ginzburg-Landau theory.

6.1 Bardeen-Stephen (BS) Model

Bardeen and Stephen (1965) have assumed a vortex line with zero order parameter in the core region (local model). For $r > \xi_c$ (ξ_c = core radius) the order parameter increases gradually with the radial distance r from the vortex axis to its equilibrium value far away from the core. The transport current density is assumed small but uniform. More specifically, the transport current density and the normal current density generated by the flux motion are taken as small compared to the supercurrent density circulating around the vortex core. This amounts to the criterion that

$$v_f \cdot \tau \ll \xi_c \quad , \tag{61}$$

the distance which the vortex line moves during the quasi-particle scattering time τ being small compared with the core radius ξ_c. Finally, free flux flow without influence of flux pinning is assumed.

Outside the core we apply London's equations and inside the core Ohm's law. From the London equation for the local electric field $\underline{e}$

$$\underline{e} = \frac{m}{n_s \cdot e^2} \dot{\underline{J}}_s \tag{62}$$

with $\underline{J}_s = n_s \cdot e \cdot \underline{v}_s$ we find outside the vortex core

$$\underline{e} = \frac{m}{e} \frac{\partial \underline{v}_s}{\partial t} = - \frac{m}{e} \underline{v}_f \cdot \text{grad}\ \underline{v}_s \ . \tag{63}$$

We introduce cylinder coordinates around the vortex axis according to Fig. 17. From the London model describing the supercurrent density around an isolated vortex line we find in the regime $\xi_c < r < \lambda$

$$\underline{v}_s = \frac{\hbar}{2mr} \underline{\Theta} \quad . \tag{64}$$

For flux motion along the x-direction (63) and (64) then yield

$$\underline{e} = - \frac{\Phi_o}{2\pi c} v_{fx} \cdot \frac{\partial}{\partial x} \left(\frac{\underline{\Theta}}{r}\right) = \frac{v_{fx} \cdot \Phi_o}{2\pi c\ r^2} (\underline{\Theta} \cos\Theta - \underline{r} \sin \Theta) . \tag{65}$$

This represents the field of a series of dipoles as shown in Fig. 18.

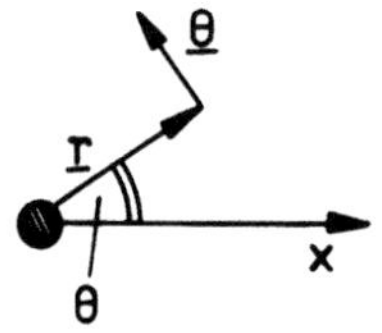

Fig. 17. Coordinates x and Θ and vectors $\underline{r}$ and $\underline{\Theta}$ for discussing vortex motion.

At the surface $r = \xi_c$ the tangential component of $\underline{e}$ must be continuous. From (65) we find a uniform field inside the core in y-direction

$$\underline{e}_{core} = \frac{v_{fx} \cdot \Phi_o}{2\pi c\ \xi_c^2} \hat{\underline{y}} = \frac{v_{fx} \cdot H_{c2}}{c} \hat{\underline{y}} \ , \tag{66}$$

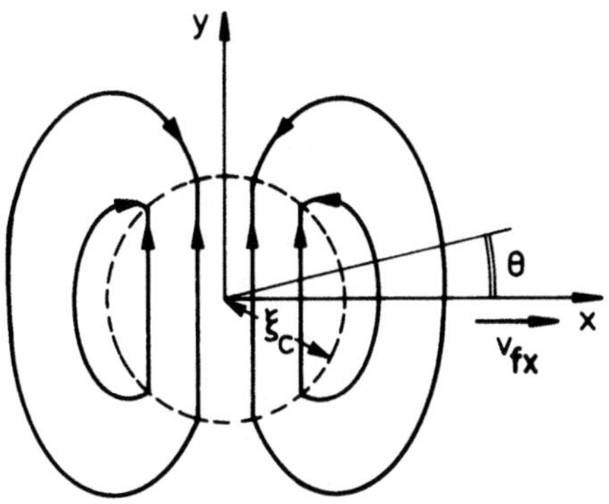

Fig. 18. Dipole field according to (65).

where we have inserted $H_{c2} = \Phi_o / 2\pi\xi_c^2$. The radial component of $\underline{e}$ shows a discontinuity at the surface $r = \xi_c$ arising from a surface charge at the core boundary. (In a more exact non-local theory this discontinuity is smeared out).

The energy dissipation in the core is per unit length

$$W_{core} = \pi\, \xi_c^2 \cdot \sigma_n \cdot \underline{e}_{core}^2 = \frac{v_f^2 \cdot \Phi_o^2}{4\pi c^2\, \xi_c^2\, \rho_n} \,. \tag{67}$$

According to BS the normal currents outside the core cause an energy dissipation identical to that in (67). (Near T_c this can easily be verified by integration of (65) yielding

$$\int_{\xi_c}^{\infty} \int_{0}^{2\pi} \underline{e}^2(r) \cdot r \cdot dr \quad d\Theta = \frac{v_f^2 \cdot \Phi_o^2}{4\pi c^2\, \xi_c^2} \qquad) \,.$$

Hence, the total dissipation per unit length of vortex is

$$W = \frac{v_f^2 \cdot \Phi_o^2}{2\pi c^2\, \xi_c^2\, \rho_n} = \eta \cdot v_f^2 \,, \tag{68}$$

yielding

$$\eta = \frac{\Phi_o^2}{2\pi c^2\, \xi_c^2\, \rho_n} = \frac{\Phi_o\, H_{c2}}{c^2\, \rho_n} \,. \tag{69}$$

This agrees with the empirical result (12b) found in the range $T \ll T_c$. In (69) we have approximated $\xi_c \approx \xi$.

It is only in the case of free vortex flow that the current density within the core is just equal to the applied current density. If free flux flow is hindered because of pinning or some other mechanism, both current densities are unequal. For complete flux pinning and $v_f = 0$ the current density in the core and the energy dissipation is zero.

So far we have neglected the Hall effect and assumed flux motion perpendicular to the applied current. Incorporating cyclotron motion into their model, BS obtained the same Hall angle as in the normal state. It appears that the detailed behavior of the Hall angle, such as shown in Fig. 5, cannot be explained by a model as simple as that adopted by BS.

The local BS model has been extended to a non-local superconductor (Bardeen and Sherman, 1975; Larkin and Ovchinnikov, 1976; Kopnin and Kravtsov, 1976, Bardeen, 1978). The order parameter now gradually approaches zero at the vortex axis, and the effective core volume must be obtained from microscopic theory. The viscosity coefficient η found from these non-local models does not differ much from the local result.

Nonlinear effects arising during vortex motion, if condition (61) is strongly violated and a redistribution of the quasiparticles must be taken into account, have been discussed by Larkin and Ovchinnikov (1977). Experimentally, so far this regime seems to be unexplored. Clearly, such effects will be difficult to observe because of other nonlinearities arising from flux pinning.

6.2 Time-Dependent Ginzburg-Landau (TDGL) Theory

Time-dependent generalizations of the Ginzburg-Landau theory have been investigated by a number of authors since about 1966. Near the transition temperature TDGL equations were first derived by Schmid (1966). In the construction of a time-dependent theory serious complications arise from the existence of an energy gap and from the singularity in the density of states at the gap edge. TDGL equations can be justified rigorously only for the gapless regime (Gor'kov and Eliashberg, 1968). A detailed discussion of these questions can be found in a review by Cyrot (1973).

Starting from the general concepts of nonequilibrium thermodynamics one can write the transport equation

$$\frac{d\psi}{dt} + \gamma \cdot \frac{df}{d\psi} = 0 \quad . \qquad (70)$$

Here the derivative $df/d\psi$ of the free-energy density with respect to the order parameter is introduced as a generalized force acting upon the order parameter ψ. The quantity γ is the corresponding transport coefficient. Expanding f in the usual way in powers of $|\psi|$, one obtains

$$\left(\frac{\partial}{\partial t} + \frac{2ie}{\hbar}\tilde{V}\right)\psi = -2\gamma[\alpha\psi+\beta|\psi|^2\psi + \frac{1}{4m}\left(\frac{\hbar}{i}\underline{\nabla} - \frac{2e}{c}\underline{A}\right)^2\psi] \quad . (71)$$

Here the second term on the left containing the effective potential $\tilde{V}$ has been added for obtaining gauge invariance. It has been proposed by Schmid (1966) that $\tilde{V}$ be written as

$$\tilde{V} = V - \frac{\mu}{e} \quad , \tag{72}$$

where V is the electric and μ the chemical potential. Setting as usual $|\psi_o|^2 = -\alpha/\beta$, $\xi^2 = -\hbar^2/4m\alpha$, and identifying the factor $\gamma\hbar^2/2m$ as the normal-state diffusion constant D

$$\gamma\hbar^2/2m \; = \; D = \frac{1}{3} \cdot v_F \cdot \ell \quad , \tag{73}$$

we obtain from (71)

$$\left(\frac{\partial}{\partial t} + \frac{2ie}{\hbar}\tilde{V}\right)\frac{\psi}{\psi_o} + \frac{D}{\xi^2}\left[\left(\frac{\psi}{\psi_o}\right)^2 - 1\right]\frac{\psi}{\psi_o} + D\cdot\left(\frac{1}{i}\nabla - \frac{2e}{\hbar c}\underline{A}\right)^2 \frac{\psi}{\psi_o} = 0 \; . \tag{74}$$

Equation (74) has been derived by Gor'kov and Eliashberg (1968) from microscopic theory for a gapless superconductor. For the current density they obtained

$$\underline{J}_s = \sigma_n\left(-\underline{\nabla}\tilde{V} - \frac{1}{c}\frac{\partial \underline{A}}{\partial t}\right) + \mathrm{Re}\left[\frac{\psi^*}{\psi_o}\left(\frac{1}{i}\underline{\nabla} - \frac{2e}{\hbar c}\underline{A}\right)\frac{\psi}{\psi_o}\right]\cdot \frac{1}{8\pi e\,\lambda^2} \tag{75}$$

and for the charge density

$$\tilde{\rho} = \frac{\tilde{V} - V}{4\pi\,\lambda_{TF}^2} \quad . \tag{76}$$

Here the temperature-dependent coherence length is $\xi = \hbar(6D/\tau_{so})^{1/2}/\psi_o$, where τ_{so} is the spin-flip scattering time, and the temperature-dependent magnetic penetration depth is $\lambda = \hbar c/(8\pi\sigma_n\tau_{so})^{1/2}\psi_o$. σ_n is the normal-state conductivity, and λ_{TF} the Thomas-Fermi static charge screening length. The three equations (74) - (76) are supplemented by Maxwell's equations coupling the scalar and vector potentials V and $\underline{A}$ to the charge and current densities $\tilde{\rho}$ and $\underline{J}_s$, respectively.

For describing flux flow we need solutions of (74) with the order parameter moving at some uniform velocity. Solutions of this type have been investigated by many authors. A summary has been given in a recent monograph (Huebener, 1979). The calculations of the various transport effects in type-II superconductors have concentrated on the regime near the upper critical field H_{c2}. Near H_{c2} the various theories yield for the flux-flow resistivity ρ_f of a triangular vortex lattice

$$\frac{\rho_f}{\rho_n} = 1 - \alpha^*\left(1 - \frac{H}{H_{c2}}\right) \quad , \tag{77}$$

where the normalized slope α^* is

$$\alpha^* = \frac{H_{c2}}{\rho_n}\left(\frac{\partial\rho_f}{\partial H}\right)_{H_{c2}} = \frac{4\,\kappa_1^2(0)}{1.16[2\kappa_2^2(T)-1]+1}\,\{\text{extensions}\}\ . \qquad (78)$$

This expression refers to the geometry of a thin superconducting plate in a perpendicular magnetic field. Here $\kappa_1(T)$ and $\kappa_2(T)$ are the temperature dependent Ginzburg-Landau parameters introduced by Maki (1964) with the property $\kappa_1(T) \to \kappa$ and $\kappa_2(T) \to \kappa$ for $T \to T_c$. The factor {extensions} indicates various temperature-dependent functions applying to the different cases regarding the strength of coupling and pair breaking.

From the Abrikosov theory of the magnetization of a type-II superconducting plate the fraction of the material remaining superconducting in the perpendicular field H near H_{c2} is

$$\text{S-fraction} = \frac{1-\dfrac{H}{H_{c2}}}{1.16[2\kappa^2-1]+1}\ . \qquad (79)$$

Using this result for a rough estimate of ρ_f we obtain

$$\frac{\rho_f}{\rho_n} \approx 1 - \frac{1-\dfrac{H}{H_{c2}}}{1.16[2\kappa^2-1]+1}\ . \qquad (80)$$

Experimental measurements of the slope α^* with high accuracy are not easy because of appreciable rounding of the $\rho_f(H)$ curves near H_{c2}. It appears that the TDGL theory can reasonably account for the experimental results if the question of the coupling strength is carefully treated (Imai, 1975; Hagmann et al., 1977).

TDGL calculations of the Hall angle for dirty type-II superconductors predict a strong increase of the Hall angle with decreasing magnetic field just below H_{c2} in agreement with the experimental results (see Fig. 5; Huebener, 1979).

7. CURRENT-INDUCED DISSIPATIVE STATE

So far we have dealt with magnetic flux structures generated by an external magnetic field applied to the superconductor and their response to an external force. In this section we treat the case of zero applied magnetic field. What happens now when we apply an electric transport current? At a sufficiently high current value flux-containing normal domains penetrate into the superconductor, and eventually electrical resistance will appear. Clearly, these current-induced normal domains represent a highly nonlinear response in a situation far from equilibrium. No simple transport coefficients exist in this case. The details of the response depend upon the geometry of the current-carrying superconductor (Huebener, 1979). In the following we restrict ourselves to a flat strip. Here a dynamic resistive state is attained with trains of flux lines or flux tubes traversing the strip perpendicular to the direction of the transport current. We will concentrate on strips of type-I superconductors with dimensions such that the moving flux tubes contain several hundred flux quanta and the dynamic details can be studied using various experimental techniques.

Figure 19 shows the current-induced dissipative state (CIDS) observed magneto-optically in a long Pb strip. As the current is gradually increased from zero in such a geometry, flux containing "channels" of normal material appear with increasing number. Above the critical current the number of these channels increases about linearly with current, as can be seen from Fig. 20. This results in a linear increase of the electric resistance and therefore a quadratic increase of the resistive voltage with increasing current (Huebener and Kampwirth, 1972). Of course, at sufficiently high currents, where the total sample length becomes normal, we have ohmic behavior.

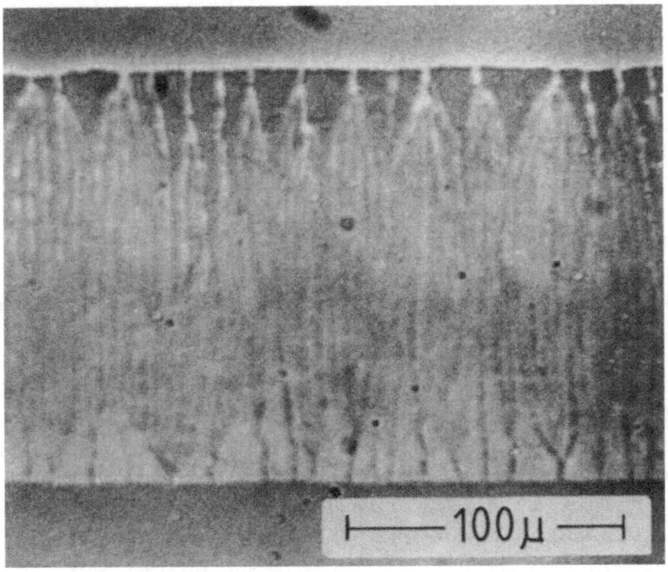

Fig. 19. Magnetic flux structure in a Pb strip of 4.0 μm thickness and 160 μm width at 4.2 K; applied current density = 5.9x 10^5 A/cm^2. At the top and the bottom the sample edges can be seen. The bright and dark "channels" indicate the current-induced flux structure detected magneto-optically (Huebener and Kampwirth, 1972).

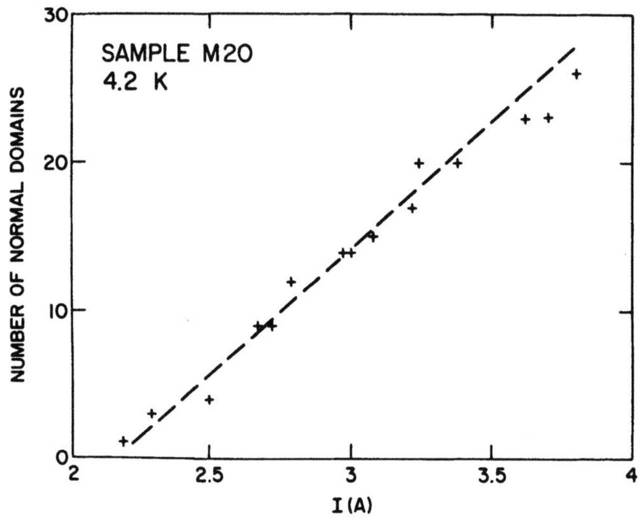

Fig. 20. Number of current-induced resistive domains versus current in a Pb strip of 4.0 μm thickness and 160 μm width at 4.2 K; length of observed section = 320 μm (Huebener and Gallus, 1973).

For investigating further details of the CIDS it is advantageous to restrict the dissipation to a single resistive "channel". This can easily be achieved in a constricted geometry where a short strip of only a few μm length is imbedded between wider and thicker end sections. The length of the constriction is chosen such that here only a single channel can be generated. Various experiments have shown that the channels of normal phase, seen in Fig. 19, actually consist of trains of flux tubes nucleated at the edge and moving rapidly to the center of the strip. In the center, opposite tubes arriving from the opposite edges annihilate each other. The most direct experimental evidence for this dynamic behavior has been obtained from stroboscopic magneto-optical flux detection at a time resolution of better than 0.1 μs (Chimenti and Huebener, 1977). The result of such an experiment performed with a constricted Pb film is shown in Fig. 21.

In a phenomenological model of the CIDS we have to take into account the following three contributions to the Gibbs free energy of a flux tube in the strip (Clem et al., 1973): a) the energy ΔG_{in} from the region within the superconductor arising from the loss of condensation energy, b) the energy ΔG_{out} contained in the magnetic field of the flux tube in the region outside the superconductor, and c) the interaction energy ΔW_T with the applied current,

$$\Delta G = \Delta G_{in} + \Delta G_{out} - \Delta W_T \quad . \tag{81}$$

For mathematical purposes it is convenient to approximate the cross section of the strip of width w and thickness d by an ellipse. (In

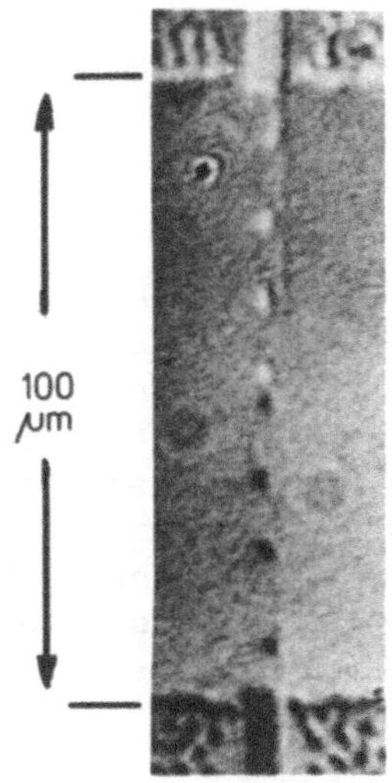

Fig. 21. Flux-tube train in the CIDS of a constricted Pb film of 6 μm thickness at 4.2 K observed by stroboscopic magneto-optic flux detection. Sample orientation same as in Fig. 19; current flows from left to right. Note black and bright flux tubes entering the sample from opposite edges.

the case $w/d >> 1$ we are interested in, this approximation only affects the immediate neighborhood of the edges). Further, we adopt a two-dimensional model, where all results are expressed per unit length in longitudinal direction. We then use this model for our constricted geometry by multiplying the results with a length to be specified. (A rigorous extension of this model to the three dimensions of the flux-tube nucleation process presents severe mathematical difficulties). In the energy summation of (81) we refer to the penetration of the first flux tube into a superconductor whose interior is completely free of flux. The three contributions in (81) to the energy of a flux tube containing the flux Φ are (Clem et al., 1973)

$$\Delta G_{in} = \frac{\Phi \cdot H_c \cdot d}{4\pi} \left(1 - \frac{4x^2}{w^2}\right)^{1/2} , \quad (82)$$

$$\Delta G_{out} = \frac{\Phi^{3/2} \cdot H_c^{1/2}}{4\pi^2} \left[\ell n\left\{\frac{w}{2r}\left(1 - \frac{4x^2}{w^2}\right)\right\} + \frac{3}{2}\right] , \quad (83)$$

$$\Delta W_T = \frac{\Phi \cdot I}{\pi c} \arccos\left(\frac{2x}{w}\right) . \quad (84)$$

In (82) - (84) the coordinate x is measured from the center of the strip; r is the radius of the flux tube. A plot of these terms versus the coordinate x across the sample for a typical experimental situation is shown in Fig. 22. In expression (82) we have ignored the wall energy of the flux tube. Of course, the wall energy becomes important for determining the actual flux-tube size during the dynamic dissipative process. Here we only look at the function $\Delta G(x)$ for a given tube of flux Φ. The particular coordinate dependence of ΔG_{in} results from the assumed elliptical cross section of the strip. The interaction energy ΔW_T in (84) is obtained by incorporating the current-density distribution $J(x)$ across the strip valid for the Meissner state.

The most important consequence of the terms in (81) is the appearance of an energy barrier against flux entry near the sample edge. Irreversible flux penetration can only take place when this energy barrier has been removed through the term ΔW_T. The smallest current value at which the barrier at the edge just reaches zero then defines the critical current I_c. (Figure 22 shows the situation where such a barrier still exists). Because of the energy barrier, I_c is enhanced above the value expected from Silsbee's rule (stating that the critical current is reached just when the current-induced field attains the value H_c at the sample edge).

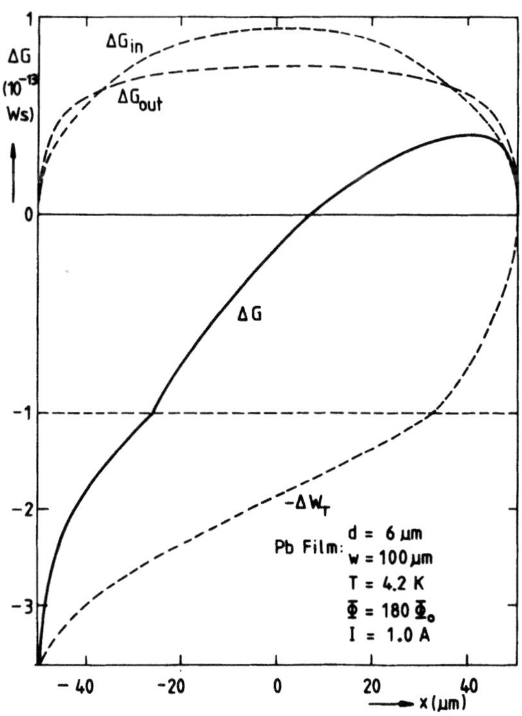

Fig. 22. Dependence of the three contributions to the flux-tube energy in (81), given explicitly in (82)-(84), upon the sample coordinate (sample parameters are given in the inset; note change in scale on vertical axis).

The model expressed in (81) - (84) and displayed in Fig. 22 so far excludes the presence of any flux tubes within the superconductor from previous nucleation events. The interaction between flux tubes is repulsive if they have the same orientation and attractive if they have opposite orientation. This flux-tube interaction in a thin-film superconductor can be treated conveniently in terms of magnetic monopoles located at the top and bottom of the film where the flux tube intersects the film surface (Chimenti and

Clem, 1978). This leads to the interaction energy

$$\Delta G_{int} = \pm \Phi^2/4\pi^2 s_{12}^2 \tag{85}$$

between two tubes of flux Φ, where s_{12} is the distance between their axes. The positive and negative sign refers to the repulsive and attractive mutual orientation, respectively. Equation (85) can be interpreted in terms of Coulomb's law for "magnetic charges" $\Phi/2\pi$. From (85) the magnetic field produced on the surface by flux tube 1 at the location of flux tube 2 is given by

$$B_{12} = \Phi/2\pi\, s_{12}^2 \quad . \tag{86}$$

From the spatial variation of the Gibbs free energy $\Delta G(x)$ of a flux tube we find the force acting upon the tube from the derivative $-\Delta G(x)/\partial x$. By equating this force with the damping force $\eta\, v_f$ we obtain the spatial variation of v_f. Such a calculation can be done analytically using (81) - (84) for the case where no other flux tubes exist within the constriction and where the function $J(x)$ valid for the Meissner state can be applied. If other tubes exist within the constriction, we must add the proper sum of all interaction terms of type (85) to the sum in (81), find the forces through the spatial derivative of the Gibbs free energy, and numerically solve the force-balance equation including the damping term $\eta\, v_f$ for the stationary flux-tube configuration. The dynamics of the CIDS and in particular the dependence of v_f upon the coordinate within the constriction can well be accounted for with the model we have outlined (Chimenti and Clem, 1978; Selig, 1980; Buck et al., 1980).

Experimentally, a systematic study of the repulsive influence of the previous flux tubes in the sample interior upon the flux nucleation process at the sample edge is possible by externally varying the flux-tube nucleation frequency ν^*. The controlled variation of ν^* can be achieved by biasing the constriction with a current smaller than I_c and superimposing short current pulses such that the total current during each pulse reaches a value larger than I_c. The parameters are taken such that just a single flux tube is nucleated during each pulse. In order to keep the time-averaged current nearly constant during the variation of the pulse frequency, the ratio of the pulse amplitude to the bias current should be as small as possible. Typical experimental values for this ratio are 1 %. (Selig, 1980; Selig and Huebener, 1980).

If the nucleation frequency ν^* is small, the flux nucleation process is not affected by any previous flux tubes since they have all disappeared. The resistive voltage is proportional to ν^*, following the Josephson relation

$$V = \frac{1}{c}\,\Phi \cdot \nu^* = \frac{1}{c}\, N_t \cdot \Phi_o \cdot \nu^* \quad , \tag{87}$$

where N_t is the number of flux quanta per flux tube. With increasing frequency we expect a discontinuity in the V-ν^* relation each time an additional pair of the flux tubes nucleated previously still exists within the constriction when the nucleation of the new pair takes place at the edges. Since the return flux of the preceding tubes at the sample edges is directed opposite to the magnetic field of the applied current, the flux Φ per tube will be reduced causing a change in the slope of the $V(\nu^*)$ curve (see Fig. 23). Here we have assumed symmetric behavior with simultaneous nucleation of two flux tubes of opposite sign at the opposite edges. For the nonsymmetric case (where due to the detailed sample structure annihilation occurs near one edge) these considerations need only little modification. Typical experimental results confirming our expectation are shown in Fig. 24.

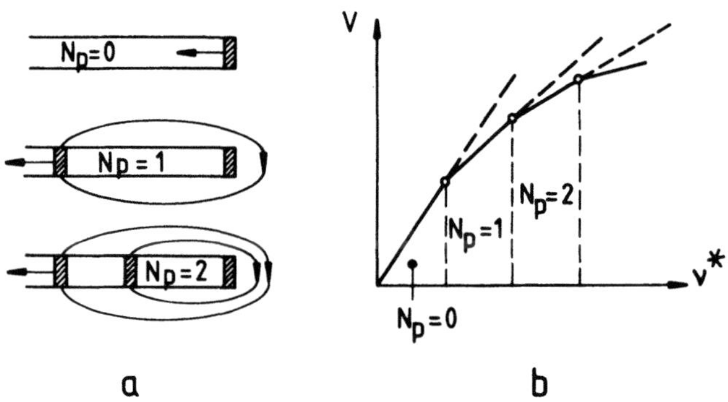

Fig. 23. Return field of the preceding flux tubes at the sample edge (a) and resistive voltage versus the nucleation frequency ν^* (b); N_p indicates the number of the preceding flux-tube pairs existing during the nucleation process.

The reduction of the nucleated flux per tube due to the compensating effect of the previous tubes can be expressed in the form

$$\Phi(N_p + 1) = \Phi(N_p) - \frac{d\Phi}{dB_e} \Delta B_e \quad . \qquad (88)$$

Here N_p is the number of flux-tube pairs existing already within the constriction during the nucleation process; ΔB_e is the increment in the compensating field at the edge caused by the additional pair. The magnetic field at the sample edge due to a flux tube located at the distance s_{12} from the edge is approximately given by (86). The increment ΔB_e can then be obtained by linear superposition of this contribution from the different tubes. Here the coordinates of each tube can be found from the stationary solution of the force-balance equation as indicated above.

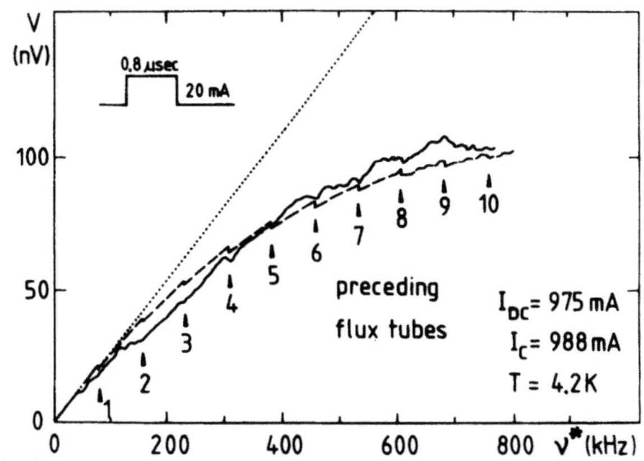

Fig. 24. Voltage V versus nucleation frequency ν^* for a Pb constriction (thickness = 7.0 μm; width = 100 μm; length = 6.5 μm). The arrows indicate the frequencies at which the time-averaged number of the preceding flux tubes within the constriction reaches the values $\bar{N}_f$= 1 , 2, ..., etc.; nonsymmetric nucleation. Solid line: experimental result; dashed line: theoretical curve obtained as indicated in the text; dotted line: behavior if preceding tubes are neglected.

The derivative $d\Phi/dB_e$ in (88) can be determined in another experiment, namely through the variation of the bias current at constant pulse frequency. The observed V-I characteristic with and without the superimposed pulses is shown schematically in Fig. 25. The application of the current pulses results in a linear precursor region, where the flux Φ per tube increases linearly with I (see (87)). By calculating the current-induced field at the edge, we obtain the derivative $d\Phi/dB_e$ directly from the slope of the V(I) curve in the precursor region. A theoretical expression for this derivative can be obtained by considering the compensation at the sample edge between the field of the applied current and the time-averaged return flux of the tubes within the sample such that the total field remains constant and equal to the critical entry field. This theoretical value of the derivative $d\Phi/dB_e$ agrees reasonably with the experimental results (Selig and Huebener, 1980).

Inserting the flux Φ from (88) into (87) we can explain the $V(\nu^*)$ dependence observed experimentally (Fig. 24). Saturation of this dependence is to be expected when the nucleation frequency approaches the inverse flux-tube nucleation time valid for free-running flux nucleation and an applied direct current whose magnitude is equal to the maximum total current reached by the superposition of the current pulses.

The ideas we have outlined provide a satisfactory understanding of the dynamic features of the CIDS in a constricted superconducting

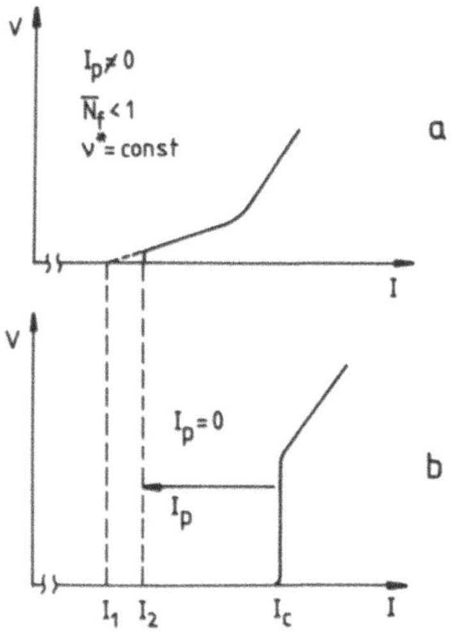

Fig. 25. V-I characteristic observed for a Pb constriction with (a) and without (b) superimposed current pulses of amplitude I_p and constant frequency, schematically.

strip (Chimenti and Clem, 1978; Selig and Huebener, 1980; Buck et al., 1980). Theoretical lines similar to those presented here have been followed by Likharev (1971) and by Aslamazov and Larkin (1975) for treating the resistive behavior in wide microbridges of a type-II superconductor.

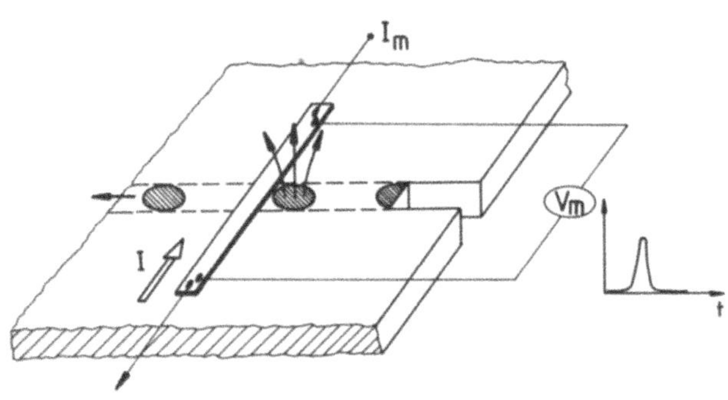

Fig. 26. Magneto-resistive micro-field probe for counting the flux tubes traversing the constricted superconductor. The probe is electrically insulated from the sample and consists of a narrow film of O_2-doped Al. It is biased at the current I_m. Each flux tube causes a voltage pulse $V_m(t)$ in the field probe (Selig et al., 1978).

Direct measurements of the flux-tube nucleation frequency ν^* in the CIDS of a constricted film of a type-I superconductor have been performed using a magneto-resistive microprobe for detecting the flux carried by each single tube. Here the micro-field probe acts as a practically passive flux-tube counter. The principle is shown in Fig. 26. Knowing the frequency ν^* and the resistive voltage, the number N_t of flux quanta per flux tube can then be obtained with high accuracy from (87). The sensitivity of this technique is exceptionally high, since the method essentially represents a counting technique. The method has been used for studying the temporal fluctuations of the nucleation frequency ν^* and for determining the power spectrum of these fluctuations (Selig et al., 1978; Hurm et al., 1979).

ACKNOWLEDGMENT

It is a pleasure to acknowledge stimulating and clarifying conversations with Prof. J. R. Clem and Dipl. Phys. K. P. Selig.

REFERENCES

Alden, T. H. and Livingston, J. D., 1966, Ferromagnetic particles in a type-II superconductor, J. Appl. Phys., 37:3551.

Aslamazov, L. G. and Larkin, A. I., 1975, Josephson effect in wide superconducting bridges, Zh. Eksp. Teor. Fiz. 68, 766, Sov. Phys. JETP, 41:381.

Bardeen, J., 1978, Motion of flux lines in nearly pure superconductors, Phys. Rev. B, 17:1472.

Bardeen, J. and Sherman, R. D., 1975, Flux flow in nearly pure low-κ superconductors, Phys. Rev. B, 12:2634.

Bardeen, J. and Stephen, M. J., 1965, Theory of the motion of vortices in superconductors, Phys. Rev. A, 140:1197.

Brandt, E. H., 1980, Flux flow and pinning, to be published.

Buck, W. and Selig, K. P. and Parisi, J., 1980, Model calculations of flux-tube nucleation in thin-film type-I superconductors, to be published.

Campbell, A. M. and Evetts, J. E., 1972, Flux vortices and transport currents in type-II superconductors, Adv. in Phys., 21:199.

Chandrasekhar, B. S., 1962, A note on the maximum critical field of high-field superconductors, Appl. Phys. Lett., 1:7.

Chimenti, D. E. and Clem, J. R., 1978, Current-induced flux motion in type-I superconducting films studied at 100-ns time resolution, Phil. Mag. B, 38:635.

Chimenti, D. E. and Huebener, R. P., 1977, Stroboscopic study of flux dynamics in superconductors during current-induced breakdown, Solid State Commun., 21:467.

Clem, J. R., 1970, Theory of flux-flow noise voltage in superconductors, Phys. Rev. B, 1:2140.

Clem, J. R., 1974, Theory of magnetically coupled type-II superconducting films, Phys. Rev. B, 9:898.

Clem, J. R., 1975, Theory of the coupling force in magnetically coupled type-II superconducting films, Phys. Rev. B, 12:1742.

Clem, J. R., Huebener, R. P. and Gallus, D. E., 1973, Gibbs free-energy barrier against irreversible magnetic flux entry into a superconductor, J. Low Temp. Phys., 12:449.

Clogston, A. M., 1962, Upper limit for the critical field in hard superconductors, Phys. Rev. Lett., 9:266.

Cyrot, M., 1973, Ginzburg-Landau theory for superconductors, Rep. Prog. Phys., 36:103.

Ekin, J. W. and Clem, J. R., 1975, Magnetic coupling force of the superconducting dc transformer, Phys. Rev. B, 12:1753.

Ekin, J. W., Serin, B. and Clem, J. R., 1974, Magnetic coupling in superposed type-II superconducting films, Phys. Rev. B, 9:912.

Giaever, I., 1965, Magnetic coupling between two adjacent type-II superconductors, Phys. Rev. Lett., 15:825.

Giaever, I., 1966, Flux pinning and flux-flow resistivity in magnetically coupled superconducting films, Phys. Rev. Lett., 16:460.

Good, J. A. and Kramer, E. J., 1970, Deformation-induced anisotropy of the critical current in single crystal niobium, Phil. Mag., 22:329.

Gor'kov, L. P. and Eliashberg, G. M., 1968, Generalization of the Ginzburg-Landau equations for non-stationary problems in the case of alloys with paramagnetic impurities, Zh. Eksp. Teor. Fiz., 54, 612, Sov. Phys. JETP, 27:328.

Haenssler, F. and Rinderer, L., 1967, Statique et dynamique de l'état intermédiaire des supraconducteurs du type I, Helv. Phys. Acta, 40:659.

Hagmann, P., Meier-Hirmer, R. and Winter, H., 1977, Strong coupling effects in the temperature dependence of the flux-flow resistance near H_{c2}, Z. Physik B, 26:233.

Van Hove, L., 1954a, Correlations in space and time and born approximation scattering in systems of interacting particles, Phys. Rev., 95:249.

Van Hove, L., 1954b, Time-dependent correlations between spins and neutron scattering in ferromagnetic crystals, Phys. Rev., 95:1374.

Huebener, R. P., 1972, Thermoelectricity in metals and alloys in: "Solid State Physics," Vol. 27, H. Ehrenreich, F. Seitz and D. Turnbull, eds., Academic Press, New York.

Huebener, R. P., 1974, Dynamics of magnetic flux structures in superconductors, Phys. Reports, 13:143.

Huebener, R. P., 1979, "Magnetic Flux Structures in Superconductors," Springer Verlag, Berlin.

Huebener, R. P. and Gallus, D. E., 1973, Current-induced intermediate state in thin-film type-I superconductors: electrical resistance and noise, Phys. Rev. B, 7:4089.

Huebener, R. P. and Kampwirth, R. T., 1972, Magneto-optical studies of superconductors at high transport current densities in: "Proc. 1972 Appl. Supercond. Conf.," H. M. Long and W. F. Gauser, eds., IEEE, New York, p. 422.

Huebener, R. P. and Seher, A., 1969, Nernst effect and flux flow in superconductors. II. Lead films, Phys. Rev., 181:710.

Hurm, V., Selig, K. P. and Huebener, R. P., 1979, Fluctuations in the flux nucleation process of the current-induced resistive state in type-I superconductors, Z Physik B, 32:175.

Imai, S., 1975, Fluctuation conductivity and vortex flow resistivity of a dirty strong-coupling superconductor, Prog. Theor. Phys. (Japan) 54:624.

Josephson, B. D., 1962, Possible new effects in superconductive tunnelling, Phys. Lett., 1:251.

Kim, Y. B. and Stephen, M. J., 1969, Flux flow and irreversible effects in: "Superconductivity," R. D. Parks, ed., Marcel Dekker, New York, p. 1107.

Kirchner, H., 1971, Ein hochauflösendes magnetooptisches verfahren zur untersuchung der kinematik magnetischer strukturen in supraleitern, Phys. Stat. Sol. (a), 4:531.

König, B. and Kirchner, H., 1975, Magnetooptische beobachtungen und elektrische messungen an stromdurchflossenen, supraleitenden bleischichten im zwischenzustand, Phys. Stat. Sol. (a), 28:467.

Kopnin, N. B. and Kravtsov, V. E., 1976, Conductivity and Hall effect of pure type-II superconductors at low temperatures, Pis'ma Zh. Eksp. Teor. Fiz., 23, 631, JETP Letters, 23:578.

Larkin, A. I. and Ovchinnikov, Yu. N., 1976, Viscosity of vortices in pure superconductors, Pis'ma Zh. Eksp. Teor. Fiz., 23, 210, JETP Letters, 23:187.

Larkin, A. I. and Ovchinnikov, Yu. N., 1977, Nonlinear effects during the motion of vortices in superconductors, Zh. Eksp. Teor. Fiz., 73, 299, Sov. Phys. JETP, 46:155.

Larkin, A. I. and Ovchinnikov, Yu. N., 1979, Pinning in type II superconductors, J. Low Temp. Phys., 34:409.

Likharev, K. K., 1972, Vortex motion and the Josephson effect in superconducting thin bridges, Zh. Eksp. Teor. Fiz., 61, 1700, Sov. Phys. JETP, 34:906.

Lowell, J., 1970, The frictional force on a moving fluxon, J. Physics C: Solid State Phys., 3:712.

Maki, K., 1964, The magnetic properties of superconducting alloys. II, Physics , 1:127

Morrison, D. D. and Rose, R. M., 1970, Controlled pinning in superconducting foils by surface microgrooves, Phys. Rev. Lett., 25:356.

Nicolis, G. and Prigogine, I., 1977, "Self-Organization in Non-equilibrium Systems," John Wiley, New York.

Schmid, A., 1966, A time dependent Ginzburg-Landau equation and its application to the problem of resistivity in the mixed state, Phys. Kond. Materie, 5:302.

Selig, K. P., 1980, Flux-tube motion in the current-induced resistive state of superconductors, Thesis, University of Tubingen (unpublished).

Selig, K. P., Chimenti, D. E. and Huebener, R. P., 1978, Passive measurement of flux nucleation in the current-induced resistive state of type I superconductors, Z. Physik B, 29:33.

Selig, K. P. and Huebener, R. P., 1980, Flux nucleation in the current-induced resistive state of a constricted type-I superconductor, to be published.

Shoenberg, D., 1952, "Superconductivity," Cambridge University Press, Cambridge.

Swartz, P. S. and Hart, H. R., 1965, Surface effects in the mixed superconducting state, Phys. Rev., 137A:818.

Ullmaier, H., 1975, Irreversible properties of type-II superconductors, "Springer Tracts in Modern Physics," Vol. 76, Springer Verlag, Berlin.

Usui, N., Ogasawara, T., Yusukochi, K. and Tomoda, S., 1969, Magnetization, flux flow resistance and Hall effect in superconducting vanadium, J. Phys. Soc., Japan, 27:574.

Van Vijfeijken, A. G., 1968, On the theory of vortices in type-II superconductors, Philips Res. Rep. Suppl., 8:1.

Yamafuji, K. and Irie, F., 1967, On the concept of pinning force in type II superconductors, Phys. Lett., 25A:387.

PHONON REFLECTIONS FROM SOLID-SOLID INTERFACES

D. Goodstein, G. Paternò, F. Scaramuzzi and P. Taborek

Laboratori Nazionali del CNEN
Frascati, Italy
and
Department of Physics
California Institute of Technology
Pasadena, CA 91125

The first two chapters of this book separate the Kapitza boundary resistance problem into two parts: classical solid-solid interfaces, which obey classical theory, and solid-liquid helium interfaces which behave in an anomalous way yet to be understood. In this note we present evidence which tends to a surprising conclusion: that the anomalous behavior is a very widespread--perhaps even universal--property of imperfect interfaces between any two media. If this view is correct, then the subjects covered in Ch. 1 and Ch. 2 differ only in the relative importance of the normal and the anomalous channels for transmitting heat at an interface.

The experiments described in this paper all concern heat pulses reflected internally from an optically polished sapphire interface. This interface is decidely imperfect, and therefore exhibits the Kapitza anomaly in liquid helium. (Weber et.al. 1978). In our experiments, one observes time of flight spectra basically similar to those in Figs. 15 and 17 of Ch. 2. If a phase transition in the medium on the other side of the interface alters the probability of phonon reflections, the change is detected in the observed spectrum. One of the fundamental purposes of this work was to see whether the reflection of phonons depends on the thermodynamic state of the helium on the surface.

This question has a somewhat confused history. There is evidence that the Kapitza conductance to liquid helium differs from that to solid helium (see Ch. 2 Fig. 1) and evidence to the contrary (Ch. 2 Fig. 22). There is evidence that the Kapitza conductance does not depend on whether the helium is superfluid or normal (Guo and Maris, 1974) but also many experiments indicating a sharp change in passing the lambda transition. These experiments are often difficult to judge because the effect of Kapitza resistance is hard to separate from the very large changes in other forms of thermal behavior in going from normal fluid to superfluid to solid. The experiments reported here overcome these difficulties and others, as we shall see below. Above all, the question is a very important one because many theories of boundary resistance certainly depend on whether the receiving medium is solid, liquid or superfluid.

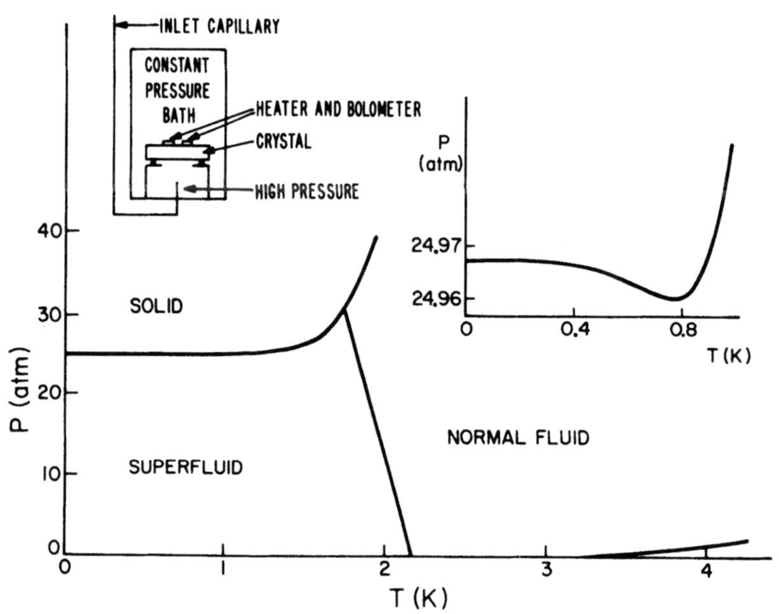

Fig. 1. Pressure - Temperature phase diagram of ^{4}He. Left inset: schematic design of the apparatus. Right inset: expanded view of the minimum in the melting curve.

The experimental arrangement is shown in the left inset of Fig. 1. Heater and bolometer films are evaporated on one side of a sapphire crystal and are kept in a constant pressure superfluid bath whose temperature is kept constant typically to within $10^{-5} - 10^{-6}$ K. Phase transitions are caused to occur on the reflecting surface by changing the pressure there. A glance at the phase diagram for

^{4}He, in Fig. 1 shows that both the lambda transition and solidification can be crossed isothermally in this way. As a result, the thermal properties of the heater bolometer system are not affected by the phase transition.

To summarize briefly the results of our early experiments, it was found that passing through the lambda transition did not affect any part of the reflection spectrum, with an uncertainty of 2 - 3% (Cole et.al., 1979). This degree of sensitivity was achieved by studying each part of the spectrum separately while drifting slowly through the transition. We also found that the relaxation time of the entire crystal after being heated by a long (500 nsec) pulse did not depend on the phase of the fluid in the pressure cell. This latter result was in disagreement with a report (Fox and Strandberg 1977) that such an experiment had sharply different results above and below the lambda transition.

Similar experiments near the solidification curve yielded similar results. However, in analyzing the data, we came to doubt that we (or, for that matter, earlier phonon reflection investigators) had ever observed solidification at the interface. The reason, briefly, is that pressurization of the high pressure cell inevitably produces gradients in both temperature and pressure which cause solidification to occur first in the inlet capillary. Once a plug forms in the capillary, no further communication with the cell is possible. Morever, even if it were possible to cause solidification to occur first in the cell, the crystal surface, heated by the heat pulses of the experiment, is the least likely place in the cell for the solid to form. We therefore found it necessary to work out a novel way to do this experiment.

The new technique takes advantage of the fact that there is a minimum in the P - T melting curve of ^{4}He, as shown in the second inset of Fig. 1. On the solidification curve below the minimum--say at 0.7K--freezing is promoted by heating at constant pressure. In these conditions, the heated crystal interface is the most likely place in the cell for solidification to occur. To assure that solid forms in the reflection cell, the entire system is raised to 35 atm. and 2K (so that no solid forms anywhere) then cooled. A plug forms in the capillary sealing into the cell helium at a sufficently high density to assure that partial solidification occurs at the working temperature. A simple calculation using standard values of the He-sapphire Kapitza resistance shows that solidification by heating at the crystal interface is an order of magnitude more effective than by pressure due to the gravitational potential difference to the bottom of the cell, thus assuring that solid forms at the interface. After recording a spectrum with solid at the interface, the pressure in the system is reduced isothermally until the plug melts and a new spectrum is taken for comparison, this time with liquid at the interface. The techinique of recording a single point of the spectrum on a chart recorder during melting is also used to increase the sensitivity of the measurement. The entire procedure may be checked by repeating it just above the minimum, where no solid is ever formed at the interface. As will be seen in the data presented below, the result of these experiments is that melting of the helium at the interface has no effect on the reflection signal.

Before presenting our most recent data, it is necessary to describe some recent technical innovations in heat pulse spectroscopy which affect both how the experiment was done, and our understanding of the results.

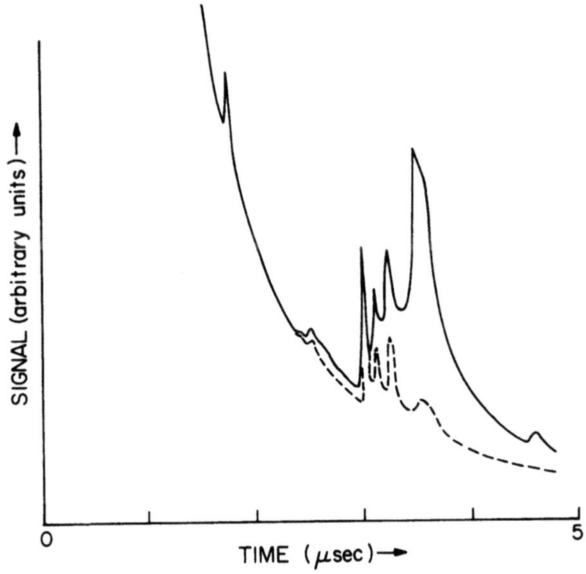

Fig. 2. Phonon reflection spectrum 7mm thick sapphire crystal in "big bump" geometry. Solid curve: vacuum interface. Dashed curve: Argon film on the interface.

In a series of experiments and analyses at Caltech (Taborek and Goodstein, 1979; 1980 a; 1980 b) techniques have been developed to produce extremely high time resolution reflection spectra, and in addition, to make use of phonon focusing caustics to "design" heater - bolometer - crystal geometres particularly suitable for these experiments. An example is shown in the vacuum spectrum of Fig. 2.

A series of sharp peaks due to specular reflection is followed by a large bump due to diffusely reflected phonons which are carried from heater to surface to bolometer along paths of singular focusing intensitity. When helium is admitted to the interface, it is found that the specular peaks are not measureably changed, but the diffuse background and bump are drastically reduced. Thus, the diffuse part of the spectrum corresponds to the "background channel" discussed in Ch. 2. It is in this geometry, where the crystal has been used as a lens to focus the important diffuse part of the spectrum, that our solidification experiments have been done.

In Fig. 3, we see a vacuum spectrum taken at Caltech in this geometry at 2K, using a tin bolometer. Just below it is a spectrum from the same crystal, in the same geometry at Frascati, at 1K using an Al bolometer. The time resolution of the Al bolometer is lower than that of the tin, probably because as discussed in Ch. 5, the characteristic electron-phonon relaxation time τ_0 is 90 nsec in Al, compared with the 50 nsec resolution in the tin spectrum (τ_0 is 2 nsec in tin). Nevertheless, the principal features of the spectrum, including the big diffuse bump and the change from vacuum to He at the interface are clearly seen in the Al bolometer spectrum.

Finally, in Fig. 4, we see spectra with solid and liquid at the interface, at 0.7K, taken as described above. Using chart recorder tracings taken during melting in the region of the focusing bump, we can put an upper limit of 1% on any change taking place due to melting of the solid.

These experiments are not yet finished. A number of refinements are called for and will be carried out. However, recent developments at Caltech have caused us to see our results in a new light. Experiments have been done in which, using the same "big bump" geometry, instead of placing helium at the interface, films of each of the other noble gases were evaporated there in turn. The results of all of these experiments are indistinguishable from that of putting liquid helium at the interface. For example, while Fig. 3b shows the effect of helium, the dashed curve in Fig. 2 was actually reproduced by placing argon at the interface. Exactly the same result is obtained using Ne, Kr and Xe.

The difference between the solid and dashed curves in Fig. 2, or the upper and lower spectra in Fig. 3b, _is_ the anomalous Kapitza effect. Apparently, all or nearly all of the diffusely reflected phonons are removed from the signal when any medium is present at the interface, whether it be liquid helium or solid Xenon. Under the circumstances, it is hardly surprising that no difference is observed between liquid and solid helium.

Our observations lead us to suggest that the anomalous Kapitza

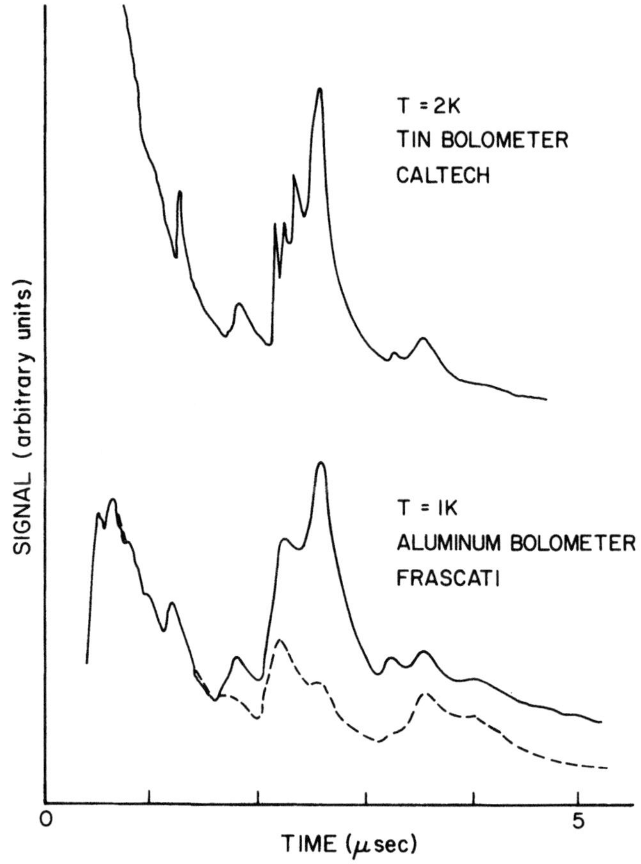

Fig. 3. Upper solid curve: phonon reflection spectrum, sapphire-vacuum interface in "big bump" geometry 7mm thick x 22 mm dia. crystal. Lower solid curve: same crystal, sapphire-vacuum interface. Dashed curve: liquid helium at the interface.

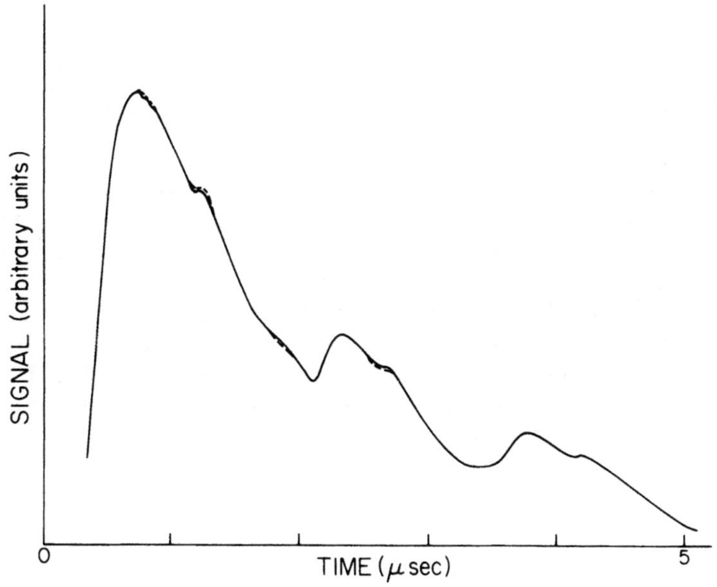

Fig. 4. Phonon reflection spectrum, same crystal as in Fig. 3. Solid curve: sapphire-solid helium interface. Dashed curve: sapphire liquid helium interface.

effect is a universal property of all interfaces, excluding only those so perfect that only specular processes occur. It is relatively less important in heat transmission in the cases discussed in Ch. 1 where acoustic mismatch is not severe, and it dominates heat transmission in the cases discussed in Ch. 2 where acoustic mismatch is severe. If further work confirms our hypothesis, however, the anomaly is always present at imperfect interfaces of all kinds, and future theories will have to yield that result.

ACKNOWLEDGEMENT

We wish to acknowledge the invaluable assistance of Luciano Mori and Lucia Cavalieri in performing these measurements.

REFERENCES

Cole, M., Goodstein, D., Mobilio, S., Paternó, G., Savoia, A., Scaramuzzi, F., and Taborek, P., 1979, Phonon Reflection at the Sapphire-Liquid Helium Interface, C.N.E.N. Edizioni Scientifiche, Frascati Rome, Italy.

Guo, C-J and Maris, H. J., 1974, Experimental Study of the Reflection of Phonons at an Interface Between Dielectic Crystals and Liquid Helium, Gaseous Helium and Solid Neon, Physical Review A, 10:960

Fox, L. R. and Strandberg, M. N. P., 1977, Measurement of Kapitza Resistance by Use of Transient Temperature Response, Physical Review Letters, 38:1295

Taborek, P. and Goodstein, D., 1979, Phonon Reflection at Sapphire-Vacuum Interface, J. Physics C,: Solid State Physics 12:4737

Taborek, P. and Goodstein, D., 1980 a, Phonon Focusing Catastrophes, Solid State Communications, 33:1191

Taborek, P. and Goodstein, D., 1980 b, Diffuse Reflection of Phonons and the Anomalous Kapitza Resistance, Physical Review B (to be pub.)

Weber, J., Sandemann, W., Dietsche, W. and Kinder, H., 1978, Absence of Anomalous Kapitza Conductance on Freshly Cleaved Surfaces, Physical Review Letters 40:1469

MEASUREMENTS OF THE MAXIMUM JOSEPHSON SUPERCURRENT VS. MAGNETIC FIELD AS A PROBE OF TUNNEL BARRIER UNIFORMITY

A. Barone
Istituto di Cibernetica, CNR,
80072 Arco Felice, and
Istituto di Fisica
Universitá di Napoli, Italy

G. Paternó
Laboratori Nazionali, CNEN
Frascati, Italy

M. Russo
Istituto di Cibernetica, CNR
80072 Arco Felice, Italy

R. Vaglio
Istituto di Fisica
Universitá di Salerno I
84100 Salerno, Italy

INTRODUCTION

Nonequilibrium quasiparticle distributions can be studied as a result of elastic single-particle electron tunneling in superconducting junctions. The use of a tunnel junction both for creation and detection of nonequilibrium states is widely reported in the literature. Quasiparticle, phonon and photon generation by tunneling have been in fact investigated, as well as phonon, and microwave detection, etc. For instance, a simple symmetric (i.e., equal superconducting gaps $\Delta_1=\Delta_2=\Delta$) junction can be used in the voltage region $V<2\Delta/e$ for the detection of phonons of energy $\hbar\omega>2\Delta$ (see Chapter 3). Another example is a structure consisting of two superimposed tunnel junctions with one superconducting layer in common (i.e., a S-I-S-I-S structure), which has been successfully employed in a variety of experiments of nonequilibrium superconductivity (see Chapter 5). In this case the role of the one junction is the creation of the nonequilibrium superconducting state, whereas the second serves as a probe of such a state. The possibility of phonon detection by using Josephson junctions has also been demonstrated (e.g., Tredwell and Jacobsen, 1974).

As emphasized in several lectures throughout this course, the uniformity of the tunneling barrier in a junction can be of paramount importance in various kinds of experiments besides those belonging to the Josephson context.

The aim of this lecture is to present a rather comprehensive discussion on the possibilities offered by measurements of the magnetic field dependence of the critical supercurrent to infer variations of the junction barrier thickness, which can very likely occur in actual samples. A number of theoretical considerations will be outlined and experimental results discussed which will provide the basic tool for the interpretation of any deviation of barrier uniformity by a careful investigation of the critical current-magnetic field patterns. To this purpose we shall resort also to the use of light-sensitive barrier junctions whose properties, besides their intrinsic physical interest, provide a tutorial overview of the Josephson structures. It is in fact possible, by varying the sample illumination, to realize a "simulation" of various situations which can occur in the "usual" oxide barrier junctions.

GENERAL

The well-known constitutive relations of the Josephson effect can be written as (Josephson 1965)

$$J = J_c \sin\phi \qquad \frac{\partial\phi}{\partial t} = \frac{2e}{\hbar} V$$

$$\frac{\partial\phi}{\partial x} = \frac{2\pi d}{\Phi_o} H_y \qquad \frac{\partial\phi}{\partial y} = \frac{-2\pi d}{\Phi_o} H_x \qquad (1,a\text{-}d)$$

Let us consider junction dimensions smaller than the Josephson penetration depth so that the magnetic field we are dealing with is just the external one. Furthermore, let us assume that such a field H is applied along the y direction ($H=H_y$) as in Fig. 1a; then the resulting dependence $I_c(H)$ gives the usual Fraunhofer-like pattern.

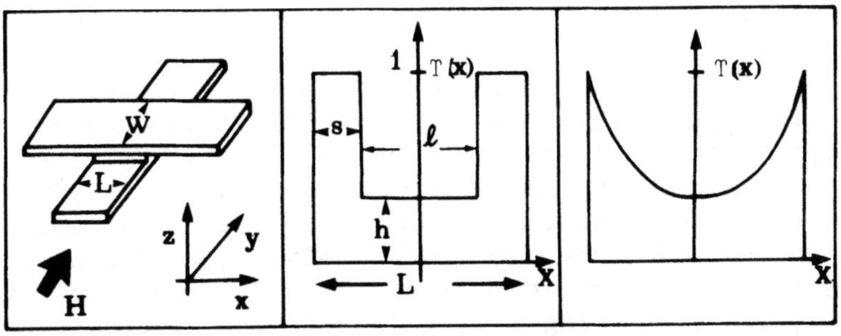

Fig. 1. a) Sketch of the junction.
b) Step-like current density distribution.
c) One-parameter current density profile.

Let us now consider a nonuniform current density distribution in the barrier so that we can assume (Dynes and Fulton, 1971)

$$J = J_c(x,y)\sin(kx + \phi_o) \quad ,$$

where

$$k = \frac{2\pi d}{\Phi_o} H_y = \frac{2\pi}{L}\frac{\Phi}{\Phi_o}$$

and ϕ_o a constant resulting from integration of 1c. The total current will be

$$I = \int_0^W \int_0^L dxdyJ_c(x,y)\sin(kx + \phi_o) \quad ,$$

which can be written

$$I(k,\phi_o) = \text{Im}\ [e^{i\phi_o} \int_0^L dx\ T(x)e^{ikx}]$$

where $T(x) = \int dyJ_c(x,y)$ represents the current density profile along the y direction. Taking the maximum with respect to ϕ_o and assuming $T(x) = 0$ for $x < 0$ and $x > L$ we get finally

$$I_c(k) = \left| \int_{-\infty}^{+\infty} dx\ T(x)e^{ikx} \right| \doteq \left| \mathcal{T}(k) \right| .$$

Hence, the maximum critical current vs. the applied magnetic field can be expressed as T(x), the modulus of the Fourier transform of the current density profile along the y direction. It is clear that, going back from the Fourier transform to the $\mathcal{T}(x)$, the information on the phase term is lost, that is: the procedure to recover the barrier shape from the $I_c(H)$ pattern is equivocal unless independent physical hypotheses are supplied (Zappe, 1975). If J_c is uniform (i.e., $\mathcal{T}(x)$ constant) the Fraunhofer-like dependence is immediately recovered. Let us observe that the $I_c(H)$ dependence is essentially determined by $\mathcal{T}(x)$. On the other hand a given $\mathcal{T}(x)$ can be obtained either by a rectangular junction with a nonuniform $J_c(x,y)$ or by a uniform $J_c(x,y)$ but adopting a suitable electrode geometry (Balsamo et al., 1976). The possibilities offered by such a "shaping" of the junctions is a relevant point (e.g., Broom et al., 1980).

LOCALIZED NONUNIFORMITIES

Assuming for $\mathcal{T}(x)$ the step-like shape sketched in Fig. 1b, it is easy to obtain the expression

$$I_c(\phi) = I_c(0) \left| \frac{2s}{h\ell+2s} \frac{h\ell}{2s} \frac{\sin \pi \frac{\ell}{L}\phi}{\pi \frac{\ell}{L}} + \frac{\sin \pi \frac{s}{L}\phi}{\pi \frac{s}{L}\phi} \cos\left(\pi \frac{(\ell+s)}{L}\phi\right) \right| \tag{2}$$

where $\phi = \Phi/\Phi_0$ and the meaning of the other symbols is clearly indicated in the figure. By inspection of Eq. 2, we can see that depending on the values of the parameters (i.e., the specific current density distribution), a diffraction behavior or a pure interferential behavior can occur as limiting cases as well as all possible intermediate situations as shown by Barone et al. (1977) (see also Richter and Seidel, 1978 and Köhler et al., 1979). In Fig. 2a is reported the experimental I_c vs. H dependence of an $Sn-Sn_xO_y-Sn$ Josephson junction. This pattern shows the presence of a modulation of an interference component due to the occurrence of current density enhancement at the edges of the sample. Such a peaking of J_c can be related to a thinning of the oxide layer at the edges of the barrier where the top film overlays the bottom one. The plausibility of this interpretation is also supported by technological considerations. This circumstance, as well as other situations corresponding to typical localized nonuniformities into the barrier, can be investigated in conjunction with the use of light sensitive Josephson junctions in which spatial modulation of the supercurrent density can be deliberately created by optical input and a suitable shading of the sample (Barone et al., 1975).

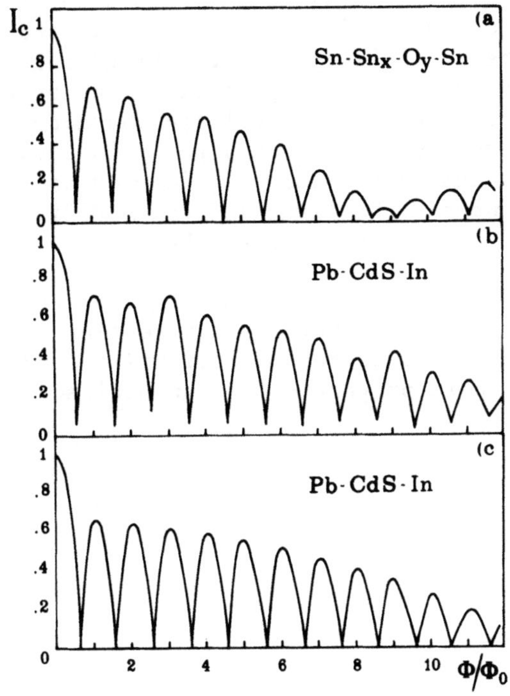

Fig. 2. Critical current vs. externally applied magnetic field:
a) experimental pattern for a dielectric barrier junction;
b) experimental pattern of a light sensitive junction; and
c) theoretical curve calculated assuming for T(x) the step-like profile of Fig. 1b.

Before embarking into the study of some examples of the diagnostic aspects of these measurements, let us first outline some of the properties of light sensitive junctions to which we shall refer.

Among the various possibilities explored in the last decade, or even before, for realizing new systems of weakly coupled superconductors, semiconducting barrier junctions have been widely investigated both for quasiparticles and Josephson tunneling studies. Particular attention is paid to junctions employing light sensitive barriers (namely CdS), since such devices exhibit a variety of behaviors depending on the optical input. The main interest lies in the possibility of "adjusting" the effective barrier, in a given sample, by varying the light exposure. Indeed, in junctions whose current-voltage characteristics I(V) in "dark" conditions exhibit a clear quasiparticle tunneling structure, a Josephson current can be subsequently induced and continuously enhanced by light input. The light induced modification of the I(V) is "permanent", and this storage-like response, as well as the possibility of "resetting", are discussed in the above quoted reference. Light sensitive superconducting devices appear to be difficult to the degree that they are interesting and can be prepared today with a good degree of reliability and reproducibility. The basic optoelectronic properties are summarized in Fig. 3.

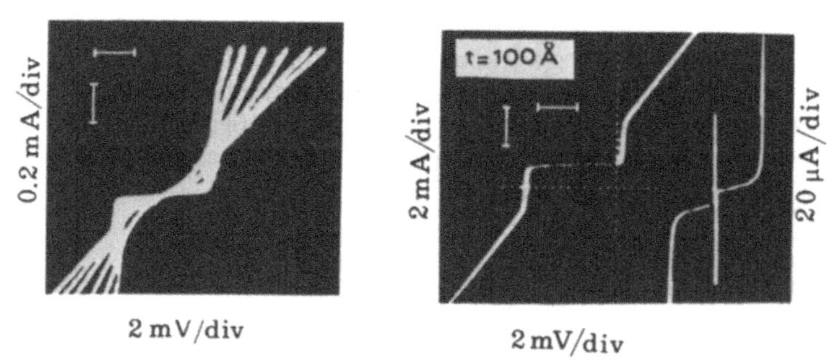

Fig. 3. a) I(V) characteristics Pb-CdS-Pb junctions. Storage-like response to optical inputs.
b) I(V) curve in dark conditions and after illumination, exhibiting light induced Josephson current.

We have observed that varying the transparency of the top electrode (i.e., film thickness) in combination with the junction dimensions, it is possible to range over all cases, from the uniform light-induced current density to a limiting situation of current peaking at the edges of the junction with a resulting double junction-like behavior. In Fig. 2b is reported an experimental $I_c(H)$ dependence for a Pb-CdS-In junction and in Fig. 2c the corresponding fitting curve calculated on the basis of the step-like current density distribution of Fig. 1b with $s/L = 0.11$ and $h = 0.06$.

Finally, we observe that, among other possibilities, it can be reasonable to resort to the one-parameter model for $J_c(x,y)$ as sketched in Fig. 1c. This has been adopted, for instance, by Russo and Vaglio (1978) to account for the magnetic field dependence of I_c as well as the light-induced step amplitude.

STRUCTURAL FLUCTUATIONS

In the previous section we have discussed the influence of nonuniformities which are "localized" in the junction barrier and whose size is comparable with the junction dimensions. As Yanson (1970) first pointed out, it is of great interest to consider also the role of the intrinsic thickness fluctuations of the barrier that can be depicted as randomly distributed nonuniformities whose size is small compared with the junction dimensions. Such unavoidable nonuniformities can be related to irregular oxide growth or, in the case of semiconductor junctions, to structural defects of the barrier. As for the "localized" nonuniformities, also in this case useful information can be deduced by measuring the magnetic field dependence of the maximum Josephson current. In fact, the effect of structural fluctuations results in the occurrence of a background current in the pattern whose amplitude is simply connected to the average properties of the nonuniformities. To deepen this matter we shall follow, in a somewhat modified form, the statistical approach by Yanson (1970).

The presence of a structural fluctuations can be studied by introducing a stochastic function which describes the fluctuations in the current density determined by the structural nonuniformities of the barrier. The current density can be written in the form:

$$J_c = J_i(x) + J_f(x) \qquad (3)$$

For simplicity we have considered a one-dimensional problem, that is: all the current densities are averaged along the y direction. $J_i(x)$ describes the localized nonuniformities ($J_i(x) \equiv \mathcal{T}(x)$ of the previous section. $J_f(x)$ describes the structural fluctuations and we shall assume the autocorrelation function has the form:

$$\langle J_f(x_1)J_f(x_2)\rangle = \langle J_f^2\rangle \exp(- \quad x_1 - x_2 \quad /r) \tag{4}$$

We further assume that

$$\langle J_f(x)\rangle = 0 \quad , \tag{5}$$

the angular brackets indicate the average over the junction length, r is the autocorrelation radius which gives the average size of the nonuniformities and $\langle J_f^2\rangle = I_f^2/2rLW^2$ is the mean square amplitude of the current density fluctuations. Inserting Eq. 3 into Eq. la and using the properties of the stochastic function $J_f(x)$ described by Eqs. 4 and 5, we get, after rather tedious calculations, the following form for the magnetic field dependence of the maximum super current:

$$I_c(\phi) = \left\{ \mathfrak{F}^2(\phi) + I_f^2 \left[a(r\ ,\phi)+b(r\ ,\phi)\right] \quad \frac{\sin^2\pi\phi}{\pi^2\phi^2} \right\}^{1/2} . \tag{6}$$

The explicit expressions for a and b can be found in Barone et al. (1978). The usual dependence $I_c(\phi) = T(\phi)$ is found if the autocorrelation radius r or the mean square amplitude of the fluctuations $\langle J_f^2\rangle$ goes to zero. The expression (6) takes a rather simpler form in the limit $2\pi\ r\phi/L \ll 1$ (small fluctuation limit). In fact, in that limit:

$$I_c(\phi) = \left\{ \mathfrak{F}^2(\phi) + I_f^2\left(1 - \frac{\sin^2\pi\phi}{\pi^2\phi^2}\right) \right\}^{1/2} . \tag{7}$$

It is clear that, besides the oscillating terms $\mathfrak{F}(\phi)$ and $I_f \sin\pi\phi/\pi\phi$ there is also a background current of constant amplitude I_f. If the complete expression (6) is considered, the magnetic field pattern can be shown to present a background current of amplitude I_f at $\phi = 0$ that slowly decreases toward zero for $\phi \approx L/2\pi r$. This effect has been observed by Yanson on $Sn\text{-}Sn_x\ O_y\text{-}Sn$ junctions in connection with a prolonged aging of the samples.

We have performed some experiments (Barone et al. 1978) using the photoconductive CdS semiconductor barrier junctions described above. There are two main reasons for using such samples in investigating the effect of fluctuations: the CdS barrier has an intrinsic inhomogeneous character; moreover, in these structures, the absence of any Josephson current in dark conditions make sure of the absence of small short-circuits that can also produce background currents in the magnetic field pattern.

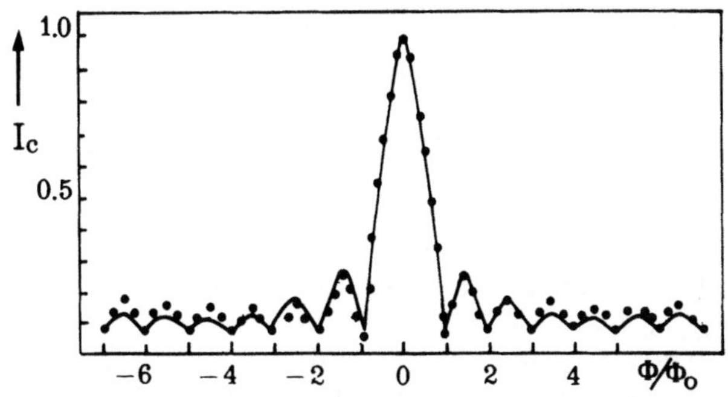

Fig. 4. Critical current vs. externally applied field in a Pb-CdS-In junction in the presence of "small" fluctuations. Experimental (dots), theoretical (full line).

In Fig. 4 the experimental dependence of the $I_c(H)$ is reported for a Pb-CdS-In junction. The current is normalized to the maximum current and the field is given in terms of flux in the junction normalized to the flux quantum. The presence of rather strong barrier fluctuations is clearly apparent from the background current in the pattern. Only small "localized" nonuniformities are present, which leads to a slight double modulation effect. The theoretical curve is obtained by Eq. 7 (small fluctuation limit) for $\gamma = I_f/I_c(0) = 0.07$. The double modulation effect is accounted assuming $J_i(x)$ as a step-like barrier with $h = 0.9$ and $s/L = 0.22$. No noticeable decrease of the background current is observed in the field range considered. From the analysis of the experimental results obtained on Pb-CdS-In junctions, typical values of the correlation radius for the nonuniformities result to be of the order of 1-2 μm.

In conclusion we hope that the arguments outlined in this lecture can give at least the flavor of the potential of $I_c(H)$ measurements in the diagnostics of the junction barrier uniformity.

This work was partially supported by the Consiglio Nazionale delle Ricerche under the "Progetto Finalizzato Superconduttività.

REFERENCES

Balsamo, E. P., Paternó, G., Barone, A., Russo, M. and Vaglio, R., 1976, Supercurrent interference patterns in a tunneling junction structure, Proc. IC-SQUID, 193, H. D. Hahlbohm and H. Lübbig eds., W. de Gruyter, Berlin.

Barone, A., Paternó, G., Russo, M. and Vaglio, R., 1977, Diffraction and interference phenomena in single Josephson junctions, Phys. Status Sol., 41a:393.

Barone, A., Paternó, G., Russo, M. and Vaglio, R., 1975, Experimental results on temperature and magnetic field dependence of light induced Josephson current, Proc. 14th Int. Conf., Low Temp. Phys., 4:88.

Barone, A., Paternó, G., Russo, M. and Vaglio, R., 1978, Experimental results and analysis of structural fluctuations in photosensitive Josephson junctions, Zh. Eksp. Teor. Fiz., 74:1483, Sov. Phys. JETP, 47:776.

Broom, R. F., Kotyczka, W. and Moser, A., 1980, Modeling of characteristics for Josephson junctions having non-uniform width or Josephson current density, IBM J. Res. Develop., 24:178.

Dynes, R. C. and Fulton, T. A., 1971, Supercurrent density distribution in Josephson junctions, Phys. Rev., 3B:3015.

Josephson, B. D., 1965, Supercurrents through barriers, Adv. Phys., 14:479.

Köhler, H. J., Müller, U., Weber, P. and Blüthner, K., 1970, Investigations of the current density distribution in Nb-NbO_x-Pb tunnel junctions, Phys. Status Sol., 56a:521.

Richter, J. and Seidel, P., 1978, Theoretical studies of the influence of barrier inhomogeneities on the behavior of Josephson junctions, Exp. Tech. der Physk, 26:217.

Tredwell, T. J. and Jacobsen, E. H., 1974, Phonon detection by Josephson junctions, Phys. Lett., 50A:281.

Yanson, I. K., 1970, Effect of fluctuations on the dependence of the Josephson current on the magnetic field, Zh. Eksp. Teor. Fiz., 58:1497; Sov. Phys. JETP, 31:800.

Zappe, H. H., 1975, Determination of the current density distribution in Josephson tunnel junctions, Phys. Rev., 11B:2535.

APPENDIX: List of Symbols

Any additional symbols are defined in the chapter in which they appear. All symbols are generalized to include dependence on other variables or parameters. If such a parameter has a continuous set of values, a parenthesis is used, e.g., $\Delta(T)$ or $\Delta(V_i)$ referring to the temperature or injector voltage dependence of the energy gap. If the parameter is fixed, a subscript may be used: Δ_a or Δ_o, referring to the energy gap in film "a" or at some stated initial conditions. If there is a distribution in values of a parameter about some average, the mean value is made from the symbol with a bar on top: $\bar{\Delta}$ or $\overline{\Delta(T)}$.

a, a^+ = phonon annihilation, creation operator (also a_λ, a_λ^+)

A = area

$\underline{A}$ = magnetic vector potential

B = magnetic induction field

c = speed of light

c, c^+ = electron annihilation, creation operator

C = capacitance

d = thickness (e.g., of film)

D = quasiparticle diffusion constant

e = absolute value of electronic charge

E = quasiparticle energy = $(\xi_k + \Delta_k^2)^{1/2}$ (also use E_k)

$\underline{E}$ = electric field

E_F = Fermi energy

f(E) = quasiparticle occupation probability at energy E

f_k = quasiparticle occupation probability at momentum k

f_L = Lorentz force per unit length of flux line

f° = Fermi distribution function [also f_k° and $f^\circ(E)$]

f_p = pinning force per unit length of flux line

f_{th} = thermal force per unit length of flux line

$f_>$ = quasiparticle occupation probability in momentum $k>k_s$

$f_<$ = quasiparticle occupation probability in momentum $k<k_s$

$\delta f_>$ = quasielectron branch excess = $f_> - f°$

$\delta f_<$ = quasihole branch excess = $f_< - f°$

F = free energy

F_ω = phonon trapping factor

$F(\Omega)$ = phonon density of states at frequency Ω

g_{ns} = normalized conductance of normal-superconducting tunnel junction = G_{ns}/G_{nn}

G = Gibbs free energy

G_{nn} = conductance of normal-normal tunnel junction

G_{ns} = conductance of normal-superconductor tunnel junction

$\hbar$ = Planck constant

$\hbar\omega$ = photon energy (ω = photon frequency)

$\hbar\Omega$ = phonon energy (Ω = phonon frequency)

$\hbar\Omega_D$ = Debye energy = $k_B\Theta_D$ (Ω_D = Debye frequency)

H = applied field

H_c = thermodynamic critical field

H_{c1} = flux entry critical field

H_{c2} = superconducting critical field for type II

I = current

I_c = critical current

I_d = detector current

I_i = injector current

I(V) = current-voltage characteristic

J = current density

J_c = critical current density

J_n = normal current density

J_s = supercurrent density

k = quasiparticle wave number

k_B = Boltzman constant (as a combination kT is acceptable)

k_F = Fermi wave number

k_s = pair wave number

K = thermal conductivity

ℓ = electron or quasiparticle mean free path

L = inductance

L_T = thermoelectric transport coefficient

m = electron rest mass

M = magnetization

n = reduced number density of quasiparticles = $N/4N(0)\Delta(0)$

n_e = conduction electron number density

n_n = number density of normal electrons

n_s = number density of superconducting electrons

n° = Bose distribution function [also $n^\circ(\Omega)$]

$n(\Omega)$ = phonon occupation probability at frequence Ω

$n(q)$ = phonon occupation probability at momentum q

N = number density of quasiparticles

$N(E)$ = single spin density of electron states at E

$N(0)$ = single spin density of electron states at E_F

$N(T)$ = number density of quasiparticles in equilibrium at T

N_ω = number density of phonons with $\hbar\Omega > 2\Delta$

$N_\omega(T)$ = number density of phonons with $\hbar\Omega > 2\Delta$ at temperature T

p_D = Debye momentum = $\hbar q_D$

p_F = Fermi momentum = $\hbar k_F$

p_k = quasiparticle momentum = $\hbar k$

p_q = phonon momentum = $\hbar q$

p_s = pair momentum = $\hbar k_s$

P = power

q = phonon wave number

q_D = Debye wave number

Q = branch imbalance quasiparticle number density

Q^* = charge imbalance quasiparticle number density

Q = heat flux per unit area

R = resistance

R_K = Kapitza boundary resistance

R_n = normal resistance

s = speed of sound (also s_λ, where λ = polarization L or T)

S = entropy

S, S^+ = Cooper pair annihilation, creation operator

S_T = normal state thermoelectric power

t = T/T_c = reduced temperature

T = temperature

T_c = superconducting transition temperature

$T_{k,q}$ = tunneling matrix element

T_o = ambient temperature (substrate or bath)

T^* = effective distribution function temperature

∇T = thermal gradient

u_k = BCS coherence factor

v_f = flux flow velocity

v_F = Fermi velocity

v_k = BCS coherence factor = $(1 - u_k^2)^{1/2}$

v_n = normal fluid velocity (helium or superconductor)

v_s = superfluid velocity (helium or superconductor)

V = voltage

V_{BCS} = BCS pair interaction potential

V_d = detector voltage

V_i = injector voltage

w = width

$w(\omega)$ = noise power spectrum

Y_K = Kapitza conductance

$Y(T)$ = Yosida function

$Z(E)$ = quasiparticle renormalization = $Z_1(E) + i\, Z_2(E)$

$Z(T)$ = $2 \int_\Delta^\infty (\frac{-\partial f^\circ}{\partial E}\; \rho^{-1}(E)dE$

α = coefficient of Ginzburg-Landau equation

$\alpha(\Theta)$ = phonon transmission probability at boundary; also $\alpha(\theta,\phi)$

$\bar{\alpha}$ = average phonon transmission probability at boundary

$\alpha^2(\Omega)$ = electron phonon coupling strength

β = coefficient of Ginzburg-Landau equation

γ, γ^+ = quasiparticle annihilation, creation operator

Γ_x = τ_x^{-1} (any x)

Δ = energy gap

$\Delta(E)$ = BCS energy gap of energy state E

Δ_k = BCS energy gap of momentum state k

$\Delta(T)$ = BCS energy gap at temperature T

$\Delta(0)$ = BCS energy gap at zero temperature

ε_F = Fermi energy

ε_k = Block energy of momentum state k

ζ = $Lg_{ns}V_d/I_i$

η = flux flow viscosity

η_{th} = thermal healing length

$\theta(x)$ = Heaviside function

θ_D = Debye temperature

κ = Ginzburg-Landau parameter = λ/ξ

λ = magnetic field penetration depth

λ_{ep} = electron phonon coupling constant

λ_L = London penetration depth

Λ = phonon mean free path

Λ_N = quasiparticle number diffusion length = $(D\tau_R)^{1/2}$

Λ_{Q*} = quasiparticle charge imbalance diffusion length = $(D\tau_{Q*})^{1/2}$

μ = chemical potential

μ_n = quasiparticle chemical potential

μ_s = pair chemical potential

$\mu*$ = effective shift in quasiparticle chemical potential with respect to ε_F

ν = Nernst coefficient

ξ = coherence length

ξ_c = flux quanta core radius

ξ_k = Block energy referred to pair chemical potential = $\varepsilon_k-\mu_s$

ξ_o = electromagnetic coherence length

ρ = resistivity

$\rho(E)$ = reduced superconducting density of states

ρ_f = flux flow resistivity

ρ_m = mass density

ρ_n = normal resistivity

σ = conductivity

σ_n = normal conductivity (junction or superconductor)

$\sigma(\omega)$ = dynamic conductivity of superconductor = $\sigma_1(\omega) + i\sigma_2(\omega)$

τ_B = phonon - Cooper pair breaking time

τ_{eff} = effective quasiparticle recombination time

τ_{es} = phonon escape time

τ_E = total quasiparticle inelastic collision time at T_c and E_F

$\tau(E)$ = energy E dependent scattering time

τ_f = minimum flux passage time

τ_{GL} = Ginzburg-Landau relaxation time

τ_{in} = inelastic scattering time

τ_J = acceleration time for supercurrent

τ_o = characteristic quasiparticle scattering time = 8.4 τ_E

τ_o^{ph} = characteristic electron-phonon scattering time

τ_Q = branch imbalance relaxation time

τ_{Q^*} = charge imbalance relaxation time

τ_R = quasiparticle recombination time

τ_{R^*} = effective quasiparticle recombination time (also τ_{eff})

τ_s = magnetic field pair breaking time

τ_{so} = spin flip scattering time

τ_S = quasiparticle inelastic scattering (not incl. recombination time)

τ_S^{ph} = phonon-quasiparticle scattering time

τ_{tr} = elastic scattering time

τ_T = tunneling time

τ_{Th} = thermalization time

τ_Δ = energy gap relaxation time

ϕ = phase of order parameter

Φ = magnetic flux

Φ_o = magnetic flux quanta

$\chi(\tau)$ = Fourier transform of $w(\omega)$

ψ = order parameter

ω = photon frequency

ω_D = Debye frequency

Ω = phonon frequency

INDEX

GPSR Compliance

The European Union's (EU) General Product Safety Regulation (GPSR) is a set of rules that requires consumer products to be safe and our obligations to ensure this.

If you have any concerns about our products, you can contact us on ProductSafety@springernature.com

In case Publisher is established outside the EU, the EU authorized representative is:

Springer Nature Customer Service Center GmbH
Europaplatz 3
69115 Heidelberg, Germany

Batch number: 10684222

Printed by Printforce, the Netherlands